DISCRETE MATHEMATICS AND FUNCTIONAL PROGRAMMING

Thomas VanDrunen

Franklin, Beedle & Associates Inc.
22462 SW Washington St.
Sherwood, Oregon 97140
503/625-4445
www.fbeedle.com

— *Dedication* —

To Esther.
Instead of the brier shall come up the myrtle.
Isaiah 55:13

President and Publisher	Jim Leisy (jimleisy@fbeedle.com)
Production	TomSumner
Associate	Jaron Ayres

Printed in the U.S.A.

Library of Congress Cataloging-in-Publication data is available from the publisher.

Contents

Chapter 9 Complexity Class 461

Chapter 10 Lattice 513

Chapter 11 Group 561

Chapter 12 Automaton 597

Preface (to the instructor)

This text's approach

This text provides a distinct way of teaching discrete mathematics. Since discrete mathematics is crucial for rigorous study in computer science, many texts include applications of mathematical topics to computer science or have selected topics of particular interest to computer science. This text fully integrates discrete mathematics with programming and other foundational ideas in computer science.

In fact, this text serves not only the purpose of teaching discrete math. It is also an introduction to programming, although a non-traditional one. *Functional programming* is a paradigm in which the primary language construct is the function—and *function* here is essentially the same as what it is in mathematics. In the functional paradigm we conceive the program as a collection of composed functions, as opposed to a sequence of instructions (in the imperative paradigm) or a set of interacting objects (in the object-oriented paradigm). Dominant computer science curricula emphasize object-oriented and imperative programming, but competence in all paradigms is important for serious programmers—and functional programming in particular may be appropriate for many casual programmers, too. For our purposes, the concepts underlying functional programming are especially grounded in those of discrete mathematics.

Discrete mathematics and *functional programming* are equal partners in this endeavor, with the programming topics giving concrete applications and illustrations of the mathematical topics, and the mathematics providing the scaffolding for explaining the programming concepts. The two work together in mutual illumination. (The time spent on the math and programming content streams, however, is closer to a 60/40 split.)

Discrete math courses and texts come in many flavors. This one emphasizes sets, using the set concept as the building block for discussions on propositional logic, relations, recursion, functions, graphs, complexity classes, lattices, groups, and automata. Of course the specific topics themselves are subservient to our real object: teaching how to write proofs.

The presentation of functional programming is based on the classic approach embodied in Abelson and Sussman's *Structure and Interpretation of Computer Programs* (SICP) [1]. In short, it is SICP-light, modernized and adapted to the discrete math context, and translated into the ML programming language.

Rationale

Why teach discrete mathematics—or functional programming—this way? This approach is proposed as a solution to a variety of curricular problems.

Serving the constituencies. As a rookie professor, the author taught a discrete math course designed for math majors but filled with math-averse computer science majors fulfilling their math requirement and, in the same semester, an introductory programming course designed for computer science majors but filled with math majors fulfilling their computing requirement. Needless to say, much effort went into convincing each population that the course was relevant to them.

If the supporting courses really are so important, than why do these need to be separate tasks? Why not offer a single course intertwining programming and proof-based discrete mathematics, making their mutual relevancy explicit? The math content will motivate the math majors and make the programming more familiar and palatable. The programming content will sweeten the math pill for the computer science majors.

Moreover, bringing the two populations together will allow them to learn from and help each other. This text's premise is

Math majors should learn to write programs
and
computer science majors should learn to write proofs
together.

Further still, this text is appropriate for non-majors who are interested in a compact experience in both fields. Accordingly, liberal arts colleges will find this approach particularly useful for interdisciplinary cross-pollination.

Theory vs practice. Where do the foundations of computing belong in undergraduate curriculum? Some departments have faced battles between, on one hand, those wanting to use the foundations and theory of computer science as a gateway for incoming majors and, on the other, those who want to ensure students develop programming skills immediately.

With this text, you do not have to choose. While this is not explicitly a course on the theory of computation or other foundations of computer science, discrete mathematics provides the right context for previewing and introducing these ideas. Moreover, the programming content keeps the theory grounded in practice.

Coverage of functional programming. Despite the paradigm's importance, finding the right place for functional programming in an undergraduate curriculum has been tricky. From the 1980s through the early 2000s, several good texts were produced for a *functional-first* approach to teaching programming, but the approach never became dominant. Most students receive a brief exposure to functional programming in a sophomore-level programming languages course, if at all. As the undergraduate curriculum becomes crowded with new topics vying for attention and old, elective topic being pushed earlier (mobile computing and concurrency, for two examples), the programming languages course absorbs much of the pressure. The temptation to jettison functional programming altogether is strong.

This would be a great mistake. Many design patterns in object-oriented programming are imports of ideas from functional programming. Features from functional languages are appearing in languages not traditionally thought of as functional (Python, C#, and Java 8). And functional programming is a handy way to reason about questions in the foundations of computer science.

The solution proposed here is to put functional programming in the discrete math course. This way functional programming is seen early (ideally freshman or sophomore year) and, for computer science majors, is taught in parallel with a traditional introduction to programming course covering imperative, object-oriented, and concurrent programming.

Why ML? The ML programming language—to be specific, Standard ML or SML—is chosen as the language for this book, but the approach is not inherently tied to ML. In choosing a language, our first goal should be to do no harm—the language, especially its syntax, should not get in the way. ML was chosen for its simplicity, particularly its similarity to mathematical notation. Students used to mathematics should find the syntax natural. For students with prior programming experience, it is not much of a transition syntactically from, say, Java, either.

With a little effort, this text could be used with programming examples and exercises in F#, OCaml, or Haskell. With a little *more* effort, a Lisp dialect or even Python could be used.

Themes

A diverse set of topics fall under the discrete mathematics umbrella, but certain themes come up throughout the book.

Proof and program. This text strives to make skill in proof-writing and programming transferable from one to the other. The interplay between the two content streams shows how the same thinking patterns for writing a rigorous proof should be employed to write a correct and useful program.

Thinking recursively. Many students have found recursion to be one of the most mind-boggling parts of learning programming. The functional paradigm and the discrete math context make recursion much more natural. Recursive thinking in both algorithms and data structures is a stepping stone to proofs using structural induction and mathematical induction.

Formal definitions. Precision in proofs hangs on precision in the definitions of things the proof is about. Informal definitions are useful for building intuition, but this text also calls students' attention to the careful definitions of the terms.

Analysis and synthesis. *Analysis* is taking something apart. *Synthesis* is putting things together. Whether we are proving a proposition or writing a program, thinking through the problem often comes into two phases: breaking it down into its components and assembling them into a result. In a proof this manifests itself in a first section of the proof where the given information is dissected using formal definitions and a second section where the known information is assembled into the proposition being proven, again using formal definitions. (We call that the analytical and synthetic use of the definitions.) A similar pattern is seen in operations on lists—analytical operations that take lists apart and synthetic operations that construct them.

How to use this text

Audience. Ideally this text can be used by first-year undergraduates who are well-prepared mathematically—that is, able to reason abstractly and willing to think in new ways. Weaker students mathematically may be better served by taking such a course in their sophomore or junior years. The only hard prerequisites are high school algebra and pre-calculus. Occasionally the text uses examples from differential or integral calculus, but these can be skipped without harming the general flow the material.

The structure of each chapter. The names of the chapters are singular nouns: Set, Proof, Function, Graph, etc. This is to draw attention to each chapter's primary object of study. The opening sections of a chapter provide foundational definitions and examples, followed by the most important properties, propositions (and proofs) about the object of study, and finally applications of them, especially computational applications.

All sections except special topics or those providing a very general introduction end with a selection of exercises. Almost all chapters end with an extended programming example (including a project) and a special topic that condenses an advanced mathematical or computational idea to an accessible level based on the contents of the chapter.

The structure of the book. The chapters are collected into three parts. Set, List, Proposition, and Proof constitute the Foundations—ideas and skills for a student to master before seeing the other material which is built on them. The Core part of the text is the chapters on Relation, Self Reference, and Function. The Foundations and Core chapters build on each other in sequence.

The Elective chapters—Graph, Complexity Class, Lattice, Group, and Automaton—represent more advanced topics, building on both the foundations and the core. These chapters are almost completely independent of each other. The instructor may pick and choose from these and even reorder them with very little difficulty. A new recurring theme emerges in Graph, Lattice, and Group, that of *isomorphism*, structural equivalence between mathematical objects.

Pacing and planning. There is a wide range of difficulty in this material, and accordingly it can be adapted to courses at several levels. The Electives part naturally contains more advanced material, but particularly challenging sections pop up at various points in the book. These include Sections 4.9, 5.9, 6.7–6.10, 7.9, and 10.4. These (and any of the extended examples or special topics) can be omitted without harming the flow of the text.

For average to strong students in their freshman or sophomore year, most of Chapters 1–7 can be covered in one semester—the instructor is encouraged to choose some of the challenging or extended example sections to cover, but not all. The chapters in the Elective part can be used in a second semester, although there may be time at the end of a first semester to cover select topics from the Elective part.

The author has used this material for six years in a one-semester course with this approximate outline (mapping class days to sections, assuming a four-hour course meeting three days a week):

Day 1: 1.1–1.3
Day 2: 1.4–1.6
Day 3: 1.7–1.9
Day 4: 1.10–1.13
Day 5: 2.1–2.2
Day 6: 2.3–2.4
Day 7: 2.5–2.6
Day 8: 3.1–3.4
Day 9: 3.5–3.7
Day 10: 3.8–3.9
Day 11: 3.10–3.13
Day 12: 3.14
Day 13: Review
Day 14: Test
Day 15: 4.1–4.2
Day 16: 4.3–4.4
Day 17: 4.5–4.8
Day 18: 4.9
Day 19: 4.10–4.11
Day 20: 5.1–5.3
Day 21: 5.4
Day 22: 5.5
Day 23: 5.6–5.7
Day 24: 5.8–5.9
Day 25: Review; 4.12
Day 26: Test
Day 27: 6.1–6.2
Day 28: 6.4
Day 29: 6.5–6.6
Day 30: 6.9–6.10
Day 31: 6.10–6.11
Day 32: 6.12
Day 33: 7.1–7.3
Day 34: 7.4–7.5
Day 35: 7.6–7.8
Day 36: 7.9
Day 37: Review; 7.15
Day 38: Test

This schedule leaves about two weeks at the end of the semester for topics chosen from later chapters.

For a class of students who are less mathematically prepared, this text can be used in a more slowly paced course that covers Chapters 1–3 and roughly the first half of each of Chapters 4–7.

For a class of very experienced students, this text can be adapted to an advanced course where Chapters 1–4 are treated very quickly, leaving time for significant coverage of material from Chapters 8–12.

Any students unfamiliar with the citation system should note that numbers in square brackets refer to bibliography entries. Supplemental materials and other related resources can be found at **http://cs.wheaton.edu/~tvandrun/dmfp**.

Introduction (to the student)

So you have a book entitled *Discrete Mathematics and Functional Programming* on your shelf. That has to be a hard subject to explain to your roommates. Have they made any snide remarks about *indiscreet* math or *dysfunctional* programming?

First, what these words mean. Most of the mathematics taught in high school and college curricula revolve around preparing students for calculus, teaching them calculus, showing them applications of calculus, and teaching them to prove things, mostly about calculus. The backbone of all this is the real number line, which you have learned can be broken down into smaller and smaller segments, all of which are still filled with real numbers. There are no gaps between real numbers, no matter how close together, which are not filled with an infinite number of real numbers. Thus the mathematics of real-number land is continuous mathematics.

We love calculus, but there are other important branches of mathematics that get short changed, things like set theory, symbolic logic, number theory (the study of natural numbers and integers), graph theory, the study of algorithms with their analysis and correctness, matrices, sequences and recurrence relations, counting and combinatorics, probability, and languages and automata. Any of these could become a course of their own, but to make a survey of them accessible to all students, college math curricula have made courses out of a sampling of these topics and called the courses discrete mathematics. That term has two main explanations: These topics always have discrete—that is, separable, indivisible, or quantized—objects in view; and these topics stand in contrast to continuous mathematics. You might say the difference between discrete and continuous is like that between digital and analog.

Next time you are in the kitchen, look around for the discrete/continuous distinction. Some ingredients can be measured in amounts of any size: flour and milk are continuous. But eggs are discrete—they must be used in whole egg units. What kind of stove do you have? An electric stove can have discrete settings (in my house growing up, ours was push-button), but the burner knobs on a gas stove are continuous. You can even see it in how you serve dessert. In some families the server asks everyone how big a slice they want; in others the pie is sliced in equal portions and you take it or leave it. Do you consider pie slices to be continuous or discrete?

About the other half of the title: Computer science education almost always begins with programming. Two of the major brands ("paradigms") of programming are imperative programming, where a computer program is series of commands, and object-oriented programming, where a program is a collection of interacting objects. A typical introductory programming course will be a mixture of the two. But there is an alternative. In the functional paradigm, a program is built from a collection of functions, each of which is a little program in its own right. We will not venture into a debate on the relative merits of these paradigms. It is safer just to say that a serious programmer needs to be competent in all of them.

This text, *Discrete Mathematics and Functional Programming*, is an effort to bring together the material from discrete mathematics most useful to computer science and the programming material most accessible through mathematics. Learning these two content streams does not have to be separate tasks, but rather the insights and thought patterns you gain through writing mathematical proofs can transfer into those you need to write computer programs and vice versa. Uniting these two also illuminates the ideas that fall at their intersection.

The goal of this book is to train your mind. It is to teach you to reason at the level of precision necessary to write rigorous proofs and correct programs. It is to teach you to think recursively, that is, to reason about things with self-referential definitions. It is to teach you to think formally about abstract structure. It is to teach you to organize your thoughts when writing a proof or program into analysis (taking things apart) and synthesis (putting things together).

As you embark on this journey, make every effort not to be intimidated. The material contained here spans a wide range of difficulty. If some of what you see when flipping through seems well beyond your reach, do not think that for that reason you won't be able to get the foundational material. Some of the harder parts, especially in Chapters 8–12 are "elective" and aren't crucial for learning the main points of this study. On the other hand, if you learn the basics well and persist with hard work, you will be able to tame even the advanced parts.

On that note, a special word to self-directed students, reading this book outside of a course for which it is a required text (this goes also for students who, having taken a course covering the first half of the book, are working through the second half on their own). The chapters in the "Elective" part are dense. Their topics—graph theory, computational complexity, lattices and related structures, group theory, and theory of computation—each deserve a course in themselves. The presentation here is to get you acquainted with what these fields hold and help you decide if you want to pursue them further. I have not confined the presentation in those chapters only to the fields' most introductory material. In each chapter you will find some serious results of that field made accessible to readers with no more background than is covered by this book. They will take serious work to digest but will repay study.

Acknowledgments

The people who most deserve my gratitude in this project are my family—my wife Esther, our daughter Annika, and our son Isaac. Their support, encouragement, patience, and love made this work possible. The writing of this book is closely tied to our family's milestones. I churned out the first cut of the manuscript (for use in my own classes) during the summer Esther and I became engaged. The first major rewriting was done the summer when Annika was an infant—I particularly remember pecking out a few paragraphs at Annika's nap times during my three weeks as a stay-at-home dad. I worked on finishing the text and preparing it for publication during Isaac's "first two summers" (one summer on the inside, the other on the outside). Thanks, too, for the encouragement of my and Esther's extended families—especially for not ribbing me too much about how long it was taking.

Thanks to Jim Leisy and Tom Sumner at Franklin, Beedle who patiently shepherded a first-time author through the process of publishing a book. Thanks also to the reviewers for their valuable suggestions, not only improving the presentation but also catching technical errors. Any remaining mistakes are mine.

Two of my teachers played an indirect but important role in this work. The exploration of number sets in Section 1.1 is based on the presentation my high school math instructor, Dick Zylstra, would make on the first day of class each year. I also credit him for helping me find math interesting for the first time. During graduate school Jens Palsberg (then of Purdue, now of UCLA) taught me the role of proofs in computer science. Later as a rookie professor I had a conversation with him about prioritizing my time among various potential projects. He sensed my interest in writing this book and encouraged me to do it. His approval gave me confidence that I was not running my race in vain.

Thanks to all my department colleagues at Wheaton College (IL) for their encouragement and support. In particular Robert Brabenec, Cary Gray, Don Josephson, Steve Lovett, Terry Perciante, and Mary Vanderschoot each gave useful suggestions at various points along the way.

Much thanks is owed to the many students on whom this material was classroom-tested over the course of six years. Their feedback has been crucial to its improvement. The following either served as teaching assistants, read drafts, were grand champion mistake-finders, or proposed an important example or exercise: Jeremy Woods, Ben Ashworth, Chet Mancini, John Charles Bauschatz, Daniel Turkington, Andrew Wolfe, Sarah Cobb, Neile Havens, Kendall Park, Drew Hannay, Jacqui Carrington Goertz, Tim MacDonald, Ben Torell, and Amy Nussbaum.

The automata in Chapter 12 were rendered using the VAUCANSON-G package by Sylvain Lombardy and Jacques Sakarovitch.

I also greatly appreciate the following reviewers; they provided expert help and offered many suggestions for improvement:

Chuck Allison, Utah Valley State University
Robert Easton, California Polytechnic State University, San Luis Obispo
Lester McCann, University of Arizona
Brian Palmer, Depauw University
Marc Pomplun, University of Massachusetts, Boston
Michael Stiber, University of Washington
My T. Thai, University of Florida

Finally, thanks be to God for his grace and mercy. *Soli Deo Gloria*. Διὰ τῶν οἰκτιρμῶν τοῦ θεοῦ . . . τὴν λογικὴν λατρείαν ὑμῶν. “In light of the mercies of God. . . your rational service.” Rom 12:1.

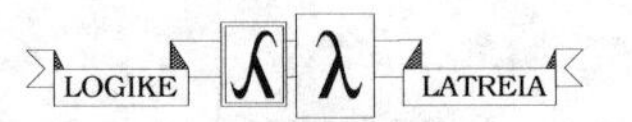

Part I

Foundations

Chapter 1 Set

Our starting point in our exploration of discrete mathematics is the *set*. This is not the only place where we could begin—some courses, for example, put *formal logic* first, since it is the most fundamental, simple system. Logic, however, is about sentences. We start with sets because it allows you to start by learning nouns. Sets also make for a familiar stepping-off point from continuous mathematics to discrete mathematics.

This chapter and the next form an important unit. In this chapter, in addition to set concepts you also will learn the basics of the ML programming language, and by the end of it you will be able to write simple programs yourself. The most important groundwork for ML, though, begins in the next chapter. Some of the more complicated set concepts are reserved until then as well.

Chapter goals. The student should be able to

- describe sets and relationships among them, both informally and using mathematical notation.
- evaluate operations on sets.
- perform simple arithmetic operations in ML.
- verify propositions about sets using Venn diagrams.
- compose types in ML using the datatype construct.
- compose simple functions in ML.

1.1 Your mathematical biography

You may be wondering what discrete mathematics is. To get there, we will retrace all the math you have learned until now. That might sound like a tall order for a first section, but we can expedite the process by highlighting some major events—say, every time you learned about a new kind of number.

Try to put yourself back to when you first realized what a number was. At some point you began to conceptualize differences in quantities (10 Cheerios is more than 3 Cheerios), and eventually you learned to count: 1, 2, 3, ... up to the highest number you knew. We call these numbers the *natural numbers*, and we will represent all of them with the symbol $\mathbb{N}$.

natural numbers

We are using this to symbolize *all* of them, not *each* of them, and that is an important distinction. $\mathbb{N}$ is not a variable that can stand for *any* natural number. It is a constant that stands for *all* natural numbers as a collection.

As time went on, you learned more natural numbers—for example, learning that 100 comes after 99. Since there are an infinite number of natural numbers, you can line up the natural numbers starting at one (which is essentially what you are doing when you count) and keep going forever and still stay within the realm of natural numbers. But it is not as *natural* to go the other direction. What comes before one? At some point early in grade school, you came to accept *zero* as a number, just a different kind of number. We will designate all natural numbers together with 0 to be *whole numbers*, and represent them collectively by $\mathbb{W}$.

whole numbers

0 is a whole number but not a natural number. 5 is both. In fact, anything that is a natural number is also a whole number, but only some things (specifically, everything but zero) that are whole numbers are also natural numbers. Thus $\mathbb{W}$ is a broader category than $\mathbb{N}$.

Tangent: What is the ability to count?

In this story, I have suggested that learning the names of new numbers constitutes expanding mathematical knowledge. Many two-year-olds, however, can recite numbers in order, even appearing to count objects, without understanding what the words mean.

Following an experiment described by Barbara Sarnecka and Susan Carey, I poured a pile of buttons in front of my two-year-old daughter, who can recite numbers up to ten very well. When I asked for one button, she placed one button in my palm. Similarly she was able to give me two buttons when asked. When I asked for three buttons, however, she again gave me two. When I told her that she had given me two although I asked for three, she took away the two she had given me and in place of them gave me two others. "Three" did not have a meaning to her different from "two."

Sarnecka and Carey distinguish "one"-knowers, "two"-knowers, "three"-knowers, and even "four"-knowers among toddlers, before the children can generalize the concept of named quantities. The transitions between understanding individual numbers may be months apart.[29]

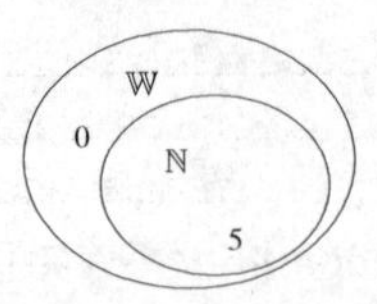

Of course in math class you did not merely meet new numbers. You also learned things we can do with numbers, like addition and subtraction.

Any time you added two whole numbers together, you got another whole number as a result. Subtraction, however, forced you to face a dilemma: what happens if you subtract a larger number from a smaller number?

The problem is, $\mathbb{W}$ is insufficient to answer this question. No whole number is the difference between 3 and 20; in technical terms, $\mathbb{W}$ is not *closed* under subtraction. To fix this problem, we invented negative numbers. We call all whole numbers together with their *opposites* (that is, their negative counterparts) *integers*, and we use $\mathbb{Z}$ (from *Zahlen*, the German word for "numbers") to symbolize the integers. Now we could do addition and subtraction on integers and stay within the realm of $\mathbb{Z}$. Multiplication worked fine, too.

integers

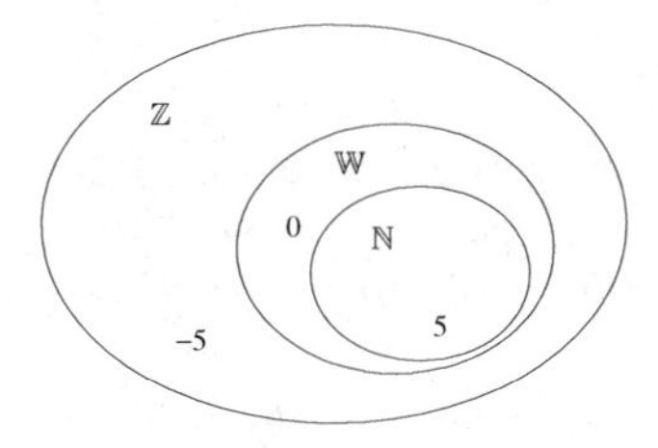

Division was a different story. What happens if we divide 5 by 2? (In the history of mathematical thinking, we can imagine two cavemen arguing over how to split five apples. Physically, they could chop one of the apples into two equal parts and each get one part, but how can you describe the resulting quantity that each caveman would get?)

Human languages handle this with words like "half." Mathematics handles this with fractions, like $\frac{5}{2}$ or the equivalent $2\frac{1}{2}$, which is shorthand for $2+\frac{1}{2}$. We call numbers that can be written as fractions (that is, ratios of integers) *rational numbers*, symbolized by $\mathbb{Q}$ (for *quotient*). Since a number like 5 can be written as $\frac{5}{1}$, all integers are rational numbers.

rational numbers

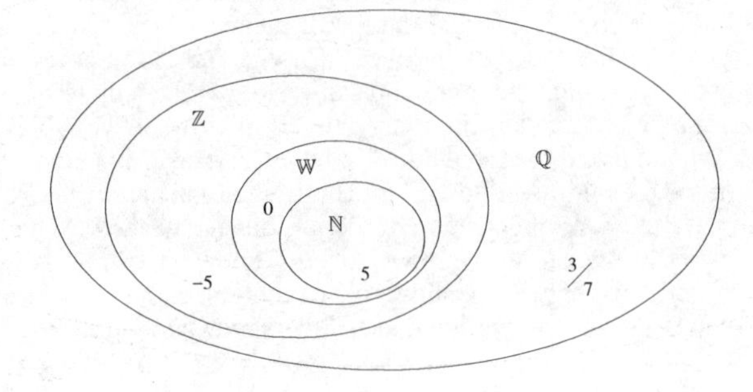

Just as whole numbers could not handle subtraction and integers could not handle division, in a similar fashion geometry exposed the insufficiency of rational numbers. For example, an isosceles right triangle with sides of length 1 inch has a hypotenuse of length $\sqrt{2}$ by the Pythagorean theorem. A circle with diameter 1 inch has a circumference of π.

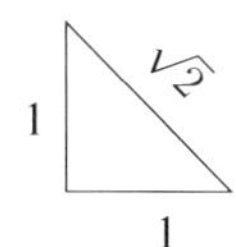

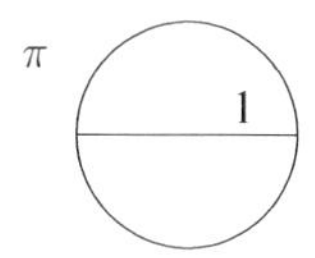

$\sqrt{2}$ and π cannot be written as fractions, and hence they are not rational numbers. However, these illustrations on paper prove that $\sqrt{2}$ and π are measurable quantities. We call all of these "possible real world quantities" *real numbers*, and symbolize them by $\mathbb{R}$.

real numbers

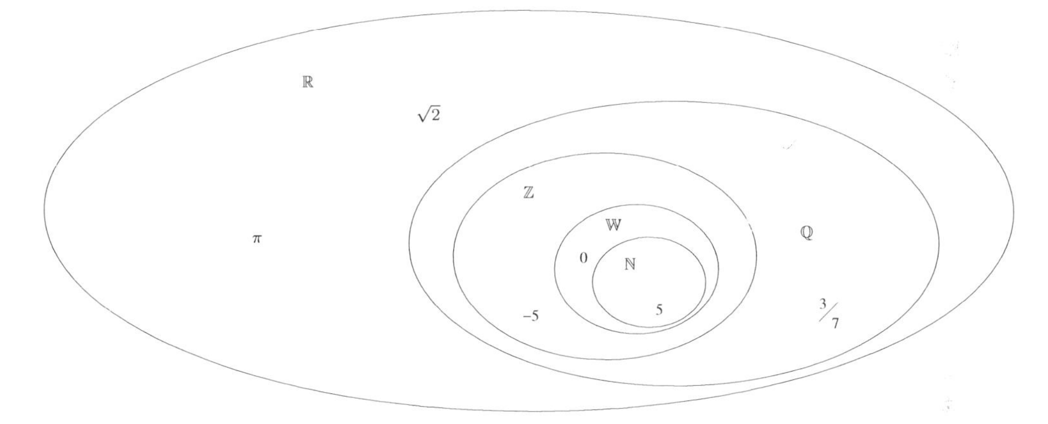

Here is a category you might not remember learning: real numbers can be split up into two camps—*algebraic numbers* ($\mathbb{A}$), each of which is a root to some polynomial function with integer coefficients ($\sqrt{2}$ and all the rationals are algebraic); and *transcendental numbers* ($\mathbb{T}$), which are not.

algebraic numbers

transcendental numbers

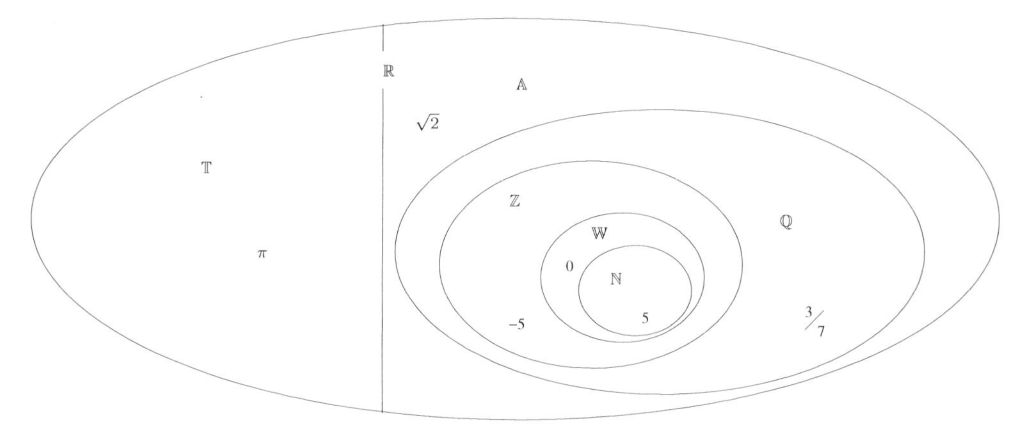

To play this game a little more, let us go back to talking about negative numbers. We first considered negative numbers when we invented the integers. However, as we expanded to rationals and reals, we introduced both new negative numbers and new positive numbers. Thus negative (real) numbers considered as a collection ($\mathbb{R}^-$) cut across all of these other collections, except $\mathbb{W}$ and $\mathbb{N}$.

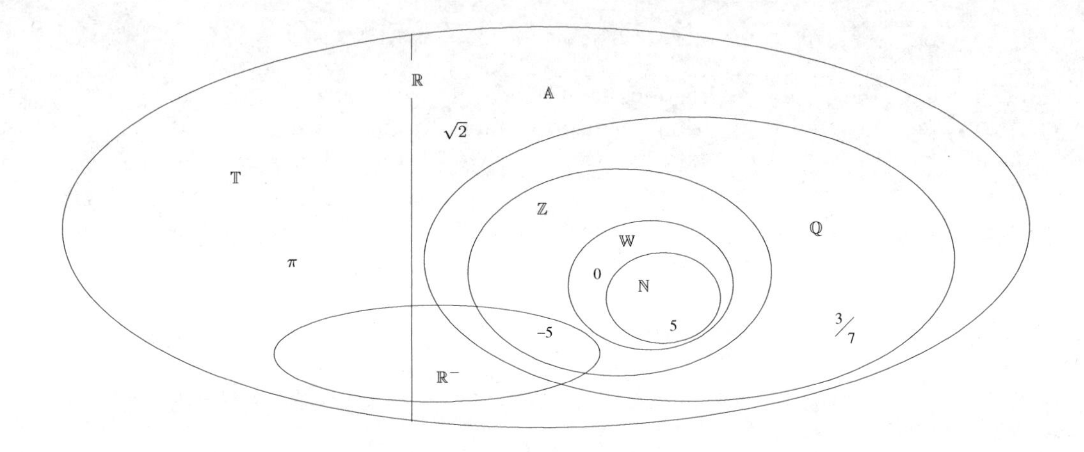

To finish off the picture, remember how $\mathbb{N}$, $\mathbb{Z}$, and $\mathbb{Q}$ each in turn proved to be inadequate because of operations we wished to perform on them. Likewise $\mathbb{R}$ is inadequate for operations like square root. What is $\sqrt{-1}$? Not a real number, at any rate. To handle that, we have *complex numbers*, $\mathbb{C}$.

complex numbers

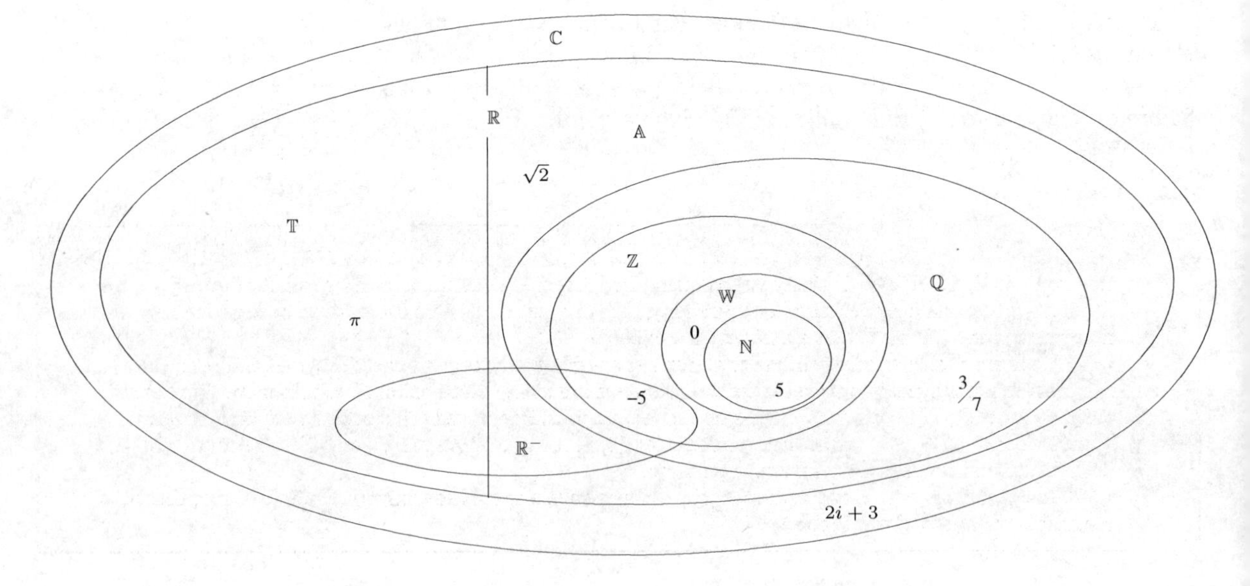

1.2 Reasoning about items collectively

These circle diagrams and symbols representing various collections of numbers, like all concepts and their representations, are tools for communicating ideas and reasoning about them. The following table lists various sentences that express ideas using the terminology introduced in the preceding exercise. In the right column we write these statements symbolically. We will introduce the symbols formally later.

5 is a natural number (*or* the collection of natural numbers contains 5).	$5 \in \mathbb{N}$
All integers are rational numbers.	$\mathbb{Z} \subseteq \mathbb{Q}$
Merging the algebraic numbers and the transcendental numbers makes the real numbers.	$\mathbb{R} = \mathbb{A} \cup \mathbb{T}$
Negative integers are both negative and integers.	$\mathbb{Z}^- = \mathbb{R}^- \cap \mathbb{Z}$
Transcendental numbers are those real numbers that are not algebraic numbers.	$\mathbb{T} = \mathbb{R} - \mathbb{A}$
Nothing is both transcendental and algebraic, *or* the collection of things both transcendental and algebraic is empty.	$\mathbb{T} \cap \mathbb{A} = \emptyset$
Adding 0 to the collection of natural numbers makes the collection of whole numbers.	$\mathbb{W} = \{0\} \cup \mathbb{N}$
Since all rational numbers are algebraic numbers and all algebraic numbers are real numbers, it follows that all rational numbers are real numbers.	$\mathbb{Q} \subseteq \mathbb{A}$ $\mathbb{A} \subseteq \mathbb{R}$ $\therefore\ \mathbb{Q} \subseteq \mathbb{R}$

Notation Note: A slight break from convention

The symbols $\mathbb{Z}$, $\mathbb{Q}$, $\mathbb{R}$, and $\mathbb{C}$ are conventionally used just as we use them in this book, for integers, rationals, reals, and complex numbers. Similarly, the way that we denote the positive or negative numbers in a set ($\mathbb{R}^+$ or $\mathbb{Q}^-$, as examples) is standard.

The terms whole number and natural number are not used as consistently, however. Some authors use them interchangeably or reverse the definition we are using. The symbol $\mathbb{N}$ usually stands for the set $\{0, 1, 2, 3, \ldots\}$ and $\mathbb{N}^*$ stands for the same set except with 0 removed. This book strictly uses natural numbers for $\{1, 2, 3, \ldots\}$ and whole numbers for $\{0, 1, 2, 3, \ldots\}$. We use $\mathbb{W}$ and $\mathbb{N}$ to be consistent with this, even though it differs from convention.

The symbols $\mathbb{A}$ and $\mathbb{T}$ are not standard. ($\mathbb{A}$ does have a special meaning in the field of ring theory, but that is well beyond our scope.)

In all these sentences, we are dealing with two different kinds of nouns: on one hand we have 5, $\frac{3}{7}$, $\sqrt{2}$, π, $2i + 3$, and the like; on the other hand we have things like $\mathbb{N}, \mathbb{Z}, \mathbb{Q}, \mathbb{R}$, and $\mathbb{C}$. We have been referring to the former by terms such as "number" or "item," but the standard mathematical term is *element*. We have called any of the latter category a "collection," but the mathematical term is *set*. Informally, a set is a collection of items categorized together because of perceived common properties.

element

set

This is the first shift in your thinking from *continuous* mathematics (such as pre-calculus and calculus) to *discrete* mathematics. Up till now, you have concerned yourself with the *contents* of these sets. In discrete mathematics, we will be reasoning about *sets themselves*.

There is nothing so special about the sets we mentioned earlier. We can declare sets arbitrarily—such as the set of even whole numbers, or simply the set containing only 1, 15, and 23. We can have sets of things other than numbers: a set of geometric points, a set of matrices, or a set of functions. But we are not limited to mathematical objects, either. Grammatically, anything that is a noun can be discussed in terms of sets.

Since *set* is a noun, we can even have a set of sets—for example, the set of number sets completely enclosed in $\mathbb{R}$, such as $\mathbb{Q}, \mathbb{Z}, \mathbb{W}$, and $\mathbb{N}$.

Nevertheless, sets themselves are abstract ideas. According to Hrbacek and Jech,

> Sets are not objects of the real world, like tables or stars; they are created by our mind, not by our hands. A heap of potatoes is not a set of potatoes, the set of all molecules in a drop of water is not the same object as that drop of water[17].

It is legitimate, though, to speak of the set of molecules in the drop and of the set of potatoes in the heap.

This course emphasizes precise definitions. Not only does mathematical rigor rely on them, but in all areas of study, correct use of vocabulary reflects full understanding of the concepts. However, one thing we will not define for you are the terms *set* and *element*, and that is because they are *primitive terms*. Primitive terms are the simplest building blocks we use in defining other things and are therefore undefinable themselves.

primitive terms

Think back to high school geometry. You met terms like *point*, *line*, and *plane* through their analogy in the physical word: a point is like a dot (a small mark such as made on paper) but infinitely small, a line is like a taut string but extending forever, and a plain is like a flat surface as on a table but, again, infinite. These are not definitions, though.

Instead, geometry is founded on a set of axioms (also called *postulates*) that describe the relationships among points, lines, and planes—for example, that exactly one line exists connecting two distinct points. They are the starting assumptions from which geometry is constructed.

Set theory is likewise grounded in a series of axioms. Here are two, to give a flavor of what these look like.

Axiom 1 (Existence.) *There is a set with no elements.*

Axiom 2 (Extensionality.) *If every element of a set X is an element of a set Y and every element of Y is an element of X, then $X = Y$.*

How do we know there is such a thing as an empty set? Because we say so. It is just a rule of the game.

empty set

The second axiom captures what we mean for two sets to be equal. Notice that this means that sets are unordered, or that the order in which we present the elements of a set is unimportant. We also can combine these two axioms to obtain a new result, that there exists only one empty set, because if we had two sets each with no elements, Axiom 2 implies they are in fact the same set.

A complete axiomatic formulation of set theory is intricate and beyond our scope. We will introduce primitive concepts as we go along as needed. The important point is that we start with presupposed propositions about undefined terms and derive other facts from these.

1.3 Set notation

We can describe a set explicitly by listing the elements of the set inside curly braces. Order does not matter in a set, so, for example, the set of primary colors is

$$\{\text{Red}, \text{Green}, \text{Blue}\} = \{\text{Green}, \text{Blue}, \text{Red}\} = \{\text{Blue}, \text{Red}, \text{Green}\} = \{\text{Blue}, \text{Green}, \text{Red}\}$$

When we use the phrase *set notation*, we mean this explicit listing of elements in curly braces.

It is possible for a set to contain only one item, for example $\{\text{Red}\}$, but we still distinguish that set from the item itself, that is $\text{Red} \neq \{\text{Red}\}$. Moreover $\{\}$ stands for a set with no elements, that is, the empty set, but we also have a special symbol for that, $\emptyset$.

The symbol $\in$ stands for set membership and should be read "an element of" or "is an element of", depending on the grammatical context (sometimes just "in" works if you are reading quickly).

Tangent: The nature of definition

What makes a good definition? Consider a typical definition from a dictionary:

definition • a statement expressing the essential nature of something. [20]

To define a noun, the dictionary-writers began with a more general noun (*statement*) and then narrowed it (*expressing the essential nature . . .*). In other words, to define a noun, we need to put a boundary around the set of things we can label with that noun (etymologically, *to define* means to set limits around). Compare this with $\mathbb{N} = \{x \in \mathbb{Z} \mid x > 0\}$: $\mathbb{Z}$ is the more general set, $x > 0$ is the qualifier.

$$\text{Red} \in \{\text{Green}, \text{Red}\}$$

The curly braces can be used more flexibly if you want to specify the elements of a set by property rather than listing them explicitly. Begin an expression like this by giving a variable to stand for an arbitrary element of the set being defined, a vertical bar (read as "such that"), a statement that the element is already a member in another set, and finally a statement that some other property is true for these elements. For example, one way to define the set of natural numbers is

$$\mathbb{N} = \{x \in \mathbb{Z} \mid x > 0\}$$

which reads "the set of natural numbers is the set of all x such that x is an integer and x is greater than 0." But there are many ways to describe or name sets. Recall from pre-calculus that you can specify a range on the real number line (say, all numbers starting from one, exclusive, to 5, inclusive) by the notation $(1, 5]$. A range is a set, the collections of numbers in that range.

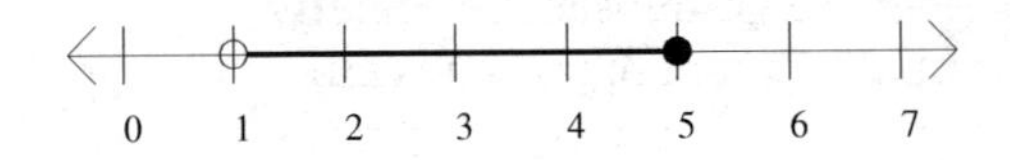

$$(1, 5] = \{x \in \mathbb{R} \mid 1 < x \leq 5\}$$

Exercises

Determine whether each statement is true or false.

1.3.1 $-12 \in \mathbb{N}$.

1.3.2 $-12 \in \mathbb{W}$.

1.3.3 $-12 \in \mathbb{Z}$.

1.3.4 $-12 \in \mathbb{Q}$.

1.3.5 $-12 \in \mathbb{R}$.

1.3.6 $\frac{1}{56} \in \mathbb{N}$.

1.3.7 $\frac{1}{56} \in \mathbb{W}$.

1.3.8 $\frac{1}{56} \in \mathbb{Z}$.

1.3.9 $\frac{1}{56} \in \mathbb{Q}$.

1.3.10 $\frac{1}{56} \in \mathbb{R}$.

Use set notation to denote the following sets (look up information as necessary).

1.3.11 The set of living species of the genus *canis*.

1.3.12 The set of United States Presidents from the Whig party.

1.3.13 The set of entrees at the dining hall tonight.

1.3.14 The set of pages in this book that the index refers to under "cardinality."

Describe each of the following sets using set notation, in terms of the other number sets (that is, do not use the given set itself).

1.3.15 $\mathbb{W}$.

1.3.16 $[12, 13.5)$.

1.4 Set operations

In elementary arithmetic, we have two kinds of operations. The first kind, including $+$, $-$, $\cdot$, and $\div$, all operate on numbers and produce another number. Their result can appear any place where a number can appear, including as the operand to another operation. Thus we have

$$5+3 \qquad\qquad 12-7 \qquad\qquad (18\cdot 13)\div 21$$

The other kind of operation also operates on numbers but produces something which grammatically can be considered a sentence. In other words, it does not produce another number, but a true or false value. $=$, $\leq$, $>$, $\geq$, $>$, and $\neq$ are all of this kind. The operator acts as the verb of the sentence.

$$5+3=8 \qquad\qquad 17>18\div 6 \qquad\qquad (15+4)\cdot 21\leq 3-2$$

The last of these happens to be false, but it still is a (grammatically) valid use of the operation.

Similarly we have operations on sets, both those that make sentences about sets and those that simply make new sets. Of the sentence-making variety, there are two that we make frequent use of, $=$ and $\subseteq$. If one set is completely contained in another, that is, every element of the one set is also an element of the other, then we say that the first is a *subset* of the second and use the symbol $\subseteq$.

subset

Suppose A and B are sets. (As with other mathematical objects, we can use variables to stand for sets. By convention, we use capitals.) Picture the subset relation using a diagram like those from Section 1.1.

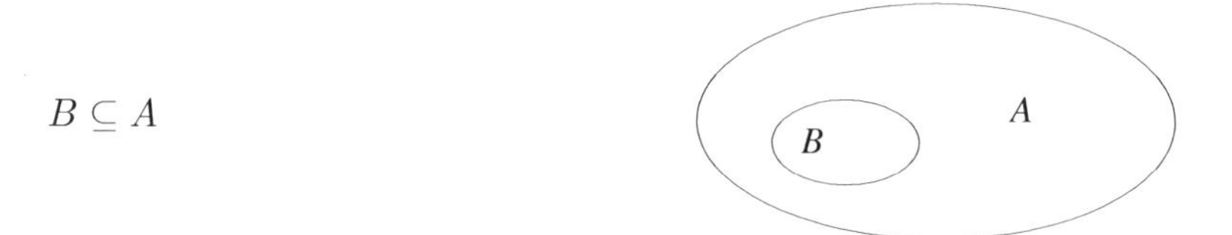

We already declared what it means for two sets to be equal in Axiom 2. Notice that $A=B$ means the same thing as $A\subseteq B$ and $B\subseteq A$ taken together, and this also implies that for any set A, $A\subseteq A$. If we mean $B\subseteq A$ and $B\neq A$—in other words, A has everything in B, but there is at least one element in A that is not in B—then we say that B is a *proper subset* of A and write $B\subset A$. This is rarely an important observation.

proper subset

In these cases we also call A a *superset* (or, if appropriate, a *proper superset*) of B and write $A\supseteq B$ or $A\supset B$, but we will not use these symbols very often, either. Thus the main sentence-making set operations we are interested in are $\subseteq$ and $=$.

superset

There are three basic operations to compute a new set given two other sets, just as we have arithmetic operations for computing numbers from two other numbers: *union* ($\cup$), *intersection* ($\cap$), and *difference* ($-$). Informally, the union of two sets, $A \cup B$, is the set containing all elements of A and all elements of B. The intersection of two sets, $A \cap B$, is the set containing only those elements that are both in A and in B. The difference of two sets, $A - B$, is the set of elements in A that are not elements of B.

Union

The set of elements that are in ***either*** set.

$$A \cup B = \{ x \mid x \in A \text{ or } x \in B \}$$

$$\begin{aligned} \{1,2,3\} \cup \{2,3,4\} &= \{1,2,3,4\} \\ \{1,2\} \cup \{3,4\} &= \{1,2,3,4\} \\ \{1,2\} \cup \{1,2,3\} &= \{1,2,3\} \end{aligned}$$

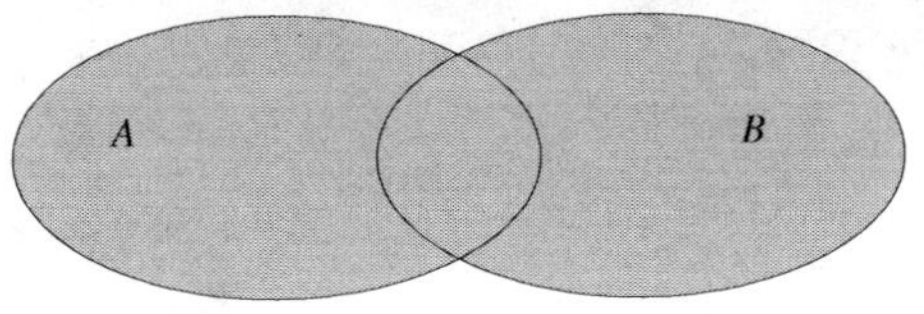

Intersection

The set of elements that are in ***both*** sets.

$$A \cap B = \{ x \mid x \in A \text{ and } x \in B \}$$

$$\begin{aligned} \{1,2,3\} \cap \{2,3,4\} &= \{2,3\} \\ \{1,2\} \cap \{3,4\} &= \emptyset \\ \{1,2\} \cap \{1,2,3\} &= \{1,2\} \end{aligned}$$

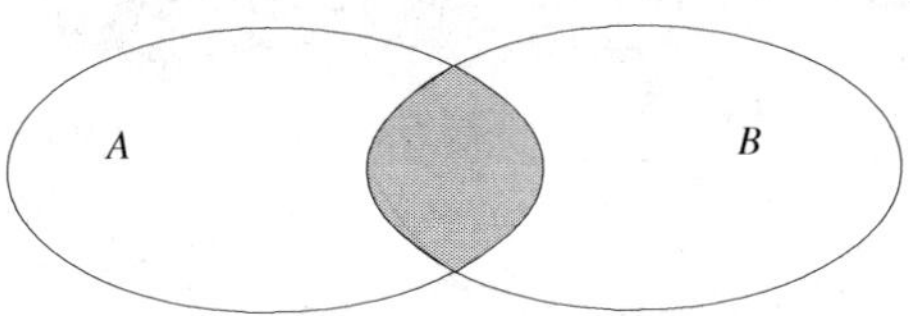

Difference

The set of elements that are in the ***first*** set but ***not*** in the ***second***.

$$A - B = \{ x \mid x \in A \text{ and } x \notin B \}$$

$$\begin{aligned} \{1,2,3\} - \{2,3,4\} &= \{1\} \\ \{1,2\} - \{3,4\} &= \{1,2\} \\ \{1,2\} - \{1,2,3\} &= \emptyset \end{aligned}$$

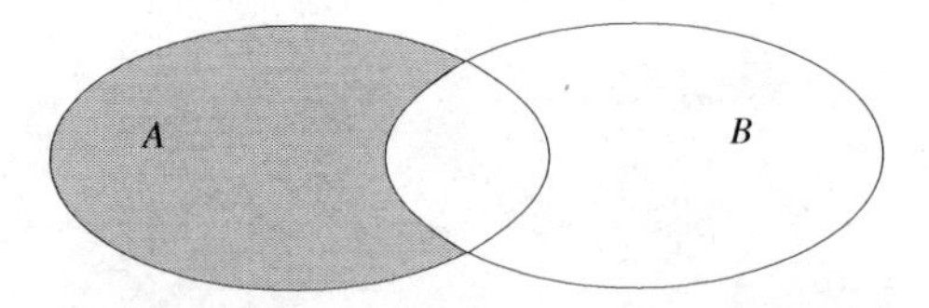

It is crucial to understand that these are not sentence-making operations but set-making operations. A beginning student might take $A \cap B$ to mean "sets A and B overlap at some point." $A \cap B$ is a noun phrase that describes the set that is the overlap of A and B, and it may very well be empty. To say that A and B overlap, write $A \cap B \neq \emptyset$.

Venn diagrams

These diagrams with overlapping ovals that we have been using are called *Venn diagrams*, named after English logician John Venn.

universal set

When sets are used to describe ideas, the context usually assumes a *universal set*, a set from which all elements under discussion come. For example, if the sets being discussed are $\{1,3,5\}$ and $\{2,3,4,7,10\}$, the universal set can be assumed to be one of the number sets, perhaps $\mathbb{Z}$ or $\mathbb{W}$. If the sets are things like $\{a,b,m,q\}$, then the universal set may be the set of letters in the alphabet. If we are talking about the set of students registered for one course or another in a given semester, the universal set is the set of students enrolled at the school.

Generically we use $\mathcal{U}$ to stand for the universal set, and it appears in a Venn diagram as a box framing the rest of the picture.

complement

The universal set concept allows us to define the *complement* of a set, which is the set of everything that is (in the universal set but) not in the set. Formally,

$$\overline{X} = \{x \in \mathcal{U} \mid x \notin X\}$$

One could also say $\overline{X} = \mathcal{U} - X$. Some authors use X' instead of $\overline{X}$ to denote complement.

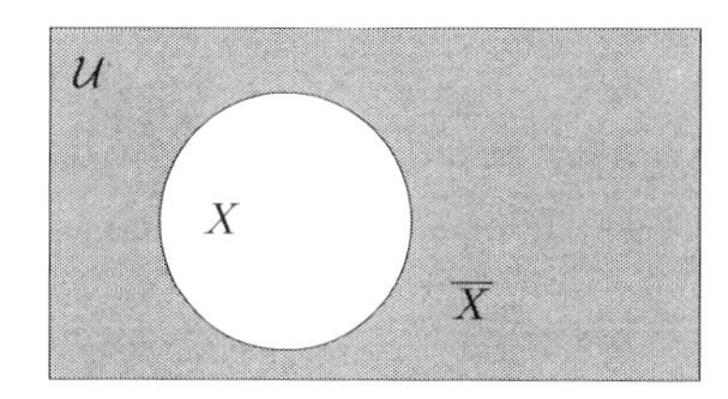

Complement is a set operation just like union, intersection, and difference, except that it is a unary (one-operand) operation, not a binary (two-operand) operation. This is just how a negative sign takes only one operand, whereas a minus sign takes two. As with arithmetic operations, these can be combined. The shaded area below is $\overline{X - (Y \cup Z)}$.

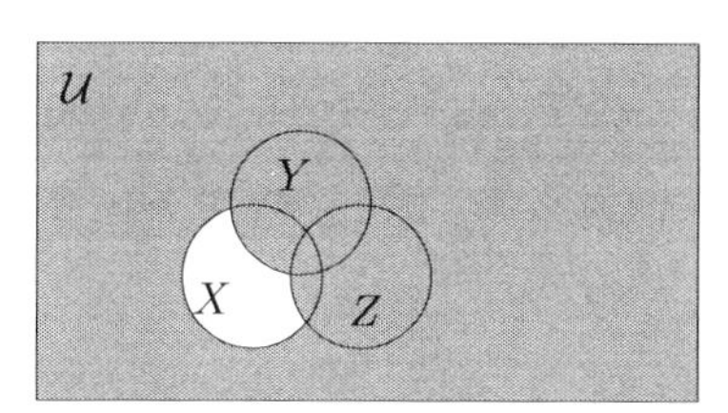

Biography: John Venn, 1834–1923

John Venn was born in Hull, England. His mother died when he was very young, and his father was a prominent clergyman. Venn followed the family profession of ordained ministry, being educated at Caius College of Cambridge University, where he also spent his career. During his life he was known most notably for his work on moral philosophy and logic. He also worked with his son to compile historical data on Caius College and Cambridge University as a whole. The diagrams that bear his name were introduced in an 1880 paper, although they are similar to diagrams used earlier by Leonhard Euler and others. They are important tools in any field that makes use of formal logic.[36]

Exercises

Let T be the set of trees, B be the set of broadleaf trees, and C be the set of coniferous trees. In Exercises 1.4.1–1.4.6, write each sentence symbolically.

1.4.1 Oak is a broadleaf tree.

1.4.2 Pine is not a broadleaf tree.

1.4.3 All coniferous trees are trees.

1.4.4 Broadleaf trees are those that are trees but are not coniferous.

1.4.5 Broadleaf trees and coniferous trees together make all trees.

1.4.6 There is no tree that is both broadleaf and coniferous.

In Exercises 1.4.7–1.4.10, let A, B, and C be sets, subsets of the universal set $\mathcal{U}$. Illustrate the following with Venn diagrams (do not draw C in the cases where it is not used).

1.4.7 $(A \cap B) - A$.

1.4.8 $(A - B) \cup (B - A)$.

1.4.9 $(A \cup B) \cap (A \cup C)$.

1.4.10 $\overline{(A \cap B)} \cap (A \cup C)$.

In Exercises 1.4.11–1.4.18, determine whether each statement is true or false.

1.4.11 $-12 \in \mathbb{R}^-$.

1.4.12 $\mathbb{A} \subseteq \mathbb{C}$.

1.4.13 $\mathbb{R} \subseteq \mathbb{C} \cap \mathbb{R}^-$

1.4.14 $4 \in \mathbb{C}$.

1.4.15 $\mathbb{Q} \cap \mathbb{T} = \emptyset$.

1.4.16 $\frac{1}{63} \in \mathbb{Q} - \mathbb{R}$.

1.4.17 $\mathbb{Z} - \mathbb{R}^- = \mathbb{W}$.

1.4.18 $\mathbb{T} \cup \mathbb{Z} \subseteq \mathbb{A}$.

1.4.19 All of the labeled sets we considered in Section 1.1 have an infinite number of elements, even though some are completely contained in others. (We will later consider whether all infinities should be considered equal.) However, two regions have a finite number of elements.

a. Describe (both verbally and using the notation in this chapter) the region shaded ▇. How many elements does it have?

b. Describe the region shaded ▢. How many elements does it have?

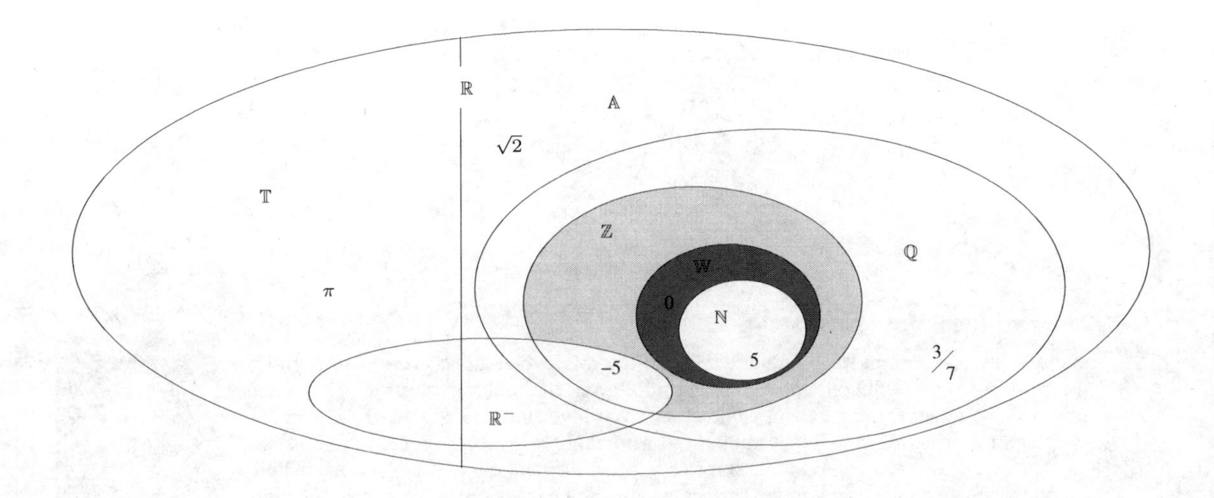

1.5 Verifying facts about sets

Let $A = \{1, 2, 3\}$, $B = \{3, 4, 5\}$, and $C = \{5, 6, 7\}$. Notice

$$\begin{aligned} A \cup (B \cap C) &= \{1,2,3\} \cup (\{3,4,5\} \cap \{5,6,7\}) \\ &= \{1,2,3\} \cup \{5\} \\ &= \{1,2,3,5\} \end{aligned}$$

and

$$\begin{aligned} (A \cup B) \cap (A \cup C) &= (\{1,2,3\} \cup \{3,4,5\}) \cap (\{1,2,3\} \cup \{5,6,7\}) \\ &= \{1,2,3,4,5\} \cap \{1,2,3,5,6,7\} \\ &= \{1,2,3,5\} \end{aligned}$$

In other words,

$$A \cup (B \cap C) = (A \cup B) \cap (A \cup C)$$

Since there does not seem to be anything special about how we chose A, B, and C, we might suspect that this formula holds true for any three sets. It turns out that it is, and this is called the *distributive law*. Notice how similar it is to the distributive law you learned in grade school: $x \cdot (y + z) = x \cdot y + x \cdot z$ for $x, y, z \in \mathbb{R}$. The union operation is analogous—in this context, at least—to multiplication and the intersection operation to addition. Much of higher mathematics is about finding patterns that recur in ostensibly dissimilar domains.

Later in this course we will prove this and similar propositions about sets. Informally we can demonstrate this using a series of Venn diagrams. First we draw a diagram with three regions.

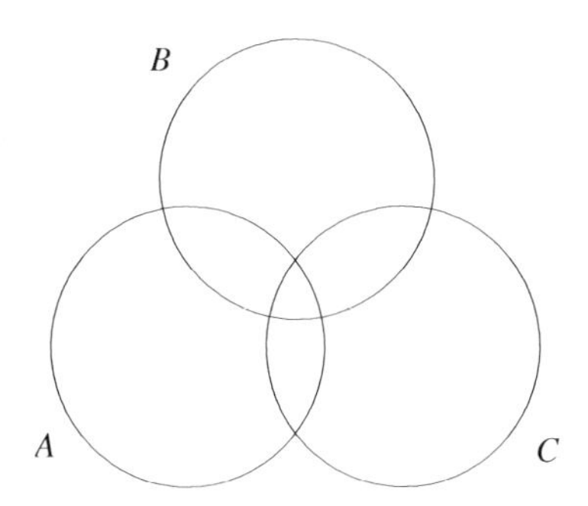

Then we shade it according to the left side of the equation and, separately, according to the right and compare the two drawings. First, shade A with ▒ and $B \cap C$ with ▓.

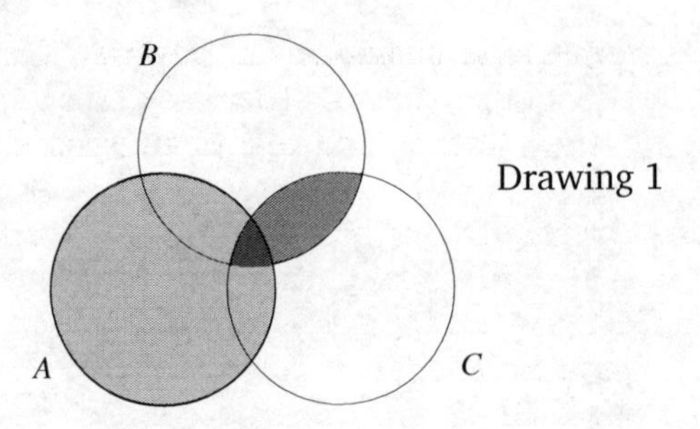

Drawing 1

Note that $A \cap (B \cap C)$ has the darkest tint, indicating that it has been shaded twice. With pencil and paper, you can shade one region with lines going in one direction and another region with lines in the other direction. The intersection of those two regions will then have a criss-cross pattern. The double-shaded region happens not to be important in this picture, though. We are interested in the union operation, indicated by all the shaded regions together.

Now, in separate pictures, shade $A \cup B$ with and $A \cup C$ with .

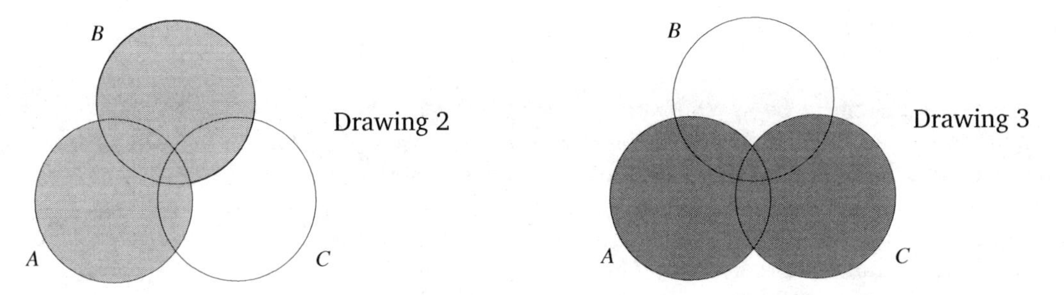

Drawing 2

Drawing 3

Put these two on top of each other, and the intersection of these two sets is the region that is double-shaded, .

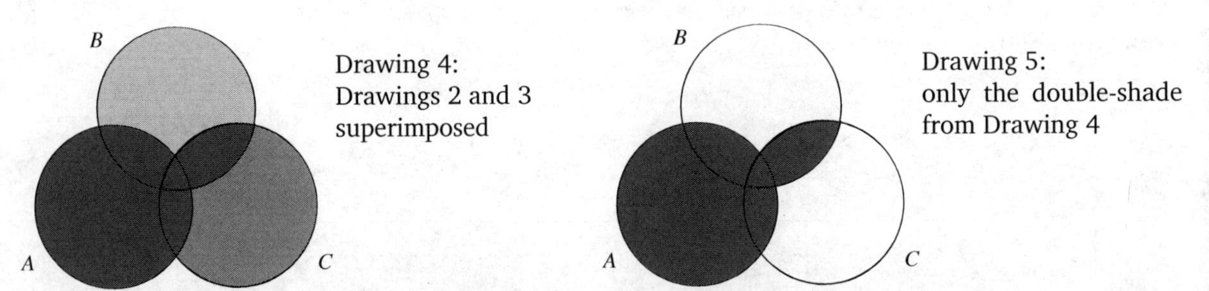

Drawing 4: Drawings 2 and 3 superimposed

Drawing 5: only the double-shade from Drawing 4

Since the total shaded region in the first picture is the same as the double-shaded region in the combined picture, we have verified the proposition. Graphics, with a little narration to help, can be used for an informal, intuitive proof.

As a second example, consider the claim

$$\overline{A} \cup B = \overline{A - B}$$

Suppose a farmer has some cattle named Alvin, Beverley, Camus, Daisy, Eddie, and Gladys. Let A be the set that are cows (Beverley, Daisy, and Gladys), the rest ($\overline{A}$) bulls. Alvin, Beverley, Camus, and Gladys are spotted; designate them as set B.

The cattle that are bulls or spotted (or both) are Alvin, Beverley, Camus, Eddie, and Gladys; that is $\overline{A} \cup B$. The cattle that are cows but not spotted ($A - B$) are just Daisy. The rest ($\overline{A - B}$) are Alvin, Beverley, Camus, Eddie, and Gladys. In this example, it seems to work.

Now we will draw it. This time we must include the universal set in the drawing.

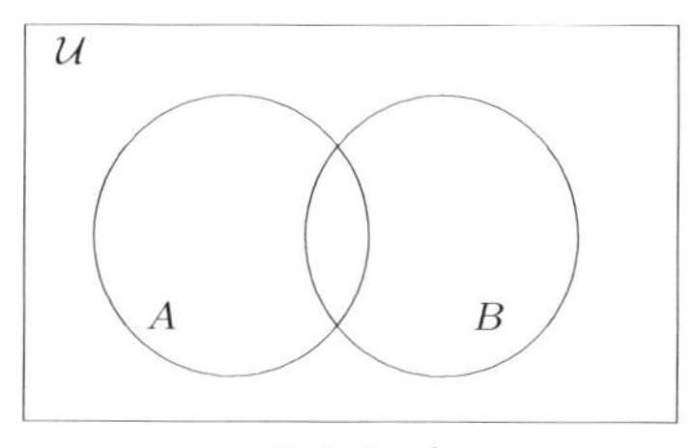

Original

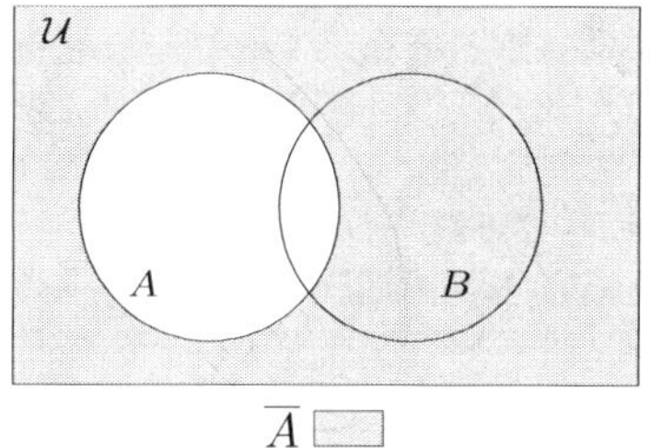

$\overline{A}$

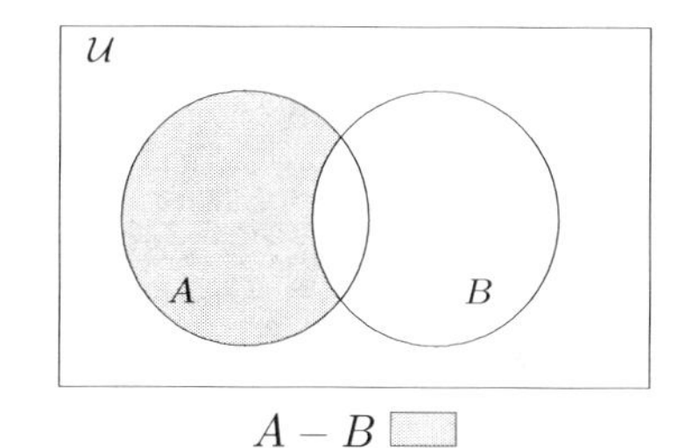

$A - B$

B

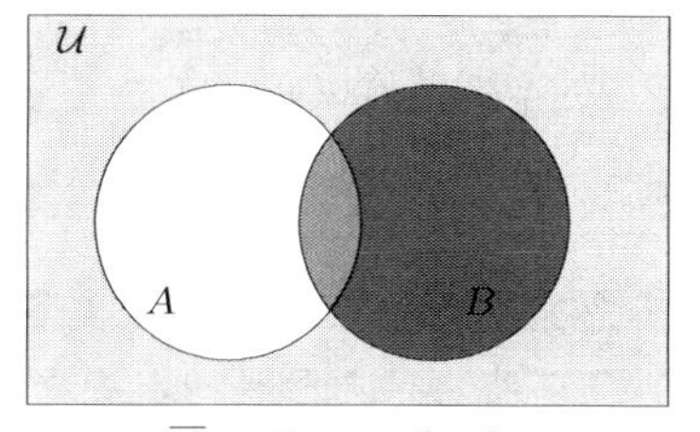

$\overline{A} \cup B$ any shade

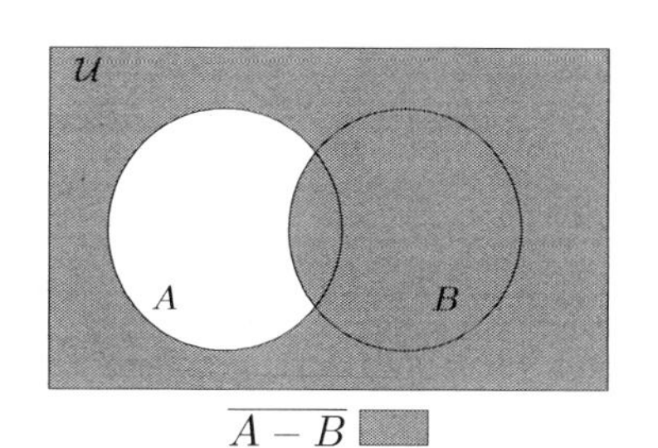

$\overline{A - B}$

Exercises

Verify using Venn diagrams (with verbal explanations as necessary).

1.5.1 $A \cup (A \cap B) = A$.

1.5.2 $A \cup \overline{A} = \mathcal{U}$.

1.5.3 $A \cup (B \cup C) = (A \cup B) \cup C$.

1.5.4 $A \cap (B \cup C) = (A \cap B) \cup (A \cap C)$.

1.5.5 $A \cap B = A - (A - B)$.

1.5.6 $(A \cap C) - (C - B) = A \cap B \cap C$.

1.5.7 $A \cup (A - B) = A$.

1.5.8 $A \cup (B - A) = A \cup B$

1.5.9 $(A - B) \cap (A \cap B) = \emptyset$.

1.5.10 $A \cap (B - (A \cap B)) = \emptyset$.

1.5.11 $\overline{(A \cup B)} = \overline{A} \cap \overline{B}$.

1.5.12 $(A \cup (B - C)) \cap \overline{B} = A - B$.

1.5.13 $(B \cup C) - A = (B - A) \cup (C - A)$

1.6 Values, expressions, and types in ML

This text teaches discrete mathematics and functional programming together for two reasons: so that you can transfer skills for writing proofs to writing programs and vice versa, and so that we can model mathematical concepts on a computer. The programming language ML is chosen for its usefulness towards these goals. The discrete math topics and the programming topics are as tightly woven as possible, but we must lay out the basic threads of each topic stream first.

Now and then we will make some comments that are significant only to students who have previous programming experience in a language other than ML. However, no such experience is necessary to acquire ML proficiency in this study.

Although ML can be used to build stand-alone applications, we will assume that the ML interpreter is running in *interactive mode*. This means the ML interpreter has been launched without a program to run, and the ML interpreter itself gives a prompt to the user, who is also the programmer. The user enters ML code, which the interpreter executes. Thus we have a work-cycle of the user-programmer entering an expression or declaration (we will define those terms below), the interpreter evaluating it, and the interpreter displaying the result.

Biography: Robin Milner, 1934–2010

Robin Milner was born near Plymouth, England, and was educated at Eton School and the University of Cambridge. After his first experience with computer programming, he decided that programming was "inelegant" and that he "would never go near a computer" again. He later changed his mind, or he changed programming.

Milner originally developed the programming language ML as a metalanguage for his automatic theorem-proving system Logic for Computable Functions. The language proved useful enough to take on a life of its own. This and his work on concurrency led to his winning the 1991 Turing Award. He died in March 2010.[17]

It is crucial that you acquire the basic vocabulary of ML programming. The term *value* is primitive, so we will not define it, but, informally, values model mathematical ideas. Examples of values include 5, 16, and 7.5. One of the beauties of our approach to programming is that we will not be working with just numbers all the time, so we will see values of all sorts of things. However, numbers make the easiest examples to start with.

value

An *expression* is any construct in the programming language that the interpreter can evaluate to a value. On one hand, a value is the result of an expression; alternately, values themselves are trivial expressions if used in the right context. A value used as an expression (or, the simplest expression which evaluates to a certain value) is called a *literal*.

expression

literal

That basic form used when interacting with the ML interpreter is

$$\langle\text{expression}\rangle\ ;$$

Try entering a literal. Text that the user types into the prompt will be in `typewriter font`; ML's response will be in *`slanted typewriter font`*. The - at the beginning of the line is the prompt, not part of what the user enters.

```
- 5;

val it = 5 : int
```

Here is what ML's response means: `val` is short for value and indicates that the result value, not surprisingly, is `5`. `it` is a variable, and `int` is the type of the value—we will discuss all these later.

We can make more interesting expressions using mathematical operators. We can enter

```
- 7 - 2;

val it = 5 : int
```

Note that this expression itself contains two other expressions, 7 and 2. Smaller expressions that compose a larger expression are called *subexpressions* of that expression. - is an *operator*, and the subexpressions are the *operands* of that operator. + means what you would expect, * stands for multiplication, and ~ (tilde) is used as a negative sign (having one operand, to distinguish it from -, which has two); division we will discuss later. To express (and calculate) $67 + 4 \times -13$, enter

subexpression

operator

operand

```
- 67 + 4 * ~ 13;

val it = 15 : int
```

Variables in ML work just as they do in mathematics (though subtly different from most programming languages). The variable `it` is where ML stores its latest result, unless you tell it otherwise. We can use a variable as an expression.

```
- it + 3;

val it = 18 : int
```

To use a name other than `it`, imitate the interpreter's response using `val`, the desired variable, and equals, something in the form of

val <identifier> = <expression>;

```
- val x = 5;

val x = 5 : int
```

identifier

An *identifier* is a programmer-given name, such as a variable. ML has the following rules for valid identifiers:

1. The first character must be a letter. (Certain kinds of identifiers begin with an apostrophe followed by a letter.)
2. Subsequent characters must be letters, digits, underscores, or apostrophes.
3. Identifiers are case-sensitive.

keywords

Some character sequences that fit the criteria above already have meaning in the language, words like `val` and `int`. These are called *keywords*, and they may not be used as identifiers.

It is convention to use mainly lowercase letters in variables. If you use several words joined together to make a variable, capitalize the first letter of the subsequent words.

```
- val minutesInHour = 60;

val minutesInHour = 60 : int

- val hoursInDay = 24;

val hoursInDay = 24 : int

- val daysInYear = 365;

val daysInYear = 365 : int

- val minutesInYear =
=     minutesInHour * hoursInDay * daysInYear;
```

```
val minutesInYear = 525600 : int

- val minutesInCentury = minutesInYear * 100;

val minutesInCentury = 52560000 : int
```

So far, all these values have had the type int. A *type* is a set of values that are associated because of the operations defined for them. Types also categorize values by how they are stored in computer memory, but that will not be a concern for us. We will use sans serif font for types. Understanding types is crucial for success in programming.

type

Types model the mathematical concept of a set, but imperfectly. ML does not provide a way to use the concepts of subsets, unions, or intersections on types. We will later study other ways to model sets to support these concepts. Moreover, the type int is not the same thing as the set $\mathbb{Z}$, although it corresponds to that set in terms of how we interpret it. The values (elements) of int are computer representations of integers, not the integers themselves, and since computer memory is limited, int comprises only a finite number of values. On the computer used to write this book, the largest integer ML recognizes is 1073741823. Although $1,073,741,824 \in \mathbb{Z}$, it is not a valid ML int.

```
- 1073741824;

Error: int constant too large
```

ML also has a type real corresponding to $\mathbb{R}$. We will not use real much, but it is a good example of a second type. The operators you have already seen are also defined for reals, plus / for division

```
- ~4.73;

val it = ~4.73 : real

- 7.1 - 4.8 / 63.2;

val it = 7.02405063291 : real
```

Tangent: Comprise

Comprise is one of those words commonly used in ways that annoy English purists. H. W. Fowler says, "The common use of *comprise* as a synonym of *compose* or *constitute* is a wanton and indefensible weakening of our vocabulary" [10]. Strunk and White say, "A zoo comprises mammals, reptiles, and birds. But animals do not comprise a zoo—they constitute a zoo"[32].

If you care about such things, then just remember, a set comprises its elements, and elements constitute a set.

The difference between the type real and the set $\mathbb{R}$ is even more striking than the difference between int and $\mathbb{Z}$. Since real is finite, it not only leaves out numbers too big or too negative, but also an infinite number of numbers along the way. For example, ML has a pre-defined value for π.

```
- Math.pi;

val it = 3.14159265359 : real
```

That value is certainly not an exact representation of π. Instead, it is the value of type real that is closer than any other real to the exact value of π.

For another limitation of using int and real to represent $\mathbb{Z}$ and $\mathbb{R}$, observe this result:

```
- 5.3 - 0.3;

val it = 5.0 : real
```

`5.0` has type real, not type int. int is not a subset (or subtype) of real, and `5.0` is a completely different value from `5`. It is better to think of int and real as being separate universal sets, rather than subsets of the universal set of all numbers.

type agreement

A consequence of int and real being unrelated is that you cannot mix them in arithmetic expressions. English requires that the subject of a sentence have the same number (singular or plural) as the main verb, which is why it does not allow a sentence like, "Two dogs walks down the street." This is called subject-verb agreement. In the same way, these ML operators require *type agreement*. That is, +, for example, is defined for adding two reals and for adding two ints, but not one of each. Attempting to mix them will generate an error.

```
- 7.3 + 5;

Error: operator and operand don't agree [literal]
  operator domain: real * real
  operand:         real * int
  in expression:   7.3 + 5
```

This rule guarantees that the result of an arithmetic operation will have the same type as the operands. It also complicates the division operation on ints. We expect that $5 \div 4 = 1.25$—as we noted in Section 1.1, division takes us out of the circle of integers. Actually, the / operator is not defined for ints at all.

```
- 5/4;

Error: overloaded variable not defined at type
  symbol: /      type: int
```

You will have to revert to fourth grade to understand ML's view of division on integers. Instead of producing one (real number) result ($7 \div 3 = 2.\bar{3}$), division produces two (integer) results, the quotient and the remainder ($7 \div 3 = 2$ R 1). ML computes the quotient—that is, with the result you would otherwise expect rounded down to the nearest integer, or with the remainder discarded—with the operator `div`. This is called *integer division*. The remainder is calculated by the modulus or *mod operator*, `mod`.

integer division

mod operator

```
- 5 div 3;

val it = 1 : int

- 5 mod 3;

val it = 2 : int
```

If there is need to mix reals and ints in the same computation, the following can be used to convert between the types.

Converter	converts from	to	by
`real()`	int	real	appending a 0 decimal portion
`round()`	real	int	conventional rounding
`floor()`	real	int	rounding down
`ceil()`	real	int	rounding up
`trunc()`	real	int	throwing away the decimal portion

For example,

```
- 15.3 / real(6);

val it = 2.55 : real

- trunc(15.3) div 6;

val it = 2 : int
```

Not everything in this section is crucial for you to remember. In fact, we will not use real very often in this text, and you will be hard pressed to find any occurrence of the `trunc()` converter. The point here is to understand how ML, like almost any programming language, categorizes values into types, and how that affects the way we use them. In the next sections we will learn other kinds of types and learn to create our own.

Now, without looking back, what do the terms *value, expression, literal, identifier*, and *type* mean? **Make sure you both understand what they are and can articulate a precise definition.** Do not leave this section until they are mastered.

Exercises

1.6.1 Is `ceil(15.2)` an expression? If not, why not? If so, what is its value and what is its type? Would ML accept `ceil(15)`? Why or why not?

1.6.2 What is the difference between `trunc()` and `floor()`? Give an example of an actual parameter for which they would produce different results.

1.6.3 Which of the following are valid ML identifiers?

(a) `hello`

(b) `points_scored`

(c) `points'scored`

(d) `pointsScored`

(e) `vElOcItY`

(f) `_velocity`

(g) `velocity12`

(h) `12velocity`

1.6.4 Compute the number of seconds in a year, similar to how we computed the number of minutes. You may assume 365 days in a year.

1.6.5 Redo the computation of the number of seconds in a year, taking into consideration that there are actually 365.25 days in a year, and so the variable `daysInYear` should be of type real. Your final answer should still be an int.

1.6.6 Mercury orbits the sun in 87.969 days. Calculate how old you will be in 30 Mercury-years. Your answer should depend on your birthday (that is, do not simply add a number of years to your age), but it should be an int.

1.6.7 Use ML to compute the circumference and area of a circle and the volume and surface area of a sphere, first each of radius 12, then of radius 12.75.

1.6.8 Store the values 4.5 and 6.7 in variables standing for base and height, respectively, of a rectangle, and use the variables to calculate the area. Then do the same but assuming they are the base and height of a triangle.

1.6.9 You have a distance $18\frac{1}{2}$ long that you wish to cover with tiles $\frac{5}{16}$ of an inch long. You cannot break the tiles; they must be used in whole. Use ML to determine how many tiles will fit and how much length will be left over.

1.7 Characters and strings

Processing text has always been an important domain of computer applications. Consider how much computer usage consists of using word processors, not to mention email, blogs, and Twitter. Our first examples of types other than number types are those used in manipulating text.

character

A *character* is a display symbol—a letter, a digit, a punctuation mark, etc. ML represents these as values using the char type. A char literal is made by enclosing the character in quotation marks and prepending a pound sign.

```
- #"a";

val it = #"a" : char
```

Some characters are hard to make. For example, the quotation mark symbol would confuse the interpreter because it would not differentiate between the intended char value and the opening and closing quotes.

```
- #""";
```

```
character constant not length 1
```

To make certain characters, ML has two-symbol special codes, each beginning with the \ symbol: \" for the quote sign, \t for the tab character, \n for the newline character, and \\ for the backslash character.

Every character has a numeric code associated with it (based on the ASCII or Unicode character set). You can convert a character to the int of its numeric code using ord, and you can obtain the char of a specific code using chr

```
- ord(#"q");

val it = 113 : int

- chr(114);

val it = #"r" : char
```

What we see is that alphanumeric characters, like almost all other kinds of information stored and processed by computers, are represented as numbers. One use of this is encryption—suppose we want to replace each letter in a text with the letter that comes three positions later in the alphabet.

```
- val cc = #"b";

val cc = #"b" : char

- val d = chr(ord(cc) + 3);

val d = #"e" : char
```

A *string* is a sequence of characters. ML represents them with the string type. A string literal is like a char literal except that many characters may appear between the quotes and they do not begin with a pound sign.

string

```
- "hello";

val it = "hello" : string

- it ^ " goodbye";

val it = "hello goodbye" : string
```

A string may have any number of characters, including zero or one.

```
- "";
```

```
val it = "" : string

- "a";

val it = "a" : string
```

Notice that the value #"a" of type char is a different value from "a" of type string. Values of type char can be converted to strings (of length 1) by the function str. Integers can be converted using Int.toString.

```
- str(#"q");

val it = "q" : string

- Int.toString(32);

val it = "32" : string
```

The following summarizes some operations on strings (supposing variables a and b have type string and variables n and m have type int):

`a ^ b`	*Concatenate* or chain two strings together to make a larger string.
`size(a)`	Compute the number of characters in a string.
`String.sub(a, n)`	Find the character at position n (counting from 0).
`substring(a, n, m)`	Compute the string made from the m characters starting in position n.

```
- val s = "Live Free";

val s = "Live Free" : string

- val t = "or Die";

val t = "or Die" : string

- s ^ t;

val it = "Live Freeor Die" : string
```

Oops, we forgot the space between "Free" and "or."

```
- val st = s ^ " " ^ t;

val st = "Live Free or Die" : string
```

```
- size(st);
val it = 16 : int
- String.sub(st, 5);
val it = #"F" : char
- substring(st, 5, 4);
val it = "Free" : string
- size(it);
val it = 4 : int
```

Exercises

1.7.1 Suppose the variable `cc` holds a char (and assume it is a lowercase letter), the variable `n` holds an int, and we want to produce the char that is `n` positions in the alphabet away from `cc`. The example given in the text (`chr(ord(cc) + n)`) would not always work, because we may run off the end of the alphabet. What we want in that case is to wrap around to the beginning of the alphabet—if `cc` is `y`, then one position away is `z`, but two positions away is `a`, three positions away is `b`, etc. Write an ML expression to compute the desired character. (Hint: You will need to use the `mod` operator. First, think about how you would do this if `a` had numeric code 0. However, `a` does not have code 0—fix up your solution to deal with `a`'s code as an offset.)

1.7.2 Suppose the variable `longName` is set to the string

```
Light amplification by stimulated
emission of radiation
```

Use ML string operations to produce the string `"Laser"` from it.

1.7.3 Suppose the variable `x` is set to the string

```
brother-in-law
```

Use ML string operations to produce the string `"brothers-in-law"` from it.

1.7.4 Suppose the variable `y` is set to the string

```
asdfghjkl;
```

Use ML to determine how many letters would be left over if could divide the string into segments of four letters each. (The answer is 1: the characters `l;` are left over after dividing the string into `asdf` and `ghjk`.)

1.8 Cardinality, disjointness, and partitions

Having introduced the basics of the ML programming language, we return to mathematical concepts of sets in this section and the next. As we pursue these two streams of ideas, they will illuminate each other more and more. We have already seen how sets are used to describe ML types, and in the next chapter we will see other means in ML for modeling sets. Meanwhile, we will increase the sophistication of the concepts and terminology we have to talk about sets, as well as introduce new features of the ML programming language.

cardinality

The *cardinality* of a finite set is the number of elements in the set. This is symbolized by vertical bars, like absolute value. For example, $|\{\text{violin}, \text{cello}, \text{viola}, \text{bass}\}| = 4$, or $|\mathbb{W} - \mathbb{N}| = |\{0\}| = 1$. This is an informal definition. A careful definition of cardinality is surprisingly complicated, as we will see in Section 7.9. There is no reason, however, to avoid using the term until then, since it is pretty clear intuitively.

It is important to understand that this definition applies only to finite sets. It does not allow you to say, for example, that the cardinality of $\mathbb{Z}$ is infinity. Instead, say that a set like $\mathbb{Z}$ *is infinite*. Cardinality is defined only for finite sets for our current purposes. At a later point, we will explore how to extend the notion of cardinality to infinite sets—it is not an easy task, bringing up certain interesting problems.

disjoint

If we have two sets from the same universal set, we say they are *disjoint* if they have no elements in common. Formally, sets A and B are disjoint if their intersection is empty,

$$A \cap B = \emptyset$$

Some examples of disjoint sets:

$\mathbb{R}^-$	$\mathbb{N}$
U.S. Presidents from the Whig party	U.S. Presidents who later became Chief Justice of the Supreme Court
Students enrolled in Discrete Mathematics at 9:15	Students enrolled in Art Survey at 9:15

Tangent: Cardinal and ordinal numbers

Cardinal numbers are the words one, two, three, . . . , which are used to express quantity (from Latin *cardo*, "hinge," think "something fundamentally important"). Ordinal numbers are the words *first, second, third,* . . . , which are used to express position, such as in a line or sequence (from Latin *ordo*, "line" or "row").

To keep these straight, just remember that sets are unordered—there is no "third" element—but finite sets do have a cardinality.

In a Venn diagram, if sets A and B are drawn so that they do not overlap (as seen on the left), then one can be certain that they are disjoint. However, just because two sets are drawn with an overlap does not mean that they are necessarily *not* disjoint. Regions of a Venn diagram do not indicate size of the represented set, and the region of their overlap could simply be empty.

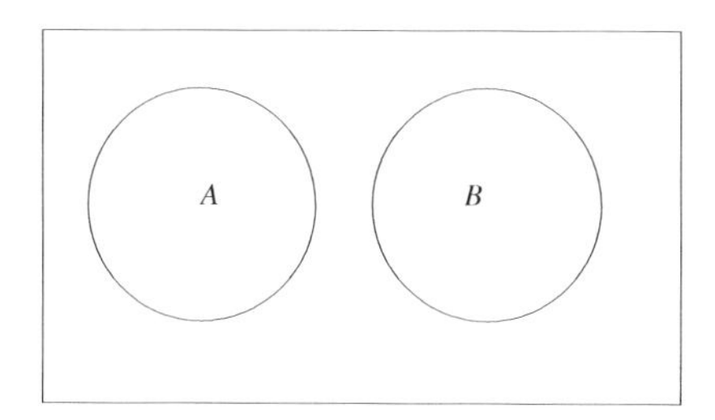

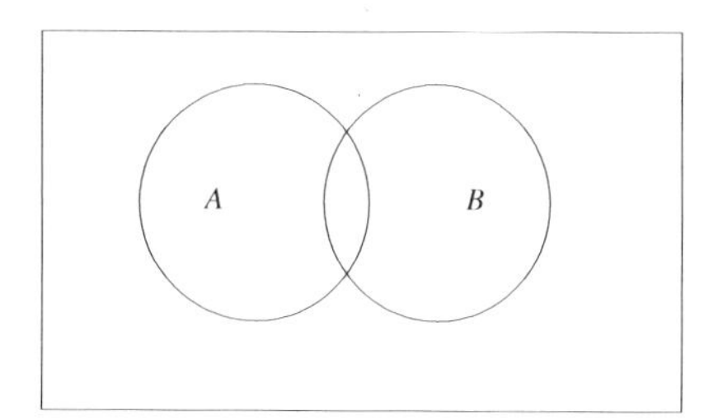

Disjointness will come up frequently. We also have an extended version of this concept that is useful when we are dealing with a larger number of sets. Consider the set S of single-serving items at a bakery:

$S = \{$ muffin, cupcake, scone, cookie, brownie, tart, donut, bagel $\}$

From this we could have a variety of sets:

A	Items that are round, viewed from above	{ muffin, cupcake, cookie, donut, bagel}
B	Items with exposed filling	{ tart }
C	Items considered relatively healthy	{ muffin, bagel }

Are these sets disjoint? Notice $A \cap B = \emptyset$, $B \cap C = \emptyset$, and most importantly $A \cap B \cap C = \emptyset$. However $A \cap C \neq \emptyset$. What does it mean for three sets to be disjoint? These three are disjoint if taken together, but not if we consider all pairs.

To clarify this terminology, we say that a set of n sets from a universal set are *pairwise disjoint* if no two of them have any overlapping elements; you can pick any two distinct sets from the set of sets, and they will be disjoint. The following set of sets is pairwise disjoint:

pairwise disjoint

Items roughly-mushroom shaped	{ muffin, cupcake }
Items roughly cylindrical	{ cookie, tart }
Items torus-shaped	{ donut, bagel }

The concept of *pairwise disjoint* naturally leads to that of a *partition*. If X is a set, then a partition of X is a set of sets $\{X_1, X_2, \ldots, X_n\}$ such that $X_1, X_2, \ldots, X_n$ are pairwise disjoint and $X_1 \cup X_2 \cup \ldots \cup X_n = X$. Intuitively, a partition of a set is a bunch of non-overlapping subsets that constitute the

partition

entire set. If we add { scone, brownie } to the three sets mentioned above, then those four sets make up a partition of the set of single-serving bakery items. Putting that in the universe of all bakery items, we can illustrate these sets this way:

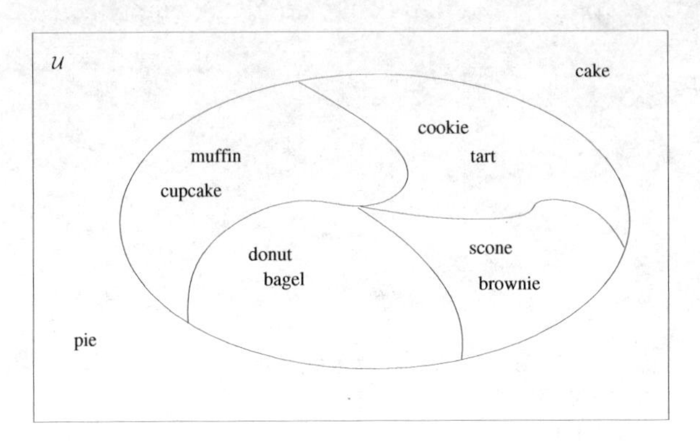

From Section 1.1, $\mathbb{T}$ and $\mathbb{A}$ make up a partition of $\mathbb{R}$. One partition of $\mathbb{Z}$ is the set of evens and the set of odds (these *two* sets together make *one* partition). $\{\mathbb{Z}^-, \{0\}, \mathbb{N}\}$ is also a partition. Or consider the set of students in your class. If we grouped students together by major (say, the set of computer science majors, the set of math majors, the set of physics majors, the set of music majors), would these sets constitute a partition? Not if the class included double majors or undecided students. A double major would be a member of more than one set, violating the fact that the sets must be pairwise disjoint. An undecided student would not be a member of any set; that would violate the fact that the union of all the sets must be the original set.

Exercises

1.8.1 What is the cardinality of $\{0, 1, 2, \ldots, n\}$?

1.8.2 Use Venn diagrams (supplemented with verbal explanations, as necessary) to demonstrate that $(A - B) \cap (B - A) = \emptyset$; that is, $A - B$ and $B - A$ are disjoint.

1.8.3 One might be tempted to think $|A \cup B| = |A| + |B|$, but this is not true in general. Why not? (In mathematics, "in general" means "always.") Under what special circumstances is it true? (Assume A and B are finite.)

1.8.4 Is $|A - B| = |A| - |B|$ true in general? If so, explain why. If not, under what special circumstances is it true? (Assume A and B are finite.)

1.8.5 Consider the sets $\{1, 2, 3\}$, $\{2, 3, 4\}$, $\{3, 4, 5\}$, and $\{4, 5, 6\}$. Notice that

$$\{1, 2, 3\} \cap \{2, 3, 4\} \cap \{3, 4, 5\} \cap \{4, 5, 6\} = \emptyset$$

Are these sets pairwise disjoint?

1.8.6 Describe three distinct partitions of the set $\mathbb{Z}$, apart from the partitions given in this section.

1.9 Cartesian products

In earlier mathematical experiences, you used the number sets we talked about at the beginning of this section not only to enumerate and measure, but also for labeling. Consider a classroom full of desks, with each position named by what column and row it is in, or consider how points in the real plane are identified.

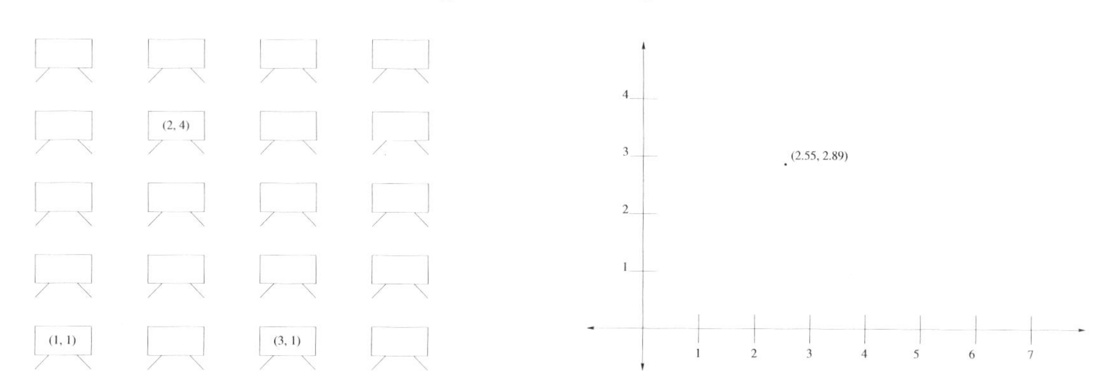

The set of desks (or names for the desk positions) and the set of points (or labels for those points) are built from other sets. Specifically, each element has two components, each component drawn from a set we have already talked about.

ordered pair

An *ordered pair* is two elements (not necessarily of the same set) written in a specific order. Suppose X and Y are sets, and say $x \in X$ and $y \in Y$. Then we say that (x, y) is an ordered pair *over* X and Y. Two ordered pairs are equal, say $(x, y) = (w, z)$ if $x = w$ and $y = z$. An ordered *pair* is different from a *set* of cardinality 2 in that it is ordered—and that the items in a pair may be drawn from different universes. Moreover, the *Cartesian product* (named after Descartes) of two sets, X and Y, written $X \times Y$ and read "X cross Y," is the set of all ordered pairs over X and Y. Formally,

Cartesian product

$$X \times Y = \{(x, y) \mid x \in X \text{ and } y \in Y\}$$

The real plane from analytic geometry is doubtless the example of Cartesian product most familiar to you, and it is written $\mathbb{R} \times \mathbb{R}$. Applying our methods of set theory, we recognize that the real plane is really just a set, like the big number sets, whose elements are points, or pairs of real numbers, depending on whether it is more convenient to think geometrically or algebraically.

Generally we will be interested in more restricted sets of pairs, that is, subsets of the Cartesian product. Thus a Cartesian product will be taken as a universal set in most contexts, and the component sets of the product will usually be universal sets in their contexts.

We can extend the idea of pairs to that of triples, quadruples, or more generally, n-tuples, or just *tuples* for short. The notation for Cartesian products

tuples

extends naturally to higher orders: For sets A, B, and C, tuples of elements from them form the set $A \times B \times C$.

Cartesian products are not only for sets of numbers, however. Consider how sets and tuples can be used to describe a paper doll set.

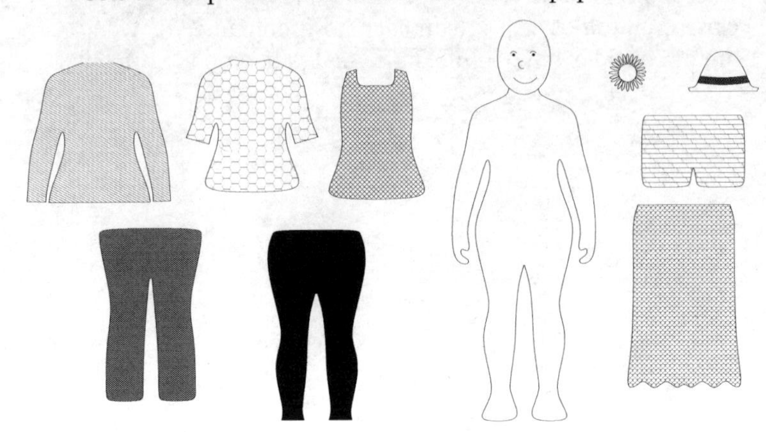

We have a set of tops, { , , }; a set of bottoms, { , , , , }; and a set of headwear, { , }. Picking one from each set makes an outfit, which is an element of { , , } × { , , , , } × { , }.

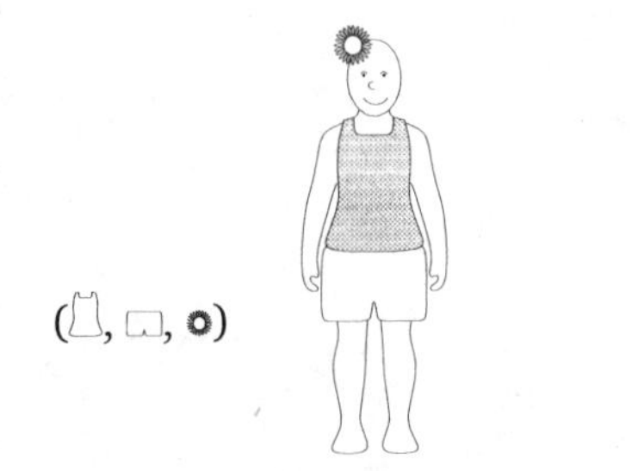

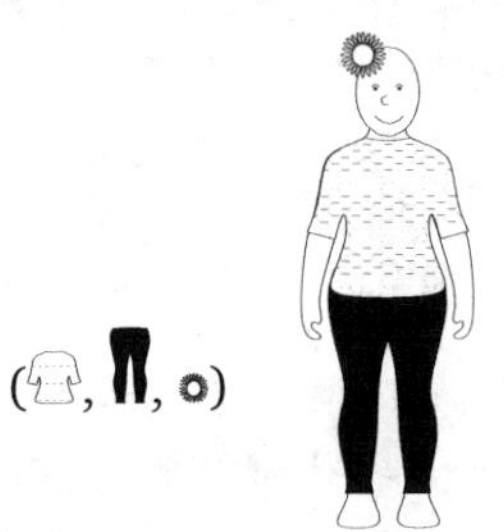

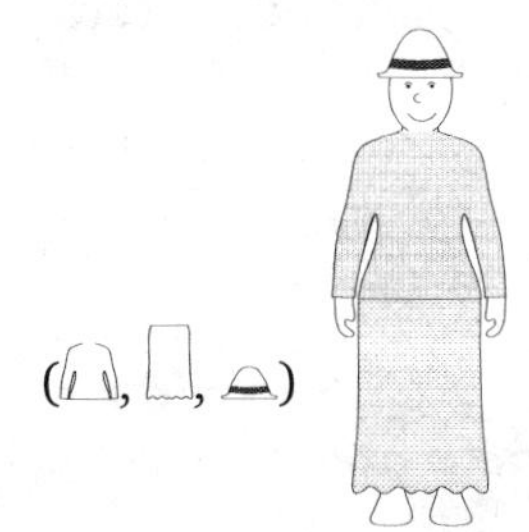

Biography: René Descartes, 1596–1650

Descartes was a French mathematician, philosopher, and mercenary soldier. His greatest contribution to mathematics is analytic geometry, by which geometric entities are described algebraically—hence the familiar coordinate system named for him.

Do you hate morning classes? Descartes is your patron saint. He did some of his most productive thinking while lying in bed until late in the morning—in fact, an unconfirmed story suggests he invented the coordinate system by watching a fly on his ceiling and considering how its two-dimensional position could be described by horizontal and vertical coordinates. He died while in the service of Catherine of Sweden, a young queen who demanded philosophy tutoring at 5:00 in the morning. [2]

Tuples are an important class of types in ML. Tuples appear just as they do in mathematical notation, and the components may be of any type, including lists and other tuples.

```
- (5, 4);

val it = (5,4) : int * int

- (2,"discrete", 8.3);

val it = (2,"discrete", 8.3) : int * string * real
```

The type int * int corresponds to the set $\mathbb{Z} \times \mathbb{Z}$.

You can make tuples of any other type—even tuple types. This next one is a tuple whose last component is itself a tuple.

```
- (5.3, "functional", (5, #"x"));

val it = (5.3,"functional",(5,#"x"))
    : real * string * (int * char)
```

To understand complicated types like this, we can analyze them by subexpressions. This demonstrates why the expression has this type.

$$\underbrace{(\underbrace{5.3}_{\text{real}}, \underbrace{\texttt{"functional"}}_{\text{string}}, \underbrace{(\underbrace{5}_{\text{int}}, \underbrace{\#\texttt{"x"}}_{\text{char}})}_{\text{int * char}})}_{\text{real * string * (int * char)}}$$

The operations #1, #2, etc, retrieve the respective element from the tuple. Continuing the above example,

```
- #1(#3(it));

val it = 5 : int
```

As we will see later, these operations are rarely necessary. Context will provide other, more convenient mechanisms for retrieving the components of a tuple.

Exercises

In Exercises 1.9.1–1.9.3, describe the given Cartesian product (by listing all the elements).

1.9.1 $\{1, 2\} \times \{1, 2\}$.

1.9.2 $\{a, b, c, d\} \times \{a, b, c\}$.

1.9.3 $\{a, b, c, d\} \times \{1, 2\}$.

1.9.4 If A and B are finite sets, what is $|A \times B|$ in terms of $|A|$ and $|B|$? Give an intuitive argument why you think your answer is correct.

1.9.5 Based on our description of the real number plane as a Cartesian product, explain how a line can be interpreted as a set.

1.9.6 Explain how $\mathbb{C}$, the set of complex numbers, can be thought of as a Cartesian product.

1.9.7 Any rational number (an element of set $\mathbb{Q}$) has two integers as components. Why not rewrite fractions as ordered pairs (for example, $\frac{1}{2}$ as $(1, 2)$ and $\frac{3}{4}$ as $(3, 4)$) and claim that $\mathbb{Q}$ can be thought of as $\mathbb{Z} \times \mathbb{Z}$? Explain why these two sets *cannot* be thought of as two different ways to write the same set. (There are at least two reasons.)

Which of the following are a partition of $\mathbb{R} \times \mathbb{R}$? For those that are not, why not?

1.9.8 $\{A, B\}$ where A is the set of points above the line $y = 3x + 2$, and B is the set of points below the line.

1.9.9 $\{A, B\}$ where A is the set of points in the first (upper right) quadrant, and B the set of points not in the first quadrant.

1.9.10 $\{A, B, C\}$ where A is the set of points that are on the x-axis, B is the set of points that are on the y-axis, and C is the set of points that are on neither axis.

1.9.11 Consider entering a cafeteria with a variety of food lines: A salad line, an entrée line, a beverage line, etc, each having several options. Suppose you plan to take one item from each line. Explain how your meal is a tuple, and that the set of possible meals is a Cartesian product.

Analyze the type of the following expressions.

1.9.12 `(3, 5.4)`

1.9.13 `(3, (#"x", #"y"), 12)`

1.9.14 `("discrete", #"m", #"a", #"t", #"h")`

1.9.15 `("discrete", (#"m", #"a", #"t", #"h"))`

1.9.16 `(3, (2, (1, (8, 4)), 13), 7, 9)`

1.10 Making your own types

We have seen types that model the sets of integers, real numbers, display characters, and sequences of display characters, and tuples combining any of these. We now turn to making our own, arbitrary types. Suppose we want to model the set of tree genera. We could define the following *datatype*:

datatype

```
- datatype tree = Oak | Elm | Maple | Spruce | Fir |
=                 Pine | Willow;

datatype tree = Elm | Fir | Maple | Oak | Pine | Spruce | Willo
```

What we are saying is that we have seven items (*oak, elm, maple, spruce, fir, pine, willow*) that constitute a set, *tree*, expressed in ML by naming a type and listing the values of that type. The name of the type (in this case tree) and the values of that type (in this case there are exactly seven values) can be any identifiers, but by convention the name of the type is lowercase like an ordinary variable and the first letter of each value is capitalized.

Since this type has no operations defined for it (until you learn to compose your own), there is little we can do with datatypes yet. However, observe that their values can be stored in variables, as with any other type.

```
- val t = Maple;

val t = Maple : tree

- t;

val it = Maple : tree
```

Here are two more examples.

```
- datatype vegetable = Carrot | Zucchini | Tomato |
=                      Cucumber | Lettuce;

datatype vegetable = Carrot | Cucumber | Lettuce | Tomato |
                     Zucchini

- datatype grain = Wheat | Oat | Barley | Maize;

datatype grain = Barley | Maize | Oat | Wheat
```

To extend this, suppose we want to model plots of ground. These plots can be used to grow trees, vegetables, or grains, or they may be empty (we will call such plots *groves*, *gardens*, *fields*, and *vacant plots*, respectively). We want to model this with a datatype as well, but we need to capture the fact that a garden is not just a garden, it is a garden *of something*. For this, you can specify the values of a datatype to be containers for another value. In that case, we say that the datatype value is carrying "extra information."

```
- datatype plot = Grove of tree | Garden of vegetable |
=                 Field of grain | Vacant;

datatype plot = Field of grain | Garden of vegetable |
                Grove of tree | Vacant;
```

This indicates that Garden by itself is not a value, but rather we need a vegetable value to complete it.

```
- Garden(Lettuce);

val it = Garden Lettuce : plot
```

```
- Garden(Zucchini);

val it = Garden Zucchini : plot

- Field(Oat);

val it = Field Oat : plot
```

Garden(Lettuce) and Garden(Zucchini) are different values (which the ML interpreter displays as *Garden Lettuce* and *Garden Zucchini*—we think the parenthesized version is clearer). Now that we have seen this, we can describe formally that one makes a datatype by giving the name of the datatype followed by one or more *constructor expressions*:

constructor expression

datatype <identifier> = <constructor expression> | . . . | <constructor expression> ;

A constructor expression indicates the form that a value of the datatype can take. The form is an identifier followed optionally by the type of any extra information, introduced by the keyword of:

<identifier> of <type>

One way to extract the extra information from a datatype value is to declare a variable for it implicitly.

```
- val glen = Grove(Oak);

val glen = Grove Oak : plot

- val Grove(whatTree) = glen;

val whatTree = Oak : tree
```

The second val declaration may look strange (but perhaps all of ML still looks strange to you). The variable we are declaring there is whatTree. The declaration is implicit in the sense that the variable is put in the context of the constructor expression Grove. We are saying, "Assuming glen is a Grove of some tree, let whatTree be that tree."

The type of the variable glen is plot, which means that in theory it could have been a Field, a Garden, or Vacant. You will get into trouble if you do an implicit declaration using a form that happens to be wrong:

```
- val Field(whatGrain) = glen;
```

```
uncaught exception Bind [nonexhaustive binding failure]
```

Unfortunately, the error messages given by typical ML implementations are not very helpful for beginners.

When we combine datatypes with tuple types, we can make datatype values that contain several pieces of extra information, and from there we can make types to capture complicated real world information. Suppose a restaurant offers value meals for which customers can pick an entrée, a side, and a beverage. The entrée can be either make-your-own sandwich or make-your-own pasta. These datatypes model the basic components (we omit the interpreter's response for brevity).

```
datatype bread = White | MultiGrain | Rye | Kaiser;
datatype spread = Mayo | Mustard;
datatype vegetable = Cucumber | Lettuce | Tomato;
datatype deliMeat = Ham | Turkey | RoastBeef;
datatype noodle = Spaghetti | Penne | Fusilli | Gemelli;
datatype sauce = Pesto | Marinara | Creamy;
datatype protein = MeatBalls | Sausage | Chicken | Tofu;
```

An entrée, then, is a sandwich or a pasta, and each of those options needs a tuple of elements chosen from the ingredients. Finally, a meal is a tuple over the set of entrées, sides, and beverages.

```
datatype entree =
     Sandwich of bread * spread * vegetable * deliMeat
   | Pasta of noodle * sauce * protein;
datatype salad = Caesar | Garden;
datatype side = Chips | CarrotSticks | Salad of salad;
datatype beverage = Water | Coffee | Pop[1] | Lemonade | IceTea;
datatype meal = Meal of entree * side * beverage;
```

Now we can make meals.

```
- Meal(Sandwich(MultiGrain, Mustard, Lettuce, Ham),
=      Chips, IceTea);

val it = Meal (Sandwich (MultiGrain,Mustard,Lettuce,Ham),
              Chips,IceTea) : meal

- Meal(Pasta(Fusilli, Creamy, Chicken), Salad(Caesar), Pop);

val it = Meal (Pasta (Fusilli,Creamy,Chicken),
              Salad Caesar,Pop) : meal
```

[1] I'm from the Midwest, and I'll call it *pop* if I want to.

As a final example, we can use a datatype to do something like a union of int and real:

```
- datatype number = Int of int | Real of real;

datatype number = Int of int | Real of real

- Int(5);

val it = Int 5 : number

- Real(5.3);

val it = Real 5.3 : number
```

This is not a perfect solution, though. `Int(5)` and `Real(5.0)` are still different values. Nevertheless, students who have programmed in an object-oriented language like Java should notice the similarity between this and subtyping.

One note on terminology. In ML, `datatype`s constitute a certain kind of programmer-defined types. The term *datatype*, however, is sometimes used in computer science as a synonym for *type*. The kinds of types made using ML's `datatype` construct are sometimes called *enumerated types* or *variant types*. In F# and OCaml, these types are made using the keyword `type` instead of `datatype`.

Exercises

1.10.1 Finish the following datatype book so that it has data constructors for various kinds of books (fiction, reference book, ...), each with string extra information, used for the title and other information (such as author, number of pages, etc) where appropriate.

```
datatype book =
            (* author, title, pages *)
       Novel of string * string * int
     | ...
```

1.10.2 Make a datatype periodical that has data constructors for various kinds of periodicals (newspaper, magazine, academic journal, ...), each also with extra information for the title.

1.10.3 Make a datatype recording that has data constructors for various recording media or formats (CD, DVD, BluRay, ...), each also with extra information for the title and running time.

1.10.4 Finish the following datatype libraryItem so that it has data constructors for various items in the library, using your solutions in Exercises 1.10.1–1.10.3 for extra information.

```
datatype libraryItem = Book of book | ...
```

1.10.5 Make a datatype element that has data constructors for certain chemical elements.

1.10.6 Make a datatype ion that has data constructors for cations and anions, using your solution to Exercise 1.10.5 for extra information; also include the charge (resulting from the number of electrons gained or lost).

1.10.7 Make a datatype molecule that has extra information about elements and their number of occurrences in the empirical formula. Restrict this to molecules with only two elements, such as ammonia (NH_3), methane (CH_3) and water (H_2O).

1.11 Making your own operations

Modeling information by making your own types is only half of programming. You must also do things with the values of those types. In these next two sections we will learn how to write new operations, both for types that already exist in ML and for our own types.

ML will reject an expression containing an unbound variable.

```
- x + 3;

Error: unbound variable or constructor: x
```

However, `x + 3` is not completely without meaning. It could denote the process of adding 3 to a given x, whatever it may be. In that case, we should encapsulate the expression in a function that indicates that x is an independent variable or *parameter*. We will study functions carefully in Chapter 7, but from your previous mathematical experience, you know we write

parameter

$$f(x) = x + 3$$

to define such a function, if we wish to call it f.

Another way to look at this is that we are generalizing from a specific case, say 5 + 3, by replacing a specific value, say 5, with a parameter—we are *parameterizing* the expression. One way to look at a function, then, is that it is a parameterized expression.

In ML, you define a function much like the way you declare a variable. In either case, we are associating an expression with a name (the name of the function or the name of the variable), but in the case of the function, the expression is parameterized. For functions, we use the keyword `fun` in place of `val`.

fun <identifier>(<identifier>) = <expression>;

The *application* of a function in ML looks exactly as it does in mathematical notation.

application

```
- fun f(x) = x + 3;

val f = fn : int -> int

- f(5);

val it = 8 : int
```

actual parameter

formal parameter

In the application above, 5 is called an *actual parameter* since it is the value substituted in for the parameter x in the function. To distinguish from actual parameters, we call x a *formal parameter*. Notice that a formal parameter is a kind of variable; an actual parameter is an expression. A function application is an expression, too.

The interpreter's response to the function declaration was "val f = fn : int -> int." Parts of this will not make sense until a later chapter, but int -> int means that this function takes an int and returns an int. Functions may take any number of parameters.

```
- fun mul(x, y) = x * y;

val mul = fn : int * int -> int

- mul(3, 16);

val it = 48 : int
```

ML has an interesting facility which allows you to give a different function body for specific parameters. For example, the following function is undefined at $x = 1$—it has point discontinuity because the denominator is 0.

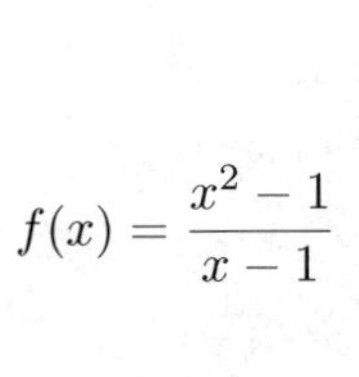

$$f(x) = \frac{x^2 - 1}{x - 1}$$

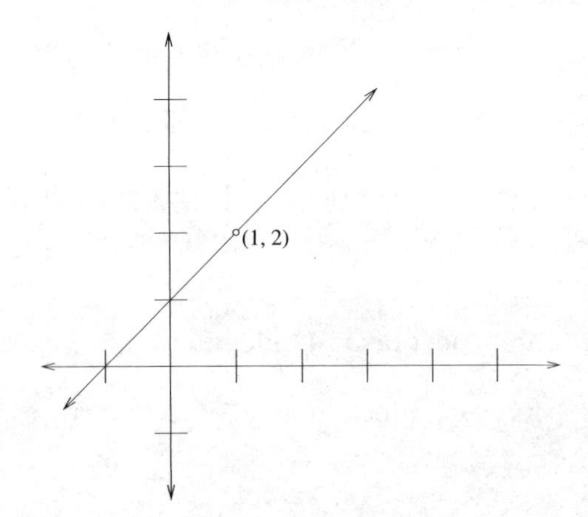

We can "fill-in" that point discontinuity by giving an explicit, different definition for the case of $x = 1$:

$$f(x) = \begin{cases} 0 & \text{if } x = 1 \\ \frac{x^2-1}{x-1} & \text{otherwise} \end{cases}$$

(Actually, the way we have done this, the function still has discontinuity at that point, only now the function is at least defined everywhere.) In ML,

```
- fun f(1) = 0
=    | f(x) = (x * x - 1) div (x - 1);

val f = fn : int -> int

- f(0);

val it = 1 : int

- f(1);

val it = 0 : int

- f(2);

val it = 3 : int
```

This style is called *pattern-matching* because we give a list of patterns describing what the actual parameters may look like (1, a very specific pattern, and x, a completely wild card pattern), and when the function is applied, the function definition of the first pattern that matches is used.

pattern-matching

When we saw datatypes earlier, we noted that we did not yet have any operations for them. We now can define our own operations using functions. Pattern-matching is a necessary tool for writing functions on datatypes.

Suppose we want to write a program that will predict the height of a tree after a given number of years. The prediction is calculated based on three pieces of information: the variety of tree (assume that this determines the tree's growth rate), the number of years, and the initial height of the tree. Obviously this assumes tree growth is linear, as well as making other assumptions, with nothing based on any real botany. So, supposing that a tree's genus can be used to determine growth rate (say, per month), we write

```
- fun growthRate(Elm) = 1.3
=    | growthRate(Maple) = 2.7
=    | growthRate(Oak) = 2.4
=    | growthRate(Pine) = 0.4
=    | growthRate(Spruce) = 2.9
=    | growthRate(Fir) = 1.1
=    | growthRate(Willow) = 5.3;

val growthRate = fn : treeGenus -> real
```

How much growing the tree does depends on how much of the year it grows. Let us assume that broadleaf (usually deciduous) trees grow for half a year and conifers (almost all evergreen) grow the entire year. Thus we need a function to associate each genus as either broadleaf (actually division *angiospermae*, which includes all flowering plants) or coniferous (division *pinophyta*).

```
- datatype treeDivision = Broadleaf | Coniferous;

datatype treeDivision = Broadleaf | Coniferous

- fun growingMonths(Broadleaf) = 6.0
=   | growingMonths(Coniferous) = 12.0;

val growingMonths = fn : treeDivision -> real

- fun division(Pine) = Coniferous
=   | division(Spruce) = Coniferous
=   | division(Fir) = Coniferous
=   | division(x) = Broadleaf;

val division = fn : tree -> treeDivision
```

Now, the function we are interested in:

```
- fun predictHeight(genus, initial, years) =
=      initial + (years * growingMonths(division(genus))
=                      * growthRate(genus));

val predictHeight = fn : tree * real * real -> real
```

Notice that a function can take any number of parameters. Actually, that is not true. A function takes exactly one parameter, but that parameter may be a tuple type. The function above has tree * real * real as its parameter type.

For our next example, suppose a farmer has a system for crop rotation. Most plots go through a cycle of being a wheat field one year, followed by a tomato garden, then a carrot garden, and then a year lying fallow as a vacant plot. He has a few plots which alternate between being maize fields and lettuce gardens. His one cucumber garden does not need to be rotated. He rarely grows barley or oats, but when he does, they are rotated to zucchini, and then vacant (after which they would go into the wheat-tomato-carrot cycle). Any groves, of course, do not rotate.

Picture this system with the following diagram describing how plots rotate.

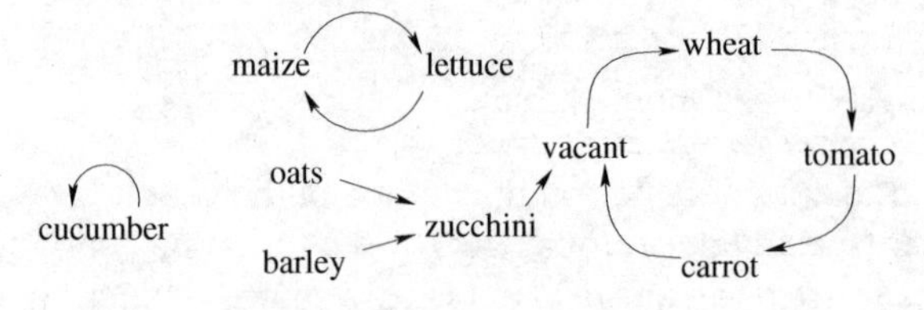

This function, when given a current use for a plot of land, determines the use of that plot for next year.

```
- fun rotate(Vacant) = Field(Wheat)
=     | rotate(Garden(Cucumber)) = Garden(Cucumber)
=     | rotate(Garden(Lettuce)) = Field(Maize)
=     | rotate(Garden(Tomato)) = Garden(Carrot)
=     | rotate(Garden(x)) = Vacant
=     | rotate(Field(Wheat)) = Garden(Tomato)
=     | rotate(Field(Maize)) = Garden(Lettuce)
=     | rotate(Field(y)) = Garden(Zucchini)
=     | rotate(z) = z;

val rotate = fn : plot -> plot

- rotate(Field(Oat));

val it = Garden Zucchini : plot

- rotate(it);

val it = Vacant : plot

- rotate(it);

val it = Field Wheat : plot

- rotate(Grove(Fir));

val it = Grove Fir : plot
```

Notice that wildcard variables can appear anywhere in the parameter. This is why this is called pattern-matching—`Garden(x)` and `Field(y)` are separate patterns. The value `Garden(Tomato)`, for example, actually fits two patterns, the `Garden(Tomato)` and `Garden(x)`. When the function is applied, the *first* matching pattern for the actual parameter is the one that is executed. If a more specific pattern follows a more general pattern, then the more specific pattern is unreachable, and ML will give an error.

```
- fun badFunction(x) = 5
=     | badFunction(Garden(Zucchini)) = 10;

stdIn:82.5-83.39 Error: match redundant
          x => ...
    -->   Garden Zucchini => ...
```

Functions can return values of any type, even tuple types. The following function takes a real and returns its floor and ceiling.

```
- fun floorAndCeiling(x) = (floor(x), ceil(x));

val floorAndCeiling = fn : real -> int * int

- floorAndCeiling(12.8);

val it = (12,13) : int * int
```

We can make functions for strings, too, by using some of the string operations already available. Suppose we want a function that does the opposite of finding a substring—that is, returning the parts of the string that are not the substring as a tuple.

```
- fun antisubstr(s, n, m) =
=  (substring(s, 0, n), substring(s, n+m, size(s) - m - n));

val antisubstr = fn : string * int * int -> string * string

- antisubstr("aaabbbccc", 3, 3);

val it = ("aaa","ccc") : string * string
```

For a more complicated example, we will write a function that takes a string and capitalizes the first character (assumed to be a letter). We need two new pieces: The function `Char.toUpper` that converts a char to an uppercase letter (if it is lowercase), and the function `str` that converts a char to a length-one string.

```
- fun capitalize(s) =
=     str(Char.toUpper(String.sub(s, 0))) ^
=     substring(s, 1, size(s)-1);

val capitalize = fn : string -> string

- capitalize("franklin");

val it = "Franklin" : string
```

Next, we define a new type along with its operations. You may have noticed that ML does not have a type corresponding to $\mathbb{Q}$. We can write our own:

```
- datatype rational = Fraction of int * int;

datatype rational = Fraction of int * int
```

```
- Fraction(3,5);

val it = Fraction (3,5) : rational
```

It might look strange to have a datatype with only one constructor expression, but do not let that confuse you. The name of the new type is rational, and values of this type have a pattern with the identifier `Fraction` as a husk and a tuple of type int * int as a core. The ints in the tuple correspond to the numerator and denominator. Here is a function for adding two fractions. It uses a function `gcd` for computing the greatest common divisor of two integers, which we will define in Section 4.10.

```
- fun add(Fraction(a, b), Fraction(c, d)) =
=       Fraction((a * d + c * b) div gcd(a * d + c * b, b * d),
=                (b * d) div gcd(a * d + c * b, b * d));

val add = fn : rational * rational -> rational
```

The body of the add function is hard to read and inefficient; it computes the original numerator and denominator three times each and their GCD twice. We could write this more efficiently (and readably) if we could store our intermediate results in variables and reuse them. In ML, this is done using a *let expression*, which has the form

let expression

`let` <var declaration>$_1$; <var declaration>$_2$; ... <var declaration>$_n$ `in`
<expression> `end`

The "var declarations" appear exactly as the variable declarations we have seen earlier. A variable's *scope* is the range of expressions in which it is valid. A variable declared in a let expression has scope beginning at its declaration and ending at the end, including the expressions used to assign values to subsequent variables declared in the let expression. The value of the entire let expression is the value of the expression occurring between `in` and `end`. Rewriting `add`:

scope

```
- fun add(Fraction(a, b), Fraction(c, d)) =
=  let val numerator = a * d + c * b;
=      val denominator = b * d;
=      val divisor = gcd(numerator, denominator);
=  in Fraction(numerator div divisor, denominator div divisor)
=  end;

val add = fn : rational * rational -> rational

- add(Fraction(1,2), Fraction(1,6));

val it = Fraction (2,3) : rational
```

Exercises

1.11.1 Write a function `double()` that takes an int and multiplies it by 2.

1.11.2 Write a function `divide13by()` that takes an int and returns the real result of dividing 13 by that number, unless the parameter is 0, in which case it should return `~1`.

1.11.3 Write a function `add()` for the datatype number defined in Section 1.10. The function should take two numbers and return their sum. If both the numbers are integers, the result should be an integer, but if either or both of the parameters are real, the result should be real. For example, `add(Int(5), Int(12))` should return `Int(17)`, but `add(Int(5), Real(12.72))` should return `Real(17.72)`.

1.11.4 Write a function `divideAll` that takes two ints and returns the result of dividing the first by the second. Specifically, it should return a tuple with three components: The (integer) quotient, the (integer) remainder, and the (real) quotient. For example, `divideAll(13, 4)` should return `(3,1,3.25)`. If the divisor is 0, then it should return `(0, 0, 0.0)`.

1.11.5 Write a function `replaceFries()` that takes a value of the meal datatype from Section 1.10 and replaces `Fries`, if that is the side, with `Chips`. If `Fries` is not the side, then the original meal is returned unchanged. Pattern-matching can be used to make this simple.

1.11.6 Write a function `replaceChar()` that takes a string, an int position in that string and a char and returns a new string like the original but with the given char replacing what had been in the given position. For example, `replaceChar("rock", 2, #"o")` should return `"rook"`.

1.11.7 Write a function `switchSides()` that takes a string and an int and splits the given string into two substrings at the given position, switches the two substrings, and returns them concatenated into a new string. For example, `switchSides("abcdefg", 3)` should return `"efgabcd"`.

1.11.8 Write a function `subtract` for the rational type which puts the result in simplest form.

1.11.9 Write a function `multiply` for the rational type which puts the result in simplest form.

1.11.10 Write a function `divide` for the rational type which puts the result in simplest form.

The set of complex numbers is assumed to have a single infinity, $\infty_{\mathbb{C}}$, as opposed to $\mathbb{R}$ which has both a positive and negative infinity. Thus if a complex number is represented as $n + mi$ (where $i = \sqrt{-1}$), then $\lim_{n\to\infty} n + mi = \infty_{\mathbb{C}}$, $\lim_{n\to-\infty} n + mi = \infty_{\mathbb{C}}$, $\lim_{m\to\infty} n + mi = \infty_{\mathbb{C}}$, etc. Thus we can design a complex number type with the following datatype:

```
datatype complex = RealImg of real * real |
Infinity;
```

This means that a complex number is either the coupling of a real part and an imaginary part (which is a real number times i) or it is $\infty_{\mathbb{C}}$.

1.11.11 Write a function `add` for the complex type.

1.11.12 Write a function `subtract` for the complex type.

1.11.13 Write a function `multiply` for the complex type.

1.12 Recursive functions

Most interesting functions—and by "interesting" we mean "for which you would bother to use a computer"—require some sort of repetitive process. The most natural way to do this in ML is for a function to apply itself to slightly different data. For example, the factorial of a number, $n!$ is the product of n and all the natural numbers less than it, with the special case that $0! = 1$. So, $1! = 1$, $2! = 2 \cdot 1$, $3! = 3 \cdot 2 \cdot 1$, and $4! = 4 \cdot 3 \cdot 2 \cdot \cdot 1$. Now we notice that $4! = 4 \cdot 3!$, and observe that the same pattern is true for the other numbers. Formally,

$$n! = \begin{cases} 1 & \text{if } n = 0 \\ n \cdot (n-1)! & \text{otherwise} \end{cases}$$

In ML,

```
- fun factorial(0) = 1
=   | factorial(n) = n * factorial(n-1);

val factorial = fn : int -> int

- factorial(0);

val it = 1 : int

- factorial(2);

val it = 2 : int

- factorial(5);

val it = 120 : int
```

Functions that apply themselves are called *recursive*, that is, self-referential. We can use recursion also to take apart strings and process them. In Section 1.11 we made a function called `capitalize()` that converted the first character in a string to be uppercase. Suppose we want to convert the entire string to uppercase. If we cared about writing a function only on strings of length 5, we could write the ridiculous function

recursive

```
- fun capitalize5(s) =
=     str(Char.toUpper(String.sub(s, 0))) ^
=     str(Char.toUpper(String.sub(s, 1))) ^
=     str(Char.toUpper(String.sub(s, 2))) ^
=     str(Char.toUpper(String.sub(s, 3))) ^
=     str(Char.toUpper(String.sub(s, 4)));
```

Notation Note: Should we parenthesize parameters?

The parentheses around the formal parameter in a function declaration and around the actual parameter in a function application are unnecessary if there is only one parameter. We could have written

```
fun factorial 0 = 1
  | factorial n = n * factorial(n-1);

factorial 3;
```

In fact, ML convention is to omit the parentheses. We use them in this book, however, because it makes ML look more like mathematical notation.

```
val capitalize5 = fn : string -> string

- capitalize5("hello");

val it = "HELLO" : string
```

To write a better function—one that will work on a string of any size, we need to find a way to break a problem down into a smaller version of itself. We know that an empty string does not need to be converted—it already has all of its (nonexistent) characters as uppercase. If a string is not empty, then we will convert the first character, use the function we are writing to convert the rest of the string, and then concatenate the results.

```
- fun capitalizeAll("") = ""
=     | capitalizeAll(s) =
=         str(Char.toUpper(String.sub(s, 0))) ^
=         capitalizeAll(substring(s, 1, size(s) - 1));

val capitalizeAll = fn : string -> string

- capitalizeAll("discrete math");

val it = "DISCRETE MATH" : string
```

Writing recursive functions is the most radical concept we have introduced in this chapter, and it will take time for you to master. Try the following exercises, but do not be discouraged if you have trouble. Recursion will be reinforced in the next two chapters and throughout this text.

Exercises

1.12.1 Write a function `charCount` that takes a string and returns the number of characters. In other words, write your own version of `size` for strings—but do not use `size`, write your own using pattern-matching and recursion. (Hint: The string `""` has 0 characters, and any other string has one more character than its substring if you take off the first character.)

1.12.2 Write a function `reverse` that takes a string and reverses it. For example, `reverse("hello")` should return `"olleh"`. (Hint: The string `""` is its own reverse. For any other string, remove the first character, reverse the rest, and tack the first character on the back.

1.12.3 Notice that the div operator can be used to chop a number up. `1275 div 10` is 127, `127 div 10` is 12, `12 div 10` is 1, and `1 div 10` is 0. Write a function `digitCount` that takes an integer—assume nonnegative—and returns the number of digits the number has. (Do not forget that `digitCount(0) = 1`.)

1.12.4 Suppose ML did not have a multiplication operator defined for integers. Write a function `multiply` that takes two integers and produces their product. Your function needs to work only on non-negative integers. Notice that `multiply(x, 0) = 0` for any `x`.

Exercises, continued

1.12.5 Using the same principle as in the previous exercise, write a function `exp` that performs exponentiation on integers; that is, for integers `x` and `n`, `exp(x, n)` would compute x^n.

1.12.6 Write a function that will compute the sum $\sum_{i=1}^{n} i$, given n. (There is a fast way to do this, which does not require repetition. If you know it, do not use it; do it the "brute force" way.)

1.12.7 A series in the form $\sum_{i=0}^{n} ar^i$ is called a *geometric series*. That is, given a value r and a coefficient a, find the sum of $a + ar + ar^2 + ar^3 + \ldots + ar^n$. Write a function that computes a geometric series, taking n, a, and r as parameters.

1.12.8 The Fibonacci sequence is defined by repeatedly adding the two previous numbers in the sequence (starting with 0 and 1) to obtain the next number, that is

$$0, 1, 1, 2, 3, 5, 8, 13, 21, 34, \ldots$$

Write a function to compute the nth Fibonacci number, given n.

1.12.9 A *Caesar cipher* is an encryption scheme where each letter in the text is shifted up or down the alphabet. The shift amount is the key. For example, the word "functional" encrypted with key 3 is "ixqfwlrqdo", since "i" comes three letters after "f," "x" comes three letters after "u", etc. At the end of the alphabet, we wrap around to the beginning, so "y" shifted 3 is "b." Write a function `encodeString` that takes a string and a key and produces the string encrypted with the given key (so, `encodeString("functional", 3);` should produce `"ixqfwlrqdo"`). Assume that the string contains only lower case letters and spaces, and do not encode spaces. You will need to write a couple of helper functions first. Your function should also work with negative numbers (any key from -25 to 25); note that `encodeString(encodeString("functional", 3), ~3)` results in `"functional"`.

1.13 Statements and exceptions

If there is ever a need to display an intermediate result to the screen, this can be done using the `print` function. `print` works only on strings, so values of any other type would need to be converted to a string.

```
- print("hello");

helloval it = () : unit
```

Notice that that since the string `"hello"` has no end-of-line marker (`\n`), the interpreter's response appears on the same line as the printing of the string. The call of the `print` function is an example of a *statement*. A statement is like an expression except that it does not have a value to return; statements are executed for their *side effect*—in this case, printing to the screen. In ML, statements return a dummy value, `()`, which is the same as an empty tuple. The unit type is a special type whose only value is `()`.

statement

side effect

We can yoke several expressions together by enclosing them in parentheses and separating them by semicolons, a construct called a *statement list*. The interpreter evaluates the expressions in order, discarding their results (thus effectively making them statements, if they are not already), except for the last expression, whose result becomes the result of the entire expression. A statement list itself is not necessarily a statement.

statement list

```
- (3 + 5; 3 < 5; real(3) / real (5));

val it = 0.6 : real
```

The only reason one would want to do this is if the earlier expressions were useful for their side effects—again, such as printing to the screen.

It is possible to write a function that does not handle all possible cases for its parameters, for example, this function to replace tomatoes with another vegetable in a garden (but leaves gardens of other vegetable unchanged):

```
- fun replaceTomato(subst, Garden(Tomato)) = Garden(subst)
=    | replaceTomato(subst, Garden(x)) = Garden(x);

Warning: match nonexhaustive
          (subst,Garden Tomato) => ...
          (subst,Garden x) => ...

val replaceTomato = fn : vegetable * plot -> plot

- replaceTomato(Zucchini, Garden(Tomato));

val it = Garden Zucchini : plot

- replaceTomato(Zucchini, Garden(Carrot));

val it = Garden Carrot : plot

- replaceTomato(Zucchini, Grove(Maple));

uncaught exception Match [nonexhaustive match failure]
```

Notice that the interpreter gives a warning when we enter the function, indicating that the function as we have written it does not address all possible input. It works fine on input for which we have defined it, but when we have something other than a garden as the second parameter, the ML interpreter raises an *exception*, a special value indicating some error has occurred. (Perhaps the term *exception* is less prejudicial than *error*—rather than judging what happened to be wrong, just say it it was "unusual." This particular exception, though, was a "match *failure*.")

exception

If you anticipate an unusual situation—such as input to a function for which there is no clear answer—you can define your own variety of exception. This is very much like declaring a datatype; the exception may have extra information.

```
- exception NotAGarden of plot;
```

```
exception NotAGarden of plot
```

In this case an exception value indicates the nature of the problem (the input is not a garden as we expected) and gives the offending value (of type plot) as extra information. Now we rewrite the function so that it has a case for this kind of input and, in place of an expression, we raise the exception using the keyword `raise`.

```
- fun replaceTomato(subst, Garden(Tomato)) = Garden(subst)
=    | replaceTomato(subst, Garden(x)) = Garden(x)
=    | replaceTomato(subst, y) = raise NotAGarden(y);

val replaceTomato = fn : vegetable * plot -> plot

- replaceTomato(Zucchini, Grove(Maple));

uncaught exception NotAGarden
```

The advantage is that the result indicates more specifically what exception circumstance caused the function's abnormal behavior. Moreover, we can indicate what plan-B action should be taken in the event of an exception by adding a `handle` clause after an expression where an exception can come from. If we think of an exception as something like a datatype, then handle clauses perform pattern-matching on the exception. In this case, we turn any field of grain into a lettuce garden and leave everything else (vacant plots and groves) as they are:

```
- replaceTomato(Cucumber, Field(Maize))
=    handle NotAGarden(Field(x)) => Garden(Lettuce)
=          | NotAGarden(y) => y;

val it = Garden Lettuce : plot
```

1.14 Extended example: A cumulative song

Each chapter will end with an extended example. Since we still have met very little of the ML programming language, our example here will be simple and whimsical, yet it will tie together several of the things we have seen in the chapter.

We will write a set of functions to produce the words to the song "The Old Woman Who Swallowed a Fly." The structure of the song is interesting because for each new animal the woman eats, we give a description of eating that particular animal, and then list the previous animals eaten, and conclude the woman will die—except for when we hit the spider in the list, which requires us to describe the spider every time, and when the woman eats the horse, in which case the song ends without listing the animals. The reason the woman eats each animal is dependent on the previous animal eaten—except the fly, which is the base case.

```
- datatype animal =
=    Fly | Spider | Bird | Cat | Dog | Cow | Horse;

datatype animal =
   Bird | Cat | Cow | Dog | Fly | Horse | Spider

- exception WillDie;

exception WillDie

- exception DidDie;

exception DidDie
```

The exceptions will be used to determine how to end the particular stanza, since they do not all end the same way.

```
- fun animalToString(Fly) = "fly"
=   | animalToString(Spider) = "spider"
=   | animalToString(Bird) = "bird"
=   | animalToString(Cat) = "cat"
=   | animalToString(Dog) = "dog"
=   | animalToString(Cow) = "cow"
=   | animalToString(Horse) = "horse";

val animalToString = fn : animal -> string

- fun next(Spider) = Fly
=   | next(Bird) = Spider
=   | next(Cat) = Bird
=   | next(Dog) = Cat
=   | next(Cow) = Dog
=   | next(Horse) = Cow;

Warning: match nonexhaustive
val next = fn : animal -> animal

- fun reason(x) =
=  print("She swallowed the " ^ animalToString(x) ^
=         " to catch the " ^ animalToString(next(x)) ^ "\n");
```

```
val reason = fn : animal -> unit

- fun whyEat(Horse) = raise DidDie
=    | whyEat(Fly) =
=          (print("I don't know why she swallowed the fly\n");
=           raise WillDie):unit
=    | whyEat(Spider) =
=          (print("that wiggled and jiggled and tickled " ^
=                 "inside her\n");
=           reason(Spider); whyEat(next(Spider)))
=    | whyEat(x) = (reason(x); whyEat(next(x)));

val whyEat = fn : animal -> unit

- fun description(Bird) = "How absurd, she swallowed a bird.\n"
=    | description(Cat) = "Imagine that, she swallowed a cat.\n"
=    | description(Dog) = "What a hog, she swallowed a dog.\n"
=    | description(Cow) = "I don't know how she swallowed " ^
=                         "a cow.\n"
=    | description(x) = "";

val description = fn : animal -> string

- fun didEat(x) =
=           (print("There was an old lady who swallowed a " ^
=                  animalToString(x) ^ "\n");
=            print(description(x)); whyEat(x));

val didEat = fn : animal -> unit

- fun stanza(x) =
=      didEat(x)
=      handle WillDie => print("I guess she'll die\n")
=           | DidDie => print("And she died\n");

val stanza = fn : animal -> unit

- stanza(Dog);

There was an old lady who swallowed a dog
What a hog, she swallowed a dog.
She swallowed the dog to catch the cat
She swallowed the cat to catch the bird
She swallowed the bird to catch the spider
```

```
that wiggled and jiggled and tickled inside her
She swallowed the spider to catch the fly
I don't know why she swallowed the fly
I guess she'll die
val it = () : unit

- stanza(Horse);

There was an old lady who swallowed a horse
And she died
val it = () : unit
```

Project

"The Green Grass Grew All Around" is a cumulative song with the following first, second, and last verses (you can figure the other verses out by the pattern):

> There was a hole (there was a hole).
> 'Twas the prettiest hole ('twas the prettiest hole)
> that you ever did see.
> Oh, the hole in the ground
> and the green grass grew all around, all around,
> the green grass grew all around.
>
> And in that hole (and in that hole)
> there was a root (there was a root).
> 'Twas the prettiest root ('twas the prettiest root)
> that you ever did see.
> Oh, the root in the hole
> and the hole in the ground
> and the green grass grew all around, all around,
> the green grass grew all around.
>
> ...
>
> And on that flea (and on that flea)
> there was a bacterium (there was a bacterium)
> 'Twas the prettiest bacterium ('twas the prettiest bacterium)
> that you ever did see (that you ever did see).
> Oh the bacterium on the flea
> and the flea on the feather,
> and the feather on the wing,
> and the wing on the chick,
> and the chick in the egg,
> and the egg under the bird,
> and the bird in the nest,
> and the nest on the leaf,
> and the leaf on the twig,
> and the twig on the branch,
> and the branch on the bough,
> and the bough on the limb,
> and the limb on the trunk,
> and the trunk on the stump,
> and the stump on the root,
> and the root in the hole,
> and the hole in the ground,
> and the green grass grew all around, all around,
> the green grass grew all around.

(How did the flea get inside the egg, or how did the leaf support the weight of the nest? You need to suspend belief a little.)

1.A Given the datatype and exception below, write appropriate functions `next`, `pieceToString`, and `preposition` so that the following function `refrain` will print the text of the refrain starting with a certain piece of the tree (we omit the repeated phrases).

Project, continued

```
datatype piece = Ground | Hole | Root |
         Stump | Trunk | Limb | Bough |
         Branch | Twig | Leaf | Nest |
         Bird | Egg | Chick | Wing |
         Feather | Flea | Bacterium;

exception Done;

fun refrain(x) =
  (print("\nand the " ^ pieceToString(x) ^
         preposition(x) ^ "the " ^
         pieceToString(next(x)) ^ ",");
   refrain(next(x)))
  handle Done =>
   print("\nand the green grass grew all" ^
         "around.\n");
```

1.B Write a function `stanza` that, given a piece, will print the stanza for that piece followed by the appropriate refrain (since the text generated by `refrain` begins with "and the...", the function `stanza` will be responsible for the line beginning with "Oh the...").

1.15 Special topic: Comparison with object-oriented programming

This section is for students who have taken a prior programming course using an object-oriented language such as Java. If you plan to take such a course in the future, you are recommended to come back and read this section after you have learned the fundamentals of class design, subtyping, and polymorphism.

You have no doubt noticed the striking difference in flavor between functional programming and object-oriented programming. A functional language views a program as a set of interacting functions, whereas an object-oriented program is a set of interacting objects. A quick sampling of their similarities will illuminate these differences. Here are two principles, cutting across all styles of programming, which are relevant for our purposes:

- A program or system is composed of data structures and functionality.
- A well-designed language encourages writing code that is modular (it is made up of small, semi-autonomous parts), reusable (those parts can be plugged into other systems), and extensible (the system can be modified by adding new parts with minimal changes to the old).

Notice the way functional and object-oriented styles each address the first point (at least in the way they are taught to beginners). In an object-oriented language, data structures and the functionality defined on them are packaged *together*; a class represents the organization of data (the instance variables) and operations (the instance methods) in a single unit. In a functional language, the data (defined, for example, by a datatype) is less tightly coupled to the functionality (the functions written for that datatype). This touches on the second point as well: In a functional language, the primary unit of modularity is the function, and in an object-oriented language, the primary unit of modularity is the class.

To see this illustrated, consider this system in ML to model animals and the noises they make.

```
- datatype Animal = Dog | Cat ;

- fun happyNoise(Dog) = "pant pant"
=    | happyNoise(Cat) = "purrrr";

- fun excitedNoise(Dog) = "bark"
=    | excitedNoise(Cat) = "meow";
```

It is easy to extend the functionality (that is, add an operation). We need only write a new function, without any change to the datatype or other functions.

```
- fun angryNoise(Dog) = "grrrrr"
=    | angryNoise(Cat) = "hisssss";
```

It is difficult, on the other hand, to extend the data. Adding a new kind of animal requires changing the datatype and every function that operates on it. From the ML interpreter's perspective, this is rewriting the whole system from scratch.

```
- datatype Animal = Dog | Cat | Chicken;

- fun happyNoise(Dog) = "pant pant"
=       | happyNoise(Cat) = "purrrr"
=       | happyNoise(Chicken) = "cluck cluck";

- fun excitedNoise(Dog) = "bark"
=       | excitedNoise(Cat) = "meow"
=       | excitedNoise(Chicken) = "cockadoodledoo";

- fun angryNoise(Dog) = "grrrrr"
=       | angryNoise(Cat) = "hisssss"
=       | angryNoise(Chicken) = "squaaaack";
```

In an object-oriented setting, we have the opposite situation. A Java system equivalent to our original ML example would be

```
interface Animal {
    String happyNoise();
    String excitedNoise();
}

class Dog implements Animal {
```

```
    String happyNoise() { return "pant pant"; }
    String excitedNoise() { return "bark"; }
}

class Cat implements Animal {
    String happyNoise() { return "purrrr"; }
    String excitedNoise() { return "meow"; }
}
```

Although the interface demands that everything of type Animal will have methods `happyNoise` and `excitedNoise` defined for it, the code for an operation like `happyNoise` is distributed among the classes. The result is a system where it is very easy to extend the data; you simply write a new class, without changing the other classes or the interface.

```
class Chicken implements Animal {
    String happyNoise() { return "cluck cluck"; }
    String excitedNoise() { return "cockadoodledoo"; }
}
```

The price is that we have made extending the functionality difficult. Adding a new operation now requires a change to the interface and to every class.

```
interface Animal {
    String happyNoise();
    String excitedNoise();
    String angryNoise();
}

class Dog implements Animal {
    String happyNoise() { return "pant pant"; }
    String excitedNoise() { return "bark"; }
    String angryNoise() { return "grrrrr"; }
}

class Cat implements Animal {
    String happyNoise() { return "purrrr"; }
    String excitedNoise() { return "meow"; }
    String angryNoise() { return "hissss"; }
}

class Chicken implements Animal {
    String happyNoise() { return "cluck cluck"; }
    String excitedNoise() { return "cockadoodledoo"; }
    String angryNoise() { return "squaaaack"; }
}
```

We can lay out the types and the operations in a table to represent the system in a way independent of either paradigm.

	Dog	Cat	Chicken
happyNoise	pant pant	purrrr	cluck cluck
excitedNoise	bark	meow	cockadoodledoo
angryNoise	grrrrr	hisssss	squaaaaack

Relative to this table, functional programming packages things by rows, and adding a row to the table is convenient. Adding a column is easy in object-oriented programming, since a column is encapsulated by a class.

It is worth noting that object-oriented programming's most touted feature, inheritance, does not touch this problem. Adding a new operation like angryNoise by subclassing may allow us to leave the old classes untouched, but it does require writing three new classes and a new interface.

```
interface AnimalWithAngryNoise extends Animal {
    String angryNoise();
}

class DogWithAngryNoise extends Dog
                  implements AnimalWithAngryNoise {
    String angryNoise() { return "grrrrr"; }
}

class CatWithAngryNoise extends Cat
                  implements AnimalWithAngryNoise {
    String angryNoise() { return "hissss"; }
}

class ChickenWithAngryNoise extends Chicken
                  implements AnimalWithAngryNoise {
    String angryNoise() { return "squaaaack"; }
}
```

Worse yet, all the code that depends on the old classes and interfaces must be changed to refer to the new (and oddly named) types.

This exemplifies a fundamental trade-off in designing a system. The Visitor pattern[11] is a way of importing the advantages (and liabilities) of the functional paradigm to object-oriented languages. A system that is easily extended by both data and functionality is a perennial riddle in software design, although the Common LISP Object System (CLOS)—the object-oriented programming facility of the LISP programming language—allows methods to be added externally to a class.

Chapter summary

Sets are the building blocks of everything we will do in this course. They are a tool we can use to articulate careful definitions of the other ideas we will. We use them to define not only mathematical concepts, but also to describe programming.

The most important programming concept in this chapter is the *type*, and we see sets at work here, too. A type is a set of values, and we can define new types to model a variety of sets, including sets of items that themselves have components. The next steps to writing useful programs is to write code to operate on the values of types, which we have begun in this chapter and will build on further in the chapters to come.

Key ideas:

Sets are unordered collections of objects. We can use sets to group or categorize ideas.

We compare sets using the relations of subset and equality. One set is a subset of another if all of its elements are also elements of the other.

New sets are composed from old using the union, intersection, and difference operations, much like how applying arithmetic operations to numbers produces new numbers. The union of two sets is the set with all the members of either set. The intersection of two sets is the set with only the members that are in both of the other sets. The difference between two sets is all the members of the first set that are not also members of the second set.

Venn diagrams help to visualize sets and their relations. We can verify facts about sets informally using series of Venn diagrams.

Types are sets of values associated together because of the operations defined for them. We organize the information we work with in a programming language into types.

A tuple is an ordered sequence of values; a Cartesian product over sets is the set of tuples with elements drawn from those sets.

The programmer defines new types in ML using the datatype construct. We can use these to model sets in the real world.

The programmer defines new operations in ML using functions. A recursive function induces repetition by calling itself on a smaller version of the same problem.

Chapter 2 List

The previous chapter introduced you to sets—and provided some foundations for functional programming on the side. This chapter addresses the crucial concept in ML programming: The list.

All programming languages and paradigms provide ways to store an arbitrary amount of information—in fact, one challenge in writing good code is to choose the right storage model for the task at hand. However, a language will usually have a native data structure integrated with the other main concepts in its design. Lists play that role in ML.

Lists are important for the other goals of this course as well. Lists exercise your ability to think recursively, and, along with datatypes, they are a primary means for us to represent sets.

Along the way in this chapter, we will meet a few advanced topics in the study of sets.

Chapter goals. The student should be able to

- use operations on ML lists.
- write functions that operate on lists, especially functions that are recursive in the structure of the list.
- represent mathematics sets and their operations with lists and list functions.
- use powersets to describe relationships among sets.

2.1 Lists

One of the most important varieties of type and the most important data container in ML is the *list*. A list is a value that contains zero or more values in a definite order, all of the same type. One creates a list in ML by listing the values, separated by commas and enclosed by square brackets.

list

```
- [4, 6, 7, 3];

val it = [4,6,7,3] : int list
```

Notice that this is an expression, and ML infers its type to be int list. Like tuples and strings, this is an example of a *composite type*—a type made up from components. Unlike tuples, all components here must have the same type, and the number of components does not affect the type of the whole expression—a list of size 2 has the same type as a list of size 28. In this case int is the *base type* of the type int list. Any type may be a base type for a list type; even list types may be base types.

composite type

base type

```
- [2.4, 8.7];

val it = [2.4,8.7] : real list

- [Garden(Zucchini), Vacant, Garden(Carrot), Grove(Oak)];

val it = [Garden Zucchini,Vacant,Garden Carrot,Grove Oak] :
     plot list

- [#"a", #"b", #"c"];

val it = [#"a",#"b",#"c"] : char list

- [[1, 4], [3, 6, 7], []];

val it = [[1,4],[3,6,7],[]] : int list list

- [];

val it = [] : 'a list
```

type variable

explicit typing

In this last case, ML cannot infer the base type of an empty list because it contains no elements to give evidence of the intended type. This is not an error; ML simply considers it to be an 'a list, where 'a is a *type variable*, a reference to a type which is, in this case, unknown. The empty list contained within the previous list must be an empty list of ints, since the context demands an int list. If for some reason we wanted to tell ML the intended type of an empty list which could not be typed from context, we could use *explicit typing*.

```
- []:int list;

val it = [] : int list
```

Explicit typing in an ML program is rarely necessary, and only in sufficiently complex programs. However, students may find explicit typing useful as an error-finding tool. ML's error messages are not always easy to follow, and many errors in a program occur because the type of an expression differs from what fits in context. If the ML interpreter is indicating a type error in your program but you think the types are right, then putting explicit types on subexpressions will often improve the error-reporting and help you find what is wrong.

head

tail

ML views a non-empty list as being composed of a *head*—the first element—and a *tail*—everything that follows, which is itself a list. Recursively defined, a value of type 'a list, for some type 'a, is

- an empty list, or
- an 'a followed by an 'a list

It is possible to define something in terms of itself, despite what you may have been taught in grade school. Note that the tail is not the the last element alone; everything that is not the head is the tail. A snake is a possible analogy, except that a snake's tail is not itself a snake. Also, think about people standing in line. There is one person who is at the front or head of the line, and that person is in the line. However, we have the phrase "there is a line behind me," which indicates that all the people behind the front person make a line (a sub-line) themselves.

Tangent: Self-referential definitions

Merriam-Webster defines *coral* in terms of itself.

1 : the calcareous or horny skeletal deposit produced by anthozoan polyps
2 : a piece of coral and especially of red coral[20]

Were you told in grade school that you could not use a word in its definition? Well, as long as there is a base case, a recursive definition is perfectly logical and gets the dictionary's approval. We define words and concepts in terms of themselves a lot in this book, especially in Chapter 6, "Self reference."

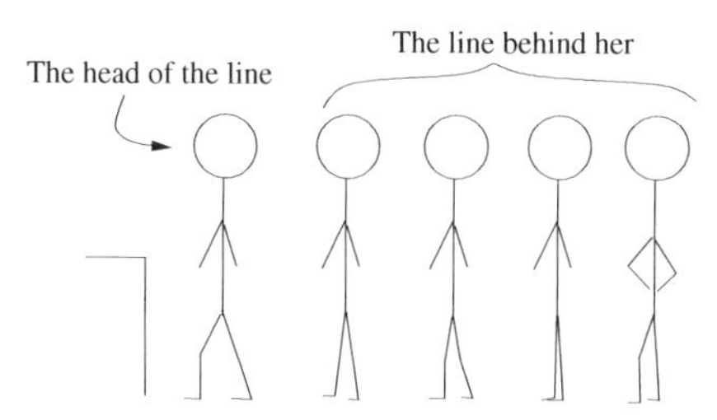

Corresponding to the defined parts of a list, ML has two *analytic list operations*, that is, operations that take lists apart: hd for retrieving the head and tl for the tail. They may be used on non-empty lists only.

analytic list operations

```
- val xx = [1,2];

val xx = [1,2] : int list

- hd(xx);

val it = 1 : int

- tl(xx);

val it = [2] : int list

- tl(it);

val it = [] : int list

- tl(it);

uncaught exception Empty
```

Compare these to the operations #1, #2, etc, for tuples. In the next section, we will see that pattern-matching can be used to analyze lists implicitly. As with the tuple operations, hd and tl are rarely used in practice.

ML has two *synthetic list operations*, that is, operations that put together new lists (from old ones). The most frequently used operation is the *cons operator*, ::, so called because it *cons*tructs a list from an element, which becomes the new list's head, and another list, which becomes the new list's tail.

synthetic list operations

cons operator

```
- 5::[];

val it = [5] : int list

- 7::it;
```

```
val it = [7,5] : int list
```

The other synthetic operation is in some ways less pure (because it can be defined in terms of other operations, see Exercise 2.2.9), but nevertheless quite useful. It is the *cat* operator, `@`, which takes two lists and concatenates them (analogous to the ^ operator for strings).

```
- [5, 6]@[1, 2];

val it = [5,6,1,2] : int list

- it@[];

val it = [5,6,1,2] : int list
```

We can use the operations, plus the explicit making of lists with square braces, to perform more complicated manipulations of lists. For example, here is one way to remove the first two elements of a list and paste them to the back.

```
- [1, 2, 3, 4, 5];

val it = [1,2,3,4,5] : int list

- tl(tl(it))@[hd(it), hd(tl(it))];

val it = [3,4,5,1,2] : int list
```

The use of such an exercise is for practice before attempting more meaningful list operations in the next section.

Expressions and their types are becoming complicated. Many mistakes that beginning programmers make manifest themselves as type errors, which happen when an expression does not have a type that ML can infer or when an operation is applied to an expression for a type other than the one for which the operation is defined. To aid in thinking about expressions—and in fixing your programs when they contain type errors—we have a method for analyzing the type of an expression by looking at the types of its subexpressions. The last expression above was found to have type int list, which we can demonstrate:

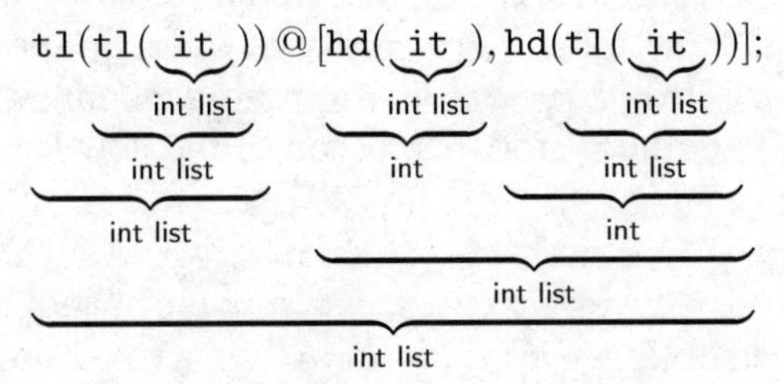

We can also intermix lists and tuples, making lists of tuples and tuples with list components.

```
- ([(Grove(Elm), 12), (Vacant, 4), (Field(Oat), 61)], 17.3);

val it = ([(Grove Elm,12),(Vacant,4),(Field Oat,61)],17.3)
  : (plot * int) list * real
```

That was

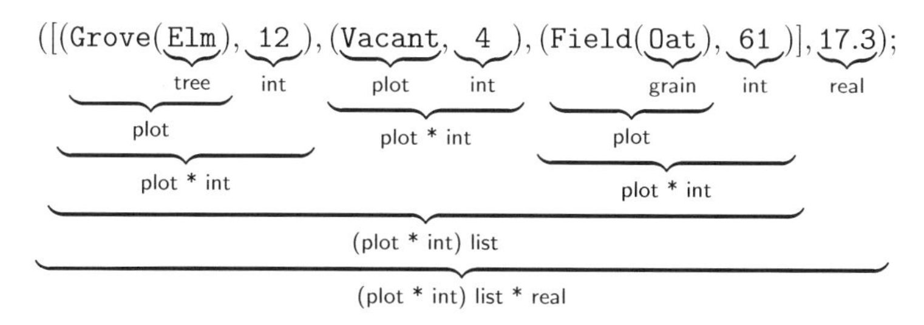

You may notice that the string type is conceptually similar to a list of chars. In fact, often the easiest way to work on strings is to convert them to lists of characters using the function `explode` and back again to lists using `implode`.

```
- explode("hi");

val it = [#"h",#"i"] : char list

- it@[#" ", #"t", #"h", #"e", #"r", #"e"];

val it = [#"h",#"i",#" ",#"t",#"h",#"e",#"r",#"e"] : char list

- implode(it);

val it = "hi there" : string
```

To make sure you understand lists, ask yourself, what differences are there between lists and tuples? The best way to address the question is to think in terms of types. In a list, all components must be of the same type, whereas the elements in a tuple may be drawn from different types. Second, the type of a list is not dependent on the list's length: `[4]`, `[4,1]`, and `[4,1,34,5,2,7,53,44]` all have type int list. The tuples `(4, 1)` and `(4, 1, 34)`, however, are values of different types. Put in other terms, a list must have a single type for its elements, but its length is arbitrary. A tuple has a fixed number of elements, and, although those elements may be of different types, the elements must be in the right order.

Exercises

Analyze the type of the expression, or indicate that there is a type error.

2.1.1 `[tl([5, 12, 6])@[8, 9]]`

2.1.2 `[14, 5]@tl([18, 91, 2])`

2.1.3 `[14, 5]::tl([3, 2, 1])`

2.1.4 `[14, 5]::[tl([1, 2, 3])]`

2.1.5 `hd([12, 5, 6])::[2, 7]`

2.1.6 `[Field(Maize), Garden(Lettuce), Vacant]`

2.1.7 `implode(#"o"::explode("kay"))`

2.1.8 `[[(2.3, 5), (8.1, 6)],[]]`

2.1.9 `([1.7, 2.9], [2, 4])`

2.1.10 `[(1.7, 2.9), (2, 4)]`

2.1.11 `[(1.3, 8), (2.6, 12), (1.7, 2)]`

2.1.12 `[(1.3, 8), (2.6, 12), (7, 2)]`

2.1.13 `([1, 12, 81], ["a", "bc"])`

2.1.14 `[5, 12, ceil(7.3 * 2.1), #2(15.2, 17)]`

Assume yy is a list of ints with sufficient length for the given problem. Write an ML expression to produce a list as indicated.

2.1.15 Remove the first item of yy.

2.1.16 Make a list containing only the first item of yy.

2.1.17 Tack 5 to the front of yy.

2.1.18 Tack [2, 5] to the front of yy.

2.1.19 Tack [2, 5] to the back of yy.

2.1.20 Remove the first item from yy and tack its double (that is, multiply it by 2) to the front.

2.1.21 Remove the third item of yy.

2.1.22 Remove the second item of yy and tack it on the front.

2.1.23 Remove the third item of yy and tack it on the back.

2.2 Functions on lists

We can encapsulate any of the list-manipulating operations we wrote in the previous section into a function and make it reusable.

```
- fun moveTwo(xx) = tl(tl(xx))@[hd(xx), hd(tl(xx))];

val moveTwo = fn : 'a list -> 'a list
```

(What does 'a mean, and why did we get it here?)

```
- moveTwo([7,6,5,4,3,2]);

val it = [5,4,3,2,7,6] : int list
```

However, pattern-matching shows its true flexibility when used with lists. We can use the cons operator in the pattern to give names to the head—or even several of the first elements.

```
- fun moveTwo2(a::b::rest) = rest@[a,b];

Warning: match nonexhaustive
          a :: b :: rest => ...

val moveTwo2 = fn : 'a list -> 'a list
```

```
- moveTwo2[5,6,7,8,9,10];

val it = [7,8,9,10,5,6] : int list
```

The pattern `a::b::rest` means, "expect a list with at least two elements; call the first element a, the second element b, and the tail `rest`." More precisely, the tail of the list is `b::rest`, which in turn has `b` as a head and `rest` as a tail. The warning indicates that the patterns listed do not cover all possible input. Indeed, this function is undefined for empty and single-element lists. Very rarely will it be acceptable for you to write functions with this sort of warning.

The cat operation is not allowed in patterns since it would be ambiguous how a list could be split to fit the pattern. A list like `[1, 2, 3]` could fit the pattern `a@b` as `[]@[1,2,3]`, `[1]@[2,3]`, `[1,2]@[3]`, or `[1,2,3]@[]`.

Here is a new problem. Suppose we want to write a function that will take a list and repeat every element. For example, given `[3,4,5]` it will produce `[3,3,4,4,5,5]`. Remember recursive functions from Section 1.12? Lists themselves are defined recursively, so recursion is the natural approach for list operations. Any problem like this must be divided into two questions:

- What is the solution to this problem for an empty list?
- How can I solve this problem for a non-empty list in terms of the solution itself?

In this case, an empty list has no elements to repeat, so the answer is just an empty list. A pattern case that does not include a call to the function itself is called a *base case*. Base cases often are trivial, and for functions that manipulate lists, almost always there is a base case for the empty list.

base case

For a list with at least one element, this problem requires us to repeat that element; then we need to repeat every element, if any, in the tail—in other words, apply our solution to the tail. Pattern cases that include a call to the function itself are called *recursive cases*. The function's calling of itself is a *recursive call*, and the technique of having a function be self-referential is called *recursion*.

recursive case

```
- fun repeatEach([]) = []
=    | repeatEach(a::rest) = a::a::repeatEach(rest);

val repeatEach = fn : 'a list -> 'a list

- repeatEach([3,4,5]);

val it = [3,3,4,4,5,5] : int list
```

Another example: Suppose we wish to sum a list of integers. If the list is empty, then the sum of all those non-existent integers is just 0. For anything else, we add the first number to the sum of the numbers in the tail.

```
- fun sum([]) = 0
=    | sum(a::rest) = a + sum(rest);

val sum = fn : int list -> int

- sum([2,4,6]);

val it = 12 : int
```

These two problems look unrelated, but their solutions are similar. The structure is the same, the contents are different. Identify the analogy between them in this summary of their differences:

	repeatEach	sum
base case result	[]	0
recursive call	repeatEach(rest)	sum(rest)
use of head item	a::a	a
combiner	::	+

Now consider a problem that mixes in tuples. This function takes a list of pairs and interchanges the elements of each pair:

```
- fun switchPair([]) = []
=    | switchPair((a,b)::rest) = (b,a)::switchPair(rest);

val switchPair = fn : ('a * 'b) list -> ('b * 'a) list
```

ML's type inference contains two type variables this time, since the component types are independent. The order changes between the input and output. It is possible that 'a and 'b are the same type, but they may also be different.

```
- switchPair([(3, 5), (21, 7)]);

val it = [(5,3),(7,21)] : (int * int) list

- switchPair([(1, 2.6), (5, 3.3)]);

val it = [(2.6,1),(3.3,5)] : (real * int) list
```

Sometimes working with lists and tuples together is more complicated than this. Suppose we wanted a function that takes a list of ints and returns the sum and product of that list as two components in a tuple. What is different here is that although the parameter to the function does not involve tuples, the result does, which means that the result of the recursive call will also be a tuple. The base case is easy:

```
- fun sumProduct([]) = (0, 1)
```

(Why is the product of an empty list 1?) But to do the recursive case, we cannot simply add and multiply the next item to the recursive result.

```
- fun sumProduct([]) = (0, 1)
=   | sumProduct(x::rest) = (x + sumProduct(rest),
=                            x * sumProduct(rest));

Error: operator and operand don't agree [overload]
  operator domain: 'Z * 'Z
  operand:         'Z * (int * int)
  in expression: x + sumProduct rest
Error: operator and operand don't agree [overload]
  operator domain: 'Z * 'Z
  operand:         'Z * (int * int)
  in expression: x * sumProduct rest
```

The ML interpreter's beef is that `sumProduct(rest)` has type (int * int), not just int. The following would work.

```
- fun sumProduct([]) = (0, 1)
=   | sumProduct(x::rest) = (x + #1(sumProduct(rest)),
=                            x * #2(sumProduct(rest)));
```

But the two calls to `sumProduct` are inefficient—we are doing the work twice. Remember let expressions? We can use one to improve this example.

```
- fun sumProduct([]) = (0, 1)
=   | sumProduct(x::rest) =
=       let val (s, p) = sumProduct(rest);
=       in (x + s, x * p) end;

val sumProduct = fn : int list -> int * int

- sumProduct([3, 5, 2, 8]);

val it = (18,240) : int * int
```

Exercises

2.2.1 Analyze the type of this expression (assuming `repeatEach` is defined as it is in this section):

```
a::a::repeatEach(rest)
```

2.2.2 Analyze the type of this expression (assuming `sum` is defined as it is in this section):

```
a + sum(rest)
```

2.2.3 Write a function `doubler` that takes a list of integers and returns the list with each element doubled. For example, `doubler([2,5,6])` would return `[4, 10, 12]`.

2.2.4 Write a function `count` that takes a list and returns the number of elements in the list.

2.2.5 Write a function `findNth` that takes a list and an integer n and returns the nth element, counting from 1. For example, `findNth([5,6,7,8], 3)` would return 7. If the nth does not exist, return `~1` as a default value. (Hint: Use pattern-matching on both parameters. What is `findNth([], n)`? What is `findNth(a::rest, 1)`?)

2.2.6 Write a function `reverse` that takes a list and reverses it. For example,

```
reverse([#"h",#"e",#"l",#"l",#"o"])
```

would return `[#"o",#"l",#"l",#"e",#"h"]`.

2.2.7 Use `implode`, `explode`, and the function you wrote for Exercise 2.2.6 to write a function `reverseString` that takes a string and returns another string, which is the reverse of the first one.

2.2.8 Write a function `digify` that takes an `int` (assume non-negative) and returns a list of the digits, from least to most significant. For example, `digify(5281)` would return `[1, 8, 2, 5]`. (Hint: Review your function `digitCount` from Exercise 1.12.3 . The digits are reversed in the list simply because it is easier.)

2.2.9 Suppose ML did not have a cat operation. Implement it yourself by writing a function `cat` that takes two lists and returns their concatenation, using only cons. (Hint: Use pattern-matching only on one parameter, say, the first. What is `cat([], xx)`?)

2.2.10 Write a function `listify` that takes a list and returns a list of lists, each element in the first list becoming its own single-element list. For example, `listify([1,2,3])` would return `[[1], [2], [3]]`.

2.2.11 Write a function `flatlist` that takes a list of lists and "flattens" it by returning the concatenation of all the lists in the original list. For example, `flatlist([[1,2,3], [4,5], [], [6,7], [8]])` would return `[1,2,3,4,5,6,7,8]`.

2.2.12 Write a function `splitlist` that takes a list of pairs and returns a pair of lists, separating the firsts of the pairs into one list and the seconds of the pair into the other list.

2.2.13 Write a function `tuplify` which takes two lists and makes a list of tuples of the corresponding elements of the two lists. If either list runs out of elements before the other, use 0 for the corresponding value. For example,

```
tuplify([4,2,1], [9,3,1])
```

would result in `[(4,9), (2,3), (1,1)]`, and

```
tuplify([8,4,2,1], [9,3])
```

would make `[(8,9), (4,3), (2,0), (1,0)]`.

2.2.14 Write a function `sumPairs` which takes a list of pairs of `int`s and makes a list of the sum of each pair. For example, `sumPairs([(2,6), (45,3), (2,7)])` would result in `[8,48,9]`.

2.2.15 Write a function `sumPairList` which takes a list of pairs of `int`s and returns a tuple containing the sum of all the first elements and the sum of all the second elements. For example, `sumPairList([(2,6), (45,3), (2,7)])` would result in `(49,16)`.

2.2.16 Rewrite your solution to the *Caesar cipher* problem (Exercise 1.12.9) using a helper function (or two) to process a list of chars. Your new `encodeString` function should use `implode` and `explode` to convert from `string` to `char list` and back to `string`.

2.3 Datatypes that use lists

The extra information carried by a datatype value may be a list type. Revisiting our field and garden example, there is no reason why a garden should have only one kind of vegetable. We can rewrite the datatype so that a garden has a list of vegetables.

```
- datatype plot = Field of grain | Garden of vegetable list |
=                 Grove of tree | Vacant;

datatype plot  = Field of grain | Garden of vegetable list |
                 Grove of tree | Vacant

- Garden([Carrot, Zucchini, Lettuce]);

val it = Garden [Carrot,Zucchini,Lettuce] : plot

- Garden([Carrot]);

val it = Garden [Carrot] : plot
```

Functions can manipulate lists in many ways. Suppose we wished to replace all the tomatoes in a garden with another plant. We can do this in two steps, first a function that replaces every tomato in a list with a given substitute, then a function that takes a garden and applies the first function to the list contained in the garden.

```
- fun replaceTomato(subst, []) = []
=    | replaceTomato(subst, Tomato::rest) =
=               subst::replaceTomato(subst, rest)
=    | replaceTomato(subst, other::rest) =
=               other::replaceTomato(subst, rest);

val replaceTomato = fn : vegetable * vegetable list
                          -> vegetable list

- fun replaceTomatoGarden(subst, Garden(crops)) =
=         Garden(replaceTomato(subst, crops));

stdIn:110.1-115.44 Warning: match nonexhaustive
          (subst,Garden crops) => ...

val replaceTomatoGarden = fn : vegetable * plot -> plot
```

```
- replaceTomatoGarden(Lettuce,
=                       Garden([Carrot,Tomato,Lettuce,Tomato,
=                              Tomato,Zucchini,Tomato,Carrot]));

val it =
  Garden [Carrot,Lettuce,Lettuce,Lettuce,Lettuce,
        Zucchini,Lettuce,Carrot]  : plot
```

We could do without the replaceTomatoGarden helper function if we used a let expression, although in this case it actually makes the code longer.

```
- fun replaceTomato(subst, Garden([])) = Garden([])
=    | replaceTomato(subst, Garden(Tomato::rest)) =
=      let val Garden(replRest) =
=              replaceTomato(subst, Garden(rest));
=      in Garden(subst::replRest)
=      end
=    | replaceTomato(subst, Garden(other::rest)) =
=      let val Garden(replRest) =
=              replaceTomato(subst, Garden(rest));
=      in Garden(other::replRest)
=      end;

Warning: match nonexhaustive
val replaceTomato = fn : vegetable * plot -> plot
```

Revisiting our meal example, there is no reason why a deli sandwich must have exactly one spread, one vegetable, and one meat. Here is a revised version in which a sandwich is a kind of bread together with a list of ingredients. (In principle, that list can be empty. Does bread by itself count as a trivial sandwich? Perhaps that would be a discussion question on the first day of Philosophy 101.)

```
datatype bread = White | MultiGrain | Rye | Kaiser;
datatype ingredient =
        Mayo | Mustard | Cucumber | Lettuce |
        Tomato | Ham | Turkey | RoastBeef;
datatype entree = Sandwich of bread * ingredient list ;

- val lunch = Sandwich(Kaiser, [Lettuce,Turkey,Turkey,Ham]);

val lunch = Sandwich(Kaiser,[Lettuce,Turkey,Turkey,Ham])
     : entree
```

How can we count the ingredients of a sandwich? A sandwich has one more ingredient than it would have if you took one ingredient away. An empty sandwich has no ingredients.

```
- fun countIngredients(Sandwich(b, [])) = 0
=   | countIngredients(Sandwich(b, i::rest)) =
=     1 + countIngredients(Sandwich(b, rest));

val countIngredients = fn : entree -> int

- countIngredients(lunch);

val it = 4 : int
```

Suppose we want to transform a sandwich, say to make it healthier: replace mayo with mustard, substitute turkey for any other kind of meat, and add a lettuce leaf. This is a bit of a pain, but we can do it:

```
- fun makeHealthy(Sandwich(b, [])) = Sandwich(b, [Lettuce])
=     | makeHealthy(Sandwich(b, RoastBeef::rest)) =
=       let val Sandwich(b, healthyRest) =
=                   makeHealthy(Sandwich(b, rest))
=       in Sandwich(b, Turkey::healthyRest) end
=     | makeHealthy(Sandwich(b, Ham::rest)) =
=       let val Sandwich(b, healthyRest) =
=                   makeHealthy(Sandwich(b, rest))
=       in Sandwich(b, Turkey::healthyRest) end
=     | makeHealthy(Sandwich(b, Mayo::rest)) =
=       let val Sandwich(b, healthyRest) =
=                   makeHealthy(Sandwich(b, rest))
=       in Sandwich(b, Mustard::healthyRest) end
=     | makeHealthy(Sandwich(b, x::rest)) =
=       let val Sandwich(b, healthyRest) =
=                   makeHealthy(Sandwich(b, rest))
=       in Sandwich(b, x::healthyRest) end;

val makeHealthy = fn : entree -> entree

- makeHealthy(lunch);

val it =
Sandwich (Kaiser,[Lettuce,Turkey,Turkey,Turkey,Lettuce])
    : entree
```

Exercises

2.3.1 Rewrite your solution to Exercise 1.10.6 so that it has two datatypes, one for cations and one for anions Each of them should have extra information about the element and the number of electrons gained or lost. Then write a datatype ionicCompound whose values have both a cation and an anion.

2.3.2 Rewrite your solution to Exercise 1.10.7 so that it is no long restricted to molecules with only two elements.

2.3.3 Suppose we rotate the crops in a garden according to the following cycle: tomato, lettuce, zucchini, carrot, cucumber. Thus a garden containing `[Tomato, Cucumber, Lettuce, Lettuce, Carrot, Cucumber, Carrot]` one year would have `[Lettuce, Tomato, Zucchini, Zucchini, Cucumber, Tomato, Cucumber]` the next. Write a function `rotateGarden` that takes a list of vegetables and returns another list of vegetables showing how a garden having the first list should be rotated. (You should first write a helper function `nextCrop` that takes a vegetable and returns the next `vegetable` in the rotation.

2.3.4 Write a function `removeHam` that takes an entree (a sandwich) and returns a similar sandwich except with the ham removed.

2.3.5 Write a function `makeVegetarian` that takes an entree and returns a similar sandwich but with all meat ingredients replaced by vegetable ingredients (the exact replacement scheme is up to you).

2.3.6 Suppose we have the following functions to price bread and ingredients (in cents)

```
fun breadPrice(White) = 75
  | breadPrice(MultiGrain) = 125
  | breadPrice(Rye) = 130
  | breadPrice(Kaiser) = 100;

fun ingPrice(Mayo) = 25
  | ingPrice(Mustard) = 25
  | ingPrice(Cucumber) = 50
  | ingPrice(Lettuce) = 60
  | ingPrice(Tomato) = 40
  | ingPrice(Ham) = 150
  | ingPrice(Turkey) = 125
  | ingPrice(RoastBeef) = 200;
```

Write a function `sandwichPrice` that takes an entree and determines the price of the sandwich by summing the bread and ingredients.

2.4 Powersets

For this course, one of the most important uses of lists is to model sets. We can represent a set of, say, certain integers by putting those ints in a list. The cat operation acts like the union operation on sets. A list is not a perfect model, however. The same value may appear more than once in a list, which does not make sense for a set; lists are inherently ordered, sets unordered. Likewise, if the cat operation is used on two lists interpreted as sets, any element appearing in both lists will appear twice in the concatenated list. Much of our programming for lists will be writing operations on them that perform set operations to solve these difficulties.

To write those operations, however, will require us to make comparisons between items (is this item an element in both sets?) and make decisions based on those comparisons. We will not learn how to do such comparisons until introducing propositional logic—in both the math side of this course and in ML. Nevertheless, using lists to think about sets will be useful as we introduce a new set concept in this section. This new concept, *powersets*, will come up in some of the most difficult problems in the text, and the ML exercises at the end of this section will help you in reasoning about building and analyzing a powerset.

Andrew, Bella, Carla, and Dirk went to visit their grandmother. On the last day of their visit, Gramma whipped out her digital camera and asked to take a few pictures. It started innocently enough, with the four siblings lined up. Then she asked to take a picture of just the brothers. Then just the sisters. Then each grandchild individually. Then the two oldest. Then the three youngest. Soon (though not soon enough), she had taken pictures of every combination of grandchildren

{ Andrew, Bella, Carla, Dirk } { Bella, Carla, Dirk } { Carla, Dirk }
{ Andrew, Bella, Carla } { Andrew, Carla , Dirk } { Andrew, Dirk }
{ Andrew, Bella, Dirk } { Bella, Dirk } { Dirk }
{ Andrew, Carla } { Bella, Carla } { Carla }
{ Andrew, Bella } { Bella } { Andrew }

"And for good measure," said Gramma at last, "let me take a picture of the background with none of you in it, so that when I'm photoshopping the results I can start from scratch if I need to."

{ }

The *powerset* of a set is the set of all subsets of the original set. Beginning students often find it a challenge to comprehend powersets, and one reason we will return to powersets so frequently in this course is the sheer experience of addressing the difficulty. The powerset of X is symbolized by $\mathscr{P}(X)$. Formally,

powerset

$$\mathscr{P}(X) = \{Y \mid Y \subseteq X\}$$

For example,

$$\mathscr{P}(\{1,2,3\}) = \{ \quad \{1,2,3\}, \\ \{1,2\}, \{2,3\}, \{1,3\}, \\ \{1\}, \{2\}, \{3\}, \\ \emptyset \quad \}$$

Two peculiarities about powersets contribute to the difficulty in understanding them. First, the elements of a powerset are themselves sets. It is not true, for example, that $1 \in \mathscr{P}(\{1,2,3\})$; rather, $\{1\} \in \mathscr{P}(\{1,2,3\})$. Make sure you understand the difference between the element 1 and $\{1\}$, the set containing only 1. It is the same as the difference between the ML value `1` and the list containing only that value.

```
- [1];


val it = [1] : int list
```

The second peculiarity is that a set and its powerset are drawn from different universal sets. If X is a set and $\mathcal{U}$ is the universal set in this context, then $X \subseteq \mathcal{U}$, $X \in \mathscr{P}(\mathcal{U})$, and $\mathscr{P}(X) \subseteq \mathscr{P}(\mathcal{U})$. The last of these suggests that $\mathscr{P}(\mathcal{U})$ is the universal set when we talk about $\mathscr{P}(X)$. X is an *element* in the universe of $\mathscr{P}(\mathcal{U})$ and a *set* in the universe of $\mathcal{U}$.

The fact that we have two universes excludes the use of Venn diagrams and makes useful visualization unlikely. You are better off thinking in terms of a story, like the one above. The grandchildren are members of the original set. Just as pictures of the children are distinct objects from the children themselves, so sets of the elements, or subsets of the original set, are distinct mathematical objects from the elements of the original sets.

Convince yourself of the following powerset facts, which follow immediately from the definition (*iff* stands for "if and only if," as we will see in Chapter 3). Let A and B be any sets, and let $S = \{\text{Andrew}, \text{Bella}, \text{Carla}, \text{Dirk}\}$.

$a \in A$ iff $\{a\} \in \mathscr{P}(A)$.	For any element in a set, the set containing just that element is a member of the powerset.	$\{\text{Bella}\} \in \mathscr{P}(S)$, the picture of jus Bella is one of the pictures taken.
$A \subseteq B$ iff $A \in \mathscr{P}(B)$.	Any subset of a set is an element of the powerset of that set.	$\{\text{Carla}, \text{Dirk}\} \subseteq S$, $\{\text{Carla}, \text{Dirk}\} \in \mathscr{P}(S)$
$A \subseteq B$ iff $\mathscr{P}(A) \subseteq \mathscr{P}(B)$.	If one set is a subset of another, then the powerset of the first is a subset of the powerset of the other. This one does not really "follow immediately." We will prove it as Theorem 4.7 and Exercise 4.9.1.	Since $\{\text{Bella}, \text{Carla}, \text{Dirk}\} \subseteq S$, the the set of pictures of all combination of Bella, Carla, and Dirk is a subset o the pictures of all the siblings.
For any set A, $A \in \mathscr{P}(A)$.	The set as a whole is a member of its own powerset.	One of the pictures Gramma took wa of all four siblings together, { Andrew Bella, Carla, Dirk }.
For any set A, $\emptyset \in \mathscr{P}(A)$.	The empty set is an element of any powerset. Make sure you understand the difference between $\emptyset$ and $\{\emptyset\}$.	In the end, Gramma took a pictur of the background with no grandchil dren.

Our modeling of sets in ML helps in comprehending powersets and the two "universes." If we model a set with a list, then a powerset is a list of lists. In Exercise 2.4.15 you will write a function `powerset` to compute the powerset of a given set. The type of that function will be 'a list -> 'a list list.

```
- powerset([1,2,3]);

val it = [[1,2,3],[1,2],[1,3],[1],[2,3],[2],[3],[]]
```

```
: int list list
```

Now make another observation. We have seen the powerset of $\{1,2,3\}$:

$$\mathscr{P}(\{1,2,3\}) = \{\emptyset, \{1\}, \{2\}, \{3\}, \{1,2\}, \{1,3\}, \{2,3\}, \{1,2,3\}\}$$

Rewrite this so that the sets without 1 in them appear first.

$$= \{\emptyset, \{2\}, \{3\}, \{2,3\}, \{1\}, \{1,2\}, \{1,3\}, \{1,2,3\}\}$$

Split those up into two sets (a partition!). 1 is something of a pivot used to divide the set into two subsets.

$$\begin{array}{ll} = & \{\emptyset, \{2\}, \{3\}, \{2,3\}\} \\ & \cup \ \{\{1\}, \{1,2\}, \{1,3\}, \{1,2,3\}\} \end{array}$$

Since each set in the second set has 1 in it, each of those sets is the result of unioning $\{1\}$ to another set:

$$\begin{array}{llllll} = & \{ & \emptyset, & \{2\}, & \{3\}, & \{2,3\} \quad \} \\ \cup & \{ & \{1\} \cup \emptyset, & \{1\} \cup \{2\}, & \{1\} \cup \{3\}, & \{1\} \cup \{2,3\} \quad \} \end{array}$$

The first of those two sets is $\mathscr{P}(\{2,3\})$. The second is the set of things we get from unioning $\{1\}$ to each element of $\mathscr{P}(\{2,3\})$

$$= \mathscr{P}(\{2,3\}) \cup \{\{1\} \cup A \mid A \in \mathscr{P}(\{2,3\})\}$$

The set $\{\{1\} \cup A \mid A \in \mathscr{P}(\{2,3\})\}$ has some hefty notation, but do not let it scare you off. It stands for the set of all things that fit the pattern "$\{1\} \cup A$," that is, sets formed by unioning $\{\ 1\ \}$ to some other set A, where A can be anything in the set $\mathscr{P}(\{2,3\})$. $A \in \mathscr{P}(\{2,3\})$ means A is either $\emptyset$ or $\{2\}$ or $\{3\}$ or $\{2,3\}$, and so $\{1\} \cup A$ is either $\{1\}$ or $\{1,2\}$ or $\{1,3\}$ or $\{1,2,3\}$.

So what? First, this provides a way to *analyze* a powerset, that is, to take one apart. When we work on proofs later in this text, this will be a handy way to reason about a powerset, namely to pick one element, break the powerset down into the sets that have that one element on one hand and the sets that do not have that element on the other hand, and consider each half of the powerset separately.

Second, this provides a way to *synthesize* a powerset, that is, to put one together. A function to compute the powerset of a given set A will

1. Pick an element $a \in A$.
2. Find the powerset of the set without a, that is, $A - \{a\}$.
3. Find the set formed by adding a to every element in the set found in part 2.
4. Combine the sets found in parts 2 and 3.

Exercises

Find the powerset of each of the following sets by listing all the elements.

2.4.1 $\{1\}$

2.4.2 $\{1, 2\}$

2.4.3 $\{1, 2, 3, 4\}$

2.4.4 $\mathscr{P}(\{1, 2\})$. (That is, find $\mathscr{P}(\mathscr{P}(\{1, 2\}))$.)

Determine which of the following propositions are true and which are false. For those that are false, find a similar proposition that is true.

2.4.5 $\emptyset = \mathscr{P}(\emptyset)$.

2.4.6 $\mathbb{Z} \in \mathscr{P}(\mathbb{R})$.

2.4.7 For any set A, $\{A\} \in \mathscr{P}(A)$.

Explain the meanings of the following propositions and give an example for each of them.

2.4.8 $\mathscr{P}(A) \cap \mathscr{P}(B) = \mathscr{P}(A \cap B)$.

2.4.9 If $B \in \mathscr{P}(A - C)$, then $B \in \mathscr{P}(A)$.

2.4.10 If $B \subseteq A$, then $\mathscr{P}(B) - \mathscr{P}(A) = \emptyset$.

Give a counterexample for each of the following propositions. That is, find example sets A and B for which each proposition is false.

2.4.11 $\mathscr{P}(A) \cup \mathscr{P}(B) = \mathscr{P}(A \cup B)$.

2.4.12 $\mathscr{P}(A) - \mathscr{P}(B) = \mathscr{P}(A - B)$.

2.4.13 Count the number of elements in your answers for Exercises 2.4.1–2.4.4 and the powerset of $\{1, 2, 3\}$ given in the text. If X is a finite set, what is $|\mathscr{P}(X)|$ in terms of $|X|$?

2.4.14 Write a function `addToEach` that takes an item and a list of lists and returns a list of lists like the one given except that the given item is prepended to each list. For example, `addToEach(5, [[1, 3], [], [8]])` would return `[[5, 1, 3], [5], [5, 8]]`.

2.4.15 Review our observations about the set $\{1, 2, 3\}$ and how it can be partitioned around 1 as a pivot. Then, using this insight and your `addToEach` function from the previous exercise, write a function `powerset` which will take a list standing for a set and return a list of lists standing for the powerset of that set.

2.5 Case expressions and option types

Briefly we show a few ML language features that are handy on occasion.

case expression

The *case expression* provides pattern-matching apart from function application. The form is

case <expression> of <pattern list>

For example, we may rewrite the factorial function as

```
- fun factorial(n) =
=    case n of
=       0 => 1
=     | n => n * factorial(n-1);
```

option type

The `option` family of types can be used in situations where an operation might not have a value to return. For example, if we write our own version of the `hd` operation, we get a match nonexhaustive error:

```
- fun homemadeHd(x::rest) = x;
```

```
Warning: match nonexhaustive
          x :: rest => ...
val homemadeHd = fn : 'a list -> 'a
```

Instead, using an option type, we can indicate if given a nonempty list that there is some head item, and none otherwise.

```
- fun optionHd(x::rest) = SOME x
=     | optionHd([]) = NONE;

val optionHd = fn : 'a list -> 'a option

- optionHd([1, 2, 3]);

val it = SOME 1 : int option

- optionHd(tl(tl(tl([1,2,3]))));

val it = NONE : int option
```

The option type and its values may be hard to read at first. Keep in mind that this is used in a context where a value is optional or when it is possible that no appropriate value exists. The result `SOME 1` above indicates that the application of function `optionHd` had a result, namely 1, and the result `NONE` is a special value indicating that the optional value is missing. The extended example in the next section will illustrate the usefulness of option types.

Option types are really a pre-defined datatype. We could have defined our own version of int option with

```
- datatype homemadeIntOption = NONE | SOME of int
```

To retrieve the value of a `SOME`, use the function `valOf`.

```
- optionHd([1,2]);

val it = SOME 1 : int option

- valOf(it);

val it = 1 : int
```

Using `valOf` brings back the possibility of error, and it usually can be avoided using pattern-matching.

2.6 Extended example: A language processor

The extended examples at the end of chapters are, naturally, more complicated and therefore harder. Do not let the size intimidate you. Instead, work through them patiently, confer with your classmates, and take notice of the power of a programming language like ML—especially considering we as yet have seen only the first principles. Stick with programming, and one day you will be able to write systems like this, and better, on your own.

Since the earliest days of modern computing machinery, users have put computers to the task of processing text. For example, in 1964 an analysis of word frequencies was used to determine authorship of various essays in *The Federalist*. Progress in these fields is evident today in the grammatical advice your word processor gives you and in online tools for automatically translating the text of websites.

In this section we will see how useful ML datatypes, lists, tuples, and options are for describing the structure of a human language. We will write a program that will take a sentence in a (very small) subset of English, analyze it grammatically, and make transformations on it (such as changing the mood or tense).

First, we define the parts of speech included in our language. We represent these with datatypes, each with a `string` for extra information. The string is the word itself; the datatype is used to indicate the part of speech. (In big code examples like this, we omit the prompts and responses of the interpreter.)

```
datatype noun = Noun of string;
datatype article = Art of string;
datatype adjective = Adj of string;
datatype preposition = Prep of string;
datatype transitiveVerb = TV of string;
datatype intransitiveVerb = IV of string;
datatype linkingVerb = LV of string;
datatype adverb = Adv of string;
```

For review, a *transitive verb* is a verb that requires a direct object. In "The man hit the ball," *hit* is a transitive verb and *the ball* is its direct object. *Intransitive verbs* do not take direct objects: "The dog ran." *Linking verbs* are complemented by an adjective: "The woman felt smart."

So much for individual words. We next define phrases, also using datatypes. We will require a noun phrase to have an article. An adjective is optional, which makes a perfect use of an option type.

```
datatype nounPhrase = NounPhrase of (article * adjective option
                                     * noun);
```

Earlier we defined transitive, intransitive, and linking verbs as distinct parts of speech, for convenience. We can comprehend all of these under the idea of a verb phrase.

```
datatype verbPhrase = TVP of (transitiveVerb * nounPhrase)
                    | IVP of (intransitiveVerb)
                    | LVP of (linkingVerb * adjective);
```

The predicate of a sentence is the main verb phrase, or the sentence itself without its subject. In our language, we take a verb phrase and add an optional adverb, such as "*happily* greeted the unicorn."

```
datatype predicate = Predicate of (adverb option * verbPhrase);
```

A prepositional phrase is a preposition plus the preposition's object, which is a noun phrase.

```
datatype prepPhrase = PrepPhrase of (preposition * nounPhrase);
```

Finally, a sentence is a noun phrase (the subject), a predicate, and optionally a prepositional phrase modifying the predicate.

```
datatype sentence = Sentence of (nounPhrase * predicate *
                                 prepPhrase option);
```

The big dog quickly chased a red ball through the bright field.

parsing

Now for the difficult task of *parsing* a sentence. To parse is to analyze the syntax (that is, grammar or structure) of a string in a language. Because we are still somewhat limited by the amount of ML we have learned so far, we will also use the code for parsing to define the (very small) vocabulary for our language—this would more naturally be separated into its own step.

Ultimately we want a function `parseSentence` that takes a string and returns a sentence. We will make a set of helper functions first: `parseArticle`, `parseAdjective`, and others to identify words and their parts of speech; then `parseNounPhrase`, `parseVerbPhrase` and others to parse larger portions; finally `parseSentence`.

The parsing functions (except for `parseSentence`) will operate under the following assumptions:

- They are given a list of words—that is, their parameter has type string list.
- The first portion of words on that list together constitute the part of speech or phrase that the parsing function is looking for.
- The parsing function "consumes" those words.
- The parsing function returns two things: the datatype value that it parses for and the rest of the list it was given, after consuming the words making up the portion of the sentence it was parsing.

Here is what that means. We call `parseNounPhrase` with a list like `["the", "big", "dog", "quickly", "chased", "a", "red", "ball", "through", "the" "bright", "field"]`. The first three items in that list constitute a noun phrase, and we would not call this function on a list that did not start with one. The function will return a tuple, the first item being a nounPhrase, in this case

```
NounPhrase(Art("the"), SOME Adj("big"), Noun("dog"))
```

The other item in the tuple will be the rest of the list: `["quickly", "chased", "a", "red", "ball", "through", "the", "bright", "field"]`.

Thus each function will process a part of the list and will return the result of the processing and the remainder of the list still needing processing.

The function `parseArticle` is simple enough, but it will fail if the list does not have an article at the beginning.

```
fun parseArticle("the"::rest) = (Art("the"), rest)
  | parseArticle("a"::rest) = (Art("a"), rest);
```

We use `parseAdjective` to specify which adjectives are in our vocabulary (you can add to this easily if you like). Since adjectives are optional, the first item in the tuple is adjective option, and if the list does not begin with one of our recognized adjectives, then we return `NONE` and the unchanged list.

```
fun parseAdjective("big"::rest) = (SOME (Adj("big")), rest)
  | parseAdjective("bright"::rest) =
                (SOME (Adj("bright")), rest)
  | parseAdjective("fast"::rest) = (SOME (Adj("fast")), rest)
  | parseAdjective("beautiful"::rest) =
                (SOME (Adj("beautiful")), rest)
  | parseAdjective("smart"::rest) = (SOME (Adj("smart")), rest)
  | parseAdjective("red"::rest) = (SOME (Adj("red")), rest)
  | parseAdjective("smelly"::rest) =
                (SOME (Adj("smelly")), rest)
  | parseAdjective(wordList) = (NONE, wordList);
```

Parsing nouns, prepositions, and adverbs are similar, adverbs being optional:

```
fun parseNoun("man"::rest) = (Noun("man"), rest)
  | parseNoun("woman"::rest) = (Noun("woman"), rest)
  | parseNoun("dog"::rest) = (Noun("dog"), rest)
  | parseNoun("unicorn"::rest) = (Noun("unicorn"), rest)
  | parseNoun("ball"::rest) = (Noun("ball"), rest)
  | parseNoun("field"::rest) = (Noun("field"), rest)
  | parseNoun("flea"::rest) = (Noun("flea"), rest)
  | parseNoun("tree"::rest) = (Noun("tree"), rest)
  | parseNoun("sky"::rest) = (Noun("sky"), rest);
```

```
fun parsePreposition("in"::rest) = (Prep("in"), rest)
  | parsePreposition("on"::rest) = (Prep("on"), rest)
  | parsePreposition("through"::rest) = (Prep("through"), rest)
  | parsePreposition("with"::rest) = (Prep("with"), rest);

fun parseAdverb("quickly"::rest) = (SOME (Adv("quickly")), rest)
  | parseAdverb("slowly"::rest) = (SOME (Adv("slowly")), rest)
  | parseAdverb("dreamily"::rest) = (SOME (Adv("dreamily")), rest)
  | parseAdverb("happily"::rest) = (SOME (Adv("happily")), rest)
  | parseAdverb(wordList) = (NONE, wordList);
```

Now to parse a complete noun phrase, we grab an article, an adjective, and a noun, in sequence. Each of these will generate a new rest-of-the-list, which we feed into the next function. The last rest-of-the-list must be returned with the nounPhrase.

```
fun parseNounPhrase(wordList) =
    let val (art, rest1) = parseArticle(wordList);
        val (adj, rest2) = parseAdjective(rest1);
        val (nn, rest3) = parseNoun(rest2);
    in
        (NounPhrase(art, adj, nn), rest3)
    end;
```

For parsing verbs, we need to do something different depending on what kind of verb it is. Our `parseVerbPhrase` function will grab the verb itself and determine whether it is transitive, intransitive, or linking. It will then pass the verb and what is left of the word list to appropriate helper functions. For transitive verbs, we need a helper function to grab a noun phrase and package the verb and direct object together; for linking verbs, the helper function must grab and package a predicate adjective.

```
fun parseTransVerb(vb, wordList) =
    let val (dirObj, rest) = parseNounPhrase(wordList);
    in (TVP(vb, dirObj), rest)
    end;

fun parseLinkingVerb(vb, wordList) =
    let val (adj, rest) = parseAdjective(wordList);
    in (LVP(vb, valOf(adj)), rest)
    end;
```

A helper function for intransitive verbs is unnecessary—`parseVerbPhrase` can handle them on the spot. Notice that our verbs are assumed to be simple past tense.

```
fun parseVerbPhrase("chased"::rest) =
            parseTransVerb(TV("chased"), rest)
  | parseVerbPhrase("saw"::rest) =
            parseTransVerb(TV("saw"), rest)
  | parseVerbPhrase("greeted"::rest) =
            parseTransVerb(TV("greeted"), rest)
  | parseVerbPhrase("bit"::rest) =
            parseTransVerb(TV("bit"), rest)
  | parseVerbPhrase("loved"::rest) =
            parseTransVerb(TV("loved"), rest)
  | parseVerbPhrase("ran"::rest) = (IVP(IV("ran")), rest)
  | parseVerbPhrase("slept"::rest) = (IVP(IV("slept")), rest)
  | parseVerbPhrase("sang"::rest) = (IVP(IV("sang")), rest)
  | parseVerbPhrase("was"::rest) =
            parseLinkingVerb(LV("was"), rest)
  | parseVerbPhrase("felt"::rest) =
            parseLinkingVerb(LV("felt"), rest)
  | parseVerbPhrase("seemed"::rest) =
            parseLinkingVerb(LV("seemed"), rest);
```

By now you should be able to understand parsePredicate and parsePrepPhra
for yourself, though parseSentence will require some explanation.

```
fun parsePredicate(wordList) =
    let val (adv, rest1) = parseAdverb(wordList);
        val (vPh, rest2) = parseVerbPhrase(rest1);
    in (Predicate(adv, vPh), rest2)
    end;

fun parsePrepPhrase(wordList) =
    let val (prep, rest1) = parsePreposition(wordList);
        val (nPh, rest2) = parseNounPhrase(rest1);
    in (PrepPhrase(prep, nPh), rest2)
    end;

fun parseSentence(message) =
   let val wordList = String.tokens(fn(x) =>
                          not(Char.isAlpha(x)))(message);
       val (subj, rest1) = parseNounPhrase(wordList);
       val (pred, rest2) = parsePredicate(rest1);
    in
      case rest2 of
        [] => Sentence(subj, pred, NONE)
      | prPh => Sentence(subj, pred,
                         SOME(#1(parsePrepPhrase(prPh))))
    end;
```

The part about `String.tokens(fn(x) ...` uses more ML than we have covered so far. All you need to know is that it turns a string containing a sentence into a list of strings, each string holding one word. It discards spaces and punctuation. Then we parse the subject and the predicate. What is left is the list `rest2`. If there is no prepositional phrase at the end, then `rest2` is empty, and the sentence is `Sentence(subj, pred, NONE)`—the first pattern in the case expressions handles this. Notice that the second pattern in the case expression effectively renames `rest2` to `prPh` (as well as guaranteeing that it is not empty). The call to `parsePrepPhrase` returns a tuple—the prepPhrase and the (now empty) list—of which we want only the first. We package that into an option (`SOME(#1(...))`), and the sentence is complete.

(What about the rest of the list returned from `parsePrepPhrase`? We completely ignored `#2(parsePrepPhrase(prPh))`. If the sentence fits our grammar, there should be nothing else after the prepositional phrase, so in that sense it should be harmless to ignore the presumably empty list—but see Project 2.F.)

```
- val dog = parseSentence("the big dog quickly chased a red ball
= through the bright field");

val dog =
  Sentence
    (NounPhrase (Art #,SOME #,Noun #),Predicate (SOME #,TVP #),
     SOME (PrepPhrase (#,#))) : sentence
```

Unfortunately, since the ML interpreter abbreviates long responses like this, we still have little to see for all the effort we have put in. Although parsing itself is an interesting problem, the fact is that so far we have only built values for our datatypes—we have not yet implemented any operations on the datatypes themselves.

To see how powerful this system is, we will use it to transform the sentences. The sentences that can be parsed so far must be declarative, meaning the sentences state a simple fact. In English, interrogative sentences (for example, "why" questions) require the verb to change form:

Why did the big dog quickly chase a red ball through the bright field?

Chased became *did chase*, and the two words fall to different places in the sentence. To make a function `makeInterrogative` that will print the sentence after transforming it to a "why" question, we first make helper functions to transform individual verbs. Note that these must return tuples, string * string:

```
fun interrogativeT(TV("chased")) = ("did", "chase")
  | interrogativeT(TV("saw")) = ("did", "see")
  | interrogativeT(TV("greeted")) = ("did", "greet")
  | interrogativeT(TV("bit")) = ("did", "bite")
  | interrogativeT(TV("loved")) = ("did", "love");
```

```
fun interrogativeI(IV("ran")) = ("did", "run")
  | interrogativeI(IV("slept")) = ("did", "sleep")
  | interrogativeI(IV("sang")) = ("did", "sing");

fun interrogativeL(LV("was")) = ("was", "")
  | interrogativeL(LV("felt")) = ("did", "feel")
  | interrogativeL(LV("seemed")) = ("did", "seem");
```

Since our transformed sentences are going to be printed as regular strings, we need some helper functions to undo the parsing of parts of the sentence, that is, to take values in our data types and turn them back into strings.

```
fun printNounPhrase(NounPhrase(Art(a), adj, Noun(n))) =
    a ^ " " ^ (case adj of SOME(Adj(aa)) => aa ^ " "
                          | NONE => "") ^ n;

fun printPrepPhrase(SOME(PrepPhrase(Prep(p), nPh))) =
      " " ^ p ^ " " ^ printNounPhrase(nPh)
   | printPrepPhrase(NONE) = "";

fun printAdverb(SOME(Adv(aa))) =  " " ^ aa
   | printAdverb(NONE) = "";
```

Now for the real work. The function `makeInterrogative` uses pattern-matching based on the the kind of verb phrase in the predicate.

```
fun makeInterrogative(Sentence(subj,
                               Predicate(adv, TVP(v, dObj)),
                               prPh)) =
    let val (v1, v2) = interrogativeT(v)
    in "why " ^ v1 ^ " " ^ printNounPhrase(subj) ^ " " ^ v2
       ^ " " ^ printNounPhrase(dObj) ^
       printAdverb(adv) ^ printPrepPhrase(prPh) ^ "?"
    end
  | makeInterrogative(Sentence(subj, Predicate(adv, IVP(v)),
                               prPh)) =
    let val (v1, v2) = interrogativeI(v)
    in "why " ^ v1 ^ " " ^ printNounPhrase(subj) ^ " " ^ v2 ^
       printAdverb(adv) ^ printPrepPhrase(prPh) ^ "?"
    end
  | makeInterrogative(Sentence(subj,
                               Predicate(adv, LVP(v, Adj(a))),
                               prPh)) =
    let val (v1, v2) = interrogativeL(v)
    in "why " ^ v1 ^ " " ^ printNounPhrase(subj) ^ " " ^ v2 ^
```

```
      printAdverb(adv) ^ " " ^ a ^ printPrepPhrase(prPh) ^ "?"
   end;
```

Trying this out:

```
- makeInterrogative(dog);

val it = "why did the big dog chase a red ball quickly
through the  bright field?"  : string

- makeInterrogative(parseSentence("the beautiful woman felt
= smart"));

val it = "why did the beautiful woman feel smart?" : string

- makeInterrogative(parseSentence("the flea dreamily slept on
= the smelly unicorn"));

val it = "why did the flea sleep dreamily on the smelly
unicorn?" : string
```

Project

2.A Write a function `diagram` (and appropriate helper functions, say `diagramNP`, `diagramVP`, etc) that takes a sentence and produces a string that shows the structure of the sentence in a readable format, equivalent to diagramming a sentence. The diagram of our sample sentence would be something like `(dog, the, big)(chased:(ball, a, red), quickly(through(field, the, bright)))`.

2.B Following the pattern for `makeInterrogative`, write a function `makeImperative` (and appropriate helper functions) that transforms and prints the sentence into the imperative mood. The verb should be in an appropriate place, and the subject should be vocative. Our sample sentence would be transformed into something like, *Chase a red ball quickly through the bright field, o big dog!"*

2.C Add to the datatypes and parsing functions so that our system will allow present active participles such as *chasing*. Participial phrases have the grammatical value of adjectives. They also need to be differentiated by whether the participle is of a transitive, intransitive, or linking verb, and have a direct object or predicate adjective if appropriate. Once present active participles are implemented, our language has the imperfect tense for free, since sentences can be constructed like *The big dog was chasing the red ball.*

2.D So far in our system, prepositional phrases have been only with an adverbial value, meaning they modify the verb phrase (*through the bright field* describes the action of chasing). However, prepositional phrases can also have an adjectival value, where they would modify a noun: "The dog *with the smart flea* chased the ball." Modify the datatypes and parsing functions so that prepositional phrases can follow a noun and modify it. One thing that makes this difficult is that if a prepositional phrase occurs at the end of a sentence that has a transitive verb, it is ambiguous whether it modifies the direct object or the verb. You can solve this by differentiating between prepositions that form adverbial prepositional phrases (such as *through*) and those that form adjective prepositional phrases (such as *with*).

Project, continued

2.E The system presented here will crash if given any string that is not in the language. Modify the parsing functions so that an exception is thrown when input that is not parsable is given. The parsing functions should display an error message and return a default value.

2.F Suppose we change the grammar to allow an arbitrary number of prepositional phrases at the end. Then sentences would have a list of `prepPhrases` rather than a single, optional one:

```
datatype sentence =
    Sentence of (nounPhrase * predicate *
                      prepPhrase list);
```

Rewrite `parseSentence` (and other functions, as appropriate) to handle this change.

2.7 Special topic: Lists vs. tuples vs. arrays

If you need to store a collection of values, should you use a list or a tuple? It depends on the nature of what you want to store and how you want to use them. A list requires all values in the collection to be of the same type. Using tuples requires uniform length of all similar tuples.

For most purposes we will use lists, and this decision affects how we can use our collections. By using a list instead of a tuple, we surrender the ability to extract an element from an arbitrary position in one step which can be done on a tuple using #1, #2, etc. This feature of tuples is called *random access*. The way we consider items in a list is called *sequential access*.

random access

sequential access

ML also provides another sort of container called an array, something we will touch on only briefly here, but it is familiar to those who have programmed in other languages. An *array* is a finite, uniformly-typed, ordered sequence of values, each value indicated by an integer *index*. Since it is a non-primitive part of ML, the programmer must load a special package that allows the use of arrays.

array

index

```
- open Array;
```

Tangent: Books and scrolls

The modern book format, with pages bound together, is called the codex (Latin for “block of wood”), as opposed to a scroll, where writing continues on a long, rolled-up piece of paper. Western civilization switched from scrolls to codices in the late Roman period.

Scrolls and codices illustrate the difference between sequential and random access. To find something in a scroll, one must roll and unroll the scroll until it is found (compare the idea of scrolling through text on a computer screen). To find a page in a codex, one can open to it almost immediately by approximating the location and flipping until the page is found. Page numbers also allow for the indirection found in a table of contents or an index.

The random access of a codex revolutionized how people used books. As we witness a new revolution in reading with various electronic formats, search functions have replaced much of the need for random-access dependent actions like indexing. It is interesting that in a format like a web page, we have largely reverted to a scroll format.

One creates a new array by typing `array(n, v)`, which will evaluate to an array of size n, with each position of the array initialized to the value v. The value at position i in an array is produced by `sub(A, i)`, and the value at that position is modified to contain v by `update(A, i, v)`.

```
- val A = array(10, 0);

val A = [|0,0,0,0,0,0,0,0,0,0|] : int array

- update(A, 2, 16);

val it = () : unit

- update(A, 3, 21);

val it = () : unit

- A;

val it = [|0,0,16,21,0,0,0,0,0,0|] : int array

- sub(A, 3);

val it = 21 : int
```

The function `update` returns the `unit` value, so it is a statement. This function changes the array—arrays are *mutable*. Although we can generate new tuples and lists, we cannot *change* the value of a tuple or list. However, unlike lists, new arrays cannot be made by concatenating two arrays together.

mutable

Much of the work of programming is weighing trade-offs among options. In this case, we are considering the appropriateness of various data structures, each of which has its advantages and liabilities. The following table summarizes the differences among tuples, lists, and arrays.

	Access	Concatenation	Length	Element types	Mutability
Tuples	random	unsupported	fixed	unrelated	immutable
Lists	sequential	supported	indefinite	uniform	immutable
Arrays	random	unsupported	indefinite	uniform	mutable

Chapter summary

As sets make the foundation of the mathematical concepts in this text, so lists are the primary feature of ML for storing arbitrary amounts of information. They are simple data structures, with access to a list made by breaking it into its first element and the list of its other elements. In this chapter, we used lists to reinforce the pieces of ML programming we have seen before and provide practice thinking recursively and writing recursive functions. We also use lists to model set concepts; in this chapter we used it to introduce and reason about powersets.

In the coming chapters, lists will be vital to almost all of our ML programming. We will also use them to model sets. The next chapter introduces formal logic, and in particular we will see how propositions that are quantified over a set parallel algorithms that operate over a list.

Key ideas:

Lists are finite, ordered sequences of values, all of the same type. They are used to store and organize arbitrary-lengthed sequences of values.

Lists differ from tuples in that their values have uniform type and the length of the list is not reflected in the type of the list.

A list is analyzed (decomposed or taken apart) into its first value (the head) and the list comprising the rest of the values (the tail). We call these *analytic* list operations, `hd` and `tl`. These are rarely used since we can analyze a list implicitly using pattern-matching.

New lists are synthesized (composed or put together) from old using the cons (`::`) and cat (`@`) operators. Cons takes an item and a list and makes a new list with the given item as the head and the given list as the tail. Cat takes two lists and joins them together in a new list. Only cons can be used in pattern-matching.

Recursive functions have at least one recursive case, which calls the function itself, and at least one base case, which gives an immediate answer. A function on a list will almost always have a base case for an empty list and at least one recursive case with a pattern using the cons operator.

The powerset of a set is the set of all subsets of that set.

Chapter 3 Proposition

This text is about constructing proofs and programs. Sets (along with types and lists and such) provide the raw materials. In this chapter, we study the simple toolkit of formal logic.

Logic is the set of rules for making deductions, that is, for producing new bits of information by knitting together other known bits of information. It not only is the foundation for mathematics and computing. It also trains the mind for any field, whether it be natural science, rhetoric, philosophy, or theology. Three clarifications before we begin:

First, we are studying what may be called *symbolic* or *formal* logic. Other approaches to the study of argumentation exist, especially in the field of philosophy.

Second, formal logic in particular studies the *form* of arguments, not their contents. The argument

> *If the moon is made of blue cheese, then alligators wear slippers.*
> *The moon is made of blue cheese.*
> *Therefore, alligators wear slippers.*

is perfectly logical. Its absurdity lies in the content of its premises. Similarly, the argument

> *If an integer is a multiple of 3, then its digits sum to a multiple of 3.*
> *The digits of 216 sum to 9, a multiple of 3.*
> *Therefore 216 is a multiple of 3.*

is illogical, even though everything (except the word *therefore*) is true.

Third, logic is a *tool* for thinking; it is not thinking itself, nor is it a substitute for thinking. Common sense, observation, and intuition are necessary to find the right context to apply logic. G. K. Chesterton observed that a madman "is the more logical for losing certain sane affections... The madman is not the man who has lost his reason. The madman is the man who has lost everything except his reason" [4].

Chapter goals. The student should be able to

- evaluate symbolic propositions using the operations of the propositional calculus and using a truth table.
- use the ML bool type.
- verify new logical equivalences by simplifying expressions with known logical equivalences.
- write ML functions using conditional expressions.
- verify the validity of simple argument forms using truth tables.
- verify the validity of larger argument forms using simple argument forms.
- write ML functions whose specifications include single or multiple quantification.
- verify the validity of larger quantified argument forms using simple quantified argument forms.

3.1 Forms

Consider these two arguments:

> If it is Wednesday or spinach is on sale, then I go to the store.
> So, if I do not go to the store, then it is not Wednesday and spinach is not on sale.
>
> If $x < -5$ or $x > 5$, then $|x| > 5$.
> So, if $|x| \not> 5$, then $x \not< -5$ and $x \not> 5$.

What do these have in common? If we strip out everything except for the words *if, or, then, so, not* and *and*—that is, if we strip out all the content but leave the logical connectors—we are left in either case with

> If p or q, then r.
> So, if not r, then not p and not q.

We replaced the content with the variables p, q, and r. This allowed us to abstract from the two arguments to find a form common to both.

What sorts of things do these variables represent? Not numbers. Grammatically we would substitute in for them *independent clauses*, grammatical items that have a complete meaning and can be true or false. In the terminology of logic, we say a *proposition* is a sentence that is true or false, but not both. The words *true, false*, and *sentence* we leave undefined, like *set* and *element*.

proposition

Sentences that are not propositions include "sentences" that are ungrammatical, or sentences that are grammatical but still meaningless; sentences with insufficient information (in mathematics, the missing information would be unbound variables; questions or commands; and paradoxical sentences, which are either simultaneously true and false or neither true nor false).

Since a proposition can be true or false, the following qualify as propositions:

> $7 - 3 = 4$
> $3 - 7 = 4$
> Springfield is the capital of Massachusetts.

Something that is like a proposition except that it has variables (for true or false values) instead of content is a *propositional form* or simply a *form*.

form

> ***Tangent: The benevolent wishbone paradox***
>
> Fred and Frank are pulling a wishbone. Fred is a benevolent wisher—his only wish is that Frank's wish comes true. But Frank is also a benevolent wisher, and he's wishing that Fred's wish comes true.
>
> The sentence "Fred's wish and Frank's wish both came true" is both true and false—at least, supposing it either to be true or false is consistent with the facts.

Exercises

Determine which of the following are propositions.

3.1.1 Spinach is on sale.

3.1.2 Spinach on sale.

3.1.3 This sentence is true.

3.1.4 This sentence is false.

3.1.5 $3 > 5$.

3.1.6 $3 + 5$.

3.1.7 If $3 > 5$, then spinach is on sale.

3.1.8 $x > 5$

3.1.9 Why is spinach on sale?

3.1.10 If p, then spinach is on sale.

3.2 Symbols

Let $\mathbb{B}$ be a set with elements T and F.

$$\mathbb{B} = \{T, F\}$$

For this set, we define three operations, $\sim$, $\wedge$, and $\vee$. They all take operands from $\mathbb{B}$ ($\sim$ takes one operand, the others take two) and result in an element from $\mathbb{B}$, defined in this way (compare with multiplication and addition tables used to teach arithmetic):

"AND"

"OR"

p	$\sim p$
T	F
F	T

p	q	$p \wedge q$
T	T	T
T	F	F
F	T	F
F	F	F

p	q	$p \vee q$
T	T	T
T	F	T
F	T	T
F	F	F

negation

conjunction

disjunction

We call these operations *negation*, *conjunction*, and *disjunction*, respectively. Negation has a higher precedence than conjunction and disjunction; $\sim p \vee q$ means $(\sim p) \vee q$, not $\sim (p \vee q)$. Conjunction and disjunction are performed from left to right, with conjunction having higher precedence than disjunction (that is, it gets executed first). For example

$$\begin{aligned} T \vee \sim (F \vee T) \wedge \sim F &= T \vee \sim T \wedge \sim F && \text{since } F \vee T = T \\ &= T \vee F \wedge T && \text{since } \sim T = F \\ &= T \vee F && \text{since } F \wedge T = F \\ &= T && \text{since } T \vee F = T \end{aligned}$$

propositional calculus

We have presented this as a mathematical system—the *propositional calculus*—but you have probably guessed that T and F stand for *true* and *false*, and our operations are *not*, *and*, and *or*. The adverb *not* can modify an entire independent clause, and so the unary operator $\sim$ takes a proposition—in this case, something that evaluates to either T or F—as an operand. The conjunctions *and* and *or* join two independent clauses, just as the binary operators $\wedge$ and $\vee$ take two propositions as operands. (We regret to use the word *conjunction* in two separate meanings: *conjunction* the operator and *conjunction* the part of speech. The English conjunction *and* corresponds to the conjunction logical operator, but the English conjunction *or* corresponds to the disjunction logical operator.)

From this observation we might draw a correspondence between propositions in English and a symbolic representation. In the following examples, assume r stands for "it is raining" and s stands for "it is snowing."

It is neither raining nor snowing.	$\sim r \wedge \sim s$
It is raining and not snowing.	$r \wedge \sim s$
It is both raining and snowing.	$r \wedge s$
It is raining but not snowing.	$r \wedge \sim s$
It is either raining or snowing but not both.	$(r \vee s) \wedge \sim (r \wedge s)$.

The problem is that our formal system and the corresponding elements of natural language are not an exact match, and your experience can mislead you.

First, the subtlety of natural language packs meaning into simple phrases. Notice that "and" and "but" have the same symbolic translation. This is because both conjunctions have the same denotational, logical value—they each assert that both of two things are true. That does not mean that the differences in their *connotations* are not important. "a but b" spins the statement to imply something like "a and b are both true, and b is surprising in light of a."

Moreover, the logical operators $\wedge$ and $\vee$ are less flexible than their corresponding English words. English conjunctions can yoke almost any part of speech: nouns ("He needs a car or truck."), verbs ("Yesterday I ran and swam."), or adjectives ("My dog is big and hungry."). Logical operators can operate only on propositions or their substitutionary equivalents—that is, independent clauses ("He needs a truck, and my dog is hungry.").

The gulf between natural language and formal logic will widen as our formal system grows. Natural language is a slippery creature, full of tricks and ambiguities. The propositional calculus is tidy and clear, and that is why we use it. Some have argued that the very values T and F mislead introductory students to connect them too closely to everyday speech. Edsger Dijkstra proposed replacing them with neutral values like *black* and *white*, or 1 and 0, or $\top$ and $\bot$, to "further sever the links to intuition" [7].

Tangent: And _and_ but

If it is hard to swallow the idea that "and" and "but" have the same literal meaning, observe how another language differentiates things more finely. Greek has three words to cover the same semantic range as our "and" and "but": *kai,* meaning "and"; *alla,* meaning "but"; and *de,* meaning something halfway between "and" and "but," joining two things together in contrast but not as sharply as *alla.*

Exercises

Evaluate each logical expression

3.2.1 $T \wedge (\sim F \vee F) \wedge (T \vee T)$

3.2.2 $(T \vee F) \wedge \sim (F \wedge T)$

3.2.3 $\sim T \vee F \wedge T \vee T$

3.2.4 $(F \vee F \vee T) \wedge (\sim T \vee F)$

3.2.5 $\sim (T \vee (F \wedge T))$

3.2.6 $(T \vee T) \wedge (F \vee F)$

Let s stand for "spinach is on sale" and k stand for "kale is on sale." Write the following using logical symbols.

3.2.7 Kale is on sale, but spinach is not on sale.

3.2.8 Either kale is on sale and spinach is not on sale or kale and spinach are both on sale.

3.2.9 Kale is on sale, but spinach and kale are not both on sale.

3.2.10 Spinach is on sale and spinach is not on sale.

3.3 Boolean values

ML provides a type for logical values, bool, short for *boolean,* named after George Boole. The two values of bool are `true` and `false`. Expressions and variables, accordingly, can have the type bool.

```
- true;

val it = true : bool

- val p = false;

val p = false : bool
```

The basic boolean operators are named `not`, `andalso`, and `orelse`.

```
- p andalso true;

val it = false : bool

- true orelse not p;

val it = true : bool
```

Biography: George Boole, 1815–1864

Boole was an English mathematician who brought logic under the domain of mathematics. Born in Lincoln, England, he taught himself mathematics while serving as a school teacher. When he was 34 he became a professor at a newly established university in Cork, Ireland. Much of his ideas about logic and mathematics were systematized in his book *An Investigation of the Laws of Thought*, which he published when he was 39, a late age to publish a first work of such originality. [2]

Comparison operators, which test for equality and related ideas, are different from others we have seen in that their results have types different from their operands. If we compare two ints, we do not get another int, but a bool. The ML comparison operators are =, <, >, <=, >=, and <>, the last three for $\leq$, $\geq$, and $\neq$.

```
- 5 <= 4;

val it = false : bool
```

ML does not allow testing reals for equality and inequality because real numbers are not stored on a computer exactly—computers, being finite, can store only a finite representation of a number. Instead, test if the difference between two real values is less than a certain small value. For example, if you want to know if `x` is 5, test whether it is "close enough" to 5.0 with `abs(x - 5.0) < 0.0001`, where `abs` computes absolute value.

= and <> are defined for datatypes, and they can be used on lists and tuples that contain only types on which = and <> are defined.

When using recursion to write a boolean-valued function, remember that there are two possible values for it to return. This often manifests itself in two base cases. Consider the problem of determining whether a whole number is even. Zero is even, and any number has the same even-or-odd-ness as the number two less than it. Thus we might try

```
- fun isEven(0) = true
=   | isEven(n) = isEven(n-2);    (* bad! *)

val isEven = fn : int -> bool
```

This has no mechanism for returning `false` and will tailspin into infinite recursion when given an odd number. Instead,

```
- fun isEven(0) = true
=   | isEven(1) = false
=   | isEven(n) = isEven(n-2);
```

As you work through problems like this, you probably will find it easiest to think of them in terms of two base cases. However, it is not always necessary to make those two outcomes explicit. In the present example, if we subtract one from `n` then we should *negate* the result and so we need only a `true` base case.

```
- fun isEven(0) = true
=   | isEven(n) = not(isEven(n-1));
```

For this problem there is also a non-recursive solution.

```
- fun isEven(n) = n mod 2 = 0;
```

Exercises

Analyze the type of the expression, or indicate that there is a type error.

3.3.1 `3 + 4 <= 7`

3.3.2 `[(5, 8) = (5, 7), 3.4 < 7.8]`

3.3.3 `5 < 8 andalso not ([4, 5] = [])`

3.3.4 Write a function `turnToInts` that, given a list of bools, produces a similar list of ints, with 1s and 0s replacing the `true`s and `false`s.

3.3.5 Write a function `areEven` that, given a list of ints, produces a list of bools indicating whether the list elements are even or odd. For example, `areEven([5,6,7,10])` should return `[false,true,false,true]`. (Hint: x is even if $x \bmod 2$ is 0.

3.3.6 Write a function `containsOne` that, given a list of ints, determines whether the list contains 1.

3.4 Logical equivalence

truth tables

The tables used in Section 3.2 to define the logical operators are called *truth tables*. They can be used not only for defining those operators but also for analyzing logical formulas with variables. The left columns of a truth table show the variables in the formulas, and the rows enumerate the possible combinations of values assigned to the variables. The other columns are for formulas.

Suppose we wanted to determine the value of $(p \wedge q) \vee \sim (p \vee q)$ for the various assignments to p and q. We do this by making a truth table with columns on the left for the arguments; a set of columns in the middle for intermediate propositions, each one the result of applying the basic operators to the arguments or to other intermediate propositions; and a column on the right for the final answer.

p	q	$p \wedge q$	$p \vee q$	$\sim (p \vee q)$	$(p \wedge q) \vee \sim (p \vee q)$
T	T	T	T	F	T
T	F	F	T	F	F
F	T	F	T	F	F
F	F	F	F	T	T

Truth tables are a "brute force" way of attacking logical problems. For complicated propositions and arguments, they can be tedious, and we will learn more expeditious ways of exploring logical properties. However, truth tables are a sure method of evaluation to which we can fall back when necessary.

Consider the forms $\sim (p \wedge q)$ and $\sim p \vee \sim q$. This truth table evaluates them for all possible assignments to the variables.

p	q	$\sim p$	$\sim q$	$p \wedge q$	$\sim (p \wedge q)$	$\sim p \vee \sim q$
T	T	F	F	T	F	F
T	F	F	T	F	T	T
F	T	T	F	F	T	T
F	F	T	T	F	T	T

logically equivalent

The two rightmost columns are identical. This is because the forms $\sim (p \wedge q)$ and $\sim p \vee \sim q$ are *logically equivalent*, that is, they have the same truth value as each other for all assignments of their arguments. We use $\equiv$ to indicate that two forms are logically equivalent. For example, similar to the equivalence demonstrated above, it is true that $\sim (p \vee q) \equiv \sim p \wedge \sim q$. These two equivalences are called De Morgan's laws, after Augustus De Morgan.

It is important to remember that the operators $\vee$ and $\wedge$ flip when they are negated. For example, take the sentence

x is even and prime.

We do not call this a proposition, because x is an unknown, but by supplying a value for x (taking $\mathbb{N}$ as the universal set) we would make it a proposition. The set of values that make this a true proposition is $\{2\}$. The set of values that make this proposition false needs to be the complement of that set—that is, the set of all natural numbers besides 2. It may be tempting to negate the not-quite-a-proposition as

x is not even and not prime.

But this is wrong. The set of values that makes this a true proposition is the set of all numbers except evens and primes—a much different set from what is required. The correct negation is

x is not even or not prime.

A form that is logically equivalent with the constant value T (something always true, no matter what the assignments are to the variables) is called a *tautology*. A form that is logically equivalent to F (something necessarily false) is called a *contradiction*. Obviously all tautologies are logically equivalent to each other, and similarly for contradictions. The following truth table explores a few tautologies and contradictions and related forms.

tautology

contradiction

p	$\sim p$	$p \vee \sim p$	$p \wedge \sim p$	$p \wedge T$	$p \vee T$	$p \wedge F$	$p \vee F$
T	F	T	F	T	T	F	T
F	T	T	F	F	T	F	F

Hence $p \vee \sim p$ and $p \vee T$ are tautologies, $p \wedge \sim p$ and $p \wedge F$ are contradictions, $p \wedge T \equiv p$, and $p \vee F \equiv p$. Compare these facts with laws you learned for

Biography: Augustus De Morgan, 1806–1871

Augustus De Morgan was a British mathematician and logician, though he was the third generation of his family born in India. He spent most of his career at the newly formed University College of London. He was known as a witty controversialist. De Morgan made contributions to many areas of mathematics, including calculus and algebra. He coined the term *mathematical induction*—see Section 6.5 His work on logic prepared the way for the work of George Boole a few years later. [19]

arithmetic and algebra: $x+0=x$, $x\cdot 1=x$, $x+-x=0$, $x/x=1$. The predicate calculus is its own sort of arithmetic with its own values, operators, and laws.

The following Theorem displays common logical equivalences.

Theorem 3.1 (Logical equivalences.) *Given logical variables p, q, and r, the following equivalences hold.*

Commutative laws:	$p \wedge q$	$\equiv$	$q \wedge p$	$p \vee q$	$\equiv$	$q \vee p$
Associative laws:	$(p \wedge q) \wedge r$	$\equiv$	$p \wedge (q \wedge r)$	$(p \vee q) \vee r$	$\equiv$	$p \vee (q \vee r)$
Distributive laws:	$p \wedge (q \vee r)$	$\equiv$	$(p \wedge q) \vee (p \wedge r)$	$p \vee (q \wedge r)$	$\equiv$	$(p \vee q) \wedge (p \vee r)$
Absorption laws:	$p \wedge (p \vee q)$	$\equiv$	p	$p \vee (p \wedge q)$	$\equiv$	p
Idempotent laws:	$p \wedge p$	$\equiv$	p	$p \vee p$	$\equiv$	p
Double negative law:	$\sim\sim p$	$\equiv$	p			
DeMorgan's laws:	$\sim (p \wedge q)$	$\equiv$	$\sim p \vee \sim q$	$\sim (p \vee q)$	$\equiv$	$\sim p \wedge \sim q$
Negation laws:	$p \vee \sim p$	$\equiv$	T	$p \wedge \sim p$	$\equiv$	F
Universal bound laws:	$p \vee T$	$\equiv$	T	$p \wedge F$	$\equiv$	F
Identity laws:	$p \wedge T$	$\equiv$	p	$p \vee F$	$\equiv$	p
Tautology and contradiction laws:	$\sim T$	$\equiv$	F	$\sim F$	$\equiv$	T

Do not memorize this table, but take a little time deciphering it. The first three are not difficult, since they are analogous to properties of addition and multiplication (as well as to set operations $\cup$ and $\cap$). Most of the others are intuitive; you merely need to learn their names.

These can be verified using truth tables. They also can be used to prove other equivalences without using truth tables but instead reducing them by a series of steps to a simpler form. For example, $q \wedge (p \vee T) \wedge (p \vee \sim (\sim p \vee \sim q))$ is equivalent to $p \wedge q$:

$q \wedge (p \vee T) \wedge (p \vee \sim (\sim p \vee \sim q))$	
$\equiv\ q \wedge T \wedge (p \vee \sim (\sim p \vee \sim q))$	by universal bounds
$\equiv\ q \wedge (p \vee \sim (\sim p \vee \sim q))$	by identity
$\equiv\ q \wedge (p \vee \sim\sim (p \wedge q))$	by DeMorgan's
$\equiv\ q \wedge (p \vee (p \wedge q))$	by double negative
$\equiv\ q \wedge p$	by absorption
$\equiv\ p \wedge q$	by commutativity.

Problems like these constitute the first of three logical games we play in this chapter. The games are designed to prepare you for the proof-writing you will do in Chapter 4. Remember from high school algebra that there are "simplify" problems and "solve" problems.

■ Simplify $3x(2+3x)^2+1$.

$$\begin{aligned} & 3x(2+3x)^2+1 \\ = \quad & 3x(4+12x+9x^2)+1 \\ = \quad & 12x+36x^2+27x^3+1 \\ = \quad & 27x^3+36x^2+12x+1 \end{aligned}$$

■ Solve $12x = 57 - 7x$ for x.

$$\begin{aligned} 12x &= 57-7x \\ 19x &= 57 \\ x &= 3 \end{aligned}$$

On the left is a chain of expressions all equal to each other. On the right is a sequence of equations, each complete, but each following logically from the previous. It is important to recognize that Game 1 problems are "simplify" problems, not "solve" problems—even though we give you the simplified expression. Your task is to list the intermediate steps. Suppose we were to show that $\sim(\sim p \wedge q) \vee (p \vee \sim p) \equiv \sim p \vee \sim q$.

Do this:

$$\begin{aligned} & \sim(\sim p \wedge q) \vee (p \wedge \sim p) & \\ \equiv \quad & \sim(\sim p \wedge q) \vee F & \text{by negation law} \\ \equiv \quad & \sim(\sim p \wedge q) & \text{by identity law} \\ \equiv \quad & p \vee \sim q & \text{by De Morgan's} \end{aligned}$$

Don't do this:

$$\begin{aligned} \sim(\sim p \wedge q) \vee (p \wedge \sim p) &\equiv \sim p \vee \sim q & \\ \sim(\sim p \wedge q) \vee F &\equiv \sim p \vee \sim q & \text{by negation law} \\ \sim(\sim p \wedge q) &\equiv \sim p \vee \sim q & \text{by identity law} \\ p \vee \sim q &\equiv p \vee \sim q & \text{by De Morgan's} \end{aligned}$$

The problem with the "solve" format is that it reads as if we assumed the first line were true and then validated the assumption by deriving the last line, which we know is true—that would be a "proof by tautology," a fallacy exposed in Section 4.4. By contrast, in the algebraic "solve" problem above, the equation $12x = 57 - 7x$ *is* a given fact, and $x = 3$ is deduced from it.

Exercises

Verify the following equivalences using a truth table. Then verify them using ML (that is, type in the left and right sides for all four possible assignments to p, q, and r checking that they agree each time).

3.4.1 $\sim(p \wedge q) \equiv \sim p \vee \sim q$.

3.4.2 $p \wedge (p \vee q) \equiv p$.

3.4.3 $(p \wedge q) \wedge r \equiv p \wedge (q \wedge r)$.

3.4.4 $p \wedge (q \vee r) \equiv (p \wedge q) \vee (p \wedge r)$.

Verify the following equivalences by applying known equivalences from Theorem 3.1.

3.4.5 $\sim(\sim p \vee (\sim p \wedge \sim q)) \vee \sim p \equiv T$.

3.4.6 $p \wedge (\sim q \vee (p \wedge \sim p)) \equiv p \wedge \sim q$.

3.4.7 $(q \wedge p) \vee \sim(p \vee \sim q) \equiv q$.

3.4.8 $((q \wedge (p \wedge (p \vee q))) \vee (q \wedge \sim p)) \wedge \sim q \equiv F$.

3.4.9 $\sim(\sim(p \wedge p) \vee (\sim q \wedge T)) \equiv p \wedge q$.

3.4.10 $\sim(\sim(q \vee q) \vee (\sim p \vee F)) \equiv p \wedge q$

3.4.11 $\sim((p \wedge (p \vee r)) \vee \sim(p \vee (p \wedge q))) \equiv F$

3.4.12 $((p \wedge T) \vee \sim(\sim r \vee \sim q)) \wedge \sim((\sim r \wedge \sim r) \vee (F \wedge \sim q)) \equiv ((p \vee q) \wedge r)$

3.4.13 $\sim((\sim p \vee \sim r) \wedge (\sim q \vee r)) \vee (p \wedge \sim r) \vee (q \wedge r) \equiv p \vee q$

3.4.14 $\sim((\sim p \wedge \sim q) \wedge \sim(p \wedge \sim r)) \wedge (\sim r \vee F) \equiv (p \vee q) \wedge \sim r$

3.5 Conditional propositions

The formula $\sim p \vee q$ is so commonly used that we have a special symbol for it, $\rightarrow$. By definition, $p \rightarrow q \equiv \sim p \vee q$, that is

p	q	$\sim p \vee q$	$p \rightarrow q$
T	T	T	T
T	F	F	F
F	T	T	T
F	F	T	T

conditional

hypothesis

conclusion

This operator is the *conditional* operator, and propositions or forms in which the operator has lowest precedence (that is, it is done last) are called *conditional propositions* and *conditional forms*. p is called the *hypothesis* and q the *conclusion*. $p \rightarrow q$ is normally read "if p then q." However, do not rely on your understanding of *if* and *then* in natural language to reason about the conditional operator.

To see what the conditional operator means (and how it differs from what we would expect from everyday English), consider some examples. The subpropositions—those independent clauses we substitute for p and q—must be full-fledged propositions with definite, known values—no missing information or ambiguity. In common sentences, that is rarer than you might think. We will restrict ourself with mathematical facts (and lies).

	p		q
If	12 divides 36 evenly,	**then**	3 divides 72 evenly.

This sounds pretty reasonable.Since both p and q are true, the conditional operator indicates that the entire proposition is true.

If	$3 < 72$,	**then**	3 divides 72 evenly.

What does the hypothesis have to do with the conclusion? Nothing. But according to the logical operator $\rightarrow$, we consider this proposition to be true formally because both p and q are true.

If	12 divides 36 evenly,	**then**	$72 < 3$.

Our expectations and the conditional operator happen to agree that this is a false proposition. The hypothesis is true, but the conclusion is false.

If	$72 < 3$,	**then**	3 divides 72 evenly.

In English, this sentence does not mean much of anything. But in formal logic, a false hypothesis and true conclusion mean it is true.

If	$72 < 3$,	**then**	12 divides 3 evenly.

Similar to the previous example, anytime the hypothesis is false, no matter what the conclusion is, the proposition is counted as true formally.

Most offensive to our intuition are the cases where the hypothesis is false and yet we define the entire proposition to be true. In English we might consider them to be neither true nor false but irrelevant. We say that such propositions are *vacuously true*.

vacuously true

There are three common forms that are variations on the conditional. Switching the order of the hypothesis and the conclusion, $q \to p$, is the *converse* of $p \to q$.

converse

If 12 divides 36 evenly, then 3 divides 72 evenly.
If 3 divides 72 evenly, then 12 divides 36 evenly.

The converse is not logically equivalent to the proposition.

p	q	$p \to q$	$q \to p$
T	T	T	T
T	F	F	T
F	T	T	F
F	F	T	T

Many common errors in reasoning come down to a failure to recognize this. p being correlated to q is not the same thing as q being correlated to p.

If spinach is on sale, then I go to the store.

is not the same as

If I go to the store, then spinach is on sale.

The *inverse* is formed by negating each of the hypothesis and conclusion separately (*not* negating the entire conditional), $\sim p \to \sim q$.

inverse

If 12 divides 36 evenly, then 3 divides 72 evenly.
If 12 does not divide 36 evenly, then 3 does not divide 72 evenly.

For the same reason as the converse, the inverse is not logically equivalent to the proposition either, but a proposition's inverse and converse are logically equivalent to each other.

p	q	$p \to q$	$\sim p \to \sim q$	$q \to p$
T	T	T	T	T
T	F	F	T	T
F	T	T	F	F
F	F	T	T	T

The *contrapositive* is formed by negating and switching the components of a conditional, $\sim q \to \sim p$.

contrapositive

If 12 divides 36 evenly, then 3 divides 72 evenly.
If 3 does not divide 72 evenly, then 12 does not divide 36 evenly.

The contrapositive of a proposition is logically equivalent to the original proposition.

p	q	$p \to q$	$\sim q \to \sim p$
T	T	T	T
T	F	F	F
F	T	T	T
F	F	T	T

Notice that the converse and inverse are contrapositives of each other.

To *negate* the conditional, we want to find a formula that is equivalent to $\sim (p \to q)$. One trying to guess such a formula might try propagating the negation to the hypothesis or conclusion or both, but all of these fail as logical equivalences:

p	q	$p \to q$	$\sim (p \to q)$	$\sim p \to q$	$p \to \sim q$	$\sim p \to \sim q$
T	T	T	F	$\boxed{\text{T}}$	F	$\boxed{\text{T}}$
T	F	F	T	T	T	T
F	T	T	F	$\boxed{\text{T}}$	$\boxed{\text{T}}$	F
F	F	T	F	F	$\boxed{\text{T}}$	$\boxed{\text{T}}$

A smarter approach is to observe that the negation of the conditional is true only in the "T F" row, that is, only when p is true and q is false: $p \wedge \sim q$. We also could deduce with Game 1 and the logical equivalence laws:

$$\begin{array}{lll} \sim (p \to q) & & \\ \equiv & \sim (\sim p \vee q) & \text{since } p \to q \equiv \sim p \vee q \\ \equiv & \sim\sim p \wedge \sim q & \text{by DeMorgan's} \\ \equiv & p \wedge \sim q & \text{by double negative} \end{array}$$

Exercises

3.5.1 Use a truth table to show that $(p \vee q) \to r \equiv \sim r \to (\sim p \wedge \sim q)$.

3.5.2 The equivalence $p \to q \equiv \sim q \to \sim p$ is called the contrapositive law. Use this and known equivalences from Theorem 3.1 to show that $(p \vee q) \to r \equiv \sim r \to (\sim p \wedge \sim q)$.

Write the negation, converse, inverse, and contrapositive of each of the following propositions.

3.5.3 If it is not snowing, then spinach is on sale.

3.5.4 If spinach is on sale and it is raining, then it is not snowing.

3.5.5 If it is raining, then spinach is not on sale or it is snowing.

3.5.6 If spinach is on sale, then if it is not raining, then it is snowing.

3.6 Conditionals and natural language

Wendy's functional programming class is held in Classroom 3 on the first floor of the science building, which has four classrooms, a lab, a lecture hall, the dean's office, bathrooms, and a display area with a reconstructed mastodon.

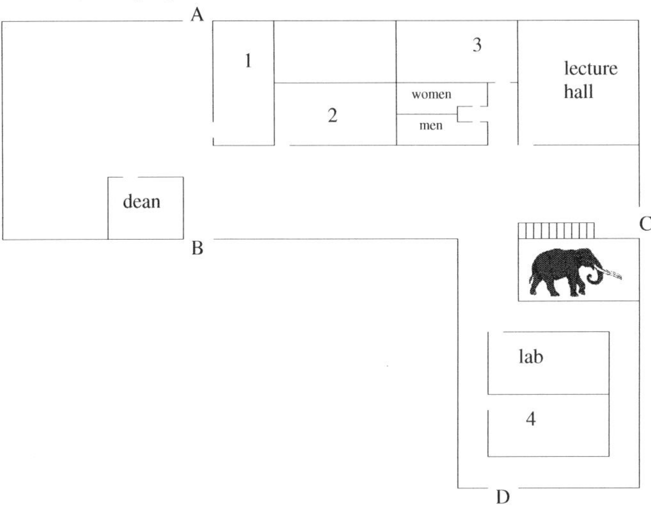

One day Wendy entered the building through one of the four entrances and went to her classroom without going into any other room. From this diagram, determine which propositions are true or false, or whether you do not have enough information to tell.

Wendy walked past the bathrooms.	This is *true*, since we know Wendy entered classroom 3, and she had to talk past the bathrooms to get there.
Wendy walked past the lab.	We *cannot tell* whether this is true or false with the given information. There are possible paths that would take her past the lab and paths that would not.
If Wendy walked past the mastodon, then Wendy went into the lecture hall.	This is *false* because she had to walk past the mastodon to get to Classroom 3, but we are told she entered no other rooms.
If Wendy entered through entrance A, then she walked past the dean's office.	This is *true*, since the dean's office lies on the way from entrance A to classroom 3.
If Wendy walked past the dean's office, then she entered through entrance A.	This is *false* because she could have walked past the dean's office by entering through entrance B, right? No—because what we did not tell you was that Wendy actually entered through entrance C and went straight to classroom 3, so this sentence turns out to be vacuously true! (The correct answer, then, was that you *cannot tell* with the given information.)

If you think this is ridiculous, then well done. Of course the last sentence is false. It is true that in the fictional world where Wendy walked straight from entrance C to classroom 3,

$$\text{Wendy walked past the dean's office} \rightarrow \text{Wendy entered through entrance A}$$

is vacuously true, but that is not what the English sentence means. The sentence is expressing something about all hypothetical paths, not a logical operation on two specific facts. This is to show that in almost all conditional-structured sentences in plain speech (and almost all interesting ones in mathematics and computer science, too) deeper meaning lurks beneath the surface. In this case, the sentence has implicit *quantification*, the topic of Section 3.11.

Consider the variety of conditions that can appear in everyday speech:

Past conditionals:

"If Fred was at the dock at midnight, then he's the murderer."
(This is making a deduction that Fred's presence at the scene of th crime at the time of the crime is evidence of his guilt. It holds if a possible—or reasonable—scenarios involving Fred at the dock als involve Fred committing the murder.)

Present conditionals:

"If it's raining at home and the windows are still open, then wate is coming in."
(This is similar to a past conditional. The speaker is making a clai about any state of the universe that includes his or her window being open during a rain storm.)

Conditions contrary to fact:

"If I were John and John were me, then he'd be six and I'd b three." — A. A. Milne
(If we blindly applied the conditional operator, then all condition contrary to fact would be vacuously true. In reality, the speake reasons about what would be true *in a hypothetical world*—just a we will be doing when we write proofs of conditional proposition in Section 4.5.)

Conditional commands:

"If the dryer is finished, then unload it."
(This is not a proposition, but "the dryer is finished" is.)

Promises:

"If you finish your spinach, then I will give you some cake.'
(Some texts make the idea of vacuous truth more intuitive by using promises as examples. What turn of events would show the promise to be a true one? Suppose the child does not finish the spinach, but the parent gives in and gives cake anyway. The promise still vacuously holds, since the parent never said that the child would *not* get cake unless he or she finished the spinach—although that was probably the intent. The promise is only broken if the child finishes the spinach but is then denied cake.)

Other future conditionals:

"If it rains tomorrow, the zucchini will come up."
(This is similar to a promise. But since no one can predict the future, what does it mean for a future conditional to be true at the time when it is spoken?)

Do not conclude from this that the conditional operator is useless. It is in fact the right tool for making deductions in our formal system. Moreover, it is a core element for capturing the logic of real sentences, when used together with other formal tools.

Finally, take note that there are several ways to phrase conditionals in natural language. Here are a few phrasings common in mathematics (all examples involve some kind of quantification).

A shape is a square only if it is a rectangle.

This certainly does not say that if a shape is a rectangle, then it is a square. It leaves open the possibility that a shape is a rectangle even though it is something other than a square. That is, it says that if a shape is a square, then it is a rectangle. In other words, *only if* is a way of phrasing the converse of a proposition. "p only if q" means $p \rightarrow q$, the converse of "p if q" ($q \rightarrow p$).

Along the same lines, the phrase "if and only if," often abbreviated "iff," is of immense use in mathematics. If we expand

$x \texttt{ mod } 2 = 0$ iff x is even.

we get

If $x \texttt{ mod } 2 = 0$, then x is even, and if x is even, then $x \texttt{ mod } 2 = 0$.

That is, we get both the conditional and its converse. The connector *iff* is sometimes called the *biconditional*, or *biimplication* $p \leftrightarrow q$, and is defined to be equivalent to $(p \rightarrow q) \wedge (q \rightarrow p)$. (How many English words can you think of with two consecutive *i*s? *Wii* does not count.)

biconditional

p	q	$p \rightarrow q$	$q \rightarrow p$	$p \leftrightarrow q$
T	T	T	T	T
T	F	F	T	F
F	T	T	F	F
F	F	T	T	T

necessary conditions

sufficient conditions

We also sometimes speak of *necessary conditions* and *sufficient conditions*, which refer to converse conditional and conditional propositions, respectively.

> An even degree is a necessary condition for a polynomial to have no real roots
>
> *means*
>
> If a polynomial function has no real roots, then it has an even degree.

> A positive global minimum is a sufficient condition for a polynomial to have no real roots
>
> *means*
>
> If a polynomial function has a positive global minimum, then it has no real roots.

> Values all of the same sign is a necessary and sufficient condition for a polynomial to have no real roots.
>
> *means*
>
> A polynomial function has values all of the same sign if and only if the function has no real roots.

Note that in the first two examples the converse does not hold, but since the last example is equivalent to a biconditional, it means that both a conditional and its converse are true.

3.7 Conditional expressions

Suppose we wish to write a function which determines whether or not a given item occurs in a list. Here is a try:

```
- fun contains(x, []) = false
=   | contains(x, x::rest) = true
=   | contains(x, y::rest) = contains(x, rest);
```

ML, however, does not allow this. A variable like x cannot appear twice in the same pattern. The intent is to compare the parameter x with the first item of the list, but ML will not perform the implicit comparison.

Pattern-matching is a decision-making mechanism in ML. Here is another decision-making mechanism, which will solve the problem above. *Conditional expressions* have the form

conditional expression

$$\text{if } <\text{expression}>_1 \text{ then } <\text{expression}>_2 \text{ else } <\text{expression}>_3$$

Make sure you notice the difference between ML's conditional expressions and the conditional operator from Section 3.5. It is also a little different from conditional statements in programming languages like Java. The first subexpression is called the *test*. The second two are called the then-clause and else-clause. The test must have type bool. The other expressions must have the same type as each other, but that type can be anything. If the condition is true, then the value of the entire expression is the value of the then-clause; if it is false, then the entire expression's value is that of the else-clause. Thus the type of the second two expressions is also the type of the entire expression.

test

```
- fun contains(x, []) = false
=    | contains(x, y::rest) = if x = y then true
=                                     else contains(x, rest);

Warning: calling polyEqual
val contains = fn : ''a * ''a list -> bool

- contains(5, [7, 6, 5, 4]);

val it = true : bool

- contains(5, [9, 8, 7, 6]);

val it = false : bool
```

The warning indicates that in writing this function we are assuming equals is defined for the type of elements in the list. We could not use this function on a list of reals.

We could use conditional expressions wherever we would use pattern-matching, but pattern-matching is more elegant. Some functions, on the other hand, that cannot be written without conditionals. Notice that in our use of lists to model sets, the function shown above implements the $\in$ operation.

Now we can generalize our `replaceTomato` function so that it replaces any kind of vegetable with any other:

```
- fun replaceVegetable(orig, subst, Garden([])) = Garden([])
=    | replaceVegetable(orig, subst, Garden(first::rest)) =
=        let val Garden(newRest) =
=                replaceVegetable(orig, subst, Garden(rest));
```

```
=           in Garden((if first = orig then subst else first)
=                       :: newRest)
=           end
=       | replaceVegetable(orig, subst, x) = x;

val replaceVegetable = fn : vegetable * vegetable * plot -> plo

- replaceVegetable(Tomato, Carrot,
=                       Garden([Zucchini, Tomato, Carrot, Lettuce,
=                               Tomato]));

val it = Garden [Zucchini,Carrot,Carrot,Lettuce,Carrot] : plot
```

Conditional expressions also greatly simplify the language processor example from Section 2.6. Instead of having a separate pattern for every word in the vocabulary, we can maintain vocabulary lists and check a work against the list.

```
- exception WordNotFound of string

- val adjectives = ["big", "bright", "fast", "beautiful",
=                   "smart", "red", "smelly"];

-  val transitiveVerbs = ["chased", "saw", "greeted", "bit",
=                         "loved"];
=

-  val intransitiveVerbs = ["ran", "slept", "sang"];
=

-  val linkingVerbs = ["was", "felt", "seemed"];
=

- val nouns = ["man", "woman", "dog", "unicorn", "ball",
=              "field", "flea", "tree", "sky"];

- fun parseAdjective(next::rest) =
=     if contains(next, adjectives)
=     then (SOME(Adj(next)), rest)
=     else (NONE, next::rest);

- fun parseVerbPhrase(next::rest) =
=     if contains(next, transitiveVerbs)
=     then parseTransVerb(TV(next), rest)
=     else if contains(next, intransitiveVerbs)
=     then (IVP(IV(next)), rest)
=     else if contains(next, linkingVerbs)
=     then parseLinkingVerb(LV(next), rest)
=     else raise WordNotFound(next);
```

```
- fun parseNoun(next::rest) = if contains(next, nouns)
=       then (Noun(next), rest) else raise WordNotFound(next);
```

Exercises

3.7.1 Type-analyze the following expression. Make an inference about variable x that makes this expression type correct.

```
if x < 3 then 5.8 else if x > 10 then
9.3 else real(x)
```

3.7.2 Suppose you wanted to divide x by some value y, but only if y was non-zero; you would like the result simply to be x otherwise. Write an ML expression for this calculation, but let x appear in the expression only once.

3.7.3 Consider the following ML expression form:

if <boolean expression>$_1$ then true else <boolean expression>$_2$

Find a simpler, equivalent form in terms of the two boolean expressions. If you do not see the answer immediately, then experiment on some specific expressions.

3.7.4 Write a function `factorPow` that takes integers n and i and determines how many times i is a factor of n—that is, what is the largest exponent k such that i^k divides n? For example, consider 360: 2 is a factor 3 times ($360 = 2^3 \cdot 45$), 6 is a factor 2 times ($360 = 6^2 \cdot 10$), 15 is a factor once ($360 = 15^1 \cdot 24$), and 7 is not a factor at all, that is, 0 times.

3.7.5 Using `contains` from this section, write a function `hasNoRepeats` that takes a list and determines whether or not all the items in the list are unique.

3.7.6 Using `contains`, write a function `isSubsetOf` that takes two lists (modeling sets) and determines whether or not the first list is a subset of the second.

3.7.7 Using `contains`, write a function `intersection` which takes two lists (modeling sets) and returns a list modeling the intersection of those two sets.

3.7.8 Using `contains`, write a function `difference` which takes two lists (modeling sets A and B) and returns a list modeling the difference of the first set from the second ($A - B$). Once you have written this, the following function will compute the union of two sets:

```
fun union(a, b) = intersection(a, b)@
   difference(a, b)@difference(b,a);
```

3.7.9 Using `contains` write a function `makeNoRepeats` which takes a list and removes any duplicate entries. For example, `makeNoRepeats([4, 6, 7, 4, 3, 6])` would return `[4, 6, 7, 3]` or `[7, 4, 3, 6]` (whether your function keeps the first or last occurrence does not matter). Once you have written this, we will have the following alternate definition of the `union` function:

```
fun union(a, b) = makeNoRepeats(a@b)
```

3.7.10 Write a function `numEvens` that takes a list of integers and returns the number of even items. (Hint: x is even iff x `mod` $2 = 0$.)

3.7.11 Write a function `removeFirst` that takes an item and a list and returns a list like the given one except that the first occurrence, if any, of the given item is removed. For example, `removeFirst(5, [1, 2, 6, 5, 2, 4, 5, 9])` would return `[1, 2, 6, 2, 4, 5, 9]`.

3.7.12 Write a function `listMin` that takes a list of integers and returns the smallest item in the list. (Hint: The parameter of your base case should be a list with one element, not an empty list. A "match non-exhaustive" warning is acceptable for this function.)

3.7.13 Using `removeFirst` and `listMin` from the previous two exercises, write a function `sort` that takes a list of integers and returns a list of the same integers, but sorted from smallest to greatest. The sorting algorithm that these exercises are leading you to is called *selection sort* because it sorts a list by repeatedly selecting the next smallest item from it.

3.8 Arguments

To use logic either for writing mathematical proofs or engaging in any other sort of discourse we need to work in units larger than propositions. For example,

> During the full moon, spinach is on sale.
> The moon is full.
> Therefore, spinach is on sale.

argument

argument form

premises

conclusion

valid

contains several propositions, and they are not connected to become a single proposition. This, instead, is an *argument*, a sequence of propositions, with the last proposition beginning with the word "therefore"—or "so" or "hence" or some other such word, and possibly in a postpositive position, as in "Spinach, therefore, is on sale." Similarly, an *argument form* is a sequence of propositional forms, with the last prefixed by the symbol $\therefore$. All except the last in the sequence are called *premises*; the last proposition is called the *conclusion*.

Propositions are true or false. Arguments are *valid* or invalid. We say that an argument form is valid if, whenever all the premises are true (depending on the truth values of their variables), the conclusion is also true.

Consider another argument:

> If my crystal wards off alligators, then there will be no alligators around.
> There are no alligators around.
> Therefore, my crystal wards off alligators.

The previous argument and this one have the following argument forms, respectively (rephrasing "During the full moon..." as "If the moon is full, then..."):

$$\begin{array}{l} p \to q \\ p \\ \therefore q \end{array} \qquad\qquad \begin{array}{l} p \to q \\ q \\ \therefore p \end{array}$$

It is to be hoped that you readily identify the left argument form as valid and the right as invalid. The truth table verifies this.

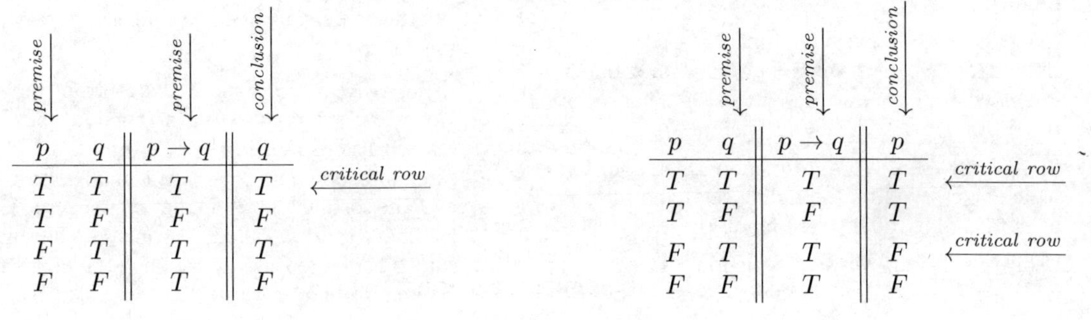

p	q	$p \to q$	q
T	T	T	T
T	F	F	F
F	T	T	T
F	F	T	F

p	q	$p \to q$	p
T	T	T	T
T	F	F	T
F	T	T	F
F	F	T	F

The first argument has only one case where both premises are true, and we see there that the conclusion is also true. The rest of the truth table does not matter—only the rows where all premises are true count. We call these *critical rows*, and when you are evaluating large argument forms, it is acceptable to leave the entries in non-critical rows blank. Moreover, once you have found a critical row where the conclusion is false, nothing more needs to be done. The second argument has a critical row where the conclusion is false; hence it is an invalid argument.

critical rows

There is an alternate, equivalent definition of validity. Suppose we have an argument with premises $p_1, p_2, \ldots p_n$ (these might themselves be formulas, not just variables) and conclusion s. If $(p_1 \wedge p_2 \wedge \ldots \wedge p_n) \rightarrow s$ is a tautology, then the argument is valid. Notice that $(p_1 \wedge p_2 \wedge \ldots \wedge p_n)$ will be true only in the critical rows. In other rows the conditional will be vacuously true.

p	q	$p \rightarrow q$	$(p \wedge (p \rightarrow q)) \rightarrow q$
T	T	T	T
T	F	F	T
F	T	T	T
F	F	T	T

p	q	$p \rightarrow q$	$(q \wedge (p \rightarrow q)) \rightarrow p$
T	T	T	T
T	F	F	T
F	T	T	F
F	F	T	T

A *syllogism* is a short argument. Usually the definition restricts the term to arguments having exactly two premises, but we will include one- and three-premised arguments as well in our discussion. The correct argument form above is the most famous, and is called *modus ponens*, Latin for "method of affirming," or, more literally, "method of placing" (or, more literally still, "placing method," since *ponens* is a participle, not a gerund).

syllogism

modus ponens

$p \rightarrow q$	If Socrates is a human, then he is mortal.
p	Socrates is a human.
$\therefore q$	Therefore, Socrates is mortal.

Since the contrapositive of a conditional is logically equivalent to the conditional itself, a truth table from Section 3.7 proves

$p \rightarrow q$
$\therefore \sim q \rightarrow \sim p$

If we substitute "$\sim q$" for "p" and "$\sim p$" for "q" in modus ponens, we get

$\sim q \rightarrow \sim p$
$\sim q$
$\therefore \sim p$

Putting these two together results in the second most famous syllogism, *modus tollens*, "lifting [that is, denying] method."

modus tollens

$p \rightarrow q$	If the moon is full, then the werewolves are out.
$\sim q$	The werewolves are not out.
$\therefore \sim p$	Therefore, the moon is not full.

We can also prove this directly with a truth table, with only the critical row completed for the conclusion column.

p	q	$p \rightarrow q$	$\sim q$	$\sim p$
T	T	T	F	
T	F	F	T	
F	T	T	F	
F	F	T	T	T

generalization

Now some other common syllogisms. The form *generalization* may seem trivial and useless, but, in fact, it captures a reasoning technique we often use subconsciously. It relies on the fact that a true proposition *or*'ed to any other proposition makes a true proposition.

p	We are in Pittsburgh.
$\therefore p \vee q$	Therefore, we are in Pittsburgh or Mozart wrote *The Nutcracker*.

specialization

Turn this argument form around and use *and* instead of *or*, and you have *specialization*.

$p \wedge q$	$x > 5$ and x is even.
$\therefore p$	Therefore, $x > 5$.

elimination

Sherlock Holmes describes the process of elimination as, "... when you have eliminated the impossible, whatever remains, however improbable, must be the truth." But you do not need to be a master of deduction to use it; even a bear of little brain said, "If anyone knows anything about anything, it's owl who knows something about something, or my name isn't Winnie-the-Pooh. Which it is. So there you have it." Formally, *elimination* is

$p \vee q$	x is even or x is odd.
$\sim p$	x is not even.
$\therefore q$	Therefore, x is odd.

transitivity

We will later prove that, given sets A, B, and C, if $A \subseteq B$ and $B \subseteq C$, then $A \subseteq C$. This means that the $\subseteq$ relation is transitive. The logical form *transitivity* is analogous to this.

$p \rightarrow q$	If $x > 5$, then $x > 3$.
$q \rightarrow r$	If $x > 3$, then $x > 0$.
$\therefore p \rightarrow r$	Therefore, if $x > 5$, then $x > 0$.

In *Alice's Adventures in Wonderland*, the heroine finds a key to a door leading into a beautiful garden. She wants to enter the garden, but the door is too small for her. She then finds a bottle labeled "DRINK ME" which causes her to shrink. Although she is now small enough, she discovers she left the key to the door on a table which she no longer can reach. Next, she notices a box labelled "EAT ME," and reasons,

> Well, I'll eat it, and if it makes me grow larger, I can reach the key; and if it makes me grow smaller, I can creep under the door; so either way I'll get into the garden, and I don't care which happens!

We call this form *division into cases*.

division into cases

$p \vee q$	It will make me grow or it will make me shrink.
$p \to r$	If it makes me grow, I'll get into the garden.
$q \to r$	If it makes me shrink, I'll get into the garden.
$\therefore r$	Either way, I'll get into the garden.

Finally, we have proof by *contradiction*.

contradiction

$p \to F$	If $x - 2 = x$ then $-2 = 0$ and $-2 \neq 0$.
$\therefore \sim p$	Therefore, $x - 2 \neq x$.

Because of certain paradoxes that have arisen in the study of the foundations of mathematics, some mathematicians call into question the validity of proof by contradiction. If p leads to a contradiction, it might not be that p is false; it could be that p is neither true nor false, that is, p might not be a proposition at all. For our purposes, however, we can rely on proof by contradiction.

Exercises

Verify the following syllogisms using truth tables.

3.8.1 Generalization.

3.8.2 Specialization.

3.8.3 Elimination.

3.8.4 Transitivity.

3.8.5 Division into cases.

3.8.6 Contradiction.

3.9 Using argument forms for deduction

Non-trivial argument forms would require huge truth tables to verify them. However, we can use known argument forms to verify larger argument forms in a step-wise fashion. Take the argument form

(a)	$\sim p \wedge \sim r \rightarrow s$
(b)	$p \rightarrow \sim q$
(c)	$\sim t$
(d)	$t \vee \sim s$
(e)	$r \rightarrow \sim q$
(f)	$\therefore \sim q$

This does not follow immediately from the argument forms we have given. However, we can deduce $\sim s$ immediately by elimination, using (c) and (d). Our goal is to generate new propositions from known argument forms until we have verified proposition (f). (We will use lowercase Roman numerals for derived propositions to distinguish them from the given propositions.)

(i)	$\sim s$	by (c), (d), and elimination.
(ii)	$\sim (\sim p \wedge \sim r)$	by (a), (i), and modus tollens
(iii)	$p \vee r$	by (ii) and De Morgan's laws
(iv)	$\sim q$	by (iii), (b), (e), and division into cases.

Notice that in step (iii) we used De Morgan's laws to produce $p \vee r$ from $\sim (\sim p \wedge \sim r)$—we also used the double negative law twice, but that is simple enough to go uncited. Logical equivalences from Theorem 3.1 can be considered little syllogisms, and you may use them in simplifying formulas when verifying argument forms.

This is our second logical game. Just as in Game 1 you used previously known logical equivalences to verify more complicated logical equivalences, in Game 2 you use known argument forms to verify more complicated argument forms. The premises and conclusions in the form to be verified are labeled with letters; you are to generate new propositions, leading up to the conclusion, and label them with lowercase Roman numerals, justifying each one.

Mastering this takes a little practice, looking for patterns and applying them. This should pay off when we begin writing proofs in the next chapter. Observe this example.

(a)	u
(b)	$s \rightarrow F$
(c)	$t \vee p$
(d)	$u \wedge p \rightarrow q \wedge r$
(e)	$t \rightarrow s$
(f)	$\therefore r$

There are two ways to approach this sort of problem: You can look at what new propositions can be produced easily from what is given, or you can consider what you want to prove and work your way backwards from there. Work both ends of the problem to get the advantages of both approaches.

Usually there is some low-lying fruit . In this case the premise $s \to F$ is all we need to apply contradiction.

(i)	$\sim s$	by (b) and contradiction.

Now that $\sim s$ is on the table, we can use $t \to s$ to get $\sim t$, and knocking out t gives us p.

(ii)	$\sim t$	by (e), (i), and modus tollens.
(iii)	p	by (c), (ii), and elimination.

At this point we should consider how we will get to r. Clearly $u \wedge p \to q \vee r$ will be necessary, since that is the only way to get a proposition involving r. We have both u and p, so $u \wedge p$ must be true. Why? Just the definition of conjunction.

(iv)	$u \wedge p$	by (a), (iii), and conjunction.
(v)	$q \wedge r$	by (d), (iv), and modus ponens.

Finally, we can pick r out of that last one by specialization.

(vi)	r	by (v) and specialization.

Here is another example, where the first step is not as immediately obvious.

(a) $(s \vee t) \to r$
(b) $r \to (p \wedge q)$
(c) s
(d) $\therefore q$

Solution:

(i)	$(s \vee t) \to (p \wedge q)$	by (a), (b), and transitivity.
(ii)	$s \vee t$	by (c) and generalization.
(iii)	$p \wedge q$	by (ii), (i), and modus ponens.
(iv)	$\therefore q$	by (iii) and specialization.

For many arguments, there is more than one way to verify it. In the problem above, we could have used two applications of modus ponens rather than transitivity and modus ponens.

Exercises

Use known syllogisms and logical equivalences to verify the following arguments.

3.9.1 (a) $t \to u$

(b) $p \vee \sim q$

(c) $p \to (u \to r)$

(d) q

(e) $\therefore t \to r$

3.9.2 (a) $p \to t$

(b) $\sim (q \to t) \to w$

(c) $p \vee q$

(d) $\sim w$

(e) $\therefore t$

3.9.3 (a) $\sim u$

(b) $p \to (r \to q)$

(c) $\sim t \vee u$

(d) $s \to q$

(e) $(\sim s \wedge \sim r) \to t$

(f) $\sim u \to p$

(g) $\therefore q$

3.9.4 (a) w

(b) $t \vee r$

(c) $p \to \sim q$

(d) $r \to q$

(e) $w \to \sim t$

(f) $\therefore \sim p \vee s$

3.9.5 (a) s

(b) $(p \wedge q) \to F$

(c) p

(d) $\sim q \to (r \vee t)$

(e) $s \to \sim t$

(f) $\therefore r \vee w$

3.9.6 (a) $\sim u$

(b) $r \to q$

(c) $s \to q$

(d) $t \to (r \vee s)$

(e) $p \to t$

(f) $\sim p \to u$

(g) $\therefore q$

3.9.7 (a) $t \to q$

(b) $q \to r$

(c) $(t \to r) \to w$

(d) $p \vee t$

(e) $\therefore w$

3.9.8 (a) w

(b) $q \to r$

(c) $t \to s$

(d) $u \to s$

(e) $(\sim t \wedge \sim u) \to \sim w$

(f) $(s \vee y) \to (p \to q)$

(g) $\sim (p \to r) \vee x$

(h) $\therefore x$

3.9.9 (a) $p \to q$

(b) x

(c) $\sim (p \vee w) \to r$

(d) $q \to u$

(e) $x \to t$

(f) $w \to u$

(g) $r \vee s$

(h) $r \to F$

(i) $\therefore t \wedge s \wedge u$

3.9.10 (a) $u \to \sim p$

(b) $(\sim p \vee q) \to (r \to s)$

(c) $u \wedge \sim w$

(d) $t \to s$

(e) $(\sim t \wedge \sim r) \to w$

(f) $\therefore s$

3.10 Predicates

When we introduced conditionals, we moved from the specific case to the general case by replacing parts of a conditional sentence with variables.

If Socrates is a human, then he is mortal. If p then q.

These variables are *parameters*, or independent variables, just as we have seen when writing ML functions. When we replace parts of a mathematical expression with independent variables, we are parameterizing that expression. We see the same process here:

parameters

Socrates is a human x is a human.

This makes a proposition with a hole in it. A sentence that would be a proposition but for an independent variable is called a *predicate*. More formally, a predicate is a function whose value is true or false (technically, whose *codomain* is $\mathbb{B}$; we will define "codomain" and other function terminology in Chapter 7). Thus we can use function notation for predicates:

predicate

$$P(x) = x \text{ is mortal}$$

You should remember learning the term *predicate* in grammar school. A predicate is the part of a clause that expresses something about the subject.

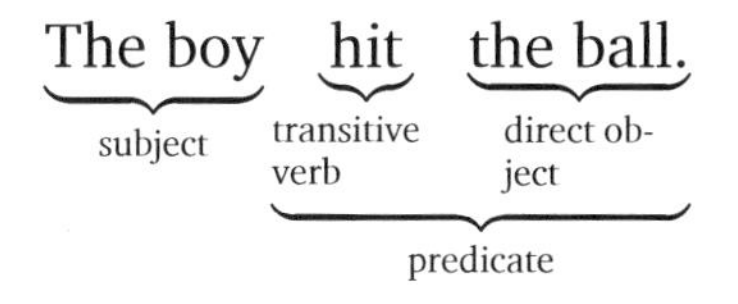

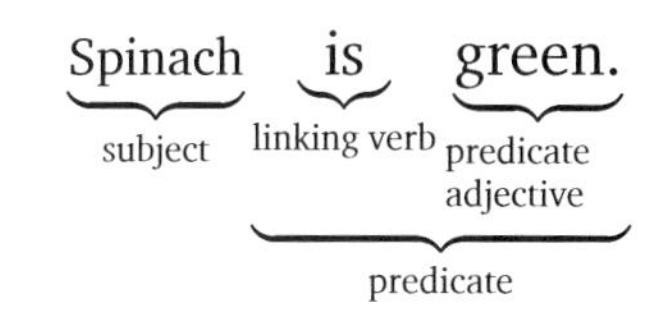

If you look carefully, you will see that these two concepts (the grammatical predicate and the logical predicate) are the same thing. A predicate is something that is like a sentence, except that it is missing a subject. To play with the two sentences above, let

$P(x) = x$ hit the ball.
$Q(x) = x$ is green.

And so we can note P(the boy), P(the bat), Q(spinach), Q(Kermit the Frog), and $\sim Q$(ruby). These are symbolic or mathematical ways to capture the logical sense of various English sentences.

Here is a mathematical example. Let $P(x) = x^2 > x$. What is $P(x)$ for various values of x, if we assume $\mathbb{R}$ as the domain?

x	5	π	2	1	$\frac{1}{2}$	0	$-\frac{1}{2}$	-1
$P(x)$	T	T	T	F	F	F	T	T

truth set

The *truth set* of a predicate $P(x)$ with domain D is the set of all elements in D that make $P(x)$ true when substituted for x. We can denote this using set notation as $\{x \in D \mid P(x)\}$. In this case,

$$\{x \in \mathbb{R} \mid P(x)\} = (-\infty, 0) \cup (1, \infty)$$

3.11 Quantification

Recall the most famous syllogism, modus ponens,

> If Socrates is a human, then he is mortal.
> Socrates is a human.
> Therefore, Socrates is mortal.

It is unlikely that we would either presume or prove such a narrow premise as "If Socrates is a human, then he is mortal." What is so special about Socrates that this conditional mortality accrues to him? Rather, we would be more likely to say

> All humans are mortal.
> Socrates is a human.
> Therefore, Socrates is mortal.

We have not yet talked about how to capture words like "all" in formalisms. Could we express the first premise using a conditional?

> If someone is a human, then he is mortal.

This is equivalent, but now we have introduced the pronouns "someone" and "he," which is English's way of referring to the same but unknown value. If we simplistically replace the pronouns with pseudo-mathematical notation, we get

> If x is a human, then x is mortal.

This allows us to avoid gender-specific language, but the variables, like pronouns with uncertain antecedents, mean that the sentence is no longer a proposition. If we define a predicate

> $P(x) =$ if x is a human, then x is mortal.

then we still are not capturing the original sentence, because "All humans are mortal" is a proposition, no mere predicate. The original sentence truly does make a claim that is either true or false.

The problem is that the variable x is not free. Instead, we want to remark on the extent of x, the values for which we are asserting this predicate. Words that express this are called *quantifiers*. Here is a rephrasing of "all humans are mortal" that uses variables correctly:

quantifiers

For all humans x, x is mortal.

The symbol $\forall$ stands for "for all." Then, if we let $\mathbb{H}$ stand for the set of all humans and $M(x) = x$ is mortal, we have

$$\forall\, x \in \mathbb{H}, M(x)$$

$\forall$ is called the *universal quantifier*. Unfortunately, defining the meaning of a universally quantified proposition cannot be done simply with a truth table. Instead we say, almost banally, that the proposition is true if $P(x)$ is true for every element in the set we are quantifying over. For example, let $D = \{3, 54, 219, 318, 471\}$. Which of the following are true?

universal quantifier

$\forall\, x \in D, x^2 \leq x$	No, this is actually false for all of them.
$\forall\, x \in D, x$ is even	No, this fails for 3.
$\forall\, x \in D, x$ is a multiple of 3	Yes: $3 = 3 \cdot 1, 54 = 3 \cdot 18, 219 = 3 \cdot 73, 318 = 3 \cdot 106$, and $471 = 3 \cdot 157$.

What we used on the last proposition was the *method of exhaustion*, that is, we tried all possible values for x until we exhausted the domain, demonstrating that each element made the predicate true. Obviously this method of proof is possible only with finite sets, and it is impractical for any set much larger than the one in this example (unless you use a computer). On the other hand, *disproving* a universally quantified proposition is easy, since it takes only one hole to sink the ship. If for any element of D, $P(x)$ is false, then the entire proposition is false. Having found 3, not an even number, we disproved the second proposition, without even noting that the predicate fails also for 219 and 471. An element of the quantified set for which the predicate is false is called a *counterexample*.

method of exhaustion

counterexample

If the set we are quantifying over is empty, then a universally quantified proposition is considered to be vacuously true. That is, for any predicate P,

$$\forall\, x \in \emptyset, P(x) \equiv T$$

Why is this true? No counterexamples.

Next, it is not true that

$$x^2 = 16 \text{ for all } x \in \mathbb{R}.$$

It is true, however, that

$$x^2 = 16 \text{ for some } x \in \mathbb{R}.$$

namely, for 4 and -4. While it is true that

$$\sim\ (\forall\, x \in \mathbb{R}, x^2 = 16)$$

it is not true that

$$\forall\, x \in \mathbb{R}, \sim (x^2 = 16)$$

The word "some" expresses the situation that falls between being true for all and being true for none—in other words, the predicate is true for at least one, perhaps more. It is an *existential quantifier*, because it asserts that at least one thing *exists* with the given predicate as a property. We can rephrase the second proposition of this section to get

existential quantifier

There exists an $x \in \mathbb{R}$ such that $x^2 = 16$.

The symbol $\exists$ means "there exists." Hence we have

$$\exists\, x \in \mathbb{R} \mid x^2 = 16$$

The symbol $\mid$ is read "such that." It is the same symbol we use in set notation, for example $A \cup B = \{x \mid x \in A \vee x \in B\}$. Formally, an existential proposition is a proposition of the form $\exists\, x \in D \mid P(x)$ for some predicate $P(x)$ with domain (or domain subset) D.

Revisiting our earlier example with $D = \{3, 54, 219, 318, 471\}$, which of the following are true?

$\exists\, x \in D \mid x^2 \leq x$	No: $3 \cdot 3 = 9 \not\leq 3$; $54 \cdot 54 = 2916 \not\leq 54$; $291 \cdot 219 = 47961 \not\leq 219$; $318 \cdot 318 = 101124 \not\leq 318$; $471 \cdot 471 = 221841 \not\leq 471$.
$\exists\, x \in D \mid x$ is even	Yes, $54 = 2 \cdot 27$.
$\exists\, x \in D \mid x$ is a multiple of 3	Yes: $3 = 3 \cdot 1$.

With existentially quantified propositions, we must use the method of exhaustion to *disprove* it. Only one specimen is needed to show that it is true.

Existential quantification on an empty set is always false, that is

$$\exists\, x \in \emptyset \mid P(x) \equiv F$$

This is because there indeed does not exist such an x. There is no room for vacuous truth in this case.

Finally, we consider how to negate propositions with universal or existential quantifiers. We saw earlier that

$$\sim\ (\forall\, x \in \mathbb{R}, x^2 = 16) \not\equiv \forall\, x \in \mathbb{R}, \sim (x^2 = 16)$$

So negation is not a simple matter of propagating the negation symbol through the quantifier. What, then, is the negation of

All humans are mortal.
There exists a bag of spinach that is on sale.

What will help us here is to think about what the quantifiers are actually saying. If a predicate is true for all elements in the set, then if we could order those elements, it would be true for the first one, and the next one, and one after that. In other words, we can think of universal quantification as an extension of conjunction. Likewise, existential quantification is an extension of disjunction. Hence if $D = \{d_1, d_2, d_3, \ldots\}$, then

$$\begin{aligned} \forall\, x \in D, P(x) &\equiv P(d_1) \wedge P(d_2) \wedge P(d_3) \ldots \\ \exists\, x \in D \mid Q(x) &\equiv Q(d_1) \vee Q(d_2) \vee Q(d_3) \ldots \end{aligned}$$

Now, we can apply an extended version of DeMorgan's laws.

$$\begin{aligned} &\sim (\forall\, x \in D, P(x)) \\ &\qquad \equiv\ \sim (P(d_1) \wedge P(d_2) \wedge P(d_3) \ldots) \\ &\qquad \equiv\ \sim P(d_1) \vee \sim P(d_2) \vee \sim P(d_3) \ldots \\ &\qquad \equiv\ \exists\, x \in D \mid\ \sim P(x) \end{aligned}$$

The same process evaluates the negation of an existentially quantified proposition:

$$\begin{aligned} &\sim (\exists\, x \in D \mid P(x)) \\ &\qquad \equiv\ \sim (P(d_1) \vee P(d_2) \vee P(d_3) \ldots) \\ &\qquad \equiv\ \sim P(d_1) \wedge \sim P(d_2) \wedge \sim P(d_3) \ldots \\ &\qquad \equiv\ \forall\, x \in D, \sim P(x) \end{aligned}$$

Hence the negation of a universal proposition is an existential proposition, and the negation of an existential proposition is a universal proposition. To negate our examples above, we would say

> There exists a human who is not mortal.
> All bags of spinach are not on sale.

Having seen quantification, reconsider the sentence

> If Wendy walked past the dean's office, then she entered through entrance A

In natural language, this sentence is false. If our logical system is any good, then it should reflect this. It does, if we recognize that this sentence contains *implicit quantification*. Let R be all simple routes Wendy could take to get to classroom 3.

implicit quantification

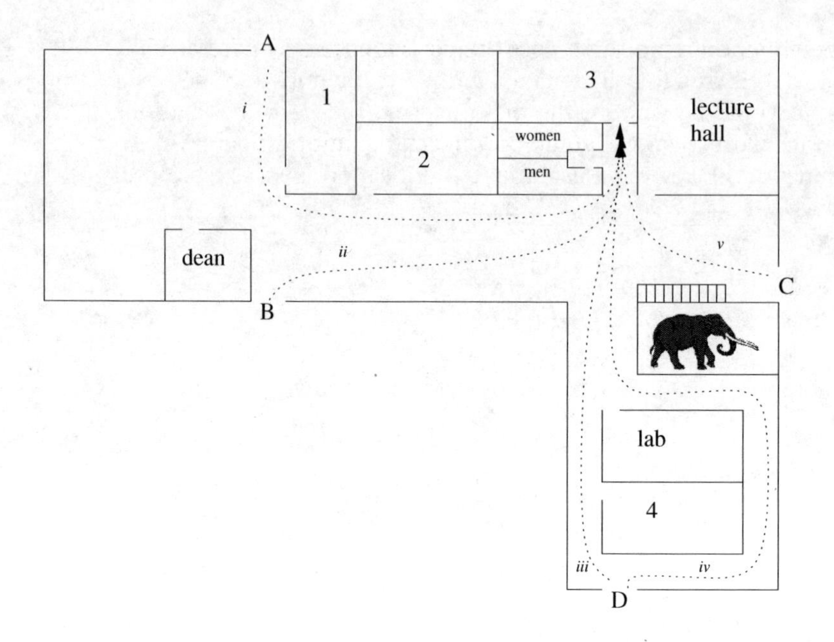

$R = \{i, ii, iii, iv, v\}$. Moreover, let

$$P(x) = x \text{ takes Wendy past the dean's office}$$
$$Q(x) = x \text{ takes Wendy through entrance A}$$

What the sentence really says is

$$\forall\, a \in R,\ P(a) \rightarrow Q(a)$$

$$= \quad (P(i) \rightarrow Q(i)) \wedge (P(ii) \rightarrow Q(ii)) \wedge (P(iii) \rightarrow Q(iii)) \wedge (P(iv) \rightarrow Q(iv)) \wedge (P(v) \rightarrow Q(v))$$

$$= \quad (T \rightarrow T) \wedge (T \rightarrow F) \wedge (F \rightarrow F) \wedge (F \rightarrow F) \wedge (F \rightarrow F)$$

$$= \quad T \wedge F \wedge T \wedge T \wedge T$$

$$= \quad F$$

Good. The idea of vacuous truth at least does not do any harm (the last three Ts in the second to last line turn out to be irrelevant). But what is it good for? Consider "All even integers are divisible by two." Teasing out the implicit quantification, we have

If an integer is even, then it is divisible by two.

For all integers, if that integer is even, then it is divisible by two.

$\forall x \in \mathbb{Z}, P(x) \rightarrow Q(x)$

where $P(x)$ is "x is even" and $Q(x)$ is "x is divisible by two." Try it out.

x	$P(x)$	$Q(x)$	$P(x) \rightarrow Q(x)$
-1	F	F	?
0	T	T	T
1	F	F	?
2	T	T	T
3	F	F	?
4	T	T	T

What should the question marks be? Intuitively one might say they are irrelevant since the hypothesis is false. However, for the universally quantified proposition to be true, all the instances must be true. Hence, we count all propositions that do not contradict the universally quantified proposition to be vacuously true.

Exercises

Let S be the set of bags of spinach, $g(x)$ be the predicate where x is green, and $s(x)$ be the predicate where x is on sale. Write the following symbolically, then negate them, then express the negations in English.

3.11.1 All bags of spinach are on sale.

3.11.2 Some bags of spinach are on sale.

3.11.3 All bags of spinach that are not green are on sale.

3.11.4 Every bag of spinach is green.

3.11.5 Some bags of spinach are not green.

3.11.6 Any bag of spinach is green and on sale.

In ML, a predicate is any function whose return type is bool. The following exercises ask you to write functions that take a list modeling a set D of integers. Do not make use of solutions to previous exercises; compute these directly.

3.11.7 Write a function `allEven` which computes $\forall\, x \in D, x$ is even. For example, `allEven([2, 12, 14, 76])` would return `true`.

3.11.8 Write a function `hasEven` which computes $\exists\, x \in D \mid x$ is even.

3.11.9 Write a function `notAllEven` which computes $\exists\, x \in D \mid x$ is not even.

3.11.10 Write a function `noneEven` which computes $\forall\, x \in D, x$ is not even.

3.11.11 Generalize what you learned from the previous four exercises. How does quantification influence the form of the function, or the strategy you use?

3.12 Multiple quantification

How would we represent the following proposition symbolically?

> Every integer has an additive inverse.

Formal representation is about making something unambiguous, so that it is perfectly clear how to go about proving it. So, in translating this to symbols, we should think about how one would go about proving it. Consider it a game between one person, the doubter, and you, the prover. What kind of challenge would you expect from the doubter? The idea here is that the integer 5 has for its additive inverse -5, -10 has 10, 0 has 0, and so on. If you chose a few integers and showed that the pattern works for the ones you have chosen, this would not convince the doubter: perhaps you have chosen one or two of the few for which it works. Instead, the challenger should get to pick the integer on which to argue. (Pick an integer, any integer.)

However, once that integer is picked, how is the rest of the game played? You, the prover, must come up with an additive inverse to match that integer. Hence the game has two steps: the doubter picks an item to challenge you, and you counter that challenge with another item. The two steps correspond to two levels of quantification.

First, you are claiming that some predicate is true for all integers, so we have something in the form

$$\forall\, x \in \mathbb{Z}, P(x)$$

But what is $P(x)$? What are we claiming about all integers? We claim that something exists corresponding to it, namely an additive inverse. $P(x) = \exists\, y \in \mathbb{Z} \mid x + y = 0$. All together,

$$\forall\, x \in \mathbb{Z}, \exists\, y \in \mathbb{Z} \mid x + y = 0$$

multiply quantified

This is a *multiply quantified* proposition, meaning that the predicate of the proposition is itself quantified. This definition does not merely say that the proposition must have more than one quantifier. The proposition "Every frog is green, and there exists a brown toad" has more than one quantifier, but this is not what we have in mind by multiple quantification, because one quantified subproposition is not nested within the other— they are just conjoined.

Quantifiers are not commutative. We would have a very different (and false) proposition if we were to say

$$\exists\, y \in \mathbb{Z} \mid \forall\, x \in \mathbb{Z}, x + y = 0$$

or

> There is an integer such that every integer is its additive inverse.

Also notice that the innermost predicate ($y = -x$) has two independent variables. The general form is

$$\forall\, x \in D, \exists\, y \in E \mid P(x, y)$$

where P is a two-argument predicate, with arguments of domains D and E, respectively; or, equivalently, P is a single-argument predicate with domain $D \times E$.

Let us try another example, to translate into symbols the proposition

> There is no greatest prime number.

To make this process easier, let us temporarily ignore the negation.

> There is a greatest prime number.

Obviously the outer quantification is existential. Let $\mathbb{P}$ stand for the set of prime numbers; now, if we write half-symbolically, we have

$$\exists\, x \in \mathbb{P} \mid x \text{ is the greatest prime number}$$

Next, focus in on the inner chunk: "x is the greatest prime number." What does it mean to be the greatest prime number? It means that all other prime numbers must be less than x.

$$\forall\, y \in \mathbb{P}, y \leq x$$

(Why did we say "$\leq$" rather than "$<$"?) Insert that back into our earlier symbolism:

$$\exists\, x \in \mathbb{P} \mid \forall\, y \in \mathbb{P}, y \leq x$$

Now we negate this.

$$\sim \exists\, x \in \mathbb{P} \mid \forall\, y \in \mathbb{P}, y \leq x$$

Evaluate the negation of the existential quantifier.

$$\forall\, x \in \mathbb{P}, \sim \forall\, y \in \mathbb{P}, y \leq x$$

Evaluate the negation of the universal quantifier.

$$\forall\, x \in \mathbb{P}, \exists\, y \in \mathbb{P} \mid y > x$$

or

> For every prime, there exists a greater prime.

Which, when you think about it, means the same thing as "there is no greatest prime number."

Recall what it means for a set of sets to be *pairwise disjoint*. If you pick any two distinct sets from the set of sets, they must be disjoint. When picking two sets, we pick them independently and sequentially. So, $\{A_1, A_2, \ldots, A_n\}$ are pairwise disjoint if

$$\forall\, A_i \in \{A_1, A_2, \ldots, A_n\},\ \text{any other set in the set of sets is disjoint with it}$$

The "any" means we have another level of quantification, choosing another set. Since these sets are chosen independently, it is possible that the first and second set are in fact the same set, except that the word "other" further specifies that they must be distinct. This requirement is captured by a conditional.

$$\forall\, A_i \in \{A_1, A_2, \ldots, A_n\}, \forall\, A_j \in \{A_1, A_2, \ldots, A_n\},\ A_i \neq A_j \rightarrow A_i \cap A_j = \emptyset$$

We can abbreviate this idea with

$$\forall\, A_i, A_j \in \{A_1, A_2, \ldots, A_n\},\ \text{where } A_i \neq A_j,\ \text{we have } A_i \cap A_j = \emptyset$$

Understand that when there are two variables in a single quantifier, this is short hand for nested quantification. Also, a *where* clause is usually shorthand for a conditional.

Since the sets in this example are labeled with numbers, we could have quantified this over $\mathbb{N}$:

$$\forall\, i, j \in \mathbb{N},\ \text{where} 1 \leq i, j \leq n \text{ and } i \neq j, \text{we have} A_i \cap A_j = \emptyset$$

or

$$\forall\, i, j \in \mathbb{N},\ 1 \leq i, j \leq n \wedge i \neq j \rightarrow A_i \cap A_j = \emptyset$$

If you have had calculus, you may recall the *formal definition of a limit*. The definition is hard to understand because it is multiply quantified. Consider the function

$$f(x) = \frac{x^2 - 1}{x - 1} = \frac{(x-1)(x+1)}{x - 1}$$

This function is almost exactly like $h(x) = x + 1$ except that it is undefined at $x = 1$. $f(1) \neq 2$, but $\lim_{x \to 1} f(x) = 2$. What does that mean? It means that f gets as close to 2 as we want. Consider a graph of the function, and suppose we zoom in to get within .001 units of 2.

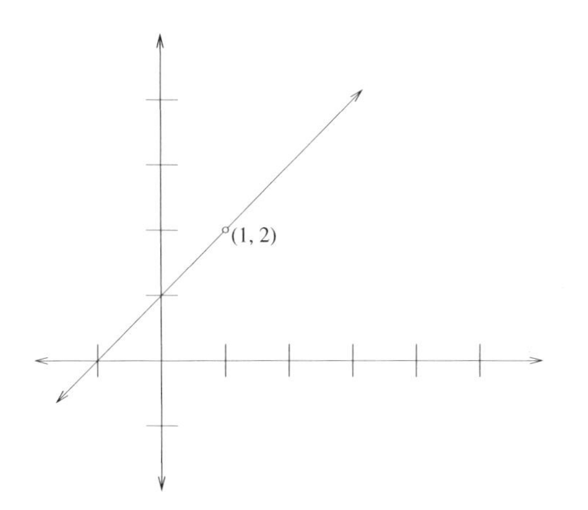

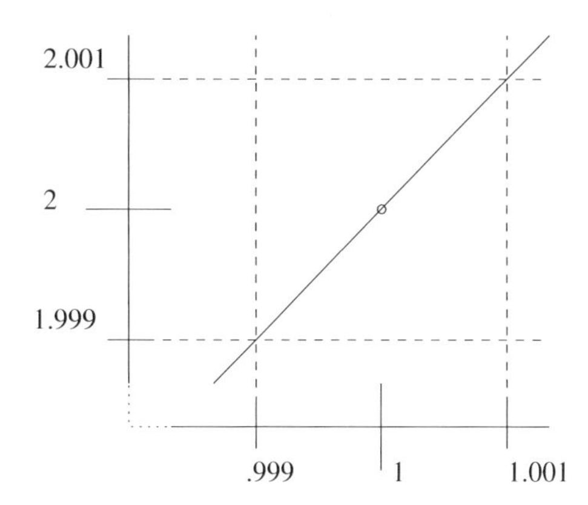

By drawing the horizontal dashed lines at $y = 2.001$ and $y = 1.999$—in mathematics, we call these *bounds*—we picked how close we wanted the function to get to 2. What we have found is that it is also possible to draw vertical dashed lines, in this case at $x = .999$ and $x = 1.001$, so that whenever the function is within the vertical lines, it is also within the horizontal lines. Formally, $\lim_{x \to a} f(x) = L$ means

$$\underbrace{\forall\, \epsilon \in \mathbb{R}^+}_{\text{For any vertical distance from } L}, \quad \underbrace{\exists\, \delta \in \mathbb{R}^+}_{\text{we can find a horizontal distance from } a} \quad \mid \forall\, x \in \mathbb{R} \ \underbrace{0 < |x - a| < \delta}_{\text{so that whenever } x \text{ is within the horizontal distance from } a,} \quad \to \quad \underbrace{|f(x) - L| < \epsilon}_{\text{then } f(x) \text{ is within the vertical distance from } L}$$

Exercises

Evaluate (that is, simplify) the negation.

3.12.1 $\sim \forall\, x \in D, \exists\, y \in E \mid P(x, y)$.

3.12.2 $\sim \exists\, x \in D \mid \forall\, y \in E, P(x, y)$.

Write the following propositions symbolically. If there is any ambiguity then write symbolic versions of all possible readings. (None are intended to be ambiguous.) Do not concern yourself with whether the propositions are true or false.

3.12.3 There is a whole number that is less than all others.

3.12.4 Any whole number is greater than all real numbers.

3.12.5 If a whole number is less than five, then it is greater than all real numbers.

3.12.6 Any natural number can be divided evenly by 4. (Think of what it means *to divide evenly*. It involves a level of quantification.)

3.12.7 For any natural number, if 12 divides it evenly, then 3 also divides it evenly.

3.12.8 If a natural number is evenly divisible by 13, then it is greater than all rational numbers.

3.13 Quantification and algorithms

Sometimes we will need to transform multiply quantified propositions or predicates into algorithms. For example, suppose we wanted to test whether every element in one list of ints had a divisor in a second list of ints. Let us call it `allHaveDivisor`. For example, `allHaveDivisor([12, 14, 15], [3, 7])` would return `true` since $12 \texttt{ mod } 3 = 0$, $14 \texttt{ mod } 7 = 0$, and $15 \texttt{ mod } 3 = 0$.

The task is multiply quantified: We want to test if *for all* elements of the first list it is true that *there exists* an element of the second list which is a divisor of the first-list element. We might reason

- A. If the first list is empty, then it is true, since universal quantification on an empty set is vacuously true.
- B. If the second list is empty, then it is false, since there would not exist a divisor for anything in the first list.
- C. If each list has at least one element,
  - C1. and if the first element of the second list is a divisor of the first element in the first list, then it is true for the first element in the first list; recur on the rest of the first list; the first case serves as a base case;
  - C2. otherwise, test to see if there is a divisor for this item somewhere else in the second list.

In ML,

```
- fun allHaveDivisor([], yy) = true
=   | allHaveDivisor(xx, []) = false
=   | allHaveDivisor(x::xRest, y::yRest) =
=         if x mod y = 0
=         then allHaveDivisor(xRest, y::yRest)
=         else allHaveDivisor(x::xRest, yRest);

val allHaveDivisor = fn : int list * int list -> bool

- allHaveDivisor([12, 14, 15], [3, 7]);

val it = false : bool
```

Can you tell what the problem is? Here is a hint: look at the the recursion of C1. We need a helper function. These can be nested in let expressions just as local variables can. Parameters of the outer function are in scope.

```
- fun allHaveDivisor([], yy) = true
=   | allHaveDivisor(x::xRest, yy) =
=       let fun hasDivisor([]) = false
=             | hasDivisor(y::yRest) = if x mod y = 0
=                                      then true
=                                      else hasDivisor(yRest);
=       in
=          hasDivisor(yy) andalso allHaveDivisor(xRest, yy)
=       end;

val allHaveDivisor = fn : int list * int list -> bool

- allHaveDivisor([12, 14, 15], [3, 7]);

val it = true : bool

- allHaveDivisor([12, 13, 15], [3, 7]);

val it = false : bool
```

This approach allows us to refer to the original list in the recursive call `allHaveDivisor(xRest, yy)`.

Exercises

In the following exercises, do not make use of solutions to previous exercises; compute them directly.

3.13.1 Write a function `allLessThan` that takes two lists of ints and tests if all the elements of the first list are less than all the elements of the second. For example, `allLessThan([1, 13, 55], [74, 56])` would return `true`.

3.13.2 Write a function `allHaveDouble` that takes two lists of ints and tests to see if the double of each element in the first list appears in the second (other elements may also appear in the second). For example, `allHaveDouble([5, 3, 17], [2, 34, 10, 9, 6])` would return `true`.

3.13.3 Write a function `allHaveAddInv` which takes a list of ints and tests whether each item in the list has its additive in the list. For example, `allHaveAddInv([4,~6, 0, 6, 5, ~5, ~4])` would return `true`. Note that 0 is its own additive inverse; the specification does not require there to be a second 0 occurring in the list.

3.13.4 Write a function `hasDivisorOfAll` that takes a list of ints and tests to see if the list contains an element that is a divisor of all the elements in the list. For example, `hasDivisorOfAll([15, 18, 3, 33, 12])` would return `true` since 3 is a divisor of each element (including itself).

3.13.5 Write a function `hasCommonElement` which takes two lists and determines whether they have at least one element in common. For example, `hasCommonElement([5, 2, 3, 1], [6, 7, 2, 8])` would return `true` since they both contain 2.

3.14 Quantification and arguments

In Game 2, we saw arguments consisting of lists of propositions, each list comprising premises and a conclusion. Most interesting mathematical propositions have quantification. Here we will extend Game 2 into a third game by introducing quantified arguments. The important thing in all this is that we are making claims about sets. Reconsidering the argument from Section 3.11,

> If Socrates is a human, then he is mortal.
> Socrates is a human.
> Therefore, Socrates is mortal.

The form is

$$\forall\, x \in A,\ P(x)$$
$$a \in A$$
$$\therefore P(a)$$

universal instantiation

We call this *universal instantiation* because we are given something predicated universally on a set ($P(x)$ on the set A) and applying that to a specific instance of the set (a). In these forms and examples, we will use variables like x, y, and z for quantified variables, that is, variables that range over sets. Variables like a, b, and c will be free variables, meaning that they stand in for a specific, though unknown, value.

Another example:

> Any biscuit that is green is moldy.
> The object in my hand is a biscuit.
> It is green.
> Therefore it is moldy.

universal modus ponens

Clearly we have a version of modus ponens here—a realistic one, in fact, and by "realistic" we mean "quantified." It is *universal modus ponens*.

$$\forall\, x \in A,\ P(x) \rightarrow Q(x)$$
$$a \in A$$
$$P(a)$$
$$\therefore Q(a)$$

What universal instantiation and universal modus ponens have in common is that they both *assume* something to be true for every element in a set—they have a universally quantified premise. What if we want to *prove* something is true for any element? What would an argument with a universally quantified conclusion look like?

Recall the doubter-prover game suggested in Section 3.12. We must allow the doubter to pick an element from the set. The word *suppose* is our invitation to the doubter, as seen the following form:

Suppose $a \in A$
    $P(a)$
$\therefore \forall\, x \in A,\ P(x)$

It is difficult to give a verbal example of an argument of this form in isolation, because it makes sense only in the context of other parts of the arguments. The indented portion ($P(a)$) contains propositions we find to be true once the doubter has picked an element from A, which we are calling a. The premises are simulating the scenario where the doubter picks and examines all elements of A one by one, hence this is really an extended version of generalization, *universal generalization*. You also may think of it as a reversed form of universal instantiation, which is itself an extended form of specialization.

universal generalization

Now we have enough tools for our first example of Game 3.

	Argument:		Solution:	
(a)	$\forall\, x \in A, P(x) \to Q(x)$		Suppose $a \in A$	
(b)	$\forall\, x \in A, P(x)$	(i)	$P(a)$	by supposition, (b), and UI
(c)	$\therefore \forall\, x \in A, Q(x)$	(ii)	$Q(a)$	by supposition, (i), (a), and UMP
		(iii)	$\therefore \forall\, x \in A,\ Q(x)$	by (ii) and UG

The supposition is not a proposition, so we do not label it. We cite it with "by supposition." When we cite universal generalization, we undo the indentation.

This table summarizes five more argument forms.

Universal modus tollens
$\forall\, x \in A,\ P(x) \to Q(x)$
$a \in A$
$\sim Q(a)$
$\therefore \sim P(a)$

Existential generalization
$a \in A$
$P(a)$
$\therefore \exists\, x \in A \mid P(x)$

Hypothetical division into cases
$p \vee q$
Suppose p
    r
Suppose q
    r
$\therefore r$

Existential instantiation
$\exists\, x \in A \mid P(x)$
Let $a \in A \mid P(a)$
$\therefore a \in A \wedge P(a)$

Hypothetical conditional
Suppose p
    q
$\therefore p \to q$

Universal modus tollens is similar to universal modus ponens. Consider *existential instantiation*. The first premise is something like "There is a green crayon somewhere." The second premise has something new, the word *let*—similar to *suppose* except that while *suppose* assumes a new fact *let* merely gives a new name to something we already know exists. Next, recall that proving existentially quantified propositions is easy—in *existential generalization*, all you need to do is find one element that works. Finally, consider the two argument forms that do not involve quantification. Earlier we saw that universal generalization supposes a fact ($a \in A$) and uses that fact hypothetically to prove a quantified proposition. The form *hypothetical conditional* uses the same approach to prove a conditional proposition. If we use two such suppositions in sequence, we are arguing from *hypothetical division into cases*.

Try them out. First, proving something existentially quantified:

Argument:

(a)	$\forall\, x \in A,\ Q(x) \to R(x)$
(b)	$\exists\, x \in A \mid P(x) \wedge S(x)$
(c)	$\forall\, x \in A,\ S(x) \to \sim R(x)$
(d)	$\therefore \exists\, x \in A \mid P(x) \wedge \sim Q(x)$

Solution:

	Let $a \in A \mid P(a) \wedge S(a)$	
(i)	$a \in A \wedge P(a) \wedge S(a)$	by (b) and EI
(ii)	$a \in A$	by (i) and specialization
(iii)	$P(a)$	"
(iv)	$S(a)$	"
(v)	$\sim R(a)$	by (iv), (c), and UMP
(vi)	$\sim Q(a)$	by (v), (a), and UMT
(vii)	$P(a) \wedge \sim Q(a)$	by (iii), (iv), and conjunction
(viii)	$\therefore \exists\, x \in A \mid P(x) \wedge \sim Q(x)$	by (ii), (vii), and EG

Notice that we also use syllogisms from Section 3.8 when appropriate. Now, proving a conditional:

Argument:

(a)	$\forall\, x \in A, x \in C$
(b)	$\forall\, x \in B,\ \exists\, y \in A \mid R(y) \wedge S(x, y)$
(c)	$\forall\, x \in C,\ P(x) \wedge R(x) \to T(x)$
(d)	$\forall\, x \in B, y \in A,\ T(y) \wedge S(x, y) \to Q(x)$
(e)	$\therefore (\forall\, x \in A,\ P(x)) \to (\forall\, y \in B,\ Q(y))$

Solution:

	Suppose $\forall\, x \in A,\ P(x)$	
	Suppose $b \in B$	
(i)	$\exists\, y \in A \mid R(y) \wedge S(b, y)$	by supposition, (b), and UI
	Let $a \in A \mid R(a) \wedge S(b, a)$	
(ii)	$a \in A \wedge R(a) \wedge S(b, a)$	by (i), and EI
(iii)	$a \in A$	by (ii) and specialization
(iv)	$a \in C$	by (iii), (a), and UI
(v)	$P(a)$	by (iii), supposition, and UI
(vi)	$R(a)$	by (ii) and specialization
(vii)	$P(a) \wedge R(a)$	by (v), (vi), and conjunction
(viii)	$T(a)$	by (vii), (iv), (c), and UMP
(ix)	$S(b, a)$	by (ii) and specialization
(x)	$T(a) \wedge S(b, a)$	by (viii), (ix), and conjunction
(xi)	$Q(b)$	by supposition, (iv), (x), (d), and UMP
(xii)	$\forall\, y \in B,\ Q(y)$	by (xi) and UG
(xiii)	$\therefore (\forall\, x \in A,\ P(x)) \to (\forall\, y \in B,\ Q(y))$	by (xii) and HC

Exercises

Some of the argument forms given in this section can be derived from others. Verify the following forms using simpler ones (especially universal instantiation and hypothetical conditional) and syllogisms from Section 3.8.

3.14.1 Universal modus ponens.

3.14.2 Universal modus tollens.

3.14.3 Hypothetical division into cases.

Verify the following arguments.

3.14.4 (a) $\forall\, x \in A,\ P(x) \wedge \sim Q(x)$
(b) $\forall\, x \in A,\ x \in B$
(c) $\forall\, x \in B,\ \sim Q(x) \rightarrow R(x)$
(d) $\therefore \forall\, x \in A,\ R(x)$

3.14.5 (a) $\forall\, x \in A,\ x \in B$
(b) $\forall\, x \in B,\ \sim P(x)$
(c) $\forall\, x \in A,\ Q(x) \rightarrow P(x)$
(d) $\therefore \forall\, x \in A,\ \sim Q(x)$

3.14.6 (a) $\forall\, x \in A,\ P(x) \rightarrow Q(x)$
(b) $\forall\, x \in B,\ x \in A \rightarrow \sim Q(x)$
(c) $\therefore \forall\, x \in A,\ x \in B \rightarrow \sim P(x)$

3.14.7 (a) $\forall\, x \in A, \forall\, y \in B,\ P(x,y) \rightarrow Q(x)$
(b) $\exists\, x \in A \mid \forall\, y \in B,\ P(x,y)$
(c) $\forall\, x \in A,\ R(x) \rightarrow \sim Q(x)$
(d) $\therefore \exists\, x \in A \mid\ \sim R(x)$

3.14.8 (a) $\forall\, x \in A,\ P(x) \vee Q(x)$
(b) $\forall\, x \in A,\ x \in B \vee \sim P(x)$
(c) $\forall\, x \in B,\ P(x) \rightarrow R(x)$
(d) $\forall\, x \in A,\ Q(x) \rightarrow R(x)$
(e) $\therefore \forall\, x \in A,\ R(x)$

3.14.9 (a) $\forall\, x \in A,\ \exists\, y \in B \mid P(x) \wedge Q(x,y)$
(b) $\forall\, x \in B,\ x \in C$
(c) $\forall\, x \in A,\ \forall\, y \in C,\ Q(x,y) \rightarrow R(y)$
(d) $\forall\, x \in D,\ x \in A$
(e) $\exists\, x \in D$
(f) $\therefore \exists\, x \in C \mid R(x)$

3.14.10 (a) $\forall\, x \in A,\ \exists\, y \in B \mid P(x,y)$
(b) $\forall\, y \in B,\ Q(y) \vee R(y)$
(c) $\forall\, x \in A,\ \forall\, y \in B,\ P(x,y) \rightarrow \sim Q(y)$
(d) $\exists\, x \in A \mid S(x)$
(e) $\therefore \exists\, y \in B \mid R(y)$

(Hint: You will use premise 10(d), even though you will not use $S(x)$.)

3.14.11 (a) $\forall\, x \in A,\ x \in B \wedge x \in C$
(b) $\forall\, x \in C,\ x \in D \vee x \in E$
(c) $\forall\, x \in B,\ x \in D \rightarrow P(x)$
(d) $\forall\, x \in B,\ x \in E \rightarrow Q(x)$
(e) $\forall\, x \in B,\ P(x) \vee Q(x) \rightarrow R(x)$
(f) $\therefore \forall\, x \in A, R(x)$

Tangent: "The Princess and the Pea"

A recent retelling of "The Princess and the Pea" has the queen saying, "If she is a princess, she'll get no comfort out of this bed. For only the delicate nature of a true princess will be able to feel the pea under all these layers." Let *W* be the set of women, *P* be the predicate "is a princess", and *Q* be the predicate "can feel the pea." The queen reasons

$$\forall\, x \in W, Q(x) \rightarrow P(x)$$
$$\therefore \forall\, x \in W, P(x) \rightarrow Q(x)$$

The queen commits the converse fallacy—the premise denies the existence of pea-sensitive peasants, but it does not make pea-sensitivity a necessary condition for princesshood. Of course, a sufficient condition for princesshood is all that matters in the story, so eliminating the "If she is a princess. . . " sentence would clear this up. The Hans Christian Anderson version does not contain that line. In the original, a true queen knows her discrete math.

3.15 Extended example: Verifying arguments automatically

The ML programming language was originally developed for writing and extending software for automatic theorem proving. When you observe the formal, almost mechanical process we have used in the games of this chapter, you can see why people would reason that if a computer can do arithmetic, it should also be able to produce or at least verify arguments.

In this section, we will build an ML program to play Game 2, or the equivalent—that is, a program that will determine whether or not an argument is valid. In our program, however, we will approach this verification at a different angle from what we used in Game 2.

Recall the alternative definition of a valid argument. If p_1, p_2, $\ldots p_n$ are premises and s is a conclusion, then the argument is valid if $(p_1 \wedge p_2 \wedge \ldots \wedge p_n) \rightarrow s$ is a tautology. For example, if you took the truth table you made for verifying division into cases in Exercise 3.8.5 and added an extra column for

$$((p \vee q) \wedge (p \rightarrow r) \wedge (q \rightarrow r)) \rightarrow r$$

you would find that that column is true for every assignment to p, q, and r. For our automatic argument checker, we will convert the argument into a formula like this and then check to see that it is a tautology.

First, we need a way to represent logical formulas—we will simply call them *propositions*—in ML. The propositions we have used can be

- Variables, say p, q, r, $\ldots$
- Negations of propositions, say $\sim p$ or $\sim (q \vee (p \wedge r))$.
- Conjunctions of propositions, say $(p \vee q) \wedge r$.
- Disjunctions of propositions, say $\sim q \vee r$.
- Conditionals of propositions, say $p \rightarrow \sim (q \vee r)$.

A proposition can contain smaller propositions—our definition of a proposition (by listing the different kinds) is self-referential. This means that if we make an ML datatype proposition, it will need to be self-referential as well. Fortunately ML datatypes, like ML functions, can be defined recursively.

To simplify things a bit, we first eliminate conditionals, because any conditional can be expressed with negation and disjunction: $p \rightarrow q \equiv \sim p \vee q$. Also, we will consider conjunctions and disjunctions to be two kinds of binary operations.

```
datatype binLogOp = Conj | Disj;
datatype proposition = Var of string
        | Neg of proposition
        | BinOp of binLogOp * proposition * proposition;
```

When a value of this datatype represents a proposition, for example $p \wedge \sim (q \vee r)$, it will have a branching structure like this

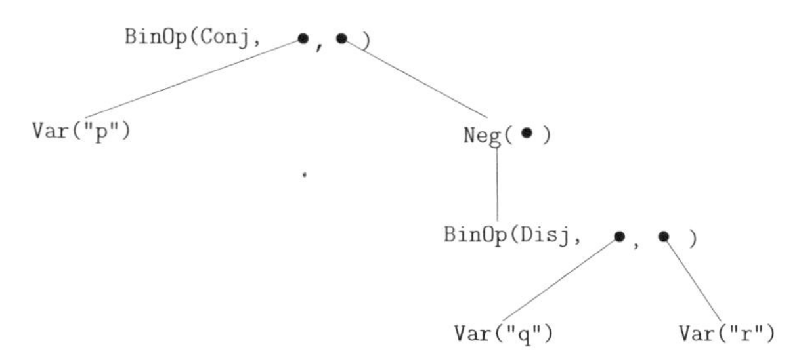

To make it a little easier to make conjunctions, disjunctions, and conditionals, we add a few functions:

```
fun conj(p, q) = BinOp(Conj, p , q);
fun disj(p, q) = BinOp(Disj, p, q);
fun cond(p, q) = BinOp(Disj, Neg(p), q);
```

It is always helpful to write a function that produces a string representation of a datatype. In this case, adding appropriate parentheses is the tricky part

```
fun display(Var(s)) = s
  | display(Neg(Var(s))) = "~" ^ s
  | display(Neg(p)) = "~(" ^ display(p) ^ ")"
  | display(BinOp(oper, p, q)) =
    (case p of
         BinOp(subOp, pp, qq) => if subOp = oper
                                 then display(p)
                                 else "(" ^ display(p) ^ ")"
       | _ => display(p)) ^
    (case oper of
         Conj => "^"
       | Disj => "v") ^
    (case q of
         BinOp(subOp, pp, qq) => if subOp = oper
                                 then display(q)
                                 else "(" ^ display(q) ^ ")"
       | _ => display(q));
```

The underscore character (_) acts like a special variable for values in patterns that we want to ignore. Trying it out:

```
- conj(Var("p"), Neg(disj(Var("q"), Var("r"))));


val it = BinOp (Conj,Var "p",Neg (BinOp (#,#,#))) : proposition
```

```
- display(it);

val it = "p^~(qvr)" : string
```

The arguments we want to verify will have long sequences of conjunctions since we want to conjoin all our premises. To make this easier, we write a function that will turn a list of propositions into a proposition that is the equivalent of anding all those propositions together. While we are at it, we may as well do the same for disjunctions.

```
exception EmptyBigConjOrDisj;

fun bigConj([]) = raise EmptyBigConjOrDisj
  | bigConj([p]) = p
  | bigConj(p::rest) = conj(p, bigConj(rest));

fun bigDisj([]) = raise EmptyBigConjOrDisj
  | bigDisj([p]) = p
  | bigDisj(p::rest) = disj(p, bigDisj(rest));
```

We can now use this to encode an argument. Take the argument in Exercise 3.9.1

(a) $t \rightarrow u$
(b) $p \vee \sim q$
(c) $p \rightarrow (u \rightarrow r)$
(d) q
(e) $\therefore t \rightarrow r$

```
- val orig =
=   cond(bigConj([cond(Var("t"), Var("u")),
=                 disj(Var("p"), Neg(Var("q"))),
=                 cond(Var("p"), cond(Var("u"), Var("r"))),
=                 Var("q")]),
=          cond(Var("t"), Var("r")));

val orig = BinOp (Disj,Neg (BinOp (#,#,#)),
                  BinOp (Disj,Neg #,Var #)) : proposition

-  print(display(orig) ^ "\n");

~((~tvu)^(pv~q)^(~pv~uvr)^q)v~tvr
val it = () : unit
```

Now for the real work. We have a formula equivalent to our argument, and we want to check that it is true for any assignment to the variables. We could take a "brute force" approach, equivalent to making a truth table: enumerate every possible combination of true and false values and evaluate the formula for each one of them. While this would make an interesting exercise (see Project 3.C), there is a more efficient way.

Consider the proposition

$$(p \vee q \vee r) \wedge (q \vee \sim t \vee \sim u) \wedge (u \vee \sim p \vee t)$$

This is not a tautology—just pick all of p, q, and r to be false. That makes the first subproposition $(p \vee q \vee r)$ false, and since we are joining several propositions together with conjunctions, that makes the entire proposition false. On the other hand,

$$(p \vee q \vee r \vee \sim q) \wedge (q \vee \sim t \vee \sim u \vee t) \wedge (u \vee \sim p \vee t \vee p)$$

is a tautology, and this is how we tell: since we again are anding several subpropositions together, this entire proposition will be true exactly under those situations where every subproposition is true. So, this proposition is a tautology (always true) if and only if all of its subpropositions are tautologies.

The first subproposition is always true because no matter how you assign the variables, either q or $\sim q$ must be true, and either of them will make the whole subproposition true. If a chain of disjunctions contains both a variable and its negation, then that chain of disjunctions is a tautology. The other two subpropositions in this example are tautologies because they contain both $\sim t$ and t and both $\sim p$ and p.

We picked an easy one, however: If the proposition is a big conjunction of subpropositions, each of which is a big disjunction of variables or negations of variables, then all we need to do is see if each subproposition contains some variable and its negation. But this is the core of our strategy: To determine if something is a tautology, we will first transform it into an equivalent "easy one," and then use a simple test to see if it is a tautology.

We will do this in two steps. First, we want to make sure that negations are applied only to variables—not to binary operations, not to other negations. A proposition is in *negation normal form* if the only negations in it are applied directly to variables. Transforming a proposition to negation normal form is straightforward: work from the outside, top-level proposition into the subpropositions; if you find a double negation, remove both because they cancel each other out; if the negation is applied to a binary operation, then push it in and flip the operator, following De Morgan's laws. For example,

negation normal form

$$\begin{aligned} & \sim ((p \vee \sim q) \wedge \sim r) \\ \equiv \quad & \sim (p \vee \sim q) \vee \sim\sim r \\ \equiv \quad & (\sim p \wedge \sim\sim q) \vee r \\ \equiv \quad & (\sim p \wedge q) \vee r \end{aligned}$$

Writing a function for this is intuitive when you use pattern-matching. Specifically, the interesting cases are the various things a negation can be applied to:

```
fun flip(Conj) = Disj
  | flip(Disj) = Conj;

fun negNormForm(Var(s)) = Var(s)
  | negNormForm(Neg(Var(s))) = Neg(Var(s))
  | negNormForm(Neg(Neg(p))) = negNormForm(p)
  | negNormForm(Neg(BinOp(oper, p, q))) =
      BinOp(flip(oper), negNormForm(Neg(p)),
            negNormForm(Neg(q)))
  | negNormForm(BinOp(oper, p, q)) =
      BinOp(oper, negNormForm(p), negNormForm(q));
```

Try it:

```
- val nnf = negNormForm(orig);

val nnf = BinOp (Disj,BinOp (Disj,BinOp #,BinOp #),
                 BinOp (Disj,Neg #,Var #))  : proposition

- print(display(nnf) ^ "\n");

(t^~u)v(~p^q)v(p^u^~r)v~qv~tvr
val it = () : unit
```

conjunctive normal form

What we had been calling "easy ones" are actually propositions in *conjunctive normal form*, which means that the proposition is the conjunction of subpropositions, each of which is the disjunction of variables and negations of variables. Just as the transformation to negation normal form uses De Morgan's law, the transformation to conjunctive normal form uses the distributive law, specifically $p \vee (q \wedge r) \equiv (p \vee q) \wedge (p \vee r)$. For example,

$$\begin{array}{ll} & \sim q \vee (r \vee (\sim p \wedge q)) \\ \equiv & \sim q \vee ((r \vee \sim p) \wedge (r \vee q)) \\ \equiv & (\sim q \vee r \vee \sim p) \wedge (\sim q \vee r \vee q) \end{array}$$

The code for conversion to conjunctive normal form is harder to follow. We start with a function that takes two propositions and distributes the first over the second using disjunction. This is only interesting if the second proposition is a conjunction—otherwise we merely create a disjunction. The function for converting to conjunctive normal form assumes the proposition is already in negation normal form and raises an exception if it is not. The interesting case is if the position being transformed is a conjunction: then we transform the two subpropositions and distribute.

```
fun distribute(p, BinOp(Conj, q, r)) =
        BinOp(Conj, distribute(p, q), distribute(p, r))
  | distribute(BinOp(Conj, p, q), r) =
        BinOp(Conj, distribute(p, r), distribute(q, r))
  | distribute(p, q) = BinOp(Disj, p, q);

exception NotInNNF of string;

fun conjNormForm(Var(s)) = Var(s)
  | conjNormForm(Neg(Var(s))) = Neg(Var(s))
  | conjNormForm(Neg(p)) = raise NotInNNF(display(Neg(p)))
  | conjNormForm(BinOp(Conj, p, q)) =
    BinOp(Conj, conjNormForm(p), conjNormForm(q))
  | conjNormForm(BinOp(Disj, p, q)) =
    distribute(conjNormForm(p), conjNormForm(q));
```

Continuing our example:

```
- val cnf = conjNormForm(nnf);

val cnf =
  BinOp (Conj,BinOp (Conj,BinOp #,BinOp #),
              BinOp (Conj,BinOp #,BinOp #))
  : proposition

- print(display(cnf) ^ "\n");

(tv~pvpv~qv~tvr)^(~uv~pvpv~qv~tvr)^(tvqvpv~qv~tvr)^
(~uvqvpv~qv~tvr)^(tv~pvuv~qv~tvr)^(~uv~pvuv~qv~tvr)^
(tvqvuv~qv~tvr)^(~uvqvuv~qv~tvr)^(tv~pv~rv~qv~tvr)^
(~uv~pv~rv~qv~tvr)^(tvqv~rv~qv~tvr)^(~uvqv~rv~qv~tvr)
val it = () : unit
```

The output is getting hard to read, but it does not need to be human readable at this point—only computer readable.

Now for the actual tautology testing. At this point we assume that every proposition is either a disjunction of variables and negations of variables or the conjunction of such variables. If it is a top-level conjunction, then we check that every subproposition is a tautology. If it is a subproposition (disjunction), we check that it is a tautology by seeing if it has both a variable and its negation; we do that by collecting all the plain variables and all the variables that are negated and seeing if they have any overlap. To do all this, we write functions `positives` and `negatives` to collect those variables. We also rely on the functions `contains` from Section 3.7 and `intersection` from Exercise 3.7.7. We raise an exception if the proposition is not in conjunctive normal form.

```
exception NotInCNF of string;

fun positives(Var(s)) = [s]
  | positives(Neg(Var(s))) = []
  | positives(BinOp(Disj, p, q)) = positives(p)@positives(q)
  | positives(x) = raise NotInCNF(display(x));

fun negatives(Var(s)) = []
  | negatives(Neg(Var(s))) = [s]
  | negatives(BinOp(Disj, p, q)) = negatives(p)@negatives(q)
  | negatives(x) = raise NotInCNF(display(x));

fun tautCNF(BinOp(Conj,p,q)) = tautCNF(p) andalso tautCNF(q)
  | tautCNF(p) =
      intersection(positives(p), negatives(p)) <> [];
```

Try it:

```
- tautCNF(cnf);
val it = true : bool
```

To make future use easier, we write a function that will merely take a list of premises and a conclusion and will do all the other work for us. For example,

(a) $p \wedge t$
(b) $\sim (q \rightarrow t) \rightarrow w$
(c) $p \vee q$
(d) $\sim w$
(e) $\therefore t$

```
- fun validArgument(premises, conclus) =
=    tautCNF(conjNormForm(negNormForm(cond(bigConj(premises),
=                                          conclus))));
val validArgument = fn : proposition list * proposition
                                        -> bool

- validArgument(
=     [cond(Var("p"), Var("t")),
=       cond(Neg(cond(Var("q"), Var ("t"))), Var("w")),
=       disj(Var("p"), Var("q")), Neg(Var("w"))],   Var("t"));

val it = true : bool
```

This example and the project were adapted from Paulson [25, pg 164–170].

Project

3.A The software in this section plays Game 2. It can be adapted to play Game 1 as well, if we make the observation that two propositions are logically equivalent ($p \equiv q$) if and only if their biconditional is a tautology ($p \leftrightarrow q \equiv T$). Write a helper function `bicond` which takes two propositions and creates a proposition in our system equivalent to $p \leftrightarrow q$. Then write a function `logEquiv` that takes two propositions and uses `bicond` and functions from this section to determine if the two propositions are logically equivalent.

3.B The "propositions" in this section of course are not truly propositions because they have variables in them. In order to evaluate a proposition (tell whether it is true or false), we need to have an *assignment* to the variables—that is, a specific true or false value assigned to each variable. Each row in a truth table represents an assignment to the variables. Consider this datatype:

```
datatype assignment =
   Assign of (string * bool) list}
```

Such an assignment associates variable names (strings) with truth values. Write a function `lookup` that takes an assignment and a string standing for a variable and returns a truth value assigned to that variable. Define an exception for it to raise if the string is not a bound variable. Then write a function `evaluateProposition` that takes a proposition and an assignment and determines its value.

3.C With `evaluateProposition`, it is possible to test if two propositions are logically equivalent by taking a "brute force" approach, similar to using a truth table. We would generate all possible assignments for a list of variables, evaluate both propositions for each assignment, and verify that the two propositions are always equal. Write a function `logEquiv2` that tests for logical equivalence in this way. Recommended helper functions are

- `enumerateAssignments`, which takes a list of variable names and produces a list of assignments, one for each possible combination of values. For example (and result):

```
enumerateAssignments(["p", "q"])

[Assign [("p",true),("q",true)],
 Assign [("p",false),("q",true)],
 Assign [("p",true),("q",false)],
 Assign [("p",false),("q",false)]]
```

- `testAllAssignments`, which takes two propositions and a list of assignments and evaluates each proposition for each assignment and verifies that the propositions are equal for each assignment.
- `collectVars`, which takes a proposition and produces a list of variable names. So that no variable occurs in the resulting list more than once, use the `union` operation from Exercise 3.7.7 or 3.7.9.

3.D A problem similar to the one in this section is that of determining whether a proposition is a *contradiction*—always false. The easy propositions for contradiction testing are those in *disjunctive normal form*—big disjunctions of subpropositions that each are conjunctions of variables or negations of variables. Write a set of functions that will turn a proposition into disjunctive normal form and then test whether the proposition is a contradiction.

Biography: Lawrence Paulson, 1955–

Lawrence Paulson is a computer scientist who has advanced the field of automatic theorem proving and logic. He has spent most of his career at the University of Cambridge. He has contributed to several software systems that produce proofs of theorems by automated reasoning. His research has included foundations of mathematics, such as set theory and symbolic logic. He also has developed a system for interactive verification of cryptographic protocols.

3.16 Special topic: Quantification and natural language

Quantification is inherently difficult to reason about—even advanced mathematics students commonly make quantification mistakes. One impediment is the various forms in which quantifiers appear in English. A proposition may have several mathematical forms that are equivalent, just as "some" and "there exists ... such that" have the same logical sense. Likewise $\forall$ can stand in for "for any," "for every," "for each," and "given any."

implicit quantification

Worse yet, the quantification may be *implicit*. Consider the proposition

A positive integer evenly divides 121.

While this sentence is awkward and ambiguous (is this supposed to define the term *positive integer*?), the most reasonable reading is "there exists a positive integer that evenly divides 121," or,

$$\exists\, x \in \mathbb{Z}^+ \mid x \text{ divides } 121$$

which is true, letting $x = 11$. The existential quantification is implicit, and the indefinite article here means "there is some such that..." However, the indefinite article can also imply universal quantification, depending on the context:

If a number is a rational number, then it is a real number.

becomes

$$\forall\, x \in \mathbb{Q}, x \in \mathbb{R}$$

Sometimes ambiguity can be cleared up only by context, or, in the case of spoken language, by voice inflection. Adverbs (besides *not*) usually do not affect the logical meaning of a sentence, but notice how *just* turns

I wouldn't give that talk to any audience.

into

I wouldn't give that talk to just any audience.

The difference is between "there does not exist an audience to which I would give this talk" or "there exists an audience to which I would not give this talk." The second sentence can have only the second meaning, but the first sentence can have either, depending on how the person says "any."

The word *anyone* can mean either universal quantification or existential quantification. Compare

Anyone would want a Junior Mint.

with

Does anyone still think math majors don't need to program?

The use of symbols becomes increasingly critical when multiple quantification is involved because natural language becomes desperately ambiguous. Consider the sentence

There is a professor teaching every class.

This could mean

$$\exists\ x \in \text{(The set of professors)} \mid \forall\ y \in \text{(The set of classes)}, x \text{ teaches } y$$

A very busy professor indeed. More likely, this is meant to indicate an adequate staffing situation, that is,

$$\forall\ y \in \text{(The set of classes)}, \exists\ x \in \text{(The set of professors)} \mid x \text{ teaches } y$$

Similarly, when someone says

A man loves every woman.

does it indicate a particularly promiscuous fellow

$$\exists\ x \in \text{(Men)} \mid \forall\ y \in \text{(Women)}, x \text{ loves } y$$

or a Hollywood ending?

$$\forall\ y \in \text{(Women)}, \exists\ x \in \text{(Men)} \mid x \text{ loves } y$$

If we say

Every man loves a woman.

do we mean

$$\forall\ x \in \text{(Men)}, \exists\ y \in \text{(Women)} \mid x \text{ loves } y$$

that every guy has a star after which he pines, or

$$\exists\ y \in \text{(Women)} \mid \forall\ x \in \text{(Men)}, x \text{ loves } y$$

some gal will be a shoo-in for homecoming queen?

Chapter summary

Symbolic logic is a formalization of one piece of the way we reason. It is important for our studies because it forms the basis for decision-making in computer programming and for proving propositions in rigorous mathematics. Most of the exercises in this chapter have been practice—practice to prepare you for real proofs and more intricate programs later in this course. We have also observed how quantification is used in charting the course for a proof and for determining an algorithm.

Key ideas:

Propositions are sentences that are either true or false, but not both. *True* and *false* are primitive terms, two elements in a set.

Propositional forms are like propositions except that they have free variables.

The basic operations of the propositional calculus are negation ($\sim$), conjunction ($\wedge$), and disjunction ($\vee$). Combine propositions (or propositional forms) to make new propositions, just as we use arithmetic operations on numbers and union, intersection, and difference on sets.

Truth values are modeled in ML using the bool type.

Two propositional forms are logically equivalent if they have the same truth value as each other for all possible assignments of their variables. We can determine logical equivalence either using a truth table or by deducing it from known equivalences.

The conditional operator ($\rightarrow$) has two propositions (or forms) as operands: the hypothesis and the conclusion. Moreover, $p \rightarrow q \equiv \sim p \vee q$.

The conditional expression in ML contains a bool-valued test expression and a *then* clause and an *else* clause. The result of the conditional expression is the result of the *then* clause if the test is true; if the test if false, the result is that of the *else* clause.

An argument form is a sequence of propositional forms, the last being the conclusion, the others being the premises. An argument is valid if, whenever all the premises are true, so is the conclusion.

A predicate is a function whose result is true or false.

A universally quantified proposition asserts a predicate to be true for all elements of a set. An existentially quantified proposition asserts a predicate to be true for at least one element of a set.

A proposition is multiply quantified if one quantified propositional form is nested inside another quantified proposition.

Chapter 4 Proof

This text is about constructing proofs and programs. Writing proofs is a skill you will practice over and over, as a figure skater does a triple axel, for the entire course. It is therefore difficult to overstate the importance of this chapter.

You may have written simple two-column proofs in high school geometry: succeeding propositions in the left column and corresponding justifications in the right. Proofs at the level of this text are to be more professional than that. Proofs should be written as paragraphs of complete English sentences, augmented with mathematical symbols.

Writing proofs is writing. Your English instructors have long forewarned that good writing skills are essential no matter what field you study. The day of reckoning has arrived. That is not to say you are expected to write good proofs immediately. Proof-writing is a fine skill that takes patience and practice to acquire—just like any other task of writing. Do not be discouraged when you write pieces of rubbish along the way.

Mathematical proofs have a character of their own among other kinds of persuasive writing. We are not about the business of amassing evidence or showing that something is probable—mathematics stands out even among the other sciences in this regard. A mathematical proof has a level of precision that no other discourse community can approach. This is the community standard to which you must live up; when you write proofs, *justify everything*.

This book has an almost militaristic rigidity in its proof-writing instruction. If you go on in mathematical study, you will see that the structure and phrasing of proofs can be quite varied, creative, and even colorful. This is not yet the place for creativity, however, but for fundamental training. We teach you to march now. Learn it well, and someday you will dance.

Chapter goals. The student should be able to

- write proofs of propositions asserting one set is a subset of another.
- write proofs of propositions asserting two sets are equal.
- write proofs of propositions asserting a set is empty.
- write proofs of conditional propositions about sets.
- write proofs of propositions about integers, particularly about being even or odd and about divisibility.
- write proofs of biconditional propositions about sets and about integers.
- write proofs of propositions about powersets.
- write algorithms based on theorems.

4.1 General outline

A *theorem* is a proposition that is proven to be true. Thus a paradox (or any other non-proposition) cannot be a theorem, because a non-proposition is neither true nor false (or it is both). A false proposition also cannot be a theorem because it is false. Like a theorem, an axiom is true (in the postulated world, at least), but unlike a theorem, an axiom is assumed to be true rather than proven true. Finally, a theorem is different from a conjecture, even one that happens to be true, in that a theorem is proven to be true, whereas a conjecture is not proven (and therefore not known for certain) to be true.

theorem

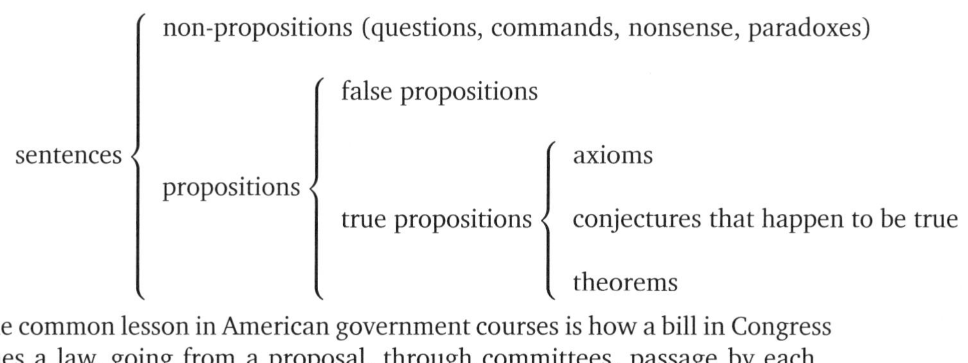

One common lesson in American government courses is how a bill in Congress becomes a law, going from a proposal, through committees, passage by each house, resolution, and presidential signing. How a conjecture becomes a theorem is a bit simpler: we write proofs for them. (We will refer to this as "proving a theorem," although one could argue that "proving a conjecture" is more accurate.)

Basic theorems take on one of three General Forms:

1. Facts. p
2. Conditionals. If p then q.
3. Biconditionals. p iff q.

Of these, General Form 2 is the most important, since facts can often be restated as conditionals, and biconditionals are just two separate conditionals. Since set theory is the setting for our proofs, we will further organize basic facts in set theory into three Set Proposition Forms:

1. Subset. $X \subseteq Y$.
2. Set equality. $X = Y$.
3. Set empty. $X = \emptyset$.

Thus our plan is to cover proofs of the subset relation, proofs of set equality, proofs that a set is empty, proofs of conditional propositions, and proofs of biconditional propositions.

4.2 Subset proofs

Let A and B be sets, subsets of the universal set $\mathcal{U}$.

Theorem 4.1 $A \cap B \subseteq B$

Our task is to prove that this is always the case, no matter what A and B are. George Pólya says,

> If you have to prove a theorem, do not rush. First of all, understand fully what the theorem says, try to see clearly what it means. ...When you have satisfied yourself that the theorem is true, you start proving it. [27, pg 76]

This particular proposition is chosen because one can readily be convinced of it. A visual proof, as in Section 1.5, will help:

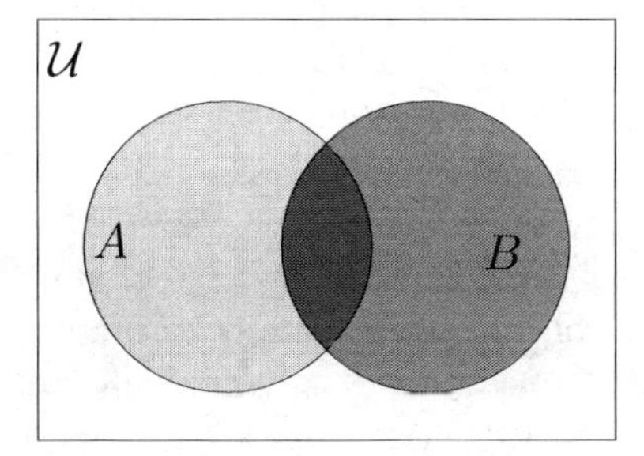

That the double-shaded area ($A \cap B$) is a subset of B cannot be doubted. But we need to transform this intuition into a formal proof. We need to ask ourselves.

What does it mean for one set to be a subset of another?

and

Why is it the case that these two sets are in that relationship?

The first question appeals to the definition of subset. Formal, precise definitions are the tools with which we build proofs. Section 1.4 gave an informal definition of the subset relation. We delayed giving a formal definition until we had talked about quantification. Here it is:

$$X \subseteq Y \text{ if } \forall\, x \in X,\ x \in Y$$

Observe how this fact now breaks down into a conditional. "$A \cap B \subseteq B$" is equivalent to "if $a \in A \cap B$ then $a \in B$." This observation will make proving conditional propositions more familiar when the time comes. More importantly, you should notice that definitions, although expressed merely as conditionals, really are biconditionals; in any definition, the "if" is an implied "iff."

The burden of a proof to show $X \subseteq Y$ is, then, to show that

$$\forall\, x \in X,\ x \in Y$$

How would you persuade someone that this is the case? We have seen this already in Game 3: Allow the doubter to pick an element of X and then show that that element makes the predicate true. The way we invite the doubter to pick an element is with the word *suppose*. "Suppose $x \in X \ldots$" is math-speak for "choose for yourself an element of X, and I will tell you what to do with it, which will persuade you of my point." Recall that this argument form is *universal generalization*.

In our case, $X = A \cap B$. We need a formal definition of intersection:

$$X \cap Y = \{z \mid z \in X \wedge z \in Y\}$$

Now follow the proof:

> **Proof.** Suppose $a \in A \cap B$.
>
> By definition of intersection, $a \in A$ and $a \in B$.
>
> *$a \in B$ by specialization.*
>
> Therefore, by definition of subset, $A \cap B \subseteq B$. □

The italicized line could be omitted for the sake of brevity. Occasionally we will add such sentences as clarifications. Specialization is the sort of logical step that it is fair to assume your audience will perform automatically, as long as you recognize that a real logical operation is indeed happening.

Notice that

- Our proof begins with *Suppose...*
- Every other sentence is a proposition joined with a prepositional phrase governed by *by*.
- The last sentence begins with *therefore* and, except for the *by* part, is the proposition we are proving.
- The proof is terminated by the symbol □, an end-of-proof marker. You will sometimes see proofs terminated with *QED*, an older convention from the Latin *quod erat demonstrandum*, which means "which was to be proven."

The strategy we used in this proof is called the *element argument* for proving facts of Set Form 1:

element argument

To prove $A \subseteq B$

say Suppose $a \in A$.

followed by ... a sequence of propositions that logically follow each other ...

with second-to-last sentence $a \in B$ by ...

and last sentence Therefore, $A \subseteq B$ by the definition of subset. □

Again, recognize this as a use of universal generalization. One way to describe the steps in proof is to categorize them into *analysis* and *synthesis*. Analysis is the taking apart of something. Break down the assumed or proven propositions by applying definitions, going from *term* to *meaning*. Synthesis is the putting together of something. Assemble the proposition to be proven by applying the definition in the other direction, going from *meaning* to *term*.

To do either of these, you must be using precise definitions. Here is a summary of the formal definitions of set operations.

$$\begin{aligned} X \cup Y &= \{z \mid z \in X \vee z \in Y\} \\ X \cap Y &= \{z \mid z \in X \wedge z \in Y\} \\ \overline{X} &= \{z \mid z \notin X\} \end{aligned} \qquad \begin{aligned} X - Y &= \{z \mid z \in X \wedge z \notin Y\} \\ X \times Y &= \{(x, y) \mid x \in X \wedge y \in Y\} \end{aligned}$$

Now a bigger example. Let A, B, and C be sets, subsets of $\mathcal{U}$. Prove

Theorem 4.2 $A \times (B \cup C) \subseteq (A \times B) \cup (A \times C)$

Immediately we know our proof will begin with

Proof. Suppose $x \in A \times (B \cup C)$.

and end with

$x \in (A \times B) \cup (A \times C)$ by .... Therefore, $A \times (B \cup C) \subseteq (A \times B) \cup (A \times C)$. □

Tangent: The nature of definition, revisited

Recall from an earlier tangent that when defining a noun, we start with a broader noun, a superset of set of things labeled with that noun. However, the objects that can be labeled with the nouns we are defining here—the set operations—are themselves sets. So with $\{z \mid z \in X \wedge z \in Y\}$, we are not using set notation to describe the set of things that are intersections, but rather giving a pattern for each element of the set of intersections.

Good mathematical definitions must be usable both analytically and synthetically. If we know something is an intersection, then what does that tell us about it? And what facts must we produce to prove that something is an intersection?

Our proof is going to be a journey from $x \in A \times (B \cup C)$ to $x \in (A \times B) \cup (A \times C)$. Most steps will be the application of a definition. To trace out a route, we need to consider what paths lead from $x \in A \times (B \cup C)$ and which paths lead to $x \in (A \times B) \cup (A \times C)$, and to find a way to connect them in the middle.

What has the definition of Cartesian product to say about $x \in A \times (B \cup C)$? x must be an ordered pair, having the form, say, (a, d). Having picked the symbols a and d, we know that $a \in A$ and $d \in B \cup C$. What we know about d can be broken down further, observing that $d \in B$ or $d \in C$. This is the analysis of our supposition, broken down to individual facts.

What about $x \in (A \times B) \cup (A \times C)$? What would this mean, if it were true? x would have to be an element of $A \times B$ or of $A \times C$. In the first case, it would have to have the form (a, d) where $a \in A$ and $d \in B$; in the other case, it would still have to have the form (a, d) where $a \in A$, but instead where $d \in C$. This is a hypothetical analysis of our destination. Notice our use of the subjunctive throughout, since we do not know that these parts are true, only that one of them would be true if (and only if) the destination were true. However, if we proved all these parts by some other means, then retracing our steps would lead to a synthesis of our destination. In this instance, the glue that connects these with the pieces of the previous analysis is the argument form we have learned called *division into cases.*

> **Proof.** Suppose $x \in A \times (B \cup C)$. By the definition of Cartesian product, $x = (a, d)$ for some $a \in A$ and some $d \in B \cup C$. Then $d \in B$ or $d \in C$, by definition of union.
>
> **Case 1:** Suppose $d \in B$. Then, by definition of Cartesian product, $(a, d) \in A \times B$. Moreover, $x \in A \times B$ by substitution. Finally, by the definition of union, $x \in (A \times B) \cup (A \times C)$.
>
> **Case 2:** Suppose $d \in C$. Then, by definition of Cartesian product, $(a, d) \in A \times C$. Moreover, $x \in A \times C$ by substitution. Finally, by the definition of union, $x \in (A \times B) \cup (A \times C)$.
>
> So $x \in (A \times B) \cup (A \times C)$ by division into cases. Therefore, $A \times (B \cup C) \subseteq (A \times B) \cup (A \times C)$ by definition of subset. □

Notice several things. First, we combined several sentences into paragraphs. This manifests the general divisions in our proof technique. The first paragraph did the analysis. The next two each dealt with one case and began the synthesis. The last paragraph completed the synthesis.

Second, division into cases was turned into prose and paragraph form by highlighting each case, with each case coming from a clause of a disjunction ("$d \in B$ or $d \in C$"), and each case requiring another supposition. Students who particularly enjoyed the logical equivalence problems in Section 3.4 often are inclined to work with longer propositions as disjunctions rather than use division into cases, in this case:

> ... $d \in B$ or $d \in C$, by definition of union. Then $(a, d) \in A \times B$ or $(a, d) \in A \times C$ by definition of Cartesian product. By substitution,

$x \in A \times B$ or $x \in A \times C$. Finally, by the definition of union, $x \in (A \times B) \cup (A \times C)$.

This is a legitimate proof (and shorter than the proof using division into cases), but it is not recommended. Writing proofs like this is more prone to logical errors and sloppiness. For example, it sounds plain in English to abbreviate the third sentence to, "By substitution, $x \in A \times B$ or $A \times C$," but that overlooks the fact that logical *or* can join only two complete propositions (independent clauses), not merely noun phrases, as is attempted here. The definition of union implies that $x \in (A \times B) \cup (A \times C)$ is equivalent to $x \in (A \times B) \vee x \in (A \times C)$, but one cannot simply change the "roundies" into "pointies": $x \in (A \times B) \vee (A \times C)$ is meaningless.

Third, we have peppered this proof with little words like "then," "moreover," "finally," and "so." These do not add meaning, but they make the proof more readable.

substitution

Finally, we have made use of one extra but very important mathematical tool, that of *substitution*. If two expressions are assumed or shown to be equal, we may substitute one for the other in another expression. In this case, we assumed x and (a, d) were equal, and so we substituted x for (a, d) in $(a, d) \in A \times B$.

Exercises

Let A, B, and C be sets, subsets of the universal set, $\mathcal{U}$. Prove.

4.2.1 $A \subseteq A \cup B$.

4.2.2 $A - B \subseteq \overline{B}$.

4.2.3 $A \cap \overline{B} \subseteq A - B$.

4.2.4 $\overline{(A \cup B)} \subseteq \overline{A} \cap \overline{B}$.

4.2.5 $A \cup (B \cap C) \subseteq (A \cup B) \cap (A \cup C)$.

4.2.6 $(A \times C) \cup (B \times C) \subseteq (A \cup B) \times C$.

4.2.7 $(A \times B) \cup (A \times C) \subseteq A \times (B \cup C)$.

4.2.8 $A \times (B - C) \subseteq (A \times B) - (A \times C)$.

Tangent: Substitute and replace

". . . we substituted x for (a, d)." Alternately, we could have said, ". . . we replaced (a, d) with x." But do not say, ". . . we substituted (a, d) with x." H.W. Fowler gives the following examples:

Correct	**Incorrect**
We had to substitute margarine for butter.	We had to substitute butter by margarine.
The substitution of margarine for butter is having bad effects.	The substitution of butter by margarine is having bad effects.
Its substitution for butter is lamentable.	Its substitution by margarine is lamentable.

Fowler goes on: "In the incorrect set the words *replace* or *replacement* would have done" [10]. I suppose, though, that the nouns are interchangeable: A substitute is also a replacement.

4.3 Set equality

We now turn to facts of Set Form 2. As with subsets, proofs of set equality must be based on the definition, what it means for two sets to be equal. Informally we understand that two sets are equal if they are made up of all the same elements, that is, if they completely overlap. In other words, two sets are equal if they are subsets of each other. Formally:

$$X = Y \text{ if } X \subseteq Y \wedge Y \subseteq X$$

This means you already know how to prove propositions of set equality—it is the same as proving subsets, only it needs to be done twice (once in each direction of the equality). Observe this proof of $A - B = A \cap \overline{B}$:

> **Proof.** First, suppose $x \in A - B$. By the definition of set difference, $x \in A$ and $x \notin B$. By the definition of complement, $x \in \overline{B}$. Then, by the definition of intersection, $x \in A \cap \overline{B}$. Hence, by the definition of subset, $A - B \subseteq A \cap \overline{B}$.
>
> Next, suppose $x \in A \cap \overline{B}$. ... *Fill in your proof from Exercise 4.2.3...* Hence, by the definition of subset, $A \cap \overline{B} \subseteq A - B$.
>
> Therefore, by the definition of set equality, $A - B = A \cap \overline{B}$. □

Notice that this proof required two parts, highlighted by "first" and "next," each part with its own supposition and arriving at its own conclusion. We used the word "hence" to mark the conclusion of a part of the proof and "therefore" to mark the end of the entire proof, but they mean the same thing. Notice also the general pattern: suppose an element in the left side and show that it is in the right; suppose an element in the right side and show that it is in the left.

To avoid redoing work (and making proofs unreasonably long), you may use previously proven propositions as justifications. in a proof. A theorem that is proven for the purpose only to be used as justification in the proof of another theorem is called a *lemma*—a "supporting theorem." Lemmas and theorems are either identified by name or by number—or exercise number, for our purposes. Here is a re-writing of the proof, lemma first:

lemma

Lemma 4.3 $A - B \subseteq A \cap \overline{B}$.

> **Proof.** Suppose $x \in A - B$. By the definition of set difference, $x \in A$ and $x \notin B$. By the definition of complement, $x \in \overline{B}$. Then, by the definition of intersection, $x \in A \cap \overline{B}$. Therefore, by the definition of subset, $A - B \subseteq A \cap \overline{B}$. □

Now the proof of our theorem becomes a one-liner (incomplete sentences are countenanced when things get this simple):

> **Proof.** By Lemma 4.3, Exercise 4.2.3, and the definition of set equality. □

Exercises

Let A, B, and C be sets, subsets of the universal set $\mathcal{U}$. Prove. You may use exercises from the previous section in your proofs.

4.3.1 $A \cup \emptyset = A$.

4.3.2 $A \cup (A \cap B) = A$.

4.3.3 $A \times (B \cup C) = (A \times B) \cup (A \times C)$.

4.3.4 $A \cup (B \cap C) = (A \cup B) \cap (A \cup C)$.

4.3.5 $A \cup \overline{A} = \mathcal{U}$.

4.3.6 $A - \emptyset = A$.

4.3.7 $A \times (B - C) = (A \times B) - (A \times C)$.

4.3.8 $(A \cup B) \cup C = A \cup (B \cup C)$.

4.3.9 $(A \cap B) \cap C = A \cap (B \cap C)$.

4.3.10 $\overline{\overline{A}} = A$.

4.3.11 $A \cup \mathcal{U} = \mathcal{U}$.

4.3.12 $\overline{A} \cup B = \overline{A - B}$.

4.3.13 $\overline{(A \cup B)} = \overline{A} \cap \overline{B}$.

4.3.14 $\overline{(A \cap B)} = \overline{A} \cup \overline{B}$.

The results of Exercises 4.3.13 and 4.3.14 are known as *DeMorgan's laws* for sets.

4.3.15 $A \cup B = A \cup (B - (A \cap B))$.

4.3.16 $(A \cap C) - (C - B) = A \cap B \cap C$.

4.3.17 $(A \cup (B - C)) \cap \overline{B} = A - B$.

4.3.18 $(A - B) \cup (C - B) = (A \cup C) - B$.

4.3.19 $(A - B) \cap (C - B) = (A \cap C) - B$.

4.4 Set emptiness

Suppose we are to prove

$$A \cap \overline{A} = \emptyset$$

To address Set Form 3, we must consider what it means for a set to be empty; though this may seem obvious, we cannot write a precise proof if we do not have a precise definition at which to aim. A set X is empty if

$$\sim \exists\, x \in \mathcal{U} \mid x \in X$$

This is the same thing as

$$\forall\, x \in \mathcal{U}, x \notin X$$

but we prefer the former.

You will frequently find the need to prove that something *does not exist*. Here the doubter might object, "Just because you have not found one does not mean they do not exist." This is a high hurdle for the prover indeed. We do, however, have a weapon for propositions of this sort—the proof by contradiction syllogism we learned in Section 3.8. We suppose the opposite of what we are trying to prove

$$\exists\, x \in \mathcal{U} \mid x \in X$$

show that this leads to a contradiction, and then conclude what we were trying to prove. This is indeed one of the most profound techniques in mathematics. G.H. Hardy remarked, "It is a far finer gambit than any chess gambit: a chess player may offer the sacrifice of a pawn or even a piece, but the mathematician offers the game" [12].

> **Proof.** Suppose $a \in A \cap \overline{A}$. Then, by definition of intersection, $a \in A$ and $a \in \overline{A}$. By the definition of complement, $a \notin A$. $a \in A$ and $a \notin A$ is a contradiction. *This proves that the supposition $a \in A \cap \overline{A}$ is false.* Therefore, by contradiction, $A \cap \overline{A} = \emptyset$. $\square$

So it is not true that $a \in A \cap \overline{A}$. Since we assumed a was *any* element in $A \cap \overline{A}$, this means that there is no element in $A \cap \overline{A}$. This is our argument pattern for proving that a set is empty, or that it equals the empty set.

The more powerful the tool, the more easily it can be misused. So it is with proof by contradiction. Since we can prove that a proposition is *false* by deriving a known false proposition from it, the novice prover is often tempted to prove that a proposition is *true* by deriving a known true proposition from it. Let this example, to prove $A = \emptyset$ (that is, all sets are empty), demonstrate the folly to would-be trespassers:

> **Proof.** Suppose $A = \emptyset$. Then $A \cup \mathcal{U} = \emptyset \cup \mathcal{U}$ by substitution. By Exercise 4.3.11, $A \cup \mathcal{U} = \mathcal{U}$ and also $\emptyset \cup \mathcal{U} = \mathcal{U}$ by Exercise 4.3.1. By substitution, $\mathcal{U} = \mathcal{U}$, an obvious fact. Hence $A = \emptyset$. $\square$

In other words, there is no *proof by tautology*. The truth table below presents another way to see why not—in the second critical row, the conclusion is false.

$p \to T$
$\therefore p$

p	T	$p \to T$	p	
T	T	T	T	← *critical row*
F	T	T	F	← *critical row*

Exercises

Let A, B, and C be sets, subsets of the universal set U. Prove.

4.4.1 $A \cap \emptyset = \emptyset$.

4.4.2 $A \times \emptyset = \emptyset$.

4.4.3 $A - A = \emptyset$.

4.4.4 $(A - B) \cap (A \cap B) = \emptyset$.

4.4.5 $(B - A) \cap A = \emptyset$.

4.4.6 $A \cap (B - (A \cap B)) = \emptyset$.

4.5 Conditional proofs

Now consider propositions in General Form 2.

Theorem 4.4 *If $A \subseteq B$, then $A \cap B = A$.*

Proof. Suppose $A \subseteq B$.

Further suppose that $x \in A \cap B$. By definition of intersection, $x \in A$. Hence $A \cap B \subseteq A$ by definition of subset.	Proof of $A \cap B \subseteq A$.	Proof of $A \cap B = A$.	Entire proof.
Now suppose that $x \in A$. Since $A \subseteq B$, then $x \in B$ as well by definition of subset. Then by definition of intersection, $x \in A \cap B$. Hence $A \subseteq A \cap B$ by definition of subset.	Proof of $A \subseteq A \cap B$.		
Therefore, by definition of set equality, $A \cap B = A$. □			

This proof is composed of smaller proofs with which we are already familiar. At the innermost levels, we have subset proofs, two of them. Together, they constitute a proof of set equality, as we saw in Section 4.3. We are now wrapping that proof in one more layer to get a proof of a conditional proposition. That "one more layer" is another supposition. Take careful stock in how the word *suppose* is used in proof above, and recall the hypothetical conditional argument form from Game 3.

The sub-proof of $A \subseteq A \cap B$ makes no sense out of context. Certainly a set A is not in general a subset of its intersection with another set. What makes that true (as we say in the proof) is that we have supposed a restriction, namely $A \subseteq B$. The wrapper provides a context that makes the proposition $A \subseteq A \cap B$ true.

When we say *suppose*, we are boarding Mister Rogers's trolley to the Neighborhood of Make-Believe. We are creating a fantasy world in which our supposition is true, and then demonstrating that something else happens to be true in the world we are imagining. The imaginary world must obey all mathematical laws plus the laws we postulate in our supposition. Sometimes it is useful to make another supposition, in which case we enter a fantasy world *within* the first fantasy world—that world must obey all the laws of the outer world, plus whatever is now supposed.

Exercises

Let A, B, and C be sets, subsets of the universal set $\mathcal{U}$. Prove.

4.5.1 If $A \subseteq B$ and $B \subseteq C$, then $A \subseteq C$.

4.5.2 If $A \cap B = A \cap C$ and $A \cup B = A \cup C$, then $B = C$.

4.5.3 If $a \notin A$, then $A \cap \{a\} = \emptyset$. (This is the same as saying that if $a \notin A$, then A and $\{a\}$ are disjoint.)

4.5.4 If $A \subseteq B$, then $A - C \subseteq B - C$.

4.5.5 If $C \subseteq A$ and $C \subseteq B$, then $C \subseteq A \cap B$.

4.6 Integers

For variety, let us try out these proof techniques in another realm of mathematics. Here we will prove various propositions about integers, particularly about what facts depend upon them being even or odd. These proofs will rely on formal definitions of even and odd: An integer x is *even* if

even

$$\exists\, k \in \mathbb{Z} \mid x = 2k$$

and an integer x is *odd* if

odd

$$\exists\, k \in \mathbb{Z} \mid x = 2k + 1$$

We will take as axioms the facts that integers are closed under addition and multiplication, and that all integers are either even or odd and not both.

Axiom 3 *If $x, y \in \mathbb{Z}$, then $x + y \in \mathbb{Z}$.*

Axiom 4 *If $x, y \in \mathbb{Z}$, then $x \cdot y \in \mathbb{Z}$.*

Axiom 5 *If $x \in \mathbb{Z}$, then x is even iff x is not odd.*

You may also use basic properties of arithmetic and algebra in your proofs. Cite them as "by rules of arithmetic" or "by rules of algebra," although if you recall the name of the specific rule or rules being used, your proof will be better if you cite them. We begin with

Theorem 4.5 *If x and y are even integers, then $x + y$ is even.*

> **Proof.** Suppose x and y are even integers. By the definition of even, there exist $j, k \in \mathbb{Z}$ such that $x = 2j$ and $y = 2k$. Then
>
> $$\begin{aligned} x + y &= 2j + 2k && \text{by substitution} \\ &= 2(j + k) && \text{by distribution} \end{aligned}$$
>
> Further, $j + k \in \mathbb{Z}$ because integers are closed under addition. *Hence there is an integer, namely $j + k$, such that $x + y = 2(j + k)$.* Therefore $x + y$ is an even integer by the definition of even. □

Notice how we brought in the variables j and k. By saying "...there exist j, k...", we have made an implicit supposition about what j and k are. The definition of even establishes that this supposition is legal. Notice also how we structured the steps from $x + y$ to $2(j + k)$. This is a convenient shorthand for dealing with long chains of equations. Finally, the second to last sentence is italicized because it merely rephrases what the previous two sentences gave us.

Exercises

Prove

4.6.1 If x and y are odd integers, then $x + y$ is even.

4.6.2 If x and y are consecutive integers, then $x + y$ is odd. (*Consecutive* means that $y = x + 1$.)

4.6.3 If n^2 is odd, then n is odd. (Hint: try a proof by contradiction using Axiom 5.)

For any $a, b \in \mathbb{Z}$, we say that a *divides* b written $a|b$, if there exists $c \in \mathbb{Z}$ such that $a \cdot c = b$. Prove

4.6.4 For all $x \in \mathbb{N}$, $x|x$.

4.6.5 For all $x, y, z \in \mathbb{Z}$, if $x|y$ and $y|z$, then $x|z$.

4.6.6 For all $x, y \in \mathbb{N}$, if $x|y$ and $y|x$, then $x = y$. (This is equivalent to saying if $x|y$ and $x \neq y$, then $y \not| x$.)

4.7 Biconditionals

Biconditional propositions (those of General Form 3) stand toward conditional propositions in the same relationship as proofs of set equality stand toward subset proofs. A biconditional is simply two conditionals written as one proposition; one merely needs to prove each of them.

Theorem 4.6 $A - B = \emptyset$ *iff* $A \subseteq B$.

> **Proof.** First suppose $A - B = \emptyset$. *We will prove that* $A \subseteq B$. Suppose $x \in A$. $x \notin \emptyset$, by definition of empty set. So, by substitution, $x \notin A - B$. Then either $x \notin A$ or $x \in B$, by DeMorgan's laws. Since $x \in A$, then by elimination, $x \in B$. Hence $A \subseteq B$ by definition of subset.
>
> Conversely, suppose $A \subseteq B$. *We will prove that* $A - B = \emptyset$. Suppose $x \in A - B$. By definition of set difference, $x \in A$ and $x \notin B$. By definition of subset, $x \in B$, contradiction. Hence $A - B = \emptyset$. □

You might have thought that the step $x \notin \emptyset$ came out of nowhere. What we needed to do was get from $x \in A$ to $x \in B$ using the fact that $A - B = \emptyset$. That fact (our supposition) helped us by showing that x could not be in A but not in B. Noting that $x \notin \emptyset$ was our way of saying this.

Also, make sure you follow how we applied DeMorgan's laws. We had

$$x \notin A - B$$

or

$$\sim (x \in A - B)$$

which means (by definition of difference)

$$\sim (x \in A \land x \notin B)$$

Here is where we apply De Morgan's:

$$\sim (x \in A) \lor \sim (x \notin B)$$

That is,

$$x \notin A \vee x \in B$$

Now, concerning the biconditional itself, notice how many suppositions we have scattered all over the proof—and we are still proving fairly simple propositions. To avoid confusion you should use paragraph structure and transition words to guide the reader around the worlds you are moving in and out of. The word *conversely*, for example, indicates that we are now proving the second direction of a biconditional proposition (which is the *converse* of the first direction).

Exercises

Let A and B be sets, subsets of the universal set $\mathcal{U}$. Let x and y be integers. Prove.

4.7.1 $A \subseteq B$ iff $(B - A) \cup A = B$.

4.7.2 $A \subseteq B$ iff $(B - A)$ and A are a partition of B. (Hint: Use Exercises 4.4.5 and 4.7.1.)

4.7.3 $x \cdot y$ is odd iff x and y are both odd.

4.7.4 $C \subseteq A$ and $C \subseteq B$ iff $C \subseteq A \cap B$.

4.7.5 $(A - B) \cup C = (A \cup C) - (B - C)$. (This one does not contain a biconditional, but it is a harder problem, requiring an application of DeMorgan's laws similarly to our proof of Theorem 4.6.)

4.7.6 $B \subseteq A - C$ iff $B \subseteq A$ and $B \cap C = \emptyset$ (that is, B and C are disjoint).

4.8 Warnings

Many logical errors seduce the young and simple prover. Let not your feet go near their houses.

Arguing from example.

> $2 + 6 = 8$. Therefore, the sum of two even integers is an even integer. □

If this reasoning were sound, then this would also prove that the sum of any two even integers is 8.

Reusing variables.

> Suppose x and y are even integers. By the definition of even, there exists $k \in \mathbb{Z}$ such that $x = 2k$ and $y = 2k$.

Since x and y are even, each of them is twice some other integer—but those are *different* integers. Otherwise, we would be proving that all even integers are equal to each other. What is confusing about the correct way we wrote this earlier, "there exist $j, k \in \mathbb{Z}$ such that $x = 2j$ and $y = 2k$," is that we were contracting the longer phrasing, "there exists $j \in \mathbb{Z}$ such that $x = 2j$ and there exists $k \in \mathbb{Z}$ such that $y = 2k$." Had we reused the variable k in the longer version, it would be clear that we were trying to reuse a variable we had already defined. This kind of mistake is extremely common for beginners.

Begging the question.

> Suppose x and y are even. Then $x = 2j$ and $y = 2k$ for some $j, k \in \mathbb{Z}$. Suppose $x + y$ is even. Then $x + y = 2m$ for some $m \in \mathbb{Z}$. By substitution, $2j + 2k = 2m$, which is even by definition, and which we were to show. □

This nonsense proof tries to postulate a world in which the proposition to be proven is already true, followed by irrelevant manipulation of symbols. You cannot make any progress by supposing what you mean to prove—this is merely a subtle form of the "proof by tautology" we repudiated earlier.

Substituting "if" for "because" or "since."

> First suppose $A - B = \emptyset$. Suppose $x \in A$. If $A - B = \emptyset$, then it cannot be that $x \in A - B$. By definition of set difference, it is not true that $x \in A$ and $x \notin B$ ...

This is more a matter of style and readability than logic. Remember that each step in a proof should yield a new known proposition, justified by previously known facts. "If $A - B = \emptyset$, then it cannot be that $x \in A - B$" does no such thing, only informing us that $x \notin A - B$, contingent upon $A - B = \emptyset$ being true. Instead, this part of the proof should assert that $x \notin A - B$ *because* $A - B = \emptyset$.

4.9 Case study: Powersets

In this section we take our skills to a truly difficult problem. The main purpose here is just that—the experience of toughing out a proof with inherent complexity. It will involve meandering around several lemmas and wrapping our heads around thick notation. The important thing is not to let the task intimidate you. It may take a few reads to follow this proof and a couple tries to get the exercises. Stick with it! If you have come this far, you can get this section, too, though it may take a little effort, patience, and determination.

Recall our previous study of powersets. Imagine Gramma at her computer organizing photos of her grandchildren in all possible combinations.

{ Andrew, Bella, Carla, Dirk }
{ Andrew, Bella, Carla }
{ Andrew, Bella, Dirk }
{ Andrew, Carla }
{ Andrew, Bella }

{ Bella, Carla, Dirk }
{ Andrew, Carla , Dirk }
{ Bella, Dirk }
{ Bella, Carla }
{ Bella }
{ }

{ Carla, Dirk }
{ Andrew, Dirk }
{ Dirk }
{ Carla }
{ Andrew }

How could she organize these? One way is to shake these out into the photos that include Andrew and those that do not.

{ Andrew, Bella, Carla, Dirk }	{ Bella, Carla, Dirk }
{ Andrew, Bella, Carla }	{ Bella, Carla }
{ Andrew, Carla , Dirk }	{ Carla, Dirk }
{ Andrew, Bella, Dirk }	{ Bella, Dirk }
{ Andrew, Bella } { Andrew, Carla }	{ Bella } { Carla }
{ Andrew, Dirk } { Andrew }	{ Dirk } { }

This whole section is about proving a claim that seems pretty obvious now: If you take all the photos of all the grandchildren, you can divide them up into ones with Andrew and ones without, because there is no photo that both has Andrew and does not have Andrew, and any photo must either have Andrew or not have Andrew. Moreover, each of the photos with Andrew is just like one of the photos without him, just with Andrew added.

Why bother proving this? Well, in Exercise 2.4.15, you wrote a program for synthesizing the powerset of a set using this very observation. Another way to put it is that a powerset—say, $\mathscr{P}(\{1,2,3\})$—can be analyzed as

$$\begin{array}{lll} \{ & \{1,2,3\}, & \{2,3\}, \\ & \{1,2\}, & \{2\}, \\ & \{1,3\}, & \{3\}, \\ & \{1\}, & \emptyset \quad \} \end{array}$$

That is, we can pick any element from the set—say, 1—and the powerset splits into those sets containing 1 and those lacking 1. This means we have a partition, two disjoint subsets that together make up the whole set. Moreover, note that each set containing 1 is like one of the 1-lacking sets but with 1 added to it.

So here is our goal in this section:

> *For any (non-empty) set A, we can pick one element, call it the pivot, a. Consider the set with the pivot removed, D, and consider the powerset of that set, $\mathscr{P}(D)$. Make a new set by adding a to every set in $\mathscr{P}(D)$, call that new set E. D and E together make a partition of the powerset of A, $\mathscr{P}(A)$.*

To get technical, we will have to use some notation. We will start assuming that we are talking about $\{1,2,3\}$, and then generalize. The set of sets lacking 1 is $\mathscr{P}(\{2,3\}) = \mathscr{P}(\{1,2,3\} - \{1\})$. But how can we describe the set of sets that contain 1? Suppose B is one of the sets that contains 1. We have said it is the union of a set from the 1-lacking sets and $\{1\}$, so,

$$B = C \cup \{1\} \text{ where } C \text{ is some set such that } C \in \mathscr{P}(\{1,2,3\} - \{1\})$$

Take note—we are not saying that B either is in C or is $\{1\}$. We are saying that B is just like set C (whatever it is), but with $\{1\}$ added to it. If C is the set $\{2\}$, then B is the set $\{1, 2\}$.

Now, we want the set of all such Bs, so using set notation, we describe that set as

$$\{ C \cup \{1\} \mid C \in \mathscr{P}(\{1, 2, 3\} - \{1\}) \}$$

In general, if A is a set and a is any element in A, we can partition $\mathscr{P}(A)$ into

$$\mathscr{P}(A - \{a\})$$

and

$$\{ C \cup \{a\} \mid C \in \mathscr{P}(A - \{a\}) \}$$

Do not be scared by that mass of symbols. Think about what it means in terms of Gramma's photos.

A	All the grandchildren (or a photo of them all).
a	Andrew.
$\{a\}$	The set (or photo) of just Andrew.
$A - \{a\}$	All the grandchildren, except for Andrew.
$\mathscr{P}(A - \{a\})$	All the photos of all combinations of grandchildren, besides Andrew.
C	Any photo of any combination of grandchildren, besides Andrew.
$C \cup \{a\}$	A photo just like one of the photos without Andrew, except this one has Andrew.
$\{ C \cup \{a\} \mid C \in \mathscr{P}(A - \{a\}) \}$	All the photos made by taking a photo without Andrew and putting him in.

What we want to prove is that these two sets, $\mathscr{P}(A - \{a\})$ and $\{ C \cup \{a\} \mid C \in \mathscr{P}(A - \{a\}) \}$ are a partition of $\mathscr{P}(A)$. Understanding the notation is a third of the battle. Another third is recognizing the demands of the definition of *partition*. The labor in between is the remaining third.

Here is a warm-up theorem to get us used to working with powersets:

Theorem 4.7 *If* $\mathscr{P}(A) \subseteq \mathscr{P}(B)$ *then* $A \subseteq B$.

> Suppose $\mathscr{P}(A) \subseteq \mathscr{P}(B)$.
>
> Suppose further that $a \in A$. Then, by definition of powerset (see Section 2.4), $\{a\} \in \mathscr{P}(A)$. Then $\{a\} \in \mathscr{P}(B)$ by definition of subset. Again by definition of powerset, $a \in B$.
>
> Therefore, by definition of subset, $A \subseteq B$. □

Not so bad. The converse is also true; its proof is left as an exercise. Now, to show that the two sets above form a partition, we need to show

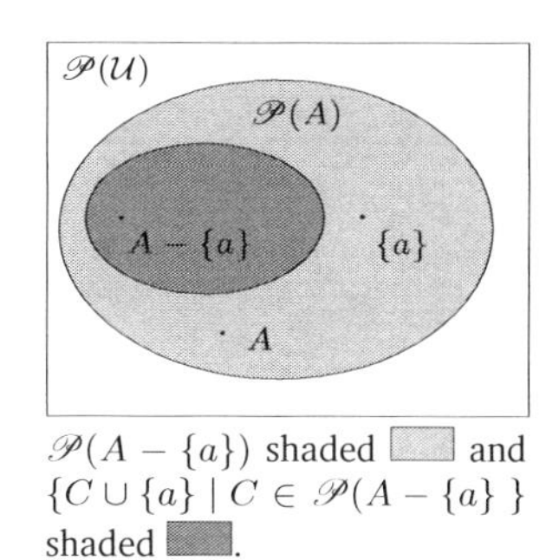

$\mathscr{P}(A - \{a\})$ shaded and $\{C \cup \{a\} \mid C \in \mathscr{P}(A - \{a\}\}$ shaded .

- They are disjoint, that is, $\mathscr{P}(A - \{a\}) \cap \{ C \cup \{a\} \mid C \in \mathscr{P}(A - \{a\}) \} = \emptyset$, and
- together they make the whole set, that is, $\mathscr{P}(A - \{a\}) \cup \{ C \cup \{a\} \mid C \in \mathscr{P}(A - \{a\}) \} = \mathscr{P}(A)$, which means
  - $\mathscr{P}(A - \{a\}) \cup \{ C \cup \{a\} \mid C \in \mathscr{P}(A - \{a\}) \} \subseteq \mathscr{P}(A)$, and
  - $\mathscr{P}(A) \subseteq \mathscr{P}(A - \{a\}) \cup \{ C \cup \{a\} \mid C \in \mathscr{P}(A - \{a\}) \}$

We will do these in the reverse order in which they are listed.

Lemma 4.8 *If $a \in A$, then $\mathscr{P}(A) \subseteq \mathscr{P}(A - \{a\}) \cup \{ C \cup \{a\} \mid C \in \mathscr{P}(A - \{a\}) \}$*

The way to start is to ignore the scary part and concentrate on what is familiar. It is a General Form 2 proposition (conditional) whose conclusion is a Set Form 1 proposition (subset).

> **Proof.** Suppose $a \in A$. Further suppose $X \in \mathscr{P}(A)$. By definition of powerset, $X \subseteq A$.

We chose a capital X because it is a set. Now, we have stuck our thumb into $\mathscr{P}(A)$ and pulled out X. We need to show it is also in the right hand side. The right hand side is a union, so it means we need to show that X is found in $\mathscr{P}(A - \{a\})$ or that it is found in $\{ C \cup \{a\} \mid C \in \mathscr{P}(A - \{a\}) \}$. Back in the story, suppose we grab a photo at random. We want to show that whatever we grab, it must either not contain Andrew or be like a picture without Andrew but with Andrew added in.

Which one is X in? On which side of the partition does X fall? That depends on whether or not X contains a. This calls for a division into cases. What is unusual here is that we use division into cases mainly for analysis, but now we are using it to synthesize a union operation. We still need an either/or to begin the division into cases. Here it is:

> Either $a \in X$ or $a \notin X$ (by the negation law).

The following diagrams summarize the preamble to the proof. Note that some diagrams use the universe $\mathcal{U}$ and others the universe $\mathscr{P}(\mathcal{U})$.

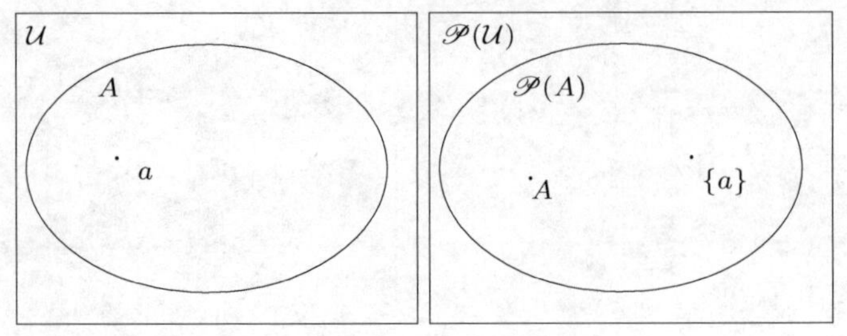

Suppose $a \in A$.

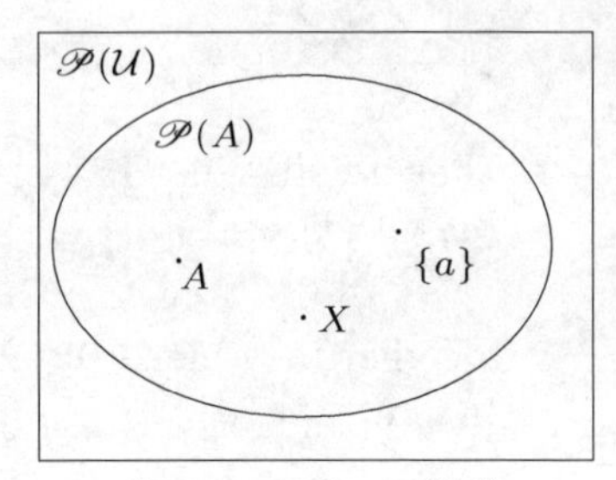

Suppose $X \in \mathscr{P}(A)$

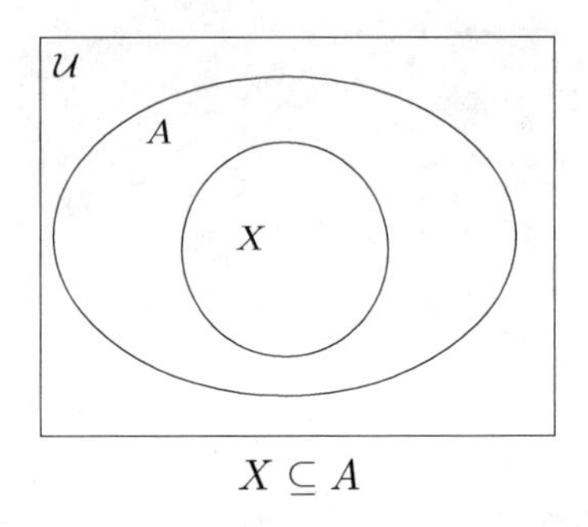

$X \subseteq A$

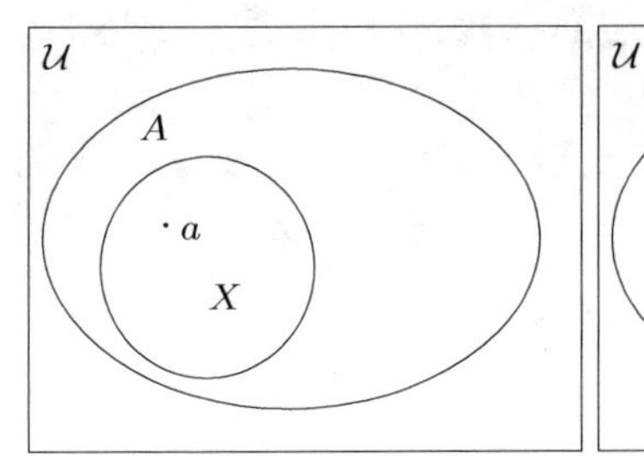

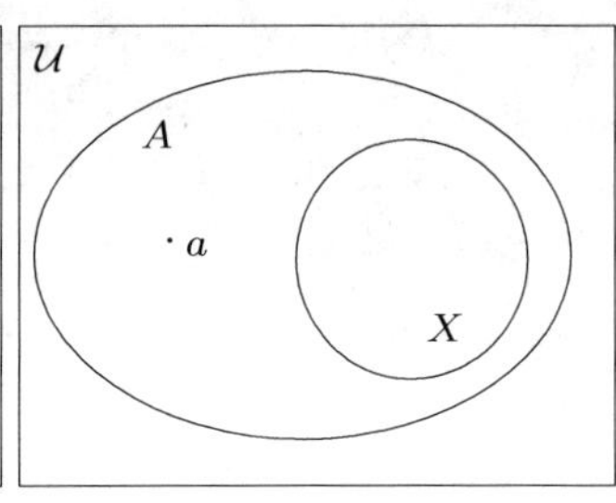

$a \in X$ or $a \notin X$

Okay, time to dive in.

Case 1: Suppose $a \in X$.

In this case, we are going to show $X \in \{ C \cup \{a\} \mid C \in \mathscr{P}(A - \{a\}) \}$, that it is one of the photos with Andrew. This means we need to show X fits the pattern $C \cup \{a\}$, that is, that $X = Y \cup \{a\}$ for some $Y \in \mathscr{P}(A - \{a\})$. To show that, we need a suitable Y, say a picture like X but without Andrew. We can find this Y by simply taking a out.

Let $Y = X - \{a\}$. By Exercise 4.7.1, $(X - \{a\}) \cup \{a\} = X$.

No, you are not expected to remember Exercise 4.7.1. But this is why it helps here:

Exercise 4.7.1 says	$B \subseteq A$	implies	$(A - B) \cup B = A$
so since	$\{a\} \subseteq X$	we know	$(X - \{a\}) \cup \{a\} = X$.

Similarly, we will use Exercise 4.5.4, which says that if $Q \subseteq R$, then $Q - S \subseteq R - S$.

By substitution, $Y \cup \{a\} = X$. By Exercise 4.5.4, $X - \{a\} \subseteq A - \{a\}$. By substitution, $Y \subseteq A - \{a\}$. By definition of powerset, $Y \in \mathscr{P}(A - \{a\})$.

Now this case is ours.

> Thus, by set notation, $Y \cup \{a\} \in \{ C \cup \{a\} \mid C \in \mathscr{P}(A - \{a\}) \}$. By substitution, $X \in \{ C \cup \{a\} \mid C \in \mathscr{P}(A - \{a\}) \}$. Then, by definition of union, $X \in \mathscr{P}(A - \{a\}) \cup \{ C \cup \{a\} \mid C \in \mathscr{P}(A - \{a\}) \}$

Here is Case 1, comic-book style:

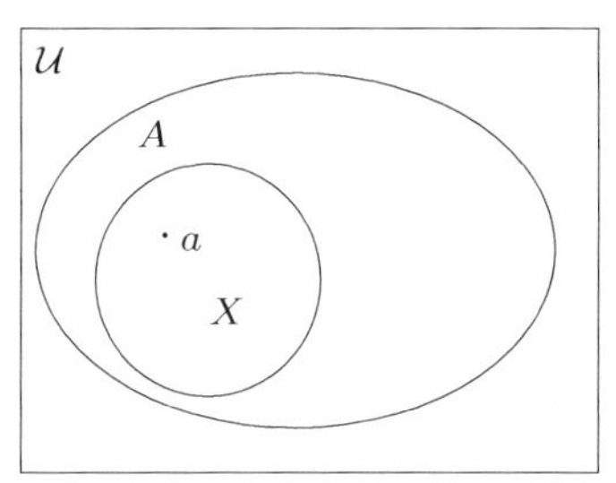

Suppose $a \in X$

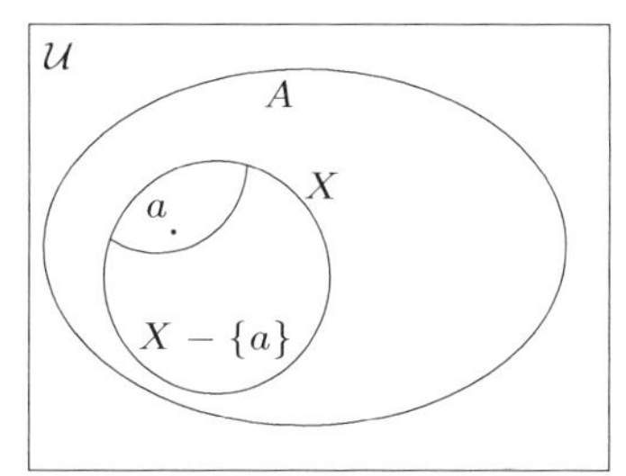

Since $\{a\} \subseteq X$, $(X - \{a\}) \cup \{a\} = X$.

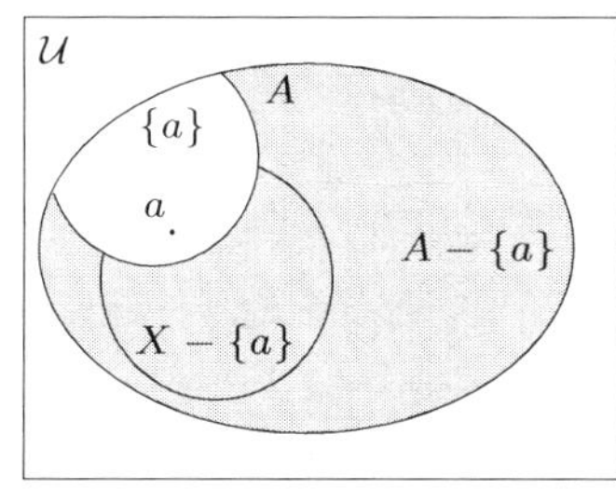

Since $X \subseteq A$, $X - \{a\} \subseteq A - \{a\}$. ($A - \{a\}$ is shaded.)

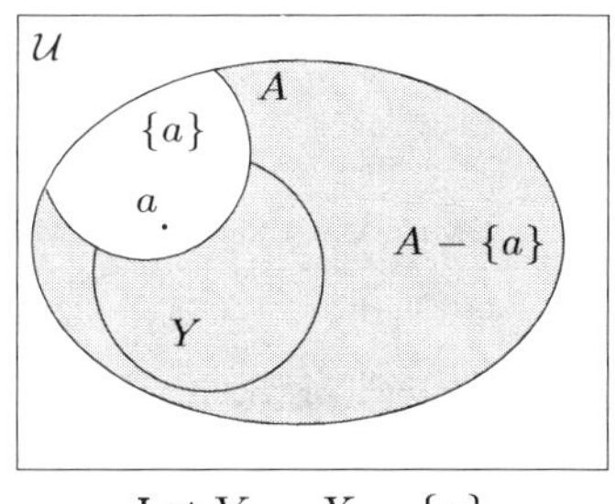

Let $Y = X - \{a\}$

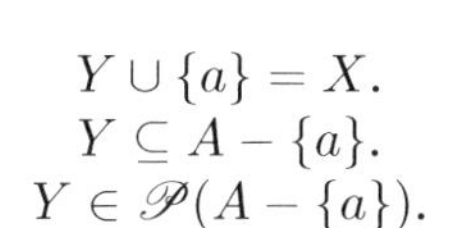

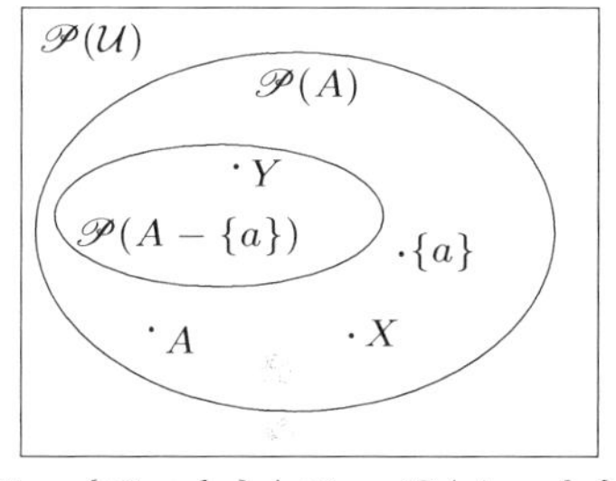

$X \in \{C \cup \{a\} \mid C \in \mathscr{P}(A - \{a\})$

Now, if a is not in X, then X must come from the other side of the union.

> **Case 2:** Suppose $a \notin X$.

We are going to show $X \in \mathscr{P}(A - \{a\})$, which requires us to show $X \subseteq A - \{a\}$.

> Suppose $b \in X$. Since $a \notin X$, we know $b \neq a$. By set notation, $b \notin \{a\}$.
>
> By definition of subset, $b \in A$. By definition of set difference, $b \in A - \{a\}$. By definition of subset, $X \subseteq A - \{a\}$. By definition of powerset, $X \in \mathscr{P}(A - \{a\})$.
>
> By definition of union, $X \in \mathscr{P}(A - \{a\}) \cup \{ C \cup \{a\} \mid C \in \mathscr{P}(A - \{a\}) \}$.

To illustrate:

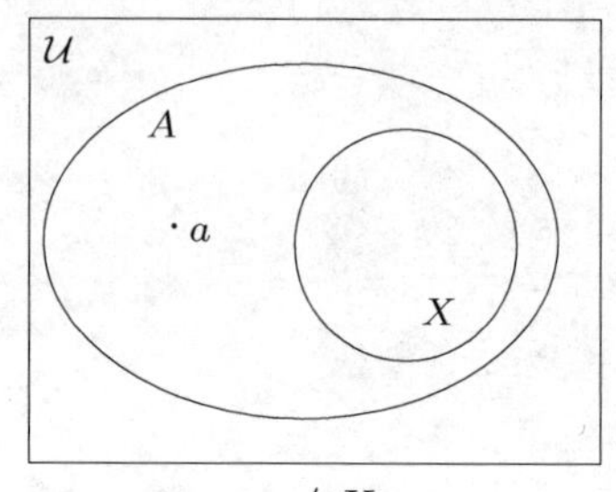

$a \notin X$

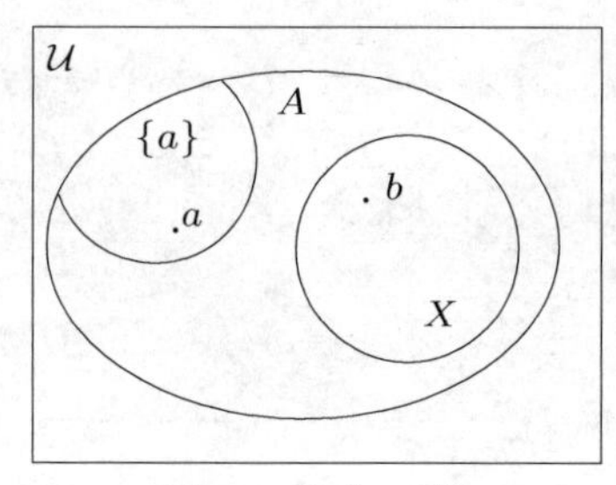

Suppose $b \in X$

$$b \neq a$$
$$b \notin \{a\}$$
$$b \in A$$
$$b \in A - \{a\}$$

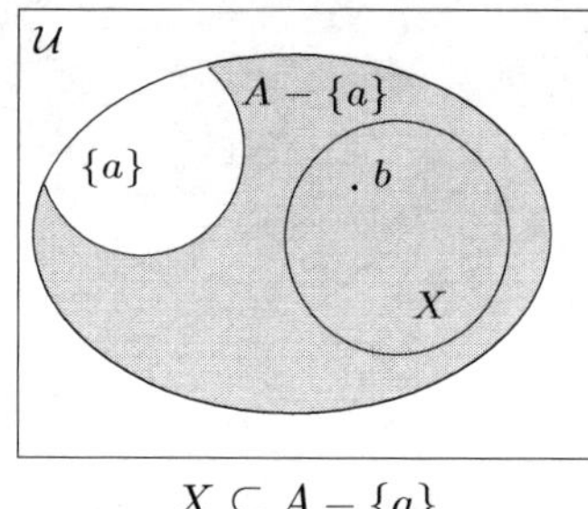
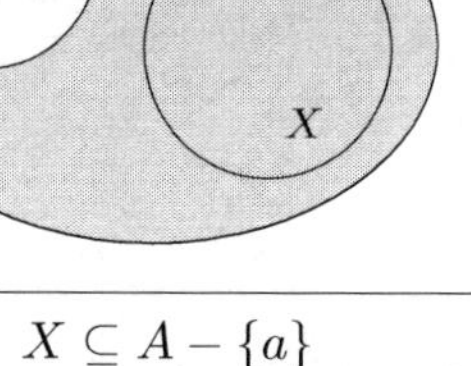

$X \subseteq A - \{a\}$

$\mathscr{P}(\mathcal{U})$
$\mathscr{P}(A)$
$\cdot X$
$\mathscr{P}(A - \{a\})$
$\cdot \{a\}$
$\cdot A$

$X \in \mathscr{P}(A - \{a\})$

Now it is all over.

In either case, $X \in \mathscr{P}(A - \{a\}) \cup \{ C \cup \{a\} \mid C \in \mathscr{P}(A - \{a\}) \}$. Therefore, by definition of subset, $\mathscr{P}(A) \subseteq \mathscr{P}(A - \{a\}) \cup \{ C \cup \{a\} \mid C \in \mathscr{P}(A - \{a\}) \}$. □

Taking it this slowly has the liability that you can lose sight of the big picture. Here is the whole proof again, together with an outline.

Pick a pivot, a. Pick a set X.

Maybe a is in X. Then we can find Y such that $Y \in \mathscr{P}(A - \{a\})$ and $X = Y \cup \{a\}$.

So $X \in \{ C \cup \{a\} \mid C \in \mathscr{P}(A - \{a\}) \}$ and so X is in the union.

On the other hand, maybe a is not in X.

Then we can show $X \subseteq A - \{a\}$, and so $X \in \mathscr{P}(A - \{a\})$. Again, X is in the union.

Either way, X is in the union, $\mathscr{P}(A)$ is a subset of the union.

Proof. Suppose $a \in A$. Further suppose $X \in \mathscr{P}(A)$. By definition of powerset, $X \subseteq A$. Either $a \in X$ or $a \notin X$ (by the negation law).

Case 1: Suppose $a \in X$. Let $Y = X - \{a\}$. By Exercise 4.7.1, $(X - \{a\}) \cup \{a\} = X$. By substitution, $Y \cup \{a\} = X$. By Exercise 4.5.4, $X - \{a\} \subseteq A - \{a\}$. By substitution, $Y \subseteq A - \{a\}$. By definition of powerset, $Y \in \mathscr{P}(A - \{a\})$.

Thus, by set notation, $Y \cup \{a\} \in \{ C \cup \{a\} \mid C \in \mathscr{P}(A - \{a\}) \}$. By substitution, $X \in \{ C \cup \{a\} \mid C \in \mathscr{P}(A - \{a\}) \}$. Then, by definition of union, $X \in \mathscr{P}(A - \{a\}) \cup \{ C \cup \{a\} \mid C \in \mathscr{P}(A - \{a\}) \}$

Case 2: Suppose $a \notin X$. Suppose $b \in X$. Since $a \notin X$, we know $b \neq a$. By set notation, $b \notin \{a\}$.

By definition of subset, $b \in A$. By definition of set difference, $b \in A - \{a\}$. By definition of subset, $X \subseteq A - \{a\}$. By definition of powerset, $X \in \mathscr{P}(A - \{a\})$.

By definition of union, $X \in \mathscr{P}(A - \{a\}) \cup \{ C \cup \{a\} \mid C \in \mathscr{P}(A - \{a\}) \}$.

In either case, $X \in \mathscr{P}(A - \{a\}) \cup \{ C \cup \{a\} \mid C \in \mathscr{P}(A - \{a\}) \}$. Therefore, by definition of subset, $\mathscr{P}(A) \subseteq \mathscr{P}(A - \{a\}) \cup \{ C \cup \{a\} \mid C \in \mathscr{P}(A - \{a\}) \}$. □

Take a deep breath. Fortunately, the second lemma is a little shorter.

Lemma 4.9 *If* $a \in A$, *then* $\mathscr{P}(A-\{a\}) \cup \{ C \cup \{a\} \mid C \in \mathscr{P}(A-\{a\}) \} \subseteq \mathscr{P}(A)$.

This will require a division into cases again. What is easier is that since the union is on the left hand side, the division into cases comes in the analysis of the union, not in the synthesis. However, the concept is the same. We pick something from the union, and it either contains a or it does not. Either way, we will find it to be in $\mathscr{P}(A)$.

> **Proof.** Suppose $a \in A$. Further suppose $X \in \mathscr{P}(A - \{a\}) \cup \{ C \cup \{a\} \mid C \in \mathscr{P}(A - \{a\}) \}$ By the definition of union, either $X \in \mathscr{P}(A - \{a\})$ or $X \in \{ C \cup \{a\} \mid C \in \mathscr{P}(A - \{a\}) \}$.
>
> **Case 1.** Suppose $X \in \mathscr{P}(A - \{a\})$. By Exercise 4.9.4, $X \in \mathscr{P}(A)$.

This is all we need for this case. Really, we have thrown the work to Exercise 4.9.4, observing that

if	$B \in \mathscr{P}(D - E)$	then	$B \in \mathscr{P}(D)$,
so since	$X \in \mathscr{P}(A - \{a\})$	we know	$X \in \mathscr{P}(A)$.

Now, for the other case, we look at the structure of the set from what set notation tells us.

> **Case 2.** Suppose $X \in \{ C \cup \{a\} \mid C \in \mathscr{P}(A-\{a\}) \}$. By set notation, there exists a $Y \in \mathscr{P}(A - \{a\})$ such that $X = Y \cup \{a\}$.

So, X is like a photo that we make by grabbing a photo that does not have Andrew and then adding Andrew in. We want to show $X \subseteq A$, that is, this photo is a part of the full collection of photos.

> Suppose $x \in X$. By the definition of union, $x \in \{a\}$ or $x \in Y$.
>
> > **Case 2a.** Suppose $x \in \{a\}$. By set notation, $x = a$. By substitution, $x \in A$.
> >
> > **Case 2b.** Suppose $x \in Y$. By the definition of powerset, $Y \subseteq A - \{a\}$. By the definition of subset, $x \in A - \{a\}$. By the definition of difference, $x \in A$.
> >
> > Either way, $x \in A$. By the definition of subset, $X \subseteq A$. By the definition of powerset, $X \in \mathscr{P}(A)$.
>
> Either way, $X \in \mathscr{P}(A)$. Therefore, by definition of subset $\mathscr{P}(A - \{a\}) \cup \{ C \cup \{a\} \mid C \in \mathscr{P}(A - \{a\}) \} \subseteq \mathscr{P}(A)$. □

Now we can say that these two sets together make up the entire powerset.

Theorem 4.10 *If* $a \in A$, *then* $\mathscr{P}(A - \{a\}) \cup \{ C \cup \{a\} \mid C \in \mathscr{P}(A - \{a\}) \} = \mathscr{P}(A)$.

Proof. From Lemmas 4.8 and 4.9. □

The other half of being a partition is that these two sets are disjoint. The proof is yours to do in Exercise 4.9.6.

Theorem 4.11 *If* $a \in A$*, then* $\mathscr{P}(A-\{a\}) \cap \{\, C \cup \{a\} \mid C \in \mathscr{P}(A-\{a\}) \,\} = \emptyset$.

Now this is the sum of the matter:

Corollary 4.12 *If* $a \in A$*, then* $\mathscr{P}(A-\{a\})$ *and* $\{\, C \cup \{a\} \mid C \in \mathscr{P}(A-\{a\}) \,\}$ *make a partition of* $\mathscr{P}(A)$.

Proof. By Theorem 4.10, the two sets together make up the entire powerset. By Theorem 4.11 and definition of disjoint, the two sets are disjoint. Therefore, by definition of partition, these two sets make a partition of the powerset. □

Exercises

Let A, B, and C be sets, subsets of the universal set $\mathcal{U}$. (In some exercises, C has a local scope within a formula.) Prove.

4.9.1 If $A \subseteq B$, then $\mathscr{P}(A) \subseteq \mathscr{P}(B)$.

4.9.2 If $B \subseteq A$, then $\mathscr{P}(B) - \mathscr{P}(A) = \emptyset$.

4.9.3 $\mathscr{P}(A) \cap \mathscr{P}(B) = \mathscr{P}(A \cap B)$.

4.9.4 If $B \in \mathscr{P}(A - C)$, then $B \in \mathscr{P}(A)$.

4.9.5 $C \in \mathscr{P}(A) \cap \mathscr{P}(B)$ iff $C \in \mathscr{P}(A \cap B)$. (You may want to cite Exercise 4.7.4.)

4.9.6 If $a \in A$, then $\mathscr{P}(A - \{a\}) \cap \{\, C \cup \{a\} \mid C \in \mathscr{P}(A-\{a\}) \,\} = \emptyset$. (This is Theorem 4.11.) (Hint: The overall structure of this is $X \cap Y = \emptyset$. So, pick an element in $X \cap Y$ (which in this case will be a set). You need to derive a contradiction, that is, the element cannot be in both sets. In this case, does the set you picked have a or not?)

4.9.7 If $x \in \mathcal{U}$, then $\{B \cup C \mid C \in \mathscr{P}(A)\} \cup \{B \cup C \cup \{x\} \mid C \in \mathscr{P}(A)\} = \{B \cup C \mid C \in \mathscr{P}(A \cup \{x\})\}$. (This is a challenging question—even compared to the others in this section. To take this one on, first spend time dissecting what is being stated and determine what needs to be shown. Note that A and B are specific sets, but C ranges over all the sets in a powerset in the various places it is used.)

4.10 From theorems to algorithms

The programming in this section is review, more or less. We will not be learning any new ML constructs or even any new programming techniques. However, this section does exercise your problem-solving skills in a new way, and it builds a bridge between programming and our work in this chapter on proving propositions.

greatest common divisor

For any $a, b \in \mathbb{Z}$, the *greatest common divisor* (GCD) of a and b is a positive integer c such that c is the greatest integer that divides both a and b. The GCD is useful for such things as putting fractions in simplest form. Formally,

- $c|a$ and $c|b$ (that is, c is a common divisor), and
- $\forall\, d \in \mathbb{Z}$ such that $d|a$ and $d|b$, $d \leq c$ (that is, c is the greatest such integer).

We said $d \leq c$ rather than $d < c$ because the "$\forall\ d$" does not mean d and c have to be distinct. By the same token c might be the same number as a or b. Also, we could have said $d|c$ rather than $d \leq c$; it would have been an equivalent definition.

We could apply this definition blindly to get the following set of functions which compute the GCD. The first one takes two ints and makes a list of all positive int that are divisors of the given two. It does this by counting down from the smaller of the two given numbers and adding each number to the list only if it evenly divides both.

```
- fun commonDivisors(a, b) =
=    let fun comDiv(1) = [1]
=          | comDiv(n) = if a mod n = 0 andalso b mod n = 0
=                        then n::comDiv(n-1)
=                        else comDiv(n-1);
=    in comDiv(if a < b then a else b)
=    end;

val commonDivisors = fn : int * int -> int list

- commonDivisors(12, 42);

val it = [6,3,2,1] : int list

- fun greatest([x]) = x
=   | greatest(y::rest) =
=        let val z = greatest(rest)
=        in if y > z then y else z
=        end;

Warning: match nonexhaustive
val greatest = fn : int list -> int

- greatest([5, 17, 2, 4, 82, 9, 22]);

val it = 82 : int

- fun gcd(a, b) = greatest(commonDivisors(a, b));

val gcd = fn : int * int -> int

- gcd(574, 392);

val it = 14 : int
```

A pair of theorems about the GCD, however, lead to a more efficient algorithm.

Lemma 4.13 *If* $a \in \mathbb{N}$, *then* $\gcd(a, 0) = a$.

Lemma 4.14 *If* $a, b \in \mathbb{Z}$, $q, r \in \mathbb{W}$, *and* $a = b \cdot q + r$, *then* $\gcd(a, b) = \gcd(b, r)$.

Lemma 4.13 is straightforward and is an exercise. Lemma 4.14 is more complicated and can be found in any discrete mathematics text with a significant number theory component. To make sense of it, note that q and r are named as they are because they are the quotient and remainder. We essentially throw q away; $r = a \texttt{ mod } b$. But anything that divides both a and b must also divide r, since

$$r = a - b \cdot q$$

If d is a divisor of a and b, then $a = d \cdot i$ and $b = d \cdot j$, for some i and j. By substitution,

$$\begin{aligned} r &= (d \cdot i) - (d \cdot j) \cdot q \\ &= d(i - j \cdot q) \end{aligned}$$

Tracing this out for the GCD of 574 and 392,

$$\begin{array}{rclll} 574 & = & 392 \cdot 1 & +182 & \text{now find GCD of 392 and 182} \\ 392 & = & 182 \cdot 2 & +28 & \text{now find GCD of 182 and 28} \\ 182 & = & 28 \cdot 6 & +14 & \text{now find GCD of 28 and 14} \\ 28 & = & 14 \cdot 2 & +0 & \text{the answer, then is 14} \end{array}$$

How does this help us write a more efficient algorithm? Notice that the first lemma gives us a "final answer"—it tells us the what the GCD is, in a simple number, though in a restricted case. The second lemma reduces the GCD to another problem which has the same answer. This begins to look like a base case and a recursive case. In fact, we could unify these two propositions to

$$\textit{If } a, b \in \mathbb{Z}, \textit{ then } \gcd(a, b) = \begin{cases} a & \textit{if } b = 0 \\ \gcd(b, a \texttt{ mod } b) & \textit{otherwise} \end{cases}$$

which becomes in ML,

```
- fun gcd(a, 0) = a
=   | gcd(a, b) = gcd(b, a mod b);

val gcd = fn : int * int -> int

- gcd(574, 392);

val it = 14 : int
```

Euclidean Algorithm

This is called the *Euclidean Algorithm*, though it was known before Euclid.

Exercises

4.10.1 Prove Lemma 4.13.

4.10.2 The following lemmas are not surprising.

Lemma 4.15 *If $a, b \in \mathbb{Z}$, then $a \cdot b^0 = a$.*

Lemma 4.16 *If $a, b, n \in \mathbb{Z}$, then $a \cdot b^n = (a \cdot b) \cdot b^{n-1}$.*

However, they can be used to generate an algorithm for exponentiation using repeated multiplication.

$$
\begin{array}{lll}
3^4 & & \\
= 1 \cdot 3^4 & & \\
= (1 \cdot 3) \cdot 3^{4-1} & = 3 \cdot 3^3 & \\
= (3 \cdot 3) \cdot 3^{3-1} & = 9 \cdot 3^2 & \text{by Lemma 4.16} \\
= (9 \cdot 3) \cdot 3^{2-1} & = 27 \cdot 3^1 & \\
= (27 \cdot 3) \cdot 3^{1-1} & = 81 \cdot 3^0 & \\
= 81 & & \text{by Lemma 4.15}
\end{array}
$$

Write an ML function to compute x^y; your function should contain a helper function that takes three parameters corresponding to a, b, and n; Lemma 4.15 provides a base case, Lemma 4.16 a recursive case.

4.10.3 The algorithm you used in Exercise 4.10.2 required n multiplication operations. Improve this by making use of the following additional lemma.

Lemma 4.17 *If $a, b, n \in \mathbb{Z}$ and n is even, then $a \cdot b^n = a \cdot (b^2)^{\frac{n}{2}}$.*

(If you have done this correctly, your function/algorithm should require about $\log_2 n$ multiplications. Can you tell why?)

4.10.4 In a similar way, use the following three lemmas to write a function that computes $x \cdot y$ without using multiplication. (To make it fast, you will need to use division, but only dividing by 2.)

Lemma 4.18 *If $a, b \in \mathbb{Z}$, then $a + b \cdot 0 = a$.*

Lemma 4.19 *If $a, b, c \in \mathbb{Z}$, then $a + b \cdot c = a + b + b \cdot (c - 1)$.*

Lemma 4.20 *If $a, b, c \in \mathbb{Z}$ and c is even, then $a + b \cdot c = a + (b + b) \cdot (c \div 2)$.*

Exercises 4.10.2–4.10.4 are adapted from Abelson and Sussman[1, pages 46–47].

4.10.5 The following is known as the Quotient-Remainder Theorem (QRT).

Theorem 4.21 *If $n, d \in \mathbb{N}$, then there exist unique whole numbers q and r such that $n = d \cdot q + r$ and $0 \leq r < d$.*

It guarantees that the integer division and mod operations are well defined, that there exist q and r that meet the restrictions

- $n = d \cdot q + r$
- $r \geq 0$
- $r < d$

Write a function that computes the quotient and the remainder of n and d, returning them as a tuple. Do not use `div` or `mod`. Instead, use the QRT to search for q and r. Your function should have a helper function that has parameters corresponding to q and r, initially call it with 0 for q and n for r. Thus you begin meeting the first two restrictions and adjust q and r until you also meet the third. The algorithm you come up with is called the *Division Algorithm*.

4.10.6 Let the base-b *floor logarithm* (flog) of x be the greatest integer less than $\log_b x$. Define the base-b *remainder logarithm* (rlog) of x to be the difference between x and b raised to the floor log of x. For examples,

			floor log	**remainder log**
17	$=$	$16 + 1$		
	$=$	$2^4 + 1$	$\text{flog}_2 17 = 4$	$\text{rlog}_2 17 = 1$
30	$=$	$27 + 3$		
	$=$	$3^3 + 3$	$\text{flog}_3 30 = 3$	$\text{rlog}_3 30 = 3$
68	$=$	$64 + 4$		
	$=$	$4^3 + 4$	$\text{flog}_4 68 = 3$	$\text{rlog}_4 68 = 4$

Lemma 4.22 *For all $a, b \in \mathbb{N}$, there exists unique $n, r \in \mathbb{W}$ such that $a = b^n + r$ and $0 \leq r < b^n$.*

Lemma 4.23 *For all $b, n, r \in \mathbb{W}$, $b^n + r = b^{n+1} + r - (b - 1) \cdot b^n$.*

Using these lemmas, write an ML function that takes two integers (a and b) and returns $\text{flog}_b a$ and $\text{rlog}_b a$ as a tuple.

Exercises, continued

4.10.7 This exercise presents the *Extended Euclidean Algorithm*. First consider this convoluted lemma of seemingly dubious worth:

Lemma 4.24 *If* $a, b, A, B \in \mathbb{W}$, $s, t, u, v \in \mathbb{Z}$, $a = s \cdot A + t \cdot B$, $b = u \cdot A + v \cdot B$, $gcd(a, b) = gcd(A, B)$, *and* $b \neq 0$, *then*

$$\begin{aligned} gcd(b, a \bmod b) &= gcd(A, B) \\ b &= u \cdot A + v \cdot B \\ a \bmod b &= (s - u \cdot (a \text{ div } b)) \cdot A \\ &\quad + (t - v \cdot (a \text{ div } b)) \cdot B \end{aligned}$$

What this gives us is a means of computing x and y for a given a and b such that $a \cdot x + b \cdot y = \gcd(a, b)$. Finish the following function that takes two such integers a and b and returns a triple (a 3-tuple) containing the GCD and the appropriate coefficients x and y.

```
fun extEuc(a, b) =
    let fun step(A, 0, s, t, u, v) =
                (A, ??, ??)
          | step(A, B, s, t, u, v) =
                step(B, A mod B,
                     ??, ??, ??, ??)
    in
       step(a, b, ??, ??, ??, ??)
   end;
```

(Hint: First think about how this algorithm computes the GCD, as the regular Euclidean algorithm does. Then fill in the blanks for the recursive call of `step` based on what the lemma says. When the second parameter is 0, what are the coefficients to our final answer? Finally, fill in the blanks for the original call of `step` by thinking about values for s, t, u, and v that would work when $a = A$ and $b = B$.)

4.11 Extended example: Solving games

Using computers to solve puzzle-like games illustrates the power of computing machinery, or at least an illustration of what machinery can do better than the human mind can. Games that have positions or configurations—that is, the arrangement of pieces on a board or numbers or letters in a puzzle—can be played by a computer by sifting through large amounts of data on possible outcomes of the game. The speed and abundance of computer memory gives machines an incomparable advantage over humans. An unbeatable tic-tac-toe-playing program is fairly straightforward to write. Computer chess-playing engines now regularly beat the world's top human players.

In this section, we try our hand at writing programs to solve puzzles that require searching through solutions. The first is called *bulls and cows*. One player thinks of a four-digit number with no digit appearing more than once (we also assume that the number does not begin with 0). The other player guesses the number. After each guess, the first player tells how many *bulls* and *cows* the guess has: a digit in the correct place is a bull, and a correct digit in a wrong place is a cow. If the secret number is 4721 and the guess is 5124, then this guess has one bull (2) and two cows (1 and 4).

We will represent a solution or guess in ML using a tuple of `int`s. Here is how we test if a 4-tuple is a valid guess (the function `makeNoRepeats` comes from Exercise 3.7.9, `count` from Exercise 2.2.4):

```
fun isValidGuess(a,b,c,d) =
     0 < a andalso a < 10 andalso 0 <= b andalso b < 10 andalso
     0 <= c andalso c < 10 andalso 0 <= d andalso d < 10 andalso
     count(makeNoRepeats([a, b, c, d])) = 4;
```

First consider how to make the computer play the role of the first player and evaluate the second player's guesses. (Later we will write a program that acts as the guesser.) This program will be interactive. When the human player makes a guess, he or she will call a function that will check the guess and report on bulls and cows. To do this, we need the system to remember the secret number and the number of guesses between function calls. This is not the usual mode for ML, but it still can be done using *reference variables*, which are symbols that can be updated. Reference variables are declared in the ordinary way except with the keyword `ref` prepended to the value:

reference variables

```
val soln = ref (4, 7, 2, 1);
val guesses = ref 0;
```

To set a reference variable to a new value, we use the operator `:=`. Such an operation is a statement, returning a `unit` value, like `print` does. Recall from Section 2.5 that expressions and statements can be compounded using a statement list. For example, we can reset the game using the following function (realistically we would want the computer to generate the secret number randomly, but this is just a warm-up for us):

```
fun resetGuesses() = (soln := (2, 6, 4, 9);
                      guesses := 0);
```

The function `bullsAndCows` takes a guess and the solution and returns a tuple, the number of bulls and cows. For each digit in the guess, we compare it with the digit in the same position in the solution; similarly, we see if each digit in the guess is one of the other digits in the solution. We package these as lists of booleans, which `numTrue` counts.

```
fun numTrue([]) = 0
  | numTrue(false::rest) = numTrue(rest)
  | numTrue(true::rest) = 1 + numTrue(rest);
fun bullsAndCows((a,b,c,d), (aa, bb, cc, dd)) =
     (numTrue([a = aa, b = bb, c = cc, d = dd]),
      numTrue([contains(a, [bb, cc, dd]),
               contains(b, [aa, cc, dd]),
               contains(c, [aa, bb, dd]),
               contains(d, [aa, bb, cc])]));
```

The human player submits a guess with `checkGuess`, which reports on the bulls and cows in readable format, as well as keeping track of the number of guesses. To retrieve a value from a reference variable, we apply the `!` operator.

```
fun checkGuess(a, b, c, d) =
   if isValidGuess(a, b, c, d)
   then
      let val (bulls, cows) =
               bullsAndCows((a, b, c, d), !soln);
      in
         (guesses := !guesses + 1;
          print("Guess # " ^ Int.toString(!guesses) ^ "\n");
          print("bulls: " ^ Int.toString(bulls) ^ "\n");
          print("cows: " ^ Int.toString(cows) ^ "\n");
          if bulls = 4 then print ("Correct!\n") else ())
      end
   else
      print("Invalid guess.\n");

- checkGuess(9,7,2,4);

Guess # 1
bulls: 2
cows: 1
val it = () : unit

- checkGuess(4,7,2,1);


Guess # 2
bulls: 4
cows: 0
Correct!
val it = () : unit
```

A more interesting problem is having the computer guess. Here is our strategy: We generate a giant list of all possible guesses. We pull guesses from this list. Every time we are told the bulls and cows for our previous guess, we remove all guesses from the list that are inconsistent with that result. Suppose we guess 2943 and are told that there is 1 bull and 2 cows. Then we would eliminate 5913 from the list because if 5913 were the answer, then our guess 2943 would have had 2 bulls and 0 cows, but we would keep 3742 because it is consistent with the bull and cow result—it is a possible solution, based on this information.

Our first task is generating the initial guess list. We will create a list containing all possible combinations of four element from the set of digits. This is similar to our function to generate a powerset (Exercise 2.4.15): for each combination of two elements, we want to add each digit to it, then add each digit to all of those combinations of size three. (See Exercise 2.2.10 for `listify` and Exercise 2.4.14 for `addToEach`. The variable `guessList` is initially set to an empty list as a dummy variable; we must type it explicitly.)

```
val guessList = ref ([]:(int*int*int*int) list);
fun addEachToEach([], y) = []
  | addEachToEach(a::aRest, y) =
      addToEach(a, y)@addEachToEach(aRest, y);
fun enumerateCombinations(x, 1) = listify(x)
  | enumerateCombinations(x, r) =
     addEachToEach(x, enumerateCombinations(x, r-1));
```

But this is too much. Our guesses are supposed to be tuples, not lists, and this process generates many invalid guesses. We can remove those. Also, we would like to be able to print guesses readably.

```
fun makeGuessesTuples([]) = []
  | makeGuessesTuples([a,b,c,d]::rest) =
          (a, b, c, d)::makeGuessesTuples(rest);
fun removeInvalid([]) = []
  | removeInvalid(a::rest) =
        if isValidGuess(a) then a::removeInvalid(rest)
                           else removeInvalid(rest);
fun printGuess(a, b, c, d) =
   print(Int.toString(a) ^ " " ^ Int.toString(b) ^ " " ^
         Int.toString(c) ^ " " ^ Int.toString(d) ^ "\n");
```

Now, to generate a first guess, we enumerate our guesses and pick one. We keep track of our most recent guess in a reference variable `currentGuess`.

```
val currentGuess = ref (0,0,0,0);
fun firstGuess() =
     (guessList :=
       removeInvalid(
         makeGuessesTuples(
          enumerateCombinations([0,1,2,3,4,5,6,7,8,9],
                                4)));
      currentGuess := hd(!guessList);
      guessList := tl(!guessList);
      printGuess(!currentGuess));
```

The user will play by calling the function `guessAgain`, providing the number of bulls and cows in the previous guess. Our last task, then, is to filter our guess list by removing those guesses we now know to be infeasible. For a given guess `a`, `bullsAndCows(!currentGuess, a)` determines the number of bulls and cows there would be if `a` were the right answer and `currentGuess` were guessed; keep `a` only if it matches what the user supplied.

```
fun removeInfeasible(bulls, cows, []) = []
  | removeInfeasible(bulls, cows, a::rest) =
      if bullsAndCows(!currentGuess, a) = (bulls, cows)
      then a::removeInfeasible(bulls, cows, rest)
      else removeInfeasible(bulls, cows, rest);
fun guessAgain(bulls, cows) =
     (guessList := removeInfeasible(bulls, cows, !guessList);
      currentGuess := hd(!guessList);
      guessList := tl(!guessList);
      printGuess(!currentGuess));
```

Our secret number is 2416.

```
- firstGuess();

1 0 2 3
val it = () : unit

- guessAgain(0, 2);

2 1 4 5
val it = () : unit

- guessAgain(1, 2);

2 3 5 4
val it = () : unit

- guessAgain(1, 1);

2 4 1 6
val it = () : unit
```

All told, bulls and cows is still a simple game. Your project is a program to solve Sudoku puzzles. The rest of this section will provide the infrastructure and lay out a skeleton of the algorithm.

A Sudoku puzzle contains 81 spaces laid out in a 9×9 grid, and organized not only into rows and columns but also into nine 3×3 squares. Each space must be filled with one of nine values, and each value must appear exactly once in each row, column, and square. The values for some spaces are given as part of the puzzle; to solve the puzzle, the player must fill-in the remaining spaces. Traditionally the values are the digits 1–9, but they could be any nine arbitrary values.

To model an instance of the puzzle (a Sudoku *board*), we use datatypes to define Sudoku values and Sudoku spaces. A space is either a single, given value, or it is an open space together with a list of possible values.

```
datatype sudokuValue = S1 | S2 | S3 | S4 | S5 | S6 | S7 | S8 | S9;
datatype sudokuSpace = Given of sudokuValue |
                       Open of sudokuValue list;
```

Initially any open space has all values of sudokuValue in its list of possibilities. Our program to solve the puzzle will narrow down the list for each open space until each has only one possibility. So, when we print the board, for each given space and for each open space with only one possibility left, we print that one value; for each open space with more than one possibility left, we print an underscore, indicating it is still blank; if any space has no possibilities left, the board is in an impossible state, and we print an X for that space. Hence `printSpace` and `printBoard`:

```
fun printSpace(Given(a)) = printValue(a)
  | printSpace(Open([a])) = printValue(a)
  | printSpace(Open([])) = "X"
  | printSpace(Open(aa)) = "_";
fun printBoard(board) =
  let fun printHelper([], n) = ()
        | printHelper(a::rest, n) =
          (print(printSpace(a) ^ " ");
           if n = 80 then print("\n")
           else if n mod 27 = 26
                then print("\n----------------------\n")
           else if n mod 9 = 8 then print("\n")
           else if n mod 3 = 2 then print "| " else ();
           printHelper(rest, n+1));
  in printHelper(board, 0) end;
```

See the accompanying website for the code for `printValue`, as well the code for a sample board ("Easy Puzzle 4,240,655,484" from `www.websudoku.com`). Here is what the result looks like:

```
- printBoard(sample);

2 _ _ | 7 _ 8 | _ 1 3
_ 7 3 | _ _ _ | _ 5 _
6 _ _ | 2 _ 3 | 4 _ _
----------------------
8 _ _ | _ _ 4 | _ 3 _
_ _ 1 | 8 _ 9 | 7 _ _
_ 4 _ | 1 _ _ | _ _ 9
----------------------
_ _ 9 | 5 _ 2 | _ _ 8
_ 2 _ | _ _ _ | 1 4 _
4 8 _ | 3 _ 1 | _ _ 5
val it = () : unit
```

Take the leftmost blank on the top row. Because of the other numbers in that row, we know that numbers 1, 2, 3, 7, and 8 cannot go in that space. Inspecting the column eliminates 2, 4, 7, and 8. The numbers already appearing in the square are 2, 3, 6, and 7. This reduces the possibilities for that space to just 5 and 9. Thus by eliminating some values from an open space's list of possibilities, we come closer to a solution.

Our general strategy, then, is to transform a board into a new one just like it except that the list of possible values for the open spaces is reduced by looking at the numbers appearing in that space's row, column, and square. If any open space is reduced to only one possibility, it can be treated just like a given space. If all open spaces have exactly one possibility, then the puzzle instance is solved. If any open space is reduced to zero possibilities—well, that is a problem. But if the board is solvable (and we write our program correctly), that should not happen.

Breaking this task down, for each open space we need to gather the other positions in its row, column, and square, then create a new space but with any impossible values removed. To pick the rowmates, columnmates, and square-mates out of the board, it will help to have a way to name each position. Since the board is conceptually a grid but represented as a list, each position will in fact have two names—its grid coordinates (conceptual) and its position in the list (representational). The naming scheme we choose is summarized by the following board template:

	0	1	2	3	4	5	6	7	8
0	0	1	2	3	4	5	6	7	8
1	9	10	11	12	13	14	15	16	17
2	18	19	20	21	22	23	24	25	26
3	27	28	29	30	31	32	33	34	35
4	36	37	38	39	40	41	42	43	44
5	45	46	47	48	49	50	51	52	53
6	54	55	56	57	58	59	60	61	62
7	63	64	65	66	67	68	69	70	71
8	72	73	74	75	76	77	78	79	80

So, position 31 in the list has coordinates (4, 3). Other indexing schemes are possible, but the math works out more cleanly in some than others. It might seem strange (if you have not programmed in a language like C or Java) to start naming from 0 rather than 1, but notice that in this scheme the list position n can be computed from the coordinates (x, y) by $n = 9 \cdot y + x$. To compute x and y from n,

```
fun numToCoords(n) = (n mod 9, n div 9);
```

In order to do the elimination, we need to collect the 'mates for a given position. The function `getRowMates` takes x and y coordinates and a list of sudokuSpaces—the board, or part of the board. For bookkeeping, it also takes one more parameter: our current position in the list, or, what position the head of the given list has in the original list. So, if we wanted to collect the rowmates for position $(3, 4)$ in a board called `sample`, we would call `getRow(3, 4, sample, 0)`, since the head of `sample` is position 0 in `sample`. The call would make a recursive call which would in effect be a call of `getRow(3, 4, tl(sample), 1)`. We show the interpreters response to this one because the type is important.

```
- fun getRowMates(x, y, [], n) = []
=    | getRowMates(x, y, a::rest, n) =
=      if y = n div 9 andalso x <> n mod 9
=      then a::getRowMates(x, y, rest, n+1)
=      else getRowMates(x, y, rest, n+1);


val getRowMates = fn : int * int * 'a list * int -> 'a list
```

We could use this on a list of anything that is interpreted as a 9×9 grid indexed as described above. For us, 'a is sudokuSpace. The functions `getColumnMates` and `getSquareMates` are for you to write.

The function `removeImpossible` takes a list of sudokuValues and a list of sudokuSpaces. It iterates through the second list—for every given space and every open space with one option left, it removes that value from the first list, using helper function `remove`.

```
fun remove(x, []) = []
  | remove(x, a::rest) =
     if x = a then remove(x, rest) else a::remove(x, rest);
fun removeImpossible(possible, []) = possible
  | removeImpossible(possible, Given(x)::rest) =
      removeImpossible(remove(x, possible), rest)
  | removeImpossible(possible, Open([x])::rest) =
      removeImpossible(remove(x, possible), rest)
  | removeImpossible(possible, _::rest) =
      removeImpossible(possible, rest);
```

The underscore in the last pattern for `removeImpossible` is a common way to refer to parts of the pattern we are not interested in.

With these tools, oneRound takes a board and transforms it by removing the impossible values from every open space. Writing oneRound is your task, but we will show you what it does.

```
- printBoard(oneRound(sample));
```

```
2 _ _ | 7 _ 8 | _ 1 3
_ 7 3 | _ _ 6 | _ 5 _
6 _ _ | 2 _ 3 | 4 _ 7
-----------------------
8 _ _ | 6 _ 4 | _ 3 _
_ _ 1 | 8 _ 9 | 7 _ _
_ 4 _ | 1 _ _ | _ _ 9
-----------------------
_ _ 9 | 5 _ 2 | _ _ 8
_ 2 _ | _ _ _ | 1 4 _
4 8 _ | 3 _ 1 | _ _ 5
val it = () : unit
```

Three blanks have been filled in. Although our display of the board does not show this, the other blanks have had some of their possibilities eliminated. What we need is to repeat this process (which is why it was called `oneRound`) until no more progress can be made. We can tell when we have made all the progress possible when we apply `oneRound` to a board and get the same board back as a result. This is a strategy of continuous improvement until a fixed point is found, a strategy we will see again in later chapters. The function `attemptWithoutBifurcation` performs the repetition of `oneRound` (its unusual name will make sense in a moment).

```
fun attemptWithoutBifurcation(board) =
   let val board2 = oneRound(board);
   in if board = board2
      then board
      else attemptWithoutBifurcation(board2)
   end;

- printBoard(attemptWithoutBifurcation(sample));

2 5 4 | 7 9 8 | 6 1 3
9 7 3 | 4 1 6 | 8 5 2
6 1 8 | 2 5 3 | 4 9 7
-----------------------
8 9 2 | 6 7 4 | 5 3 1
5 3 1 | 8 2 9 | 7 6 4
7 4 6 | 1 3 5 | 2 8 9
-----------------------
1 6 9 | 5 4 2 | 3 7 8
3 2 5 | 9 8 7 | 1 4 6
4 8 7 | 3 6 1 | 9 2 5
val it = () : unit
```

Before we celebrate, remember that this was an easy instance. If we try it on something harder, we will not get as good a result. Here is "Evil Puzzle 4,724,921,946" from www.websudoku.com.

```
- printBoard(evil);

2 _ 6 | _ 5 _ | _ 7 _
_ _ _ | _ _ 6 | _ _ _
4 _ _ | _ _ 2 | 1 _ _
---------------------
_ 8 _ | _ 1 _ | _ _ 5
7 _ _ | 6 _ 5 | _ _ 1
6 _ _ | _ 4 _ | _ 3 _
---------------------
_ _ 4 | 1 _ _ | _ _ 3
_ _ _ | 5 _ _ | _ _ _
_ 1 _ | _ 7 _ | 6 _ 8
val it = () : unit

- printBoard(attemptWithoutBifurcation(evil));

2 _ 6 | _ 5 _ | _ 7 _
_ _ _ | _ _ 6 | _ _ _
4 _ _ | _ _ 2 | 1 _ _
---------------------
_ 8 _ | _ 1 _ | _ _ 5
7 _ _ | 6 _ 5 | _ _ 1
6 _ _ | _ 4 _ | _ 3 _
---------------------
_ _ 4 | 1 _ _ | _ _ 3
_ _ _ | 5 _ _ | _ _ _
_ 1 _ | _ 7 _ | 6 _ 8
```

Rats. Some puzzles are too difficult to be solved this way. No matter how many "impossibles" we remove, there will still be some blanks with more than one possibility. What do we do next?

We guess. Suppose a board that cannot be improved has an open space with possible values 2, 5, and 7. We pick 2, make that the space's only option, and see if we can solve the board that way. If not—that is, if this turns out to be impossible—then we pick up where we left off and try with 5. This is called *bifurcation*: when we have several choices and not enough ready information to tell which is right, we try them all until we find one that works.

bifurcation

To bifurcate a board, we pick an open position with more than one possibility and split the board about that position: the spaces before that position, the list of possibilities at that position, and the rest of the board. The function `splitAtFirstOpen`, which you are to write, does this by choosing the first open space. For efficiency's sake one might choose the open space with the smallest number of possibilities, but this way will work just as well. The function `bifurcate` creates a list of board guesses, given the results of `splitAtFirstOpen`.

```
fun bifurcate(front, [], back) = []
  | bifurcate(front, a::rest, back) =
      (front@[Open([a])]@back)::bifurcate(front, rest, back);
```

This is as far as we will take you. Your function `solve`, which will make use of `bifurcate` and `splitAtFirstOpen` will solve even the evil instances.

```
- printBoard(solve(evil));


2 9 6 | 8 5 1 | 3 7 4
1 7 8 | 4 3 6 | 5 2 9
4 3 5 | 7 9 2 | 1 8 6
-----------------------
9 8 2 | 3 1 7 | 4 6 5
7 4 3 | 6 2 5 | 8 9 1
6 5 1 | 9 4 8 | 2 3 7
-----------------------
8 2 4 | 1 6 9 | 7 5 3
3 6 7 | 5 8 4 | 9 1 2
5 1 9 | 2 7 3 | 6 4 8
val it = () : unit
```

Project

4.A Write the functions `getColumnMates` and `getSquareMates`. The column function requires only a few changes from the row function, but the square function is trickier. Since these functions can work on any list of length 81, you may want to test your functions on plain lists of ints.

4.B Finish the code for oneRound:

```
fun oneRound(board) =
  let fun processEachPosition([], n) = ??
        | processEachPosition(Given(a)::rest,
                                          n) = ??
        | processEachPosition(Open(aa)::rest,
                                          n) =
            let val (x, y) = numToCoords(n);
                val mates =
             getRowMates(x, y, board, 0) @
             getColumnMates(x, y, board, 0) @
             getSquareMates(x, y, board, 0);
            in
               ??
            end;
  in
     processEachPosition(board, 0)
  end;
```

4.C When we are trying to solve a board that was produced by bifurcation, we need to detect when the board is solved and when it is impossible. Write a function `isSolved` that takes a board (sudokuSpace list) and returns true if every open space has exactly one option and a function `isImpossible` that takes a board and returns true if it contains an open space that has zero options. All the interesting work in these functions can be done in pattern-matching.

4.D Finish the code for `splitAtFirstOpen`. It should return a tuple of type sudokuSpace list * sudokuValue list * sudokuSpace list. (The first pattern is a dummy to avoid a match nonexhaustive warning.)

```
                                 (* shouldn't happen *)
fun splitAtFirstOpen([]) = ([], [], [])
         (* pattern for unsolved open space *)
  | splitAtFirstOpen(??::restBoard) = ??
                                        (* other case *)
  | splitAtFirstOpen(aa::restBoard) = ??
```

4.E The function `solve` takes a board and returns a solved version of that board, if there is one. It tries to solve it with our original strategy. If that does not work, it bifurcates. A bifurcation fails when it reaches an impossible board, in which case we will raise an `impossibleSudoku` exception. When a bifurcation fails, we try a different guess. The local function `processBifurcations` iterates through the possible guesses. Finish `solve`. (The `handle impossibleSudoku` indicates that the body of `processBifurcations(a::rest)` may result in the exception.)

```
exception impossibleSudoku;

fun solve(board) =
   let val fstAttempt = ??
   in if isSolved(fstAttempt) then ??
      else if isImpossible(fstAttempt)
           then ??
      else
       let fun processBifurcations([]) =
                  ??
             | processBifurcations(a::rest) =
                  ??
                  handle impossibleSudoku
                         => ??;
       in
           processBifurcations(??)
        end
    end;
```

4.12 Special topic: Russell's paradox

The usefulness of the set concept is its simpleness and its flexibility. For this reason set theory is a core component of the foundations of mathematics. An example of the concept's flexibility is that we can talk sensibly about sets of sets—for example, powersets. Reasoning becomes more subtle, however, if we speak of sets that contain themselves. For example, the set X of all sets mentioned in this book is hereby mentioned in this book, and so $X \in X$. Bertrand Russell called for caution when playing with such ideas with the following paradox:

Let X be the set of all sets that do not contain themselves, that is, $X = \{Y \mid Y \notin Y\}$. Does X contain itself?

First, suppose it does, that is $X \in X$. However, then the definition of X states that only those sets that do not contain themselves are elements of X, so $X \notin X$, which is a contradiction. Hence $X \notin X$. However, the same definition of X now tells us that $X \in X$, and we have another contradiction.

A well-known puzzle, also attributed to Russell, presents the same problem. Suppose a certain town has only one man who is a barber. That barber shaves only every man in the town who does not shave himself. Does the barber shave himself? If he does, then he doesn't; if he doesn't, then he does.

We can conclude from this only that the setup of the puzzle is an impossibility. There could not possibly be a man who shaves only every man who does not shave himself. Likewise, the set of all sets that do not contain themselves must not exist. This is why rigorous set theory must be built on axioms. Although we have not presented a complete axiomatic foundation for set theory here, we have assumed that at least one set exists (namely, the empty set), and we have assumed a notion of what it means for sets to be equal. We have not assumed that just because a set can be described that it for that reason exists.

For this reason, when we name sets in a proof ("let X be the set..."), we really are doing more than assigning a name to some concept; we are jumping to the conclusion that the concept exists. Therefore if we are defining sets in terms of a property ("let $X = \{x \mid x < 3\}$"), it is more rigorous to make that set a subset of a known (or postulated) set merely *limited* by that property ("let $X = \{x \in \mathbb{Z} \mid x < 3\}$").

This clears away the paradox nicely. Since it makes sense only to speak about things in the universal set, we will assume that $X \subset \mathcal{U}$, that is, $X = \{Y \in \mathcal{U} \mid Y \notin Y\}$. Now the question, "Does X contain itself?" becomes "Is it true that $X \in \mathcal{U}$ and $X \notin X$?" First suppose $X \in \mathcal{U}$. Then either $X \notin X$ or $X \in X$. As we saw before, both of those lead to contradictions. Hence $X \notin \mathcal{U}$. In other words, X does not exist.

Interestingly, this leaves powerset without a foundation, since we cannot define it as a subset of something else. Remember that a set and its powerset exist in different universes. Assuming $X \subseteq \mathcal{U}$, if we try to define powerset as

$$\mathscr{P}(X) = \{Y \in \ldots \mid Y \subseteq X\}$$

what would go in the ellipses? It cannot be $\mathcal{U}$: Take $\mathcal{U} = \mathbb{Z}$ and $X = \{4, 5\}$. Then $\mathscr{P}(X) = \{\emptyset, \{4\}, \{5\}, \{4, 5\}\}$, none of which are elements of $\mathbb{Z}$. Rather, they are subsets of $\mathbb{Z}$. The only thing that makes the definition of $\mathscr{P}(X)$ complete is $\mathscr{P}(\mathcal{U})$, and that would make the definition of powerset circular.

Sometimes mathematics can be like Calvinball. When things do not work out right, we can say "new rule, new rule." Since we do not want to abandon the idea of powerset nor define it circularly, we place it on firm ground with its own axiom.

Axiom 6 (Powerset.) *For any set X, there exists a set $\mathscr{P}(X)$ such that $Y \in \mathscr{P}(X)$ if and only if $Y \subseteq X$.*

We have declared it okay to speak of powersets. However, we can prove that no set contains its own powerset.

Theorem 4.25 *If A is a set, then $\mathscr{P}(A) \nsubseteq A$.*

> **Proof.** Suppose A is a set. Further suppose that $\mathscr{P}(A) \subseteq A$. Let $B = \{b \in A \mid b \notin b\}$. Since $B \subseteq A$, $B \in \mathscr{P}(A)$ by definition of powerset. By definition of subset, $B \in A$. Now we have two cases:
>
> Case 1: Suppose $B \in B$. Then $B \notin B$, contradiction; this case is impossible.
>
> Case 2: Suppose $B \notin B$. Then $B \in B$, contradiction, this case is impossible.
>
> Either case leads to a contradiction. Hence $\mathscr{P}(A) \nsubseteq A$. □

Moreover, this shows that the idea of a "set of all sets" is downright impossible in our axiomatic system.

Corollary 4.26 *The set of all sets does not exist.*

> **Proof.** Suppose A were the set of all sets. By Axiom 6, $\mathscr{P}(A)$ also exists. Since $\mathscr{P}(A)$ is a set of sets and A is the set of all sets, $\mathscr{P}(A) \subseteq A$. However, by the theorem, $\mathscr{P}(A) \nsubseteq A$, a contradiction. Hence A does not exist. □

This chapter draws heavily from Epp[8] and Hrbacek and Jech[17].

Chapter summary

Learning to write proofs is one of the two main goals of this course. You have learned the fundamentals here, and the much of the rest of this text rolls out specific areas of study on which we can practice proof-writing. When proving a theorem, focus your attention on the definitions of the terms and symbols being used. What information do the definitions of terms used in the given facts provide? What are the demands of the definitions used in the conclusion? A useful strategy is to break the proof down into *analysis* and *synthesis*: Break down the given information, compose what is to be shown.

In the chapters ahead one of the most difficult (and, therefore, most important) aspects of getting a proof right is to consider how the proposition is quantified.

Key ideas:

Propositions about one set being a subset of another are proven using the element argument: pick any element of the subset, and show that it is an element of the other set.

If two expressions are known to be equal, one can be substituted for another in a proof.

Propositions about two sets being equal are proven by two sub-proofs, each showing one set to be a subset of the other.

Propositions about a set being empty are proven by supposing an element in the set and deriving a contradiction.

Conditional propositions are proven by supposing the hypothesis and then proving the conclusion from the assumption that the hypothesis is true.

Propositions about an expression being even are proven by finding an integer such that multiplying that integer by 2 produces the original expression.

Biconditional propositions are proven by writing two subproofs, each supposing one side and proving the other side from that side.

Theorems can produce useful algorithms.

Part II

Core

Chapter 5 Relation

We have laid the foundation of writing proofs and programs. Now we build on this by expanding our range of raw material and investigate more structures of discrete mathematics and the use of these structures in programming.

Relations, the topic of this chapter, are not new to you. Most likely, you learned about relations in junior high or high school where they were presented in terms of how they differed from functions. We, too, will use relations as a building block for studying functions in Chapter 7, but we will also consider them for logical and computational interest in their own right.

The most important thing to learn in this chapter and the chapter on functions is not the details of relations and functions themselves. Instead, all of this is about advancing your proof- and program-writing skills. You will be given definitions about relations and their properties and be asked to prove propositions about those properties. This requires you to read definitions and discern from them how one should meet the demands of the proposition to be proven. One of the most important parts of the definition to study is the quantification.

Chapter goals. The student should be able to

- describe relations as sets of ordered pairs, as predicates, and as digraphs.
- represent relations in ML.
- determine the image and inverse of a relation and the composition of two relations.
- determine whether a relation is reflexive, symmetric, and transitive.
- prove results about relations, including the operations and properties mentioned above.
- prove results about equivalence relations.
- compute relation operations, properties, and property closures in ML.
- prove results about partial order relations.

5.1 Definition

To refresh your memory on relations, consider curves in the real plane. The equation $y = 4 - x^2$ represents a function because its graph passes the vertical line test. A circle, like $y^2 = 4 - x^2$, fails the test and thus is not a function, but it is still a relation.

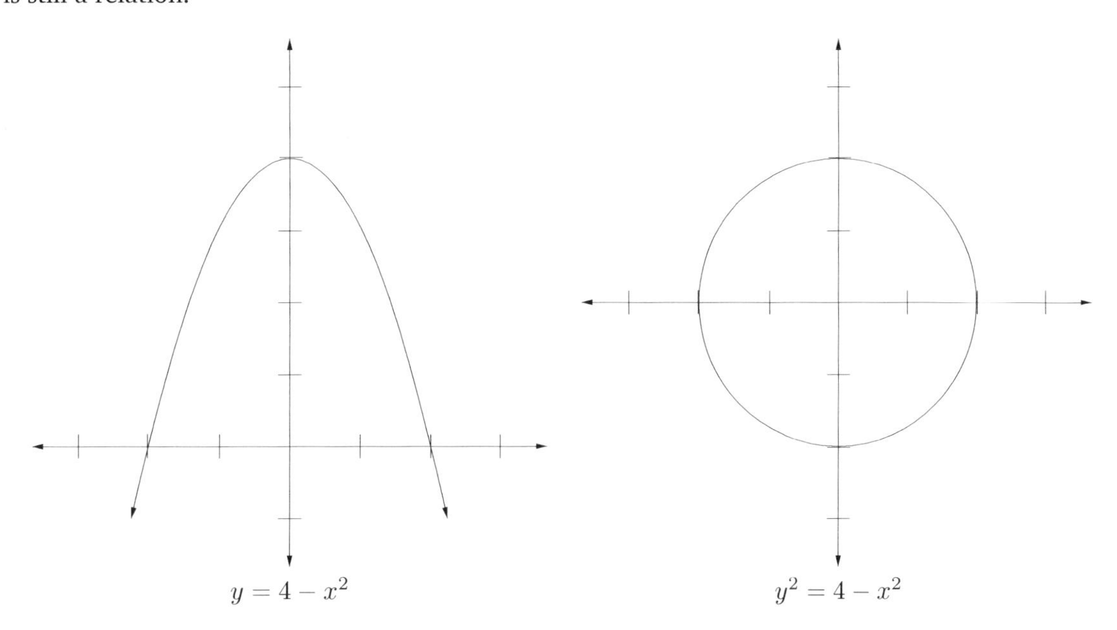

Reexamine your high school understanding of curves and graphs with the lens of set theory. Exercise 1.9.5 asked you to describe how a line can be considered a set. We hope you said something like, "A line is a set of points." The same can be said of a curve.

Since a point is just an ordered pair of real numbers, we can say that an equation is also a set, or at least that it describes a set, in the same way $\{x \in \mathbb{Z} \mid x > 0\}$ describes the set $\mathbb{N}$ or that both $(1, 5]$ and $\{x \in \mathbb{R} \mid 1 < x \leq 5\}$ describe the same set. The equation $y^2 = 4 - x^2$ defines a set that includes $(0, 2)$, $(2, 0)$, $(0, -2)$, $(-2, 0)$, $(\sqrt{2}, -\sqrt{2})$, $(-\sqrt{2}, \sqrt{2})$, etc.

The high school definition of a relation, then, is a set of ordered pairs, or a subset of $\mathbb{R} \times \mathbb{R}$. Our definition is more general: If X and Y are sets, then a binary *relation* R *from* X *to* Y is a subset of $X \times Y$. X and Y are the *domains* of R. If X and Y happen to be the same set, then we say that R is a relation *on* X. If $(x, y) \in R$, then we say that x and y are related to each other. A ternary relation is a set of ordered triples, a subset of $X \times Y \times Z$, and an n-place relation is a subset of $X_1 \times X_2 \times \ldots \times X_n$. When we just say "relation," we mean binary relation.

relation

domain

There are three ways to denote the fact that x and y are related to each other in relation R. Consider four examples of relations: Let A be the set { hawk, coyote, fox, rabbit, clover }, and *eats* be the relation of what species preys on what other species. Consider also $|$ (divides) on the set $\mathbb{N}$. Letting B be the set of logical formulas, consider $\equiv$, the relation describing what formulas are logically equivalent to each other. Finally, let $C = \{a, b, c\}$, $D = \{d, e, f, g\}$ and $R = \{(a, d), (a, e), (b, e), (b, g), (c, d)\}$. (This last one is something we would call an "arbitrary example"—it has no real meaning, we just made it up arbitrarily.)

First, since relations are sets, we can use the same notation we use for other sets.

$$\begin{array}{rcl@{\qquad}rcl@{\qquad}rcl@{\qquad}rcl}
\text{(hawk, rabbit)} & \in & \textit{eats} & (3,12) & \in & | & (p \vee T, T) & \in & \equiv & (a,d) & \in & R \\
\text{(rabbit, clover)} & \in & \textit{eats} & (4,48) & \in & | & (p \wedge q, q \wedge p) & \in & \equiv & (b,g) & \in & R \\
\text{(fox, rabbit)} & \in & \textit{eats} & (13,39) & \in & | & (\sim (p \vee q), \sim p \wedge \sim q) & \in & \equiv & (c,d) & \in & R \\
\text{(clover, coyote)} & \notin & \textit{eats} & (4,25) & \notin & | & (\sim (p \vee q), \sim p \vee \sim q) & \notin & \equiv & (a,f) & \notin & R \\
\text{(hawk, fox)} & \notin & \textit{eats} & (12,3) & \notin & | & (p \vee T, p) & \notin & \equiv & (c,f) & \notin & R
\end{array}$$

This looks strange when the relation is a predefined symbol like $|$ or $\equiv$. The following infix notation emphasizes relations as operators.

$$\begin{array}{rcl@{\qquad}rcl@{\qquad}rcl@{\qquad}rcl}
\text{hawk} & \textit{eats} & \text{rabbit} & 3 & | & 12 & p \vee T & \equiv & T & a & R & d \\
\text{rabbit} & \textit{eats} & \text{clover} & 4 & | & 48 & p \wedge q & \equiv & q \wedge p & b & R & g \\
\text{fox} & \textit{eats} & \text{rabbit} & 13 & | & 39 & \sim (p \vee q) & \equiv & \sim p \wedge \sim q & c & R & d \\
\text{clover} & \not{\textit{eats}} & \text{coyote} & 4 & \nmid & 25 & \sim (p \vee q) & \not\equiv & \sim p \vee \sim q & a & \not R & f \\
\text{hawk} & \not{\textit{eats}} & \text{fox} & 12 & \nmid & 3 & p \vee T & \not\equiv & p & c & \not R & f
\end{array}$$

From a third perspective, these operations all return boolean values, so we can think of them as predicates—and write them that way.

$$\begin{array}{l@{\qquad}l@{\qquad}l@{\qquad}l}
\textit{eats}(\text{hawk, rabbit}) & |(3,12) & \equiv (p \vee T, T) & R(a,d) \\
\textit{eats}(\text{rabbit, clover}) & |(4,48) & \equiv (p \wedge q, q \wedge p) & R(b,g) \\
\textit{eats}(\text{fox, rabbit}) & |(13,39) & \equiv (\sim (p \vee q), \sim p \wedge \sim q) & R(c,d) \\
\sim \textit{eats}(\text{clover, coyote}) & \sim |(4,25) & \sim\equiv (\sim (p \vee q), \sim p \vee \sim q) & \sim R(a,f) \\
\sim \textit{eats}(\text{hawk, fox}) & \sim |(12,3) & \sim\equiv (p \vee T, p) & \sim R(c,f)
\end{array}$$

This also looks strange on the predefined symbols, so for them, we will prefer infix notation. Since we like to emphasize the set perspective, we will prefer the first form of notation in other cases.

directed graph

If a relation is over a finite (and small) set, we can make a visual representation for it using what is called a *directed graph* or *digraph*—but this is unlike the graphs of functions you used in precalculus. We will study graphs thoroughly in Chapter 8, but for now understand that a graph is a collection of elements called *vertices* connected by *edges*. If the edges have arrow heads on them, then the graph is *directed*. Here are digraphs illustrating the four relations discussed earlier (on subsets of the domains), with one more relation: $\subseteq$ over $\mathscr{P}(\{1, 2, 3\})$.

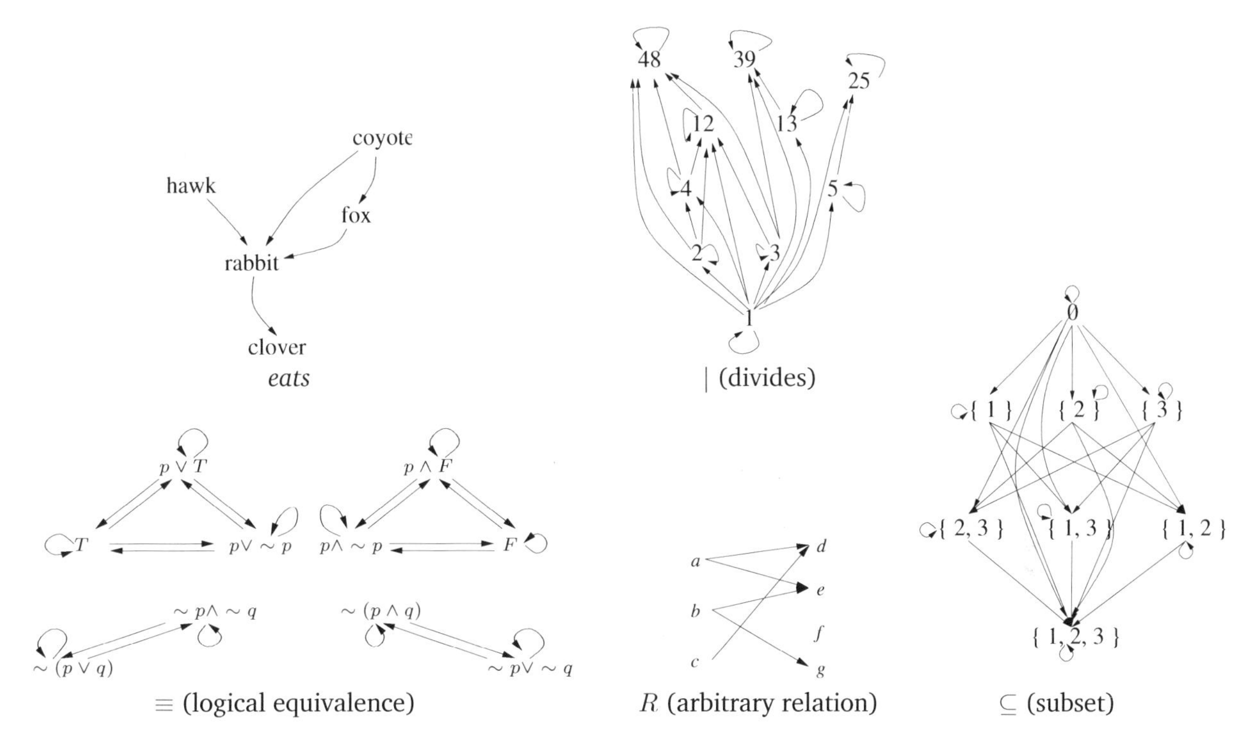

eats | (divides) ≡ (logical equivalence) R (arbitrary relation) ⊆ (subset)

Notice that in some cases, an element may be related to itself. This is represented graphically by a *self-loop*. Every natural number divides itself, every logical formula is logically equivalent to itself, and every set is a subset of itself.

self-loop

Exercises

In the following exercises, you may describe the specified relations by listing the pairs explicitly or by drawing a graph (except for the first, where a verbal description is appropriate).

5.1.1 Describe three relations on $\mathbb{Z}$ or subsets of $\mathbb{Z}$ not mentioned in this chapter. Try to think of "real" relations, like | (divides), rather than contrived ones.

5.1.2 Make a set of actors A and a set of films F, and then define a relation *hasStarredIn* from A to F.

5.1.3 Make a set of US presidents P and a set of home states S and define a relation *wasFrom* from P to S. Some presidents may have claimed more than one state as home.

5.1.4 Make a list of some of your relatives and define a relation *isRelatedTo* to indicate biological relation.

5.2 Representation

One goal of a discrete mathematics course is for you to think mathematically about things other than numbers. Some of your prior math courses did introduce you to other mathematical objects: shapes, points, matrices. Already we have treated sets and truth values as full-fledged mathematical objects. This chapter broadens the horizon to include relations.

In ML, our concept of a *value* corresponds to what we mean when we say *mathematical object*. We now consider how to represent relations in ML. As a running example for this chapter we will explore relations over two sets: a set of North American cities, and a set of rivers in the Mississippi tributary system.

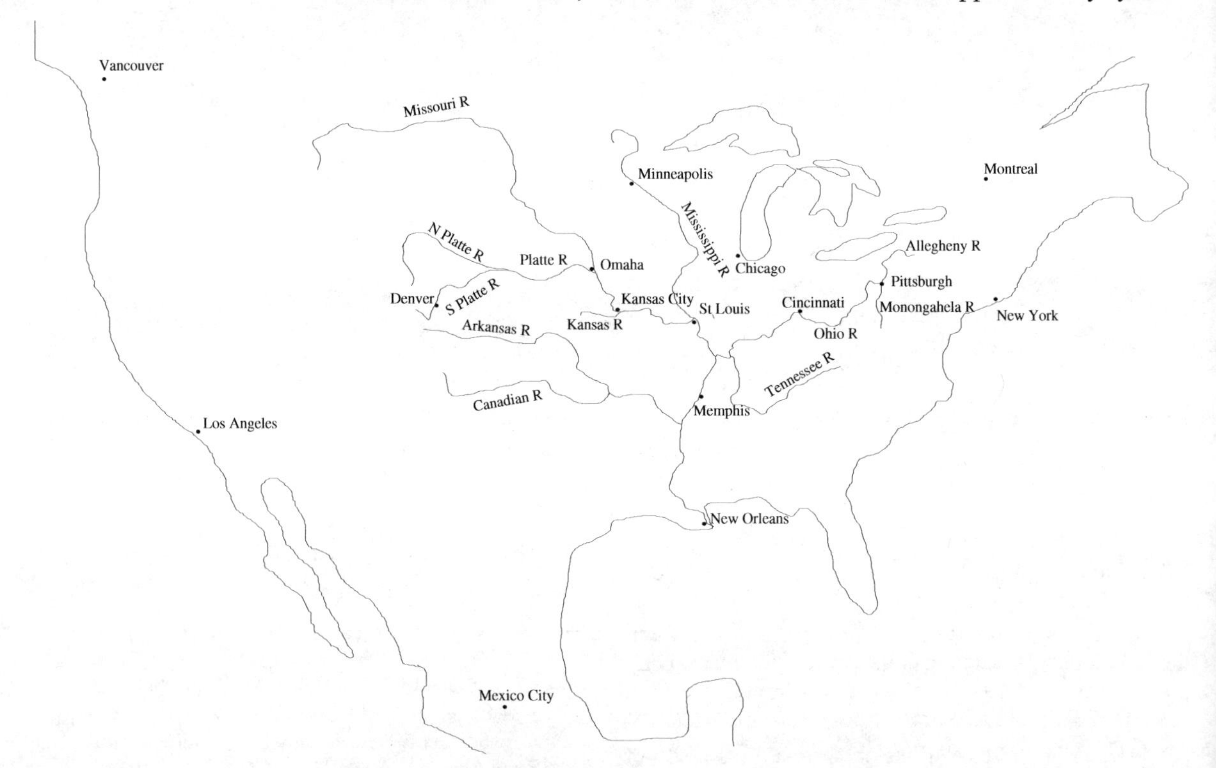

Suppose we have relations *isOnRiver* from cities to rivers and *flowsInto* over rivers. We can represent the original sets using datatypes. The relations, since they are sets of ordered pairs, can be lists of tuples. (An alternative is to define a relation as a predicate; we will explore that option later.)

```
datatype city =  Vancouver | LosAngeles | Mexico |
  Minneapolis | Omaha | KansasCity | Denver |  StLouis |
  Memphis | Chicago | NewOrleans | Cincinnati | Pittsburgh |
  Montreal | NewYork;
```

```
datatype river = Missouri | Platte | NPlatte | SPlatte |
  Arkansas | Canadian | Kansas | Mississippi | Tennessee |
  Ohio | Allegheny | Monongahela;

val isOnRiver =
  [(Denver, Platte), (Omaha, Missouri), (Omaha, Platte),
   (KansasCity, Missouri), (KansasCity, Kansas),
   (Minneapolis, Mississippi), (StLouis, Mississippi),
   (StLouis, Missouri), (Memphis, Mississippi),
   (NewOrleans, Mississippi), (Cincinnati, Ohio),
   (Pittsburgh, Ohio), (Pittsburgh, Allegheny),
   (Pittsburgh, Monongahela)];

val flowsInto =
  [(Platte, Missouri), (Kansas, Missouri),
   (Missouri, Mississippi), (Canadian, Arkansas),
   (Arkansas, Mississippi), (Allegheny, Ohio),
   (Monongahela, Ohio), (Tennessee, Ohio), (NPlatte, Platte),
   (SPlatte, Platte), (Ohio, Mississippi)];
```

(Omaha is actually about 12 miles north of the Platte, but close enough.)

The type of these relations are (city * river) list and (river * river) list. To test if two things are related to each other, we define a function `isRelatedTo` which takes two elements and a relation.

```
- fun isRelatedTo(a, b, []) = false
=     | isRelatedTo(a, b, (h1, h2)::rest) =
=         (a = h1 andalso b = h2) orelse isRelatedTo(a, b, rest);

val isRelatedTo = fn : ''a * ''b * (''a * ''b) list -> bool

- isRelatedTo(Platte, Missouri, flowsInto);

val it = true : bool

- isRelatedTo(Chicago, Mississippi, isOnRiver);

val it = false : bool
```

Alternatively, we could use `contains`, defined in Section 3.7.

Exercises

5.2.1 Describe the relation "is downstream from" over cities by listing the pairs explicitly.

5.2.2 Draw a graph of the relation "is west of" over cities. If two cities appear to be more or less vertically aligned on the given map, then consider neither to be west of the other (for example, the map given does not show much difference in longitude among St Louis, Memphis, and New Orleans).

5.3 Image, inverse, and composition

image

The *image* of an element $a \in X$ under a relation R from X to Y is the set of elements in Y to which a is related.

$$\mathcal{I}_R(a) = \{b \in Y \mid (a,b) \in R\}$$

Turning back to some of the examples we had earlier, The image of 3 under the relation | is $\mathcal{I}_|(3) = \{3, 6, 9, 12, \ldots\}$. The image of fox under *eats* is { rabbit }. The image of a under R (the arbitrary relation from Section 5.1) is $\{d, e\}$. The image of `Pittsburgh` under `isOnRiver` is { `Allegheny`, `Monongahela`, `Ohio` }.

The important thing to remember is that images are sets, so we can operate on them and reason about them as we do other sets. Take this theorem:

Theorem 5.1 *If $a, b \in \mathbb{N}$ and $a|b$, then $\mathcal{I}_|(b) \subseteq \mathcal{I}_|(a)$.*

> **Proof.** Suppose $a, b \in \mathbb{N}$ and $a|b$. By definition of divides, there exists $i \in \mathbb{N}$ such that $a \cdot i = b$.
>
> Suppose further that $c \in \mathcal{I}_|(b)$. By definition of image, $b|c$. By definition of divides, there exists $j \in \mathbb{N}$ such that $b \cdot j = c$.
>
> By substitution, $a \cdot i \cdot j = c$, and so $a|c$ by definition of divides. By definition of image, $c \in \mathcal{I}_|(a)$, and by definition of subset, $\mathcal{I}_|(b) \subseteq \mathcal{I}_|(a)$. □

The concept of image can be extended so that it can be applied to a set rather than an individual element. So, if $A \subseteq X$,

$$\mathcal{I}_R(A) = \{b \in Y \mid \exists\, a \in A \mid (a,b) \in R\}$$

Take careful notice of the quantification. If b is some element in $\mathcal{I}_R(A)$, that means there must be some $a \in A$ such that $(a,b) \in R$. Further, an equivalent definition would be, if $A = \{a_1, a_2, \ldots\}$, then

$$\mathcal{I}_R(A) = \mathcal{I}_R(a_1) \cup \mathcal{I}_R(a_2) \cup \ldots$$

inverse

The *inverse* of a relation R from X to Y is the relation

$$R^{-1} = \{(b,a) \in Y \times X \mid (a,b) \in R\}$$

The inverse of | is $isMultipleOf$, and the inverse of *eats* is $isEatenBy$. The inverse is easy to perceive graphically; all one does is reverse the direction of the arrows.

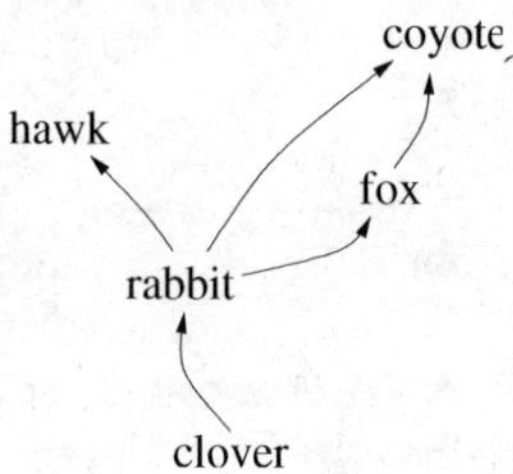

The previous theorem was a result about a specific relation, $|$. We also have some results that are true for all relations.

Theorem 5.2 *If R is a relation on a set A, $a \in A$, and $\mathcal{I}_R(a) \neq \emptyset$, then $a \in \mathcal{I}_{R^{-1}}(\mathcal{I}_R(a))$.*

> **Proof.** Suppose R is a relation on A, $a \in A$, and $\mathcal{I}_R(a) \neq \emptyset$.
>
> Let $b \in \mathcal{I}_R(a)$. By definition of image, $(a, b) \in R$. By definition of inverse, $(b, a) \in R^{-1}$. By definition of image (extended for sets), $a \in \mathcal{I}_{R^{-1}}(\mathcal{I}_R(a))$. □

The *composition* of a relation R from X to Y and a relation S from Y to Z is the relation

composition

$$S \circ R = \{(a, c) \in X \times Z \mid \exists\, b \in Y \text{ such that } (a, b) \in R \text{ and } (b, c) \in S\}$$

Suppose we had the set of cities $C = \{$ Chicago, Green Bay, Indianapolis, Minneapolis$\}$, the set of professional sports $P = \{$ baseball, football, hockey, basketball $\}$, and the set of seasons $S = \{$ summer, fall, spring, winter $\}$. Let $hasMajorProTeam$ be the relation from C to P representing whether a city has a major professional team in a given sport, and let $playsSeason$ be the relation from P to S representing whether or not a professional sport plays in a given season. The composition $playsSeason \circ hasMajorProTeam$ represents whether or not a city has a major professional team playing in a given season. In the illustration to the right, pairs in the composition are found by following two arrows.

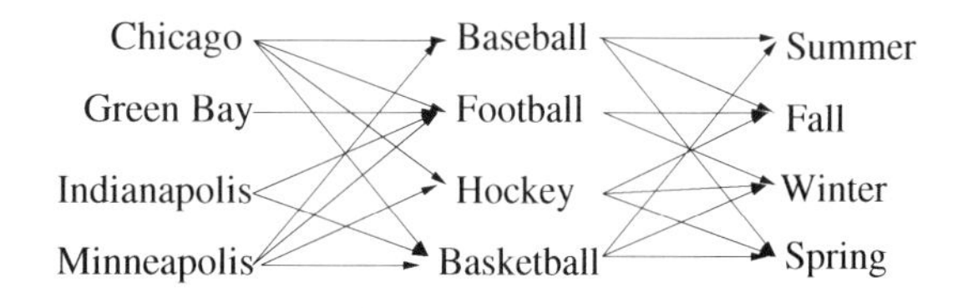

Sometimes it is interesting to compose a relation with itself. The relation `flowsInto` composed with itself relates the rivers `Allegheny`, `Monongahela`, `Tennessee`, `Platte`, `Kansas`, `Arkansas`, and `Canadian` to `Mississippi`.

The following theorem illustrates relation composition—self-composition, in
fact—but also raises the question of what it means for two relations to be equal. Since relations are sets, this is simply an application of the definition of set equality. Define $<$, a relation on $\mathbb{R}$, formally as $a < b$ if there exists a $c \in \mathbb{R}^+$ such that $a + c = b$.

Theorem 5.3 $<=< \circ <$

If that is hard to read, all it says is that $<$ composed with itself is just $<$. We will use $< \circ <$ in the infix style. Remember that $a < b$ is the same as $(a, b) \in <$.

> **Proof.** Suppose $a, b \in \mathbb{R}$ and $a < b$. *(This is the same as saying, "Suppose $(a, b) \in <$," which is how we usually would begin a proof of*

set equality.) By definition of $<$, there exists a $c \in \mathbb{R}^+$ such that $a + c = b$.

Let $d = \frac{c}{2}$. Since $d \in \mathbb{R}^+$, the definition of $<$ means that $a < a + d$ and $a + d < a + d + d = a + c = b$. Since $a < a + d$ and $a + d < b$, by definition of composition, $a < \circ < b$. *In other words,* $(a, b) \in < \circ <$.

(*Now the other direction.*) Next suppose $a < \circ < b$. By definition of function composition, there exists $c \in R$ such that $a < c$ and $c < b$. By definition of $<$, there exist $d \in \mathbb{R}^+$ such that $a + d = c$ and $e \in \mathbb{R}^+$ such that $c + e = b$.

By substitution, $a + d + e = b$. Since d and e are both elements of $\mathbb{R}^+$, $d + e \in \mathbb{R}^+$, so, by definition of $<$, $a < b$.

Therefore, by definition of relation equality, $< = < \circ <$. □

In the "other direction" of this proof, we used the definition of composition analytically, going from the fact that $a < \circ < b$ to asserting that a certain c existed, $a < c$ and $c < b$. Another theorem about arbitrary relations will illustrate this analytical use. The important things in a section like this are first, recognize how to use the proof techniques from earlier chapters and second, how to apply the new definitions.

Theorem 5.4 *If R is a relation from a set X to a set Y and S is a relation from Y to a set Z, then $\mathcal{I}_{S \circ R}(X) \subseteq \mathcal{I}_S(Y)$.*

Proof. Suppose R is a relation from a set X to a set Y and S is a relation from Y to a set Z.

This is a subset proof. We need to pick something out of $\mathcal{I}_{S \circ R}(X)$ and show that it is in $\mathcal{I}_S(Y)$. But what sort of thing would we pull out of $\mathcal{I}_{S \circ R}(X)$? Since $S \circ R$ is a relation from X to Z, any image of that composed relation must be a subset of Z. A good choice of variable name will help you think straight, not to mention make your proof more understandable.

Suppose $z \in \mathcal{I}_{S \circ R}(X)$. By definition of image, there exists $x \in X$ such that $(x, z) \in S \circ R$.

Here is the analytical use of the definition of composition. If $S \circ R$ relates x to z, there must be a go-between in Y.

By definition of composition, there exists $y \in Y$ such that $(x, y) \in R$ and $(y, z) \in S$.

By definition of image, $z \in \mathcal{I}_S(Y)$. Therefore, by definition of subset, $\mathcal{I}_{S \circ R}(X) \subseteq \mathcal{I}_S(Y)$. □

Exercises

5.3.1 What is the image of $\{2, 3\}$ under the relation $\subseteq$ on $\mathscr{P}(\{1, 2, 3\})$?

5.3.2 What is the inverse of $\in$?

5.3.3 What is the composition of the relation *eats* (on the set { hawk, coyote, rabbit, fox, clover }) with itself?

5.3.4 The hypothesis of Theorem 5.2 requires that $\mathcal{I}_R(A) \neq \emptyset$. Where did we use that in the proof?

5.3.5 Prove that if R is a relation from a set X to a set Y, S is a relation from Y to a set Z, and $A \subseteq X$, then $\mathcal{I}_{S \circ R}(A) = \mathcal{I}_S(\mathcal{I}_R(A))$.

5.3.6 Prove that $| = | \circ |$.

5.3.7 Prove that if R is a relation over a set A and $(a, b) \in R$, then $\mathcal{I}_R(b) \subseteq \mathcal{I}_{R \circ R}(a)$.

5.3.8 Suppose R is a relation from a set X to a set Y and $A \subseteq X$. Are either of the following true? $\mathcal{I}_{R^{-1}}(\mathcal{I}_R(A)) \subseteq A$. $A \subseteq \mathcal{I}_{R^{-1}}(\mathcal{I}_R(A))$. Prove or give a counterexample for each.

5.3.9 The *identity relation* on a set X is defined as $i_X = \{(x, x) \mid x \in X\}$ (this is really just a fancy way to talk about $=$ as it is usually understood). Prove that for a relation R from A to B, $i_B \circ R = R$.

5.3.10 Prove that if R is a relation from A to B, then $(R^{-1})^{-1} = R$.

5.3.11 If R is a relation from A to B, is $R^{-1} \circ R = i_A$? Prove or give a counterexample.

5.3.12 Write a function `image` that takes an element and a relation and returns the image of the element over the relation. For example, `image(2, [(2, 1), (3, 6), (2, 4), (5, 2), (3, 4), (2, 9)])` would return `[1, 4, 9]`.

5.3.13 Write a function `addImage` that takes an element (`x`) and a set (represented as a list, `L`) and returns a relation R with the property that the image of `x` under R is `L`. For example, `addImage(5, [1, 2, 3])` would return `[(5, 1), (5, 2), (5, 3)]`.

5.3.14 Write a function `compose` that takes two relations and returns the composition of those two relations. (Hint: use your `image` and `addImage` functions.)

5.4 Properties of relations

We now study several interesting properties that some relations have. These occur only on relations that relate a set to itself ($R \subseteq A \times A$, not $R \subseteq A \times B$, unless $B = A$.)

- A relation R on a set X is *reflexive* if every element is related to itself: $\forall\, x \in X, (x, x) \in R$. You can tell a reflexive relation by its graph because every element will have a self-loop. Examples of reflexive relations are $=$ on anything, $\equiv$ on propositional forms, $\leq$ and $\geq$ on $\mathbb{R}$ and its subsets, $\subseteq$ on sets, and "is acquainted with" on people. *reflexive*

- A relation R on a set X is *symmetric* if for every pair in the relation, the inverse of the pair also exists: $\forall\, x, y \in X$, if $(x, y) \in R$ then $(y, x) \in R$. In the graph of a symmetric relation, every arrow has a corresponding reverse arrow (except for self-loops, which are their own reverse arrow). Examples of symmetric relations are $=$ on anything, $\equiv$ on logical propositions, "is opposite of" on $\mathbb{Z}$, "is on same river as" on cities, and "is acquainted with" on people. *symmetric*

transitive

- A relation R on a set X is *transitive* if any time one element is related to a second and that second is related to a third, then the first is also related to the third: $\forall\, x, y, z \in X$, if $(x, y) \in R$ and $(y, z) \in R$, then $(x, z) \in R$. If you imagine "traveling" around a graph by hopping from element to element as the arrows permit, then you recognize a transitive relation because any trip that can be made in two (or any positive number of) hops can also be made in one hop. Examples of transitive relations are $=$ on anything, $\equiv$ on logical propositions, $\leq$, $\geq$, $<$, and $>$ on $\mathbb{R}$ and its subsets, $\subseteq$ on sets, "is downstream from" on cities, and "is ancestor of" on people.

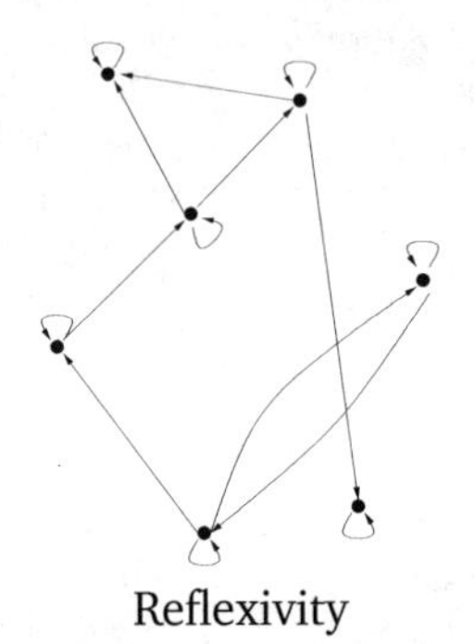

Reflexivity

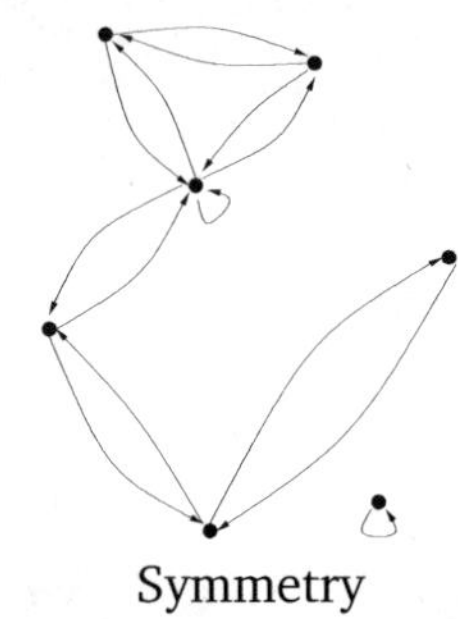

Symmetry

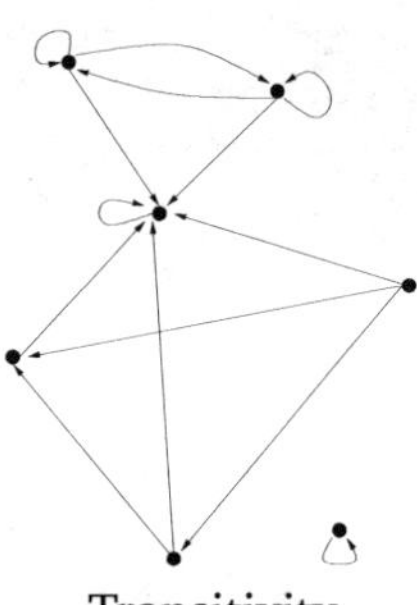

Transitivity

These properties provide opportunities to practice proof-writing because they require careful consideration of the burden of proof demanded by their definition. For example, suppose we have the theorem

Theorem 5.5 *The relation $|$ on $\mathbb{N}$ is reflexive.*

If we unpack this using the definition of "reflexive," we see that this is a "for all" proposition:

$$\forall\, x \in \mathbb{N}, (x, x) \in \,|$$

This, in turn, contains a set-membership proposition which requires the application of the definition of this particular relation.

Tangent: Friends and relations

Many human relationships among peers are symmetric or transitive or both. Friendship is symmetric, and although it is not transitive, many friendships do begin through mutual friendships. "Is sibling of" is likewise symmetric, and as long as we mean full, biological siblings, it is transitive, too—though this also requires us to consider it reflexive, that one is one's own sibling.

A relation like "is sibling-in-law of" is not transitive, and that is because in English the term encompasses three distinct relationships: sibling's spouse, spouse's sibling, and spouse's sibling's spouse. My sister's husband and my wife's brother are close friends; among all the family interaction, they sometimes mistake themselves for being brothers-in-law to each other.

Is "is in love with" a symmetric relation?

$$\forall\, x \in \mathbb{N}, \exists\, c \in \mathbb{Z} \text{ such that } x \cdot c = x.$$

Proof. Suppose $a \in \mathbb{N}$.	Let the doubter pick any one element.
By arithmetic, $a \cdot 1 = a$, and so by the definition of divides, $a \vert a$.	Apply the definition of the relation—in this case, divides. Since divides is existentially quantified, we need to find the appropriate object, in this case, 1.
Hence, by the definition of reflexive, $\vert$ is reflexive. □	Satisfy the demands of reflexivity.

Symmetry and transitivity require similar reasoning, except that their "for all" propositions require two and three picks, respectively, and they contain General Form 2 propositions.

For most symmetric relations we know, the fact of their symmetry follows trivially from their definition—it would not be interesting to prove that $\equiv$ is symmetric, for example. But here is a result involving symmetry and relation operations:

Theorem 5.6 *If R is a relation on a set A, then $R \cap R^{-1}$ is symmetric.*

Try an example by hand to see what this is saying. Let $A = \{a, b, c, d, e, f, g\}$ and let $R = \{(a,b), (a,c), (c,a), (c,d), (c,g), (e,d), (f,g), (g,c)\}$. Then $R^{-1} = \{(b,a), (c,a), (a,c), (d,c), (g,c), (d,e), (g,f), (c,g)\}$. Finally, $R \cap R^{-1} = \{(a,c), (c,a), (c,g), (g,c)\}$, which is symmetric by inspection.

Proof. Suppose R is a relation on a set A. Suppose $(a,b) \in R \cap R^{-1}$.

The second supposition also implicitly supposes $a, b \in A$.

By definition of intersection, $(a,b) \in R$ and $(a,b) \in R^{-1}$. Since $(a,b) \in R$, the definition of inverse tells us that $(b,a) \in R^{-1}$. Similarly, since $(a,b) \in R^{-1}$, by definition of inverse it is also the case that $(b,a) \in R$.

By definition of intersection, $(b,a) \in R \cap R^{-1}$. Therefore $R \cap R^{-1}$ is symmetric by definition. □

Many interesting relations are transitive.

Theorem 5.7 *The relation $\vert$ on $\mathbb{Z}$ is transitive.*

Proof. Suppose $a, b, c \in \mathbb{Z}$, and suppose $a|b$ and $b|c$. By the definition of divides, there exist $d, e \in \mathbb{Z}$ such that $a \cdot d = b$ and $b \cdot e = c$. By substitution and associativity, $a(d \cdot e) = c$. By the definition of divides, $a|c$. Hence $\vert$ is transitive. □

Notice that this proof involved two applications of the definition of the relation, the first to analyze the fact that $a|b$ and $b|c$, the second to synthesize the fact that $a|c$. As you work on increasingly sophisticated proofs, do not forget the fundamental strategies or fail to notice how proofs involving this newer material grow out of the earlier. Proofs still divide into analysis and synthesis sections, and you will find yourself making analytical and synthetic applications of the definitions.

Exercises

5.4.1 When talking about the relation $|$, why did we restrict ourselves to $\mathbb{Z}^+$ for reflexivity, but consider all of $\mathbb{Z}$ for transitivity?

5.4.2 Give a counterexample proving that $|$ is not symmetric over $\mathbb{Z}$.

For $x, y \in \mathbb{R}$, let $x \leq y$ if there exists $z \in \mathbb{R}^{\text{nonneg}}$ such that $x + z = y$.

5.4.3 Prove that $\leq$ is reflexive over $\mathbb{R}$.

5.4.4 Give a counterexample proving that $\leq$ is not symmetric over $\mathbb{R}$.

5.4.5 Prove that $\leq$ is transitive over $\mathbb{R}$.

Let R be the relation on $\mathbb{Z}$ defined that $(a, b) \in R$ if $a - b$ is even.

5.4.6 Prove that R is reflexive.

5.4.7 Prove that R is symmetric.

5.4.8 Prove that R is transitive.

5.4.9 Which of these properties would be true of a similar relation S, where $(a, b) \in S$ if $a - b$ is odd?

Let R be the relation on $\mathbb{N}$ defined that $(a, b) \in R$ if a is a power of b—for example, $2, 4, 16, 32, \ldots$ are powers of 2. Formally, $(a, b) \in R$ if there exists $i \in \mathbb{W}$ such that $b^i = a$.

5.4.10 Prove that R is reflexive.

5.4.11 Prove that R is transitive.

5.4.12 What would you conclude about a and b if $(a, b) \in R$ and $(b, a) \in R$?

Let A be a set in the universe $\mathcal{U}$. Let R be a relation on $\mathscr{P}(\mathcal{U})$ (that is R is a relation on sets in the universe) defined that $(B, C) \in R$ if $B - C = A$. Which of the following are true? Prove or give a counterexample. (If a later proposition is implied by an earlier one, then simply say so; you do not need to repeat the proof.)

5.4.13 R is symmetric.

5.4.14 R is transitive.

5.4.15 If $A = \emptyset$, then R is reflexive.

5.4.16 If $A = \emptyset$, then R is symmetric.

5.4.17 If $A = \emptyset$, then R is transitive.

5.4.18 If $A \neq \emptyset$, then R is symmetric.

5.4.19 If $A \neq \emptyset$, then R is transitive.

Let R and S be relations on a set X, and let $A \subseteq X$. Prove the following except when asked to give a counterexample.

5.4.20 $R^{-1} \circ R$ is reflexive. This is false; give a counterexample.

5.4.21 If R and S are both reflexive, then $R \cap S$ is reflexive.

5.4.22 $R \cup R^{-1}$ is symmetric.

5.4.23 If R and S are both symmetric, then $(S \circ R) \cup (R \circ S)$ is symmetric.

5.4.24 If R and S are both transitive, then $R \cap S$ is transitive.

5.4.25 If R is reflexive, then $A \subseteq \mathcal{I}_R(A)$.

5.4.26 If R is symmetric and for all $a \in A$, there exists a $b \in X$ such that $(a, b) \in R$, then $A \subseteq \mathcal{I}_R(\mathcal{I}_R(A))$. (The complicated premise is saying that every element in a is related to something.)

5.4.27 If R is transitive, then $\mathcal{I}_R(\mathcal{I}_R(A)) \subseteq \mathcal{I}_R(A)$.

5.5 Equivalence relations

An *equivalence relation* over a set is a relation that is reflexive, symmetric, and transitive.

equivalence relation

Equivalence relations are important because they group elements together into useful subsets and, as the name suggests, express some notion of equivalence among certain elements. In the previous section, we saw that the various relation properties are easy to recognize from the the graphs of the relations. The graphs of equivalence relations obviously have the visual properties of reflexivity, symmetry, and transitivity, and together the properties have the effect of grouping the vertices in the graph into cliques that all have arrows to each other and no arrows to any elements in other cliques.

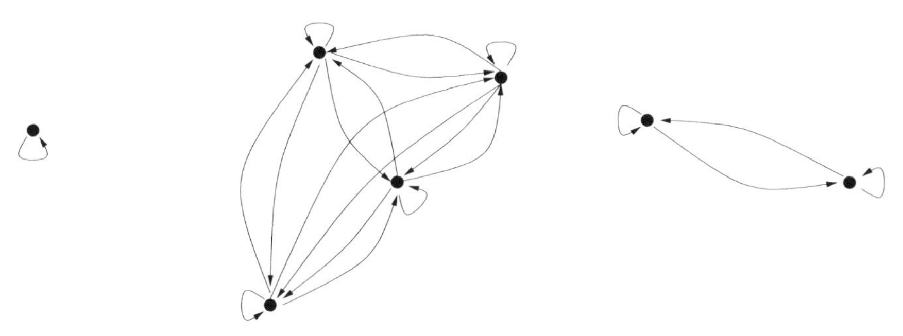

Examples of equivalence relations:

- R on $\mathbb{Z}$ where $(a, b) \in R$ if $a - b$ is even; see Exercises 5.4.6–5.4.8.
- $\equiv$ on the set of propositional formulas; see the graph on page 199.
- R_n on $\mathbb{W}$ where $(a, b) \in R_n$ if a and b are *congruent modulo* n where $n \in \mathbb{N}$. Here is what this means: Suppose $n \in \mathbb{N}$. Two numbers a and b are congruent modulo n if $a \bmod n = b \bmod n$. For example, let $n = 7$. Notice $7 \bmod 7 = 0$, $8 \bmod 7 = 1$, $9 \bmod 7 = 2$, $14 \bmod 7 = 0$, $15 \bmod 7 = 1, \ldots$. So, 0, 7, and 14 are congruent with each other, 1, 8 and 15 are congruent, etc.
- R on the set of triangles where $(a, b) \in R$ if a and b are similar.
- R on the set of students where $(a, b) \in R$ if a and b have the same major—assuming each student has exactly one major.
- R on the set of English words where $(a, b) \in R$ if a and b are inflected forms of the same lexeme—that is, they have the same dictionary entry. For example, *run*, *runs*, *ran*, *running* share a single dictionary entry; *be*, *is*, *are*, *am*, *were*, *was*, and *being* similarly are related.
- R on $\mathbb{Z} \times \mathbb{Z}$ where $((a, b), (c, d)) \in R$ if $\frac{a}{b} = \frac{c}{d}$.

- R on the set of real-valued functions where $(f(x), g(x)) \in R$ if $f'(x) = g'(x)$—that is, functions are related to each other if they have the same derivative.

Proving that a specific relation is an equivalence relation follows a fairly predictable pattern. The parts of the proof are the proving of the three individual properties.

Theorem 5.8 *Let R be a relation on $\mathbb{Z}$ defined that $(a, b) \in R$ if $a + b$ is even. R is an equivalence relation.*

pick one

Proof. Suppose $a \in \mathbb{Z}$. Then by arithmetic, $a + a = 2a$, which is even by definition. Hence $(a, a) \in R$ and R is reflexive.

pick two

Now suppose $a, b \in \mathbb{Z}$ and $(a, b) \in R$. Then, by the definition of even, $a + b = 2c$ for some $c \in \mathbb{Z}$. By the commutativity of addition, $b + a = 2c$, which is still even, and so $(b, a) \in R$. Hence R is symmetric.

pick three

Finally suppose $a, b, c \in \mathbb{Z}$, $(a, b) \in R$ and $(b, c) \in R$. By the definition of even, there exist $d, e \in \mathbb{Z}$ such that $a + b = 2d$ and $b + c = 2e$. By algebra, $a = 2d - b$. By substitution and algebra

$$\begin{aligned} a + c &= 2d - b + c &&= 2d - 2b + b + c \\ &= 2d - 2b + 2e &&= 2(d - b + e) \end{aligned}$$

which is even by definition (since $d - b + e \in \mathbb{Z}$). Hence $(a, c) \in R$, and so R is transitive.

Therefore, R is an equivalence relation by definition. □

Once one knows that a relation is an equivalence relation, there are other facts that follow.

Theorem 5.9 *If R is an equivalence relation, then $R = R^{-1}$.*

A relation is a set, so $R = R^{-1}$ is in Set Proposition Form 2 and wrapped in a General Form 2 proposition.

Proof. Suppose R is an equivalence relation.

First suppose $(a, b) \in R$. Since R is an equivalence relation, it is symmetric, so $(b, a) \in R$ by definition of symmetry. Then by the definition of inverse, $(a, b) \in R^{-1}$, and so $R \subseteq R^{-1}$ by definition of subset.

Next suppose $(a, b) \in R^{-1}$. By definition of inverse, $(b, a) \in R$. Again by symmetry, $(a, b) \in R$, and so $R^{-1} \subseteq R$.

Therefore, by definition of set equality, $R = R^{-1}$. □

In fact, the only equivalence relation property we used is symmetry, so could have stated this as *If R is symmetric ....* An equivalence relation combines the powers of the three properties.

Take a closer look at the graph of an equivalence relation. Equivalence relations have a natural connection to *partitions*. Those "cliques" are pairwise disjoint subsets, which all together make the entire set. Let X be a set, and let $P = \{X_1, X_2, \ldots, X_n\}$ be a partition of X. Let R be the relation on X defined so that $(x, y) \in R$ if there exists $X_i \in P$ such that $x, y \in X_i$. We call R the *relation induced* by the partition.

relation induced

Similarly, let R be an equivalence relation on X. Let $[x]$ be the image of a given $x \in X$ under R. We call $[x]$ the *equivalence class* of x (under R). It turns out that any relation induced by a partition is an equivalence relation, and the collection of all equivalence classes under an equivalence relation is a partition.

equivalence class

Theorem 5.10 *Let A be a set, $P = \{A_1, A_2, \ldots, A_n\}$ be a partition of A, and R be the relation induced by P. R is an equivalence relation.*

> **Proof.** Suppose $a \in A$. Since $A = A_1 \cup A_2 \cup \ldots \cup A_n$ by the definition of partition, $a \in A_1 \cup A_2 \cup \ldots \cup A_n$. By the definition of union, there exists $A_i \in P$ such that $a \in A_i$. By definition of relation induced, $(a, a) \in R$. Hence R is reflexive.
>
> The rest is left for an exercise. □

On the other hand, if we already know an equivalence relation, then we can define a partition based on what elements are related to each other. The set of (singled-majored) students with the equivalence relation "has same major as" is partitioned by the students' majors. The partition itself, then, is the set of equivalence classes of all the elements in the set, $\{[a] \mid a \in A\}$ The following theorem is essentially the converse of the the previous.

Theorem 5.11 *If A be a set and R be an equivalence relation on A, then $\{[a] \mid a \in A\}$ is a partition of A.*

> **Proof.** Exercise 5.5.3.

To prove this, think about what it means for something to be a partition. Remember that $[a]$ is a set and a partition is a set of sets. The proof requires us to show two things: first that the union of all the sets in the form $[a]$ for some $a \in A$ make up the set A—that is, $\bigcup_{a \in A} [a] = A$; second, that the elements are pairwise disjoint—that is, for all $a, b \in A$, if $[a] \neq [b]$, then $[a] \cap [b] = \emptyset$. Note in particular that even though $a \neq b$, we still could have $[a] = [b]$.

The point of all this is the we can talk about the equivalence relation or we can talk about the partition, our choice; but the meaning is the same. The sets of things related form a partition, and the partition defines an equivalence relation.

Exercises

Let R and S be relations on a set A.

5.5.1 Prove that i_A, the identity relation on set A (see Exercise 5.3.9) is an equivalence relation.

5.5.2 Finish the proof of Theorem 5.10. Proof of symmetry is straightforward. Proof of transitivity is more complicated. You will suppose $a, b, c \in A$, $(a, b) \in R$, and $(b, c) \in R$. By definition of relation induced, this means there exist $A_i, A_j \in P$ such that $a, b \in A_i$ and $b, c \in A_j$. In other words, you cannot immediately say that a, b, and c are all in the same subset; instead, part of your proof must be to show $A_i = A_j$.

5.5.3 Prove Theorem 5.11.

5.5.4 Suppose that R is reflexive and that for all $a, b, c \in A$, if $(a, b) \in R$ and $(b, c) \in R$, then $(c, a) \in R$. Prove that R is an equivalence relation. (Hint: Once you have proved either symmetry or transitivity, the other is straightforward. Proving the first of these is tricky. However, notice that in the special fact about this relation—that for all $a, b, c \in A$ etc—there is nothing saying that a, b, and c must be distinct.)

5.5.5 Prove that if R is an equivalence relation, then $R \circ R \subseteq R$.

5.5.6 Prove that if R is an equivalence relation and $(a, b) \in R$, then $\mathcal{I}_R(a) = \mathcal{I}_R(b)$. (Hint: Do not overlook the Set Proposition Form 2 of the conclusion.)

5.5.7 If R and S are both equivalence relations, is $R \cap S$ necessarily an equivalence relation? Prove or give a counterexample.

5.5.8 If R and S are both equivalence relations, is $R \circ S$ necessarily an equivalence relation? Prove or give a counterexample.

5.5.9 It is tempting to think that if a relation R is symmetric and transitive, then it is also reflexive. This, however, is false. Give a counterexample. (Hint: Try to prove that R is reflexive. Then think about what unfounded assumption you are making in your proof.)

5.5.10 Based on what you discovered in Exercise 5.5.9, fill in the blank to make this proposition true and prove it. "If a relation R on a set A is symmetric and transitive and if ________, then R is reflexive."

5.6 Computing transitivity

We have already thought about representing relations in ML. Now we consider testing such relations for some of the properties we have talked about. A test for transitivity is delicate because the definition begins with a double "for all." We can tame this by writing a modified (but equivalent) definition, that a relation R on A is transitive if

$$\forall\, (x, y) \in R, \forall\, (w, z) \in R, \text{ if } y = w \text{ then } (x, z) \in R$$

That is, we choose (x, y) and (w, z) one at a time—but (w, z) is the same thing as (y, z). This allows us to break down the problem. Assume that (x, y) is already chosen, and we want to know if transitivity holds specifically *with respect to* (x, y). In other words, is it the case that for any element to which y is related, x is also related to that element? To get a feel for what we mean by being transitive with respect to a specific pair, consider an example. Suppose we have the relation

$$\{(1, 2), (2, 3), (5, 2), (1, 5), (2, 5), (1, 3)\}$$

Is this transitive with respect to $(1, 2)$? Yes, for each pair in the form $(2, z)$ (lightly shaded), there exists a corresponding $(1, z)$ (darkly shaded).

$\{(\overset{\downarrow}{1}, 2), (2, 3), (5, 2), (1, 5), (2, 5), (1, 3)\}$

Is this transitive with respect to $(2, 3)$? Yes, since there are no pairs that begin with 3, this is vacuously true.

$\{(1, 2), (\overset{\downarrow}{2}, 3), (5, 2), (1, 5), (2, 5), (1, 3)\}$

Is this transitive with respect to $(5, 2)$? No. The pair $(2, 3)$ exists, but there is no corresponding $(5, 3)$, as transitivity demands.

$\{(1, 2), (2, 3), (\overset{\downarrow}{5}, 2), (1, 5), (2, 5), (1, 3)\}$

To start thinking in terms of ML, remember that the relation will be represented as a list of pairs. We will need to iterate through that list, testing for each pair (x, y) whether the relation is transitive with respect to (x, y). When testing a single pair, then, we must iterate through the list, using a process that makes decisions based on the following conditions:

1. The list is empty. Then *true*, the list is (vacuously) transitive with respect to (x, y).
2. The list begins with (y, z) for some z. Then the list is transitive with respect to (x, y) if (x, z) exists in the relation, and if the rest of the list is transitive with respect to (x, y).
3. The list begins with (w, z) for some $w \neq y$. Then the list is transitive with respect to (x, y) if the rest of the list is.

Expressing this in ML:

```
- fun testOnePair((a, b), []) = true
=   | testOnePair((a, b), (c, d)::rest) =
=           ((not (b = c)) orelse isRelatedTo(a, d, relation))
=           andalso testOnePair((a,b), rest);
```

Note two things. First, our symbolic notation is more expressive than ML since we can write $\forall(x, y), (y, z) \in R$, using y twice, but we cannot use a variable more than once in a pattern. That is, to write

```
...testOnePair((a, b), (b, c)::rest) = ...
```

is not allowed in ML. Hence we coalesce cases 2 and 3 as far as pattern-matching goes and test b and c separately for equality. Second, we used the undefined variable `relation`, something we will have to fix by putting it in a proper context. The reason for this is that it is crucial that in case 2 we test if (a, c) exists in *the original relation*, not just the portion of the list currently being examined. In other words, it would be a serious error (why?) to write

```
...((not (b=c)) orelse isRelatedTo(a, d, rest))
```

The difficult part is complete. Now we must cover the first "for all," which can be done by iterating over the list, testing each pair.

```
- fun test([]) = true
=   | test((a,b)::rest) =
=          testOnePair((a,b), relation) andalso test(rest);
```

Again, we must be able to refer to the original relation, not just the rest of the list. It is good style to bundle into one package pieces like `test` and `testOnePair` that will not be used independently of our test for transitivity. This will also solve our problem of making `relation` a valid variable.

```
- fun isTransitive(relation) =
=   let fun testOnePair((a, b), []) = true
=         | testOnePair((a, b), (c, d)::rest) =
=                 ((not (b = c))
=                      orelse isRelatedTo(a, d, relation))
=                 andalso testOnePair((a,b), rest);
=       fun test([]) = true
=         | test((a,b)::rest) =
=                 testOnePair((a,b), relation)
=                 andalso test(rest);
=   in
=     test(relation)
=   end;

val isTransitive = fn : (''a * ''a) list -> bool
```

Test this on a relation that ought to be transitive.

```
- val isDownstreamFrom = [(Omaha,Denver), (KansasCity,Omaha),
=    (KansasCity,Denver), (StLouis,KansasCity),
=    (StLouis,Omaha), (StLouis,Denver),
=    (StLouis,Minneapolis), (Cincinnati,Pittsburgh),
=    (Memphis,StLouis), (Memphis,KansasCity), (Memphis,Omaha),
=    (Memphis, Denver), (Memphis,Minneapolis),
=    (Memphis,Cincinnati), (Memphis,Pittsburgh),
=    (NewOrleans,Memphis), (NewOrleans,StLouis),
=    (NewOrleans,KansasCity), (NewOrleans,Omaha),
=    (NewOrleans,Denver), (NewOrleans,Minneapolis),
=    (NewOrleans,Cincinnati), (NewOrleans,Pittsburgh)];

val isDownstreamFrom =
  [(Omaha,Denver), ...] : (city * city) list
```

```
- isTransitive(isDownstreamFrom);

val it = true : bool
```

Continuing with ou1 use of the cities as an example, notice that we can partition them by time zones. The relation induced by this partition, `inSameTimeZone`, is then an equivalence relation. Here is a representation of the relation (for brevity's sake, we reduce the number of cities we consider):

```
- val inSameTimeZone = [(Montreal,NewYork), (NewYork,Montreal),
=    (NewYork,Pittsburgh), (Pittsburgh,NewYork),
=    (Montreal,Pittsburgh), (Pittsburgh,Montreal),
=    (Chicago,NewOrleans), (NewOrleans,Chicago),
=    (Chicago,Mexico), (Mexico,Chicago), (NewOrleans,Mexico),
=    (Mexico, NewOrleans), (LosAngeles, Vancouver),
=    (Vancouver, LosAngeles)];

val inSameTimeZone =
  [(Montreal,NewYork),...]
  : (city * city) list

- isTransitive(inSameTimeZone);


val it = false : bool
```

To our surprise, it reports that `inSameTimeZone`, a supposed equivalence relation, is not transitive. We must then turn our attention to debugging—why is our relation not transitive? Since the test for transitivity fails when pairs (a, b) and (c, d) are found in the list and $b = c$ but (a, d) is not found in the list, the most useful information we could get would be to ask the predicate `isTransitive`, "What pairs did you find that contradict transitivity?"

What if we modified `isTransitive` so that it returns a list of counterexamples instead of a simple true or false? This would tell us not only *that* a relation is intransitive, but *why* it is intransitive. This would require the following changes to `testOnePair`:

- Instead of returning `true` on an empty list or when `a` is related to `d`, return `[]`, that is, an empty list, indicating no contradicting pairs are found.
- Instead of returning `false` when `a` is not related to `d`, return `[(a, d)]`, that is, a list containing the pair that was expected but not found.
- Instead of anding the (boolean) value of `testOnePair` on the rest of the list with what we found for the current pair, concatenate the result for the current pair to the (list) value of `testOnePair`.

Observe the new function.

```
- fun counterTransitive(relation) =
=  let fun testOnePair((a, b), []) = []
=            | testOnePair((a, b), (c,d)::rest) =
=                (if ((not (b=c))
=                     orelse isRelatedTo(a, d, relation))
=                  then [] else [(a, d)])
=                @ testOnePair((a,b), rest);
=        fun test([]) = []
=          | test((a,b)::rest) =
=                testOnePair((a,b), relation) @ test(rest)
=  in
=     test(relation)
=  end;
```

```
val counterTransitive = fn
     : (''a * ''a) list -> (''a * ''a) list
```

Note the type. It is no trouble to write the same things again, and it is a safeguard for you.

```
- counterTransitive(inSameTimeZone);
```

```
val it =
  [(Montreal,Montreal),(NewYork,NewYork),
   (Pittsburgh,Pittsburgh), ...] : (city * city) list
```

This reveals the problem: we forgot to add self-loops. Adding them (but eliminating repeats) makes the predicate transitive.

```
- val correctedInSameTimeZone =
=   inSameTimeZone @
=   makeNoRepeats(counterTransitive(inSameTimeZone));
```

```
val correctedInSameTimeZone =
  [(Montreal,NewYork),(NewYork,Montreal),
   (NewYork,Pittsburgh), ...] : (city * city) list
```

```
- isTransitive(correctedInSameTimeZone);
```

```
val it = true : bool
```

Exercises

5.6.1 Write a predicate `isSymmetric` which tests if a relation is symmetric, similar to `isTransitive`.

5.6.2 Why do we not ask you to write a predicate `isReflexive`? What would such a predicate require that you do not know how to do?

5.6.3 Write a function `counterSymmetric` which operates on the list representation of relations.

5.7 Transitive closure

Consider a person wanting to travel by air from city D (out of a set of cities $\{A, B, C, D\}$) to city C, and suppose the available flights are represented by the relation hasFlightTo $= \{(A, B), (B, C), (C, D), (D, B)\}$, shown on the left.

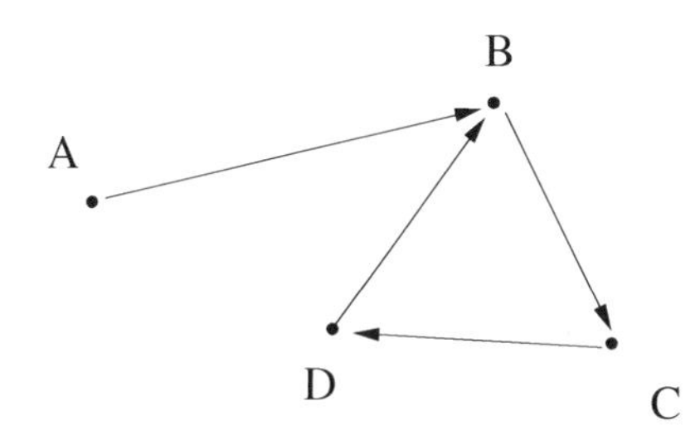

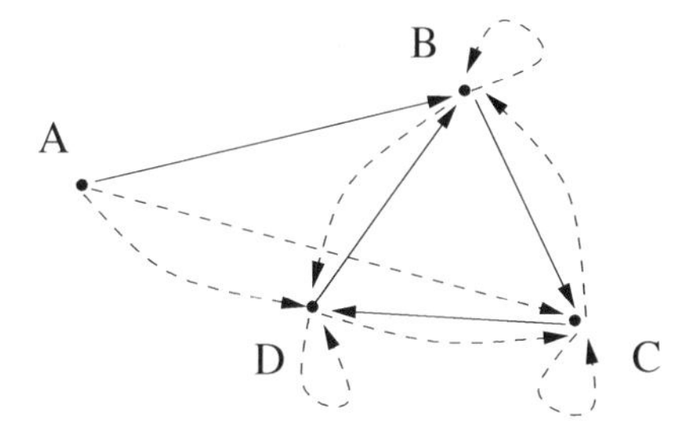

Just because there is no (direct) flight from D to C does not mean that the journey is impossible. Instead, the person can fly to B and make a connecting flight from B to C. The relation hasFlightTo does not tell us what trips between cities are possible; we need another relation for that: canFlyTo, shown on the right. (D, C) exists in this relation, that is, a journey from D to C is possible. Notice the self-loops (B, B), (C, C), and (D, D)—this means round-trips beginning and ending at these cities are possible. A is still not reachable from D, or anywhere else (presumably there is an airplane factory in city A). R is not transitive, but the relation—which we derived from hasFlightTo by adding more pairs to it—is transitive.

Look for a pattern in the following pairs of relations.

Domain	***First relation***	***Second relation***
Rivers	flows into The Platte flows into the Missouri, and the Missouri flows into the Mississippi.	is tributary to The Platte is a tributary to the Missouri; both the Platte and the Missouri are tributaries to the Mississippi.
People	is parent of Bill is Jane's parent; Jane is Leroy's parent	is ancestor of Bill is Jane's ancestor; Leroy has both Jane and Bill as ancestors.

Animals	*eats* Rabbit eats clover; coyote eats rabbit.	derives nutrients from Coyote derives nutrients from rabbit; rabbit derives nutrients from clover; both coyote and rabbit ultimately derive nutrients from clover.
$\mathbb{Z}$ (and other number sets)	is one less than 2 is one less than 3; 3 is one less than 4	$<$ $2 < 3;\ \ 3 < 4;\ \ 2 < 4.$

A similar pattern holds between our original `inSameTimeZone` and our `correctInSameTimeZone`. In all cases the second relation is a bigger, transitive version of the first. Our task now is to pin down what we mean by "bigger, transitive version." Take the relation "flows into."

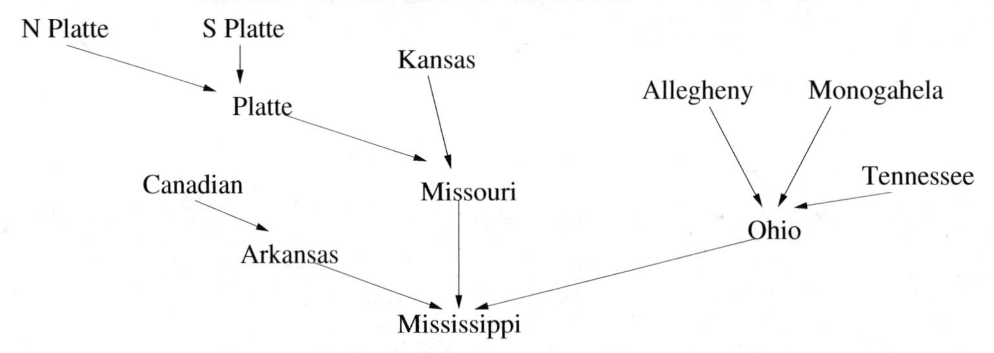

To make it transitive, we add more pairs to the relation. We could take every pair where the second element in the first pair is the same as the first element in the second pair (so, arrows where the first arrowhead touches the second arrow's tail) and add another pair relating the first of the first to the second of the second.

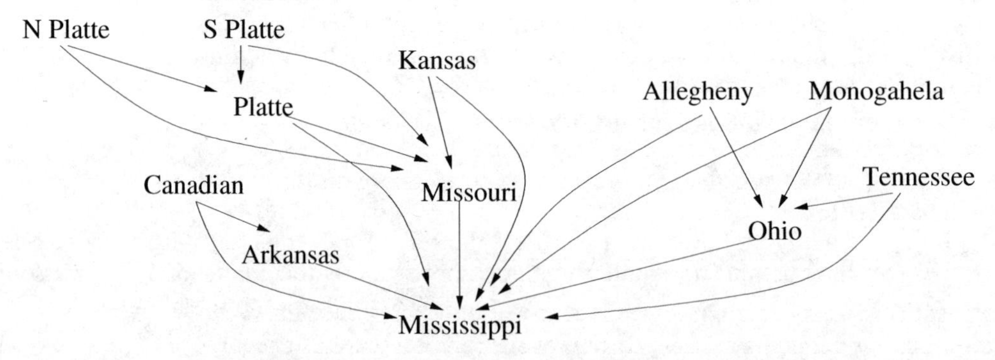

But that still is not transitive. Now `NPlatte` is related to `Missouri`, which is related to `Mississippi`, but `NPlatte` is not related to `Mississippi`. By adding an extra pair, we have added the need for yet another pair. We could simply add every possible pair, making everything related to everything else.

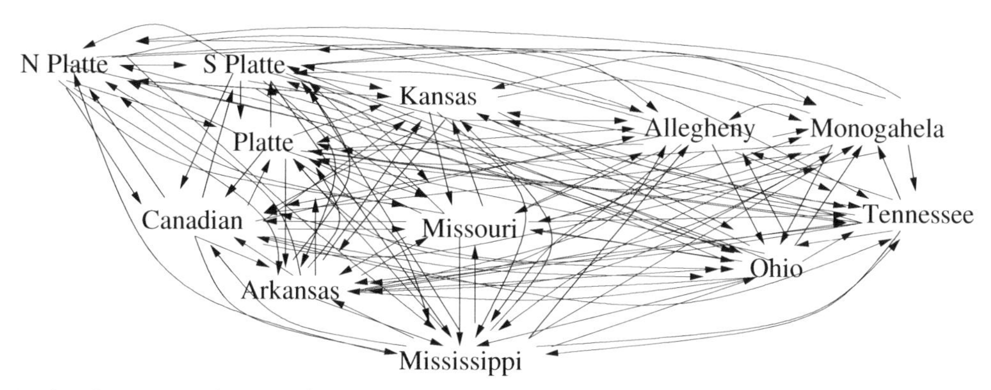

This is clearly too much, not what we had in mind. Instead, what we want is the *fewest* possible additional pairs that would make the relation transitive.

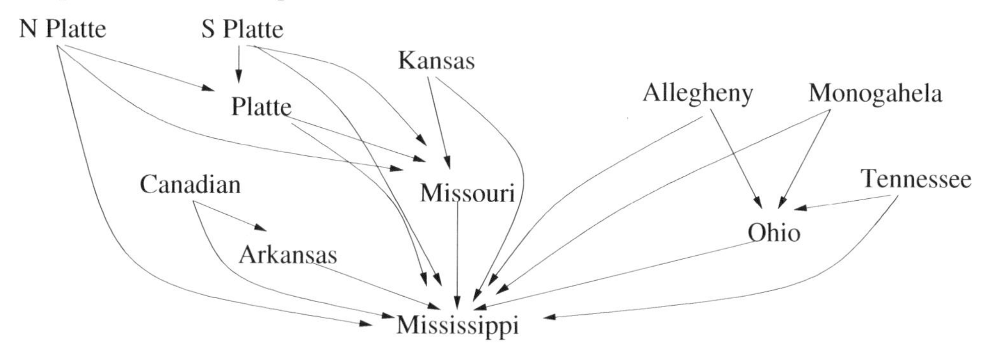

This is called the *transitive closure*. Formally, if R is a relation, then R^T is the transitive closure of R if

transitive closure

1. R^T is transitive.

2. $R \subseteq R^T$.

3. If S is a relation such that $R \subseteq S$ and S is transitive, then $R^T \subseteq S$.

In other words (reversing the order), R^T is the smallest bigger relation that is transitive. The third requirement captures what we mean when we say "the smallest relation": All others like it are bigger than it.

Here is a new kind of proposition, which we use in this case for talking about transitive closures.

Theorem 5.12 *The transitive closure of a relation R is unique.*

To say that something is unique means it is the only thing that fits a certain description. We defined transitive closure as *the* relation that has the three properties mentioned above. This suggests that there is only one such relation, something we have not yet proved. To prove a proposition of uniqueness requires that we assume two objects (possibly distinct, but not necessarily) that fit the description and then show that the two objects are in fact the same.

> **Proof.** Suppose S and T are relations fulfilling the requirements for being transitive closures of R. By items 1 and 2, S is transitive and $R \subseteq S$, so by item 3, $T \subseteq S$. By items 1 and 2, T is transitive and $R \subseteq T$, so by item 3, $S \subseteq T$. Therefore $S = T$ by the definition of set equality. □

If a relation is already transitive, it is its own transitive closure. See Exercise 5.7.1.

reflexive closure

symmetric closure

Along with the transitive closure, we similarly can define the *reflexive closure* as the relation with the fewest pairs added to make a relation reflexive, and the *symmetric closure* as the smallest symmetric superset of a relation. Formally we use the same definition as with the transitive closure; just replace "transitive" with "reflexive" or "symmetric."

Reflexive and symmetric closures are rarely used and not particularly interesting. It is difficult to come up with examples that do not seem overly contrived. The reflexive closure of $<$ is $\leq$, and in general the reflexive closure can be found by adding self-loops (see Exercise 5.7.2). A relation's symmetric closure is just the relation itself except with each arrow's reverse added (see Exercise 5.7.3).

Our success in the previous section at using `@` and `makeNoRepeats` might make us optimistically write a function

```
- fun transitiveClosure(relation) =
=     relation @ makeNoRepeats(counterTransitive(relation));
```

However, we saw that this was wrong on our first attempt to make a transitive version of "flows into."

```
- val flowsInto = [(Canadian,Arkansas), (NPlatte,Platte),
= (SPlatte,Platte), (Platte,Missouri), (Kansas,Missouri),
= (Missouri,Mississippi), (Allegheny,Ohio),
= (Monongahela,Ohio), (Tennessee,Ohio), (Ohio,Mississippi)];

val flowsInto =  [...]  : (river * river) list

- counterTransitive(transitiveClosure(flowsInto));

val it =
  [(NPlatte,Mississippi),(SPlatte,Mississippi),
   (NPlatte,Mississippi),(SPlatte,Mississippi)]
  : (river * river) list
```

To compute the transitive closure correctly, we must add missing pairs repeatedly until the relation is transitive. In other words, we must add not only the pairs of $R \circ R$, but also those of $R \circ R \circ R$, $R \circ R \circ R \circ R$, etc. (To make this notation more succinct and readable, let us say $R^2 = R \circ R$, $R^3 = R \circ R \circ R$, etc.) The following theorem informs how to calculate the transitive closure.

Theorem 5.13 *If R is a relation on a set A, then*

$$R^{\infty} = \bigcup_{i=1}^{\infty} R^i = \{(x, y) \mid \exists\, i \in \mathbb{N} \text{ such that } (x, y) \in R^i\}$$

is the transitive closure of R.

The nasty-looking part is called an infinite iterated union. Do not let it scare you. The notation we are using allows us to extend the set union operation to more than two sets at a time. An *iterated union* of sets $A_1, A_2, \ldots, A_n$ takes the form

iterated union

$$\bigcup_{i=1}^{n} A_i$$

For an *infinite* iterated union, we union together all the sets A_i—or in this case, relations R_i—where $i \in \mathbb{N}$. In Theorem 5.13 all we are saying is that the transitive closure connects elements that are connected in a relation resulting from some number (however large it may be) of compositions. To do this proof, we use the definition of transitive closure.

Proof. Suppose R is a relation on a set A.

R^{∞} is transitive.

Suppose $a, b, c \in A$, $(a, b), (b, c) \in R^{\infty}$. By the definition of R^{∞}, there exist $i, j \in \mathbb{N}$ such that $(a, b) \in R^i$ and $(b, c) \in R^j$. By the definition of relation composition and Exercise 5.7.4, $(a, c) \in R^j \circ R^i = R^{i+j}$. $R^{i+j} \subseteq R^{\infty}$ by the definition of R^{∞}. By the definition of subset, $(a, c) \in R^{\infty}$. Hence, R^{∞} is transitive by definition.

$R \subseteq R^{\infty}$.

Suppose $a, b \in A$ and $(a, b) \in R$. By the definition of R^{∞} (taking $i = 1$), $(a, b) \in R^{\infty}$, and so $R \subseteq R^{\infty}$, by definition of subset.

$\forall\, S$ such that S is transitive and $R \subseteq S$, $R^{\infty} \subseteq S$.

Suppose S is a transitive relation on A and $R \subseteq S$. Further suppose $(a, b) \in R^{\infty}$. Then, by definition of R^{∞}, there exists $i \in \mathbb{N}$ such that $(a, b) \in R^i$. By Lemma 5.14 (see below), $(a, b) \in S$.

Hence $R^{\infty} \subseteq S$ by definition of subset.

Therefore, R^{∞} is the transitive closure of R. □

We extracted one of the hard parts of this proof into the following lemma:

Lemma 5.14 *If R is a relation on a set A, S is a transitive relation on A, $R \subseteq S$, and $i \in \mathbb{N}$, then $R^i \subseteq S$.*

We did this because the straighforward way to prove this uses a technique called *mathematical induction*, which we will not learn until Chapter 6. (We can put off part of a proof until a later point as long as that proof is not dependent on any later result that is dependent on this one, since that would make our reasoning circular.) The proof of this lemma is found on page 275.

If we want this theorem to inform us how to compute the transitive closure of a relation, the infinite number of relation compositions might seem discouraging. However, this does not mean that it necessarily takes an infinity of compositions to reach closure, and since any relation on finite sets can have only a finite number of pairs, we can infer that the process implied by this theorem will terminate. Students with prior experience in other programming languages may recognize an iterative algorithm for doing this, that is, an algorithm containing a sequence of repeated steps, as opposed to a recursive algorithm (we will study iteration in Section 6.9):

- Given a relation R, first assume $R^T = R$.
- While our current estimate for R^T is not transitive,
  - Find the pairs that constitute a counterexample to R^T being transitive.
  - Union the counter example to our current estimate for R^T.
  - Consider that union to be our new current estimate.

To write a recursive algorithm, we can restate our definition of the transitive closure of a relation to be

- The relation itself, if it is already transitive, or
- The transitive closure of the union of the relation to its immediately missing pairs, otherwise.

Hence the following program

```
- fun transitiveClosure(relation) =
=   if isTransitive(relation)
=             then relation
=             else transitiveClosure(
=                  makeNoRepeats(counterTransitive(relation))
=                  @ relation);
```

Exercises

5.7.1 Prove that if R is a transitive relation on A, then $R^T = R$. As in the proof for Theorem 5.13, consider the relation formed by the definition of R^T, and then prove that that relation is identical to R (a proof of set equality).

5.7.2 Prove that if R is a relation on A, then $R \cup i_A$ is the reflexive closure of R.

5.7.3 Prove that if R is a relation on A, then $R \cup R^{-1}$ is the symmetric closure of R.

5.7.4 Our proof of Theorem 5.13 assumes that composition of relations is associative (particularly our claim that $R^j \circ R^i = R^{i+j}$). Prove that if A, B, C, and D are sets and Q, R, and S are relations from A to B, B to C, and C to D, respectively, then $(R \circ S) \circ Q = R \circ (S \circ Q)$.

5.7.5 Using your function `counterSymmetric` from Exercise 5.6.3, write a function `symmetricClosure`.

5.8 Partial orders

Study the graphs of the following relations, looking for patterns.

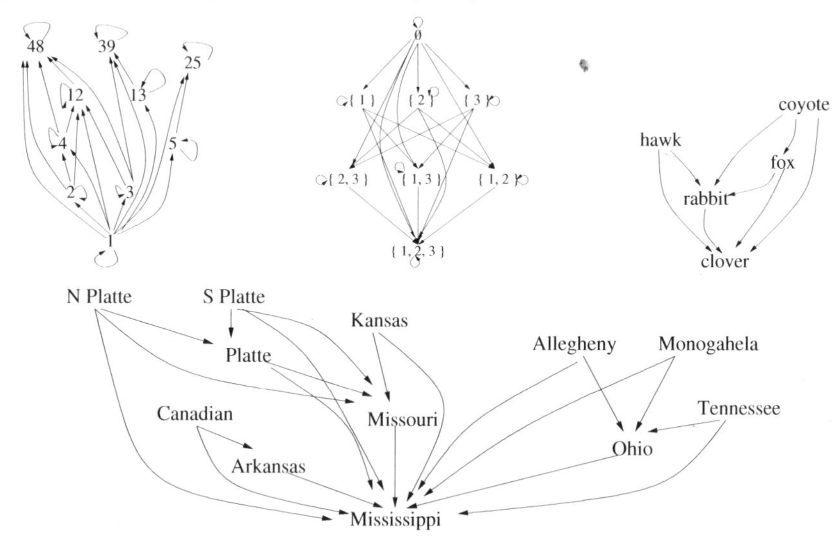

They are all transitive, but equivalence relations are transitive also, and these clearly are not equivalence relations. Moreover, just as equivalence relations organize a set by partitioning its elements up into cliques, these relations organize the sets by putting some sort of ordering to them. One difference is that equivalence relations are symmetric, whereas these are asymmetric. But even observing that they are not symmetric is not strong enough: to be asymmetric (that is, not symmetric) means there is at least one pair for which the reverse does not exist:

$$\sim \forall\, x, y \in X, \text{ if } (x, y) \in R \text{ then } (y, x) \in R$$

$$\equiv \quad \exists\, x, y \in X \mid (x, y) \in R \text{ and } (y, x) \notin R$$

Instead, these relations have the property that *no* arrow has its reverse—with the exception that self loops still have themselves as their own reverse. We call this *antisymmetry*; a relation is *antisymmetric* if

antisymmetric

$$\forall\, x, y \in X, \text{ if } (x, y) \in R \text{ and } (y, x) \in R, \text{ then } x = y.$$

Tangent: A-, in-, un-

Sometimes a word's etymology can be figured out by its negation. The negative of symmetric is *a*symmetric, but the negatives of transitive and reflexive are *in*transitive and *ir*reflexive. This is because *symmetric* comes from Greek (*sym*-, together, *metron*, measure) but *transitive* and *reflexive* come from Latin (*trans*-, across, *ire*, to go; *re*-, again, *flectere*, to bend). English words negated with *un*- usually have Anglo-Saxon origins.

Being antisymmetric, informally, means that no two distinct elements in the set are mutually related—though any single element may be (mutually) related to itself. Note carefully, though, that the definition of antisymmetric says "if two elements are mutually related, they must be equal," not "if two elements are equal, they must be mutually related." Relations like this are important because they give a sense of order to the set, in this case a hierarchy from least to greatest, with certain elements at more or less the same level.

partial order relation

A *partial order relation* (or *partial order*) is a relation R on a set X that is reflexive, transitive, and antisymmetric. A set X on which a partial order is defined is called a *partially ordered set* or a *poset*. The idea of the ordering being only partial is because not every pair of elements in the set is organized by it. In the case of $\subseteq$, for example, $\{1, 2\}$ and $\{1, 3\}$ are not *comparable*, which we will define formally in the next section.

poset

strict partial order relation

If a relation is transitive, antisymmetric, and *irreflexive*, that is, no element is related to itself, it is called a *strict partial order relation*. The partial orders we have seen are $|$ (divides), $\subseteq$, and $\leq$; strict partial orders include "is tributary to," "is downstream from," "gets nutrients from," "is ancestor of," $\subset$, and $<$. Partial orders get more attention because some of the more interesting relations mathematically (like $\subseteq$) are partial orders. However, most of the interesting results apply both to partial orders and to strict partial orders. Generic partial orders are often denoted by the symbol $\preceq$, with obvious reference to $\leq$ and $\subseteq$.

Consider how to prove that a relation is a partial order, or that a set is a poset:

Theorem 5.15 *Let A be any set of sets over a universal set $\mathcal{U}$. Then A is a poset with the relation $\subseteq$.*

> **Proof.** Suppose A is a set of sets over a universal set $\mathcal{U}$.
>
> Suppose $a \in A$. By the definition of subset, $a \subseteq a$. Hence $\subseteq$ is reflexive.
>
> Suppose $a, b, c \in A$, $a \subseteq b$, and $b \subseteq c$. Suppose further that $x \in a$. By definition of subset, $x \in b$, and similarly $x \in c$. Again by the definition of subset, $a \subseteq c$. Hence $\subseteq$ is transitive.
>
> Finally suppose $a, b \in A$, $a \subseteq b$, and $b \subseteq a$. By the definition of set equality, $a = b$. Hence, by the definition of antisymmetry, $\subseteq$ is antisymmetric.
>
> Therefore, A is a poset with $\subseteq$. □

Hasse diagram

The graphs of equivalence relations and of partial orders become very cluttered. For equivalence relations, it is more useful visually to illustrate regions representing the equivalence classes. For partial orders, we use a pared down version of a graph called a *Hasse diagram*, after German mathematician Helmut Hasse. It strips out redundant information. To transform the graph of a partial order relation into a Hasse diagram, first draw it so that all the arrows (except for self-loops) are pointing up. Antisymmetry makes this possible.

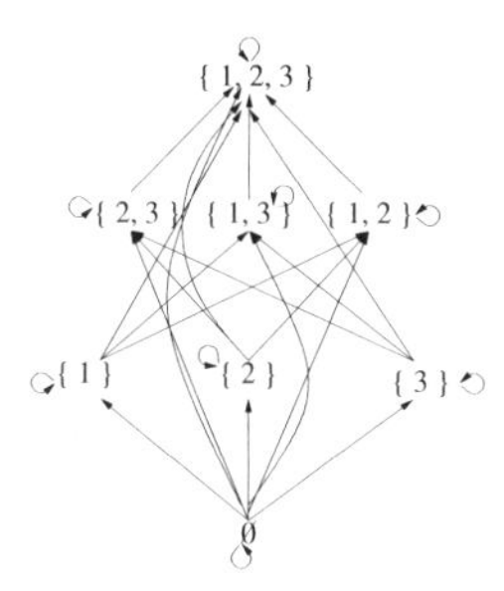

Then, since the arrangement on the page informs us what direction the arrows are going, the arrowheads themselves are redundant and can be erased.

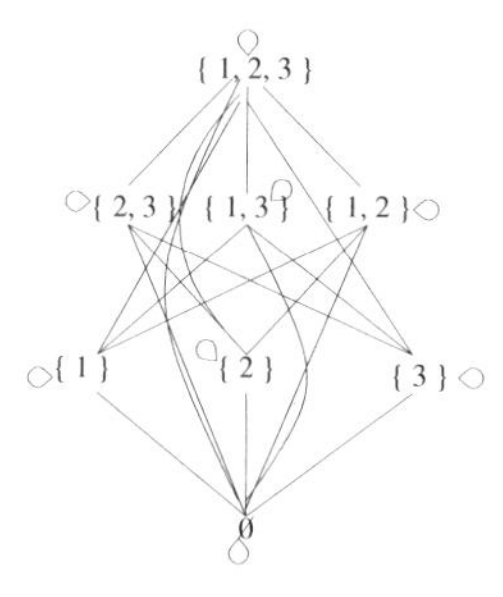

Finally, since we know that the relation is transitive and reflexive, we can remove self-loops and short-cuts. In the end we have something more readable, with the (visual) symmetry apparent.

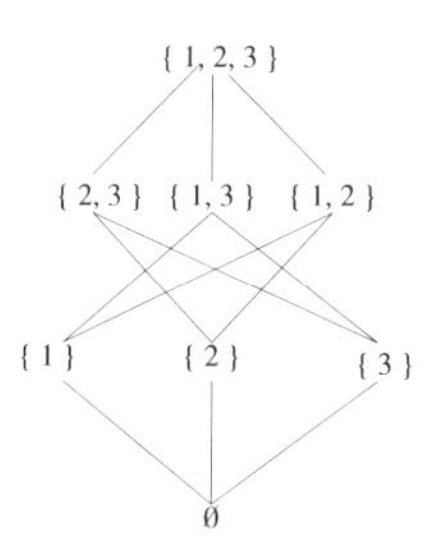

Biography: Helmut Hasse, 1898–1979

Helmut Hasse was a German mathematician who contributed to algebra and number theory. He served in the German navy during World War I. Hasse had many Jewish friends and collaborators, but he did work on ballistics for the German military during World War II, and after the war British occupying forces dismissed him from his teaching position. He was interested in the connections between mathematics and other fields. When he returned to teaching at another school in 1948, his first lecture compared the aesthetics of number theory to that of music. [21]

Exercises

In Exercises 5.8.1–5.8.5, consider the set of Jedi and Sith, $A = \{$ Yoda, Palpatine, Mace, Anakin, Obi-Wan, Maul, Qui-Gon, Dooku, Luke $\}$.Let R be a relation on A defined so that $(a, b) \in R$ if a has defeated b in a lightsaber fight (not necessarily fatally). Specifically:

Ep 1. Maul defeats Qui-Gon.
Ep 1. Obi-Wan defeats Maul.
Ep 2. Dooku defeats Obi-Wan.
Ep 2. Dooku defeats Anakin.
Ep 2. Yoda defeats Dooku.
Ep 3. Anakin defeats Dooku.
Ep 3. Mace defeats Palpatine.[a]
Ep 3. Palpatine defeats Yoda.
Ep 3. Obi-Wan defeats Anakin.
Ep 4. Anakin defeats Obi-Wan.
Ep 5. Anakin defeats Luke.
Ep 6. Luke defeats Anakin.

5.8.1 Draw a graph representing this relation.

5.8.2 Let R^T be the transitive closure of R. Which of the following are true?

(a) (Obi-Wan, Qui-Gon) $\in R^T$.

(b) (Mace, Maul) $\in R^T$.

(c) (Luke, Yoda) $\in R^T$.

(d) (Palpatine, Luke) $\in R^T$.

5.8.3 R is not antisymmetric. Give a counterexample, that is a set of tuples that show R is not antisymmetric.

5.8.4 If you consider only the *final* meeting of each Jedi/Sith pair (eliminate the pairs (Obi-Wan, Anakin), (Anakin, Luke), and (Dooku, Anakin)) and take the transitive closure, we are left with a (strict) partial order. Draw the Hasse diagram.

5.8.5 Suppose you wanted to find a relation S that associated people with others on the same side in the conflict (that is, S = "is on the same side"; assume there are only two sides). S should be an equivalence relation, and it should be that if $(a, b) \in S$ then $(a, b) \notin R$, that is, a person never fights with someone on the same side. Explain why S does not exist for the given data (you need not know the story; merely find a counterexample in the given data). If you know the movies, explain why this is the case in terms of the story.

[a]Palpatine killed Mace, but that was *after* the lightsaber fight.

5.8.6 Notice that the Hasse diagrams of `isTributaryOf` and "gets nutrients from" look very much like the graphs of `flowsInto` and *eats*, just without the arrowheads. Using this observation, we could implement in ML a minimal representation of a partial order that only contained pairs that would be connected in the Hasse diagram. Write a function `isRelatedPO` which takes two elements and a minimal list representation of a partial order and determines if the first element is related to the second in the partial order.

5.8.7 If R and S are antisymmetric relations on a set A, is $R \cup S$ antisymmetric? Prove or give a counterexample.

5.8.8 If R and S are antisymmetric relations on a set A, is $S \circ R$ antisymmetric? Prove or give a counterexample.

5.8.9 Prove that | (divides) on $\mathbb{N}$ is antisymmetric.

5.8.10 $\mathbb{C}$ stands out from the other standard number sets in that $\leq$ (and similar comparison operators) is no longer defined: Which is greater, $4 + 3i$ or $3 + 4i$? Invent a partial order for $\mathbb{C}$ based on $\leq$.

5.8.11 Write a predicate `isAntisymmetric` which tests if a relation is antisymmetric.

5.8.12 Write a function `counterAntisymmetric` which operates on the list representation of relations and constructs a list of counter examples to antisymmetry. This is different from `counterSymmetric` and `counterTransitive` because it will create a list of extraneous pairs, not missing pairs.

5.9 Comparability and topological sort

Partial orders are useful in that they describe a way of organizing a set, as one might put items in a drawer in a way that expresses the relationship the items have with each other. The previous section pointed out that partial order relations are partial in the sense that there may be elements that are not related to each other, that the relation cannot put in order. The partial order relation $\subseteq$ does not put, for example, $\{1, 3\}$ and $\{2, 3\}$ in order.

To give us terms to use about this situation, we say that for a partial order relation $\preceq$ on a set A, $a, b \in A$ are *comparable* if $a \preceq b$ or $b \preceq a$. 12 and 24 are comparable for | (since $12|24$), but 12 and 15 are not comparable. `NPlatte` and `Missouri` are comparable for "is tributary of," but `NPlatte` and `Kansas` are not comparable. For $\subseteq$, $\emptyset$ is comparable to everything; in fact, it is a subset of everything. For `Allegheny`, on the other hand, there are some things to which it is not comparable, but to all things to which it is comparable, it is a tributary—nothing in our set of rivers is a tributary to it. These last observations lead us to say that if $\preceq$ is a partial order relation on A, then $a \in A$ is

comparable

- *maximal* if $\forall\, b \in A$, $b \preceq a$ or b and a are not comparable.
- *minimal* if $\forall\, b \in A$, $a \preceq b$ or b and a are not comparable.
- *greatest* if $\forall\, b \in A$, $b \preceq a$.
- *least* if $\forall\, b \in A$, $a \preceq b$

maximal

minimal

greatest

least

From our observations below, we now can say that $\emptyset$ is the least element and A is the greatest element in the poset $\mathscr{P}(A)$ on $\subseteq$, for some set A. `Allegheny`—and `NPlatte` and `Kansas`, too—are minimal elements.

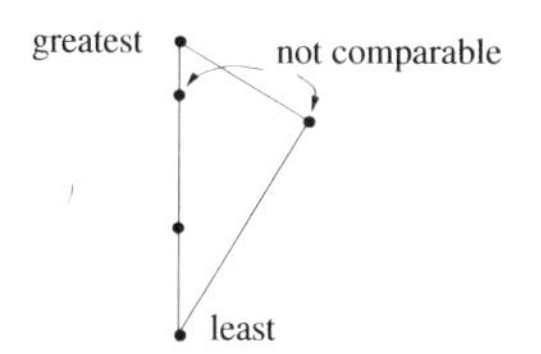

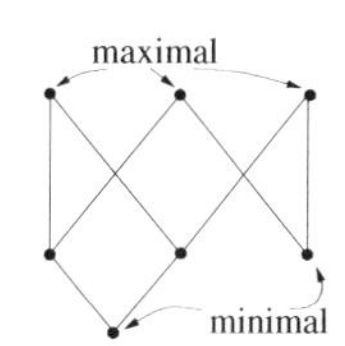

Observe that a poset may have many maximal or minimal elements, but at most one greatest or least. Must it have *any*? If the poset is an infinite set

> **Tangent: "I am the greatest."**
>
> Boxer Muhammad Ali is famous for his self-applied nickname, "The Greatest." However, he also regarded Sugar Ray Robinson as the true master of the sport and by "The Greatest" he did not mean to put himself ahead of Robinson. Since bulk is a considerable advantage in a sport like boxing, boxers are divided into weight classes so that matches can exhibit boxers' skills rather than raw size. Thus a comparison between Ali, a heavyweight, and Robinson, a middleweight, is difficult. If only Muhammad Ali had known his discrete math. . . . He should have said, "I am maximal."

then it is possible that it has no maximal or no minimal elements, as $\mathbb{R}$ with $\leq$ does not. Finite posets, on the other hand, must have at least one maximal and minimal. This is our next result, although for simplicity we will state it only for maximal elements (the statement and proof for minimal elements is symmetric).

Theorem 5.16 *A finite, non-empty poset has at least one maximal element.*

We will not give a complete proof of this theorem until Section 6.6 because it requires a proof technique we will cover later. However, we will give an informal argument for it, and, first, a lemma that supports it:

Lemma 5.17 *If A is a poset with partial order relation $\preceq$, and $a \in A$, then $A - \{a\}$ is a poset with partial order relation $\preceq - \{(b, c) \in \preceq \mid b = a$ or $c = a\}$.*

The notation is confusing, especially since the relation symbol $\preceq$ is used to stand for a set (since a relation is a set after all). Take a moment to digest what we are saying. We are deriving a new set from A and a new relation for that set. We make the new set B by removing an element a from A. Then the new relation defined in the lemma (call it $\preceq'$) is just like $\preceq$ except that all pairs that involve a are removed. It may seem especially strange to use set difference on the symbol $\preceq$ (in $\preceq - \{(b, c) \in \preceq \mid b = a \text{ or } c = a\}$). However, $\preceq$ is a relation, and a relation is a set. The lemma says that the new relation is a partial order. The proof needs to be structured around the definition of partial order relation.

Proof. Suppose A is a poset with partial order relation $\preceq$, and suppose $a \in A$. Let $B = A - \{a\}$ and $\preceq' = \preceq - \{(b, c) \in \preceq \mid b = a \text{ or } c = a\}$

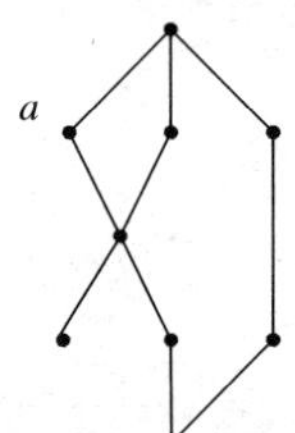

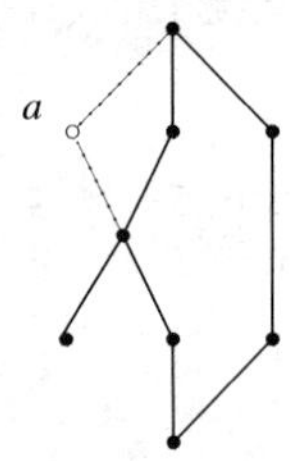

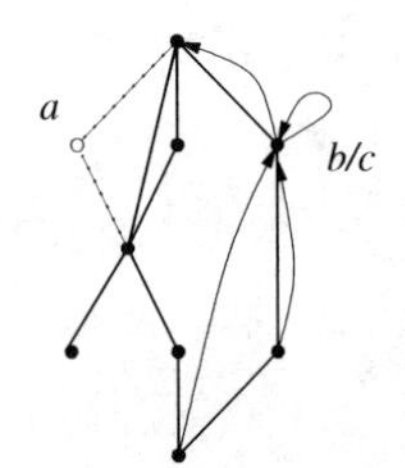

Now suppose $b \in B$. By definition of difference, $b \in A$, and since $\preceq$ is reflexive, $b \preceq b$. Also by definition of difference, $b \neq a$, so $b \preceq' b$. Hence $\preceq'$ is reflexive.

Suppose $b, c \in B$, $b \preceq' c$, and $c \preceq' b$. By definition of difference, $b, c \in A$, $b \preceq c$, and $c \preceq b$. Since $\preceq$ is antisymmetric, $b = c$. Hence $\preceq'$ is antisymmetric.

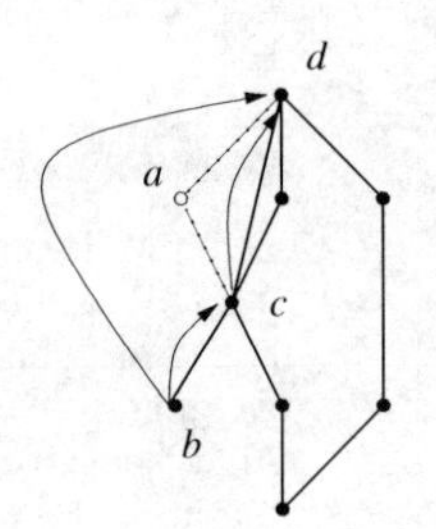

Finally, suppose $b, c, d \in B$, $b \preceq' c$, and $c \preceq' d$. By definition of difference, $b, c, d \in A$, $b \preceq c$, and $c \preceq d$. Since $\preceq$ is transitive, $b \preceq d$. Also by the definition of difference, $b \neq a$ and $d \neq a$, so $b \preceq' d$. Hence $\preceq'$ is transitive.

Therefore B is a poset with partial order relation $\preceq'$. □

In other words, the reflexivity, antisymmetry, and transitivity of $\preceq'$ depend on $\preceq$, but they do not depend on a being in the set.

This is important because it helps us show Theorem 5.16, that any nonempty, finite poset must have at least one maximal element. Suppose we had a poset A with partial order $\preceq$. First, remove the elements of A one at a time. By our lemma, each time we remove an element, we still have a poset. Eventually there will be only one element of A left (call it x), and that one element is the maximal element of that poset of cardinality 1.

Next, add back to the set the last one we removed (call it y). If $y \preceq x$, then x is still maximal. If $x \preceq y$, then y is the new maximal. If neither are true, then they are both maximal. Either way, the poset of cardinality 2 has at least one maximal element.

Keep adding elements back to the set. Whenever we add one back (call it z), either z is greater than the previous maximal (of which there is at least one), less than the previous maximal, or not comparable to the previous maximal. But no matter what, there must be at least one maximal. When we get back to the original set, it must have a maximal also.

If all pairs in a poset are comparable, then the partial order relation $\preceq$ is a *total order relation*. This simply means that all elements are placed in a definite or complete ordering in relation to one another. Such a relation could be used to sort a list of elements from the poset. The obvious total order is $\leq$ on $\mathbb{R}$.

total order relation

However, there are many situations where elements of a poset need to be put into sequence even though the relation is not a total order relation. In that case, we must disambiguate the several possibilities for sequencing noncomparable elements. This disambiguation is a matter of adding extra pairs into the relation to order otherwise noncomparable elements. If X is a poset with partial order relation $\preceq$, then the relation $\preceq'$ is a *topological sort* if $\preceq \subseteq \preceq'$ and $\preceq'$ is a total order relation. In other words, a topological sort of a partial order relation is simply that partial order relation with the ambiguities worked out. (Some students find the term *topological sort* confusing just from the words used. We do not mean a sorting *algorithm* like selection sort or insertion sort. A topological sort is a relation.)

topological sort

Suppose Grover Cleveland University has a computer science program with the following courses illustrated with their prerequisite chain in a Hasse diagram:

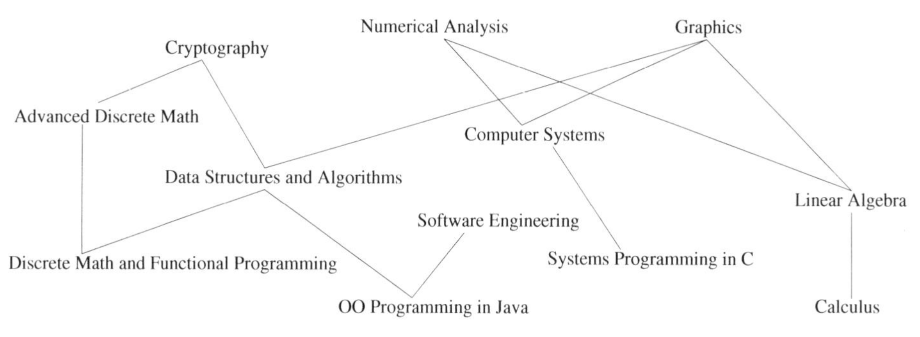

As illustrated, the relation "is prerequisite of" is a (strict) partial order. It is transitive in the sense that since Calculus is a prerequisite for Linear Algebra and Linear Algebra is a prerequisite for Numerical Analysis, then Calculus is an indirect prerequisite for Numerical Analysis.

Ralph is triple majoring in computer science, French, and tuba performance, and is also pre-vet. Accordingly, he is on the six-year plan. He plans to take one course in the computer science program each semester, but of course he needs to take them in a way that is consistent with the prerequisite chain. Here is one possible schedule:

	Fall	*Spring*		*Fall*	*Spring*
Year 1	DM and FP	OO Prog.	*Year 4*	Comp Systems	Linear Algebra
Year 2	DS and Algor.	Soft. Eng.	*Year 5*	Num. Analysis	Graphics
Year 3	Calculus	Sys. Prog.	*Year 6*	Advanced Discrete	Crypto

The order in which Ralph takes these courses is a topological sort of the prerequisite chain partial order. Keep this in perspective: Even though we have displayed it as a course schedule, the topological sort is a relation, organizing courses as $\leq$ organizes $\mathbb{R}$. It includes such pairs as (DM and FP, OO Prog) and (OO Prog, DS and Algor), not to mention (DM and FP, Comp Systems) and (DS and Algor, Crypto). The important things are, first, that this relation is a superset of the prerequisite chain relation, and second, that every two courses are comparable. Of course this schedule is not the only topological sort of the original partial order relation. In general, a partial order relation may have many topological sorts.

Even though the formal definition of a topological sort is confusing, everyday examples abound. When you write a research paper (or almost any composition, for that matter), you gather a collection of facts and identify points you wish to convey. Some points are used as evidence to assert others—there are dependencies between ideas that constrain the order in which they should be presented. As you map these things out, you may find the interrelations among ideas to be complex, like an intricate Hasse diagram. However, your paper presents ideas in a sequence. A key part of crafting a composition is finding the best way to linearize the points for making your argument.

Software that puts records in order often requires a topological sort in its design. Suppose a program collates records that include a date, a person associated with them, and some descriptor of the other contents. If the records were to be sorted by date, then records with the same date would not be comparable and so we might choose to disambiguate the ordering of such records by the name of person associated with a record in alphabetical order. If we have two records with the same date and the same person (or two people with the same name), we can disambiguate them with the record descriptor. The second two criteria make a topological sort of the original relation "has earlier date than." If you have programming experience in Java and are familiar with the `Comparable` interface, you may note that the documentation for `Comparable` in the Java API specifies that contract for the `compareTo` method mandates that it define a total order on the class implementing the interface.

One final note. The topological sorts mentioned so far are finite, but there exist infinite ones, too. The relation $\leq$ on $\mathbb{N}$ is a topological sort of $|$ (divides).

Exercises

5.9.1 Give a topological sort of the partial order in Exercise 5.8.4.

5.9.2 Alphabetical order is generalized by what is called *lexicographical order*. If $A_1, A_2, \ldots, A_n$ are posets with total order relations $\preceq_1, \preceq_2, \ldots, \preceq_n$, respectively, then the lexicographical order of $A_1 \times A_2 \times \ldots \times A_n$ is $\preceq_\ell$ defined as $(a_1, a_2, \ldots, a_n) \preceq_\ell (b_1, b_2, \ldots, b_n)$ if $a_1 \preceq_1 b_1$, $a_2 \preceq_2 b_2$, $\ldots$, $a_n \preceq_n b_n$. Prove that a lexicographical order is a total order relation.

5.9.3 The previous example was defined for tuples. We can imagine a similar concept for lists, where all the elements must come from the same set, using only one total order relation. To deal with the varying sizes of lists, we will say that if a longer list has a shorter list as a prefix, then the shorter list comes before the longer. Write a function `isRelatedLex` which takes two lists and a total order relation on the elements in those lists and determines if the first list should come after the second.

5.9.4 Give an example of an *infinite* topological sort (that is, an infinite poset and a relation that is a topological sort for that poset).

5.9.5 Suppose A is a poset with a total order $\preceq$, and suppose $a \in A$ is a maximal element. Prove that a is the greatest element.

5.9.6 What relation is reflexive, symmetric, transitive, and antisymmetric?

5.9.7 Find a total order for $\mathbb{C}$. Describe it verbally or with mathematical notation—since the set and relation are infinite, no visual description is possible. (Hint: Find a topological sort of your answer to Exercise 5.8.10.)

5.9.8 Professor Bumstead is getting dressed. He needs to put on a belt, a jacket, pants, a shirt, shoes, socks, a tie, an undershirt, undershorts, and a watch. Draw a Hasse diagram showing the order in which certain articles must be put on in relation to one another. Then give a topological sort that gives a reasonable order in which Professor Bumstead can put them on. This exercise comes from Cormen et al [5].

5.10 Extended example: Unification and resolution

Relations are an important way to organize information. They play a central role in several kinds of databases—information can be queried from collections of relations, and new information can be deduced.

Suppose we have two sets—a set of people and a set of things—and two relations, *likes* from people to things and *knows* from people to people. We do not know exhaustively all the tuples in these relations, but suppose we do know at least that the tuples (Kathy, Cars), (Maisie, Cars), and (Maisie, Oatmeal) are in the relation *likes*.

Now suppose we also know some propositions about these particular relations. For example, anyone who likes cars or likes oatmeal knows Jim, and anyone who likes cars and likes oatmeal knows Fred. From these facts we would be able to deduce that Kathy knows Jim and that Maisie knows both Jim and Fred.

In this section we develop a system to model information about a set of relations and to infer new information about those relations. Setting up the input information for such a system is called *logic programming*. The programming language Prolog is the most widely used language for this style of programming. The system we are writing is an interpreter of sorts for a very small programming language similar to Prolog. This is also how the ML interpreter infers types.

logic programming

To begin, suppose we have two kinds of facts about relations: facts that tell us a specific tuple in the relation ("Stephanie likes chocolate"), and facts that, using quantification, tell us about a large set of tuples ("Everyone likes chocolate"). In addition to facts we have queries, which can take the same two forms. If a query has the first form ("Does Stephanie like chocolate?"), then the result should be a yes or no answer; for the second form ("What does Stephanie like?") a query should result in a set of things we know fit the query. Suppose, then, this is our set of facts:

Likes(Kathy, Cars)
Likes(Maisie, Cars)
Likes(Maisie, Oatmeal)
Likes(Stephanie, Michigan)
Likes(x, Chocolate)
Likes(Harvey, x)

Sample queries:

Likes(Maisie, Oatmeal) [Answer: yes]
Likes(Kathy, Oatmeal) [Answer: no]
Likes(Stephanie, x)
  [Answer: { { x=Michigan }, { x=Chocolate } }]
Likes(x, Cars)
  [Answer: { { x=Harvey }, { x=Kathy }, { x=Maisie } }]

Notice that for facts variables are implicitly universally quantified—x means "everything." In queries, variables range over a truth set.

The main building block in the process by which queries are answered is called *unification*. The idea is to take a query and a fact and make them match, if possible, by substituting a constant in one of them for a variable in the other. Consider the following pairs of sentences. For readability, assume the first is a query and the second is a fact, but it really does not matter.

unification

Likes(Maisie, Oatmeal), Likes(Maisie, Oatmeal)	There are no variables, and this matches immediately. No substitution is necessary.
Likes(Stephanie, x), Likes(Stephanie, Michigan)	These can be unified by substituting Michigan for x.
Likes(Stephanie, y), Likes(Maisie, Oatmeal)	The constants Stephanie and Maisie are in conflict. No substitution for variables can fix this.
Likes(Stephanie, y), Likes(x, Chocolate)	These can be unified by substituting Stephanie for x and Chocolate for y.

In the last case, this works only because different variables were used. We cannot unify "Likes(Stephanie, x)" and "Likes(x, Chocolate)". However, since the name of a variable does not affect meaning, we can replace any variable with another name to avoid clashes like this.

Before deriving an algorithm for unification, consider how to represent the information. Variables and strings can appear in the same place, so they can be subsumed in the same datatype which we will call atom. A fact contains a string for the name of the relation and a list of atoms. We do not need a separate type for queries—they are essentially facts put to a different purpose.

```
datatype atom = Var of string | Const of string;
datatype fact = Fact of string * atom list;
```

Some examples:

```
Fact("Likes", [Const("Kathy"), Const("Cars")]);
Fact("Likes", [Const("Maisie"), Const("Cars")]);
Fact("Likes", [Const("Maisie"), Const("Oatmeal")]);
Fact("Likes", [Const("Stephanie"), Const("Michigan")]);
Fact("Likes", [Var("x"), Const("Chocolate")]);
Fact("Likes", [Const("Harvey"), Var("x")]);
```

substitution

As alluded to earlier, the solution to a unification is a *substitution*, a set of pairs each associating a variable with atoms that can replace it. If no such substitution exists, then the unification results in a failure. Thus we have the following datatype to model results of unifications, as well as a function to lookup a variable in a substitution. Since the first item in each pair must be a variable, we represent it with a string rather than an atom.

```
datatype substitution = Failure | Sub of (string * atom) list;
exception failedLookup;

fun lookup(v, []) = raise failedLookup
  | lookup(v, (w, x)::rest) = if v = w then x else lookup(v, rest);
```

To unify two facts, we make sure that we are testing the same relation (which we will call the facts' *operators*), and then unify the lists of parameters item by item, updating our solution substitution as we go. This breaks down to unifying individual atoms, given a "substitution so far." We do this with two mutually recursive functions, `unifyAtom` and `unifyVar`.

```
fun unifyAtom(Const(c), Const(d), subst) =
        if c = d then subst else Failure
  | unifyAtom(Var(v), y, subst) = unifyVar(v, y, subst)
  | unifyAtom(x, Var(w), subst) = unifyVar(w, x, subst)

and unifyVar(v, y, s as Sub(subs)) =
     (unifyAtom(lookup(v, subs), y, s)
      handle failedLookup =>
        (case y of
              Var(w) => (unifyAtom(Var(v), lookup(w, subs), s)
                           handle failedLookup =>
                              if v = w then Failure
                                       else Sub((v, y)::subs))
            | Const(s) => Sub((v, Const(s))::subs)))
  | unifyVar(_,_,_) = Failure;
```

as

This is a tricky piece of code. First, one new bit of ML: The third parameter to `unifyVar` is written `s as Sub(subs)`. The `as` construct allows us to define two names in one parameter: the variable `s` to stand for the entire substitution value and the variable `subs` to stand for the (string * atom) list component. It gives us the best of both worlds between parameter names and pattern-matching. Recall that the underscore (_) is a special variable for values in patterns we want to ignore. It can appear in a pattern more than once.

Now for what these functions do. If given two constants, `unifyAtom` makes sure they are equal. If given at least one variable, the action is passed off to `unifyVar`; notice that the parameter `v` has type string (the name of a variable) and parameter `y` has type atom—it could be either a variable or a constant. If the variable `v` already has a replacement in the substitution `s`, then try to unify that replacement with `y`.

Instead of first testing if `v` has a replacement and then retrieving the replacement if it exists, we optimistically retrieve the replacement and handle an exception if no replacement exists. Thus the first `handle failedLookup` clause is for the case that `v` has no replacement yet. If that happens, we look at `y`.

If `y` is a variable, then do what we did with `v`: optimistically lookup its replacement and unify. If that fails (another `handle failedLookup`), then we have two variables without replacements. If they are the same variable, that is bad, and we throw up our hands. If they are not equal, then we add `y` as a replacement for `v` to our substitution. We similarly add to our substitution if `y` is a constant.

Unifying a list of atoms is more straightforward. We apply `unifyAtom` on every pair of items in the two lists, but accumulate the substitution as we go—if the unification of two atoms results in a new replacement pair, that pair must be included in the substitution used for the later atoms in the list. If the lists are not of the same length, then we fail.

```
fun unifyList(_, _, Failure) = Failure
```

```
  | unifyList(xFst::xRest, yFst::yRest, subst) =
      unifyList(xRest, yRest, unifyAtom(xFst, yFst, subst))
  | unifyList([], [], subst) = subst
  | unifyList(_, _, _) = Failure;
```

The main unification function is `unifyFact`, which checks the operators and parameter lists.

```
fun unifyFact(Fact(xOp, xArgs), Fact(yOp, yArgs), subst) =
    if xOp = yOp then unifyList(xArgs, yArgs, subst)
                 else Failure;
```

Trying these out:

```
  - unifyFact(Fact("Likes", [Const("Stephanie"), Var("x")]),
  =           Fact("Likes", [Const("Stephanie"),
  =                          Const("Michigan")]), Sub([]));

  val it = Sub [("x",Const "Michigan")] : substitution

  - unifyFact(Fact("Likes", [Const("Stephanie"), Var("x")]),
  =           Fact("Likes", [Var("y"), Const("Chocolate")]),
  =                Sub([]));

  val it = Sub [("x",Const "Chocolate"),("y",Const "Stephanie")]
      : substitution

  - unifyFact(Fact("Likes", [Const("Stephanie"), Var("x")]),
  =           Fact("Likes", [Const("Maisie"),
  =                          Const("Oatmeal")]), Sub([]));

  val it = Failure : substitution
```

Unification is just a part of the process we are interested in. We have not yet addressed how to infer new facts using information like “Everyone who likes cars or oatmeal knows Jim” and “Everyone who likes cars and oatmeal knows Fred.” Notice that at their core these are conditionals:

$$\begin{array}{lcl} \forall x, \text{Likes}(x, \text{cars}) \vee \text{Likes}(x, \text{oatmeal}) & \rightarrow & \text{Knows}(x, \text{Jim}) \\ \forall x, \text{Likes}(x, \text{cars}) \wedge \text{Likes}(x, \text{oatmeal}) & \rightarrow & \text{Knows}(x, \text{Fred}) \end{array}$$

We will call sentences like these *rules*, to distinguish them from facts. To standardize the format, we will specify that the hypothesis of any rule is a conjunction of facts (the premises) and the hypothesis is a single fact. As before, all variables are universally quantified. We can handle disjunctions—when we would like to “or” the premises—by splitting them up into several rules. Here is our datatype modeling rules and some examples.

```
datatype rule = Rule of fact list * fact;

Rule([Fact("Likes", [Var("x"), Const("Cars")])],
     Fact("Knows", [Var("x"), Const("Jim")]));
Rule([Fact("Likes", [Var("x"), Const("Oatmeal")])],
     Fact("Knows", [Var("x"), Const("Jim")]));
Rule([Fact("Likes", [Var("x"), Const("Cars")]),
      Fact("Likes", [Var("x"), Const("Oatmeal")])],
     Fact("Knows", [Var("x"), Const("Fred")]));
```

resolution

The process of generating new facts from a set of facts and rules is called *resolution*. If a fact matches the premise of a rule under a certain substitution, then the conclusion of the rule is a known fact under that substitution. As we work through the process of resolution, we will assume a list of facts and rules; a current *goal*, the query we are currently trying to prove or find a substitution for; and a substitution or set of substitutions. As before, the result of a query is a set of substitutions that make the query true.

We will maintain a list of facts and a list of rules as reference variables in the system. This way it is easy to add to them, and they do not have to be passed to the resolution functions.

```
val factList =
  ref [Fact("Likes", [Const("Kathy"), Const("Cars")]),
   Fact("Likes", [Const("Maisie"), Const("Cars")]),
   Fact("Likes", [Const("Maisie"), Const("Oatmeal")]),
   Fact("Likes", [Const("Stephanie"), Const("Michigan")]),
   Fact("Likes", [Var("x"), Const("Chocolate")]),
   Fact("Likes", [Const("Harvey"), Var("x")])];
val ruleList =
  ref [Rule([Fact("Likes", [Var("x"), Const("Cars")])],
        Fact("Knows", [Var("x"), Const("Jim")])),
   Rule([Fact("Likes", [Var("x"), Const("Oatmeal")])],
        Fact("Knows", [Var("x"), Const("Jim")])),
   Rule([Fact("Likes", [Var("x"), Const("Cars")]),
         Fact("Likes", [Var("x"), Const("Oatmeal")])],
        Fact("Knows", [Var("x"), Const("Fred")]))];
```

Here is how resolution works.

Resolving a goal. Suppose we have a goal to resolve. This may be a sub-goal produced by the resolve functions themselves, or it may be a query that the user makes. We also have a substitution generated so far, but it may be easier to assume for now that the substitution is empty. We find all substitutions (if any) by resolving using the fact list and concatenate that list of substitutions to those found by resolving using the rules. These resolution functions are mutually recursive, but here is what the main `resolve` function looks like.

```
and resolve(goal, subst) =
    resolveFact(goal, !factList, subst)@
    resolveRule(goal, !ruleList, subst);
```

Resolving a goal using facts. The function `resolveFact` takes a goal, a list of facts, and a substitution (the replacements found so far). We attempt to unify the goal with the first fact in the list. If it fails, try the next one. If it succeeds with a refined substitution, then try the next one anyway, concatenating the refined substitution with the substitutions found by trying the rest of the list.

Resolving a goal using rules. The function `resolveRule` takes a goal, a list of rules, and a substitution. For each rule in the list, we try to unify the goal with the conclusion. If it matches (with a refined substitution), then the substitution only works if we can satisfy the premises of the rule. All the premises, then, become sub-goals which we try to resolve, starting with the substitution that unifies the goal with the conclusion. Any substitution we find to work from this rule we add to a list of other substitutions that work, using other rules.

Suppose our goal is "Who does Harvey know?"—that is, Knows(Harvey, x). Since none of the facts in our list are about the Knows relation, we start looking at the rules. The first rule has Knows(x, Jim) as a conclusion, and that would match if it were not for the reuse of variable x. We will rename the variable to get Knows(x_1, Jim), and now the two are unified with substitution { x_1 = Harvey, x = Jim }.

This does not mean we have yet found someone Harvey knows. We still need to prove the premises. Our subgoal is now Likes(Harvey, Cars). More accurately, our subgoal is Likes(x_1, Cars) with substitution { x_1 = Harvey, x = Jim }. Testing this against the facts, we find that it unifies with Likes(Harvey, x_2) (the variable x renamed to be unambiguous), with refined substitution { x_1 = Harvey, x = Jim, x_2 = x_1 } Notice that variables can be associated with other variables in a substitution. It does imply, of course, that ultimately x_2 = Harvey.

However, that substitution is not the only answer, nor is that the only route to that answer. The next rule in the list also has Knows(x_1, Jim) as a conclusion, with Likes(x_1, Oatmeal) as the premise. This also can be resolved with the fact Likes(Harvey, x_2).

Finally, the query unifies with the conclusion of the third rule, Knows(x_1, Fred). The substitution is { x_1 = Harvey, x = Fred }. Now we have a list of two sub-goals: Likes(x_1, Oatmeal) and Likes(x_1, Cars). It is important to note that both variables are x_1—that is, we already have a value for them in the substitution. We need to satisfy both of them.

As we have seen, we can resolve Likes(x_1, Oatmeal) with this substitution by unifying with Likes(Harvey, x_2). Our substitution is now { x_1 = Harvey, x = Jim, x_2 = Oatmeal }, and we need to use this substitution when resolving for the other subgoal. It can be resolved by unifying with Likes(Harvey, x_3)—same fact as before, but with a fresh variable. The final substitution in this case is { x_1 = Harvey, x = Jim, x_2 = Oatmeal, x_3 = Cars }.

Two observations in particular with all this. First, to do any of this, we need to be able to take a fact or rule and give the variables fresh, unique names. As an easy naming scheme, we will append each variable name with a unique number. The mechanism for doing this is a reference variable `idGen` to keep track of the last number used and a function `makeUnique`:

```
val idGen = ref 0;

fun makeUnique(v) = (idGen := !idGen + 1;
                     v ^ Int.toString(!idGen));
```

Making all the variables unique in parts of our data representation is handled by functions `standardizeFact`, `standardizeRule`, and others. The algorithm for doing this is simple but tedious. The code can be found on the accompanying website, `http://cs.wheaton.edu/~tvandrun/dmfp`.

The second observation involves how we resolve lists of goals. Given a list of goals and a single substitution, we try to resolve the first goal. Remember that `resolve` returns a list of substitutions that work. If that list is empty, everything fails (for the original given substitution). If it is not empty, then we need to check every other goal in the list *with every substitution* and concatenate the results together. This function is part of the process:

```
and resolveSubstList(goals, []) = []
  | resolveSubstList(goals, subst::rest) =
    resolveGoalList(goals, subst)@resolveSubstList(goals, rest)
```

As a final step, we provide a few functions to make the system easier to use. We want functions that allow the user to tell the system a new fact or rule and to ask the system a query. The tell functions are easy:

```
fun tellFact(fact) = factList := fact::!factList;
fun tellRule(rule) = ruleList := rule::!ruleList;
```

The result should be printed in a readable format. The result is a a list of substitutions, and these substitutions include variables that the user does not need to know about. We want to display the ultimate replacement for each of the variables in the original query.

```
fun deepLookup(v, subs) =
    case lookup(v, subs) of
        Const(c) => c
      | Var(vv) => deepLookup(vv, subs);

fun getVars([]) = []
  | getVars(Const(c)::rest) = getVars(rest)
```

```
  | getVars(Var(v)::rest) = v::getVars(rest);

fun printResult([], _) = print(";\n")
  | printResult(_, Failure) = print("fail")  (* shouldn't happen *)
  | printResult(x::rest, Sub(subs)) =
    (print(x ^ "=" ^ deepLookup(x, subs) ^ " ");
     printResult(rest, Sub(subs)));

fun printResults(vars, []) = print(".\n")
  | printResults(vars, sub::rest) =
    (printResult(vars, sub); printResults(vars, rest));
```

The function ask puts all this together.

```
fun ask(Fact(oper, args)) =
    let val results = resolve(Fact(oper, args), Sub([]));
        val vars = getVars(args);
    in if results = [] then print("no\n")
                         else printResults(vars, results)
    end;
```

Trying it out:

```
- ask(Fact("Likes", [Const("Maisie"), Var("x")]));

x=Cars ;
x=Oatmeal ;
x=Chocolate ;
.
val it = () : unit

- ask(Fact("Knows", [Const("Harvey"), Var("x")]));

x=Jim ;
x=Jim ;
x=Fred ;
.
val it = () : unit
```

More information about the unification and resolutions algorithms can be found in Harrison [13] and Russell and Norvig [28].

Project

5.A Finish the following function to resolve a goal using a list of facts.

```
fun resolveFact(goal,fact::rest,subst) =
    (case unifyFact(goal,
            #1(standardizeFact(fact,[])),
            subst) of
         Failure => ??
       | Sub(s) => ??
  | resolveFact(goal, [], subst) = []
```

5.B Finish the following function to resolve a goal using a list of rules.

```
and resolveRule(goal, r::rest, subst)  =
   let val Rule(premises, conclus) =
             standardizeRule(r)
   in
    case unifyFact(goal,conclus,subst) of
        Failure => ??
      | subst2 =>
        (case resolveGoalList(premises,
                                  subst2)
            of
              [] => ??
            | substs => ??
   end
 | resolveRule(goal, [], subst) = []
```

5.C Finish the following function to resolve a list of goals. This will include a call to function resolveSubstList.

```
and resolveGoalList(goal::rest, subst) =
    (case ?? of
         [] => ??
       | substs=> ??
  | resolveGoalList([], subst) = [subst]
```

5.D Write improved versions of the tell and ask functions that will parse string input and create appropriate datatype values, so the user can enter something like `tell("Likes(Maisie, ML)")` instead of `tellFact(Fact("Likes", [Const("Maisie"), Const("ML")]))`. Use as a model the parser for the language system in Section 2.6. Make reasonable assumptions like all constants begin with a capital letter and all variables begin with a lowercase letter.

5.E Make a collection of facts and rules so that you can use the system to solve the following problem.

Angie, Brad, Casey, Dora, Evert, and Fuchsia are at a conference. It is lunchtime, and they need to figure out who can eat with whom at what restaurant.

The nearby restaurants are the East Grille, Bertie's, the City Tavern, and Fish King. The East Grille serves hamburgers, tofu, and halibut. Bertie's serves hamburgers. The City Tavern serves hamburgers and tofu. Fish King serves halibut. Any place that serves hamburgers also serves French fries. Bertie's and Fish King have patios. The City Tavern and Fish King require patrons to wear shoes, and the East Grille requires (male) patrons to wear ties. Any place that requires ties also requires shoes.

Angie wants to eat halibut, and Brad wants to eat French fries. Casey will only eat a place that requires shoes, and he wants to eat tofu. Dora is more formal and wants only to eat at a place that requires ties. Evert wants to discuss his research with Angie and so will eat any place where she will eat. Fuchsia wants to eat on a patio.

Who can eat with whom, and where? (You do not need to have a solution to Project 5.D to do this one, but it would make the input more convenient.) For the output, list the possible "assignments," that is, who is eating where.

5.11 Special topic: Representing relations

Our initial problem with the relation `inSameTimeZone` was that it was missing self-loops. Hence the reflexive closure would have solved our problem just as well as the transitive closure. Unfortunately, there is no way to compute the reflexive closure given our way of representing relations (think back to Exercise 5.6.2).

This is particularly frustrating because the reflexive closure is so simple. For example, we could write a predicate like `isRelatedTo` except that it tests if two elements are related by the relation's reflexive closure.

```
- fun isRelatedToByRefClos(a, b, relation) =
=   a = b orelse isRelatedTo(a, b, relation);
```

This suggests that our situation could be remedied if we represented relations another way. At the beginning of this chapter, we alluded to the fact that relations could be considered predicates. We could use that intuition to build an alternate representation of relations in ML.

```
- fun flowsInto(Canadian, Arkansas) = true
=   | flowsInto(NPlatte, Platte) = true
=   | flowsInto(SPlatte, Platte) = true
=   | flowsInto(Platte, Missouri) = true
=   | flowsInto(Kansas, Missouri) = true
=   | flowsInto(Missouri, Mississippi) = true
=   | flowsInto(Allegheny, Ohio) = true
=   | flowsInto(Monongahela, Ohio) ==  true
=   | flowsInto(Tennessee, Ohio) = true
=   | flowsInto(Ohio, Mississippi) = true
=   | flowsInto(x, y) = false;

val flowsInto = fn : river * river -> bool
```

What we do not want to lose is the ability to treat relations as what we have been calling "mathematical objects." What we mean is that we should be able to pass a relation to a function and have a function return a relation, as `transitiveClosure` does. A value in a programming environment that can be passed to and returned from a function is a *first-class value*. A pillar of functional programming is that functions are first-class values.

first-class value

A relation represented by a predicate will have a type like ('a * 'b) $\rightarrow$ bool. This means "function that maps from an 'a $\times$ 'b pair to a bool." Our first task is to write a function that will convert a relation from list representation to predicate representation. To return a predicate from a function, simply define the predicate (locally) within the function and return it by naming it without giving any parameters.

```
- fun listToPred(oldRelation) =
=    let fun newRelation(a,b) = isRelatedTo(a,b,oldRelation);
=    in newRelation end;


val listToPred = fn : (''a * ''b) list -> ''a * ''b -> bool
```

In a similar way, a function can receive a predicate. Observe this function to compute the reflexive closure:

```
- fun reflexiveClosure(relation) =
=    let fun closure(a, b) = a = b orelse relation(a, b);
=    in closure end;


val reflexiveClosure = fn : (''a * ''a -> bool) ->
                              ''a * ''a -> bool
```

Computing the symmetric closure is left to the reader. We cannot compute the predicate-form transitive closure directly, but if the relation is originally represented as a list, we could compute the transitive closure before converting to predicate form.

Another way to represent relations is to use a matrix or table. Label the columns and rows with the elements of the set (or sets), the rows representing the first values in each pair, the columns representing the second values. The contents of the matrix are boolean values, T if the pair is in the relation, F otherwise. Here we illustrate `isOnRiver` with a matrix; to reduce clutter, we only display the T values, leaving the F values blank.

	Missouri	Platte	NPlatte	SPlatte	Arkansas	Canadian	Kansas	Mississippi	Tennessee	Ohio	Allegheny	Monongahela
Vancouver												
LosAngeles												
Mexico												
Denver				T								
Omaha	T	T										
KansasCity	T					T						
Minneapolis								T				
StLouis	T							T				
Memphis								T				
NewOrleans								T				
Chicago												
Cincinnati										T		
Pittsburgh										T	T	T
Montreal												
NewYork												

If a relation is on a single set, then the row and column labels are the same. In that case, how could you tell if the relation is reflexive or symmetric, just by looking at the matrix?

Matrix representation is not convenient in ML, but it makes sense in languages that have strong support for arrays, like C and Java. Assume each element of the set the relation is on is associated with a number which can be used as an index into an array. Then the matrix is represented using a two-dimensional array, and the numbers of elements are used for indices into both dimensions. If you know Java, then consider this method for determining whether a relation is transitive:

```
public static boolean isTransitive(boolean[][] relation) {
    // the number of elements in the set
    int n = relation.length;

    // loop over all possible pairs
    for (int i = 0; i < n; i++)
        for (int j = 0; j < n; j++)
            // if (i, j) is in the relation,
            if (relation[i][j]) {
                // then for all pairs (j, k) in the relation,
                for (int k = 0; k < n; k++)
                    // make sure (i, k) is also in the relation
                    if (relation[j][k] && ! relation[i][k])
                        // if not, then it is not transitive
                        return false;
            }
    // if we make it through the big loop, then it is transitive
    return true;
}
```

A small modification turns the same algorithm into one equivalent to our ML function `transitiveClosure`:

```
public static boolean[][] transitiveClosure(boolean[][] relation) {
    boolean[][] trClosure = relation.clone();
    int n = relation.length;

    // initially true to insure at least on iteration of the loop
    boolean changed = true;

    // repeat as long as there was a change last time
    while (changed) {
        changed = false;
        for (int i = 0; i < n; i++)
          for (int j = 0; j < n; j++)
```

```
                if (trClosure[i][j]) {
                    for (int k = 0; k < n; k++)
                      // if (i, k) is not in the relation,
                      // add it, and note the change.
                      if (trClosure[j][k] && ! trClosure[i][k]) {
                          trClosure[i][k] = true;
                          changed = true;
                      }
                }
        }

        return trClosure;
    }
```

As in Section 2.7, we are faced with the dilemma of choosing among three representations, each of which has favorable features and drawbacks—in fact, the dilemma extends to choosing between two languages. The list representation is the more intuitive, at least in terms of the formal definition of a relation; with the predicate representation, however, we can test a pair for membership directly, as opposed to relying on a function like `isRelatedTo`. Because we can iterate through all pairs, the list representation allows testing and computing of transitivity, but with the predicate representation we can compute reflexive and symmetric closures. Conversion is one-way, from lists to predicates. Neither representation allows us to test reflexivity. These aspects are summarized below.

	List	Predicate
first-class value	yes	yes
membership test	indirectly	directly
`isReflexive`	no	no
`isSymmetric`	yes	no
`isTransitive`	yes	no
`isAntiSymmetric`	yes	no
`reflexiveClosure`	no	yes
`symmetricClosure`	yes	yes
`transitiveClosure`	yes	no
convert to other	yes	no

Every "no" would become a "yes" if only we had the means to iterate through all elements of a datatype. A matrix representation would have a "yes" for all, but that is an unfair comparison, since the dimensions of the matrix implicitly enumerate all the items in the set.

Finally, a word of interest only to those who have programmed in an object-oriented language such as Java. When you read examples like these, you should be thinking about the best way to represent the concept in other programming languages. In an object-oriented language, objects are first-class values; hence we would want to write a class to represent relations.

The primary difference between the list and predicate representations presented here is that the former represents the relation as *data*, the latter as *functionality*. The primary characteristic of objects is that they encapsulate data and functionality together in one package. Since the predicate representation can be built from a list representation, one would expect that it would be strictly more powerful; however, in the conversion we lost the ability to test for transitivity, and this is because we have lost access to the list.

If instead we made the list to be an instance variable of a class, the methods could do the functional work of `reflexiveClosure` as well as the iterative work of `isTransitive`. Java's enum types provide the functionality of ML's datatypes, as well as a means of iterating over all elements of a set, something that hinders representing relations as lists or predicates in ML (ML's datatype is more powerful than Java's enum types in other ways, however). The following shows a Java implementation of relations.

```
/**
 * Class Relation to model mathematical relations.
 * The relation is assumed to be over a set modeled
 * by a Java enum.
 */

public class Relation<E extends Enum<E>> {

    private class List {
        public E first, second;
        public List tail;
        public List(E first, E second, List tail) {
            this.first = first;
            this.second = second;
            this.tail = tail;
        }

        /**
         * Concatenate a given list to the end of this list.
         * @param other The list to add.
         * POSTCONDITION: The other list is added to
         * the end of this list.
         */
        public void concatenate(List other) {
            if (tail == null) tail = other;
            else tail.concatenate(other);
        }
    }
```

```
/**
 * The set ordered pairs of the relation.
 */
private List pairs;

/**
 * Constructor to create a new relation of n pairs from an
 * n by 2 array.
 * @param input The array of pairs; essentially an array of
 * length-2 arrays of the base enum.
 */
public Relation(E[][] input) {
    for (int i = 0; i < input.length; i++) {
        assert input[i].length == 2;
        pairs = new List(input[i][0], input[i][1], pairs);
    }
}

/**
 * Constructor to create a new realtion from a list of pairs.
 */
public Relation(List pairs) {
    this.pairs = pairs;
}

/** Is a related to b in this relation? */
public boolean relatedTo(E a, E b) {
    for (List current = pairs; current != null;
         current = current.tail)
        if (current.first == a && current.second == b)
            return true;
    return false;
}

/** Is this relation reflexive? */
public boolean isReflexive() {
    // if there are no pairs, we can assume this is
    // not reflexive.
    if (pairs == null) return false;
    try {
        for (E t : (E[]) pairs.first.getClass()
                 .getMethod("values").invoke(null))
            if (! relatedTo(t, t)) return false;
    } catch (Exception e) { }  // won't happen
```

```
        return true;
    }

    /** Is this relation symmetric? */
    public boolean isSymmetric() {
        for (List current = pairs; current != null;
             current = current.tail)
            if (! relatedTo(current.second, current.first))
                return false;
        return true;
    }

    /**
     * Is it the case that everything b is related to,
     * a is also related to?
     */
    private boolean isTransitiveWRTPair(E a, E b) {
        for (List current = pairs; current != null;
             current = current.tail)
            if (b == current.first
                && ! relatedTo(a, current.second))
                return false;
        return true;
    }

    /** Is this relation transitive? */
    public boolean isTransitive() {
        for (List current = pairs; current != null;
             current = current.tail)
            if (! isTransitiveWRTPair(current.first,
                                      current.second))
                return false;
        return true;
    }

    /** Is this relation antisymmetric? */
    public boolean isAntisymmetric() {
        for (List current = pairs; current != null;
             current = current.tail)
            if (current.first != current.second &&
                relatedTo(current.second, current.first))
                return false;
        return true;
    }
```

```
/**
 * Create a new relation, the reflexive closure of this.
 */
public Relation reflexiveClosure() {
    if (isReflexive()) return this;
    else return new Relation<E>() {
        public boolean isReflexive() { return true; }
        public boolean relatedTo(E a, E b) {
            return a == b
                || Relation.this.relatedTo(a, b);
        }
    };
}

/**
 * Create a new relation, the symmetric closure of this.
 */
public Relation symmetricClosure() {
    if (isSymmetric()) return this;
    else {
       List newPairs = null;
       for (List current = pairs; current != null;
             current = current.tail)
            newPairs =
              new List(current.second,current.first,newPairs);
       newPairs.concatenate(pairs);
       return new Relation(newPairs);
    }
}

/**
 * Make a list of missing pairs (a, c) for each (b, c).
 */
private List counterTransitiveWRTPair(E a, E b) {
    List toReturn = null;
    for (List current = pairs; current != null;
          current = current.tail)
         if (b == current.first
             && ! relatedTo(a, current.second))
             toReturn = new List(a, current.second, toReturn);
    return toReturn;
}

/**
 * Make a list of pairs needed for transitivity.
```

```
     */
    private List counterTransitive() {
        List toReturn = null;
        for (List current = pairs; current != null;
             current = current.tail) {
            List currentCounter =
                counterTransitiveWRTPair(current.first,
                                         current.second);
            if (currentCounter != null) {
                currentCounter.concatenate(toReturn);
                toReturn = currentCounter;
            }
        }
        return toReturn;
    }

    /**
     * Default constructor used by transitiveClosure().
     */
    private Relation() {}

    /**
     * Create a new relation, the transitivie closure of this.
     */
    public Relation transitiveClosure() {
        if (isTransitive()) return this;
        Relation toReturn = new Relation();
        toReturn.pairs = counterTransitive();
        toReturn.pairs.concatenate(pairs);
        return toReturn.transitiveClosure();
    }

    /**  Is this an equivalence relation? */
    public boolean isEquivalenceRelation() {
        return isReflexive() && isSymmetric() && isTransitive();
    }

    /** Is this a partial order relation? */
    public boolean isPartialOrder() {
        return isReflexive() && isAntisymmetric()
            && isTransitive();
    }
}
```

Chapter summary

Relations are our first applications of sets, and they are useful in describing many phenomena in mathematics, computer science, and the world around us. To master proofs about relations, one must pay careful attention to the quantification in the definitions. Representing relations and computing their properties is also an interesting application of functional programming.

Key ideas:

Relations are sets of ordered pairs. They also can be thought of as operators or as predicates, and we can visualize them using digraphs.

We can represent relations in ML using lists of tuples.

The image of an element under a relation is the set of elements to which it is related.

The inverse of a relation is the relation formed by reversing all the tuples in the original relation.

The composition of two relations is the relation which is the set of tuples (a, c) where there exists some element b such that (a, b) is the the first relation and (b, c) is in the second.

The transitive closure of a relation is the relation that is the smallest transitive superset of the original relation.

A relation is antisymmetric if no distinct elements are mutually related.

A partial order relation is reflexive, antisymmetric, and transitive. A strict partial order relation is irreflexive, antisymmetric, and transitive. A poset is a set together with a partial order relation.

A relation is reflexive if every element is related to itself.

A relation is symmetric if for every element a that is related to element b, b is also related to a.

A relation is transitive if for all elements a, b, and c, if a is related to b and b is related to c, then a is also related to c.

An equivalence relation is reflexive, symmetric, and transitive. The set of elements that an element a is related to in an equivalence relation is the equivalence class of a. Equivalence classes make a partition of the set. Conversely, any partition of a set can be used to define an equivalence class, called the relation induced by the partition.

Two elements a and b in a poset are comparable if either (a, b) or (b, a) are related. A total order relation is a partial order relation where every two elements are comparable.

A topological sort of a partial order relation is a total order relation that is a superset of the partial order relation. Informally, a topological sort disambiguates the incomparable elements in a poset.

Chapter 6 Self Reference

The chambered nautilus is a mollusk with a shell composed of a series of chambers to which the animal adds throughout its life. Whenever it has outgrown the chamber in which it is living, it builds a new chamber at the entrance to the shell, proportional to the others but larger. (The older chambers become filled with air and give the mollusk buoyancy.) Over time, these chambers form a spiral pattern which can be seen if the shell is cut at a cross section. Many people find it beautiful.

We can define a "chambered nautilus shell" as being either

- The initial one-chambered shell of a developing chambered nautilus, or
- An outermost chamber together with the inner chambered nautilus shell.

Thus we have a recursive definition of what the shell is. The base-case is the innermost chamber, which by itself is a chambered nautilus shell. A non-initial chamber is not a shell by itself, but it can be combined with another shell to make a new shell.

This chapter explores the use of self-reference in programming and mathematics. There is nothing inherently self-referential about the chambered nautilus shell—it is not a fractal, since the chambers are not miniatures of the entire shell. It is our *description* of the shell which depends on itself.

We have seen this pattern already in algorithms which reduce the problem to a smaller version of the original. Now we will see it in types that structure data in a way so that a type can contain a component of the same type. We will also see the pattern in proofs that rely on the very proposition they are proving.

Chapter goals. The student should be able to

- write recursive datatypes and functions that operate on them.
- reason about full binary trees.
- write mutually recursive functions.
- prove results about recursively defined sets using structural induction.
- prove results about whole numbers using mathematical induction.
- prove results about constructed sets using mathematical induction.
- prove that a recursive function is correct using structural or mathematical induction.
- implement several well-known sorting algorithms and prove certain aspects of their correctness.
- write functions that use ML's reference variables and iteration constructs, as an alternative to recursion.
- prove loop invariants for iterative algorithms.

6.1 Peano numbers

Italian mathematician Giuseppe Peano introduced a way of reasoning about whole numbers that includes the following axioms:

Axiom 7 *There exists a whole number 0.*

Axiom 8 *Every whole number n has a successor,* `succ` n.

successor

Axiom 9 *No whole number has 0 as its successor.*

Axiom 10 *If $a, b \in \mathbb{W}$, then $a = b$ iff* `succ` a = `succ` b.

If we interpret *successor* to mean "one more than," then these axioms allow us to define whole numbers recursively (called *Peano numbers*); a whole number is

Peano numbers

- zero, or
- one more than another whole number.

As with the `fun` form for declaring functions, the `datatype` form may refer to the type being defined.

```
- datatype wholeNumber = Zero | OnePlus of wholeNumber;
```

When we wrote self-referential functions or algorithms, they were defined as processes which contained smaller versions of themselves. They were recursive *verbs*. A self-referential datatype is a recursive *noun*. The recursive part establishes a pattern for generating every possible whole number. For example,

5 is a whole number because it is the successor of
 4, which is a whole number because it is the successor of
  3, which is a whole number because it is the successor of
   2, which is a whole number because it is the successor of
    1, which is a whole number because it is the successor of
     0, which is a whole number by Axiom 7.

Biography: Giuseppe Peano, 1858-1932

Giuseppe Peano was raised on a farm in Piedmont, a region of northern Italy. He studied and spent his career at the University of Turin. Many of his important discoveries came about by noticing subtle flaws in widely accepted definitions and results. In doing so he contributed to set theory, mathematical logic, and foundations of mathematics. As a side project, he developed a language based on Latin vocabulary but with a very simple and regular grammar. His intent was for it to be used for international communication. [22]

In ML,

```
- val five = OnePlus(OnePlus(OnePlus(OnePlus(OnePlus(Zero)))));

val five = OnePlus (OnePlus (OnePlus (OnePlus (OnePlus Zero))))
  : wholeNumber

- val six = OnePlus(five);

val six = OnePlus (OnePlus (OnePlus (OnePlus (OnePlus
    (OnePlus Zero))))) : wholeNumber
```

Finding the successor of a number is just a matter of tacking "OnePlus" to the front, a process easily automated.

```
- fun succ(num) = OnePlus(num);
```

Conversion from an int to a wholeNumber is a recursive process—the base case, 0, can be returned immediately; for any other case, we add one to the wholeNumber representation of the int that comes before the one we are converting. (Negative ints will get us into trouble.)

```
- fun asWholeNumber(0) = Zero
=    | asWholeNumber(n) = OnePlus(asWholeNumber(n-1));

val asWholeNumber = fn : int -> wholeNumber
```

predecessor

Notice how subtracting one from the int and adding one to the resulting wholeNumber balance each other off. Opposite the successor, we define the *predecessor* of a natural number n, $\texttt{pred}(n)$, to be the number of which n is the successor. From Axiom 10 we can prove that the predecessor of a number, if it exists, is unique; Axiom 9 says that 0 has no predecessor. Pattern-matching makes stripping off a "OnePlus" easy:

```
- fun pred(OnePlus(num)) = num;

Warning: match nonexhaustive
          OnePlus num => ...

val pred = fn : wholeNumber -> wholeNumber
```

The "match nonexhaustive" warning is okay; we truly want to leave the operation undefined for Zero. Using the function on Zero, rather than failing to define it for Zero, would be the mistake.

```
- val three = asWholeNumber(3);
```

```
val three = OnePlus (OnePlus (OnePlus Zero)) : wholeNumber

- pred(three);

val it = OnePlus (OnePlus Zero) : wholeNumber

- pred(Zero);

uncaught exception nonexhaustive match failure
```

Now we can start defining arithmetic recursively. `Zero` will always be our base case. Anything we add to `Zero` is just itself. For other numbers, picture an abacus. We have two wires, each with a certain number of beads pushed up. At the end of the computation, we want one of the wires to contain our answer. Thus we push down one bead from the other wire, bring up one bead on the answer wire, and repeat until the other wire has no beads left. In other words, we define addition similarly to our recursive gcd lemmas from Section 4.10.

$$\begin{aligned} 0+b &= b \\ a+0 &= a \\ a+b &= (a+1)+(b-1) \quad \text{if } b \neq 0 \end{aligned}$$

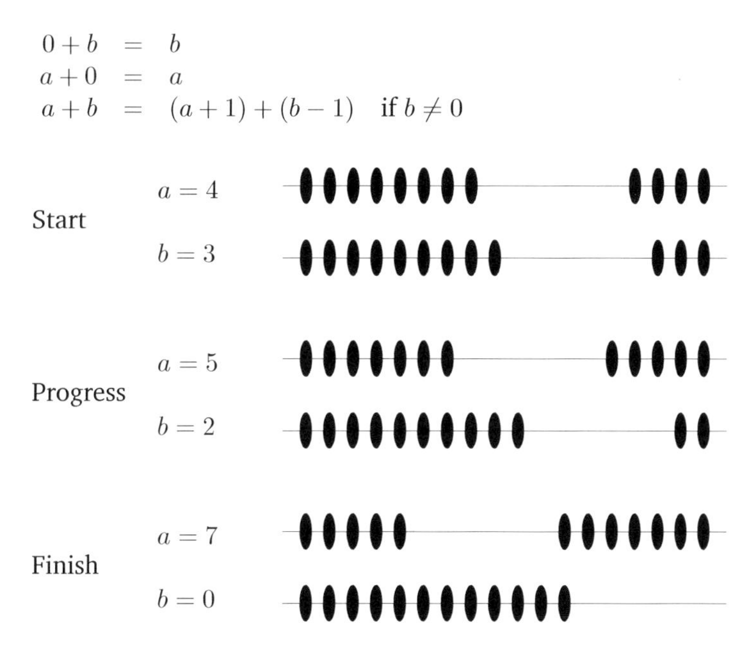

In ML,

```
- fun plus(Zero, num) = num
=   | plus(num, Zero) = num
=   | plus(num1, OnePlus(num2)) = plus(OnePlus(num1), num2);
```

Next we define `isLessThanOrEq`. Examine for yourself the similar structure. Keep in mind that recursively defined predicates have two base cases, one true and one false. Here the first and second parameters are in a survival contest; they repeatedly shed a `OnePlus`, and the first one reduced to `Zero` loses.

```
- fun isLessThanOrEq(Zero, num) = true
=     | isLessThanOrEq(num, Zero) = false
=     | isLessThanOrEq(OnePlus(num1), OnePlus(num2)) =
=         isLessThanOrEq(num1, num2);
```

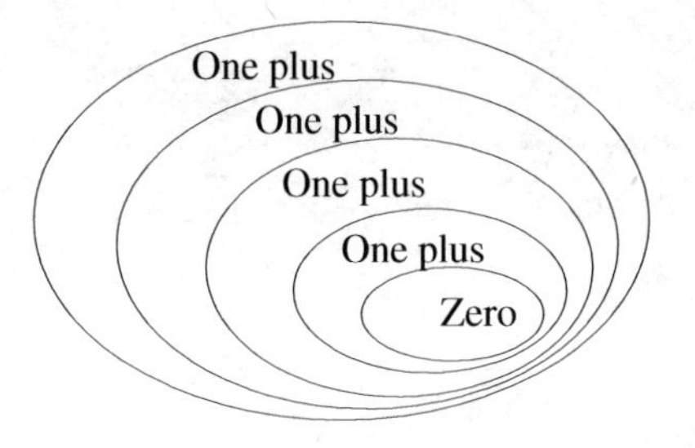

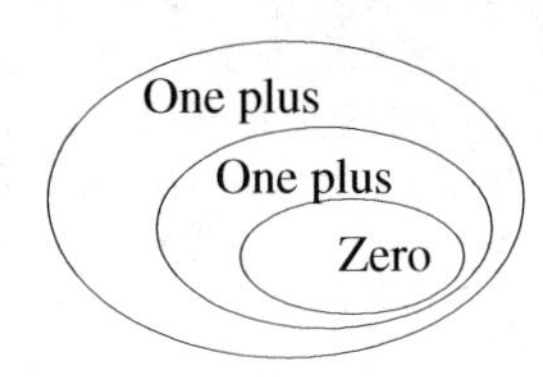

What happens if the two numbers we are comparing are equal? In other words, what if they both strip down to zero simultaneously? The result should be `true`, since it is true that $7 \leq 7$. The expression `isLessThanOrEq(Zero, Zero)` will match to the first pattern, so it will indeed return `true`. Notice how important the order of the first two patterns is. Swapping them would give us `isLessThan`:

```
- fun isLessThan(num, Zero) = false
=     | isLessThan(Zero, num) = true
=     | isLessThan(OnePlus(num1), OnePlus(num2)) =
=         isLessThan(num1, num2);
```

For subtraction, we observe

$$
\begin{aligned}
a - 0 &= a \\
a - b &= (a-1) - (b-1) \quad \text{if } a \neq 0 \text{ and } b \neq 0
\end{aligned}
$$

In ML,

Tangent: "A Sheep in the Deep"

In a 1962 animated short starring sheep-stealing Ralph Wolf and his nemesis Sam Sheepdog, Ralph walks off with a sheep, thinking Sam is sleeping. The sheep, however, turns out to be Sam, who unzips a sheep costume. No matter, Ralph unzips too, revealing himself not to be Ralph after all, but a sheep in a wolf costume. Sam counters by unzipping a sheepdog costume—it turns out he is actually Ralph. The costume shedding continues until one turns at his core to be Ralph—and the other, a stick of dynamite.

```
- fun minus(num, Zero) = num
=   | minus(OnePlus(num1),  OnePlus(num2)) =
=       minus(num1, num2);


Warning: match nonexhaustive
val minus = fn : wholeNumber * wholeNumber -> wholeNumber
```

This rightly leaves the pattern `minus(Zero, OnePlus(num))` undefined. Finally, conversion back to int is just a literal interpretation of the identifiers we gave to the constructor expressions.

```
- fun asInt(Zero) = 0
=   | asInt(OnePlus(num)) = 1 + asInt(num);
```

Exercises

6.1.1 Write a function `isEven` for the wholeNumber type.

6.1.2 Write a function `multiply` for the wholeNumber type.

6.1.3 Write a function `divide`, performing integer division, for the wholeNumber type. (Hint: The Division Algorithm from Exercise 4.10.5 is a good place to start.)

6.1.4 Write a function `modulo`, performing the `mod` operation, for the wholeNumber type.

6.1.5 Write a function `gcd`, computing the greatest common divisor, for the wholeNumber type.

Natural numbers that are powers of 2 can be defined as

- 1, or
- 2 times a power of 2.

We can represent this in ML as

```
datatype powerOfTwo =
     One | TwoTimes of powerOfTwo;
```

6.1.6 Write a function `multiply` that takes two powerOfTwos and returns another powerOfTwo, the product of the two given.

6.1.7 Write a function `asPowerOfTwo`, converting an int to the nearest powerOfTwo less than or equal to it.

6.1.8 Write a function `asInteger`, converting a powerOfTwo to an equivalent int

6.1.9 Write a function `logBase2`, computing an int base 2 logarithm from a powerOfTwo (that is, the type of this function should be powerOfTwo -> int).

If ML did not come with a list construct, we could define our own using the datatype

```
datatype  homemadeList =
        Null | Cons of int * homemadeList;
```

6.1.10 Write a function `head` for homemadeList, equivalent to the ML primitive `hd`.

6.1.11 Write a function `tail` for homemadeList, equivalent to the ML primitive `tl`.

6.1.12 Write a function `cat` to concatenate two homemadeLists. equivalent to the ML primitive `@`.

6.1.13 Write a function `contains` for homemadeList.

6.1.14 Write a function `makeNoRepeats` for homemadeList.

6.1.15 Write a function `sum` for homemadeList.

6.2 Trees

Trees are useful mechanisms for organizing information. They are used in many contexts and fields of study. Common uses include genealogical trees and the phone trees used by customer service numbers.

tree

node

link

root

leaf

internal node

parent

child

A *tree* is a collection of *nodes* and *links*. We will take *node* to be a primitive. A link is an ordered pair of nodes, and each node is the second node in exactly one link, except for one special node which is not the second node in any link; we call this special node the *root* of the tree. Nodes that are not the first node in any link are called *leaves*; all other nodes (nodes which are the first node in at least one link) are called *internal nodes*. To avoid the awkward phrasing of "being the first node in a link," we will say that there exists a link *from* one node *to* another, and we will call the first node the *parent* and the other the *child*.

If relations are still on your mind, you may notice that we could define trees as a kind of relation: the nodes are the set that the relation is on, and the links are the pairs in the relation. By a "kind" of relation, we mean a relation with a restriction like, for one special node (the root), there exists a path (a sequence of tuples matching up like $(a, b), (b, c), (c, d)$) to every other element in the set, and for all elements in the set, no path starting from that element leads back to the element. This restriction is hard to state formally, though. See Exercise 6.2.1.

We have seen trees before. Recall our proposition type from our argument verifier of Section 3.15.

```
datatype proposition = Var of string
                     | Neg of proposition
                     | BinOp of binLogOp * proposition * proposition;
```

The various proposition values are nodes; variables are leaves, negations have one child, and binary operations have two children. Our language system from Section 2.6 also had values that were technically trees, although the types were not recursive and so the trees were not of arbitrary depth. In the next section, we will make the tree-structure more explicit by extending the system to language structures that are recursive.

full binary tree

One of our main interests is in a specific kind of tree, a *full binary tree* which is either

- a single node with no links, or
- a node together with links to two other full binary trees

(Technically, the designation that it is a *binary* tree means that any node may have at most two children; *full* eliminates the case of a node having exactly one child, that is, it requires internal nodes to have the full two children possible or none at all.)

We can model this in ML with this datatype:

```
- datatype tree = Leaf of int | Internal of (int * tree * tree);
```

Here is an example, with the ML equivalent. In this case we are assuming that each node carries with it an integer as a datum.

```
        5
      /   \
     2     6
    / \   / \
   1   8 2   7
            / \
           3   1
```

```
val t =
    Internal(5, Internal(2, Leaf(1), Leaf(8)),
                Internal(6, Leaf(2),
                            Internal(7, Leaf(3), Leaf(1))));
```

Operations over a tree will be recursive in the structure of the tree. What this means is that we will have two cases (or patterns) for any function: how the function finds its answer for a leaf (the base case) and how the function finds its answer for an internal node (the recursive case). Say we want to sum the numbers in the tree:

```
- fun treeSum(Leaf(x)) = x
=     | treeSum(Internal(x, left, right)) =
=            x + treeSum(left) + treeSum(right);

val treeSum = fn : tree -> int

- treeSum(t);

val it = 35 : int
```

If we wrote a synthetic tree operation (something that returns a tree), we would have trouble seeing our result since ML would abbreviate any tree larger than a certain size; besides, ML's default display for datatypes is not very readable to begin with. Instead, we will use ML's string type. Here is one way to print such trees:

```
- fun printTree(Leaf(x)) =  Int.toString(x)
=   | printTree(Internal(x, left, right)) =
=      "(" ^ Int.toString(x) ^ ": " ^ printTree(left) ^ ", " ^
=                                     printTree(right) ^ ")";

val printTree = fn : tree -> string

- printTree(t);

val it = "(5: (2: 1, 8), (6: 2, (7: 3, 1)))" : string
```

Exercises

6.2.1 We said that defining a tree as a relation is hard, but it is not impossible. Write a definition of *tree* as a relation R on a set A. (Hint: State the restriction in terms of R^T, the transitive closure of the relation.)

6.2.2 Write a function `nodes` that takes a tree and returns the number of nodes in the tree.

6.2.3 Write a function `leaves` that takes a tree and returns the number of leaves in the tree.

6.2.4 Write a function `internals` that takes a tree and returns the number of internal nodes in the tree.

6.2.5 Write a function `contains` that takes a tree and an `int` and determines whether or not the tree contains the `int`.

6.2.6 Write a function `count` that takes a tree and an `int` and determines the number of times the `int` occurs in the tree.

6.2.7 The `height` of a tree is the largest number of links on any path from the root to a leaf. For example, a tree that is just a leaf has height 0, and the tree pictured on the previous page has height 3. Write a function `height` that takes a tree and returns the height.

6.2.8 Write a function `links` that takes a tree and computes the number of links in the tree.

6.2.9 Write functions `min` and `max` that each take a tree and return the least or greatest value contained in the tree, respectively.

6.2.10 Write a function `isInOrder` that takes a tree and determines whether or not the numbers in the tree are in order. They are in order if for any internal node, all the numbers found in the left subtree are less than or equal to it, and all the numbers found in the right subtree are greater than or equal to it. Use your `min` and `max` functions from the previous exercise.

6.2.11 Write a function `reflection` that takes a tree and return a tree that is the mirror image of the given tree.

Return tree height:
fun height (Leaf(x)) = 0
| height (Internal(x, left, right)) =
let val l = 1 + height(left)
val r = 1 + height(right)
fun max (x, y) = if x > y then x else y
in max (l, r) end;

6.2.12 Write a function `primeFactTree` that takes an `int` and returns a prime factorization tree for that number. For example, a prime factorization tree of 12 would be `(12: 2, (6 : 2, 3))` or `(12: (4: 2, 2), 3)`.

6.2.13 Write a function `halveEvens` that takes a tree and returns a tree like the one given except that all even numbers are replaced by their halves.

We can use the following datatypes to model four-function mathematical expressions:

```
datatype operation = Plus | Minus | Mul | Div;
datatype expression =
        Internal of operation *
                       expression * expression
      | Leaf of int;
```

For example, $((5-7)*((3+2)/8))$ would be

```
Internal(Mul,Internal(Minus,Leaf(5),Leaf(7)),
            Internal(Div,
                       Internal(Plus,Leaf(3),
                                         Leaf(2)),
                       Leaf(8)));
```

6.2.14 Write a function `printExpression` that takes an expression and returns a string displaying it in normal mathematical formatting. The result will have to be fully-parenthesized to avoid confusion over order of operations. See Section 2.5 to review the string type.

6.2.15 Write a function `execute` that takes an expression and evaluates it.

6.2.16 Write a function `numOperators` that takes an expression and computes the number of operators.

6.2.17 Write a function `numNumbers` that takes an expression and computes the number of literal numbers.

6.3 Mutual recursion

Recall our language processing system from Section 2.6. Suppose we wanted to extend the system to accept sentences like *The unicorn knew that the big dog was smelly.* What is different about this?

The independent clause *the big dog was smelly*, which would be a sentence by itself, is embedded in the larger sentence, and it is the direct object of the verb *knew*. What we want to introduce to the language are independent clauses (sentences or proposition prefixed with *that*) and verbs that take propositions as their direct object.

In theory, *that*- introduced propositions are full-fledged nouns, just as boolean-valued expressions are full-fledged expressions. However, since only a small number of verbs make sense with propositions as direct objects (The woman *proved* that a red dog ran through the field) or propositions as subjects (That a flea bit the unicorn *concerned* the man), we will treat those verbs specially. Also, for simplicity's sake, we restrict our discussion to verbs taking propositions as direct objects.

A sentence contains a predicate which contains a verb phrase. One such kind of verb phrase is a transitive propositional-object verb phrase, which contains a sentence. Thus a sentence is defined in terms of itself, but not directly: a sentence is defined in terms of a verb phrase, which is in turn defined in terms of a sentence. This is *mutual recursion*, where two or more things are defined in terms of each other.

mutual recursion

Consider two examples of pairs of simple functions that are mutually recursive. Suppose we want to cut down a list to make a new list containing every other element in the original list. For example, we would transform `[6, 3, 12, 8, 4, 2, 7]` to `[6, 12, 4, 7]`. We can decompose this process into two parts: taking an element to add to the new list and skipping an element. If we take, then we want to skip next time, and vice versa. (This example is from Ullman [34].)

```
- fun take([]) = []
=   | take(x::rest) = x::skip(rest)
= and skip([]) = []
=   | skip(y::rest) = take(rest);


val take = fn : 'a list -> 'a list
val skip = fn : 'a list -> 'a list

- take([6, 3, 12, 8, 4, 2, 7]);


val it = [6,12,4,7] : int list
```

There are two differences: the `fun` of the second function is replaced with `and`, and only the last function is terminated by a semicolon.

As a second example, we can rewrite isEven and isOdd to be mutually recursive:

```
- fun isEven(0) = true
=     | isEven(n) = isOdd(n-1)
= and isOdd(0) = false
=    | isOdd(n) = isEven(n-1);


val isEven = fn : int -> bool
val isOdd = fn : int -> bool

- isEven(17);


val it = false : bool

- isOdd(17);


val it = true : bool
```

Mutually recursive datatypes also are joined together using and. Much of the code in this language example is the same as from Section 2.6, but first we need to add a datatype for transitive propositional-object verbs:

```
datatype transpoVerb = TPOV of string;
```

Next, we define phrases and sentences:

```
datatype nounPhrase = NounPhrase of
                          (article * adjective option * noun)
and verbPhrase = TVP of (transitiveVerb * nounPhrase)
                     | IVP of (intransitiveVerb)
                     | LVP of (linkingVerb * adjective)
                     | TPOVP of (transpoVerb * sentence)
and predicate = Predicate of (adverb option * verbPhrase)
and prepPhrase = PrepPhrase of (preposition * nounPhrase)
and sentence = Sentence of (nounPhrase * predicate *
                            prepPhrase option);
```

The parsing functions for articles, adjectives, nouns, prepositions, and adverbs can stay the same, and similarly for parseNounPhrase, parsePrepPhrase, parseTransVerb, and parseLinkingVerb. The function parseTransPOVerb is new, and the remaining parse functions (for verb phrases, predicates, and sentences) are reorganized below. The original version of the language processor was written before we had introduced ifs, so all decision making was done by pattern-matching. We can simplify matters now by keeping a list of different kinds of verbs:

```
val transitiveVerbs = ["chased", "saw", "bit", "loved"];
val intransitiveVerbs = ["ran", "slept", "sang"];
val linkingVerbs = ["was", "felt", "seemed"];
val transitivePOVerbs = ["knew", "believed", "proved"];
exception unknownVerb;
```

For reasons to be explained soon, we also make a list of prepositions:

```
val prepositions = ["in", "on", "through", "with"];
```

Now, the remaining parsing code, all at once:

```
fun parseTransPOVerb(vb, wordList) =
    let val (dirObj, rest) = parseSentence(wordList);
    in (TPOVP(vb, dirObj), rest) end

and parseVerbPhrase(vb::rest) =
    if contains(vb, transitiveVerbs)
        then parseTransVerb(TV(vb), rest)
    else if contains(vb, intransitiveVerbs)
        then (IVP(IV(vb)), rest)
    else if contains(vb, linkingVerbs)
        then parseLinkingVerb(LV(vb), rest)
    else if contains(vb, transitivePOVerbs)
           andalso hd(rest) = "that"
        then parseTransPOVerb(TPOV(vb), tl(rest))
    else raise UnknownVerb

and parsePredicate(wordList) =
    let val (adv, rest1) = parseAdverb(wordList);
        val (vPh, rest2) = parseVerbPhrase(rest1);
    in (Predicate(adv, vPh), rest2) end

and parseSentence(wordList) =
    let val (subj, rest1) = parseNounPhrase(wordList);
      val (pred, rest2) = parsePredicate(rest1);
    in
      case rest2 of
        [] => (Sentence(subj, pred, NONE), [])
      | next::rest3 =>
        if contains(next, prepositions)
        then let val (prPh, rest4) = parsePrepPhrase(rest3);
             in (Sentence(subj, pred, SOME prPh), rest4)
             end
        else (Sentence(subj, pred, NONE), rest3)
    end
```

```
and parseString(message) =
    #1(parseSentence(String.tokens(fn(x) => not(Char.isAlpha(x)))
                              (message)));
```

We separated the old parseSentence (which took a sentence-string and chopped it up into a list of word-strings) into parseSentence and parseString because having subsentences means that we need to parse sentences from word lists.

In parseSentence, we needed to check whether rest2 (the words following the predicate) began with a preposition and parse the prepositional phrase if it did. The sentence *The man believed the unicorn slept in the field* is technically ambiguous as to whether *in the field* describes the place of the unicorn's sleeping or the place of the man's believing. Our parser associates prepositional phrases with the nearest, smallest sentence, as would be our intuition in the given example.

Now the sentences in our system can be arbitrarily nested.

```
- parseString("the man happily believed that the woman knew
= that the man proved that the man loved the woman");

val it =  Sentence (NounPhrase (Art #,NONE,Noun #),
Predicate (SOME #, TPOVP #),NONE)   : sentence
```

Exercises

6.3.1 What is wrong with the following?

```
fun isEven(0) = true
  | isEven(n) = isOdd(n-1)
and isOdd(1) = true
  | isOdd(n) = isEven(n-1);
```

6.3.2 Suppose you wanted every other item in a list but starting with the second item, not the first—for example, you wanted to extract [3, 8, 2] from [6, 3, 12, 8, 4, 2, 7]. How could you achieve this with code we have already written (that is, without writing a new function)?

A *forest* is a set of trees. We can define a type tree where each node contains an int datum and an arbitrary number of children with

```
datatype tree = Node of int * forest
and forest = Forest of tree list;
```

So, a node's children are viewed as a forest of trees.

6.3.3 Write mutually recursive functions sumTree and sumForest that determine the sum of all the data in a tree or forest.

6.3.4 Write mutually recursive functions heightTree and heightForest that determine the height of a tree or forest.

6.3.5 Write mutually recursive functions leavesTree and leavesForest that determine the number of leaves in a tree or in a forest. Notice that a leaf is a tree node with an empty forest.

6.3.6 Extend the language processor by introducing relative clauses: "The man *who chased the unicorn* saw a dog *that was big*." Restrict the system to relative clauses where the relative pronoun is the subject of the clause (*the dog that chased the ball*, not the *the ball that the dog chased* or *the field in which the dog chased the ball*). Assume relative pronouns *who*, *that*, and *which*. The relative clause should be an optional part of a noun phrase and come last in the noun phrase. Relative clauses can be arbitrarily nested (*the man who loved the woman who walked with the unicorn that chased the dog slept in the bright field*).

6.4 Structural induction

Observe the following about full binary trees:

Tree	Nodes	Links	Tree	Nodes	Links
	1	0	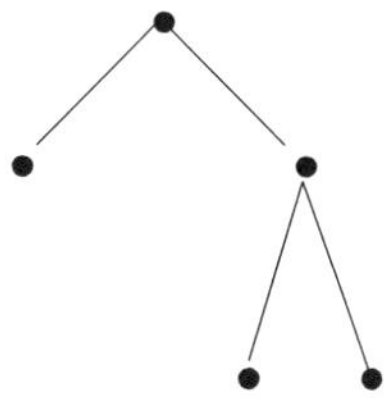	5	4
	3	2	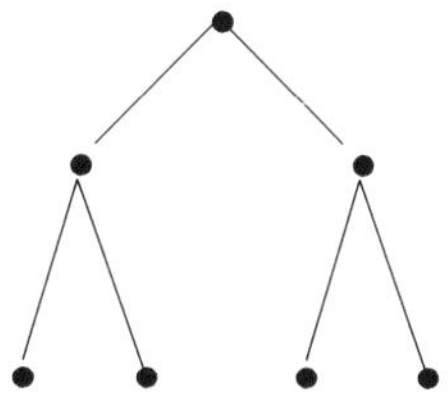		
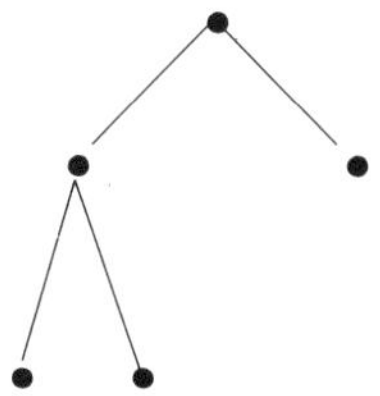	5	4		7	6

Every full binary tree has one more node than it has links. It is not hard to explain why: we can pair up each node with the link above it, and this will account for every link and also every node, except for the root. Thus there is one more node than there are links. But how can we prove this formally?

We also can state our observation by taking into account the fact that trees are built from smaller trees. Any tree (other than a single node) can be made by taking two other trees and linking them together with a new root. This adds two new links and one new node. If the two older trees already each have one more node than links, the new tree will also.

This property is an *invariant*, a proposition that does not vary through changing circumstances. In this case, the fact that the property holds for a tree depends on the property holding for its subtrees. We have a new proof strategy specifically for an invariant over a recursively defined mathematical structure or set.

invariant

Theorem 6.1 *For any full binary tree T,* $\texttt{nodes}(T) = \texttt{links}(T) + 1$.

> **Proof.** Suppose T is a full binary tree.

This is all we are given. We need to use the definition to analyze what this means.

> By definition of full binary tree, T is either a single node or a node with links to two full binary trees.

Two possibilities. This calls for division into cases. We will use special names for these cases, based on how they correspond to cases in the recursive definition of *full binary tree*.

> **Base case.** Suppose T is a single node. Then it has one node (itself) and no links, that is $\texttt{nodes}(T) = 1$ and $\texttt{links}(T) = 0$. Thus $\texttt{nodes}(T) = \texttt{links}(T) + 1$.
>
> **Inductive case.** Suppose T is a node with links to two other full binary trees, call them T_1 and T_2. Since T adds one node and two links to the subtrees, $\texttt{nodes}(T) = \texttt{nodes}(T_1) + \texttt{nodes}(T_2) + 1$ and $\texttt{links}(T) = \texttt{links}(T_1) + \texttt{links}(T_2) + 2$.

Here is the new part. From what we said earlier, we know T_1 and T_2 each have one more node than links. How do we know that, formally? The theorem itself tells us.

> By *structural induction*, $\texttt{nodes}(T_1) = \texttt{links}(T_1)+1$ and $\texttt{nodes}(T_2) = \texttt{links}(T_2) + 1$. Then

$$\begin{aligned}
\texttt{nodes}(T) &= \texttt{nodes}(T_1) + \texttt{nodes}(T_2) + 1 && \text{as stated above}\\
&= \texttt{links}(T_1) + 1 + \texttt{links}(T_2) + 1 + 1 && \text{by substitution}\\
&= \texttt{links}(T_1) + \texttt{links}(T_2) + 2 + 1 && \text{by algebra}\\
&= \texttt{links}(T) + 1 && \text{by substitution}
\end{aligned}$$

> Either way, $\texttt{nodes}(T) = \texttt{links}(T) + 1$. □

Why does that work? It is the same principle behind recursive algorithms and recursive structures. We can apply the proof of this theorem to the subtrees, which requires it to be applied to their subtrees, until we reach the leaves.

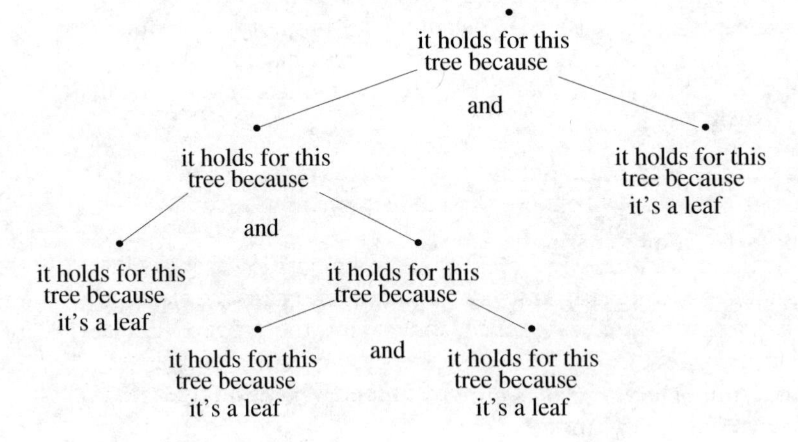

In our proof, we cite this as "by *structural induction*." This is the proof technique that performs a division into cases based on the structure of a recursively defined set. The proposition we proved can be broken down to a predicate

structural induction

$$I(T) = \quad (\texttt{nodes}(T) = \texttt{links}(T) + 1)$$

wrapped in a universally quantified proposition

$$\forall\, T \in \mathcal{T}, I(T).$$

where $\mathcal{T}$ is the set of full binary trees. Therefore, propositions which are predicates universally quantified over a recursively defined set are candidates for proof by structural induction.

When we cite our claims about $\texttt{nodes}(T_1)$ by structural induction, we are making a convenient shorthand for how this proof really works. Done more carefully, this proof really deals with that certain set of full binary tress (a subset of $\mathcal{T}$) on which the invariant I holds: $\{T \in \mathcal{T} \mid I(T)\}$. We can rephrase the proposition as saying that all full binary trees are in this set:

$$\mathcal{T} = \{T \in \mathcal{T} \mid I(T)\}$$

The proof shows how our knowledge of the set grows to include larger and larger trees, implying that *every* tree is in the set. Here is a proof again, without shorthands:

> **Proof.** Suppose T is a full binary tree. By definition of full binary tree, T is either a single node or a node with links to two full binary trees.
>
> **Base case.** Suppose T is a single node. Then it has one node (itself) and no links, that is $\texttt{nodes}(T) = 1$ and $\texttt{links}(T) = 0$. Thus $\texttt{nodes}(T) = \texttt{links}(T) + 1$, and there exists a set of sufficiently small full binary trees each of whose number of nodes is one more than its number of links.

So far we have established that that set is not empty.

> **Inductive case.** Suppose T is a node with links to two other full binary trees, call them T_1 and T_2, and that for all trees smaller than T, their number of nodes is one more than their number of links.

That last part is called our *inductive hypothesis*, the supposition that the invariant holds on elements of the set that are simpler or smaller than the element we are currently concerned about. If we have shown that the property holds for a certain nonempty set of "small" trees (and the base case has established the existence of such a set), then we can show the property holds on trees that can be built from them.

inductive hypothesis

Since T adds one node and two links to the subtrees, $\texttt{nodes}(T) = \texttt{nodes}(T_1)+\texttt{nodes}(T_2)+1$ and $\texttt{links}(T) = \texttt{links}(T_1)+\texttt{links}(T_2)+2$.

By the inductive hypothesis,

$\texttt{nodes}(T_1) = \texttt{links}(T_1) + 1$ and $\texttt{nodes}(T_2) = \texttt{links}(T_2) + 1$. Then

$$\begin{array}{rcll} \texttt{nodes}(T) & = & \texttt{nodes}(T_1) + \texttt{nodes}(T_2) + 1 & \text{as stated above} \\ & = & \texttt{links}(T_1) + 1 + \texttt{links}(T_2) + 1 + 1 & \text{by substitution} \\ & = & \texttt{links}(T_1) + \texttt{links}(T_2) + 2 + 1 & \text{by algebra} \\ & = & \texttt{links}(T) + 1 & \text{by substitution} \end{array}$$

Therefore, by structural induction, $\texttt{nodes}(T) = \texttt{links}(T) + 1$. □

Observe that formally the fact that the proposition holds for T_1 and T_2 comes from our inductive hypothesis. Structural induction is cited for combing the base and inductive cases to conclude that the proposition holds for all full binary trees—an "either way" uniting a division into cases is lurking under the surface. Our original, less formal version gave a top-down intuition, whereas the formal approach affords imagining the process from the bottom up.

It will take time to wrap your mind around structural induction and the related techniques in the next sections. It may help to think of these proofs as recipes for how to construct a specific proof for a specific member of a set. That is true not only for inductive proofs, however—the proof of any universally quantified proposition is a recipe for proving the predicate for any given element of the set.

Exercises

Prove using structural induction.

6.4.1 For any full binary tree T, $\texttt{leaves}(T) = \texttt{internals}(T) + 1$.

6.4.2 For any full binary tree T, $\texttt{leaves}(T) \leq 2^{\texttt{height}(T)}$.

6.4.3 For any full binary tree T, $\texttt{nodes}(T) \leq 2^{\texttt{height}(T)+1} - 1$.

6.4.4 For any full binary tree T, $\texttt{nodes}(T)$ is odd. (Prove this directly, using structural induction. Do not use Theorem 6.1 or an earlier exercise as a lemma.)

Following the Peano definition of the whole numbers, define a whole number as either 0 or $succ(x)$ where x is a whole number. Also assume the following axioms:

1. For all $x \in \mathbb{W}, 0 \leq x$
2. For all $x, y \in \mathbb{W}$, if $x \leq y$ then $succ(x) \leq succ(y)$.
3. For all $x \in \mathbb{W}, x + 0 = x$ and $0 + x = x$.
4. For all $x, y \in \mathbb{W}$, $x + succ(y) = succ(x) + y$

Prove using structural induction.

6.4.5 For all $x \in \mathbb{W}$, $x \leq succ(x)$.

6.4.6 For all $x, y \in \mathbb{W}$, $x + succ(y) = succ(x + y)$. (Hint: Perform structural induction on x. The proposition you want to prove for any x is that $\forall\, y \in \mathbb{W}, x + succ(y) = succ(x + y)$.)

6.4.7 Let the set S be defined so that for all $s \in S$, either

- $s = 3$, or
- $s = t + u$ for some $t, u \in S$.

Prove, using structural induction, that for all $s \in S$, $3|s$.

6.5 Mathematical induction

This section is one of our occasional forays into the world of integers. Consider numbers in the form $3^n - 1$ for $n \in \mathbb{W}$.

n	0	1	2	3	4	5	6	7	8
$3^n - 1$	0	2	8	26	80	242	728	2186	6560

From these examples, one might expect that all numbers in that form are even. Not a surprising result, in fact, since by eyeballing it we can tell that 3^n will be odd and so $3^n - 1$ will be even. Let us wrap that in a predicate

$$I(n) = 3^n - 1 \text{ is even.}$$

And our claim becomes $\forall\, n \in \mathbb{W}, I(n)$.

If we take a specific example, say $3^4 - 1$, we see

$$\begin{array}{rcll} 3^4 - 1 & = & 3 \cdot 3^3 - 1 & \text{pull out 3} \\ & = & 3 \cdot (3^3 - 1 + 1) - 1 & \text{the positive and negative 1 cancel out} \\ & = & 3 \cdot (3^3 - 1) + 3 - 1 & \text{distribute the 3} \\ & = & 3 \cdot (3^3 - 1) + 2 & \text{simplify } 3 - 1 \end{array}$$

This might seem like a haphazard rearranging of values, but what it does is relate $3^4 - 1$ to $3^3 - 1$, and moreover it relates $I(4)$ to $I(3)$. If we knew $I(3)$ were true, a simple manipulation would yield $I(4)$.

Lemma 6.2 *If $I(3)$, then $I(4)$.*

Proof. Suppose $I(3)$, that is, $3^3 - 1$ is even. By definition of even, $3^3 - 1 = 2 \cdot k$ for some $k \in \mathbb{Z}$. Then,

$$\begin{array}{rcll} 3^4 - 1 & = & 3 \cdot (3^3 - 1) + 2 & \text{as we showed above} \\ & = & 3 \cdot (2 \cdot k) + 2 & \text{by substitution} \\ & = & 2 \cdot (3 \cdot k + 1) & \text{by algebra} \end{array}$$

Since $3 \cdot k + 1 \in \mathbb{Z}$, $3^4 - 1$ is even by definition. □

In Section 6.1, we learned a recursive construction for the whole numbers. In Section 6.4, we learned a proof technique for propositions universally quantified over recursively defined sets. Let us put these things together to prove

Theorem 6.3 *For all $n \in \mathbb{W}$, $3^n - 1$ is even.*

Proof. Suppose $n \in \mathbb{W}$. By definition of whole number (from Section 6.1), either $n = 0$ or $n = m + 1$ where $m \in \mathbb{W}$.

Base case: Suppose $n = 0$. Then $3^n - 1 = 3^0 - 1 = 1 - 1 = 0 = 2 \cdot 0$, which is even by definition, and so there exists at least one $n \in \mathbb{W}$ such that $3^n - 1$ is even.

Inductive case: Suppose $n = m + 1$ where $m \in \mathbb{W}$ and $3^m - 1$ is even. By definition of even, $3^m - 1 = 2 \cdot k$ where $k \in \mathbb{Z}$. Then, by algebra and substitution,

$$\begin{aligned}
3^n - 1 &= 3^{m+1} - 1 \\
&= 3 \cdot 3^m - 1 \\
&= 3 \cdot (3^m - 1 + 1) - 1 \\
&= 3 \cdot (2 \cdot k + 1) - 1 \\
&= 3 \cdot 2 \cdot k + 3 - 1 \\
&= 3 \cdot 2 \cdot k + 2 \\
&= 2 \cdot (3 \cdot k + 1)
\end{aligned}$$

which is even by definition.

By structural induction, $3^n - 1$ is even. □

When using structural induction to prove a predicate $I(n)$ quantified over all whole numbers, the proof has the following outline:

- Show that $I(0)$.
- Show that $\forall\, n \in \mathbb{W}$ such that $n > 0, I(n-1) \rightarrow I(n)$.
- Conclude that $\forall\, n \in \mathbb{W}, I(n)$.

All this depends on the inference that the first two propositions above imply the third. It also depends on the second proposition being easier to prove than the third, which it generally is, even though it looks more intimidating. We could also use this method for proving predicates quantified over natural numbers—show that $I(1)$ and that for $n > 1$, $I(n-1) \rightarrow I(n)$.

The specialization of structural induction for whole or natural numbers is called *mathematical induction*. Seemingly a strange name, does that suggest structural induction otherwise is not mathematical? Actually, the name reflects the fact that induction over whole numbers came first historically. It is also a standard tool in mathematics, whereas general structural induction is used in only a handful of areas. Peano numbers are not necessary for understanding mathematical induction. You may leave out the line "By definition of a whole number, n is either..." in your proofs.

mathematical induction

A favorite way for instructors to introduce mathematical induction is comparing to a ladder or stairs. The base case proves you can get on the first rung. The inductive case shows that as long as you can get to a certain rung, you can get to the next one. Altogether, this means you can get to any rung. (The analogy breaks down if the ladder-climber gets tired.)

Try another example. Whole numbers are called *perfect squares* if they are the square of two integers, say $1^2 = 1$, $2^2 = 4$, $3^2 = 9$, and $4^2 = 16$. They also can be drawn as squares of dots:

If we start with $1^1 = 1$, notice that we add three dots along the bottom and right sides to get $2^2 = 4$.

Going from $2^2 = 4$ to $3^2 = 9$ is like adding five dots along the bottom and right.

And we add seven dots to make $4^2 = 16$.

Thus

$$\begin{array}{rclcl} 1^2 & = & 1 & = & 1 \\ 2^2 & = & 4 & = & 1+3 \\ 3^2 & = & 9 & = & 1+3+5 \\ 4^2 & = & 16 & = & 1+3+5+7 \end{array}$$

It seems like that for any natural number n, n^2 is equal to the sum of the first n odd numbers. We conjecture

Theorem 6.4 $\forall\, n \in \mathbb{N}, \sum_{i=1}^{n}(2i-1) = n^2$.

We will prove this in a slightly different format from our previous example, one that is more standard for math induction proofs. To indicate the structure of the proof, we begin by saying "By induction on n." This informs the reader that a base case and an inductive case are coming. For the base case, we assume n is 1, and prove the predicate. For the inductive case, we assume that the predicate works for some value (call it N to distinguish it from n) and show that it works for the next value. This assumption is safe because we already know there is at least one value for which it works—we showed that in the base case. Generally we specify that N is some number greater than or equal to 1 for clarity, but this really is not necessary. As in the formal version of structural induction, our assumption in the inductive case is called the *inductive hypothesis*, and that is how we cite it when using it to justify something.

Proof. By induction on n.

Base case. Suppose $n = 1$. Then $\sum_{i=1}^{1}(2i - 1) = 1 = 1^2$ by definition of summation and by arithmetic.

Inductive case. Suppose $\sum_{i=1}^{N}(2i - 1) = N^2$ for some $N \geq 1$. Then

$$\begin{aligned}
\sum_{i=1}^{N+1}(2i-1) &= \sum_{i=1}^{N}(2i-1) + 2(N+1) - 1 && \text{by definition of summation} \\
&= N^2 + 2(N+1) - 1 && \text{by the inductive hypothesis} \\
&= N^2 + 2N + 1 && \text{by distribution and simplifi-} \\
&= (N+1)^2 && \text{by factoring}
\end{aligned}$$

Therefore, by math induction, $\sum_{i=1}^{n}(2i - 1) = n^2$ for all $n \in \mathbb{N}$. □

Tangent: Induction

Induction is a reasoning process where one argues for a conclusion because it is consistent with or best explains the evidence. The best example of induction is the scientific method: an observation leads to a hypothesis; many controlled observations strengthen (or falsify) a theory; reproduced experiments support a law.

Mathematics is different among the sciences. Its results are proven by deduction meaning that a new proposition follows absolutely from prior propositions. *Mathematical induction* is so called because it gives a recipe for doing an indefinite number of observations with a guarantee that the results always will confirm the proposition. Math induction is still deduction, though, not induction.

Induction, in the general sense, is still useful in Mathematics. George Pólya wrote an accessible study in how mathematical observation can lead to theorems and proofs, and also explains mathematical induction in this context [27].

Exercises

6.5.1 Notice that

$$\begin{array}{lllll}\sum_{i=1}^{1} i &=& 1 &=& 1 &=& \frac{1\cdot 2}{2}\\ \sum_{i=1}^{2} i &=& 1+2 &=& 3 &=& \frac{2\cdot 3}{2}\\ \sum_{i=1}^{3} i &=& 1+2+3 &=& 6 &=& \frac{3\cdot 4}{2}\\ \sum_{i=1}^{4} i &=& 1+2+3+4 &=& 10 &=& \frac{4\cdot 5}{2}\\ \sum_{i=1}^{5} i &=& 1+2+3+4+5 &=& 15 &=& \frac{5\cdot 6}{2}\end{array}$$

Using math induction, prove that for all $n \in \mathbb{N}$, $\sum_{i=1}^{n} i = \frac{n(n+1)}{2}$. Is an example of an *arithmetic series*.

6.5.2 Notice that

$$\begin{array}{lllllll}4^0-1 &=& 1-1 &=& 0 &=& 3\cdot 0\\ 4^1-1 &=& 4-1 &=& 3 &=& 3\cdot 1\\ 4^2-1 &=& 16-1 &=& 15 &=& 3\cdot 5\\ 4^3-1 &=& 64-1 &=& 63 &=& 3\cdot 21\end{array}$$

Using math induction, prove that for all $n \in \mathbb{W}$, $3 \mid 4^n - 1$.

6.5.3 We've observed that $2 \mid 3^n - 1$ (Theorem 6.3) and $3 \mid 4^n - 1$ (Exercise 6.5.2). Generalize this. Using math induction, prove that for all $n \in \mathbb{N}$ and for all $x \in \mathbb{Z}$ such that $x > 1$, $x - 1 \mid x^n - 1$.

6.5.4 Use mathematical induction to prove that for all $n \in \mathbb{W}$, $\sum_{i=0}^{n} 3 \cdot 5^i = \frac{3(5^{n+1}-1)}{4}$. You may find it useful to try a few values of n by hand first.

6.5.5 We can generalize the previous result for all *geometric series*, series in the form $\sum_{i=0}^{n} a \cdot r^i$, for some $a, r \in \mathbb{R}$. Use mathematical induction to prove that for $n \in \mathbb{W}$, $\sum_{i=0}^{n} a \cdot r^i = \frac{a(r^{n+1}-1)}{r-1}$.

6.6 Mathematical induction on sets

Sometimes we will find mathematical induction useful for proving a result whose relationship to whole numbers is not immediately apparent. In this section we will find other opportunities for using math induction, including the proofs of some results that were left unproven earlier in the book.

Take the following observations about the size of powersets, using the `count` function from Exercise 2.2.4 and the `powerset` function from Exercise 2.4.15.

```
- count(powerset([]));

val it = 1 : int

- count(powerset([1]));

val it = 2 : int

- count(powerset([1, 2]));

val it = 4 : int

- count(powerset([1, 2, 3]));
```

```
val it = 8 : int

- count(powerset([1, 2, 3, 4]));

val it = 16 : int
```

Our conjecture is that for all finite sets A, $|\mathscr{P}(A)| = 2^{|A|}$. Since the predicate is quantified over all sets, not all whole or natural numbers, this proposition does not scream for using math induction. But think about the reasons behind this result.

We know from our previous scuffles with powersets that if we distinguish an element in the original set, say $a \in A$, we can partition the powerset $\mathscr{P}(A)$ around that element: $\mathscr{P}(A - \{a\})$, the set of sets that do not contain a; and $\{ C \cup \{a\} \mid C \in \mathscr{P}(A - \{a\}) \}$, the set of sets that do. Let us take $\{1, 2, 3\}$ and pick 3 as our pivot element. Our partition is

$$\{\{1,2\},\{1\},\{2\},\emptyset\} \cup \{\{1,2,3\},\{1,3\},\{2,3\},\{3\}\}$$

Two things to notice: First, the two sets of the partition are the same size. Second, the first set is just $\mathscr{P}(\{1,2\})$. Those observations put together imply that $\mathscr{P}(\{1,2,3\})$ is twice the size of $\mathscr{P}(\{1,2\})$. Every time we add another element to a set, we double the cardinality of the powerset.

Do you see math induction at work here? If we have an empty set, its powerset has one element. If a set of cardinality n has a powerset of cardinality m, then a set of cardinality $n+1$ has a powerset of cardinality $2 \cdot m$. By math induction (and a whole lot of details), a set of cardinality n has a powerset of cardinality 2^n. Preparing the proposition for proof is a matter of restating it so its outermost quantification is over numbers rather than over sets.

Theorem 6.5 *For all $n \in \mathbb{W}$, if A is a set such that $|A| = n$, then $|\mathscr{P}(A)| = 2^n$.*

Proof. By induction on n.

Base case. Suppose $n = 0$. Further suppose A is a set and $|A| = 0$. By definition of the empty set, $A = \emptyset$. Moreover, by definition of powerset, $\mathscr{P}(A) = \mathscr{P}(\emptyset) = \{\emptyset\}$, since $\emptyset$ is the only subset of $\emptyset$. By definition of cardinality, $|\{\emptyset\}| = 1 = 2^0$, so $|\mathscr{P}(A)| = 2^0$.

Inductive case. Suppose the result holds for some $N \in \mathbb{N}$. Now, suppose A is a set such that $|A| = N + 1$. Let $a \in A$.

Time out for a subtlety. The last sentence had a *let*, not a *suppose*. We use *suppose* to introduce a new proposition, like "the result holds for some $N \in \mathbb{N}$" or that there exists a set A such that $|A| = N + 1$. "$a \in A$" assumes no new information. Since $N \geq 0$, we have $N + 1 \geq 1$, and so we know A must have at least one element. We are merely giving that known element a new name, a.

Note that $\{a\} \subseteq A$ by definition of subset and $|\{a\}| = 1$ by definition of cardinality.

The rest is left as an exercise. □

Do you remember the following result that we needed in order to prove that the infinite iterated union of a relation is the relation's transitive closure?

Lemma 5.14 *If R is a relation on a set A, S is a transitive relation on A, $R \subseteq S$, and $i \in \mathbb{N}$, then $R^i \subseteq S$.*

To remember what this means, take an example. Let our set be $A = \{1, 2, 3, 4, 5, 6, 7\}$ and our relation be $R = \{(1, 2), (2, 3), (3, 4), (4, 5), (5, 6), (6, 7)\}$. The other relation, S, can be any superset of R that is transitive—it could be the transitive closure, or it could be something much bigger. Watch what happens as we repeatedly compose R with itself.

$$\begin{array}{lclcl}
R^1 & = & R & = & \{(1,2),(2,3),(3,4),(4,5),(5,6),(6,7)\} \\
R^2 & = & R \circ R & = & \{(1,3),(2,4),(3,5),(4,6),(5,7)\} \\
R^3 & = & R^2 \circ R & = & \{(1,4),(2,4),(3,6),(4,7)\} \\
R^4 & = & R^3 \circ R & = & \{(1,5),(2,6),(3,7)\} \\
R^5 & = & R^4 \circ R & = & \{(1,6),(2,7)\} \\
R^6 & = & R^5 \circ R & = & \{(1,7)\} \\
R^7 & = & R^6 \circ R & = & \{\}
\end{array}$$

All of these, we claim, are subsets of S. Why? Take R^4, and within it take $(3, 7)$. $(3, 7)$ must be in S because $(3, 5)$ and $(5, 7)$ are in S and S is transitive. How do we know $(3, 5)$ and $(5, 7)$ are in S? Because $(3, 4)$, $(4, 5)$, $(5, 6)$, and $(6, 7)$ each are in R, $R \subseteq S$, and S is transitive.

Do you see induction at work here? We could have said that we know $(3, 5)$ and $(5, 7)$ are in S because they are in R^2 and the rest is by induction. In that vein, here is the proof:

Proof. Suppose R is a relation on A, S is a transitive relation on A, and $R \subseteq S$. We will prove that $R^i \subseteq S$ for all $i \in \mathbb{N}$ by induction on i.

Base case. Suppose $i = 1$. Then $R^i = R^1 = R$. By substitution, $R^i \subseteq S$.

Inductive case. Next suppose $I \in \mathbb{N}$ such that $I \geq 1$ and for all $i \leq I$, $R^i \subseteq S$.

Time out. That is an unusual supposition for the inductive case. As we would expect, we are making an assumption about R^I and going to prove $R^{I+1} \subseteq S$. What is different is that we are assuming not only that $R^I \subseteq S$ but that a similar proposition holds for all natural numbers less than I. We will see the reason for this in a moment.

Suppose then that $i = I + 1$, and so $R^i = R^{I+1}$. Suppose further that $(a, b) \in R^{I+1}$.

By the definition of relation composition, there exist $j, k \in \mathbb{N}$ (where $j < I + 1$ and $k < I + 1$) and $c \in A$ such that $(a, c) \in R^j$ and $(c, b) \in R^k$. By arithmetic, $j \leq I$ and $k \leq I$.

By our inductive hypothesis, $(a, c), (c, b) \in S$. Since S is transitive, $(a, b) \in S$. Hence, by definition of subset, $R^{I+1} \subseteq S$. Moreover, for all $i \leq I + 1$, $R^i \subseteq S$.

Therefore, by mathematical induction, $R^i \subseteq S$ for all i. □

The reason we assumed our invariant was true for *all* previous steps and not only the one immediately previous was because we were going to split R^{I+1} in an arbitrary place, that is, $R^{I+1} = R^j \circ R^k$, and we needed to be sure that the invariant held for both R^j and R^k.

strong mathematical induction

This is a notable difference from our use of mathematical induction so far, and it has its own name: *strong mathematical induction*. Once again the terminology is misleading. Structural induction is no less mathematical than mathematical induction, none of them are induction in the philosophical sense, and structural and mathematical induction are no less robust than strong induction. What makes strong induction different is that when we finished the inductive case, we had to use and prove the *stronger requirement* of the inductive hypothesis, that the property holds on *all* smaller numbers.

Next, let us return to partial order relations. In Section 5.9, we said that every finite poset has a maximal element. We will prove this with math induction on the size of the poset. In particular, remember this lemma

Lemma 5.17. *If A is a poset with partial order relation $\preceq$, and $a \in A$, then $A - \{a\}$ is a poset with partial order relation $\preceq'$, defined as $\preceq - \{(b, c) \in \preceq \mid b = a$ or $c = a\}$.*

As we said in Section 5.9, the reflexivity, antisymmetry, and transitivity of $\preceq'$ depend on $\preceq$, but they do not depend on a being in the set. Now the main result about posets:

Theorem 5.16 *A finite, non-empty poset has at least one maximal element.*

Here is where the induction comes. A poset with one element has its one element as a maximal. From there we show that every time an element is added to the poset, we retain the fact that there is at least one maximal element. Rather than rephrasing the theorem to make the dependence on $\mathbb{N}$ apparent, we will let the proof find the induction.

Proof. Suppose A is a poset with partial order relation $\preceq$. We will prove it has at least one maximal element by induction on $|A|$.

Base case. Suppose $|A| = 1$. Let a be the one element of A. Trivially, suppose $b \in A$. Since $|A| = 1$, $b = a$, and since $\preceq$ is reflexive, $b \preceq a$.

Hence a is a maximal element, and so there exists an $N \geq 1$ such that for any poset of size N, it has at least one maximal element.

Inductive case. Suppose $|A| = N + 1$. Suppose $a \in A$. Let $B = A - \{a\}$ and $\preceq' = \preceq - \{(b, c) \in \preceq \mid b = a \text{ or } c = a\}$.

You should be able to recognize the next step by intuition; we will prove it in a later chapter.

$|B| = |A| - |\{a\}| = N + 1 - 1 = N$ by Exercise 7.9.1.

By Lemma 5.17, B is a partial order with $\preceq'$. By the inductive hypothesis, B has at least one maximal element. Let b be a maximal element of B.

We have three cases in light of b. It must be that $a \preceq b$, or $b \preceq a$, or neither. (It cannot be that both $a \preceq b$ and $b \preceq a$, since $a \neq b$ and by antisymmetry.) In each case, we will show that a maximal element exists.

Case 1: Suppose $a \preceq b$. Then suppose $c \in A$. If $c \neq a$, then since b is maximal in B, either $c \preceq' b$ (in which case, by definition of difference, $c \preceq b$) or c and b are noncomparable in $\preceq'$ (in which case c and b are still incomparable in $\preceq$). If $c = a$, then $c \preceq b$ by our supposition. Hence b is a maximal element in A.

Case 2: Suppose $b \preceq a$. See Exercise 6.6.6 to show a maximal element exists.

Case 3: Suppose b and a in noncomparable with $\preceq$. See Exercise 6.6.7 to show a maximal element exists.

In any case, there is a maximal element for A. Therefore, by the principle of mathematical induction, any finite, non-empty poset has at least one maximal element. □

To sum up these last three sections, here are the three versions of inductive proof techniques:

- *Structural induction* is used on propositions universally quantified over recursively defined sets.
- *Mathematical induction* is used on propositions universally quantified over whole numbers or natural numbers: $I(0)$. $\forall\, n \in \mathbb{W}, I(n) \rightarrow I(n+1)$. $\therefore \forall\, n \in \mathbb{W}, I(n)$.
- *Strong mathematical induction* also is used on propositions universally quantified over whole numbers or natural numbers: $I(0)$. $\forall\, n \in \mathbb{W}, (\forall k \leq n, I(k)) \rightarrow I(n+1)$. $\therefore \forall\, n \in \mathbb{W}, I(n)$.

Exercises

Recall that just as we have summation notation for summing a series of algebraic expressions, so we can have *iterated union* and *iterated intersection* for unioning and intersecting a series of set theoretical expressions. If $A_1, A_2, \ldots A_n$ are all sets, then

$$\bigcup_{i=1}^{n} A_i = A_1 \cup A_2 \cup \ldots \cup A_n$$

$$\bigcap_{i=1}^{n} A_i = A_1 \cap A_2 \cap \ldots \cap A_n$$

We then generalize older set properties for these iterated versions of set operations. Use mathematical induction to prove the following, for all $n \in \mathbb{N}$.

6.6.1 $\overline{\bigcup_{i=1}^{n} A_i} = \bigcap_{i=1}^{n} \overline{A_i}$. (Hint: Use Exercise 4.3.13 as a lemma.)

6.6.2 $\overline{\bigcap_{i=1}^{n} A_i} = \bigcup_{i=1}^{n} \overline{A_i}$. (Hint: Use Exercise 4.3.14 as a lemma.)

6.6.3 $\bigcap_{i=1}^{n} (A \cup B_i) = A \cup (\bigcap_{i=1}^{n} B_i)$. (Hint: Use Exercise 4.3.4 as a lemma.)

6.6.4 $\bigcup_{i=1}^{n} (A_i - B) = (\bigcup_{i=1}^{n} A_i) - B$. (Hint: Use Exercise 4.3.19 as a lemma.)

6.6.5 $\bigcap_{i=1}^{n} (A_i - B) = (\bigcap_{i=1}^{n} A_i) - B$. (Hint: Use Exercise 4.3.18 as a lemma.)

In the following exercises you will finish proofs from this chapter. You will not need to introduce any new induction; the proofs that you are finishing already rely on math induction.

6.6.6 Prove case 2 of Theorem 5.16.

6.6.7 Prove case 3 of Theorem 5.16.

6.6.8 Finish the proof of Theorem 6.5. Several results from earlier or later in this text will help tame this somewhat intricate proof:

- If $A \subseteq B$, then $|B - A| = |B| - |A|$. (Exercise 7.9.1)
- If $a \in A$, then $\mathscr{P}(A - \{a\})$ and $\{ C \cup \{a\} \mid C \in \mathscr{P}(A - \{a\}) \}$ make a partition of $\mathscr{P}(A)$. (Corollary 4.12)
- If A is a finite set and $a \in A$, then $|\{ \{a\} \cup C \mid C \in \mathscr{P}(A - \{a\})\}| = |\mathscr{P}(A - \{a\})|$. (Exercise 7.9.6)
- If A and B are finite, disjoint sets, then $|A \cup B| = |A| + |B|$. (Theorem 7.12)

6.7 Program correctness

Programs are usually verified by testing: A large number of test cases are thrown at it, and if it produces the right answer for all of these, then it is assumed to be right. Of course, testing like this can prove only that a program is *incorrect* if it fails a test. This process cannot prove that the program is correct if the tests pass. There may be a case on which the program would fail but that we failed to test.

An alternative is to write a proof that a program or part of a program is correct. In general this is no easy task, given the complexity even of small programs. Yet for some software projects, the guarantees of a correctness proof are worth the effort. It also is a good intellectual exercise to help us reason better about programs.

This section will demonstrate proofs of correctness by four examples: two functions on integers, one function on lists, and one function on a recursive datatype. Each will be proven by some kind of induction. The following section will continue proofs of correctness by investigating a specific list-processing problem.

The correctness of a program is defined by its *specification*. The specification describes what the output should be for all valid input. The valid input is defined by the *pre-conditions* of the program. To see this in practice, take a simple ML function.

specification

pre-conditions

The function `factorial` is specified to return $n!$ for any input n, where $n \geq 0$. The restriction on what n can be is the pre-condition. Now compare the mathematical definition of factorial with our function `factorial`:

$$n! = \begin{cases} 1 & \text{if } n = 0 \\ n \cdot (n-1)! & \text{otherwise} \end{cases}$$

```
fun factorial(0) = 1
  | factorial(n) = n * factorial(n-1);
```

We claim

Theorem 6.6 *For all* $n \in \mathbb{W}$, $\texttt{factorial}(n) = n!$

There is a problem with how we set up this result: we quantified the proposition over $\mathbb{W}$, but `factorial` operates on int. Not only are *integers* and *computer representations of integers* two different things, $\mathbb{W}$ contains some values for which there are no int representations because they are too large. We will ignore these difficulties by assuming we are really talking about the subset of $\mathbb{W}$ representable by values in int and that there is a one-to-one correspondence between the numbers and their representations.

This is a boring, tedious proof. But it demonstrates what a proof of program correctness looks like.

Proof. By induction on n.

Base case. Suppose $n = 0$. By definition of `factorial`, $\texttt{factorial}(0) = 1 = 0!$, by definition of !. Hence there exists an $N \geq 0$ such that $\texttt{factorial}(N) = N!$.

Inductive case. Suppose $N \geq 0$ such that $\texttt{factorial}(N) = N!$, and suppose $n = N + 1$. Then

$$\begin{array}{lcll} \texttt{factorial}(n) & = & n \cdot \texttt{factorial}(n-1) & \text{by definition of } \texttt{factorial} \\ & = & n \cdot \texttt{factorial}(N) & \text{by algebra and substitution} \\ & = & n \cdot N! & \text{by the inductive hypothesis} \\ & = & n! & \text{by definition of } ! \end{array}$$

Therefore, by math induction, `factorial` is correct for all $n \in \mathbb{W}$.
□

integer square root

Try something new. We cannot compute square roots exactly in the realm of integers, but we can define the *integer square root*: If $n \in \mathbb{W}$, then the integer square root of n is the greatest integer i such that $i^2 \leq n$. Thus the integer square root of 16 is 4, but the integer square root of 15 is 3. We claim that the following two functions compute the integer square root for whole numbers:

```
- fun increase(k, n) = if (k + 1) * (k + 1) > n
=                      then k
=                      else k+1;

val increase = fn : int * int -> int

- fun sqrt(n) = if n = 0
=               then 0
=               else increase(2 * sqrt(n div 4), n);

val sqrt = fn : int -> int
```

To see how this works, notice that if we were dealing with $\mathbb{R}$, the following equation would hold.

$$\sqrt{n} = 2 \cdot \sqrt{\frac{n}{4}}$$

The heart of our `sqrt` function is to use this formula to reduce the radicand until we reach the base case. However, this process sometimes shoots too low for integer square roots. Take 3: $3 \texttt{ div } 4 = 0$, and the integer square root of 0 is 0, but the integer square root of 3 is 1. The function `increase` fixes this up.

First, a helper lemma about the helper function `increase`. Remember that we want a number i such that $i^2 \leq n < (i+1)^2$. This lemma says that if we have a number k fulfilling the lower part of this double inequality for i (that is, $k^2 \leq n$), `increase` will do no harm ($(\texttt{increase}(k,n))^2 \leq n$).

Lemma 6.7 *If $k, n \in \mathbb{W}$ and $k^2 \leq n$, then $(\texttt{increase}(k,n))^2 \leq n$.*

> **Proof.** Suppose $k, n \in \mathbb{W}$ and $k^2 \leq n$. It must be that either $(k+1)^2 \leq n$ or $(k+1)^2 > n$.
>
> Suppose $(k+1)^2 \leq n$. Then by definition, $\texttt{increase}(k,n) = k+1$, and by substitution, $(\texttt{increase}(k,n))^2 \leq n$.
>
> On the other hand, suppose $(k+1)^2 > n$. Then by definition, $\texttt{increase}(k,n) = k$, and by substitution, $(\texttt{increase}(k,n))^2 \leq n$.
>
> Either way, $(\texttt{increase}(k,n))^2 \leq n$. □

Now, what we want to say about the function `sqrt`:

Theorem 6.8 *For all* $n \in \mathbb{W}$ $(\texttt{sqrt}(n))^2 \leq n < (\texttt{sqrt}(n) + 1)^2$.

The math gets fairly involved in this one. We will break it up into manageable pieces.

Proof. By induction on n.

Base case. Suppose $n = 0$. Then $\texttt{sqrt}(n) = 0$ by definition, and $0^2 = 0 \leq 0 < 1 = (0+1)^2$. Hence there exists $N \geq 0$ such that for all $k \leq N$, $(\texttt{sqrt}(k))^2 \leq k < (\texttt{sqrt}(k) + 1)^2$.

Why did we say "such that *for all* $k \leq N$"? Because in our inductive case, $N+1$ does not depend on N, but on some smaller number, $(N+1)$ div 4. We are using "strong induction" as in the proof of Lemma 5.14 on page 6.6.

Inductive case. Suppose $n = N + 1$. Then $n \texttt{ div } 4 \leq N$.

What we want to prove, $(\texttt{sqrt}(n))^2 \leq n < (\texttt{sqrt}(n)+1)^2$, is actually two propositions, the left inequality and the right inequality. First we show $(\texttt{sqrt}(n))^2 \leq n$:

$$\begin{array}{rcll} (\texttt{sqrt}(n \texttt{ div } 4))^2 & \leq & n \texttt{ div } 4 & \text{by the inductive hypothesis} \\ 4(\texttt{sqrt}(n \texttt{ div } 4))^2 & \leq & 4(n \texttt{ div } 4) & \text{multiplying by 4} \\ (2 \cdot \texttt{sqrt}(n \texttt{ div } 4))^2 & \leq & 4(n \texttt{ div } 4) & \text{by algebra} \end{array}$$

We cannot simplify $4(n \texttt{ div } 4) = n$ because div is integer division: $4 \cdot (3 \texttt{ div } 4) = 0$, not 3. Instead, we will use n as an upper bound.

$$\begin{array}{rcll} (2 \cdot \texttt{sqrt}(n \texttt{ div } 4))^2 & \leq & n & \text{since } 4(n \texttt{ div } 4) \leq n \\ (\texttt{increase}(2 \cdot (\texttt{sqrt}(n \texttt{ div } 4))))^2 & \leq & n & \text{by Lemma 6.7} \\ (\texttt{sqrt}(n))^2 & \leq & n & \text{by definition of } \texttt{sqrt} \end{array}$$

Now the right side, to prove $n < (\texttt{sqrt}(n) + 1)^2$.

It must be that either $n < (2 \cdot \texttt{sqrt}(n \texttt{ div } 4) + 1)^2$ or $n \geq (2 \cdot \texttt{sqrt}(n \texttt{ div } 4) + 1)^2$.

Case 1. Suppose $n < (2 \cdot \texttt{sqrt}(n \texttt{ div } 4) + 1)^2$. Then

$$\begin{array}{rcll} \texttt{sqrt}(n) & = & \texttt{increase}(2 \cdot \texttt{sqrt}(n \texttt{ div } 4), n) & \text{by definition of } \texttt{sqrt} \text{ (since } n > 0) \\ & = & 2 \cdot \texttt{sqrt}(n \texttt{ div } 4) & \text{by definition of } \texttt{increase} \end{array}$$

And so $n < (\texttt{sqrt}(n) + 1)^2$ by substitution.

We did not need the inductive hypothesis in Case 1.

Case 2. Suppose $n \geq (2 \cdot \texttt{sqrt}(n \ \texttt{div} \ 4) + 1)^2$. Then

$$\begin{array}{rcll} \texttt{sqrt}(n) & = & \texttt{increase}(2 \cdot \texttt{sqrt}(n \ div \ 4), n) & \text{by definition of } \texttt{sqrt} \text{ again} \\ & = & 2 \cdot \texttt{sqrt}(n \ \texttt{div} \ 4) + 1 & \text{by definition of } \texttt{increase} \end{array}$$

Now, the inductive hypothesis says $n \ \texttt{div} \ 4 < (\texttt{sqrt}(n \ \texttt{div} \ 4) + 1)^2$, and so by arithmetic (since all quantities are integers), $n \ \texttt{div} \ 4 + 1 \leq (\texttt{sqrt}(n \ \texttt{div} \ 4) + 1)^2$. So,

$$\begin{array}{rcll} n & < & 4(n \ \texttt{div} \ 4) + 4 & \text{by definition of } \texttt{div} \\ & = & 4(n \ \texttt{div} \ 4 + 1) & \text{by algebra} \\ & \leq & 4(\texttt{sqrt}(n \ \texttt{div} \ 4) + 1)^2 & \text{by substitution} \\ & = & (2 \cdot \texttt{sqrt}(n \ \texttt{div} \ 4) + 2)^2 & \text{by algebra} \\ & = & (2 \cdot \texttt{sqrt}(n \ \texttt{div} \ 4) + 1 + 1)^2 & \text{by arithmetic} \\ & = & (\texttt{sqrt}(n) + 1)^2 & \text{by substitution} \end{array}$$

Either way, $n < (\texttt{sqrt(n)} + 1)^2$. □

We have seen how important list-processing functions are in ML programming. We need ways to prove that functions make correct computations on lists. In Exercise 2.4.15, you wrote a function to compute the powerset of a set represented as a list. Consider this alternative for computing powersets which accumulates its answer in its second parameter.

```
- fun powerset([], yy) = [yy]
=   | powerset(x::rest, yy) =
=             powerset(rest, yy) @ powerset(rest, x::yy);

val powerset = fn : 'a list * 'a list -> 'a list list
```

The intent is that `powerset(x, [])` will return a list representing the powerset of `x`. To see how it works, consider what happens if we call `powerset([1, 2, 3], [])`. The first call makes two recursive calls, `powerset([2, 3], [])` and `powerset([2, 3], [1])`. What are the results of those?

```
- powerset([2, 3], []);

val it = [[],[3],[2],[3,2]] : int list list

- powerset([2,3], [1]);

val it = [[1],[3,1],[2,1],[3,2,1]] : int list list
```

The function computes the powerset of the first parameter, but also unions the second parameter to every element in that powerset—not unlike our way of partitioning a powerset in our proofs.

Just as the examples on integers raised concerns about the correspondence between number sets and int, with this function we need to reason both about *sets* and *lists representing sets*, and the correspondence is not as neat. Lists, unlike sets, are ordered. Let $\mathcal{L}$ be the set of ML lists values. Each element is either [] or $a :: ww$ for some base-type value a and list value ww. When we say that a set A and a list xx are equal, we mean that A is *the* set of elements in xx, but that xx is *a* list of the elements in A.

Here is what we want to say about this particular function. Although its output is a list, we mean that the list is a representation of a certain set:

Theorem 6.9 *For all* $xx, yy \in \mathcal{L}$, $\texttt{powerset}(xx, yy) = \{zz \cup yy \mid zz \in \mathscr{P}(xx)\}$

Take a moment to understand the described set. It is the set of all subsets zz of xx, each unioned with yy. The proof will use Exercise 4.9.7, that $\{B \cup C \mid C \in \mathscr{P}(A)\} \cup \{B \cup C \cup \{x\} \mid C \in \mathscr{P}(A)\} = \{B \cup C \mid C \in \mathscr{P}(A \cup \{x\})\}$. Taking $xx = a :: ww$, then this lemma applies with $B = yy$, $x = a$, and $A = ww$.

Proof. Suppose xx and yy are lists. The proof is by induction on the structure of xx. By definition of a list, either xx is [] or there is some element a and some list ww such that $xx = a :: ww$.

Base case. Suppose $x = \texttt{[]}$. Then $\mathscr{P}(x) = \texttt{[[]]}$ or $\{\emptyset\}$, so $\{zz \cup yy \mid zz \in \mathscr{P}(xx)\} = \{yy\}$. Moreover, $\texttt{powerset}(\texttt{[]}, yy) = \texttt{[}yy\texttt{]}$ by definition of powerset.

Inductive case. Suppose $xx = a :: ww$ for some element a and some list ww. Then, by definition of powerset,

$$\begin{array}{llll}
\texttt{powerset}(xx, yy) & = & \texttt{powerset}(ww, yy) @ \texttt{powerset}(ww, a :: yy) & \text{by definition of powerset} \\
& = & \{zz_1 \cup yy \mid zz_1 \in \mathscr{P}(ww)\} & \\
& & \cup \{zz_2 \cup \{a\} \cup yy \mid zz_2 \in \mathscr{P}(ww)\} & \text{by the inductive hypothesis} \\
& = & \{zz \cup yy \mid zz \in \mathscr{P}(a :: ww)\} & \text{by Exercise 4.9.7} \\
& = & \{zz \cup yy \mid zz \in \mathscr{P}(xx)\} & \text{by substitution}
\end{array}$$

Therefore, by structural induction, $\texttt{powerset}(xx, yy) = \{zz \cup yy \mid zz \in \mathscr{P}(xx)\}$. □

Functions on recursively defined datatypes are also well-suited to proofs of correctness. Recall from Section 3.15 the `proposition` datatype and the function `negNormForm` that converts propositions into negative normal form.

```
datatype binLogOp = Conj | Disj;
datatype proposition = Var of string
```

```
                          | Neg of proposition
                          | BinOp of binLogOp * proposition
                                          * proposition;

fun flip(Conj) = Disj
  | flip(Disj) = Conj;

fun negNormForm(Var(s)) = Var(s)
  | negNormForm(Neg(Var(s))) = Neg(Var(s))
  | negNormForm(Neg(Neg(p))) = negNormForm(p)
  | negNormForm(Neg(BinOp(oper, p, q))) =
    BinOp(flip(oper),negNormForm(Neg(p)),negNormForm(Neg(q)))
  | negNormForm(BinOp(oper, p, q)) =
    BinOp(oper,negNormForm(p),negNormForm(q));
```

This time around, we will take the function `negNormForm` as our definition, that it defines *negative normal form* by its output. The following mutually recursive functions also compute negative normal form, but more efficiently:

```
fun negNormFormPos(Var(s)) = Var(s)
  | negNormFormPos(Neg(p)) = negNormFormNeg(p)
  | negNormFormPos(BinOp(oper, p, q)) =
    BinOp(oper, negNormFormPos(p), negNormFormPos(q))
and negNormFormNeg(Var(s)) = Neg(Var(s))
  | negNormFormNeg(Neg(p)) = negNormFormPos(p)
  | negNormFormNeg(BinOp(oper, p, q)) =
    BinOp(flip(oper), negNormFormNeg(p), negNormFormNeg(q));
```

What is the advantage? The pattern-matching in the original function is very complicated, with several cases dealing with negations. The new functions keep track of whether we have seen a negation simply by which of the two functions currently is being called. Notice that a binary operator is flipped in `negNormFormNeg` but not `negNormFormPos`; similarly compare the how `Var` constructors are handled.

This time our correctness claim is that two functions are equal in their output.

Theorem 6.10 *For all values r of type proposition,* `negNormForm`$(r) =$ `negNormFormPos`(r).

We chose the variable r so that we can use p and q inside the proof consistently with p and q in the functions. Induction will be on the structure of r, the cases corresponding to what r could be—for example, the three data constructors in the definition of the datatype, which would mean a base case for variables and an inductive case each for negations and binary operations. For convenience, however, the cases will correspond to the patterns of `negNormForm`, which means two base cases and three inductive cases.

In each case, we want to show that $\texttt{negNormForm}(r) = \texttt{negNormFormPos}(r)$. Sometimes it will be convenient to work from one direction, sometimes the other.

Proof. By induction on the structure of r.

Base case 1. Suppose $r = \texttt{Var}(s)$ for some string value s. Then $\texttt{negNormForm}(r) = \texttt{negNormFormPos}(r)$ by their respective definitions.

Base case 2. Suppose $r = \texttt{Neg}(\texttt{Var}(s))$ for some string value s. Then

$$\begin{array}{rcl} \texttt{negNormForm}(r) & = & \texttt{Neg}(\texttt{Var}(s)) \\ & & \text{by definition of } \texttt{negNormForm} \\ \texttt{negNormFormPos}(r) & = & \texttt{negNormFormNeg}(\texttt{Var}(s)) \\ & & \text{by definition of } \texttt{negNormFormPos} \\ & = & \texttt{Neg}(\texttt{Var}(s)) \\ & & \text{by definition of } \texttt{negNormFormNeg} \\ & = & \texttt{negNormForm}(r) \\ & & \text{by substitution} \end{array}$$

Inductive case 1. Suppose $r = \texttt{Neg}(\texttt{Neg}(p))$ for some proposition value p. Then

$$\begin{array}{rcl} \texttt{negNormFormPos}(r) & = & \texttt{negNormFormNeg}(\texttt{Neg}(p)) \\ & & \text{by definition of } \texttt{negNormFormPos} \\ & = & \texttt{NegNormFormPos}(p) \\ & & \text{by definition of } \texttt{negNormFormNeg} \\ \texttt{negNormForm}(r) & = & \texttt{negNormForm}(p) \\ & & \text{by definition of } \texttt{negNormForm} \\ & = & \texttt{negNormFormPos}(p) \\ & & \text{by induction} \end{array}$$

Inductive case 2. Suppose $r = \texttt{Neg}(\texttt{BinOp}(oper, p, q))$ for some binLogOp value $oper$ and some proposition values p and q. Then

$$\begin{array}{l} \texttt{negNormFormPos}(\texttt{Neg}(\texttt{BinOp}(oper, p, q)) \\ \qquad = \texttt{negNormFormNeg}(\texttt{BinOp}(oper, p, q)) \\ \qquad\qquad \text{by definition of } \texttt{negNormFormPos} \text{ (second pattern)} \\ \qquad = \texttt{BinOp}(\texttt{flip}(oper), \\ \qquad\qquad\quad \texttt{negNormFormNeg}(p), \\ \qquad\qquad\quad \texttt{negNormFormNeg}(q)) \\ \qquad\qquad \text{by definition of } \texttt{negNormFormNeg} \\ \texttt{negNormForm}(\texttt{Neg}(p)) = \texttt{negNormFormPos}(\texttt{Neg}(p)) \\ \qquad\qquad \text{by induction} \end{array}$$

$$
\begin{aligned}
& = \texttt{negNormFormNeg}(p) \\
& \qquad \text{by definition of } \texttt{negNormFormPos} \\
\texttt{negNormForm}(\texttt{Neg}(q)) & = \texttt{negNormFormNeg}(q) \\
& \qquad \text{similarly} \\
\texttt{negNormFormPos}(\texttt{Neg}(\texttt{BinOp}(oper), p, q)) & \\
& = \texttt{BinOp}(\texttt{flip}(oper), \\
& \qquad \texttt{negNormForm}(\texttt{Neg}(p)), \\
& \qquad \texttt{negNormForm}(\texttt{Neg}(p))) \\
& \qquad \text{by substitution} \\
& = \texttt{negNormForm}(r) \\
& \qquad \text{by definition of } \texttt{negNormForm}
\end{aligned}
$$

Inductive case 3. Suppose $r = \texttt{BinOp}(oper, p, q)$ for some binLogOp value $oper$ and some proposition values p and q. Then

$$
\begin{aligned}
\texttt{negNormFormPos}(\texttt{BinOp}(oper), p, q) & = \texttt{BinOp}(oper, \\
& \qquad \texttt{negNormFormPos}(p), \\
& \qquad \texttt{negNormFormPos}(q)) \\
& \qquad \text{by definition of } \texttt{negNormFormPos} \\
& = \texttt{BinOp}(oper, \\
& \qquad \texttt{negNormForm}(p), \\
& \qquad \texttt{negNormForm}(q)) \\
& \qquad \text{by induction} \\
& = \texttt{negNormForm}(\texttt{BinOp}(oper), p, q) \\
& \qquad \text{by definition if } \texttt{negNormForm}
\end{aligned}
$$

In all cases, $\texttt{negNormForm}(r) = \texttt{negNormFormPos}(r)$. □

The material in this section, including some exercises, is adapted from Paulson [25].

Exercises

6.7.1 Consider the following function which implements an alternative approach to computing factorial:

```
- fun fact2(0, p) = p
=   | fact2(n, p) = fact2(n-1, n * p);
```

Traditional factorial decrements the n parameter as it descends the recursion and multiplies by n as it comes back up. This version accumulates the answer in its second parameter (p), and no work needs to be done on the way back up. Prove that for all $n \in \mathbb{W}$, for all $p \in \mathbb{Z}$, $\texttt{fact2}(n, p) = n! \cdot p$. The correctness of this function is actually a corollary: For all $n \in \mathbb{W}$, $\texttt{fact2}(n, 1) = n!$.

6.7.2 Prove that your solution to Exercise 4.10.2 is correct (that is, that its result is x^y).

Exercises, continued

6.7.3 Modify your proof from the previous exercise to prove that your solution to Exercise 4.10.3 is correct.

6.7.4 Prove that your solution to Exercise 4.10.4 is correct, that its result is $x \cdot y$.

6.7.5 Prove that your solution to Exercise 4.10.5 is correct, that if given n, d, it returns (q, r) such that $n = d \cdot q + r$ and $0 \leq r < d$.

6.7.6 Consider the following two functions.

```
- fun reverse1([]) = []
=   | reverse1(a::rest) =
=                 reverse(rest) @ [a];
```

```
- fun reverse2([], bb) = bb
=   | reverse2(a::rest, bb) =
=                 reverse2(rest, a::bb);
```

We will take `reverse1` to be our "definition" of the reverse of a list. Like `fact2` from Exercise 6.7.1, `reverse2` accumulates its answer in its second parameter. Prove that for all lists x, `reverse2`(x, `[]`) $=$ `reverse1`(x).

6.7.7 Demonstrate the correctness of your solution to Exercise 6.2.11 by proving that for all $t \in \mathcal{T}$, `reflection`(`reflection`(t)) $= t$.

6.8 Sorting

In Exercise 3.7.13, you were asked to write a function `sort` that takes a list of integers and returns a list of the same integers, but sorted from smallest to greatest. Sorting is an important problem in the field of computer science partly because there are many ways to do it. The variety of sorting algorithms provides us with a domain where we can demonstrate several large categories of algorithms, make analyses and comparisons about algorithmic efficiency, and, for our present purposes, practice writing correctness proofs.

In this section we will assume we are sorting lists of integers from least to greatest, as we did in Exercise 3.7.13. The same basic algorithms, however, can be used to sort different kinds of containers—not only list, but also arrays. They can sort collections having different base types, and they can use different ways to compare values, for example from greatest to least.

Two sorting algorithms of interest to us presently are examples of *divide-and-conquer algorithms*, meaning that the problem is solved by dividing it into subproblems, solving the subproblems, and unifying the result.

divide-and-conquer algorithms

Of course, almost all recursive algorithms are divide-and-conquer algorithms in a sense. However, most of the recursive functions we have written so far "divide" the problem into only one slightly-smaller subproblem—think of our list-processing functions that divide the list into head and tail and operate recursively on the tail. What we have in mind by "divide and conquer" are algorithms that divide the work into roughly equal parts, or at least into two or more non-trivial parts.

Take a list of integers—it may be easier to imagine a pile of cards, each with a number written on it. Divide the pile into two piles each with half the cards. Sort each half-pile, recursively; our base case is a pile with one or zero cards, which is already sorted. Then, having obtained two sorted half-piles, merge them into a unified sorted pile. This is called *merge sort*.

merge sort

We illustrate this on the list [6, 5, 2, 8, 4, 7, 3, 9]. The left illustration gives the "global view," showing the breakdown of the program through the various recursive calls. The right illustration gives the "local view," showing only the top recursive call but giving more detail for the merge process.

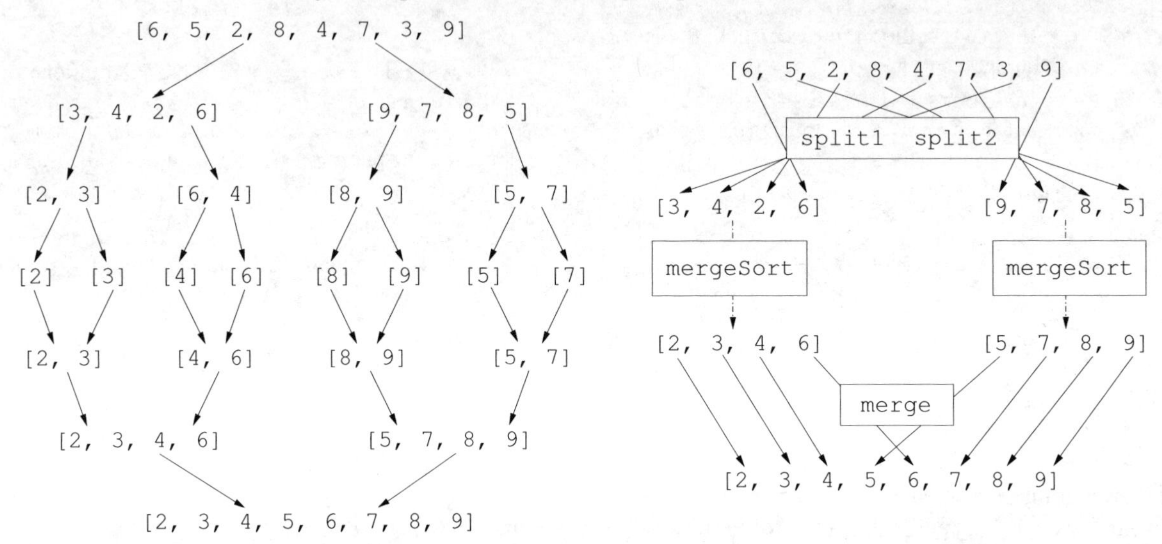

The global view shows that there are three parts to merge sort: the splitting, the recursive call, and the merging. We will consider the split first. Traditional versions of merge sort split the list in the middle—so, [6, 5, 2, 8, 4, 7, 3, 9] becomes [6, 5, 2, 8] and [4, 7, 3, 9]. In ML, however, the obvious way to split a list in the middle is inefficient for both the programmer and the computer: first we would need to compute the length of the original list, then divide it in half, and then count down halfway to find the splitting point (but see Exercise 6.8.8). It is much easier to make one pass through the list, accumulating the two half-lists as we go. This is accomplished by the mutually recursive functions split1 and split2:

```
- fun split1([], aa, bb) = (aa, bb)
=   | split1(x::rest, aa, bb) = split2(rest, x::aa, bb)
= and split2([], aa, bb) = (aa, bb)
=   | split2(x::rest, aa, bb) = split1(rest, aa, x::bb);

val split1 = fn
    : 'a list * 'a list * 'a list -> 'a list * 'a list
val split2 = fn
    : 'a list * 'a list * 'a list -> 'a list * 'a list
```

These functions have the unintended but harmless effect of also reversing the lists, as can be seen in the diagrams. For merge, we iterate through the two lists, collating them by taking the smaller of the two heads as the next item.

We have three cases: the first list is empty, the second list is empty, and neither list is empty. (The case where both lists are empty is subsumed arbitrarily by the first case.) So, if one of the lists is empty, then the other list becomes the remainder of our combined, sorted list. If they each have at least one element, then compare the first elements and take the smaller.

```
- fun merge([], bb) = bb
=    | merge(aa, []) = aa
=    | merge(a::aRest, b::bRest) =
=      if a < b then a::merge(aRest, b::bRest)
=               else b::merge(a::aRest, bRest);

val merge = fn : int list * int list -> int list
```

Now mergeSort is easy.

```
- fun mergeSort([]) = []
=    | mergeSort([a]) = [a]
=    | mergeSort(xx) =
=      let val (aa, bb) = split1(xx, [], []);
=      in merge(mergeSort(aa), mergeSort(bb))
=      end;

val mergeSort = fn : int list -> int list

- mergeSort([6, 5, 2, 8, 4, 7, 3, 9]);

val it = [2,3,4,5,6,7,8,9] : int list
```

We are going to prove that mergeSort results in a sorted list. It is worth noting, however, that this is only a piece of what it means for a sorting algorithm to be correct. The output of a sorting algorithm also must have all the same values as the input, with the same frequency. Moreover, both sortedness and value-consistency requirements of the output list assume that there is an output, that is, that the function actually terminates. A complete proof would prove all of this.

We will use the following predicate to define what it means for a list to be sorted.

```
- fun isSorted([]) = true
=    | isSorted([x]) = true
=    | isSorted(x::y::rest) =
=        x <= y andalso isSorted(y::rest);

val isSorted = fn : int list -> bool
```

First, we prove that the helper function `merge` produces a sorted list. That is true, however, only if it is given sorted lists to begin with. We need to state that as a precondition. Let $\mathcal{L}_{\text{int}}$ by the set of lists of ints.

Lemma 6.11 *For all* $aa, bb \in \mathcal{L}_{\text{int}}$, *if* $\texttt{isSorted}(aa)$ *and* $\texttt{isSorted}(bb)$, *then* $\texttt{isSorted}(\texttt{merge}(aa, bb))$.

Proof. Suppose aa and bb are sorted lists of ints. The proof is by induction on the structure of aa and bb.

Base case. Suppose $aa = \texttt{[]}$. Then $\texttt{merge}(aa, bb) = bb$ by definition of `merge`, and since we are given $\texttt{isSorted}(bb)$, then we know $\texttt{isSorted}(\texttt{merge}(aa, bb))$ by substitution. Similarly for $bb = \texttt{[]}$.

Inductive case. Suppose $aa = a :: aRest$ and $bb = b :: bRest$ for some $a, b \in \mathbb{Z}$ and $aRest, bRest \in \mathcal{L}_{\text{int}}$. Either $a < b$ or $a \geq b$.

Case 1. Suppose $a < b$. Then $\texttt{merge}(aa, bb) = a :: \texttt{merge}(aRest, bb)$ by definition of `merge`. So $\texttt{isSorted}(\texttt{merge}(aRest, bb))$ by the inductive hypothesis.

We need to show that a is less than or equal to the head of $\texttt{merge}(aRest, bb)$. To do this, break it down into more subcases: Is $aRest$ empty, does it have a head that is less than b, or does it have a head that is greater than or equal to b?

Case 1a. Suppose $\texttt{aRest} = \texttt{[]}$. Then $\texttt{merge}(aRest, bb) = bb$ and $\texttt{hd}(\texttt{merge}(aRest, bb)) = b$. Since $a < b$, then we can conclude $\texttt{isSorted}(a :: \texttt{merge}(aRest, bb))$.

Case 1b. Suppose $\texttt{aRest} = c :: cRest$ for some $c \in \mathbb{Z}$ and some $cRest \in \mathcal{L}_{\text{int}}$. Further suppose $c < b$. Then by definition of `merge`, $\texttt{hd}(\texttt{merge}(aRest, bb)) = c$. Since $\texttt{isSorted}(aa)$, we know $a \leq c$ and so $\texttt{isSorted}(a :: \texttt{merge}(c :: cRest, bb))$.

Case 1c. Again suppose $\texttt{aRest} = c :: cRest$ for some $c \in \mathbb{Z}$ and some $cRest \in \mathcal{L}_{\text{int}}$, but now further suppose $c \geq b$. Then by definition of `merge`, $\texttt{hd}(\texttt{merge}(aRest, bb)) = b$. Since $a < b$ then $\texttt{isSorted}(a :: \texttt{merge}(aRest, bb))$.

Case 2. It is similar if $a \geq b$, except that $\texttt{merge}(aa, bb) = b :: \texttt{merge}(aRest, bb)$ by definition of `merge`, and our cases would consider what is the head of $\texttt{merge}(aRest, bb)$

In any case, $\texttt{isSorted}(\texttt{merge}(aa, bb))$. □

Seeing that `merge` works as we want it, then we can state and prove our claim about `mergeSort`.

Theorem 6.12 *For all* $xx \in \mathcal{L}_{\text{int}}$, $\texttt{isSorted}(\texttt{mergeSort}(xx))$.

Proof. Suppose $xx \in \mathcal{L}_{\text{int}}$. The proof is by induction on the structure of xx.

Base case 1. Suppose $xx =$ `[]`. Then `mergeSort`$(xx) =$ `[]` by definition of `mergeSort`, and `isSorted([])` by definition of `isSorted`.

Base case 2. It is similar for $xx =$ `[`a`]`.

Inductive case. Suppose $xx = a :: b :: xRest$ for some $a, b \in \mathbb{Z}$ and $xRest \in \mathcal{L}_{\text{int}}$. By definition, `mergeSort`$(xx) =$ `merge(mergeSort`(aa), `mergeSort`(bb)`)` for some $aa, bb \in \mathcal{L}_{\text{int}}$.

By induction, we conclude both `isSorted(mergeSort`(aa)`)` and also `isSorted(mergeSort`(bb)`)`.

By Lemma 6.11, `isSorted(merge(mergeSort`(aa)`,mergeSort`(bb)`)))`.
□

Do you find it suspicious that the proof ignores `split1` and `split2`, and that the inductive case does not deal with the content of xx? This is because it is only proving the sortedness of the output—remember that a complete proof of correctness would also need to show that the output list has the same values as the input list. See Exercise 6.8.7.

In merge sort, the split step is relatively simple—at any rate, it requires no comparisons. The work of comparing items and putting them in the right order is done by the `merge` function. The *quick sort* algorithm reverses this: It divides the list into a list of lesser values and a list of greater values (called *partitioning*), it sorts each of the two sublists, and it concatenates the sorted versions of the two sublists. We determine which sublist a value should go to by comparing the value to a pivot value. We will use the the first value of the list as our pivot, but the choice is arbitrary. At any rate, the pivot value goes between the two sorted sublists when we concatenate them at the end.

quick sort

```
- fun partition(p, [], aa, bb) = (aa, bb)
=   | partition(p, x::rest, aa, bb) =
=     if x < p then partition(p, rest, x::aa, bb)
=                else partition(p, rest, aa, x::bb);

val partition = fn
  : int * int list * int list * int list -> int list * int list

- fun quickSort([]) = []
=   | quickSort(x::rest) =
=     let val (aa, bb) = partition(x, rest, [], []);
=     in  quickSort(aa) @ (x::quickSort(bb))
=     end;

val quickSort = fn : int list -> int list
```

```
- quickSort([6, 5, 2, 8, 4, 7, 3, 9]);

val it = [2,3,4,5,6,7,8,9] : int list
```

Exercises 6.8.1–6.8.6 will walk you through a proof of correctness. It requires reasoning about `@`: to show that `quickSort(aa) @ (x::quickSort(bb))` is sorted, we would need to show that the *last* element in `aa` is less than or equal to `x`. To talk about this, we introduce two new comparison operators: $x \lll yy$ if element x is less than or equal to every element in list yy, similarly for $x \ggg yy$. Formally:

$$\text{If } x \in \mathbb{Z} \text{ and } yy \in \mathcal{L}_{\text{int}}, \text{ then} \begin{cases} x \lll yy & \text{if } \forall\, z \in \mathbb{Z}, \quad \text{if } \texttt{contains}(z, yy), \text{then } x \leq z \\ x \ggg yy & \text{if } \forall\, z \in \mathbb{Z}, \quad \text{if } \texttt{contains}(z, yy), \text{then } x > z \end{cases}$$

Exercises

6.8.1 To prove that `quickSort` is correct, we must first have a grasp on what `partition` does. This lemma says that if it is given lists aa and bb that are already partitioned by p as well as an unpartitioned list xx, then it will produce two lists cc and dd that are partitioned by p.

Lemma 6.13 *For all* $p \in \mathbb{Z}$, $xx, aa, bb \in \mathcal{L}_{\text{int}}$, *if* $p \ggg aa$, $p \lll bb$, *and* $(cc, dd) = \texttt{partition}(p, xx, aa, bb)$, *then* $p \ggg cc$ *and* $p \lll dd$.

Prove this by induction on the structure of xx.

6.8.2 Prove the following lemma. Since it is a corollary to Lemma 6.13, it is almost a one-liner.

Lemma 6.14 *For all* $p \in \mathbb{Z}$, $xx \in \mathcal{L}_{\text{int}}$, *if* $(cc, dd) = \texttt{partition}(p, xx, [], [])$, *then* $p \ggg cc$ *and* $p \lll dd$.

6.8.3 Prove the following lemma about $\lll$ and `isSorted` by induction on the structure of yy.

Lemma 6.15 *For all* $x \in \mathbb{Z}$ *and* $yy \in \mathcal{L}_{\text{int}}$, *if* $\texttt{isSorted}(yy)$ *and* $x \lll yy$, *then* $\texttt{isSorted}(x :: yy)$.

6.8.4 The following "converse" is also true. Prove it by induction on the structure of yy.

Lemma 6.16 *For all* $x \in \mathbb{Z}$ *and* $yy \in \mathcal{L}_{\text{int}}$, *if* $\texttt{isSorted}(yy)$ *and* $\texttt{isSorted}(x :: yy)$, *then* $x \lll yy$.

6.8.5 This lemma gets us most of the way to our final answer. Prove this by induction on the structure of aa. The inductive case ($aa = a :: aRest$) will include a mini-proof that for all $z \in \mathbb{Z}$ such that $\texttt{contains}(z, aRest)$, $z \lll bb$. You may assume that the definition of `contains` implies that if $\texttt{contains}(z, aRest)$, then $\texttt{contains}(z, a :: aRest)$. You will use Lemmas 6.15 and 6.16.

Lemma 6.17 *For all* $aa, bb \in \mathcal{L}_{\text{int}}$, *if* $\texttt{sorted}(aa)$, $\texttt{sorted}(bb)$, *and for all* $z \in \mathbb{Z}$, *if* $\texttt{contains}(z, aa)$ *then* $z \lll bb$, *then* $\texttt{sorted}(aa@bb)$.

To clarify any confusion about the nesting in the lemma, here is the proposition formally:

$$\begin{aligned} &\forall\, aa, bb \in \mathcal{L}_{\text{int}}, \\ &\quad (\texttt{sorted}(aa) \wedge \texttt{sorted}(bb) \wedge \\ &\quad\; (\forall\, z \in \mathbb{Z}, \texttt{contains}(z, aa) \rightarrow z \lll bb)) \\ &\quad \rightarrow \texttt{sorted}(aa@bb) \end{aligned}$$

The innermost conditional simply asserts that all elements in aa are less than all elements in bb.

6.8.6 Prove the following theorem about the correctness of `quickSort` by induction on the structure of xx. Use Lemmas 6.14 and 6.17.

Theorem 6.18 *For all* $xx \in \mathcal{L}_{\text{int}}$, $\texttt{isSorted}(\texttt{quickSort}(xx))$.

Exercises, continued

6.8.7 Describe informally what would need to be shown to prove that the output of `mergeSort` has all the same values with the same frequency as its input. Break it down: What would need to be shown about `split1` and `split2`? What would need to be shown about `merge`? Given those two subresults, how would we use induction on the structure of xx to show that this holds for `mergeSort`?

6.8.8 Finish the following function `splitMid` which splits a list in the middle in one pass. To use it, apply it to the same list as both parameters and it will return the half-lists as tuples (for example, `splitMid([1,8,7,3],[1,8,7,3])` will return `([1,8],[7,3])`. It works by counting down the elements in each parameter, but counting down the first first parameter twice as fast so that it runs out of elements when the second still has half left.

```
fun splitMid([], bb) = ([], bb)
  | splitMid([a], bb) = ([], bb)
  | splitMid(a1::a2::aRest, b::bRest) =
        ??
```

Also rewrite `mergeSort` to use this.

Bubble sort moves larger items to the back of a list by comparing adjacent items and swapping them if they are out of order. Exercises 6.8.9–6.8.11 ask you to reason about the following implementation of bubble sort:

```
fun bubblePass([]) = []
  | bubblePass([a]) = [a]
  | bubblePass(a::b::rest) =
    if a <= b
    then a::bubblePass(b::rest)
    else b::bubblePass(a::rest);

fun bubbleSort([]) = []
  | bubbleSort(a::rest) =
    bubblePass(a::bubbleSort(rest));
```

6.8.9 Prove the following lemma about the helper function `bubblePass` by induction on the structure of $a :: aRest$.

Lemma 6.19 *For all* $a \in \mathbb{Z}$, $aRest \in \mathcal{L}_{\text{int}}$, *if* `isSorted`$(aRest)$, *then* `isSorted`(`bubblePass`$(a :: aRest)$).

Notice that the base case is the second pattern of `bubblePass`, when the list has one element. This lemma does not say anything about `bubblePass([])`, so you do not need to address that case in your proof.

6.8.10 Prove the following theorem about `bubbleSort` by induction on the structure of xx, making use of Lemma 6.19.

Theorem 6.20 *For all* $xx \in \mathcal{L}_{\text{int}}$, `isSorted`(`bubbleSort`$(xx)$).

6.8.11 Consider the following alternate implementation of bubble sort, which repeats `bubblePass` on the whole list until it is sorted.

```
fun bubbleSort2(aa) =
    if isSorted(aa)
    then aa
    else bubbleSort2(bubblePass(aa));
```

How does this function demonstrate that merely proving that a function's output is sorted is not sufficient truly to prove correctness of a sorting algorithm?

Exercises 6.8.12–6.8.14 walk you through a partial proof of correctness for your implementation of selection sort from Exercises 3.7.11–3.7.13.

6.8.12 Prove the following lemma about your helper function `removeFirst` from Exercise 3.7.11 by induction on the structure of aa.

Lemma 6.21 *For all* $x \in \mathbb{Z}$ *and* $aa \in \mathcal{L}_{\text{int}}$, *if* $x \ll aa$, *then* $x \ll$`removeFirst`(x, aa).

6.8.13 Prove the following lemma about your helper function `listMin` from Exercise 3.7.12 by induction on the structure of aa (the base case is a list with one element, not an empty list).

Lemma 6.22 *For all* $aa \in \mathcal{L}_{\text{int}}$ *where* $aa = a :: aRest$ *for some* $a \in \mathbb{Z}$ *and* $aRest \in \mathcal{L}_{\text{int}}$, `listMin`$(aa) \ll aa$.

Exercises, continued

6.8.14 Use Lemmas 6.21 and 6.22 and induction on the structure of yy to prove the following theorem about your function `selectionSort` from Exercise 3.7.13.

Theorem 6.23 *For all* $yy \in \mathcal{L}_{\text{int}}$, `isSorted`(`selectionSort`($yy$)).

Exercises 6.8.15–6.8.19 walk you through writing and proving part of the correctness for an implementation of *insertion sort*, which works by maintaining a sorted list and repeatedly putting new items in the right place in that sorted list.

6.8.15 Write a function `putInPlace` that takes an int and a list, assumed sorted, and returns a list like the one given except with the given item put in sorted position. For example, `putInPlace(5, [2,3,7,8])` would return `[2,3,5,7,8]`.

6.8.16 Prove the following lemma for your function by induction on the structure of aa.

Lemma 6.24 *For all* $x \in \mathbb{Z}$ *and* $aa \in \mathcal{L}_{\text{int}}$, *if* `isSorted`($aa$), *then* `isSorted`(`putInPlace`($x$, aa)).

6.8.17 Write a function `insertionSort` that takes two lists—the first is the list to be sorted, the second is the accumulated answer. The function should essentially take one element from the unsorted list (first parameter) and add it to the sorted list (second parameter). It should return a sorted version of the original list. For example, `insertionSort([4, 8, 1], [3, 6, 7])` should return `[1, 3, 4, 6, 7, 8]`, and `insertionSort([6, 2, 7, 5], [])` should return `[2, 5, 6, 7]`.

6.8.18 Prove the following lemma for your function `insertionSort` by induction on the structure of aa. Use Lemma 6.24

Lemma 6.25 *For all* $aa, bb \in \mathcal{L}_{\text{int}}$, *if* `isSorted`($bb$), *then* `isSorted`(`insertionSort`($aa$, bb)).

6.8.19 Use Lemma 6.25 to prove the following theorem about your function `insertionSort`.

Theorem 6.26 *For all* $aa \in \mathcal{L}_{\text{int}}$, `isSorted`(`insertionSort`($aa$, [])).

6.9 Iteration

For the entire course you have been writing self-referential programs. In this chapter you have seen self-referential types and self-referential proofs. This section introduces *non*-self-referential programs, which may seem out of place. However, this is a widespread programming style, and the following section presents proofs of correctness for programs in this style.

Any non-trivial program needs to repeat some steps. One mechanism for repetition is recursion, where the program applies itself to a smaller version of the problem. The alternative is *iteration*, where a program performs a sequence of steps in a loop. The latter is natural for most mainstream programming languages, but it goes against the grain of ML. Accordingly, our programs in this section will be inelegant, compared to equivalent programs in other languages and compared to the ML programs we have seen previously.

iteration

Iterative algorithms are learned in grade-school mathematics. Take multi-digit addition. The algorithm is informed by how we arrange the materials we work on. The two numbers are analyzed by "columns," which represent powers of ten. First you write the two addends, one above the other, aligning them by the ones column. The answer will be written below the two numbers, and carry digits will be written above. At every point in the process, there is a specific column that we are working on (call it the "current column").

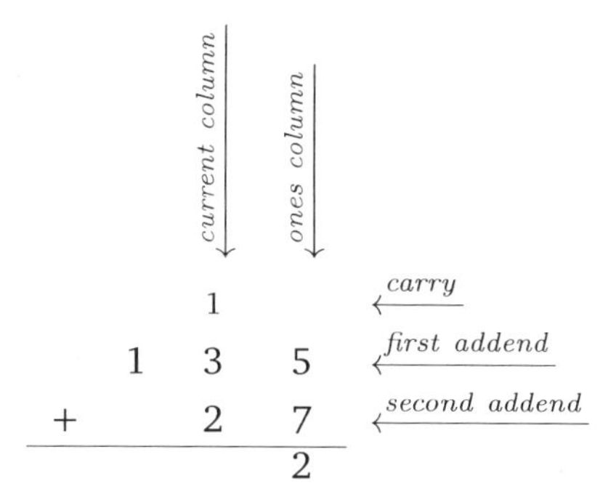

Our process is

1. Start with the ones column.
2. While the current column has a number for either of the addends, repeatedly
   (a) Add the two numbers in the current column plus any carry from last time; call this result the "sum."
   (b) Write the ones column of the sum in the current column of the answer.
   (c) Write the tens column of the sum in the carry space of the next column to the left.
   (d) Let the next column to the left become our new current column.
3. If the last round of the repetition had a carry, write it in a new column for the answer.

Notice how this algorithm takes two addends as its input, produces an answer for its output, and uses the notions of sum and current column as temporary scratch space. It is particularly important to notice that the sum and the current column keep changing.

A grammatical analysis of the algorithm is instructive. All the sentences are in the imperative mood. Contrast this with *propositions*, which are indicative. The algorithm does contain propositions, however: the independent clauses "the current column has a number for either of the addends" and "the last round of the repetition has a carry" are either true or false. Moreover, those propositions are used to guard Steps 2 and 3; the words "if" and "while" guide decisions about whether or how many times to execute a certain command. Finally, note that Step 2 is actually the repetition of four smaller steps, bound together as if they were one. This is similar to how we use simple expressions to build more complex ones.

To replicate this process in ML, we must use some features of the language that we have seen only briefly, such as reference variables and statement lists. Recall that

- to *declare* a reference variable, precede the expression assigned with the keyword `ref`.

```
- val x = ref 5;

val x = ref 5 : int ref
```

- to *use* a reference variable, precede the variable with `!`.

```
- !x;

val it = 5 : int
```

assignment statement

- to *set* a reference variable, use an *assignment statement* in the form

<identifier> := <expression>

```
- x := !x + 1;

val it = () : unit

- !x;

val it = 6 : int
```

while statement

Here is the new piece of ML. We can repeat an expression or statement with the *while* construct, which takes the form

`while` <expression> `do` <expression>

The interpreter executes the first expression (which must have type bool) and, if it is true, executes the second; it repeats this process until it finds the first expression false.

```
- val x = ref 6720;

val x = ref 6720 : int ref

- while !x mod 2 = 0 do x := !x div 2;

val it = () : unit
```

```
- !x;

val it = 105 : int
```

Or, put all together,

```
- let val x = ref 6720
= in
=   (while !x mod 2 = 0 do x := !x div 2;
=     !x)
= end;

val it = 105 : int
```

Consider our old friend factorial. Using product notation, we define

$$n! = \prod_{i=1}^{n} i$$

For instance,

$$5! = \prod_{i=1}^{5} i = 1 \times 2 \times 3 \times 4 \times 5 = 120$$

Notice the algorithmic nature of this definition (or just the product notation). It says,

keep a running product while
repeatedly multiplying by a multiplier,
incrementing the multiplier by one at each step,
starting at one,
until a limit for the multiplier is reached.

In ML,

```
- fun factorial(n) =
=   let
=      val i = ref 0;
=      val fact = ref 1;
=   in
=     (while !i < n do
=        (i := !i + 1;
=         fact := !fact * !i);
=        !fact)
=   end;
```

```
val factorial = fn : int -> int

- factorial(5);

val it = 120 : int
```

To write iterative algorithms, you must ask yourself three questions: *What are we repeating? How do we start?* and *When do we stop?* The best order in which to consider these questions might differ from problem to problem.

Suppose we want to write an iterative approach to finding the smallest element in a list of ints. First, imagine how you would do this by hand if given a list of integers on paper. You would scan the list from left to right, remembering the smallest you have seen so far. With that much, we can put together a template for this algorithm.

```
- fun listMin(L) =
=   let
=       val min = ...
=       ...
=   in
=    (while ... do
=     (if ... < !min then min := ... else ();
=      ...);
=     !min)
=   end;
```

The variable `min` stands for the smallest value we have seen so far. It is important that you do not think of it as "the smallest value over all." That would indeed be an accurate description of the variable *when the algorithm finishes,* but it does not take into account how the variable changes during the course of the execution. Initially, the smallest we have seen so far is the first item in the list. (We are gambling here that the list is not empty.)

```
- fun listMin(L) =
=   let
=       val min = ref (hd(L));
=       ...
=   in
=    (while ... do
=     (if ... < !min then min := ... else ();
=      ...);
=     !min)
=   end;
```

Now, as we go through the list, how do we mark our progress through the list, and how do we retrieve the current item? We can slice through it using `hd` and tl. Let LL represent the portion of the list we have not yet processed. Then we can find the current item at the head and mark progress by resetting LL to its tail.

```
- fun listMin(L) =
=   let
=       val min = ref (hd(L));
=       ...
=   in
=    (while ... do
=     (if hd(!LL) < !min then min := hd(!LL) else ();
=      LL := tl(!LL));
=     !min)
=   end;
```

We know we are done if there is no more LL left—that is, when LL is the empty list.

```
- fun listMin(L) =
=   let
=       val min = ref (hd(L));
=       ...
=   in
=    (while !LL <> [] do
=     (if hd(!LL) < !min then min := hd(!LL) else ();
=      LL := tl(!LL));
=     !min)
=   end;
```

Since we initialized min to the front element in the original list, we should initialize LL to tl(L).

```
- fun listMin(L) =
=   let
=       val min = ref (hd(L));
=       val LL = ref (tl(L));
=   in
=    (while !LL <> [] do
=     (if hd(!LL) < !min then min := hd(!LL) else ();
=      LL := tl(!LL));
=     !min)
=   end;

val listMin = fn : int list -> int

- listMin([54, 61, 23, 44, 89, 71, 12, 31, 22, 57, 84, 14, 33]);

val it = 12 : int
```

Exercises

6.9.1 Suppose ML did not have a multiplication operator defined for integers. Write an iterative function `multiply` that takes two integers and produces their product, without using direct multiplication (that is, using repeated addition).

6.9.2 Write an iterative function that performs integer exponentiation; that is, for integers `x` and `n`, `exp(x, n)` would return x^n.

6.9.3 Recall that series in the form $\sum_{i=0}^{n} ar^i$ is called a *geometric series*. That is, given a value r and a coefficient a, find the sum of $a+ar+ar^2+ar^3+\ldots+ar^n$. Write an iterative function that computes a geometric series, taking n, a, and r as parameters. (Do not use the result from Exercise 6.5.4; do it the "brute force" way.)

6.9.4 Write a function that takes an integer and computes its number of digits. Your function need only work on nonnegative integers—you do not need to consider negative numbers, but your function should work for zero, which has one digit. (Hint: Use `div` and `mod` to chop the number up.)

6.9.5 The Fibonacci sequence is defined by repeatedly adding the two previous numbers in the sequence (starting with 0 and 1) to obtain the next number, that is

$$0, 1, 1, 2, 3, 5, 8, 13, 21, 34, \ldots$$

Write an iterative function to compute the nth Fibonacci number, given n.

(Hint: You will need four reference variables local to the function: one to serve as a counter, one to hold the current value, one to hold the previous value, and one to hold a temporary value as you shuffle the current value to the previous:

```
temp := ...
previous := !current
current := !temp
```

Then think about how you can rewrite this so that the temporary value can be local to the while statement, and then make it a regular variable instead of a reference variable.)

6.9.6 Write an iterative function `count` that takes a list and returns the number of elements. Your solution should work for empty lists, which have size 0. Avoid making this a "special case."

6.9.7 Write an iterative function `range` that takes a list of integers and returns the range of the values.

6.9.8 Write an iterative function that takes a list of integers and returns the number of evens.

6.9.9 Write an iterative function `sum` that takes a list of integers and returns the sum of those integers.

6.9.10 Write an iterative function `doubler` that takes a list of integers and returns the list with each element doubled. For example, given `[1, 2, 5, 8]`, it would return `[2, 4, 10, 16]`.

6.9.11 Write an iterative function `reverse` that takes a list and returns the reverse of the given list.

6.9.12 Write an iterative function to compute a Caesar cipher, as in Exercise 1.12.9. The function should take a plaintext string and an int key and return an encrypted string.

6.9.13 Write an iterative version of your function `addToEach` from Exercise 2.4.14.

6.9.14 Write an iterative version of your function `powerset` from Exercise 2.4.15, using your iterative `addToEach`.

6.9.15 Write an iterative function to implement selection sort. You may find it easier to remove the maximum element repeatedly rather than the minimum because that way you can add to the front of the sorted list rather than to the back. Instead of writing helper functions to find and remove the maximum, use nested while statements.

6.10 Loop invariants

This program computes an arithmetic series from 1 to a given N, $\sum_{i=1}^{N} i$ using brute force (as opposed to using the result from Exercise 6.5.1):

```
- fun arithSum(N) =
=   let
=     val s = ref 0;
=     val i = ref 1;
=   in
=   (while !i <= N do
=     (s := !s + !i;
=      i := !i + 1);
=     !s)
= end;
```

But how do we know if this is correct? We have seen proofs of correctness for recursive functions, but how do we reason about loops and reference variables? What makes this different is that reference variables have *state*, that is, assigned values that change during the execution of the program.

state

As before, correctness needs to be defined in terms of pre-conditions, but now the pre-conditions involve not only the input but also the initial state of the reference variables. Moreover, we now speak of *post-conditions*, our expectation of what the state should be afterwards. Pre-conditions and post-conditions are sets of propositions. If all the post-conditions hold whenever the pre-conditions are met, then we say that the algorithm is correct.

post-conditions

This approach is particularly useful in that it scales down to apply to smaller portions of an algorithm; Suppose we have the pre-condition x *is a nonnegative integer* for the statement

```
y := !x + 1
```

We can think of many plausible post-conditions for this, including y *is a positive integer*, $y > x$, and $y - x = 1$. Whichever of these makes sense, they are things we can prove mathematically, as long as we implicitly take the efficacy of the assignment statement as axiomatic; propositions deduced that way are justified as *by assignment*.

Moreover, the post-conditions of one statement become the pre-conditions of the following statement; the pre-conditions of the entire program are the pre-conditions of the first statement; and the post-conditions of the final statement are the post-conditions of the entire program. Thus by inspecting how each statement in turn affects the propositions in the pre- and post-conditions, we can prove an algorithm is correct.

Consider this program for computing a `mod` b (here we are supposing that the `mod` operator does not exist and we have to write our own—but we can still use div). It does not have any reference variables, but it illustrates pre- and post-conditions.

```
- fun remainder(a, b) =
=   let
=     val q = a div b;
=     val p = q * b;
=     val r = a - p;
=   in
=     r
=   end;
```

In future examples, we will consider val-declarations merely as setting up initial pre-conditions, but since they do all the work of computation in this example, we will consider them statements to be inspected. The pre-condition for the entire program is that $a, b \in \mathbb{N}$. The result of the program should be $a \texttt{ mod } b$; equivalently, the post-condition of the declarations is $r = a \texttt{ mod } b$.

Now we intersperse the algorithm with little proofs.

Suppose $a, b \in \mathbb{Z}$.

`val q = a div b`

By assignment, $q = a \texttt{ div } b$. By the Theorem 4.21 (the Quotient-Remainder Theorem) and the definition of division, $a = b \cdot q + R$ for some $R \in \mathbb{Z}$, where $0 \leq R < b$. By algebra, $q = \frac{a-R}{b}$.

`val p = q * b`

$p = q \cdot b$ by assignment. $p = a - R$ by substitution and algebra.

`val r = a - p`

By assignment, $r = a - p$. By substitution and algebra, $r = a - (a - R) = R$. Therefore, by the definition of mod, $r = a \texttt{ mod } b$. □

So far so good, but what makes most algorithms difficult to reason about is that they have branching, repetition, and variables that change value. We will discover how to reason about the mutation of reference variables and the evaluation of conditionals along the way. First consider loops, which are more difficult.

Simplistically, a loop can be analyzed for correctness by unrolling it. If a loop were to be executed, say, five times, then we can prove correct the program that would result if we pasted the body of the loop end to end five times. The problem with that approach is that almost always the number of iterations of the loop itself depends on the input. As we see in the arithmetic series example, the guard is `i <= N`. (An *iteration* is an execution of the body of a loop; the boolean expression which we test to see if the loop should continue is called the *guard*).

iteration

guard

We want to prove a proposition that holds for an algorithm containing a loop—that is, a proposition that is true no matter how many iterations there are, as long as the number is finite. Notice:

- The proposition must be true for an arbitrary number of iterations.
- The number of iterations must be a whole number.

If this sounds like it calls for math induction, then you have figured out where all this work on iteration is heading.

The proposition in question must be flexible. It must be true before the loop starts (after 0 iterations), or the pre-condition for the entire loop. It must be true between any two iterations, or both the post-condition of the previous iteration and the pre-condition of the next iteration. And it must be true after the last iteration, or the post-condition for the entire loop. It must be parameterized by a number of iterations (or, the number of iterations "so far" in our analysis). In summary, it must state *what property the loop does not change* with respect to the number of iterations.

loop invariant

A *loop invariant* $I(n)$ is a predicate whose argument, n, is the number of iterations of the loop, chosen so that

- $I(0)$ is true (that is, the proposition is true before the loop starts; this must be proven as the base case in the proof).
- $I(n)$ implies $I(n+1)$ (that is, if the proposition is true before a given iteration, it will still be true after that iteration; this must be proven as the inductive case in the proof).
- If the loop terminates (after, say, N iterations), then $I(N)$ is true (this follows from the two previous facts and the principle of math induction).
- Also if the the loop terminates, then $I(N)$ implies the post-condition of the entire loop.

initialization

maintenance

termination

These four points correspond to steps in the proof that a loop is correct: we prove that the loop is *initialized* so as to establish the loop invariant; that a given iteration *maintains* the loop invariant; and that when/if the loop *terminates*, the loop invariant is true—and that it implies the post-condition. The first two steps constitute a proof by induction, on which we focus. The last, plus a proof that the loop *will terminate*, completes the proof of algorithm correctness.

Let us now apply this intention to the algorithm given at the beginning of this section. The post-condition for the entire algorithm is that the variable s should equal the value of the series. Since s is a reference variable, it changes value during the running of the program. Specifically, after n iterations, $s = \sum_{k=1}^{n} k$. (Why did we replace N and i with n and k?) It will also be helpful to monitor what the variable i is doing. Since i is initialized to 1 and is incremented by 1 during each iteration, then with respect to n, $i = n + 1$. Thus we have our loop invariant:

$$I(n) = \quad \text{after } n \text{ iterations, } i = n + 1 \text{ and } s = \sum_{k=1}^{n} k$$

Now we prove

Theorem 6.27 *For all $N \in \mathbb{W}$, the program* `arithSum` *computes* $\sum_{i=1}^{N} i$.

For reference, here is the algorithm again:

```
- fun arithSum(N) =
=   let
=     val s = ref 0;
=     val i = ref 1;
=   in
=   (while !i <= N do
=     (s := !s + !i;
=      i := !i + 1);
=     !s)
= end;
```

One piece we still need to fashion is reasoning about reference variables. It is difficult to reason about mathematical variables if they change mid-scope. To make this easier, we will distinguish between a variable's value before the current iteration (using the subscript "old") and the value after (using the subscript "new"). Some expressions will mix "old" and "new," so pay careful attention when that happens. For example, we interpret the statement `s := !s + !i` to imply that $s_{\text{new}} = s_{\text{old}} + i_{\text{old}}$.

Proof. We will prove $I(n)$ to be a loop invariant by induction on n.

Base case / initialization. After 0 iterations, $i = 1$ and $s = 0 = \sum_{k=1}^{0} k$ by assignment and the definition of summation. Hence $I(0)$, and so there exists an $n' \geq 0$ such that $I(n')$.

Inductive case / maintenance. Suppose $I(n')$. Let i_{old} be the value of i after the n'th iteration and before the $n' + 1$st iteration, and let i_{new} be the value of i after the $n' + 1$st iteration. Similarly define s_{old} and s_{new}. By $I(n')$, $i_{\text{old}} = n' + 1$ and $s_{\text{old}} = \sum_{k=1}^{n'} k$.

By assignment and substitution, $s_{\text{new}} = s_{\text{old}} + i_{\text{old}} = \sum_{k=1}^{n'} k + (n'+1) = \sum_{k=1}^{n'+1} k$. Similarly $i_{\text{new}} = i_{\text{old}} + 1 = (n'+1) + 1$. Hence $I(n'+1)$.

Hence by math induction, $I(n)$ for all $n \in \mathbb{W}$, and so $I(N)$.

Before we continue, be attentive to the subtle matter of our choice of variables. What is with the N, the n, and the n'? N is the input to the *program* `arithSum`. n is the parameter to (or, independent variable of) the *predicate* $I(n)$ used to analyze `arithSum`. One thing we are trying to prove is $I(N)$. n' is an arbitrary whole number such that $I(n')$, used inside of our inductive proof that $I(n)$ for all n. Properly disambiguating variables is essential for all proofs and becomes particularly hard in proofs of algorithm correctness when you are juggling similar variables and proving a property of an object which itself has variables. Failure to do this will lead to equivocal nonsense.

Termination. By $I(N)$, after N iterations, $i = N + 1 > N$, and so the guard will fail. Moreover, by $I(N)$, after N iterations, $s = \sum_{k=1}^{N} k$. By change of variable, $s = \sum_{i=1}^{N} i$, which is our post-condition for the program. Hence the program is correct. □

Our main concern for this chapter is to make you comfortable formulating and proving loop invariants. Proving termination is less important for our purposes.

Biography: C.A.R. Hoare, 1934–

Much of our understanding of program correctness comes from Charles Antony Richard (Tony) Hoare. Hoare was born in Colombo, Sri Lanka, and studied at Oxford University and the State University of Moscow. In his work on programming languages, he proposed a way to describe the semantics of a programming language using a set of axioms asserting how various statements would change the state of the computation. These axioms, then, also can be the basis of proofs of the correctness of a program. A statement together with its pre-conditions and post-conditions is called a *Hoare triple* after him.

Hoare has had his hand in many areas of computing. He developed ways to reason about concurrency and other aspects of operating systems. He performed early research on machine translation of natural languages. He is best known for the invention of the quick sort algorithm. In 2000, Queen Elizabeth knighted Hoare for his contributions to computer science. [15]

Exercises

In Exercises 6.10.1–6.10.5, prove the predicates to be loop invariants for the loops in the following programs.

6.10.1 $I(n) =$ after n iterations, x is even.

```
- fun aaa(m) =
=   let
=     val x = ref 0;
=     val i = ref 0;
=   in
=     (while !i < m do
=       (x := !x + 2 * !i;
=        i := !i + 1);
=      !x)
=   end;
```

6.10.2 $I(n) =$ after n iterations, $x + y = 100$.

```
- fun bbb(m) =
=   let
=     val x = ref 50;
=     val y = ref 50;
=     val i = ref 0;
=   in
=    (while !i < m do
=      (x := !x + 1;
=       y := !y - 1;
=       i := !i + 1);
=     !x + !y)
=   end;
```

6.10.3 $I(n) =$ after n iterations, $x + y$ is odd.

```
- fun ccc(m) =
=   let
=     val x = ref 0;
=     val y = ref 101;
=   in
=     (while !x < m do
=       (x := !x + 4;
=        y := !y - 2);
=      !x + !y)
=   end;
```

6.10.4 $I(n) =$ after n iterations, $3 \mid i$.

```
- fun ddd(a) =
=     let
=        val i = ref 0;
=        val x = ref 0;
=     in
=       (while !i < a do
=          (x := !x + !i;
=           i := !i + 3);
=        !x)
=     end;
```

6.10.5 $I(n) =$ after n iterations, $i \bmod 5 = 1$.

```
- fun eee(a) =
=     let
=        val i = ref 1;
=        val x = ref 0;
=     in
=       (while !i < a do
=          (x := !x + !i;
=           i := !i + 5);
=        !x)
=     end;
```

6.10.6 Find a useful loop invariant for your solution to Exercise 6.9.1 and prove that it is a loop invariant.

6.10.7 Find a useful loop invariant for your solution to Exercise 6.9.3 and prove that it is a loop invariant.

6.10.8 The following is the formal definition of the Fibonacci sequence:

$$F_n = \begin{cases} 0 & \text{if } n = 0 \\ 1 & \text{if } n = 1 \\ F_{n-1} + F_{n-2} & \text{otherwise} \end{cases}$$

For your solution to Exercise 6.9.5, prove the following to be a loop invariant:

$$I(n) = \begin{array}{l} \text{after } n \text{ iterations, } \textit{previous} = F_n \\ \text{and } \textit{current} = F_{n+1} \end{array}$$

You may need to adjust this slightly to fit your solution.

Exercises 6.10.2 and 6.10.3 are taken from Epp [8].

6.11 From theorems to algorithms, revisited

For better or worse, correctness proofs of algorithms are not part of most programmers' everyday life. We teach them as an intellectual exercise for two reasons: because some software should come with correctness proofs (say, software controlling the life-support system on the space station), and because loop invariants help you think about, compose, and explain your programs.

Let us suppose we want to write an iterative version of the Euclidean Algorithm from Section 4.10. Recall the Lemmas

Lemma 4.13 *If $a \in \mathbb{Z}$, then gcd$(a, 0) = a$.*

Lemma 4.14 *If $a, b \in \mathbb{Z}$, $q, r \in \mathbb{Z}^{nonneg}$, and $a = b \cdot q + r$, then gcd$(a, b) =$ gcd(b, r).*

They both provide equivalent expressions for $\gcd(a, b)$, but Lemma 4.13 addresses the special case of $b = 0$. The equivalence in Lemma 4.14 is simply a new GCD problem; Lemma 4.13 gives a final answer. Thus we have our termination condition: $b = 0$. Lemma 4.14 then helps us distinguish between what does not change and what does: The parameters to the GCD change; the answer does not. If we distinguish the changing variables a and b from their original values a_0 and b_0, we have

- Initial conditions: $a = a_0$ and $b = b_0$.
- Loop invariant: $\gcd(a, b) = \gcd(a_0, b_0)$.
- Termination condition: $b = 0$.

All that remains is how to mutate the variables a and b. a takes on the value of the old b. b takes on the value of r from Lemma 4.14.

$$\begin{array}{lllll} a_{\text{new}} & = & b_{\text{old}} & & \\ b_{\text{new}} & = & r & = & a_{\text{old}} \texttt{ mod } b_{\text{old}} \end{array}$$

In ML,

```
- fun gcd(a0, b0) =
=   let
=       val a = ref a0;
=       val b = ref b0;
=   in
=     (while !b <> 0 do
=       (a := !b;
=        b := !a mod !b);
=       !a)
=   end;

val gcd = fn : int * int -> int
```

```
- gcd(21,36);

val it = 36 : int
```

36 clearly is not the GCD of 21 and 36. What went wrong? Our proof-amenable approach has lulled us into a false sense of security, but it also will help us identify the problem speedily. The line `b := !a mod !b` is not $b_{\text{new}} = a_{\text{old}} \bmod b_{\text{old}}$ like we meant but rather $b_{\text{new}} = a_{\text{new}} \bmod b_{\text{old}}$. We must calculate r before we change a. This can be done with a let expression:

```
- fun gcd(a0, b0) =
=   let
=       val a = ref a0;
=       val b = ref b0;
=   in
=     (while !b <> 0 do
=       let val r = !a mod !b
=       in
=         (a := !b;
=          b := r)
=       end;
=      !a)
=   end;

val gcd = fn : int * int -> int

- gcd(21, 36);

val it = 3 : int
```

Exercises

6.11.1 Write an iterative version of your solution to Exercise 4.10.2.

6.11.2 Write an iterative version of your solution to Exercise 4.10.3.

6.11.3 Write an iterative version of your solution to Exercise 4.10.4.

6.11.4 Write an iterative version of your solution to Exercise 4.10.5 (the Division Algorithm).

6.11.5 Write an iterative version of your solution to Exercise 4.10.6.

6.11.6 Write an iterative version of your solution to Exercise 4.10.7 (the Extended Euclidean Algorithm).

6.12 Extended example: Huffman encoding

Computers and other digital devices store and transmit information as sequences of bits. Characters, for example, are represented either using the 7-bit ASCII code or the 16-bit Unicode standard.

In those standard encoding systems, each character takes up the same amount of memory. In the case of memory or bandwidth limitations, however, information can be represented more efficiently with *variable-length codes*, where characters that are used more frequently have shorter codes than those infrequently used. This section describes *Huffman encoding*, a way to encode information that uses variable-length bit sequences to stand for characters.

Take the string RIFFRAFF. Since only four distinct characters appear, we could use an encoding scheme with two bits per character. The entire string would require 16 bits.

A	00
F	01
I	10
R	11

11	10	01	01	11	00	01	01
R	I	F	F	R	A	F	F

Since R and F occur more frequently, we could make encode the same string with fewer bits if we adopt and encoding scheme with codes of variable length.

A	01
F	0
I	10
R	1

1	10	0	0	1	01	0	0
R	I	F	F	R	A	F	F

The message now takes only ten bits—or so it seems. But the boundaries between the letters are not part of the information sent or stored. The bit sequence is ambiguous, alternately interpreted as

A	01
F	0
I	10
R	1

1	1	0	0	01	0	10	0
R	R	F	F	A	F	I	F

To make an unambiguous scheme, we stipulate that no character can have a code that is a *prefix* for the code of another character, as the code for F is a prefix for the code for A and the code for R is a prefix for the code for I above. Here is a prefix-free variable-length encoding scheme for this example:

A	111
F	0
I	110
R	10

10	110	0	0	10	111	0	0
R	I	F	F	R	A	F	F

This takes 14 bits—only a slight win over the fixed-length version of this example, but the benefits increase with larger messages containing more distinct characters.

Before we go on, it is important to distinguish between the *code*—information describing the mapping from bit sequences to characters—and the information that is encoded using the code, which we will call the *message*. It also should be mentioned that this is not cryptographic encoding: our goal is efficiency, not secrecy.

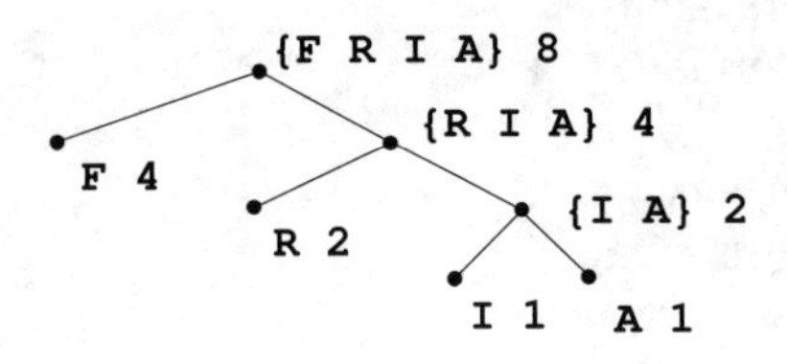

The key to understanding how variable-length prefix-free binary code works and computing an optimal one is to see the code itself represented as a full binary tree, as on the left. The tree can be searched to discover the correct character, given a bit sequence, and, given a character, to discover the bit sequence encoding that character. Each leaf stores a character and its frequency in the message being encoded. Each internal node stores (besides two subtrees) the string of all characters contained under it and the sum of all frequencies contained under it. The frequency itself is not used in encoding or decoding the message, but we will use it in building the tree.

```
- datatype huffTree =
=         Leaf of char * int
=       | Internal of huffTree * huffTree * string * int;
```

Recall that char is the ML type for values of a single character. A character literal, for example the character `A`, is written `#"A"`.

The code for each symbol is based on the path taken from the root to its leaf. Each left link is a 0, and each right link is a 1. Take the letter `I`. To get the appropriate node from the root, we would need to take a right link, another right link, and then a left link. The route is described as 110, which is the code for `I`.

Now we will implement a system to build a tree based on letter frequencies in a text, encode a text using the tree, and decode the text, also using the tree. First, a few convenience functions for analyzing and synthesizing trees. We wish to be able to find the frequency of a node, whatever it is; similarly to find all the symbols at a node; to make a tree from two nodes; and to print the tree in a way somewhat readable.

```
- fun freq(Leaf(sym, f)) = f
=   | freq(Internal(left, right, sym, f)) = f;

val freq = fn : huffTree -> int

- fun symbols(Leaf(sym, f)) = str(sym)
=   | symbols(Internal(left, right, sym, f)) = sym;

val symbols = fn : huffTree -> string
```

```
- fun makeTree(left, right) =
=       Internal(left, right, symbols(left) ^ symbols(right),
=                freq(left) + freq(right));

val makeTree = fn : huffTree * huffTree -> huffTree

- fun printTree(Leaf(sym, f)) =
=           print("Leaf(\"" ^ str(sym) ^ "\", " ^
=                  Int.toString(f) ^ ")")
=   | printTree(Internal(left, right, syms, f)) =
=           (print("makeTree("); printTree(left);
=            print(", "); printTree(right); print(")"));

val printTree = fn : huffTree -> unit
```

The above makes use of the standard ML function `str` which converts a char to a string of length 1.

Suppose we want to encode the text `ANNIKA GRACE VANDRUNEN`. Spaces count as characters. We want to count up the frequency of the various letters and make a list of pairs containing a character and its frequency. We will need to iterate over the characters in the string and add them to such a list.

First, suppose we have such a list and want to account for another character. In one way of thinking of the problem, either the character already appears in the list (in which case we would increment its frequency), or the character does not yet appear (in which case we would add it, with frequency 1). In building

Tangent: Talking with the eyes

In Alexandre Dumas's novel *The Count of Monte Cristo*, a character named Noirtier de Villefort suffers a stroke and is left completely paralyzed, able to control only his eyes and eyelids. With these alone he develops a means of communicating his basic needs to his granddaughter, for example closing his eyes to say "yes" and blinking repeatedly to say "no."

More complicated thoughts could be expressed through these building blocks. At one point, when his granddaughter sensed he wanted something,

> . . . she began to recite all the letters of the alphabet, starting with *a*, until she got to *n*, smiling and watching the invalid's face; at *n*, Noirtier indicated: "Yes."
>
> "So!" Valentine said. "Whatever you want starts with *n*. And what do we want after *n*? Na, ne, ni, no. . . "
>
> "Yes, yes, yes," the old man said.
>
> Valentine went to fetch a dictionary, which she put on a reading stand in front of Noirtier. She opened it and when she saw that he was looking attentively at the pages, she ran her finger up and down the columns. At the word *notary*, Noirtier signalled her to stop.

After Valentine teaches the method of communication to the notary, Noirtier is able even to rewrite his will. The principle of systems like the Huffman encoding are not so different: as a series of yeses and noes directs the search of a dictionary, so here a series of bits describes a path in a tree.

a function to do this, we break this into three cases: the list is empty, the list begins with the letter we are adding, and the list begins with a different letter.

```
- fun addCharToFreqList(c, []) = [(c, 1)]
=    | addCharToFreqList(c, (x, f)::rest) =
=      if c = x then (x,f+1)::rest
=               else (x,f)::addCharToFreqList(c, rest);

Warning: calling polyEqual
val addCharToFreqList = fn
     : ''a * (''a * int) list -> (''a * int) list
```

The ML interpreter cannot determine that we mean a list of *characters*, but that is not a problem. Now we write two functions that will apply this to every letter in a list of characters and finally convert a message to a list of pairs. Recall that `explode` takes a string and returns a list of the characters in the string.

```
- fun charListToFreqList([]) = []
=    | charListToFreqList(c::rest) =
=         addCharToFreqList(c, charListToFreqList(rest));

val charListToFreqList = fn : ''a list -> (''a * int) list

- fun msgToFreqList(msg) = charListToFreqList(explode(msg));

val msgToFreqList = fn : string -> (char * int) list

- val namePairs = msgToFreqList("ANNIKA GRACE VANDRUNEN");

 val namePairs =
   [(#"N",5),(#"E",2),(#"U",1),(#"R",2),(#"D",1),(#"A",4),
    (#"V",1),(#" ",2),(#"C",1),(#"G",1),(#"K",1),(#"I",1)]
         : (char * int) list
```

These will all become leaves eventually. The function `makeLeafSet` converts a list of pairs into a list of leaves, sorting the leaves in descending order of frequency as it goes. It makes use of the function `adjoinSet` which acts like the cons operator except that it sorts as it goes; writing `adjoinSet` is a project.

```
- fun makeLeafSet([]) = []
=    | makeLeafSet((sym, f)::morePairs) =
=        adjoinSet(Leaf(sym, f), makeLeafSet(morePairs));

val makeLeafSet = fn : (char * int) list -> huffTree list
```

```
- val leafList = makeLeafSet(namePairs);

val leafList =
  [Leaf (#"N",5),Leaf (#"A",4),Leaf (#" ",2),Leaf (#"R",2),
   Leaf (#"E",2),Leaf (#"I",1),Leaf (#"K",1),Leaf (#"G",1),
   Leaf (#"C",1),Leaf (#"V",1),Leaf (#"D",1),Leaf (#"U",1)]
       : huffTree list
```

An efficient tree will have the most frequently used letters closest to the root and the least frequently used at the end of the longest sequence of links, so our strategy is merging smallest-frequency trees. Since our list is reverse sorted the last two items will have a frequency at least as small as all other items in the list. The function `oneMerge` takes a list of huffTree nodes and merges the final two into a new tree.

```
- fun oneMerge([x, y]) = (makeTree(x, y), [])
=   | oneMerge(x::rest) =
=       let val (node, others) = oneMerge(rest);
=       in (node, x::others) end;

Warning: match nonexhaustive
          x :: y :: nil => ...
          x :: rest => ...

val oneMerge = fn : huffTree list -> huffTree * huffTree list

- oneMerge(leafList);

val it =
 (Internal (Leaf (#,#),Leaf (#,#),"DU",2),
   [Leaf (#"N",5),Leaf (#"A",4),Leaf (#" ",2),Leaf (#"R",2),
    Leaf (#"E",2),Leaf (#"I",1),Leaf (#"K",1),Leaf (#"G",1),
    Leaf (#"C",1),Leaf (#"V",1)])
  : huffTree * huffTree list
```

The function `successiveMerge` repeats this process until we have a single tree. It, too, is a project.

```
- val nameTree = successiveMerge(leafList);

val nameTree =
  Internal
    (Internal (Internal #,Leaf #,"DUCVREN",13),
     Internal (Internal #,Leaf #,"KGI A",9),"DUCVRENKGI A",22)
          : huffTree
```

```
- printTree(nameTree);

makeTree(makeTree(makeTree(makeTree(makeTree(Leaf("D", 1),
Leaf("U", 1)), makeTree(Leaf("C", 1), Leaf("V", 1))),
makeTree(Leaf("R", 2), Leaf("E", 2))), Leaf("N", 5)),
makeTree(makeTree(makeTree(makeTree(Leaf("K", 1),
Leaf("G", 1)), Leaf("I", 1)), Leaf(" ", 2)),
Leaf("A", 4)))
val it = () : unit
```

A graphic will illuminate that result. The _ stands for a space character.

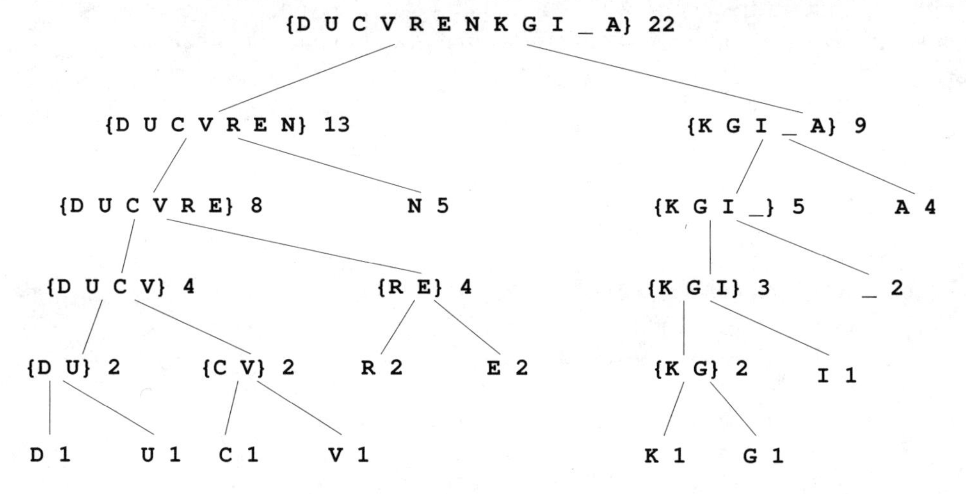

The function encodeSymbol takes a character and a Huffman tree and finds the code for that symbol by traversing the tree. It returns a list of bits (which we define below in a datatype), which serves as a record of the search. This function is left as a project. Once it is written, we can use it to encode an entire list of characters into a list of bit sequences. Recall that explode takes a string and returns a list of the characters in the string.

```
- datatype bit = One | Zero;

datatype bit = One | Zero

- fun encodeList([], tree) = []
=    | encodeList(fst::rest, tree) =
=        encodeSymbol(fst, tree)@encodeList(rest, tree);

val encodeList = fn : char list * huffTree -> bit list

- fun encode(msg, tree) = encodeList(explode(msg), tree);
```

```
val encode = fn : string * huffTree -> bit list

- val msg = encode("ANNIKA GRACE VANDRUNEN", nameTree);

val msg = [One,One,Zero,One,Zero,One,One,Zero,Zero,One,...]
: bit list
```

Notice our message is a simple list of bits, not a list of list of bits. We do not show the boundaries of the letters in our message.

Decoding the message, a final part of the project, uses the bit sequence to traverse the tree. When it hits a leaf, it adds a new character to the result and goes back to the top of the tree.

```
- decode(msg, nameTree);

val it = "ANNIKA GRACE VANDRUNEN" : string
```

This section was adapted from Abelson and Sussman[1].

Project

6.A Finish the following function adjoinSet. Remember that it takes a tree and a list of trees and places the tree in the right place in the list to maintain that list as backwards sorted by frequency.

```
fun adjoinSet(x, []) = [x]
  | adjoinSet(x, fst::rest) =
    if freq(x) > freq(fst)
    then ??
    else ??;
```

Biography: David Huffman, 1925–1999

David Huffman was a computer scientist who made important contributions to information theory. After serving in the Navy in World War II, Huffman attended graduate school at MIT. While there he took a certain course that gave students the option of writing a term paper to get out of the final exam. It was in that term paper that Huffman presented his most famous result, the encoding scheme that bears his name. Although widely used in industry, the encoding was never patented, leading some to observe that his "main compensation was dispensation from a final exam" [30].

Huffman taught at MIT and the University of California Santa Cruz, where he helped found the computer science department. His later work included research into paper folding and computational origami.

Project, continued

6.B Finish the function `successiveMerge`. It should make use of `oneMerge` to do each step of the merging, but it should also make use of your `adjoinSet` to maintain sortedness. In other words, you may assume in writing `successiveMerge` that any list your function receives is already sorted, but this means that any recursive call to `successiveMerge` must pass a list that is also sorted. Remember also that `adjoinSet` returns a list of pairs..

```
fun successiveMerge([x]) = x
  | successiveMerge(nodes) = ??;
```

6.C Finish the function `encodeSymbol`. It uses the case and option ML language features described in Section 2.5.

```
fun encodeSymbol(sym, tree) =
 let
  fun encodeSymbol1(sy, Leaf(st, f)) =
                   ??
    | encodeSymbol1(sy,Internal(left,right,
                                st,f)) =
        case (encodeSymbol1(sy, left),
              encodeSymbol1(sy, right)) of
           (NONE, NONE) => ??
         | (NONE, SOME bits) => ??
         | (SOME bits, x) => ??;
 in
   valOf(encodeSymbol1(sym,tree))
 end;
```

(This all assumes that every character in the message actually occurs somewhere in the tree. If not, `encodeSymbol1` will return `NONE`, and the line `valOf(encodeSymbol1(sym, tree))` will crash. Furthermore, the pattern `(SOME bits, x)` assumes `x` will always be `NONE`.)

6.D Finish the function `decode`. It makes use of a function called `implode`, which is a standard ML function and does the opposite of what `explode` does: it turns a list of characters into a string.

```
fun chooseBranch(Zero,
          Internal(left, right, st, f)) =
     left
  | chooseBranch(One,
          Internal(left, right, st, f)) =
     right;


fun decode(bits, tree) =
 let
  fun decode1([], currentBranch) = []
    | decode1(b::rest, currentBranch) =
      let val nextBranch = ?? in
        case nextBranch of
           Leaf(sym, w) => ??
         | Internal(left, right, syms, w)
            => ??
      end;
 in
  implode(decode1(bits, tree))
 end;
```

6.13 Special topic: Recursion vs. iteration

This chapter illustrated the difference between recursion and iteration from the programmer's perspective. In one style, steps in the algorithm are repeated by applying the algorithm to a smaller version of the same problem, represented by new parameters. Algorithms in the other style repeat steps by looping over a sequence of commands, updating mutable variables as execution proceeds.

But how do these differ concretely? How are these approaches implemented in computer hardware? This course of study mainly presents an abstract view of computing, not the lower levels of computer systems. The contrast between recursion and iteration, however, is an opportunity to peek under the hood.

At the machine level, computer programs are sequences of instructions that manipulate computer memory. Different levels of computer memory exist, differing in size and speed. For example, *registers* are few and fast, and most instructions directly manipulate them. There is more available *main memory*, but it is slower, and usually data from main memory must be moved temporarily into a register in order to use it in a computation.

registers

main memory

Each location in main memory has an address. The address is used when storing information in a memory location or looking up stored information. The instructions themselves are stored in main memory. A specially designated register called the *program counter* stores the address of the current instruction. Thus in the execution of the program we have a cycle: fetch the current instruction from main memory as indicated by the program counter, execute the instruction, update the program counter.

program counter

Machine instructions are very simple. Each instruction executes one thing such as loading some information from main memory to a register, adding the values from two registers and storing the result in a third register, or overwriting the program counter so that it jumps to another instruction. Programs written in programming languages like ML, Java, C, etc, must either be translated into machine code (by a program called a *compiler*) or interpreted by another program (an *interpreter*).

compiler

Each kind of computer hardware has its own set of instructions. The following table shows examples of instructions similar to those on many kinds of hardware.

LOAD c Rx	Store constant value c into register x
MOVE Rx Ry	Copy the value in register x to register y
ADD Rx Ry Rz	Add the values in registers x and y, storing the result in register z.
SUB Rx Ry Rz	Subtract the values in registers x and y, storing the result in register z.
MUL Rx Ry Rz	Multiply the values in registers x and y, storing the result in register z.
LESS Rx Ry Rz	Compute whether the value in register x is less than the value in register y, storing the result (0 for false, 1 for true) in register z.
BZ Rx a	"Branch zero": branch to address a if the value in register x is equal to 0.
JUMP a	Jump to address a.

"Branching" and "jumping" both refer to designating a next instruction by overwriting the program counter with an address a (the difference being that "branching" is conditional and "jumping" is unconditional). All other instructions implicitly set the program counter to the next instruction in the sequence.

It is immediately apparent that these instructions fit the iterative style. The registers are like reference variables, and the instructions modify the state, as statements do. The jump and branch instructions can be used to assemble a loop.

Accordingly, here is an example of a program in this language that is the equivalent of our iterative factorial function. Four registers have designated purposes in this program: R1 holds our input n, R2 stands for variable i, R3 stands for variable $fact$, and R4 is for temporary results. The precondition is that R1 is already set to the value of n. When this sequence of instructions reaches the end, R3 should hold our final answer. (We will refer to locations in the instruction stream using *labels* named L1, L2, etc, rather than addresses. The mark ;; indicates an explanatory comment.)

labels

```
fun factorial(n) =
  let
     val i = ref 0;
     val fact = ref 1;
  in
    (while !i < n do
      (i := !i + 1;
       fact := !fact * !i);
      !fact)
  end;
```

```
     LOAD 0 R2          ;; initialize i
     LOAD 1 R3          ;; initialize fact
L1:  LESS R1 R2 R4      ;; calculate i < n
     BZ R4 L2           ;; go to L2 if not i < n
     LOAD 1 R4          ;; set temp to 1
     ADD R2 R4 R2       ;; set i to i + 1
     MUL R3 R2 R3       ;; set fact to fact * i
     JUMP L1            ;; go back to L1
L2:                     ;; all done
```

If most computer hardware is inherently iterative, how is it possible to write recursive programs? Before reasoning about recursive functions, consider how functions in general are represented in machine language.

A function will be translated into a sequence of instructions, a subsequence of the overall program. As in our iterative factorial example, the sequence will have preconditions about where it expects its parameters to be in memory and postconditions about where in memory it will put its result. Try translating a recursive factorial function into machine language. We do not yet know how to translate a function *call* (let alone a recursive call), so we will leave that blank for now. Assume again that R1 is our input, R3 our final result, and R4 a place for temporary results. The ML function uses a conditional rather than pattern-matching to make the correspondence with machine language clearer.

```
fun factorial(n) =
  if n = 0
  then 1
  else n * factorial(n-1);
```

```
     BZ R1 L1           ;; go to L1 if n = 0
     LOAD 1 R4          ;; set temp to 1
     SUB R1 R4 R4       ;; set temp to n-1
     ???                ;; call function, somehow
     ???                ;; (expect result of call in R3)
     MUL R1 R3 R3       ;; result is n * old result
     JUMP L2            ;; jump to L2
L1:  LOAD 1 R3          ;; set result to 1
L2:                     ;; all done
```

There are a couple of tricky parts in this code. First, in our previous example the BZ instruction branched based on the result of a comparison. In this one, it branches based on R1 (n) itself, going to L1 if it is 0 (that is, if it is not equal to a nonzero value). Second, the instructions MUL R1 R3 R3 and LOAD 0 R3 both set the result (R3), for the recursive case and the base case, respectively. In either case, the program counter ends up at L2: in one case because of the JUMP, in the other case because L2 simply comes after the instruction.

Think about how to implement a call to a function using machine instructions. Before the call we must store the actual parameter in the memory location specified by the function's precondition. Then we must jump to the first instruction in the function. But that is not all—we also need to think about how to *return* from a call, that is, come back to the spot we left when we called the function.

To do this, we introduce a special purpose register like the program counter, this one called the *return address pointer*. We introduce two new instructions:

return address pointer

JAL a "Jump and link": set the program counter to the address a and the return address pointer to the address following this current instruction

RET "return": set the program counter to the value stored in the return address pointer.

With these in hand, we write this naïve recursive factorial in machine language. See if you can figure out what is wrong with it.

```
fun factorial(n) =
  if n = 0
  then 1
  else n * factorial(n-1);
```

```
L0: BZ R1 L1            ;; go to L1 if n = 0
    LOAD 1 R4           ;; set temp to 1
    SUB R1 R4 R4        ;; set temp to n-1
    MOVE R4 R1          ;; set up n for call
    JAL L0              ;; make call
    MUL R1 R3 R3        ;; result is n * old result
    JUMP L2             ;; jump to L2
L1: LOAD 1 R3           ;; set result to 1
L2: RET                 ;; all done -- return
```

First of all, when this function sets up n for the recursive call, it overwrites register R1, which it still needs for the multiplication after the call returns. Second, the JAL instruction overwrites the return address pointer. Suppose some function g called `factorial`. When `factorial` makes its recursive call, the information in the return address pointer telling how to get back to g is lost. To fix this, try saving the old value of n and the return address pointer to registers R5 and R6, respectively. (The return address pointer is R31.)

```
fun factorial(n) =
  if n = 0
  then 1
  else n * factorial(n-1);
```

```
L0: BZ R1 L1            ;; go to L1 if n = 0
    LOAD 1 R4           ;; set temp to 1
    SUB R1 R4 R4        ;; set temp to n-1
    MOVE R1 R6          ;; save old n
    MOVE R31 R7         ;; save old rap
    MOVE R4 R1          ;; set up n for call
    JAL L0              ;; make call
    MOVE R6 R1          ;; restore old n
    MOVE R7 R31         ;; restore old rap
    MUL R1 R3 R3        ;; result is n * old result
    JUMP L2             ;; jump to L2
L1: LOAD 1 R3           ;; set result to 1
L2: RET                 ;; all done -- return
```

But this has the same problem: If we call `factorial(5)`, the function will store the old n and the old return address pointer all right before calling `factorial(4)`, but that function call will also attempt to store its old n and return address pointer before calling `factorial(3)`, wiping out the records of `factorial(5)`.

activation

Now for the solution, how to implement recursion on an iterative machine. For terminological clarity, we need to distinguish between a function and an *activation* of a function. When we call a function, we create a new activation of that function. Thus the call `factorial(5)` creates an activation of function `factorial` with parameter 5. The execution of that function for that activation will result in the call `factorial(4)`, which creates another activation for the function `factorial`. Thus at this point in the execution of the program there are two activations of the same function. Eventually `factorial(0)` is called and a corresponding activation is created, but when the function returns, the activation with parameter 0 is destroyed. Similarly for the other activations until the activation with parameter 4 finishes and only the original activation is left. Finally, the activation with parameter 5 finishes and no activations of `factorial` are left.

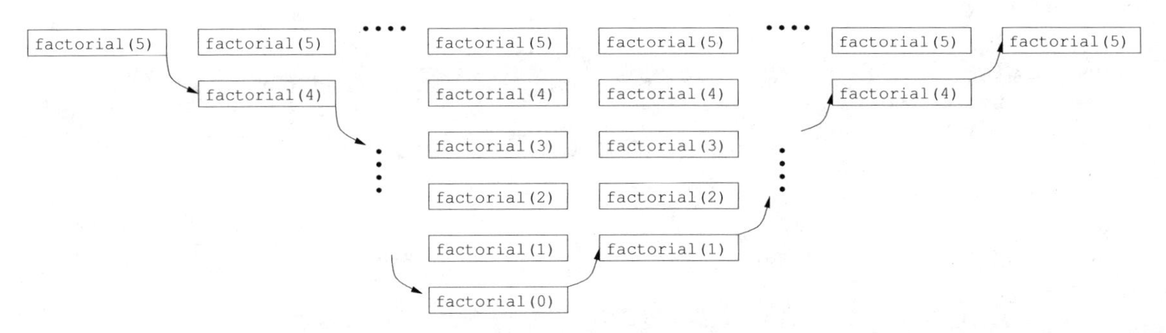

activation record

call stack

Each activation has its own parameter values and its own versions of any local variables. Information for each activation must be stored in an *activation record*. Think of the activation records as being organized like a pile of papers on a table. The top paper is the activation record of the currently executing function. If that function makes a call—whether to itself or another function—then a new activation record is created and put on top of the pile. When an activation of a function finishes, that top paper is taken away and the next paper underneath is now the top one again. We call this pile the *call stack*.

stack frame

frame pointer

Since the number of registers is limited but an arbitrary number of activations of a function may be live at any given time, we will need to store the activation records in main memory. The sequence of memory locations used for a single activation record is called a *stack frame*. A range of main memory is designated for the call stack, and the stack frames are laid out, one after another, in it. Since the call stack keeps changing as functions are called and return, we use two specially designated registers to keep track of the current state. The *frame pointer* holds the address of the beginning of the top, current stack frame, and the *stack pointer* holds the address following the current stack frame, where the next stack frame would begin.

stack pointer

Suppose function g calls `factorial(1)`. Suppose the stack frame for the activation of g is memory locations 1024–1029, and a stack frame is made for the activation of `factorial(1)` in memory locations 1030–1031. The frame pointer and stack pointer define the boundaries of the top frame, the program counter holds the address of the current instruction in the function `factorial`, and the return address pointer holds the address in the function g to be executed when `factorial` is finished.

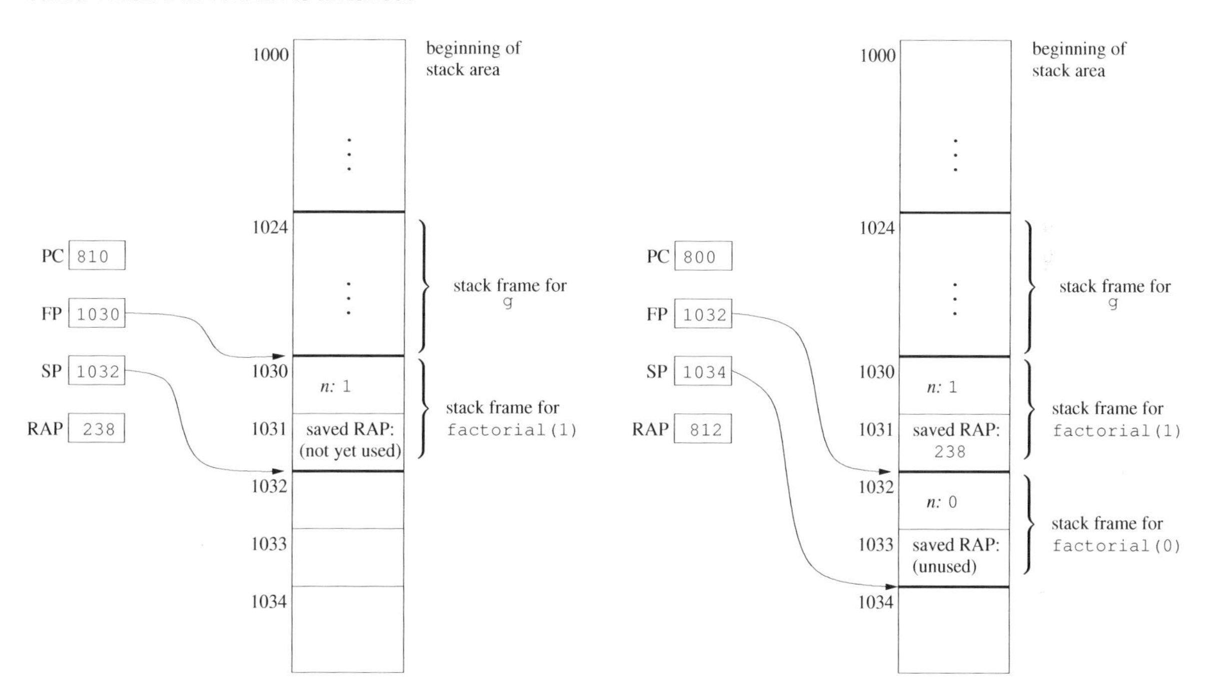

At some point, `factorial` will make a recursive call. The old return address pointer is stored in the current stack frame. A new stack frame (for the activation of `factorial(0)`) is made in locations 1032–1033, and the stack pointer and frame pointer are adjusted for this. The parameter $n = 0$ is stored in this new frame. The program counter and return address pointer are set to the beginning instruction of `factorial` and the instruction of `factorial` following the recursive call.

In order to write machine code to do all this, we need a way to read from and write to main memory. Here are two new instructions:

READ Rx c Ry — Assume register x holds a memory address. Store the value found c locations away from the address in register x into register y.

STOR Rx Ry c — Store the value in register x into the memory location c locations away from the address in register y.

Our interaction with main memory is done through *offsets*—we specify the location using a register r and a constant c and read or modify location $r + c$. This may seem unnecessarily complicated, but it makes sense when working with a stack frame. If we want to read a location within the current stack frame, then the location can be specified by an offset c from the address stored in the frame pointer.

Here then is the correct translation to machine code. Now our precondition is that a new frame has been made and that the parameter n is stored in the first memory location (offset 0) of the frame—we no longer assume that n is in register 1. This also means we do not need to save it before the recursive call—it already is saved in the current frame. We still assume register 3 is used for the returned value. Registers 4 and 5 are both used for temporary values. The stack pointer and frame pointer are registers 29 and 30, respectively.

```
fun factorial(n) =
  if n = 0
  then 1
  else n * factorial(n-1);
```

```
L0: READ R30 0 R1       ;; load n to reg 1
    BZ R1 L1            ;; go to L1 if n = 0
    LOAD 1 R4           ;; set temp4 to 1
    SUB R1 R4 R4        ;; set temp4 to n-1
    STOR R31 R30 1      ;; save old rap
    LOAD 2 R5           ;; set temp5 to 2
    ADD R29 R5 R29      ;; sp = sp+2
    ADD R30 R5 R30      ;; fp = fp+2
    STOR R4 R30 0       ;; put n-1 in new frame
    JAL L0              ;; make call
    LOAD 2 R5           ;; set temp5 to 2
    SUB R30 R5 R30      ;; fp = fp-2
    SUB R29 R5 R29      ;; sp = sp-2
    READ R30 1 R31      ;; restore old rap
    READ R30 0 R1       ;; restore old n
    MUL R1 R3 R3        ;; result is n * old result
    JUMP L2             ;; jump to L2
L1: LOAD 1 R3           ;; set result to 1
L2: RET                 ;; all done -- return
```

Do not be disheartened if the details seem overwhelming. You can learn them in a later course on computer systems. For now we are interested in one observation: Recursion is implemented using a call stack, and that has three implications. First, machine code for recursive functions is longer because of the instructions to manage the call stack. Second, recursive functions take longer to execute because of the overhead of managing the call stack. Third, recursive functions use more memory, namely the call stack.

Considering the effort we have spent learning to program in the recursive style, this assessment sounds grim. There is hope. The following factorial function uses a second parameter to accumulate its result (to compute $n!$, call `factorial(n, 1)`).

```
- factorial(n, i) =
=     if n = 0
=     then i
=     else factorial(n-1, n * i)
```

What is so different? In traditional factorial, the multiplication happens after the recursive call returns, multiplying its result by n. In this version, the multiplication happens before the call to compute the parameter, and the result of the recursive call is the final result. In other words, no more work is needed to be done in the original activation.

Suppose we call `factorial(5, 1)`. Instead of burying that activation record under a new one for `factorial(4, 5)`, we can simply *replace* the old activation with the new one. *We never need to come back.* This way a JAL can be replaced with a simple JUMP, and the call stack does not need to be used.

Here is the machine code for the accumulating version of factorial. For simplicity, we revert to the precondition that parameter n is in register 1. Further assume that parameter i is in register 2. We will also store our result in register 2.

```
fun factorial(n, i) =
   if n = 0
   then i
   else factorial(n-1, n * i)
```

```
L0: BZ R1 L1         ;; go to L1 if n = 0
    MUL R1 R2 R2     ;; i = n * i
    LOAD 1 R4        ;; set temp to 1
    SUB R1 R4 R1     ;; n = n - 1
    JUMP L0          ;; "recursive call"
L1:                  ;; all done
```

When a function call is the last thing to be done in a function, it is called a *tail call*. If all calls are tail calls, then minimal use of the call stack is necessary. However, the programmer does not need to change programming habits and rewrite all functions to use tail calls. The compiler or interpreter is able to make that transformation on the code before translating to machine code.

tail call

Chapter summary

Thinking in terms of recursion is a major goal of this course and is not confined to this chapter. However, it is here that we find the widest expression of the idea of self-reference. Of primary importance is proof by induction, that we demonstrate the truth of a proposition over a recursively defined set by proving it true for the primitive members of that set and then, for non-primitive members, showing that the proposition being true for their components implies it is true for them.

Key ideas:

Using the Peano axioms, a whole number is either 0 or one plus another whole number.

A full binary tree is either a single node with no links, or it is a node together with links to two other full binary trees. The node is the root of that tree. The roots of the subtrees are the children of that node.

Two or more functions are mutually recursive if they call each other.

Structural induction is used to prove propositions quantified over recursively defined sets. The proof involves showing that the proposition holds for the base case of the definition (which is the base case of the proof) and showing that if the proposition holds for some elements in the set, then it also holds for elements in the set built from those elements (which is the inductive case of the proof).

Mathematical induction is a special case of structural induction, where the proposition is quantified over the whole numbers or the natural numbers. The proof consists in demonstrating that the proposition holds for 0 or 1 (which is the base case of the proof) and showing that if the proposition holds for some number n, then it also holds for $n+1$ (which is the inductive case of the proof).

In a proof using math induction, when proving that the proposition holds for some number $n+1$, the inductive hypothesis is the supposition that the proposition holds for n.

So-called strong induction differs from ordinary math induction in that the inductive hypothesis is the supposition that the proposition holds for all whole numbers less than or equal to n.

Some propositions that are not quantified over whole numbers still can be proved using math induction by first rewriting the proposition so that it is quantified over whole numbers.

Many simple recursive programs can be proved correct using structural or mathematical induction. First prove that the function produces the correct answer for the base case of the recursion and then prove that the recursive cases are correct, using the correctness of the recursive calls as the inductive hypothesis.

Divide-and-conquer algorithms split the input into two or more similar, smaller problems, solve each subproblem, and combine the results to produce the final result.

Merge sort divides its input in half, sorts each half, and merges the two smaller sorted lists into a large sorted list.

Quick sort partitions its input into two smaller lists based on which items are larger or smaller than a pivot, sorts each smaller list, and combines the two smaller sorted lists by concatenating them.

Iterative algorithms compute their result by repeating a set of statements which modify mutable variables.

A loop invariant is a proposition quantified over all iterations of a loop—that is, a proposition true before a loop begins, after every iteration, and therefore after the loop finishes. Loop invariants are used in proofs of correctness for iterative algorithms.

A proposition is proved to be a loop invariant by math induction. The base case, called the initialization step, is a proof that the proposition is true before the loop starts (after 0 iterations). The recursive case, called the maintenance step, is a proof that if the proposition is true before an iteration begins, then it is also true at the end of that iteration.

The Huffman encoding is a scheme for efficient representation of data in binary. The key to the encoding is a full binary tree. The leaves of the tree stand for characters, and the code for each character is based on the path in the tree from the root to the character's node.

Chapter 7 Function

Functions pervade modern mathematics. They are essential tools for talking about algebra, analysis, calculus, analytic geometry, and trigonometry. It is hard for us today to comprehend the mathematics of earlier ages that did not have our understanding and notation of functions.

At the beginning of this text, you were told that though you had previously studied the *contents* of various number sets, you would now study sets themselves. Similarly, previous math courses have used functions to talk about other mathematical objects. Now we shall study functions as mathematical objects themselves.

There are several models or metaphors we can use to think about functions. You probably heard these in earlier math courses. First, we can think of a function as *dependence*—two phenomena or quantities are related to each other in such a way that one is dependent on the other. In high school chemistry, you may have performed an experiment where, given a certain volume of water, the change of temperature of the water was affected by how much heat was applied to the water. The number of joules applied is considered the *independent variable*, a quantity you can control. The temperature change, in kelvins, is a *dependent variable*, which changes predictably based on the independent variable. Let f be the temperature change in kelvins and x be the heat in joules. With a kilogram of water, you would discover that

$$f(x) = 4.183x$$

Next, we can think of a function as a kind of machine, one that has a slot into which you can feed raw materials and a slot where the finished product comes out, like a sausage maker.

A function is sometimes defined as a *mapping* or association between two collections of things. Think of how a telephone book associates names with phone numbers.

Finally, we can contrast functions with relations. Considered graphically, a (real-valued) function is a curve that passes the vertical line test—that is, there is no x value for which there is more than one y value. This last model best informs our formal definition of a function—a function is a restricted kind of relation. This works not only for functions in the specific world of the real-number plane, but for functions between any two sets.

Chapter goals. The student should be able to

- prove results about function equality.
- write ML programs that use anonymous functions.
- prove results about images and inverse images.
- write ML programs that use `map` and `foldl`.
- prove results about the function properties, specifically onto functions and one-to-one functions.
- prove results about inverse functions.
- prove results about composed functions.
- prove results about the cardinality of finite sets and apply those results to counting problems.

7.1 Definition

A *function* f from a set X to a set Y is a relation from X to Y such that each $x \in X$ is related to exactly one $y \in Y$. We denote that value y as $f(x)$. We call X the *domain* of f and Y the *codomain*. We write $f : X \to Y$ to mean "f is a function from X to Y."

function

domain

codomain

The definition of function is a "for all there exists unique" definition. To capture it formally, we say that $f : X \to Y$ means

$$\begin{array}{lll} \forall\, x \in X, & \exists\, y \in Y \mid (x, y) \in f & \text{(existence of } y\text{)} \\ & \wedge \quad \forall\, y_1, y_2 \in Y, ((x, y_1) \in f \wedge (x, y_2) \in f) \to y1 = y2 & \text{(uniqueness of } y\text{)} \end{array}$$

Let $A = \{1, 2, 3\}$ and $B = \{5, 6, 7\}$. Let $f = \{(1, 5), (2, 5), (1, 7)\}$. f is not a function because 1 is related to two different items, 5 and 7, and also because 3 is not related to any item. (It is not a problem that more than one item is related to 5, or that nothing is related to 6.) When a supposed function meets the requirements of the definition, we sometimes will say that it is *well defined*. Make sure you remember, however, that being well defined is the same thing as being a function. Do not say "well-defined function" unless the redundancy is warranted for emphasis.

well-defined

The term *codomain* might sound like what you remember calling the *range*. We give a specific and slightly different definition for that: for a function $f : X \to Y$, the *range* of f is the set $\{y \in Y \mid \exists\, x \in X \text{ such that } f(x) = y\}$. That is, a function may be broadly defined to a set but may actually contain pairs for only some of the elements of the codomain. An arrow diagram will illustrate.

range

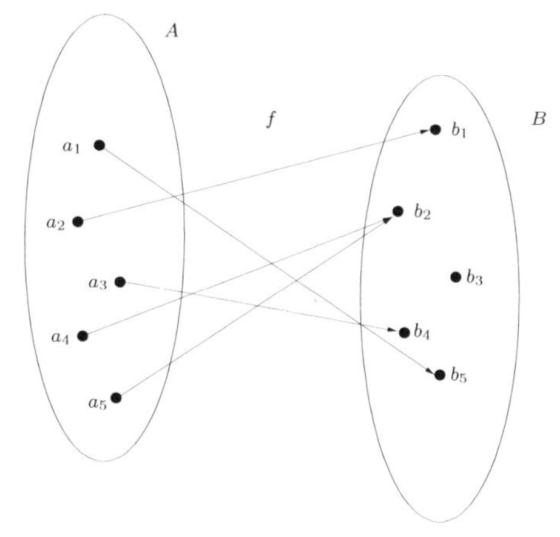

Since f is a function, each element of A is at the tail of exactly one arrow. Each element of B may be at the head of any number (including zero) of arrows. Although the codomain of f is $B = \{b_1, b_2, b_3, b_4, b_5\}$, since nothing maps to b_3, the range of f is $\{b_1, b_2, b_4, b_5\}$.

A function like $f(x) = 5$ whose range is a set with a single element is called a *constant function*. Moreover, we can redefine the term *predicate* to mean

constant function

a function whose codomain is $\mathbb{B}$. Notice that the identity relation on a set is a function. Finally, you may recall seeing some functions with more than one argument. Our simple definition of function still works in this case if we consider the domain of the function to be a Cartesian product. That is, $f(4, 12)$ is simply f applied to the tuple $(4, 12)$. This, in fact, is exactly how ML treats functions apparently with more than one argument.

7.2 Function equality

We already have a definition for what it means for two functions to be equal: functions are *relations*; relations are *sets* of ordered pairs; sets are equal if they have all the same elements—in this case, all the same ordered pairs.

However, we will use a different definition of function equality, saying that two functions are *equal* if they map all domain elements to the same things. That is, for $f : X \to Y$ and $g : X \to Y$,

$$f = g \text{ if } \forall\, x \in X, f(x) = g(x).$$

This definition is equivalent to the definition inherited from sets and relations–see Exercise 7.2.1. We use it because it makes proofs of function equality cleaner.

Theorem 7.1 *Let* $f : \mathbb{R} \to \mathbb{R}$ *as* $f(x) = x^2 - 4$ *and* $g : \mathbb{R} \to \mathbb{R}$ *as* $g(x) = \frac{(2x-4)(x+2)}{2}$. $f = g$.

Proof. Suppose $a \in \mathbb{R}$. Then, by algebra

Tangent: Functions, mappings, and metaphors

A map or mapping is a way to associate values of one domain (elements in one set) to values in another domain. A dictionary maps items from the domain of words to items in the (co-)domain of definitions. Thus a function and a mapping are really the same thing.

Fields like psychology and linguistics use the idea of a mapping to talk about metaphors. Consider the metaphor between war and love drawn up by the Roman poet Ovid in Amores I.IX:

> Every lover's in arms, and Cupid holds the fort.
> The age that's good for war, is also right for love.
> An old soldier's a disgrace, and an old lover.
> That spirit a commander looks for in a brave army,
> a lovely girl looks for in a love partner.
> Both keep watch: both sleep on the ground,
> one serves at his lady's entrance, the other his general's.
> A long road's a soldier's task: but send the girl off,
> and a restless lover will follow her to the end.[24]

War	Love
Old man as poor soldier	Old man as poor lover
The commander's approval	The crush's approval
Sleeping in a tent	Watching the crush's door
Marching	Following to the end of the earth

A metaphor is thus a kind of function.

$$\begin{aligned} f(a) &= a^2 - 4 \\ &= a^2 - 2a + 2a - 4 \\ &= (a-2)(a+2) \\ &= \frac{2(a-2)(a+2)}{2} \\ &= \frac{(2a-4)(a+2)}{2} \quad = \quad g(a) \end{aligned}$$

Hence, by definition of function equality, $f = g$. □

Notice that we chose a as the variable to work with instead of x. This is to avoid equivocation by the reuse of variables. We used x in the rules given to define f and g. a is the symbol we used to stand for an arbitrary element of $\mathbb{R}$ which we were plugging into the rule.

Exercises

7.2.1 Prove that the set theory definition of function equality and the one given in this chapter are equivalent. That is, let X and Y be sets and let f and g each be functions from X to Y. Prove that $(\forall\, x \in X, f(x) = g(x))$ iff $(f \subseteq g \wedge g \subseteq f)$.

7.2.2 Let f and g each be functions from $\mathbb{R} \times \mathbb{R}$ to $\mathbb{R}$ defined so that $f(x, y) = x + y$ and $g(x, y) = 3 \cdot x - (2 \cdot x - y)$. Prove that $f = g$.

7.2.3 Let f and g each be functions from $\mathbb{B} \times \mathbb{B}$ to $\mathbb{B}$ defined so that $f(p, q) = p \wedge (\sim q \vee (p \wedge \sim p))$ and $g(p, q) = p \wedge \sim q$. Prove $f = g$.

7.3 Functions as first-class values

Functions could be represented in ML as a special case of relations, which are a special case of sets, which are represented as a special case of lists. However, the best representation of functions in ML is of course the function, which you have been using for some time now. Our new interest is treating functions as mathematical objects.

Take the following round-about way of writing a function that takes two integers and multiplies them.

```
- fun mul(a, b) =
=    let fun m(x) = a * x;
=    in
=       m(b)
=    end;


val mul = fn : int * int -> int
```

```
- mul(2, 6);

val it = 12 : int
```

The helper function `m` is created in the local scope. But we can do more with a local function like that—we can return it as a value from a function. Values that a programming language allows to be passed to functions, stored in variables, and returned from functions are called *first-class values*, and one of the pillars of functional programming is that functions are first-class values.

first-class value

```
- fun mulMaker(a) =
=   let fun m(x) = a * x;
=   in
=     m
=   end;


val mulMaker = fn : int -> int -> int

- mulMaker(2);


val it = fn : int -> int

- it(6);


val it = 12 : int
```

A function does not even need to be stored in a variable. We can write *anonymous functions* using the following form

anonymous function

fn (<identifier>) => <expression>

Consider the following examples.

```
- fn (x) => (x + 3) mod 5;

val it = fn : int -> int

- it(15);

val it = 3 : int
```

```
- fun mulMaker(a) = fn (x) => a * x;

val mulMaker = fn : int -> int -> int

- val doubler = mulMaker(2);

val doubler = fn : int -> int

- doubler(5);

val it = 10 : int

- mulMaker(6)(12);

val it = 72 : int
```

If we take <identifier>$_1$ to be the name of the function, <identifier>$_2$ to be the name of the formal parameter, and <expression> to be the function's body, then it seems like

fun <identifier>$_1$ (<identifier>$_2$) = <expression>;

is equivalent to

val <identifier>$_1$ = fn (<identifier>$_2$) => <expression>;

but there is one crucial difference. In the latter, the scope of <identifier>$_1$ (the variable being defined) does not include the line itself. That name is valid after its definition, but it is not valid inside its definition itself. Since <identifier>$_1$ cannot be used in the body of the second form, it cannot be used to make a recursive call. Recursive functions must be written using the first form.

We will see many good uses of anonymous functions and, more generally, functions as first class values. For example, suppose we want to use quick sort to arrange integers in ways other than least to greatest or to arrange things other than integers. We can *parameterize* the function with a comparison function.

```
- fun partition(p, [], aa, bb, m) = (aa, bb)
=   | partition(p, x::rest, aa, bb, m) =
=     if m(x, p) then partition(p, rest, x::aa, bb, m)
=                else partition(p, rest, aa, x::bb, m);

val partition = fn
  : 'a * 'b list * 'b list * 'b list * ('b * 'a -> bool) ->
    'b list * 'b list

- fun quickSort([], m) = []
=   | quickSort(x::rest, m) =
=     let val (aa, bb) = partition(x, rest, [], [], m);
=     in  quickSort(aa, m) @ (x::quickSort(bb, m))
=     end;

val quickSort = fn : 'a list * ('a * 'a -> bool) -> 'a list

- quickSort([5, 2, 7, 4, 1, 9, 3], fn(a, b) => a > b);

val it = [9,7,5,4,3,2,1] : int list

- quickSort([5, ~2, ~7, 4, ~1, 9, ~3],
=                fn(a, b) => abs(a) < abs(b));

val it = [~1,~2,~3,4,5,~7,9] : int list

- fun isShorterOrEq([], _) = true
=   | isShorterOrEq(_, []) = false
=   | isShorterOrEq(a::aRest, b::bRest) =
=     isShorterOrEq(aRest, bRest);

val isShorterOrEq = fn : 'a list * 'b list -> bool

- quickSort([[1, 2], [], [6, 3, 7, 5], [5, 1, 2, 8, 6],
=                 [1, 2, 7]], isShorterOrEq);

val it = [[],[1,2],[1,2,7],[6,3,7,5],[5,1,2,8,6]]
         : int list list
```

The parameter m holds a function used as a comparison predicate. The function `quickSort` does not use it directly, but passes it on to `partition`.

Exercises

7.3.1 To convert from degrees Fahrenheit to degrees Celsius, we need to consider that a degree Fahrenheit is $\frac{5}{9}$ of a degree Celsius and that degrees Fahrenheit are "off by 32," that is, 32° F corresponds to 0° C. Write a function `fahrToCel` to convert from Fahrenheit to Celsius.

7.3.2 Redo your solution to Exercise 7.3.1 in val/fn form.

7.3.3 Finish one of the two versions of the function `makeListExtractor` which takes a list and returns a function that takes an integer n and returns the nth item in the list (or `~1` if none exists). This essentially converts the list into a function. For example, `makeListExtractor([5,6,7])(3)` would return 7. Both versions use a helper function just like your solution to Exercise 2.2.5. The second version uses an anonymous function.

```
fun makeListExtractor(xx) =
   let
     fun findNth([], x) = ~1
       | findNth(a::rest, 1) = ??
       | findNth(a::rest, x) = ??
     fun extract(x) = ??
   in
     extract
   end;
```

```
fun makeListExtractor(xx) =
   let
     fun findNth([], x) = ~1
       | findNth(a::rest, 1) = ??
       | findNth(a::rest, x) = ??
   in
     fn(x) => ??
   end;
```

7.3.4 Finish the function `makeListMul` which takes an integer and returns a function that takes a list of integers and returns a list like the original except with each element multiplied by the given integer. For example, `makeListMul(3)([5,6,7])` would return `[15, 18, 21]`.

```
fun makeListMul(x) =
   let
     fun listMul([]) = []
       | listMul(a::rest) = ??
   in
     ??
   end;
```

7.3.5 Write a function `makeInverse` that takes the predicate representation of a relation and returns the predicate representation of the inverse of that relation. For example, suppose `divides` is a predicate such that `divides(3, 9)` returns true and `divides(3, 10)` returns false. Then `makeInverse(divides)(9, 3)` should return true and `makeInverse(divides)(10, 3)` should return false.

7.3.6 Write a function `makeFindImage` that takes a list representation of a relation from a type 'a to a type 'b and returns a function which takes an element of type 'a and returns the image of that element under the relation. For example, `makeFindImage([(1, 3), (2, 5), (1, 8), (2, 6), (1, 2)])(1)` should return `[3, 8, 2]` (or any list with those numbers, any order).

7.3.7 Write a function `qsNumDigs` that takes lists of integers and sorts them from fewest digits to most digits. For example, `qsNumDigs([44, 512, 8, 6239])` would return `[8, 44, 512, 6239]`. Your function should call the parameterized `quickSort` from this section, giving it an appropriate comparison function.

7.3.8 Write a function `transformTree` that takes a tree (from Section 6.2) and a int $\rightarrow$int function and produces a new tree like the given tree but with all values replaced with the result of applying the given function to them.

7.3.9 Write a function `filter` that takes a predicate and a list and returns a list like the given one except with only the elements for which the predicate is true (that is, it *filters* out the failing elements). For example, `filter(fn(x) => x div 2 = 0, [4, 17, 3, 8, 6, 11])` would return `[4, 8, 6]`.

7.4 Images and inverse images

We can extend our idea of applying a function to an element by thinking about what it would mean to apply a function to an entire subset of the domain. In Section 5.3, we learned about the *image* of a set under a relation; the image concept is more important when studying functions, so we will restate the (equivalent) definition here. Suppose $f : X \to Y$ and $A \subseteq X$. The *image* of A under f is

image

$$F(A) = \{y \in Y \mid \exists\, x \in A \text{ such that } f(x) = y\}$$

The image of a subset of the domain is the set of elements in the codomain that are "hit" by elements in the subset. Note that the image of the entire domain is the *range*.

In the following diagram $C = \{a_3, a_4, a_5\}$. Then $F(C) = \{b_2, b_4\}$.

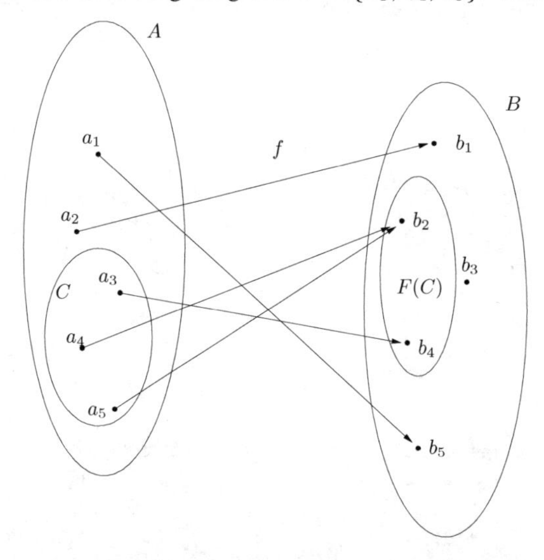

inverse image

The *inverse image* of a set $B \subseteq Y$ under a function $f : X \to Y$ is

$$F^{-1}(B) = \{x \in X \mid f(x) \in B\}$$

The inverse image of a subset of the codomain is the set of elements that hit the subset. It is vital to remember that although the image is defined by a set in the domain, it *is* a set in the *co*domain, and although the inverse image is

Notation note: Naming images

If a function is named *f*, we refer to an image of a set *A* under *f* by capitalizing the name of the function, *F*(*A*). This is unconventional. Most books would write the image as *f*(*A*). The notation used here is meant to draw our attention to how an image differs from applying a function to an element—in particular, an image is a *set*.

defined by a set in the codomain, it *is* a set in the *domain*. It is also important to be able to distinguish an inverse image from an inverse (of a) function, which we will meet later. Let $D = \{b_2, b_3\}$. Then $F^{-1}(D) = \{a_4, a_5\}$. Notice that $F^{-1}(\{b_3\}) = \emptyset$.

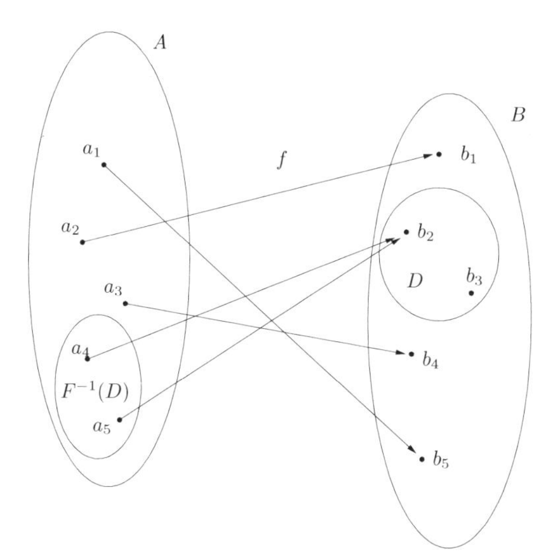

Before looking at and proving more complicated results, take note of a few things about how the empty set interacts with images and inverse images. These lemmas follow immediately from the definitions.

Lemma 7.2 *If $f : X \to Y$, then $F(\emptyset) = \emptyset$.*

Since there is nothing in the empty set, that means there is nothing in the codomain hit by anything in the empty set. The empty set's image is empty.

Lemma 7.3 *If $f : X \to Y$, $A \subseteq X$, and $A \neq \emptyset$, then $F(A) \neq \emptyset$.*

The image of a nonempty subset is never empty. If A is not the empty set, then it must contain something, so there must be something in the codomain that the elements in A hit.

Lemma 7.4 *If $f : X \to Y$, then $F^{-1}(\emptyset) = \emptyset$.*

This lemma complements the previous. Nothing maps to the empty set, so the inverse image of the empty set is the empty set.

At this point we might expect to round this out with a lemma beginning with "If $f : X \to Y$, $A \subseteq Y$, and $A \neq \emptyset$, ..." What can we say about a nonempty subset of the codomain? Nothing. It is possible that something maps to it, but it is also possible that its image is empty.

Now on to writing proofs. One could argue that proofs of propositions involving images and inverse images are straightforward. The principles from set proposition proofs and quantified reasoning are the same, we merely apply them to a few new definitions. Nevertheless, this is a common trouble spot for students. The difficulty seems to be in keeping straight whether something is a set or element. Remember that *images and inverse images are sets*. Put the techniques you learned for set proofs to work.

Theorem 7.5 *Let $f : X \to Y$ and $A, B \subseteq X$. Then $F(A \cup B) = F(A) \cup F(B)$.*

Before getting buried in the definitions, recognize that this is a proof of set equality (Set Proof Form 2). Second, recognize that $F(A \cup B) \subseteq Y$. Choose your variable names in a way that shows you understand this.

Proof. Suppose $y \in F(A \cup B)$. By the definition of image, there exists an $x \in A \cup B$ such that $f(x) = y$. By the definition of union, $x \in A$ or $x \in B$. Suppose $x \in A$. Then, again by the definition of image, $y \in F(A)$. Similarly, if $x \in B$, then $y \in F(B)$. Either way, by definition of union, $y \in F(A) \cup F(B)$. Hence $F(A \cup B) \subseteq F(A) \cup F(B)$ by definition of subset.

Next suppose $y \in F(A) \cup F(B)$. By definition of union, either $y \in F(A)$ or $F(B)$. Suppose $y \in F(A)$. By definition of image, there exists an $x \in A$ such that $f(x) = y$. By definition of union, $x \in A \cup B$. Again by definition of image, $y \in F(A \cup B)$. The argument is similar if $y \in F(B)$. Hence by definition of subset, $F(A) \cup F(B) \subseteq F(A \cup B)$.

Therefore, by definition of set equality, $F(A \cup B) = F(A) \cup F(B)$. □

It can be difficult to grasp what is going on in the proof when much of your mental effort is spent interpreting the newly introduced symbols. When you work out a proof, you may find it easier to begin by drawing a picture. We can illustrate an entire proof in a way similar to the visual verifications from Section 1.5.

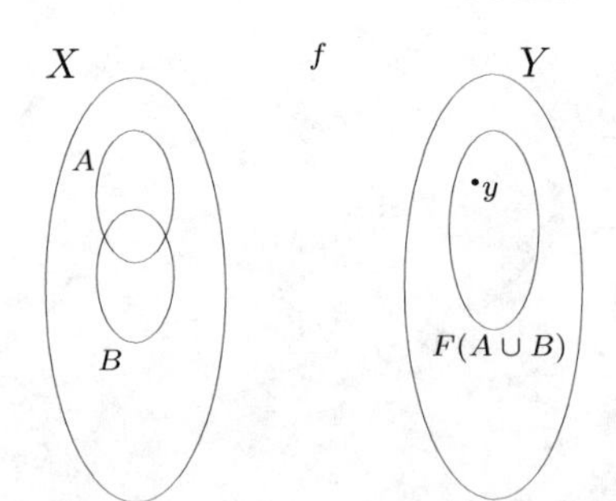

Suppose $y \in F(A \cup B)$.

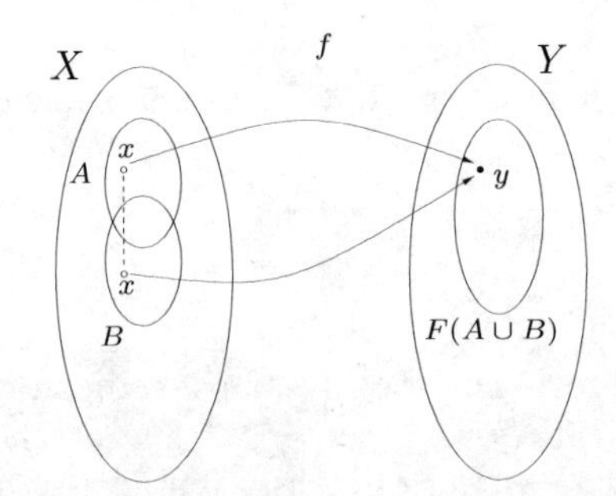

By the definition of image, there exists an $x \in A \cup B$ such that $f(x) = y$. By the definition of union, $x \in A$ or $x \in B$.

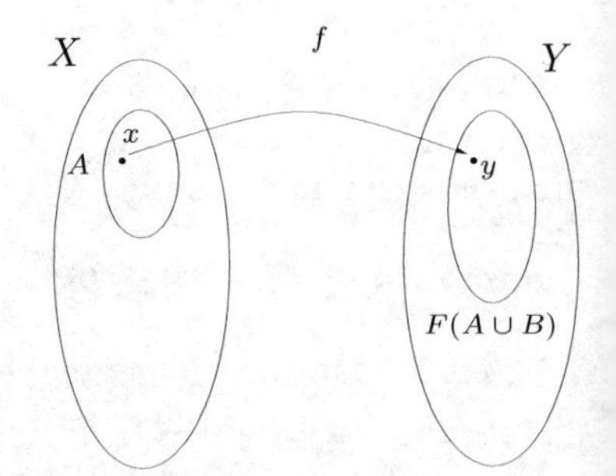

Suppose $x \in A$.

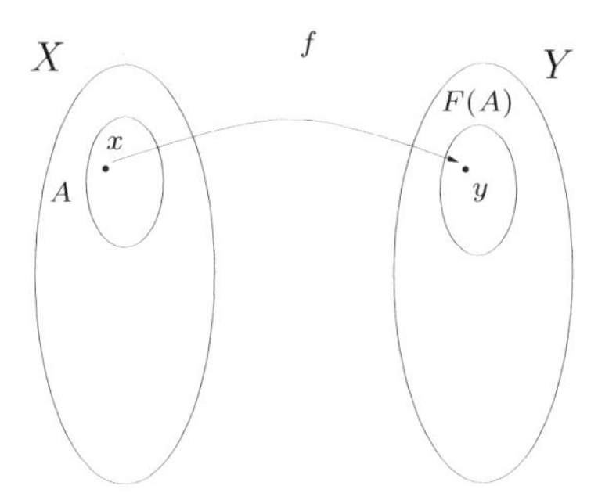

Then, again by the definition of image, $y \in F(A)$.

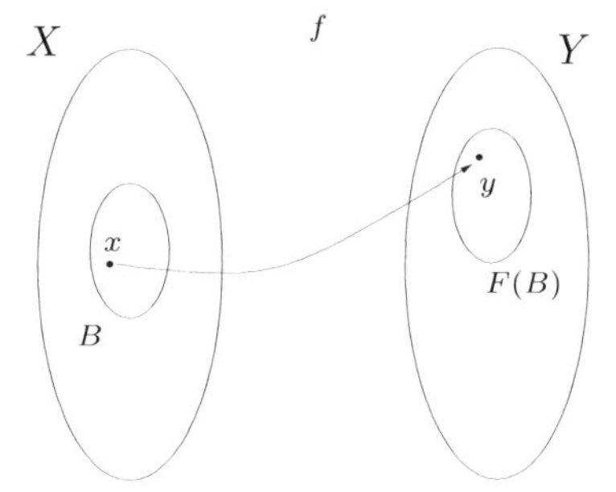

Similarly, if $x \in B$, then $y \in F(B)$.

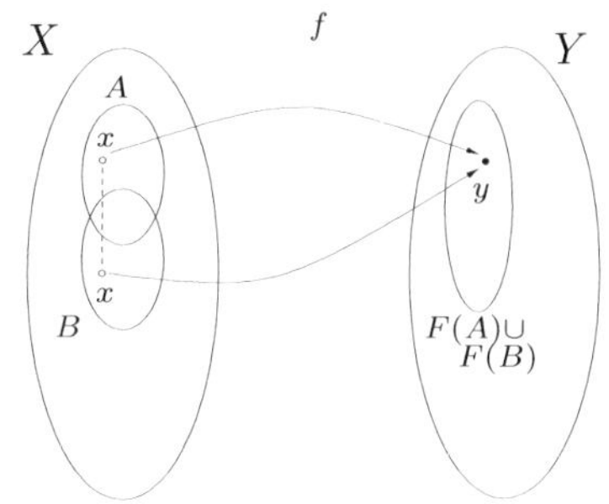

Either way, by definition of union, $y \in F(A) \cup F(B)$. Hence $F(A \cup B) \subseteq F(A) \cup F(B)$ by definition of subset.

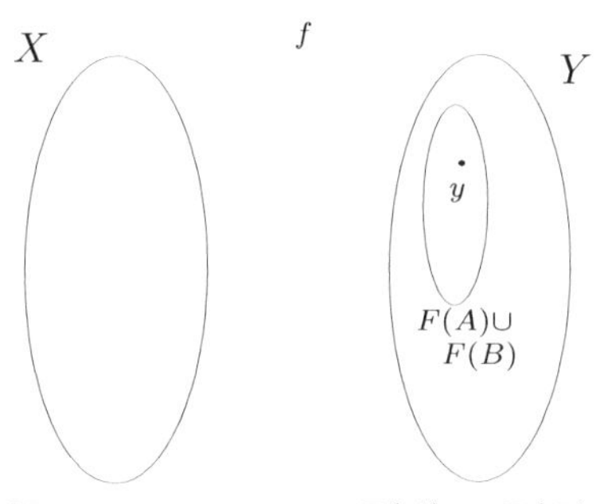

Next suppose $y \in F(A) \cup F(B)$. By definition of union, either $y \in F(A)$ or $y \in F(B)$.

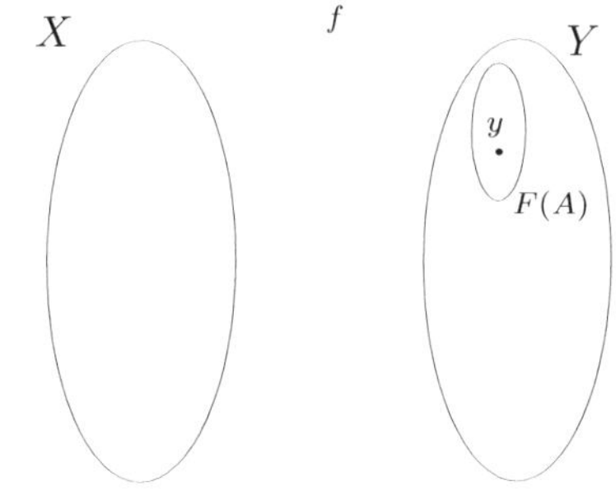

Suppose $y \in F(A)$.

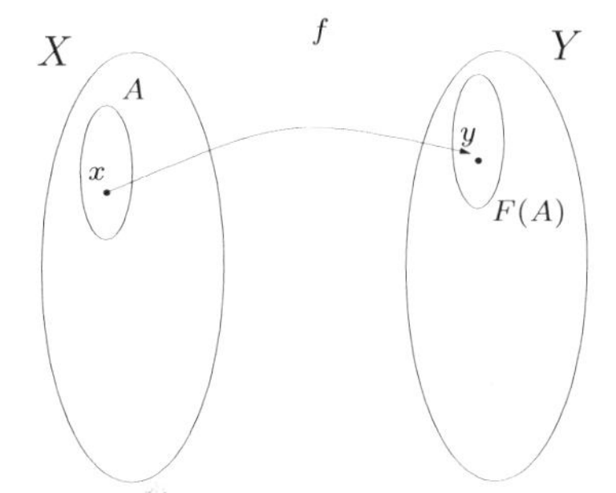

By definition of image, there exists an $x \in A$ such that $f(x) = y$.

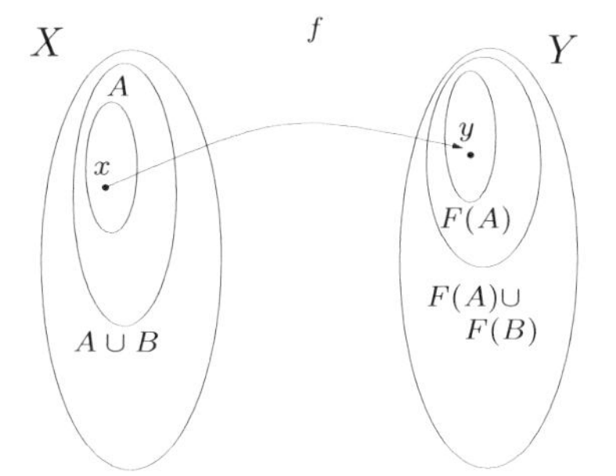

By definition of union, $x \in A \cup B$. Again by definition of image, $y \in F(A \cup B)$.

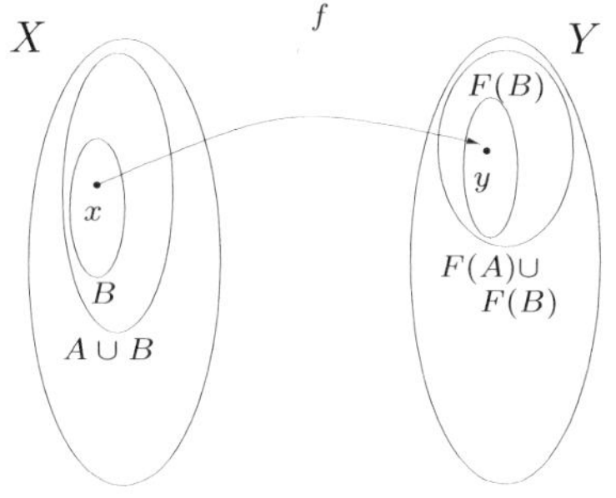

The argument is similar if $y \in F(B)$. Hence by definition of subset, $F(A) \cup F(B) \subseteq F(A \cup B)$.

Therefore, by definition of set equality, $F(A \cup B) = F(A) \cup F(B)$. □

Illustrations are also useful for showing a counterexample to a false proposition. Suppose we conjectured a proposition similar to the one we just proved, but for intersection instead of union.

$$\text{If } A, B \subseteq X, \text{ then } F(A \cap B) = F(A) \cap F(B)$$

We will use three versions of this diagram for clarity. It could be condensed to one diagram.

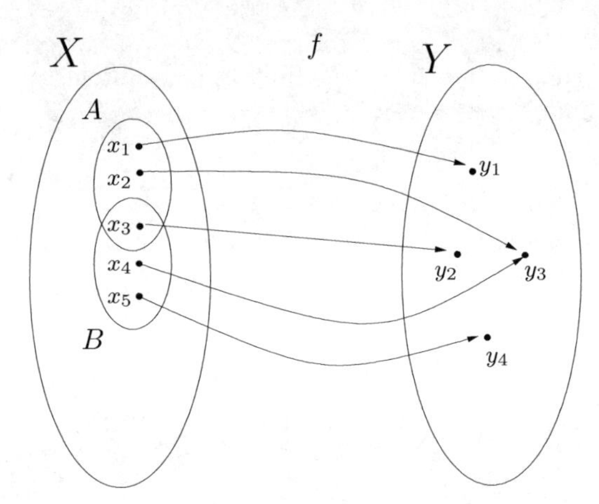

Let $A = \{x_1, x_2, x_3\}$, $B = \{x_3, x_4, x_5\}$, and $f = \{(x_1, y_1), (x_2, y_3), (x_3, y_2), (x_4, y_3), (x_5, y_4)\}$.

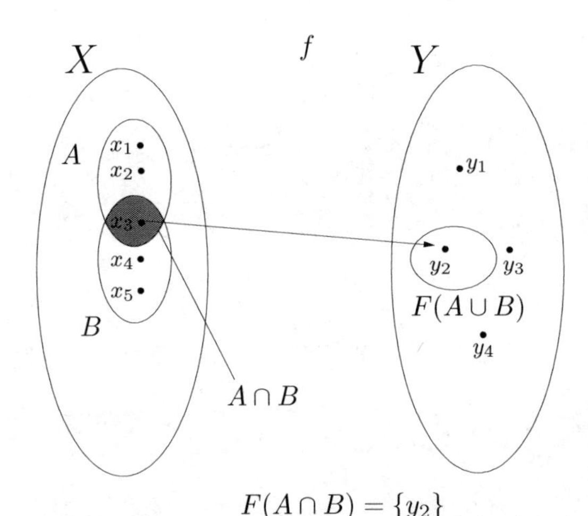

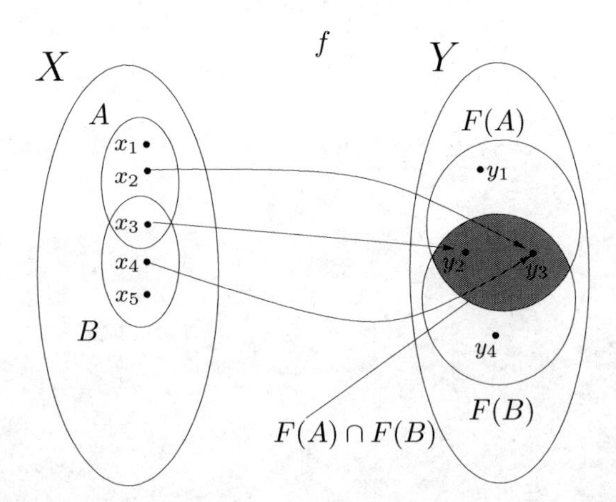

$F(A) \cap F(B) = \{y_2, y_3\}$

The entire example is more complicated than necessary. All we need is one element in A but not in B (x_2) and one element in B but not in A (x_4) to map to the same element in Y. If we let $f = \{(x_2, y_3), (x_4, y_3)\}$ and $A = \{x_2\}$ and $B = \{x_4\}$, then $A \cap B = \emptyset$ and so $F(A \cap B) = \emptyset$. Yet still $F(A) = \{y_3\}$ and $F(B) = \{y_3\}$, and so $F(A) \cap F(B) = \{y_3\} \neq F(A \cap B)$. Use this to make sure you understand images. Remember that $F(\emptyset) = \emptyset$ for all f.

Inverse images may look intimidating, but reasoning about them is still a matter of set manipulation. Just remember that an inverse image is a subset of the domain.

Theorem 7.6 *Let $f : X \to Y$ and $A, B \subseteq Y$. Then $F^{-1}(A - B) = F^{-1}(A) - F^{-1}(B)$.*

> **Proof.** Suppose $x \in F^{-1}(A - B)$. By definition of inverse image, $f(x) \in A - B$. By definition of set differencc, $f(x) \in A$ and $f(x) \notin B$. Again by definition of inverse image, $x \in F^{-1}(A)$ and $x \notin F^{-1}(B)$. Again by definition of set difference, $x \in F^{-1}(A) - F^{-1}(B)$. Hence, by definition of subset, $F^{-1}(A - B) \subseteq F^{-1}(A) - F^{-1}(B)$.
>
> Next suppose $x \in F^{-1}(A) - F^{-1}(B)$. By definition of set difference, $x \in F^{-1}(A)$ and $x \notin F^{-1}(B)$. By definition of inverse image, $f(x) \in A$ and $f(x) \notin B$. Again by definition of set difference $f(x) \in A - B$. Again by definition of inverse image, $x \in F^{-1}(A - B)$. Hence, by definition of subset, $F^{-1}(A) - F^{-1}(B) \subseteq F^{-1}(A - B)$.
>
> Therefore, by definition of set equality, $F^{-1}(A - B) = F^{-1}(A) - F^{-1}(B)$. □

Notice that this proof does not use a y variable—it is unnecessary. Instead, the expression $f(x)$ is used to refer to the only element in Y we care about. Nevertheless, it would not do any harm to introduce an extra variable, and some students find that it makes the proof easier to write. Just make sure that you declare the variable properly and in the right spot. For example, the sentence "Let $y = f(x)$" would be appropriate after the first or second sentence in the first paragraph, and then *it should reappear* in the second paragraph because then it may be referring to another element of Y. If you chose to write your proof that way, remember that y and $f(x)$ are merely two names for the same object. The following diagrams illustrate our approach to the proof, without a y variable.

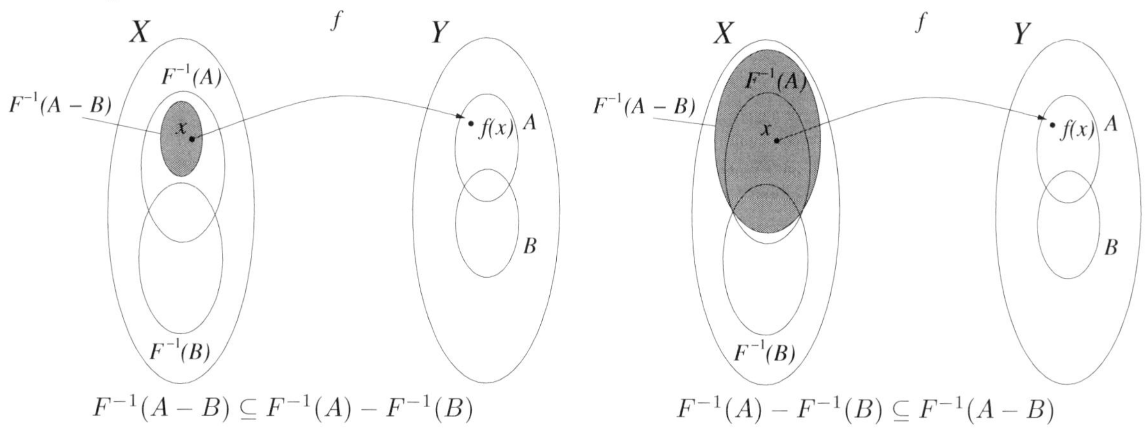

$F^{-1}(A - B) \subseteq F^{-1}(A) - F^{-1}(B)$ $\qquad$ $F^{-1}(A) - F^{-1}(B) \subseteq F^{-1}(A - B)$

Both of the proofs in this section are examples of the analysis/synthesis process. We take apart expressions using definitions, and using definitions we reconstruct other expressions.

Exercises

In Exercises 7.4.1–7.4.12, assume $f : X \to Y$. Prove, unless you are asked to give a counterexample.

7.4.1 If $A, B \subseteq X$, then $F(A \cap B) \subseteq F(A) \cap F(B)$.

7.4.2 If $A, B \subseteq X$, then $F(A) - F(B) \subseteq F(A - B)$.

7.4.3 If $A, B \subseteq X$, then $F(A - B) \subseteq F(A) - F(B)$. This is false; give a counterexample.

7.4.4 If $A \subseteq B \subseteq X$, then $F(B) = F(B - A) \cup F(A)$.

7.4.5 If $A \subseteq B \subseteq X$ and $F(A) \subseteq F(B - A)$, then $F(B) = F(B - A)$.

7.4.6 If $A \subseteq B \subseteq Y$, then $F^{-1}(A) \subseteq F^{-1}(B)$.

7.4.7 If $A, B \subseteq Y$, then $F^{-1}(A \cup B) = F^{-1}(A) \cup F^{-1}(B)$.

7.4.8 If $A, B \subseteq Y$, then $F^{-1}(A \cap B) = F^{-1}(A) \cap F^{-1}(B)$.

7.4.9 If $A \subseteq X$, then $A \subseteq F^{-1}(F(A))$.

7.4.10 If $A \subseteq X$, then $A = F^{-1}(F(A))$. This is false; give a counterexample.

7.4.11 If $A \subseteq Y$, then $F(F^{-1}(A)) \subseteq A$.

7.4.12 If $A \subseteq Y$, then $F(F^{-1}(A)) = A$. This is false; give a counterexample.

7.5 Map

The following ML functions determine which elements in a list are even, scale a list of integers by a factor of 5, convert a list of integers to reals, and convert a lists of four integers to tuples (the last taken from Section 4.11).

```
- fun testEven([]) = []
=   | testEven(x::rest) = (x mod 2 = 0) :: testEven(rest);

val testEven = fn : int list -> bool list

- fun scale([]) = []
=   | scale(x::rest) = x * 5 :: scale(rest);

val scale = fn : int list -> int list

- fun makeReals([]) = []
=   | makeReals(x::rest) = real(x)::makeReals(rest);

val makeReals = fn : int list -> real list

- fun makeGuessesTuples([]) = []
=   | makeGuessesTuples([a,b,c,d]::rest) =
=             (a, b, c, d)::makeGuessesTuples(rest);
```

```
Warning: match nonexhaustive
val makeGuessesTuples = fn : 'a list list ->
                              ('a * 'a * 'a * 'a) list
```

We have a clear pattern. When we repeat code or write code that is very similar, we must ask whether the language can do some of this work for us. Since functions are first-class values, we can write a handy function that takes a list and a function, applies that function to every element in the list, and returns the list of results. This function is traditionally called `map`.

```
- fun map(func, []) = []
=    | map(func, x::rest) = func(x) :: map(func, rest);

val map = fn : ('a -> 'b) * 'a list -> 'b list

- map(fn(x) => x mod 2 = 0, [8, 7, 6, 3, 2, 1]);

val it = [true,false,true,false,true,false] : bool list

- map(fn(x) => x * 5, [8, 7, 6, 3, 2, 1]);

val it = [40,35,30,15,10,5] : int list

- map(real, [8, 7, 6, 3, 2, 1]);

val it = [8.0,7.0,6.0,3.0,2.0,1.0] : real list

- map(fn([a,b,c,d])=>(a,b,c,d), [[5,3,7,4], [9,6,4,8]]);

Warning: match nonexhaustive
val it = [(5,3,7,4),(9,6,4,8)] : (int * int * int * int) list
```

The function `map` naturally adapts to our notion of image, using the list representation of sets.

```
- fun image(f, set) = makeNoRepeats(map(f, set));
```

Sometimes an operation we wish to perform on a list takes more than one operand and produces something other than a list. The function `foldl` takes a function, a "seed value," and a list. The given function must take two parameters, and `foldl` applies that function to the first element in the list and the seed value. Next it applies the function to the second element in the list, using the result of the first application as the second parameter. Thus, given f, s, and $x_1, x_2, \ldots, x_n$, it computes $f(x_n, \ldots f(x_2, f(x_1, s)) \ldots)$.

```
- fun foldl(f, s, []) = s
=   | foldl(f, s, x::rest) = foldl(f, f(x, s), rest);

val foldl = fn : ('a * 'b -> 'b) * 'b * 'a list -> 'b

- foldl(fn(x,y) => x + y, 0, [4, 5, 6]);

val it = 15 : int

- foldl(fn(x, y) => x * y, 1, [2, 3, 4]);

val it = 24 : int
```

Remember the function `listify` you wrote in Exercise 2.2.10, which took a list and return a list of lists, with each element now in its own list? We can rewrite it using `map`:

```
- fun listify(xx) = map(fn(a) => [a], xx);

val listify = fn : 'a list -> 'a list list

- listify([1, 2, 3]);

val it = [[1],[2],[3]] : int list list
```

Exercises

7.5.1 Rewrite your Caesar cipher function from Exercise 1.12.9 using `map`.

7.5.2 Sometimes it is useful to write a function that operates on a list but also takes some extra arguments. For example a better version of `scale` might take a list of integers and another integer and return a list of the old integers multiplied by the other integer. Write a function `mapPlus` that takes a function, a list, and an extra argument and applies the function to every element in the list and the extra argument. For example, `mapPlus(fn (x, y) => x * y, [1, 2, 3, 4], 2)` would return `[2, 4, 6, 8]`.

7.5.3 Write a function `mapPlusMaker` that takes a function and returns a function that will take a list and an extra argument and apply the given function to all the elements in the list. For example, `mapPlusMaker(fn (x, y) => x * y)` would return the `scale` program described above.

7.5.4 Use `foldl` to write a function `smallest` that takes a list of integers and returns the smallest item in the list. Assume all the integers in the list are positive.

7.5.5 Rewrite your `scale` function from Exercise 7.5.2 using `foldl`. (Hint: There is no reason why the function passed to `foldl` cannot return a list.)

7.5.6 Write a function `foldr` that is like `foldl` except that it applies the function starting at the end of the list and moving backward to the front (thus `foldr` works from right-to-left, unlike `foldl` which works left-to-right).

7.5.7 `foldr` is a more primitive operation than `map` because `map` can be written using `foldr`. Rewrite `map` using `foldr`. (Hint: Your accumulating function is cons.)

7.6 Function properties

Some functions have certain properties that imply that they behave in predictable ways. We have seen examples of functions where some elements in the codomain are hit more than once and functions where some are hit not at all. Two properties give names for the opposite of these situations. A function $f : X \to Y$ is *onto* if

onto

$$\forall\, y \in Y, \exists\, x \in X \mid f(x) = y.$$

In other words, an onto function hits every element in the codomain, possibly more than once. We may also talk about a function being onto a subset of the codomain. If $B \subseteq Y$ and for all $y \in B$ there exists an $x \in X$ such that $f(x) = y$, then we say that f is *onto* B. Notice that if f is onto, then the range of f is equal to the codomain.

A function is *one-to-one* if

one-to-one

$$\forall\, x_1, x_2 \in X, \text{ if } f(x_1) = f(x_2), \text{ then } x_1 = x_2.$$

If any two domain elements hit the same codomain element, they must be equal (compare the structure of this definition with the definition of antisymmetry). This means that no element of the codomain is hit more than once. A function that is both one-to-one and onto is called a *one-to-one correspondence*. In that case, every element of the codomain is hit exactly once.

one-to-one correspondence

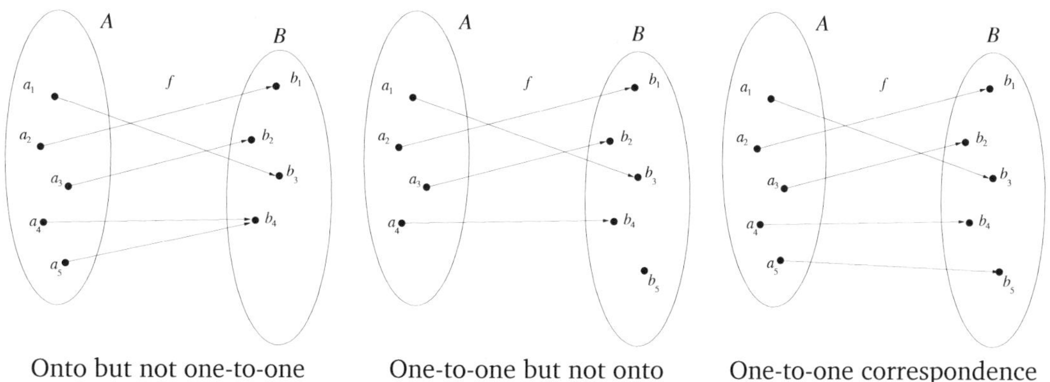

Onto but not one-to-one   One-to-one but not onto   One-to-one correspondence

Some authors refer to onto functions, one-to-one functions, and one-to-one correspondences as *surjections*, *injections*, and *bijections*.

In Exercise 7.4.1, you proved that $F(A \cap B) \subseteq F(A) \cap F(B)$. However, we saw a counterexample to $F(A) \cap F(B) = F(A \cap B)$, which really was a counterexample specifically to $F(A) \cap F(B) \subseteq F(A \cap B)$. If you turn back to that example, you will notice that we constructed it by associating two elements of the domain (x_2 and x_4) with the same element of the codomain (y_2). In other words, our counterexample depended on f not being one-to-one. If we assume f is one-to-one, on the other hand, we can prove this result.

Theorem 7.7 *If $f : X \to Y$, $A, B \subseteq X$, and f is one-to-one, then $F(A) \cap F(B) \subseteq F(A \cap B)$.*

First, we set up the proof

Proof. Suppose $f : X \to Y$, $A, B \subseteq X$, and f is one-to-one. Now suppose $y \in F(A) \cap F(B)$. Then $y \in F(A)$ and $y \in F(B)$ by definition of intersection. By the definition of image, there exist $x_1 \in A$ such that $f(x_1) = y$ and $x_2 \in B$ such that $f(x_2) = y$.

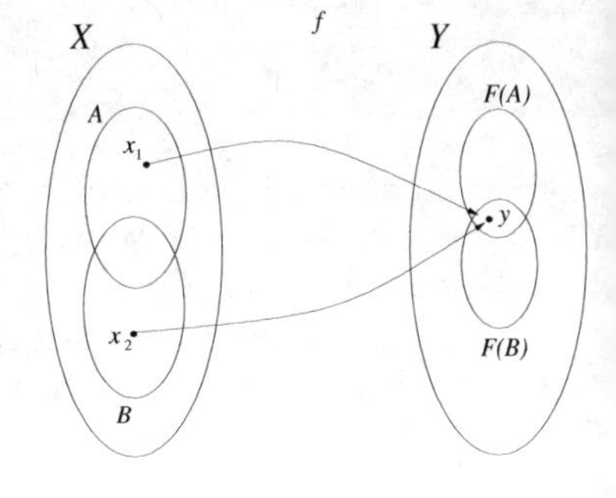

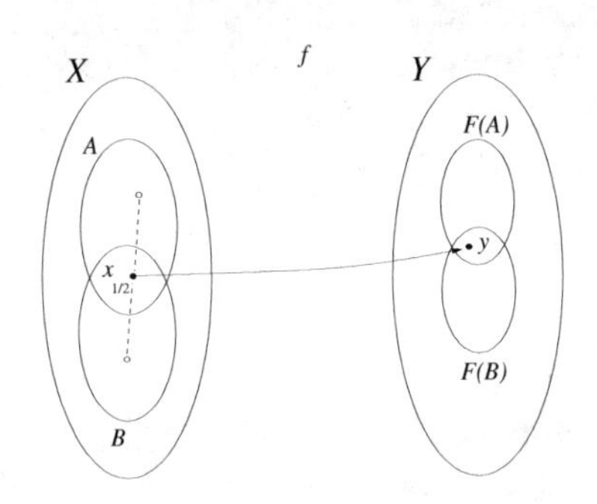

Then we can use the definition of one-to-one.

By definition of one-to-one, $x_1 = x_2$. By substitution, $x_1 \in B$. By definition of intersection, $x_1 \in A \cap B$. Therefore, by definition of image, $y \in F(A \cap B)$. □

Note that with $y \in F(A) \cap F(B)$, we had to assume *two* xs, one in A and one in B. Only then could we apply the fact that f is one-to-one and prove the two xs equal. *This is extremely important.* Notice also that by putting this result and Exercise 7.4.1 together, we conclude that if f is one-to-one, $F(A \cap B) = F(A) \cap F(B)$.

Exercises

7.6.1 Which of the following are one-to-one and/or onto? (Assume a domain and codomain of $\mathbb{R}$ unless indicated otherwise.)
(a) $f(x) = x$. (b) $f(x) = x^2$.
(c) $f(x) = \ln x$. (Use $\mathbb{R}^+$ as domain.)
(d) $f(x) = x^3$. (e) $f(x) = \frac{1}{3}x^3 - 4x$.

7.6.2 Prove that $f(x) = 3 \cdot x + 4$ is one-to-one and onto. Do this in separate proofs. Think carefully about what it takes to prove that a function is one-to-one or onto. Assume $f : \mathbb{R} \to \mathbb{R}$.

In Exercises 7.6.3–7.6.8, assume $f : X \to Y$. Prove.

7.6.3 If $A, B \subseteq X$ and f is one-to-one, then $F(A - B) \subseteq F(A) - F(B)$.

7.6.4 If $A \subseteq X$ and f is one-to-one, then $F^{-1}(F(A)) \subseteq A$.

7.6.5 If $A \subseteq Y$ and f is onto, then $A \subseteq F(F^{-1}(A))$.

7.6.6 If $f : X \to Y$, $A, B \subseteq Y$, f is onto, and $F^{-1}(A) \cap F^{-1}(B) = \emptyset$, then $A \cap B = \emptyset$.

7.6.7 If A is any set, then i_A (the identity function on A) is a one-to-one correspondence.

7.6.8 If for all non-empty $A, B \subseteq X$, $F(A \cap B) = F(A) \cap F(B)$, then f is one-to-one.

(Look carefully at how this proposition is structured. Do not begin your proof with "Suppose $A, B \subseteq X$..." but with "Suppose for all $A, B \subseteq X$..." Look back at the counterexample we found for $F(A \cap B) = F(A) \cap F(B)$ on page 340. How does the hypothesis fix things?)

7.7 Inverse functions

When we prove propositions about images and inverse images—take $A \subseteq F^{-1}(F(A))$ in Exercise 7.4.9 for example—we can imagine the function taking us on a journey from a place in domain to a place in the codomain. The inverse image addresses the complexities of the return voyage. Suppose $f : X \to Y$ is not one-to-one, specifically that for some $x_1, x_2 \in X$ and $y \in Y$, both $f(x_1) = y$ and $f(x_2) = y$. $f(x_1)$ will take us to y (or $F(\{x_1\})$ will take us to $\{y\}$), but we have in a sense lost our way: Going backwards through f might take us to x_2 instead of x_1; $F^{-1}(y) = \{x_1, x_2\} \neq \{x_1\}$. If f is one-to-one, this is not a problem because only one element will take us to y; that way, if we start in X and go to Y, we know we can always retrace our steps back to where we were in X. However, this does not help if we start in Y and want to go to X; unless f is also onto, there may not be a way from X to that place in Y.

All things work out nicely if $f : X \to Y$ is a one-to-one correspondence. Take the concept of the inverse of a relation from Section 5.3 and reapply it to functions to define the *inverse* of f:

inverse

$$f^{-1} = \{(y, x) \in Y \times X \mid f(x) = y\}$$

It is more convenient to call this the *inverse function* of f, but the title "function" does not come for free.

inverse function

Theorem 7.8 *If $f : X \to Y$ is a one-to-one correspondence, then $f^{-1} : Y \to X$ is well defined.*

We need to apply the definition of a function to show that f^{-1} is one: for all $y \in Y$, there is exactly one $x \in X$ such that $(y, x) \in f^{-1}$.

Proof. Suppose $y \in Y$. Since f is onto, there exists $x \in X$ such that $f(x) = y$. Hence $(y, x) \in f^{-1}$ or $f^{-1}(y) = x$.

Further suppose $(y, x_1), (y, x_2) \in f^{-1}$ (*That is, suppose that both $f^{-1}(y) = x_1$ and $f^{-1}(y) = x_2$.*) Then $f(x_1) = y$ and $f(x_2) = y$. Since f is one-to-one, $x_1 = x_2$.

Therefore, by definition of function, f^{-1} is well defined. □

Do not confuse the inverse of a function and the inverse image of a function. Remember that the inverse image is a set, a subset of the domain, and that it is applied to a subset of the codomain; the inverse image always exists. The inverse function exists only if the function itself is a one-to-one correspondence; it takes an element of the codomain and produces an element of the domain.

Erroneous proof. Suppose $f : X \to Y$ and $A \subseteq Y$. Suppose $y \in A$. Then $f^{-1}(y) \in F^{-1}(A)$. ...

NO! Unless f is a one-to-one correspondence, you do not know that the function f^{-1} exists.

7.8 Function composition

We have seen that two relations can be composed to form a new relation, say, given relations R from X to Y and S from Y to Z:

$$S \circ R = \{(a,c) \in X \times Z \mid \exists\, b \in Y \text{ such that } (a,b) \in R \text{ and } (b,c) \in S\}$$

This easily can be specialized for the composition of functions. Suppose $f : X \to Y$ and $g : Y \to Z$ are functions. Then the *composition* of f and g is

function composition

$$g \circ f = \{(x,z) \in X \times Z \mid z = g(f(x))\}$$

In short, $g \circ f(x) = g(f(x))$. We are just rewriting the definition of composition for relations. The interesting part is that $g \circ f$ is a function.

Theorem 7.9 *If $f : A \to B$ and $g : B \to C$ are functions, then $g \circ f : A \to C$ is well defined.*

Proof. Suppose $a \in A$. Since f is a function, there exists a $b \in B$ such that $f(a) = b$. Since g is a function, there exists a $c \in C$ such that $g(b) = c$. By definition of composition, $(a,c) \in g \circ f$, or $g \circ f(a) = c$.

Next suppose $(a,c_1),(a,c_2) \in g \circ f$, or $g \circ f(a) = c_1$ and $g \circ f(a) = c_2$. By definition of composition, there exist b_1, b_2 such that $f(a) = b_1$, $f(a) = b_2$, $g(b_1) = c_1$, and $g(b_2) = c_2$. Since f is a function, $b_1 = b_2$. Since g is a function, $c_1 = c_2$.

Therefore, by definition of function, $g \circ f$ is well defined. □

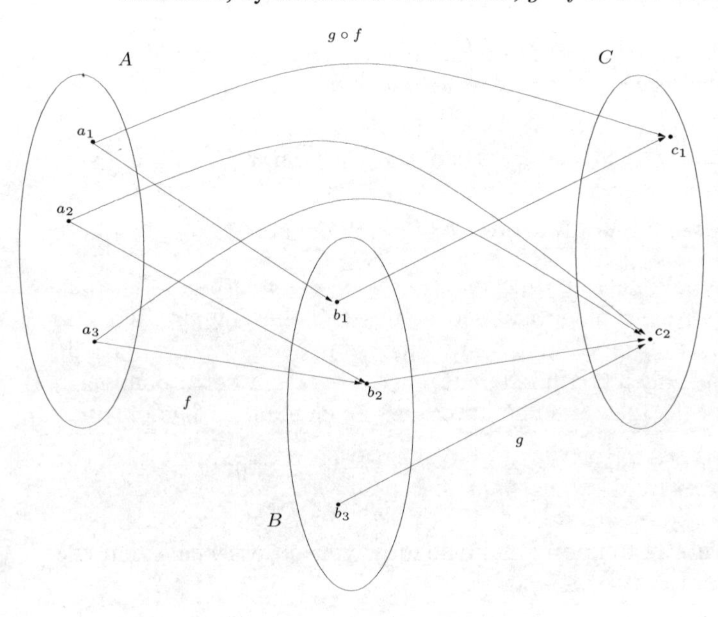

For an intuitive way to think of composition, use the machine model of functions. We simply attach two machines together, feeding the output slot of the one into the input slot of the other. To make this work, the one output slot must fit into the other input slot, and the one machine's output material must be appropriate input material for the other machine. Mathematically, we describe this by considering the domains and codomains. Given $f : A \to B$ and $g : C \to D$, $g \circ f$ is defined only for $B = C$, though it could easily be extended for the case where $B \subseteq C$.

Function composition happens quite frequently without our noticing it. For example, the real-valued function $f(x) = \sqrt{x - 12}$ can be considered the composition of the functions $g(x) = x - 12$ and $h(x) = \sqrt{x}$. The chain rule in calculus takes function composition into account.

When we examined inverse functions, we spoke of functions taking us from a spot in one set to a spot in another. Only if the function were a one-to-one correspondence could we assume a "round trip" (using an inverse function) existed. Since the net effect of a round trip is getting you nowhere, you may conjecture that if you compose a function with its inverse, you will get an identity function. The converse (if you compose two functions and get the identity, then the two are inverses of each other) is true, too.

Theorem 7.10 *Suppose $f : A \to B$ is a one-to-one correspondence and $g : B \to A$. Then $g = f^{-1}$ if and only if $g \circ f = i_A$.*

Proof. Suppose that $g = f^{-1}$. (Since f is a one-to-one correspondence, f^{-1} is well defined.) Suppose $a \in A$. Then

$$\begin{aligned} g \circ f(a) &= g(f(a)) && \text{by definition of composition} \\ &= f^{-1}(f(a)) && \text{by substitution} \\ &= a && \text{by definition of inverse function} \\ &= i_A(a) && \text{by definition of identity} \end{aligned}$$

Therefore, by function equality, $g \circ f = i_A$.

For the converse (proving that if you compose a function and get the identity function, it must be the inverse), see Exercise 7.8.7. □

Composition finally gives us enough raw material to prove some fun results on functions. Remember in all of these to apply the definitions carefully and also to follow the outlines for proving propositions in now-standard forms. For example, it is easy to verify visually that the composition of two one-to-one functions is also a one-to-one function. If no two f or g arrows collide, then no $g \circ f$ arrows have a chance of colliding. Proving this result relies on the definitions of composition and one-to-one.

Theorem 7.11 *If $f : A \to B$ and $g : B \to C$ are one-to-one, then $g \circ f : A \to C$ is one-to-one.*

> **Proof.** Suppose $f : A \to B$ and $g : B \to C$ are one-to-one. Now suppose $a_1, a_2 \in A$ and $c \in C$ such that $g \circ f(a_1) = c$ and $g \circ f(a_2) = c$. By definition of composition, $g(f(a_1)) = c$ and $g(f(a_2)) = c$. Since g is one-to-one, $f(a_1) = f(a_2)$. Since f is one-to-one, $a_1 = a_2$. Therefore, by definition of one-to-one, $g \circ f$ is one-to-one. □

In the long run, we want to prove that something is one-to-one, so we need to gather the materials to synthesize the definition. It means we can pick any two elements from the domain of $g \circ f$, and if they map to the same element, they themselves must be the same. Here is how the proof connects with a visual verification.

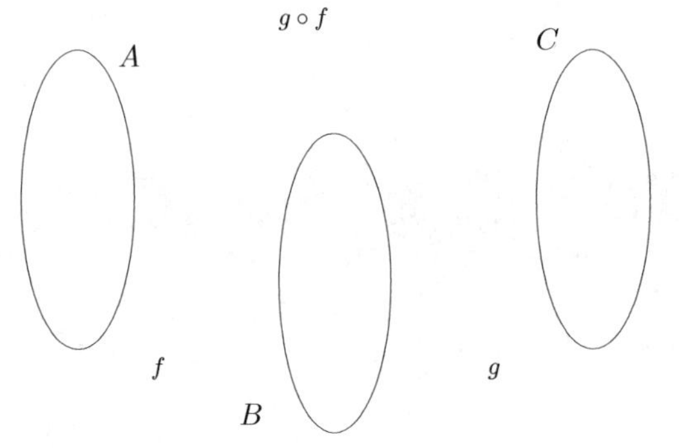

Suppose $f : A \to B$ and $g : B \to C$ are one-to-one.

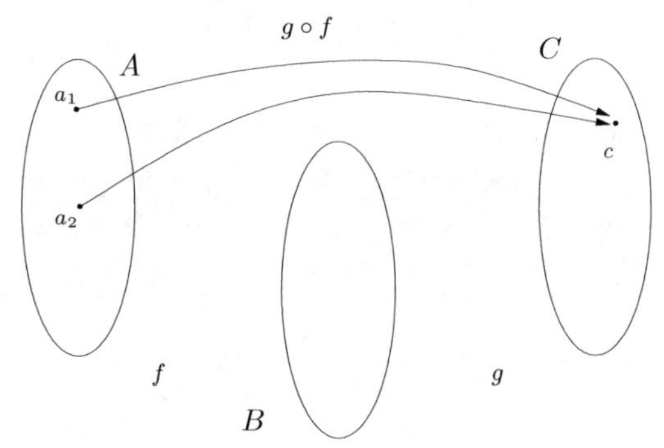

Now suppose $a_1, a_2 \in A$ and $c \in C$ such that $g \circ f(a_1) = c$ and $g \circ f(a_2) = c$.

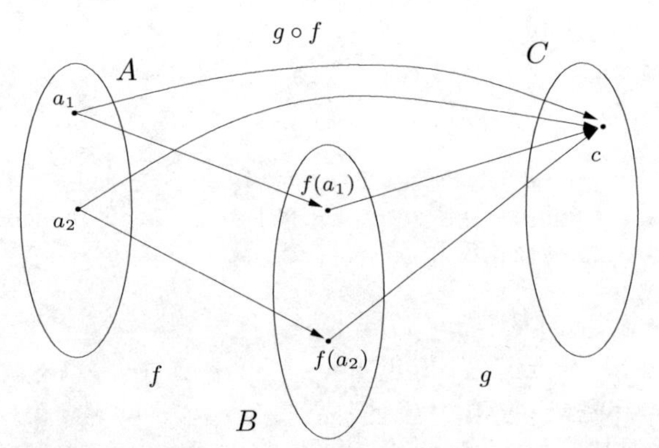

By definition of composition, $g(f(a_1)) = c$ and $g(f(a_2)) = c$.

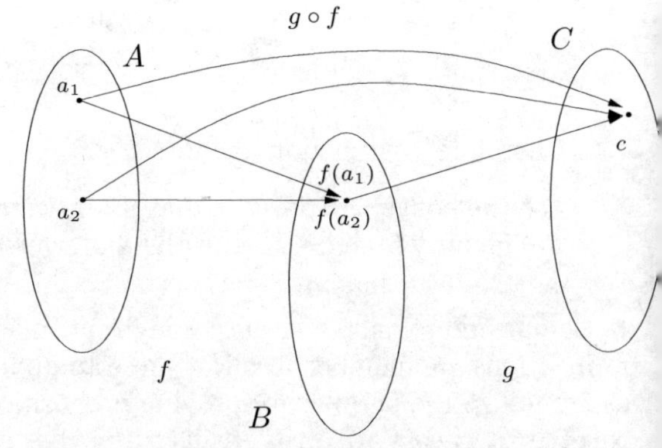

Since g is one-to-one, $f(a_1) = f(a_2)$.

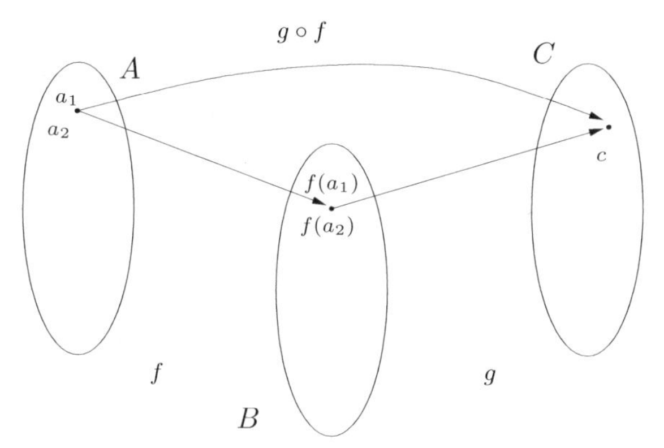

Since f is one-to-one, $a_1 = a_2$.

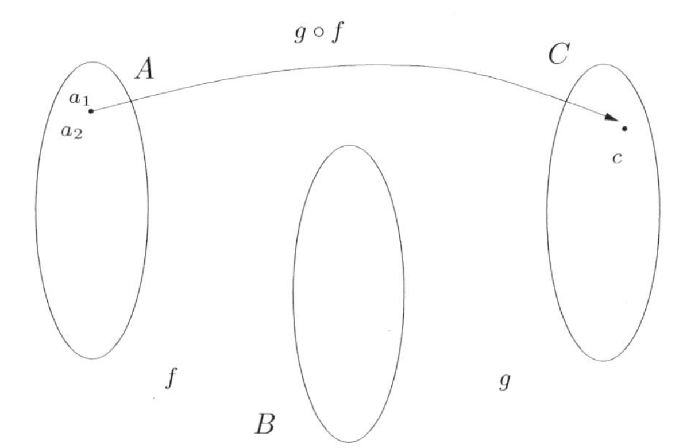

Therefore, by definition of one-to-one, $g \circ f$ is one-to-one. □

ML has a built-in composition operator, o (a lowercase letter o).

```
- map((fn(x) => 2 * x) o (fn(x) => 3 + x), [1, 8, 11]);

val it = [8,22,28] : int list
```

Exercises

Prove.

7.8.1 If $f : A \to B$, then $f \circ i_A = f$.

7.8.2 If $f : A \to B$, then $i_B \circ f = f$.

7.8.3 f $f : A \to B$, $g : B \to C$, and $h : C \to D$, then $h \circ (g \circ f) = (h \circ g) \circ f$.

7.8.4 If $f : A \to B$ and $g : B \to C$ are both onto, then $g \circ f$ is also onto.

7.8.5 If $f : A \to B$ and $g : B \to C$, and $g \circ f$ is one-to-one, then f is one-to-one.

7.8.6 If $f : A \to B$ and $g : B \to C$, and $g \circ f$ is onto, then g is onto.

7.8.7 If $f : A \to B$ is a one-to-one correspondence, $g : B \to A$, and $g \circ f = i_A$, then $g = f^{-1}$.

7.8.8 If $f : A \to B$, $g : A \to B$, $h : B \to C$, h is one-to-one, and $h \circ f = h \circ g$, then $f = g$.

7.8.9 If $f : A \to B$ and $g : B \to C$ are both one-to-one correspondences, then the inverse function $(g \circ f)^{-1}$ exists and $(g \circ f)^{-1} = f^{-1} \circ g^{-1}$. (Hint: Notice that this requires you to prove two things. However, the first thing can be proven quickly from previous exercises in this chapter.)

7.8.10 Let $G \circ F(X)$ be the image of a set X under the function $g \circ f$, for functions f and g; similarly define inverse images for composed functions. If $f : A \to B$, $g : B \to C$ and $X \subseteq C$, then $(G \circ F)^{-1}(X) = F^{-1}(G^{-1}(X))$.

7.9 Cardinality

If a function $f : X \to Y$ is onto, then every element in Y has at least one domain element seeking it, but two elements in X could be rivals for the same codomain element. If it is one-to-one, then every domain element has a codomain element all to itself, but some codomain elements may be left out. If it is a one-to-one correspondence, then everyone has a date to the dance. When we are considering finite sets, we can use this as intuition for comparing the cardinalities of the domain and codomain.

cardinality for funct. types

It seems that f could be onto only if $|X| \geq |Y|$ and one-to-one only if $|X| \leq |Y|$. If f is a one-to-one correspondence, then it must be that $|X| = |Y|$. How could we prove this, though?

Actually, we have never given a formal definition of cardinality, which is necessary for a proof like this. Rather than *proving* that the existence of a one-to-one correspondence implies sets of equal cardinality, we will simply use that idea as our *definition* of cardinality.

First, what it means for two sets to have the same cardinality *as each other*:

cardinality defn

> Two finite sets X and Y have the same *cardinality* if there exists a one-to-one correspondence from X to Y.

Now we would like to say that a finite set's cardinality *is* something, as the cardinality of $\{1, 4, 19\}$ is 3. What we mean is that we can count the elements, "one, two, three." This sort of counting is just applying a unique label to each element of the set. In other words, we are making a one-to-one correspondence from the set $\{1, 2, 3\}$ to the set we are counting. Thus

> If X is a finite set, then $|X| = n$ for some $n \in \mathbb{N}$ if there exists a one-to-one correspondence from $\{1, 2, \ldots, n\}$ to X. Moreover, $|\emptyset| = 0$.

Note that when we include the special case of the empty set, cardinality is defined over $\mathbb{W}$, not just $\mathbb{N}$.

(There is a wrinkle in this system: it lacks a formal definition of the term *finite*. Technically, we should define a set X to be finite if there exists an $n \in \mathbb{N}$ and a one-to-one correspondence from $\{1, 2, \ldots, n\}$ to X. Then, however, we would need to use this to define *cardinality*. We would prefer to keep the definition of cardinality separate from the notion of finite subsets of $\mathbb{N}$ to make it easier to extend the concepts to infinite sets later. We also are assuming that

Tangent: Respectively

Mary-Claire van Leunen says in *A Handbook for Scholars*,

> "Respective" and "respectively" are such complicated words that defining them is like a short excursion into [discrete] math. Their function is to link the members of two sets with identical cardinality in a one-to-one correspondence in the order of their enumeration. Got that? [35]

a set's cardinality is unique, or that the cardinality operator $|\ |$ is well defined as a function. This can be proven, but it is difficult.)

Now we can use cardinality formally. The proofs in this section are quite complex, but worth the work. Take them slowly and study them carefully. Here is the first:

Theorem 7.12 *If A and B are finite, disjoint sets, then $|A \cup B| = |A| + |B|$.*

What does this mean? We are claiming that if two sets are disjoint, we can count their union by counting them separately and adding the results. This is possible because there is no overlap, so there is nothing that we would count twice. Consider these two examples:

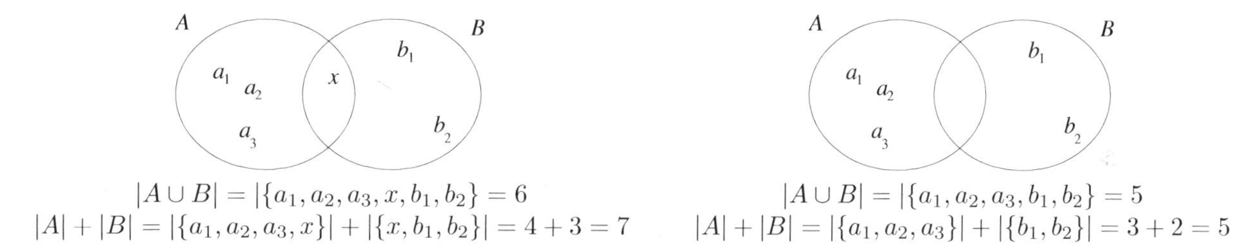

A
B
a_1
a_2
a_3
x
b_1
b_2
$|A \cup B| = |\{a_1, a_2, a_3, x, b_1, b_2\} = 6$
$|A| + |B| = |\{a_1, a_2, a_3, x\}| + |\{x, b_1, b_2\}| = 4 + 3 = 7$
$|A \cup B| = |\{a_1, a_2, a_3, b_1, b_2\} = 5$
$|A| + |B| = |\{a_1, a_2, a_3\}| + |\{b_1, b_2\}| = 3 + 2 = 5$

What is going to be difficult in this proof is dealing with the new definition, which associates cardinality with the existence of a natural number and of a one-to-one correspondence. First, analyze the meaning of $|A|$ and $|B|$ with our new definition of cardinality. (The proof implicitly assumes A and B are nonempty. It would work the same way without this assumption except that f or g below would potentially be a trivial function between empty sets.)

> **Proof.** Suppose A and B are finite, disjoint sets. By the definition of finite, there exist $i, j \in \mathbb{N}$ and one-to-one correspondences $f : \{1, 2, \ldots, i\} \to A$ and $g : \{1, 2, \ldots, j\} \to B$. Note that $|A| = i$ and $|B| = j$.

Vaguely, picture it this way (the subsets of $\mathbb{N}$ are drawn as clouds; clearly they have some overlap, though the diagrams do not show it):

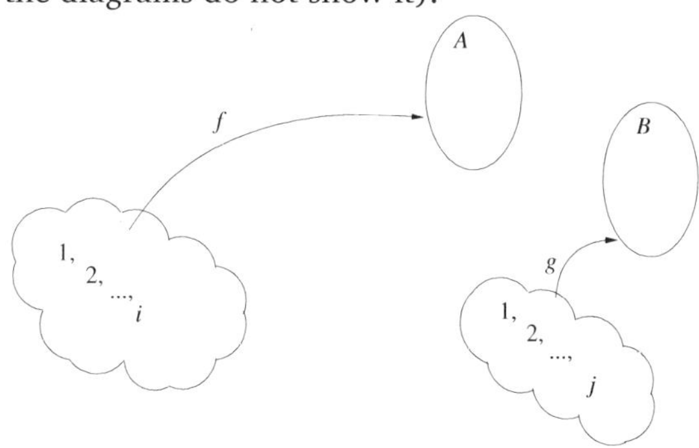

It is a simple observation that $|A| + |B| = i + j$—by substitution, really. To show $|A \cup B| = |A| + |B|$, which is our goal, means showing $|A \cup B| = i + j$. Note $i + j \in \mathbb{N}$. To show that this is its cardinality—that is, using the definition of cardinality synthetically—requires showing an appropriate one-to-one correspondence exists. We have not run into many proofs of existence, and they tend to come up in harder proofs, like this one. The natural form to prove an object exists is to

- describe an object, and
- show that it meets the requirements.

constructive proof

This is called a *constructive proof* because the proof itself gives a recipe for constructing the object that we claim exists. In our case, the "object" is a function, and the requirements are that it is a one-to-one correspondence from $\{1, 2, \ldots, i + j\}$ to $A \cup B$. We will build a function like this using f and g.

To make this understandable, suppose a family had three boys, Zed, Yelemis, and Xavier, followed by four girls, Wilhelmina, Valerie, Ursula, and Tassie. We will show that there are seven children altogether.

What does it mean that the number of boys is three? It means there is a one-to-one correspondence between the sets $\{1, 2, 3\}$ and $\{\text{Zed}, \text{Yelemis}, \text{Xavier}\}$; $f(i)$ is defined to be the ith boy in the family. Likewise "there are four girls" is proven correct by the existence of the function g, where $g(i)$ is the ith girl in the family.

Now notice that Yelemis's birth order among all the children is equal to his birth order among just the boys. Ursula's birth order in the family in general, however, is three more than her birth order among just her sisters. Let the function h map absolute birth order to child, that is, $h(i)$ is the ith child in the family. Now we observe that if $1 \leq i \leq 3$, then $h(i) = f(i)$, but if $4 \leq i \leq 7$, then $h(i) = g(i - 3)$.

i	$f(i)$
1	Zed
2	Yelemis
3	Xavier

i	$g(i)$
1	Wilhelmina
2	Valerie
3	Ursula
4	Tassie

i			$h(i)$		
1	$f(1)$	$=$	Zed		
2	$f(2)$	$=$	Yelemis		
3	$f(3)$	$=$	Xavier		
4	$g(4-3)$	$=$	$g(1)$	$=$	Wilhelmina
5	$g(5-3)$	$=$	$g(2)$	$=$	Valerie
6	$g(6-3)$	$=$	$g(3)$	$=$	Ursula
7	$g(7-3)$	$=$	$g(4)$	$=$	Tassie

One can verify that h is a one-to-one correspondence by inspection, if there is any doubt. This proves there are seven children. Back in the real proof, we similarly construct a new function:

Define a function $h : \{1, 2, \ldots, i + j\} \to A \cup B$ as

$$h(x) = \begin{cases} f(x) & \text{if } x \leq i \\ g(x - i) & \text{otherwise} \end{cases}$$

Think of "$x - i$" as an anonymous function from $\{i+1, i+2, \ldots, i+j\}$ to $\{1, 2, \ldots, j\}$, and $g(x-i)$ is the composition of g with this anonymous function. The function h is essentially a union of these two functions (remember that functions are sets). To help you picture it, we embellish our earlier diagram:

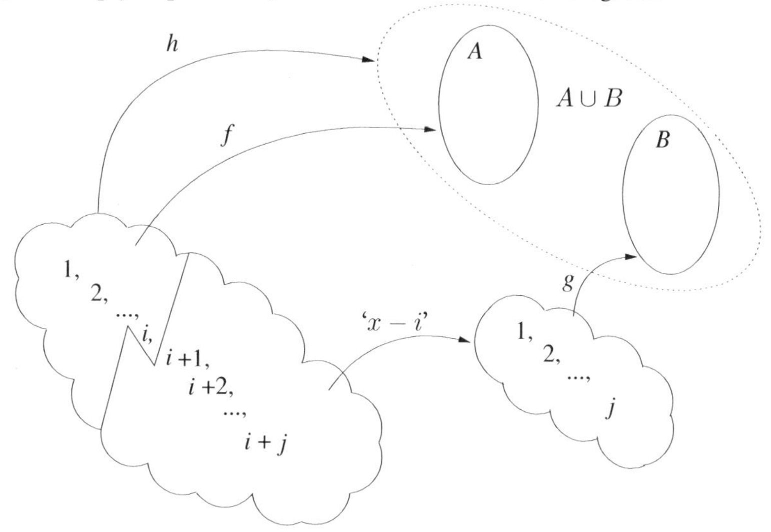

What remains is to prove that this function is a one-to-one correspondence. To do this, we pick an element in the codomain and prove it is hit exactly once. We split this into two cases, corresponding to whether the codomain element we picked is in A or B—it cannot be in both.

> Now suppose $y \in A \cup B$. Then either $y \in A$ or $y \in B$ by definition of union, and it is not true that both $y \in A$ and $y \in B$ by definition of disjoint. Hence we have two cases:
>
> Case 1: Suppose $y \in A$ and $y \notin B$. Then, since f is a onto, there exists a $k \in \{1, 2, \ldots, i\}$ such that $f(k) = y$. By our definition of h, $h(k) = y$.

Mini proof that k exists (onto).

> Further, suppose $\ell \in \{1, 2, \ldots, i+j\}$ and $h(\ell) = y$. Suppose $\ell > i$; then $y = h(\ell) = g(\ell - i) \in B$, contradiction; hence $\ell \leq i$. This implies $h(\ell) = f(\ell)$, and since f is one-to-one, $\ell = k$.

Mini proof that k is unique (one-to-one).

> Case 2: Suppose $y \in B$ and $y \notin A$. Then, since g is onto, there exists a $k \in \{1, 2, \ldots, j\}$ such that $g(k) = y$. By our definition of h, $h(k+i) = g(k) = y$. Further, suppose $\ell \in \{1, 2, \ldots, i+j\}$ and $h(\ell) = y$. Suppose $\ell \leq i$; then $y = h(\ell) = f(\ell) \in A$, contradiction; hence $\ell > i$. This implies $h(\ell) = g(\ell - i)$, and since g is one-to-one, $\ell - i = k$ or $\ell = k + i$.
>
> In either case, there exists a unique element $m \in \{1, 2, \ldots, i+j\}$ ($m = k$ and $m = k + i$, respectively) such that $h(m) = y$. Hence h is a one-to-one correspondence. Therefore, $|A \cup B| = i + j = |A| + |B|$.
> □

The phrase "there exists a unique element" is equivalent to our idea of a one-to-one correspondence, just stated in a different way. Remember, onto-ness is a matter of existence, and one-to-one-ness is a matter of uniqueness.

This proof contains three examples of proofs of existence. First, the entire proof is the existence of a one-to-one correspondence. Then, the first paragraphs each of Case 1 and Case 2 are proofs of existence since they each prove a certain function is onto. A proof of existence has two steps: finding such an object, and demonstrating that it fits the desired qualifications.

Are you ready for another hard proof? Take pride in your ability to work through these proofs which are much more sophisticated than the proofs we have done earlier.

Theorem 7.13 *If A and B are finite sets and $f : A \to B$ is one-to-one, then* $|A| \leq |B|$.

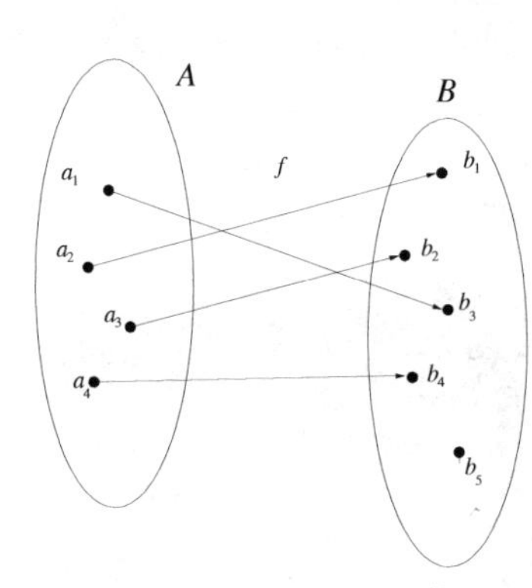

All we are claiming is what we see in this diagram. If every element in A is matched up with a unique element in B, then there might be some elements in B left over; there could not be fewer elements in B.

Here is our formal definition of "less than or equal to," something we need to prove this.

> If $x, y \in \mathbb{R}$, then $x \leq y$ if there exists $z \in \mathbb{R}^{\text{nonneg}}$ such that $x + z = y$.

For our purposes, however, we will be dealing only with whole numbers, so we could replace both $\mathbb{R}$ and $\mathbb{R}^{\text{nonneg}}$ with $\mathbb{W}$. Specifically, we want to find a number, say $k \in \mathbb{W}$ such that $|A| + k = |B|$. To do that, we will build a one-to-one correspondence from a set of size $|A| + k$ to a set of size $|B|$. We start by analyzing what is given.

Proof. Suppose $f : A \to B$ is one-to-one. Let $i = |A|$ and $j = |B|$. By the definition of cardinality, there exist one-to-one correspondences $g : \{1, 2, \ldots, i\} \to A$ and $h : \{1, 2, \ldots, j\} \to B$. (Note that $h^{-1} : B \to \{1, 2, \ldots, j\}$ exists and is a one-to-one correspondence by Theorem 7.8.) By composing these functions, we can produce the function $h^{-1} \circ f \circ g : \{1, 2, \ldots, i\} \to \{1, 2, \ldots, j\}$, which is one-to-one by Theorem 7.11.

Picture it:

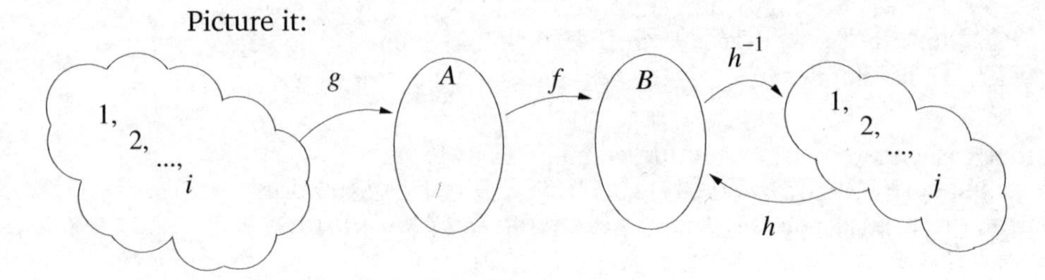

Really we want to prove that when we map A to B, we may have some elements of B left over. Equivalent to that, we will show that when we map $\{1, 2, \ldots, i\}$ to $\{1, 2, \ldots, j\}$, we likewise may have elements of $\{1, 2, \ldots, j\}$ left over. We will let k be the number of extra elements. Here is how we will count them:

> Note that $H^{-1} \circ F \circ G(\{1, 2, \ldots, i\}) \subseteq \{1, 2, \ldots, j\}$. Let $k = |\{1, 2, \ldots, j\} - H^{-1} \circ F \circ G(\{1, 2, \ldots, i\})|$. By definition of cardinality, there exists a one-to-one correspondence $\phi : \{1, 2, \ldots, k\} \rightarrow \{1, 2, \ldots, j\} - H^{-1} \circ F \circ G(\{1, 2, \ldots, i\})$.

As an example, take $i = 3$, $j = 5$, and f as shown in the following diagram. In this case, $k = 2$. You should keep in mind, however, that since the theorem says $\leq$, not $<$, it is possible that k would be 0.

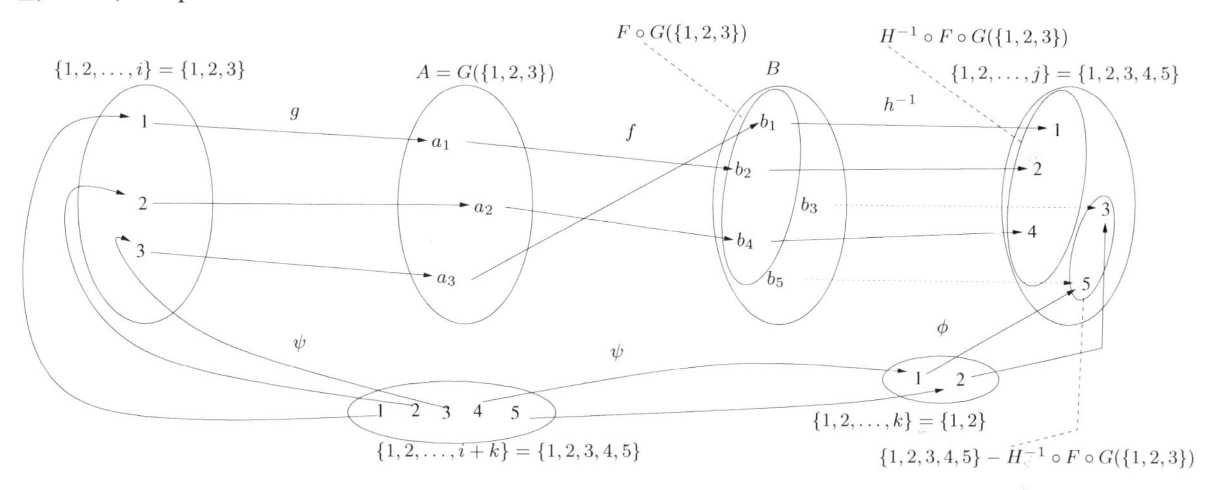

Notice that everything in $\{1, 2, \ldots, j\}$ is hit either by $h^{-1} \circ f \circ g$ or by ϕ. Now, formally to show $i + k = j$:

> Define $\psi : \{1, 2, \ldots, i + k\} \rightarrow \{1, 2, \ldots, j\}$ so that
>
> $$\psi(x) = \begin{cases} h^{-1} \circ f \circ g(x) & \text{if } x \leq i \\ \phi(x - i) & \text{if } i + 1 \leq x \leq i + k \end{cases}$$
>
> We claim that ψ is a one-to-one correspondence. Suppose $\ell \in \{1, 2, \ldots, j\}$. Either $\ell \in H^{-1} \circ F \circ G(\{1, 2, \ldots, i\})$ or $\ell \notin H^{-1} \circ F \circ G(\{1, 2, \ldots, i\})$.

We divide this up based on whether ℓ is in the set hit by $h^{-1} \circ f \circ g$ or the set hit by ϕ.

> Case 1: Suppose $\ell \in H^{-1} \circ F \circ G(\{1, 2, \ldots, i\})$. Then by definition of image, there exists $m \in \{1, 2, \ldots, i\}$ such that $h^{-1} \circ f \circ g(m) = \ell$. By how we have defined ψ, $\psi(m) = \ell$.

Further suppose $n \in \{1, 2, \ldots, i+k\}$ and $\psi(n) = \ell$. Suppose $n > i$. Then $\ell = \psi(n) = \phi(n-i) \in \{1, 2, \ldots, j\} - H^{-1} \circ F \circ G(\{1, 2, \ldots, i\})$, and so $\ell \notin H^{-1} \circ F \circ G(\{1, 2, \ldots, i\})$, which is a contradiction. So $n \leq i$.

Since $h^{-1} \circ f \circ g$ is one-to-one, this implies $n = \ell$. Hence ℓ is unique.

We have shown that in the first case, ℓ is uniquely hit by ψ, specifically from something in the set $\{1, 2, \ldots, i+k\}$.

Case 2: Suppose $\ell \notin H^{-1} \circ F \circ G(\{1, 2, \ldots, i\})$. By definition of difference, $\ell \in \{1, 2, \ldots, j\} - H^{-1} \circ F \circ G(\{1, 2, \ldots, i\})$. By definition of one-to-one correspondence, there exists $m \in \{1, 2, \ldots, k\}$ such that $\phi(m) = \ell$. By how ψ is defined, $\psi(m+i) = \phi(m+i-i) = \phi(m) = \ell$. Note that $m + i \in \{1+i, 2+i, \ldots, k+i\}$.

Further suppose $n \in \{1+i, 2+i, \ldots, k+i\}$ and $\phi(n) = \ell$. Suppose $n \leq i$. Then $\ell = \psi(n) = h^{-1} \circ f \circ g(x) \in H^{-1} \circ F \circ G(\{1, 2, \ldots, i\})$, which is a contradiction, So $n > i$, and hence $\psi(n) = \phi(n-i)$.

Since ϕ is one-to-one, $n - i = m$, and so $n = m + i$, and $m + i$ is unique.

Now we have shown that in the other case, ℓ is uniquely hit by ψ but from something in the set $\{i+1, i+2, \ldots, i+k\}$.

Either way, there exists a unique $n \in \{1, 2, \ldots, i+k\}$ such that $\psi(n) = \ell$. Hence ψ is a one-to-one correspondence, and by definition of cardinality, $|\{1, 2, \ldots, i+k\}| = |\{1, 2, \ldots, j\}|$, that is, $i+k = j$.

By definition of $\leq$, $i \leq j$. Therefore, by substitution, $|X| \leq |Y|$. □

Pigeonhole Principle

The contrapositive of this theorem is more famous and is known as the *Pigeonhole Principle*. Suppose we have n pigeons and k pigeonholes. If $k < n$, then there are not enough pigeonholes to go around for every pigeon to have its own hole. We know that some pigeons must share a hole. Consider the pigeons to stand for the elements in a finite domain and the pigeonholes to stand for the elements in a finite codomain. The function maps pigeons to their holes.

Theorem 7.14 *If A and B are finite sets, $|A| > |B|$, and $f : A \to B$, then f is not one-to-one.*

Proof. By Theorem 7.13 and contrapositive law. □

This has been a workout. Reward yourself with the exercises below, and if they do not help, then try ice cream.

Exercises

In Exercises 7.9.1–7.9.2, assume A and B are finite sets. Prove.

7.9.1 If $A \subseteq B$, then $|B - A| = |B| - |A|$. (Hint: This turns out to be a fairly short proof, by citing previous results: Exercise 4.7.1, Exercise 4.4.5, and Theorem 7.12.)

7.9.2 If $A \subseteq B$, then $|A| \leq |B|$. (Hint: Use Exercise 7.9.1.)

In Exercises 7.9.3–7.9.5, assume $f : X \to Y$ for some finite sets X and Y. Prove.

7.9.3 If $A \subseteq X$, then $|F(A)| \leq |A|$. This is a difficult proof. Try using the following outline:

- Use induction on $|A|$. That is, prove that for all $n \in \mathbb{W}$, if $|A| = n$ then $|F(A)| \leq n$.
- In your base case, $n = 0$ so $A = \emptyset$. Show (trivially) that $|A| = |F(A)|$, which generalizes to $|F(A)| \leq |A|$.
- For your inductive case, suppose this works for some n, and suppose $|A| = n + 1$.
- Pick some element $a \in A$. What do Exercise 7.9.1 and the inductive hypothesis tell you about $|A - \{a\}|$?
- Either $f(a) \in F(A - \{a\})$ or $f() \notin F(A - \{a\})$. Divide this into cases to show $|F(A)| \leq |A|$.
- In the case where $f(a) \in F(A - \{a\})$, use Exercise 7.4.5.
- In the case where $f(a) \notin F(A - \{a\})$, use Exercise 7.4.4.

7.9.4 If $|X| = |Y|$, then f is one-to-one if and only if f is onto. (Hint: Try proving the parts of this using proof by contradiction. For example, suppose that f is one-to-one but not onto. In that case, there exists $y \in Y$ such that for all $x \in X$, $f(x) \neq y$. Then construct a new function like f except with $Y - \{y\}$ as its codomain. Then apply the other theorems or exercises.)

7.9.5 If f is onto, then $|X| \geq |Y|$. (Hint: Use Exercise 7.9.3 and the fact that if f is onto, then its codomain is its range.)

7.9.6 Prove that if A is a finite set and $a \in A$, then $|\{\ \{a\} \cup C \mid C \in \mathscr{P}(A - \{a\})\}| = |\mathscr{P}(A - \{a\})|$.

7.9.7 Let B be a set of finite subsets of a set A (for example, B could be $\mathscr{P}(A)$). Prove that cardinality (that is, the relation R on B such that $(X, Y) \in R$ if $|X| = |Y|$) is an equivalence relation.

7.10 Counting

We have seen that if A and B are finite, disjoint sets, then $|A \cup B| = |A| + |B|$. For example, if we took the natural numbers 10 and under, their total would be the number of evens plus the number of odds.

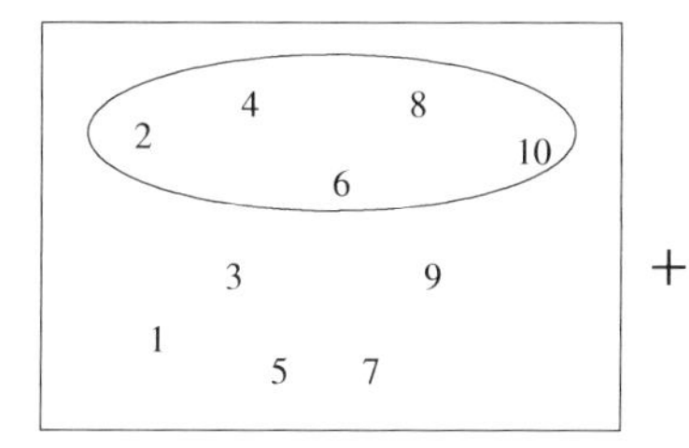

+

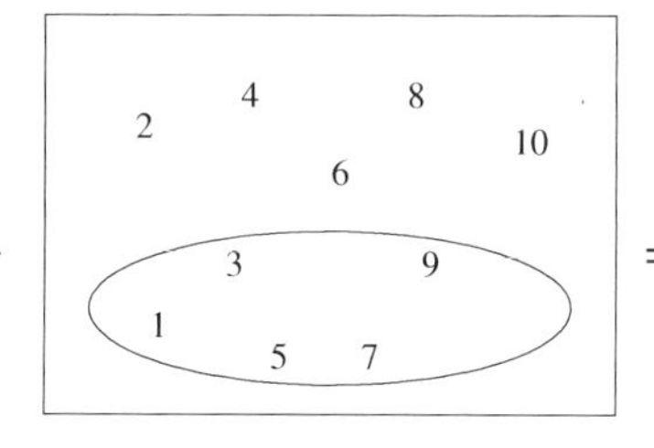

=

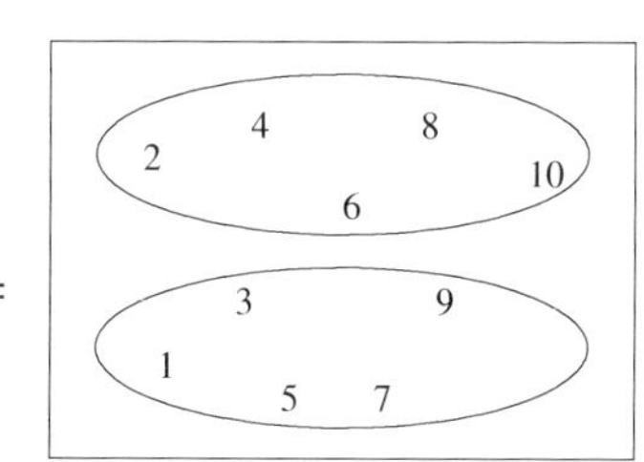

Using similar intuition, if one set is a subset of another, we can compute the cardinality of a set difference by subtracting the cardinality of the subset from the cardinality of the superset. Suppose we wanted to count the numbers that are not multiples of three.

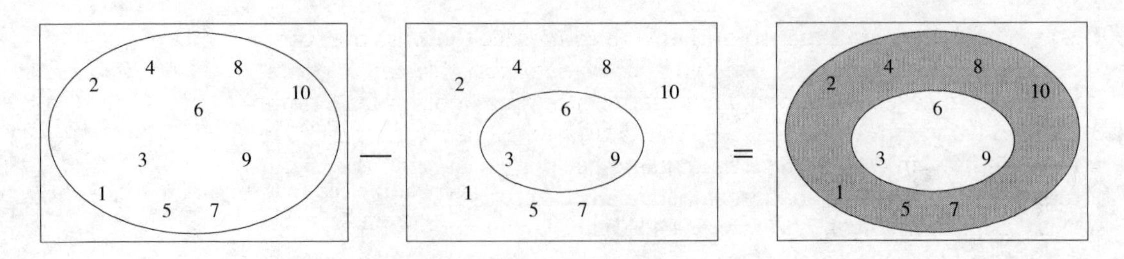

Notice that this is just Exercise 7.9.1. Now, what if we were to count the numbers that are even or multiples of three?

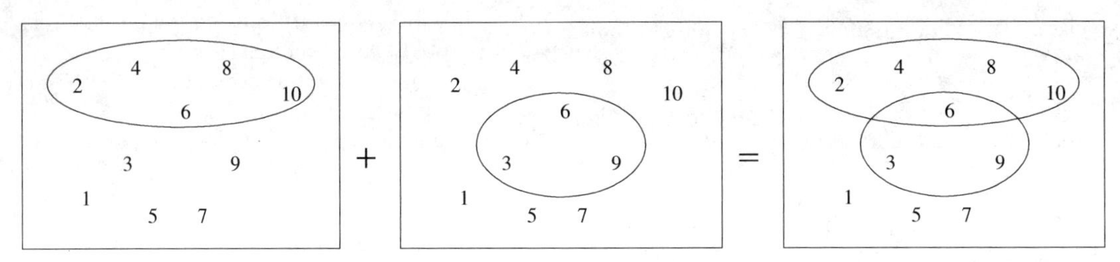

We could not simply add 5 + 3, because that would count the element 6 twice; the correct answer is 7, not 8. We can generalize the result about the union of disjoint sets by noting that all we need to do is subtract out the cardinality of the intersection (if the sets are disjoint, the intersection will be empty anyway).

Theorem 7.15 *If A and B are finite sets, then* $|A \cup B| = |A| + |B| - |A \cap B|$.

This result relies on earlier results summarized on the right.

Proof. Suppose A and B are finite sets. Note that by Exercise 4.4.6, A and $B - (A \cap B)$ are disjoint; and that by Theorem 4.1, $A \cap B \subseteq B$. Then $|A \cup B| =$

	$\lvert A \cup (B - (A \cap B))\rvert$	by Exercise 4.3.15 and substitution
$=$	$\lvert A\rvert + \lvert B - (A \cap B)\rvert$	by Theorem 7.12
$=$	$\lvert A\rvert + \lvert B\rvert - \lvert A \cap B\rvert$	by Exercise 7.9.1. □

Exercise 4.4.6	$A \cap (B - (A \cap B)) = \emptyset$.
Theorem 4.1	$A \cap B \subseteq B$.
Exercise 4.3.15	$A \cup B = A \cup (B - (A \cap B))$.
Theorem 7.12	If A and B are finite, disjoint sets, then $\lvert A \cup B\rvert = \lvert A\rvert + \lvert B\rvert$.
Exercise 7.9.1	If A and B are finite sets and $A \subseteq B$, then $\lvert B - A\rvert = \lvert B\rvert - \lvert A\rvert$.

This theorem can be interpreted as a set of directions for counting the elements of A and B. For example, if you counted the number of math majors and the number of computer science majors in your class, you may end up with a number larger than the enrollment because you counted the double math-computer science majors twice. To avoid counting the overlapping elements twice, we subtract the cardinality of the intersection.

The area of mathematics that studies how to count sets, the ways they combine, and the ways they are are ordered is *combinatorics*. (Elementary combinatorics is often called "counting," but that term invites derision from those unacquainted with higher mathematics. Who wants to tell people that they are learning "counting" in their college math class?) It plays a part in many fields of mathematics, especially probability and statistics. Exercise 7.9.1 is called the *Difference Rule* and Theorem 7.15 is called the *Inclusion/Exclusion Rule*.

combinatorics

Difference Rule

Supposing we do have disjoint sets, we generalize Theorem 7.12 to describe counting a set by counting the sets that make a partition of it. We call this the *Addition Rule*:

Inclusion/Exclusion Rule

Addition Rule

Theorem 7.16 *If A is a finite set with partition $A_1, A_2, \ldots, A_n$, then* $|A| = \sum_{i=1}^{n} |A_i|$.

This result can help us prove a generalized form of the Pigeonhole Principle. If the domain has more than k times as many elements as the codomain, some element in the codomain must be hit by more than k elements.

Theorem 7.17 *If X and Y are finite sets, $k \in \mathbb{N}$, and $|X| > k \cdot |Y|$, and $f : X \to Y$, then there exists $y \in Y$ such that $|F^{-1}(y)| \geq k + 1$.*

Proof. Suppose X and Y are finite sets, $k \in \mathbb{N}$, and $|X| > k \cdot |Y|$, and $f : X \to Y$. Further suppose that for all $y \in Y$, $|F^{-1}(y)| \leq k$.

Consider the set of inverse images of elements in Y, that is, let $A = \{F^{-1}(\{y\}) \mid y \in Y\}$. By Exercise 7.10.1, A is a partition of X, and so letting $Y = \{y_1, y_2, \ldots y_n\}$, we have

$$|X| = \sum_{i=1}^{|Y|} |F^{-1}(\{y_i\})|$$

This formula is valid only if $F^{-1}(\{y_1\}), F^{-1}(\{y_2\})$, etc, is a valid way of naming the elements in A. Even though we know that A is a partition, if for any $y_i, y_j \in Y$, $F^{-1}(\{y_i\}) = F^{-1}(\{y_j\})$, then we have counted one of the elements in the partition twice. However, by Exercise 7.10.2, all inverse images of distinct elements in Y are either distinct or $\emptyset$. It does no harm to count $\emptyset$ more than once, because $|\emptyset| = 0$.

Also, since for all $y \in Y$, $|F^{-1}(y)| \leq k$, we have by algebra

$$\sum_{i=1}^{|Y|} |F^{-1}(\{y_i\})| \leq k \cdot |Y|$$

By substitution, $|X| \leq k \cdot |Y|$, which is a contradiction.

Therefore there must exist $y \in Y$ such that $|F^{-1}(y)| \geq k + 1$.□

Suppose you had a collection of books: 12 computer science books, 8 math books, 7 physics books, 25 fiction books, and 4 telephone directories. As long as these categories are mutually exclusive, the addition rule formalizes the obvious fact that the total number of books is the sum of these, $12+8+7+25+4=56$. This views these categories as a partition of the set of books.

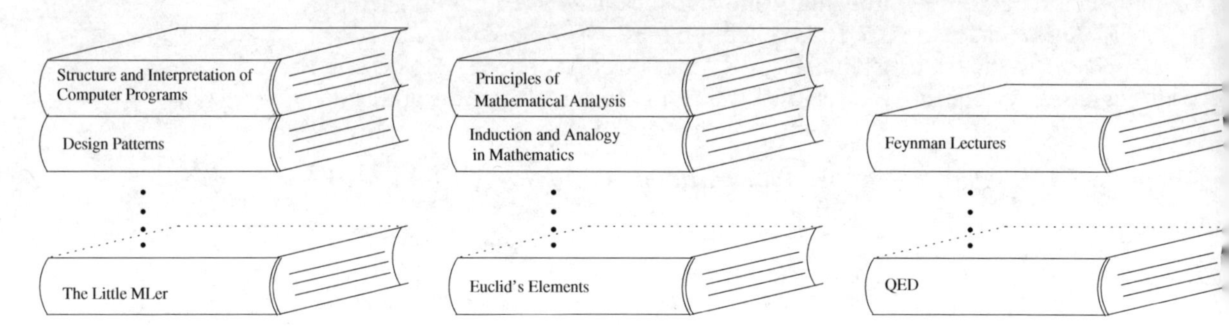

On the other hand, suppose we wanted to choose one book from each category. How many possible combinations of books could we choose? First, we have 12 CS books to choose from; for each of those, there are 8 possible math books; for each of those pairs, we need to choose one of 7 physics books; etc. This requires multiplying the number of options from each category, $12 \cdot 8 \cdot 7 \cdot 25 \cdot 4 = 67200$.

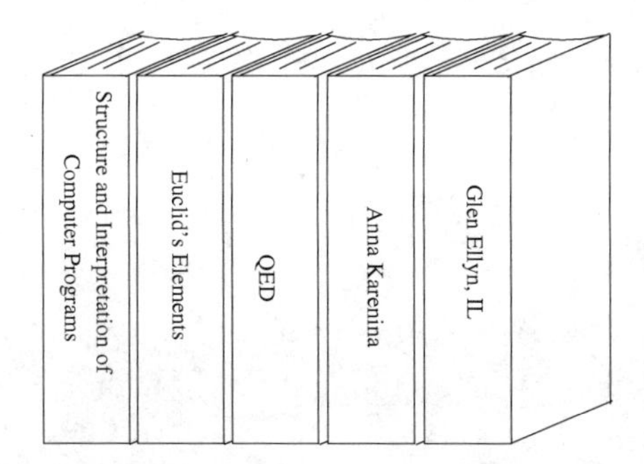

We can think of a set of books drawn this way as a 5-tuple over the book-category sets. Thus to relate the cardinality of a Cartesian product, we have the *multiplication rule*:

multiplication rule

Theorem 7.18 *If $A_1, A_2, \ldots, A_n$ are finite sets, then* $|A_1 \times A_2 \times \ldots \times A_n| = |A_1| \cdot |A_2| \cdot \ldots \cdot |A_n|$.

The proofs are left as exercises. Consider some examples that make use of these rules. ML allows identifiers to be made up of upper and lower case letters, numbers, underscores, and apostrophes, with the first character being a letter or apostrophe. Furthermore, an identifier may not be the same as a reserved word. How many possible four-character identifiers are there in ML? To solve this, consider the various sets of characters. Let A be the set of capital letters, B the set of lowercase letters, C the set { ' }, D the set { _ }, and E the set of digits. Thus

$$
\begin{array}{lll}
 & (& \overbrace{(|A| + |B| + |C|)}^{\textit{addition rule}} \\
\textit{multiplication rule} & \left\{ \begin{array}{l} \times \\ \times \\ \times \end{array} \right. & \begin{array}{l} (|A| + |B| + |C| + |D| + |E|) \\ (|A| + |B| + |C| + |D| + |E|) \\ (|A| + |B| + |C| + |D| + |E|) \;\;) \end{array} \\
\textit{difference rule} & - & |\{\text{ case, else, open, then, type, with }\}|
\end{array}
$$

The number of identifiers is $(53 \cdot 64 \cdot 64 \cdot 64) - 6 = 13893626$.

Next consider the number of possible phone numbers in an area code. Phone numbers are seven characters, chosen from the ten digits with the restrictions that no numbers begin with 0, 1, 555, 411, or 911. Notice that there are

$$10 \cdot 10 \cdot 10 \cdot 10 = 10000$$

four-digit suffixes, and

$$8 \cdot 10 \cdot 10 - 3 = 797$$

three-digit prefixes that do not start with 0 or 1 and do not include the restricted prefixes. Choosing a phone number means choosing a prefix and a suffix, hence $797 \cdot 10000 = 7970000$ possible phone numbers.

Exercises

Prove.

7.10.1 If $f : X \to Y$, then $\{F^{-1}(\{y\}) \mid y \in Y\}$ is a partition of X.

7.10.2 If $f : X \to Y$, $y_1, y_2 \in Y$, and $F^{-1}(\{y_1\}) = F^{-1}(\{y_2\})$, then either $y_1 = y_2$ or $F^{-1}(\{y_1\}) = F^{-1}(\{y_2\}) = \emptyset$.

Compare the following problems with Exercises 6.6.1–6.6.5.

7.10.3 Prove Theorem 7.16.

7.10.4 Prove Theorem 7.18.

7.11 Permutations and combinations

Now consider what happens when we try to arrange the computer science books in the earlier example on a shelf. We describe this process in terms of choosing books for each shelf position from left to right. For the first book, we have 12 options. Once we have chosen a first book, we have 11 options for the second book, etc. The possible ways to arrange or *permute* the set of books is

$$12 \times 11 \times 10 \times \ldots \times 1 = 12! = 479001600$$

permutation

In general, a *permutation* of a set is a sequence of its elements, somewhat like a tuple of the set with dimension the same as the cardinality of the entire set, except one important difference—once we choose the first element, that element is not in the set to choose from for the remaining elements, and so on. In ML, our lists which we use to represent sets are in fact permutations of their sets. Applying the multiplication rule, we see that the number of permutations of a set of size n is

$$\begin{aligned} P(n) &= n \cdot (n-1) \cdot (n-2) \cdot \ldots \cdot 1 \\ &= n! \end{aligned}$$

r-permutation

An r*-permutation* of a set is a permutation of a subset of size r. Suppose we wish to put only 5 books on the shelf. We would have still have 12 options for the first spot, 11 for the second, then 10, 9, and 8 for the remaining spots The number of r-permutations of a set of size n is

$$\begin{aligned} P(n,r) &= n \cdot (n-1) \cdot (n-2) \cdot \ldots \cdot (n-r+1) \\ &= \tfrac{n!}{(n-r)!} \end{aligned}$$

r-combination

On the other hand, suppose we were not concerned about the order of the books, only the choosing of 5 to put on the shelf. It might seem strange to sequence these problems this way—first to consider ordered collections and then to lift the restriction on order—however, it will make sense in our formulating of a solution. The five chosen books simply make a subset of the set of computer science books. When talking about sets in the context of counting, we have a special term: an r*-combination* of a set of n elements is a subset of size r. How many subsets of size r are there of a set of size n (which we write $\binom{n}{r}$ and read "n choose r")? First, we know that the number of *orderings* of subsets of that size is $\frac{n!}{(n-r)!}$. Each of those subsets can be ordered in $r!$ ways. Hence

$$\binom{n}{r} = \frac{P(n,r)}{r!} = \frac{n!}{r!(n-r)!}$$

Thus there are

$$\frac{12!}{5! \cdot (12-5)!} = \frac{12!}{5! \cdot 7!} = \frac{12 \cdot 11 \cdot 10 \cdot 9 \cdot 8}{5 \cdot 4 \cdot 3 \cdot 2 \cdot 1} = 792$$

distinct subsets of five books from a set of size 12.

Our main interest here is computing not just the number of permutations and combinations, but the permutations and combinations themselves. Suppose we want to write a function `combos(x, r)` which takes a set `x` and returns a set of sets, all the subsets of `x` of size r. We need to find a recursive strategy for this. First off, we can consider easy cases. If `x` is an empty set, there are no combinations of any size. Similarly, the only combination of size zero of any set is the empty set. Thus

```
- fun combos(x, 0) = [[]]
=   | combos([], r) = []
```

The case where r is 1 is also straightforward: Every element in the set is, by itself, a combination of size 1. Creating a set of all those little sets is accomplished by our `listify` function from Exercise 2.2.10.

```
- fun combos(x, 0) = [[]]
=   | combos([], r) = []
=   | combos(x, 1) = listify(x)
```

The strategy taking shape is odd compared to most of the recursive strategies we have seen before. The variety of base cases seems to anticipate the problem being made smaller in both the `x` argument and the `r` argument. What does this suggest? When you form a combination of size r from a list, you first must decide whether that combination will contain the first element of the list or not. Thus all the combinations are

- all the combinations of size r that do not contain the first element, plus
- all the combinations of size $r - 1$ that do not contain the first element with the first element added to all of them.

Function `addToEach` from Exercise 2.4.14 will work nicely here.

```
- fun combos(x, 0) = [[]]
=   | combos([], r) = []
=   | combos(x, 1) = listify(x)
=   | combos(head::rest, r) =
=         addToEach(head, combos(rest, r-1)) @ combos(rest, r);
```

Exercises

7.11.1 In the code for `combos` given, the pattern `combos(x, 1)` is unnecessary. Explain why.

7.11.2 Write a function `P(n, r)` which computes the number of r-permutations of a set of size n.

7.11.3 Write a function `C(n, r)` which computes the number of r-combinations of a set of size n.

The following exercises walk you through how to write a function to compute r-permutations.

7.11.4 Write a function `addEverywhere(x, y)` which takes an item `x` and a list of items of that type `y` and returns a list of lists, each like `y` but with `x` inserted into every position. For example, `addEverywhere(1, [2, 3, 4])` should return `[[1,2,3,4],[2,1,3,4],[2,3,1,4],[2,3,4,1]]`. Given an empty list, it should return the list containing just the list containing `x`.

7.11.5 Write a function `fullPermus(y)` which takes a list and returns the list of all the permutations of that list.

7.11.6 Write a function `fullPermusList(y)` which takes a list of lists and returns a united list of all the permutations of all the lists in `y`.

7.11.7 Write a function `permus(y, r)` which computes all the r-permutations of the list `y`.

7.12 Currying

arity

The *arity* of a function is the number of its parameters. We begin by introducing a common functional programming technique for reducing the arity of a function by partially evaluating it. Take for a simple example a function that takes two arguments and multiplies them.

```
- fun mul(x, y) = x * y;
```

We could specialize this by making a function that specifically doubles any given input by calling `mul` with one argument hardwired to be 2.

```
- fun double(y) = mul(2, y);
```

This idea can be generalized and automated by a function that takes any first argument and returns a function that requires only the second argument.

```
- fun makeMultiplier(x) = fn y => mul(x, y);

val makeMultiplier = fn : int -> int -> int

- makeMultiplier(2);

val it = fn : int -> int

- it(3);

val it = 6 : int
```

This process is called *currying*, after mathematician Haskell Curry, who studied this process (though he did not invent it). We can generalize this with a function that transforms any two-argument function into curried form.

currying

```
- fun curry(func) = fn x => fn y => func(x, y);

val curry = fn : ('a * 'b -> 'c) -> 'a -> 'b -> 'c

- curry(mul)(2)(3);

val it = 6 : int
```

In fact, currying is the native style for ML. The "real" versions of `map` and `foldl` which come pre-made in ML are curried.

```
- map;

val it = fn : ('a -> 'b) -> 'a list -> 'b list

- map(fn(a)=>2*a)([5, 12, 16, 30]);

val it = [10,24,32,60] : int list
```

In fact, since multiple parameters in this style are not tuples, we can dispense with the parentheses. Supposing we had your `putInPlace` method from Exercise 6.8.15, we could write a quick version of selection sort:

```
- foldl putInPlace [] [5, 30, 12, 2, 19, 4];

val it = [2,4,5,12,19,30] : int list
```

Exercises

7.12.1 Type-analyze the expression

`makeMultiplier(2)(3)`

7.12.2 Type-analyze the expression

`curry(mul)(2)(3)`

7.12.3 Write a function `makeSearcher1` which takes a list and returns a function which takes a number and returns the position of the number in the list, -1 if it is absent.

7.12.4 Write a function `makeSearcher2` which takes a number and returns a function which takes a list and returns the position of the number in the list, -1 if it is absent.

7.13 Fixed-point iteration

Past generations learned algorithms for calculating square roots. The availability of computing machinery has made that unnecessary in today's curriculum. Since we are learning how to program computing machinery, though, it is a good exercise for us to see how machines are directed to do it. We will adapt Newton's method for finding roots in general, which you may recall from calculus. Here is how Newton's method works.

root

Suppose a curve of a function f crosses the x-axis at x'. Functionally, this means $f(x') = 0$, and we say that x' is a *root* of f. Finding a decimal or fraction representation for x' may be difficult or even impossible. Obviously x' may not be rational—we are going to be looking for values like $\sqrt{x}$. If f is not a polynomial function, then x' may not even be an algebraic number (do you remember the difference between $\mathbb{A}$ and $\mathbb{T}$?). Instead, we use Newton's method to approximate the root.

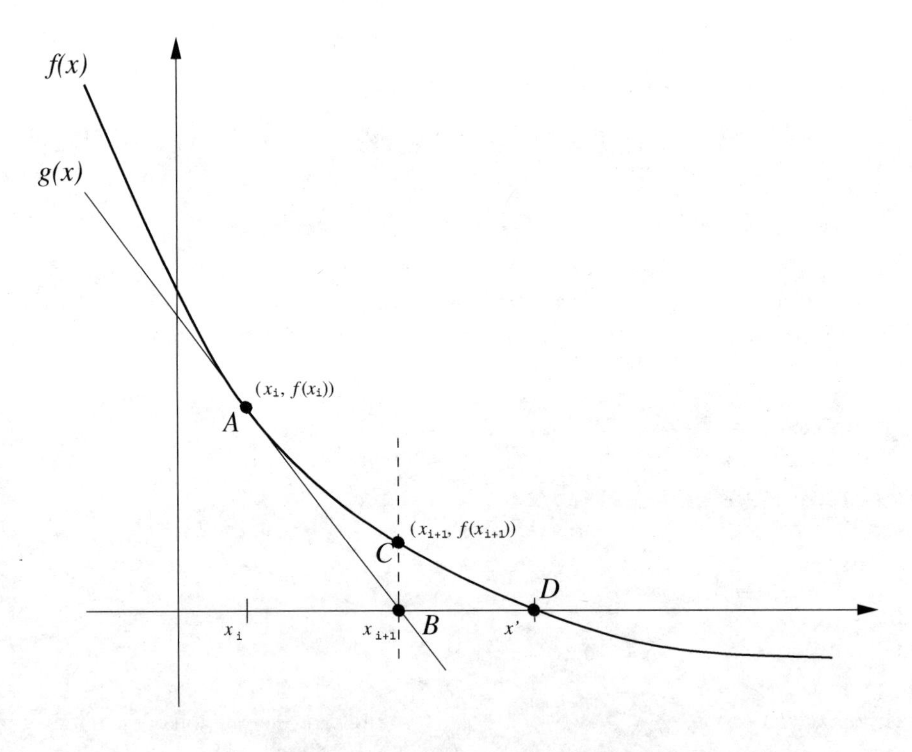

tolerance

The approximation is done by making an initial guess and then improving the guess until it is "close enough" (in technical terms, it is within a desired *tolerance* ϵ of the correct answer). Suppose x_i is a guess in this process. To improve the guess, we draw a tangent to the curve at the point A, $(x_i, f(x_i))$, and then calculate the point at which the tangent strikes the x-axis. The slope of the tangent can be calculated by evaluating the derivative at that point, $f'(x_i)$.

Recall from first-year algebra that if you have a slope m and a point (x', y'), a line through that point with that slope satisfies the equation

$$\begin{aligned} y - y' &= m \cdot (x - x') \\ y &= m \cdot (x - x') + y' \end{aligned}$$

Using point-slope form, we define a function (call it g) corresponding to the tangent,

$$g(x) = f'(x_i) \cdot (x - x_i) + f(x_i)$$

Let x_{i+1} be the x value where g strikes the x-axis. Thus we want $g(x_{i+1}) = 0$, and solving this equation for x_{i+1} lets us find point B, the intersection of the tangent and the axis.

$$\begin{aligned} 0 &= f'(x_i)(x_{i+1} - x_i) + f(x_i) \\ x_{i+1} &= \frac{x_i f'(x_i) - f(x_i)}{f'(x_i)} \\ &= x_i - \frac{f(x_i)}{f'(x_i)} \qquad (7.1) \end{aligned}$$

Drawing a vertical line through B leads us to C, the next point on the curve where we will draw a tangent. Observe how this process brings us closer to x', the actual root, at point D. x_{i+1} is thus our next guess. Equation 7.1 tells us how to generate each successive approximation from the previous one. When the absolute value of $f(x_i)$ is small enough (the function value of our guess is within the tolerance of zero), then we declare x_i to be our answer.

We can use this method to find $\sqrt{c}$ by noting that the square root is simply the positive root of the function $f(x) = x^2 - c$. In this case we find the derivative $f'(x) = 2x$ and plug this into Equation 7.1 to produce a function for improving a given guess x:

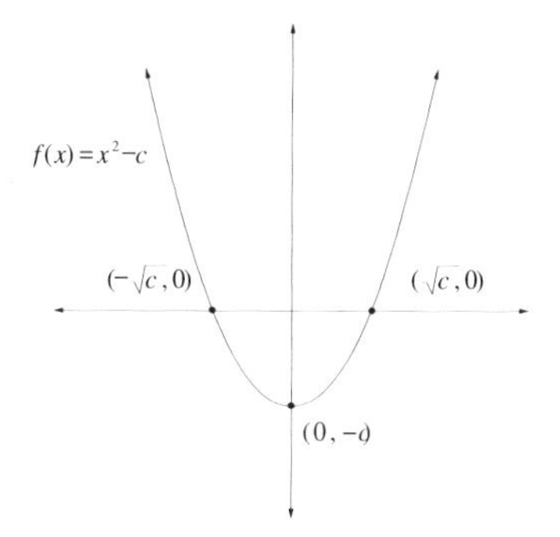

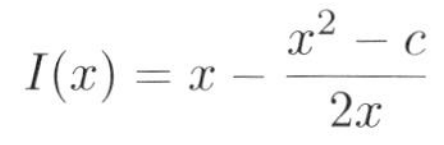

$$I(x) = x - \frac{x^2 - c}{2x}$$

In ML,

```
- fun improve(x) =
=              x - (x * x - c) / (2.0 * x);

val improve = fn : real -> real
```

Obviously this will work only if c has been given a valid definition already. Now that we have our guess-improver in place, our concern is the repetition necessary to achieve a result. Stated algorithmically, while our current approximation is not within the tolerance, simply improve the approximation. At least, that is how we might state it in iterative terms. We set up a recursive solution based on the current guess, which is the data that is being tested and updated. There are two cases, depending on whether the current guess is within the tolerance or not.

- *Base case:* If the current guess is within the tolerance, return it as the answer.
- *Recursive case:* Otherwise, improve the guess and reapply this test, returning the result as the answer.

In ML,

```
- fun sqrtBody(x) =
=             if isInTolerance(x)
=             then x
=             else sqrtBody(improve(x));

val sqrtBody = fn : real -> real
```

We called this function `sqrtBody` instead of `sqrt` because it is a function of the previous guess, `x`, *not* a function of the radicand, `c`. Two things remain: a predicate to determine if a guess is within the tolerance (say, $\epsilon = .001$; then we are in the tolerance when $|x^2 - c| < .001$), and an initial guess (say, 1). If we package this together, we have

```
- fun sqrt (c) =
=    let  fun isInTolerance(x) =
=                    abs((x * x) - c) < 0.001;
=         fun improve(x) =
=                    x - (x * x - c) / (2.0 * x);
=         fun sqrtBody(x) =
=                    if isInTolerance(x)
=                    then x
=                    else sqrtBody(improve(x));
=   in
=      sqrtBody(1.0)
=   end;

val sqrt = fn : real -> real

- sqrt(2.0);

val it = 1.41421568627 : real
```

```
- sqrt(16.0);

val it = 4.00000063669 : real

- sqrt(121.0);

val it = 11.0000000016 : real

- sqrt(121.75);

val it = 11.0340382471 : real
```

Whenever you solve a problem in mathematics or computer science, the next question to ask is whether the solution can be generalized so that it applies to a wider range of problems and thus can be reused more readily. To this end, we will first *analyze* our solution to identify its constituent parts and then *synthesize* them to a general solution.

To generalize an idea means to reduce the number of assumptions and to acknowledge more unknowns. In other words, we are replacing constants with variables. Our square root algorithm was a specialization of Newton's method. The natural next question is how to program Newton's method in general.

What assumptions or restrictions did we make on Newton's method when we specialized it? Principally, we assumed that the function for which we were finding a root was in the form $x^2 - c$ where c is a variable to the system. Let us examine how this assumption affects the segment of the solution that tests for tolerance.

```
- fun isInTolerance(x) =
=           abs((x * x) - c) < 0.001;
```

The assumed function shows itself in the expression `(x * x) - c`. By taking the absolute value of that function for a supplied `x` and comparing with .001, we are checking if the function is "within an epsilon" of zero. We know that functions can be passed as parameters to functions; here, as happens frequently, *generalization* manifests itself as *parameterization*.

```
- fun isInTolerance(function, x) =
=        abs(function(x)) < 0.001;

val isInTolerance = fn : ('a -> real) * 'a -> bool
```

Take stock of the type. The function `isInTolerance` takes a function (in turn mapping from a type 'a to real) and a value of type 'a. The given information does not allow ML to infer what type `function` would accept; hence the type variable 'a. How function's return type is inferred to be real is more subtle. abs is a special kind of function that is defined so that it can accept either reals or ints, but it must return the same type that it receives. Since we compare its result against 0.001, its result must be real; thus its parameter must also be real, and finally we conclude that function must return a real.

isInTolerance is now less easy to use because we must pass in the function whenever we want to use it, unless function is in scope already and we can eliminate it as a parameter. However, we know that functions can also return functions. To make this more general, instead of writing a function to test the tolerance, we write a function that *produces a tolerance tester*, based on a given function.

```
- fun toleranceTester(function) =
=       fn x => abs(function(x)) < 0.001;

val toleranceTester = fn : ('a -> real) -> 'a -> bool
```

Notice that the -> operator is right associative, which means it groups items on the right side unless parentheses force it to do otherwise. Fully parenthesized, the type would be ('a -> real) -> ('a -> bool), and so toleranceTester accepts something of type 'a -> real and returns something of type 'a -> bool. Now we need to call toleranceTester only once and call the function it returns whenever we want to test for tolerance. To further generalize, let us no longer assume $\epsilon = .001$, but instead parameterize it.

```
- fun toleranceTester(function, tolerance) =
=       fn x => abs(function(x)) < tolerance;

val toleranceTester = fn : ('a -> int) * int -> 'a -> bool
```

What happened to the type? Since 0.001 no longer appears, there is nothing to indicate that we are dealing with reals. Yet ML cannot simply introduce a new type variable (say, 'b in ('a -> 'b) * 'b -> 'a -> bool) because abs is not defined for all types, just int and real. Instead, ML has to guess, and when it comes between real and int, it goes with int. We will force it to choose real.

```
- fun toleranceTester(function, tolerance) =
=       fn x => abs(function(x):real) < tolerance;

val toleranceTester = fn : ('a -> real) * real -> 'a -> bool
```

Next, consider the function improve. We can generalize this by stepping back and considering how we formulated it in the first place. It comes from applying Equation 7.1 to a specific function f. Thus we can generalize it by making the function of the curve a parameter. Since we do not have a means of differentiating f, we will need f' to be supplied as well.

```
- fun nextGuess(guess, function, derivative) =
=     guess - (function(guess) / derivative(guess));
```

However, just as with tolerance testing, we would prefer to think of our next-guesser as a function only of the previous guess, not of the curve function and derivative. We can modify `nextGuess` easily so that it produces a function like `improve`:

```
- fun nextGuesser(function, derivative) =
=     fn guess => guess - (function(guess) / derivative(guess));
```

Notice that this process amounts to the partial application of a function, an example of currying. `nextGuess` has three parameters; `nextGuesser` allows us to supply values for some of the parameters, and the result is another function. `sqrtBody` also demonstrates a widely applicable technique. If we generalize our function $I(x)$ based on Equation 7.1 we have

$$G(x) = x - \frac{f(x)}{f'(x)}$$

If x is an actual root, then $f(x) = 0$, and so $G(x) = x$. In other words, a root of $f(x)$ is a solution to the equation

$$x = G(x)$$

Problems in this form are called *fixed point problems* because they seek a value which does not change when $G(x)$ is applied to it (and so it is fixed). If the fixed point is a local minimum or maximum and one starts with a good initial guess, one approach to solving (or approximating) a fixed point problem is *fixed point iteration*, the repeated application of the function $G(x)$, that is

fixed point problems

fixed point iteration

$$G(G(G(\ldots G(x')\ldots)))$$

where x' is the initial guess, until the result is "good enough." In parameterizing this process by guess-improving function, initial guess, and tester, we have this generalized version of `sqrtBody`:

```
- fun fixedPoint(improve, guess, tester) =
=    if tester(guess)
=    then guess
=    else fixedPoint(improve, improve(guess), tester);

val fixedPoint = fn : ('a -> 'a) * 'a * ('a -> bool) -> 'a
```

We have decomposed our implementation of the square root function to uncover the elements in Newton's method (and more generally, a fixed point iteration). Now we assemble these to make useful, applied functions. In the analysis, parameters proliferated; as we synthesize the components into something more useful, we will reduce the parameters, or "fill in the blanks." Simply hooking up `fixedPoint`, `nextGuesser`, and `toleranceTester`, we have

```
- fun newtonsMethod(function, derivative, guess) =
=     fixedPoint(nextGuesser(function, derivative), guess,
=                    toleranceTester(function, 0.001));

val newtonsMethod = fn : (real -> real) * (real -> real) * rea
                   -> real
```

Given a function, its derivative, and a guess, we can approximate a root. However, one parameter in particular impedes our use of `newtonsMethod`. We are required to supply the `derivative` of `function`; in fact, many curves on which we wish to use Newton's method may not be readily differentiable. In those cases, we would be better off finding a numerical approximation to the derivative. The easiest such approximation is the *secant method*, where we take a point on the curve near the point at which we want to evaluate the derivative and calculate the slope of the line between those points (which is a secant to the curve). Thus for small ϵ, $f'(x) \approx \frac{f(x+\epsilon)-f(x)}{\epsilon}$. In ML, taking $\epsilon = .001$,

```
- fun numDerivative(function) =
=    fn x => (function(x + 0.001) - function(x)) / 0.001;

val numDerivative = fn : (real -> real) -> real -> real
```

Now we make an improved version of our earlier function. Since any user of this new function is concerned only about the results and not about how the results are obtained, our name for it shall reflect *what the function does* rather than *how it does it*.

```
- fun findRoot(function, guess) =
=     newtonsMethod(function, numDerivative(function), guess);

val findRoot = fn : (real -> real) * real -> real
```

Coming full circle, we can apply these pre-packaged functions to a special case: finding the square root. We can use `newtonsMethod` directly and provide an explicit derivative or we can use `findRoot` and rely on a numerical derivative, with different levels of precision. Since for positive x, $x^2 - c$ is monotonically increasing (that is, it is always going up), 1 is a safe guess, which we provide.

```
- fun sqrt(c) =
=     newtonsMethod(fn x => x * x - c, fn x => 2.0 * x, 1.0);

val sqrt = fn : real -> real

- sqrt(2.0);
```

```
val it = 1.41421568627 : real

- sqrt(16.0);

val it = 4.00000063669 : real

- sqrt(121.0);

val it = 11.0000000016 : real

- sqrt(121.75);

val it = 11.0340382471 : real
```

There are several lessons here. First, this has been a demonstration of the interaction between mathematics and computer science. The example we used comes from an area of study called numerical analysis which straddles the two fields. Numerical analysis is concerned with the numerical approximation of calculus and other topics of mathematical analysis. More importantly, this also demonstrates the interaction between discrete and continuous mathematics. We have throughout assumed that $f(x)$ is a real-valued function like you are accustomed to seeing in calculus or analysis. However, *the functions themselves* are discrete objects. The most important lesson is how functions can be parameterized to become more general, curried to reduce parameterization, and, as discrete objects, passed and returned as values.

The running example in this chapter was developed from Abelson and Sussman [1].

Exercises

7.13.1 Write a function similar to the original `sqrt` (page 370) but to calculate cubed roots.

7.13.2 Notice that a root of f is either a local minimum or a local maximum of G. Identify under what circumstances it is a minimum and under what circumstances it is a maximum.

7.13.3 Write a function similar to the final `sqrt` (on page 374) but to calculate cubed roots.

7.13.4 The implementation of `fixedPoint` presented in this chapter differs from traditional fixed point iteration because it requires the user to provide a function which determines when the iteration should stop, based on the current guess. That approach is tied to the application of finding roots, since our criterion for termination is how close f(`guess`) is to zero. Instead, termination should depend on how close *successive guesses are to each other*, that is, if $\mid x_i - x_{i-1} \mid < \epsilon$. Rewrite `fixedPoint` so that it uses this criterion and receives a tolerance instead of a tester function.

7.14 Extended example: Modeling mathematical functions

Prior to this course, you probably associated *function* most readily with the real-valued functions from algebra and calculus. In this course, functions are a crucial element to ML programming, and they are way to reason about any sets. In this section we take up the topic of modeling real-valued functions in ML. This section requires a semester or so of calculus.

Obviously we can model mathematical functions with plain old ML functions. However, there are some operations we would like to perform on real functions—differentiation and integration, for example. For this reason we might seek other ways to model real functions.

Start with the most familiar, polynomial functions. A polynomial in x is a sum of terms with each term being x raised to a whole number power and multiplied by a real number *coefficient*. Formally, a polynomial in x of *degree* n has the form

$$\sum_{i=0}^{n} c_i \cdot x^i$$

for some real numbers $c_0, c_1, \ldots c_n$. For example

$$6x^4 + 2x^3 + 0x^2 + 12x^1 + 8x^0$$

Of course, we would normally write that as

$$6x^4 + 2x^3 + 12x + 8$$

What information do we need to store in order to model a polynomial function? It is as simple as a list of coefficients. All other information—the degree, the exponents on the individual terms—can be inferred from the structure, as long as we have reasonable assumptions about that structure. In our case, we will assume that all coefficients (up to the highest-degree term with non-zero coefficient) are present in the list, arranged from highest degree to zero degree.

```
- [6, 2, 0, 12, 8];

val it = [6,2,0,12,8] : int list
```

The simplest operation is to evaluate this function at some given x value. Instead of evaluating it in a brute-force method involving many exponentiations, we will use a more efficient approach known as Horner's rule:

$$\sum_{i=0}^{n} c_i \cdot x^i = (c_0 + x \cdot (\ldots x \cdot (c_{n-2} + x \cdot (c_{n-1} + x \cdot c_n)) \ldots))$$

Read it this way: c_n is our starting value. The first step is to multiply by x, and then we add c_{n-1}. That is our next "value so far." Going on from highest degree coefficient to lowest, we multiply our value so far by x and add the next coefficient. We can implement this using `foldl`. The seed value is 0, and the supplied function takes the next coefficient and the accumulated answer.

```
- fun evaluatePoly(coeffs, x) = foldl(fn(c, y) => c + y * x,
=                                     0.0, coeffs);

val evaluatePoly = fn : real list * real -> real

- evaluatePoly([6.0, 2.0, 0.0, 12.0, 8.0], 4.5);

val it = 2704.625 : real
```

Computing the derivative of a polynomial is also reasonably easy.

$$\frac{d}{dx}\sum_{i=0}^{n} c_i \cdot x^i = \sum_{i=1}^{n} i \cdot c_i \cdot x^{i-1}$$

In our case,

$$\frac{d}{dx}(6x^4 + 2x^3 + 12x + 8) = 24x^3 + 6x^2 + 12$$

What change would need to be computed from the list `[6.0, 2.0, 0.0, 12.0, 8.0]`? The list standing for the derivative would need to be one element shorter, specifically the last element would be removed. `12.0` would stay the same, but the other elements would need to be multiplied by factors 2, 3, 4, from back to front.

Clearly the process we are describing needs to be computed starting at the end of the list and working our way forward. In terms of recursion, this means the rest of the list would need to be processed first, then the current element. But to process the current element, we would need to know how many elements there are in the rest of the list, in order to know the factor for the current element. To handle this, we could have the function return two things: not only the transformed list, but also the factor for the next element.

```
- fun differentiatePoly([]) = ([], 0)
=   | differentiatePoly([c]) = ([], 1)
=   | differentiatePoly(c::rest) =
=       let val (diffRest, degree) = differentiatePoly(rest)
=       in (real(degree) * c ::diffRest,
=           degree + 1)
=       end;

val differentiatePoly = fn : real list -> real list * int
```

Alternately, we could use foldr (see Exercise 7.5.6). Often we will find that using of functions like map, foldl, and foldr will make our functions shorter, though less readable (unless you are used to thinking in terms of these functions). This time it does not make our program shorter, but it does make it so the final answer is just the derivative, not a derivative/next-degree tuple.

```
- fun differentiatePoly(coeffs) =
=      #1(foldr(fn(c, (derivCoeffs, degree)) =>
=                (if degree = 0
=                    then []
=                    else real(degree) * c::derivCoeffs,
=                 degree + 1),
=              ([], 0), coeffs));

val differentiatePoly = fn : real list -> real list

- differentiatePoly([6.0, 2.0, 0.0, 12.0, 8.0]);

val it = [24.0,6.0,0.0,12.0] : real list
```

The function indefIntegratePoly, to compute the coefficients of a polynomial that is an indefinite integral of a given polynomial, is left for a Project. Once computed, we can use it to compute a definite integral:

```
- fun defIntegratePoly(coeffs, min, max) =
=      let val indefIntegral = indefIntegratePoly(coeffs)
=      in evaluatePoly(indefIntegral, max) -
=         evaluatePoly(indefIntegral, min) end;

val defIntegratePoly = fn : real list * real * real -> real
```

However, our original goal was to model mathematical functions, not just the restricted class of polynomials. Different kinds of real functions would require different data storage, and the operations of evaluation, differentiation, and integration would be defined differently. Moreover, we would like to be able to use values of these different kinds of functions uniformly. In other words, we want a type function that all the kinds of functions will fit into.

Consider two other kinds of functions: Exponential functions in the form $c \cdot e^x$ for some constant c, and step functions, defined for a given step point v and step level c to be

$$f(x) = \begin{cases} 0 & \text{if } x < v \\ c & \text{otherwise} \end{cases}$$

If we take our usual approach of representing information with datatypes processed by functions with pattern-matching, we might come up with the following. (Real.Math.exp is a library function that computes powers of e.)

```
- datatype function =
=     Poly of real list | Exp of real | Step of real * real;

- fun evaluate(Poly(coeffs), x) =
=        foldl(fn(c, y) => c + y * x, 0.0, coeffs)
=    | evaluate(Exp(c), x) =
=        c * Real.Math.exp(x)
=    | evaluate(Step(v, c), x) =
=        if x < v then 0.0 else c;

val evaluate = fn : function * real -> real
```

We want to take a different approach on this problem for two reasons. First is *extensibility*. The old way of making datatypes does not allow us to add new varieties of functions later, at least not without rewriting the old code. The second reason is *data hiding*. Code where data and operations are available to the rest of the program only on a "need to know" basis is less error-prone and easier to modify because the code is less dependent on the details of other parts of the code.

extensibility

data hiding

In our current example, suppose that our system of real functions is to be used by another system, say one that draws graphs of functions or does some sort of data analysis. The system using ours (the client system) should be able to work with functions of any kind uniformly—without knowing what kind of function it is or how the information for that kind of function is stored. That way changes to the implementation of the functions can be made, including making new kinds of functions, without any change needing to be made to the client system.

The way to achieve this independence (referred to as *loose coupling* between parts of code) is to have the different parts of the code interact through stable and clearly defined operations. In our case, the available operations are evaluation, finding the derivative, and computing a definite integral. (It might seem more fundamental to make *indefinite* integration an available operation, since a definite integral always can be computed from an indefinite integral. We chose the definite integral instead of the indefinite integral because for many kinds of functions the indefinite integral is hard to find; in those cases we will compute a numerical approximation of the definite integral.)

loose coupling

Here we define the type function in terms of its *interface*, that is, the types of its available operations:

interface

```
- datatype function = Func of
=   (real -> real) * (unit -> function) *
=   (real * real -> real);
```

How then do we make values of this type? We write (ML) functions that will take the necessary information for a kind of (real) function (say, a list of coefficients for a polynomial), make appropriate (anonymous ML) functions, and return a value of type function.

```
- fun makePolynomial(coeffs) =
=  Func ((* Evaluate *)
=         fn (x) => foldl(fn(c,y) => c + y * x, 0.0, coeffs),
=         (* Find derivative *)
=         fn () =>  makePolynomial(#1(foldr(
=                      fn(c, (derivCoeffs, degree)) =>
=                          (if degree = 0
=                               then []
=                               else real(degree)*c::derivCoeffs,
=                           degree + 1),
=                      ([], 0), coeffs))),
=         (* Compute definite integral *)
=         fn (min, max) =>
=            let val Func(evaluateIntegral, _, _)  =
=                  makePolynomial(indefIntegratePoly(coeffs));
=            in evaluateIntegral(max) - evaluateIntegral(min)
=            end);

val makePolynomial = fn : real list -> function
```

This code is plenty complicated, so digest it carefully. The evaluate function is straightforward. The function for making a derivative contains our earlier code for `differentiatePoly` to make the coefficient list for the derivative. However, we no longer can return a real list—we need a value of our new function type. Accordingly, we feed that real list into a recursive call to `makePolynomial`, which will make the appropriate three operations for that new polynomial.

The function for the definite integral is similar but more complicated. As with the differentiation function (and our earlier function for integrating polynomials), we first make the indefinite integral. (We call `indefIntegratePoly` rather than put the code there directly so as not to spoil Project 6.C.) In the function returned from the recursive call to `makePolynomial`, the only component function we care about is the evaluate function, named `evaluateIntegral`. Finally, we compute the definite integral by evaluating the indefinite integral at the endpoints of the range and subtracting.

(These "recursive calls" to `makePolynomial` do not seem to have a base case. Why do we not have infinite recursion? Because those calls are not made by `makePolynomial` itself, but by the functions that `makePolynomial` returns, which in turn will not be called until after `makePolynomial` returns. Although `makePolynomial` is self-referential, it is not recursive in the conventional sense. It does not call itself. It merely makes functions that call it.)

Consider how to make a step function. Its derivative is the constant function 0, which is a special case of polynomial (we ignore the fact that the function is not differentiable at the the step point). Integration is just a matter of computing the area of a rectangle, but it depends on whether the step point falls after the max, before the min, or between the min and max. (If the min is greater than the max, switch them and return the opposite.)

```
- fun makeStep(stepPoint, stepLevel) =
=     Func (fn (x) => if x < stepPoint then 0.0 else stepLevel,
=           fn () => makePolynomial([0.0]),
=           fn (min, max) =>
=              let val (rmin, rmax, sign) =
=                    if min <= max then (min, max, 1.0)
=                                  else (max, min, ~1.0)
=              in sign * (if max < stepPoint then 0.0
=                         else if min > stepPoint
=                              then stepLevel * (max - min)
=                         else stepLevel * (max - stepPoint))
=              end);
```

Finally, we write a function to make exponential functions. You may have guessed that this kind of function was chosen as an example because of how easy the derivative and integral are.

```
- fun makeExponential(coefficient) =
=     Func (fn (x) => coefficient * Real.Math.exp(x),
=           fn () => makeExponential(coefficient),
=           fn (min, max) =>
=              coefficient * (Real.Math.exp(max) -
=                             Real.Math.exp(min)));
```

Students with experience in object-oriented programming such as in Java will recognize what we are doing in this chapter. The datatype function is equivalent to a Java interface, and the kinds of functions would be implemented in Java by classes that implement that interface. The functions `makePolynomial`, `makeStep`, and `makeExponential` stand in for classes (or constructors for those classes, depending on how you look at it). In particular, in Java the body of the derivative function of `makeExponential` would be replaced by `this`.

Project

7.A Use math induction to prove Horner's rule. The first step is to state Horner's rule so that it is quantified over whole numbers.

7.B Rewrite `evaluatePoly` to use "brute force" (not Horner's rule) in computing the value of a polynomial function for a given x. That is, given $3x^2 - 2x + 3$, start with 0 as a running sum. Then add 3. Then compute $-2x$ and add, then compute $3x^2$ and add. Use `foldr`. It is not necessary to compute the powers of x directly (such as using `Math.pow`). Instead, the anonymous function passed to `foldr` should return two things: the "answer so far" and the next or current power of x. That way each step requires only two multiplications (multiplying x by the previous power of x and multiplying the coefficient by the current power of x) and an addition, rather than an exponentiation and an addition.

Project, continued

7.C Write the function `indefIntegratePoly` which takes a list of coefficients standing for a polynomial and returns a new list of coefficients standing for an indefinite integral of that polynomial. Use `foldr`.

7.D Write a function `secantMethod` that takes an ML function (type real →real) and returns an ML function to evaluate the given function's derivative at a point. The secant method approximates a derivative at a point by choosing a nearby point on the (real) function and computing the slope between those two points. Notice that the (ML) function you are asked to write does not return a function but just the "evaluate" portion of the derivative (type real →real, like the parameter.

7.E Write a function `trapezoidMethod` that takes a function and a min and max value and computes an approximation of the definite integral of that function using the trapezoid method. The trapezoid method divides the area under the curve of the function into segments and approximates their areas with trapezoids.

7.F Write a function `makeSum` that takes two functions, say f and g, and returns a new function to stand for the function $h(x) = f(x) + g(x)$. Recall that $\frac{d}{dx}[f(x) + g(x)] = f'(x) + g'(x)$ and $\int f(x) + g(x)\, dx = \int f(x)\, dx + \int g(x)\, dx$.

7.G Write a function `makeProduct` that takes two functions, say f and g, and returns a new function to stand for the function $h(x) = f(x) \cdot g(x)$. Use the product rule (and `makeSum` and `makeProduct` for differentiation and your `trapezoidMethod` for integration).

7.H Write a function `makeQuotient` that takes two functions, say f and g, and returns a new function to stand for the function $h(x) = f(x)/g(x)$. Use the quotient rule (and various function-making functions) for differentiation and your `trapezoidMethod` for integration.

7.I Write a function `makeArbitrary` that takes an ML function (type real →real), assumed to model a real-valued mathematical function, and returns a function to model that same mathematical function. Use `secantMethod` for differentiation and `trapezoidMethod` for integration.

7.15 Special topic: Countability

Both our informal definition of cardinality in Section 1.8 and the more careful one in Chapter 7.6 were restricted to finite sets. This was in deference to an unspoken assumption that the cardinality of a set ought to be *something*, that is, a whole number. As has been mentioned already, we cannot merely say that a set like $\mathbb{Z}$ has cardinality infinity. Infinity is not a whole number—or even a number at all, if one has in mind the number sets $\mathbb{N}$, $\mathbb{W}$, $\mathbb{Z}$, $\mathbb{Q}$, $\mathbb{R}$, and $\mathbb{C}$. The definition of cardinality taken at face value, however, does not guarantee that the cardinality of a set is something; it merely inspired us to define what the operator $||$ means by comparing a set to a subset of $\mathbb{N}$. Indeed, the definition of cardinality merely gives us a way to say that two sets *have the same cardinality as each other.* What would happen if we extended this bare notion to infinite sets—that is, drop the term *finite*, which we have not defined anyway?

cardinality

Two sets X and Y have the same *cardinality* if there exists a one-to-one correspondence from X to Y. We know from Exercise 7.9.7 that this relation partitions sets into equivalence classes. (Your proof did not depend on the sets being finite, did it?) We say that a set X is *finite* if it is the empty set or if there exists an $n \in \mathbb{W}$ such that X has the same cardinality as $\{1, 2, \ldots, n\}$. Otherwise, we say the set is *infinite*.

finite

infinite

Are all infinities equal? To phrase it more rigorously, are all infinite sets in the same equivalence class? There are two conceivable ways our intuition could lead us. On one hand, one might assume that infinity is infinity without qualification. Thus

$$|\mathbb{W}| = |\mathbb{N}| = |\mathbb{Z}| = |\mathbb{Q}| = |\mathbb{R}|$$

On the other hand, every integer is a rational number, and the real numbers are all the rationals plus many more. Perhaps

$$|\mathbb{W}| < |\mathbb{N}| < |\mathbb{Z}| < |\mathbb{Q}| < |\mathbb{R}|$$

Which could it be?

Let us ask a more general question first. Is it possible for a proper subset to have the same cardinality as the whole set? Take $\mathbb{W}$ and $\mathbb{N}$ for an example. They are sets so similar that most people have trouble remembering which is which. Since $\mathbb{W}$ has one more element than $\mathbb{N}$, it seems like it should have a larger cardinality. But if $\mathbb{N}$ has cardinality "infinity," is $\mathbb{W}$'s cardinality "infinity plus one"?

We do not need to rely entirely on intuition here. We have a means of showing that these two sets have (or do not have) the same cardinality: find a one-to-one correspondence between them, or show that one cannot exist. It turns out that we can show

Theorem 7.19 *$\mathbb{W}$ and $\mathbb{N}$ have the same cardinality.*

Proof. Define a function $f : \mathbb{W} \rightarrow \mathbb{N}$ thus:

$$f(n) = n + 1$$

This simple function associates 0 with 1, 1, with 2, etc.

$$\begin{array}{cccccccccccccccccccccccc}
\mathbb{W} & 0 & & 1 & & 2 & & 3 & & 4 & & 5 & & 6 & & 7 & & 8 & & 9 & & 10 & & \\
 & & \searrow & & \searrow & & \searrow & & \searrow & & \searrow & & \searrow & & \searrow & & \searrow & & \searrow & & \searrow & & \searrow & \\
\mathbb{N} & & & 1 & & 2 & & 3 & & 4 & & 5 & & 6 & & 7 & & 8 & & 9 & & 10 & & 11
\end{array}$$

Tangent: *Infinite verbs and trains*

What is so infinite about grammatical infinitives, like *to walk*, *to compute*, *to prove*? It makes more sense in Latin. The pronoun subject of a Latin verb is given by a suffix: *ambulo*, *ambulas*, *ambulat* is the conjugation "I walk," "you walk," "he/she/it walks," and *o*, *s*, and *t* are the singulars for first, second, and third person. An infinitive—in this case, *ambulare*—has no pronominal suffix. This does not mean that it goes on forever, but that it, literally, has no proper ending.

Up through the 1980s, freight trains in North America had a special end-of-train car called a caboose, but technological advances have made them unnecessary. No longer having their proper ending, can we say that freight trains are infinite? (Actually freight trains are still terminated by something—a box called a FRED, which stands for *Flashing Rear-End Device*; no kidding.)

Now we must prove that this function is a one-to-one correspondence.

> Suppose $n \in \mathbb{N}$. (Note that $n-1 \in \mathbb{W}$: If $n = 1$ then $n-1 = 0 \in \mathbb{W}$. If $n > 1$, then $n - 1 > 0$ and so $n - 1 \in \mathbb{N}$ and $n - 1 \in \mathbb{W}$ since $\mathbb{W} \subseteq \mathbb{N}$.) Then $f(n-1) = n$, and so there exists an element in $\mathbb{W}$ that maps to n, and so f is onto by definition.
>
> Next suppose $n_1, n_2 \in \mathbb{W}$ and $f(n_1) = f(n_2)$. By the way f is defined, $f(n_1) = n_1+1$ and $f(n_2) = n_2+1$. By substitution, $n_1+1 = n_2+1$, and by algebra, $n_1 = n_2$. Hence f is one-to-one by definition.
>
> Hence f is a one-to-one correspondence, and therefore $\mathbb{W}$ has the same cardinality as $\mathbb{N}$. □

So, it is possible for a set to be the same size as one of its proper subsets because infinity always has room for one more. We can illustrate this by a story based on an idea by mathematician David Hilbert.

> Once there was a great hotel with an infinite number of rooms, numbered 1, 2, 3, etc. One day a man entered the lobby and inquired at the desk,
>
> "I need a place to stay. Do you have any vacant rooms?"
>
> "No," replied the receptionist. "All our rooms are occupied."
>
> The man thought for a moment and then said, "Perhaps I should have asked a different question. Do you have any room for one more guest?"
>
> "Of course," said the receptionist. "I'll give you Room 1. Then I'll ask the guest currently in Room 1 to move to Room 2, the guest in Room 2 to move to Room 3, and for all n, the guest in Room n to move to Room $n + 1$."

This first theorem makes it seem plausible that all the infinite sets have the same cardinality. But is it true? First, let us introduce some terminology. We

Biography: David Hilbert, 1862–1943

David Hilbert was a German mathematician and the leader of the mathematical community of his generation. At an international conference in Paris in 1900, he proposed a series of open fundamental problems (eventually published as 23 problems) in mathematics. These set the agenda for much of the mathematical research in the following decades and even century.

His influence led to the acceptance of formerly controversial ideas and methods in mathematics, such as the understanding of infinite sets explained here. He also pioneered the non-constructive proof of existence, that is, proving that something exists by deriving a contradiction from its nonexistence rather than by giving an algorithm for finding it. [23]

say that a set X is *countably infinite* if X has the same cardinality as $\mathbb{N}$. A set X is *countable* if it is finite or countably infinite. The idea is that we could count every element in the set by assigning a number 1, 2, 3, ... to each one of them, even if it took us forever. A set X is *uncountable* if it is not countable. This gives us a new question: Are all sets countable?

countably infinite

countable

uncountable

Since we have shown that $|\mathbb{W}| = |\mathbb{N}|$, let us consider $\mathbb{Z}$ next. This is a little more ambitious—$\mathbb{W}$ has only one more element than $\mathbb{N}$, whereas $\mathbb{Z}$ has twice as many elements (or, infinitely more, depending on how you look at it). First, we continue our story.

> The next day, an infinite bus pulled up to the front of the hotel with an infinite number of passengers. A tour guide got out of the bus and came up to the desk.
>
> "Hello," he said. "I need to find accommodations for my passengers. They will each need their own room. Your sign said there was no vacancy, but is there any way you could squeeze us in?"
>
> "Well," said the receptionist, "I suppose they could line up in the lobby, and I'll be able to check each one of them in one at a time, but they'll have to be prepared to change rooms a lot."
>
> "That would take too long," said the tour guide. "Isn't there a way I could check them in all at once?"
>
> "We do have a group special," the receptionist advised. "First, I'll ask each of our current guests to change to a new room. The guest in Room 1 will move to Room 2, the guest in Room 2 will move to Room 4, and for all n, the guest in Room n will move to Room $2n$. Then your passengers can find their rooms by looking at the seat number they have in your bus. The passenger in Seat 1 will have Room 1, the passenger in Seat 2 will have Room 3, and for all n, the passenger in Seat n will have Room $2n - 1$."

The moral of the story is

Theorem 7.20 *$\mathbb{N}$ and $\mathbb{Z}$ have the same cardinality.*

Proof. Define a function $f : \mathbb{N} \to \mathbb{Z}$ thus:

$$f(n) = \begin{cases} n \texttt{ div } 2 & \text{if } n \text{ is even} \\ -(n \texttt{ div } 2) & \text{otherwise} \end{cases}$$

Now suppose $n' \in \mathbb{Z}$. We have three cases: $n' = 0$, $n' > 0$, or $n' < 0$. Suppose $n' = 0$; then $f(1) = -(1 \texttt{ div } 2) = 0 = n'$. Suppose $n' > 0$; then $f(2n') = 2n' \texttt{ div } 2 = n'$. Suppose $n' < 0$; then $f((-2n') + 1) = -(((-2n') + 1) \texttt{ div } 2) = -(-n') = n'$. In each case, there exists an element of $\mathbb{N}$ that maps to n' and hence f is onto.

Next suppose $n_1, n_2 \in \mathbb{N}$ and $f(n_1) = f(n_2)$. We have three cases: $f(n_1) = 0$, $f(n_1) > 0$, and $f(n_1) < 0$.

Case 1: Suppose $f(n_1) = 0$. Since $-0 = 0$, then by our formula, $n_1 \texttt{ div } 2 = 0$ whether n_1 is even or odd. Then $n_1 = 0$ or $n_1 = 1$, but since $0 \notin \mathbb{N}$, $n_1 = 1$. By substitution and similar reasoning, $n_2 = 1$.

Case 2: Suppose $f(n_1) > 0$. Then n_1 is even. Moreover, by our formula, either $n_1 = 2 \cdot f(n_1)$ or $n_1 = 2 \cdot f(n_1) + 1$. Since n_1 is even, $n_1 = 2 \cdot f(n_1)$. By substitution and similar reasoning, $n_2 = 2 \cdot f(n_1)$.

Case 3: Suppose $f(n_1) < 0$. Then n_1 is odd. Moreover, by our formula, either $n_1 = 2 \cdot (-f(n_1))$ or $n_1 = 2 \cdot (-f(n_1)) + 1$. Since n_1 is odd, $n_1 = 2 \cdot (-f(n_1)) + 1$. By substitution and similar reasoning, $n_2 = 2 \cdot (-f(n_1)) + 1$.

In each case, $n_1 = n_2$. Hence f is one-to-one.

Hence f is a one-to-one correspondence, and therefore $\mathbb{Z}$ has the same cardinality as $\mathbb{N}$. □

$\mathbb{Z}$ is countable. It looks like infinity always has room for as many more. Let us try $\mathbb{Q}$ next. The jump from $\mathbb{N}$ to $\mathbb{W}$ required only one more element. Even the jump from $\mathbb{N}$ to $\mathbb{Z}$ was not so shocking since $\mathbb{Z}$ has only about "twice" as many elements as $\mathbb{N}$. $\mathbb{Q}$, however, has an infinite number of elements between 0 and 1 alone, and an infinite number again between 0 and $\frac{1}{10}$. Now imagine...

The next day the front desk phone rang.

"Hilbert Grand Hotel," answered the receptionist. "May I help you?"

"I'm a travel agent," said the caller, "and I heard that yesterday you were able to accommodate a tour guide with an infinite bus when he was in a pinch."

"Word gets around fast," the receptionist muttered.

"Now I'm in a real pickle," the agent went on. "Because of a tiny oversight, I forgot to make accommodations ahead of time for my clients." (The receptionist rolled her eyes.) "And now I have an infinite number of those infinite buses heading to your city today..."

"Wait," said the receptionist. "What do you mean by 'an infinite number'?"

"I mean they're numbered 1, 2, 3, etc, and go on forever," the agent continued. "The passengers are all going to need accommodations tonight, and I've been calling and calling and desperately looking for a place with enough rooms for all of them..."

"Say no more, we can help," the receptionist interrupted. "But first off, it would make things easier for us if you had one fewer bus—is there anyway you could arrange it so there is no bus numbered 1?"

"I'll radio all the buses right now to add 1 to their number, that way the first bus will be bus number 2," said the agent.

"Great," said the receptionist. "Now I'm booking rooms for everyone, but they will have to do a little bit of figuring to determine their room number. We'll consider all of our current guests to be in 'Bus 1,' and their current room number will act as their seat number. Then we'll lay out the bus number / seat number pairs in a grid, with all the seats in Bus 1 going across the first row, the seats in Bus 2 going across the second row, like this:

$$\begin{array}{cccccc}
1/1 & 1/2 & 1/3 & 1/4 & 1/5 & \cdots \\
2/1 & 2/2 & 2/3 & 2/4 & 2/5 & \\
3/1 & 3/2 & 3/3 & 3/4 & 3/5 & \\
4/1 & 4/2 & 4/3 & 4/4 & 4/5 & \\
5/1 & 5/2 & 5/3 & 5/4 & 5/5 & \\
\cdots & & & & &
\end{array}$$

"Then each person can find their room by counting off rooms while searching for their bus/seat pairs in a diagonal fashion like this:"

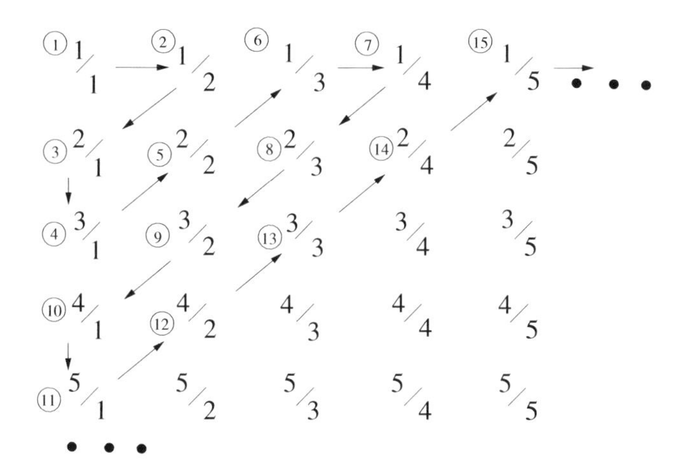

"I see," said the man on the other side of the phone line. "I'm only hesitant about asking our passengers to do so much figuring after long day in the bus."

“I’ll have our staff programmer email you an ML function to compute the room numbers,” reassured the receptionist.

An hour later, the programmer came to the desk and said,

“I’ve finished the program you asked me to write. This computes a function from bus seats to room numbers:”

```
- fun findRoom(busNum, seatNum) =
=   let
=     fun nextPair(a, b) =
=       if a = 1 andalso b mod 2 = 1 then (1, b + 1)
=       else if b = 1 andalso a mod 2 = 0
=            then (a + 1, 1)
=       else if (a + b) mod 2 = 1 then (a + 1, b - 1)
=       else (a - 1, b + 1);
=     fun findRoomHelper(i, currentPair) =
=       if currentPair <> (busNum, seatNum)
=       then findRoomHelper(i + 1, nextPair(currentPair))
=       else i;
=   in
=     findRoomHelper(1, (1, 1))
= end;

val findRoom = fn : int * int -> int

- findRoom(2, 3);

val it = 8 : int

- findRoom(1, 5);

val it = 15 : int
```

“I also wrote a program to compute the inverse, so for each room, we can tell where that guest came from.”

```
- fun findBusSeat(room) =
=   let
=     fun nextPair(a, b) =
=       if a = 1 andalso b mod 2 = 1 then (1, b + 1)
=       else if b = 1 andalso a mod 2 = 0
=            then (a + 1, 1)
=       else if (a + b) mod 2 = 1 then (a + 1, b - 1)
=       else (a - 1, b + 1);
=     fun findBusSeatHelper(i, currentPair) =
```

```
=          if i <> room
=          then findBusSeatHelper(i + 1,
=                                 nextPair(currentPair))
=          else currentPair;
=      in
=        findBusSeatHelper(1, (1, 1))
=      end;

val findBusSeat = fn : int -> int * int

- map(findBusSeat, [1,2,3,4,5,6,7]);

val it = [(1,1),(1,2),(2,1),(3,1),(2,2),(1,3),(1,4)] :
       (int * int) list
```

"It looks like your hard work in your discrete math and functional programming course has paid off," the receptionist smiled.

Later, the buses began to arrive. However, the first passenger who entered the lobby had a bit of a confused look on his face.

"It seems our travel agent forgot to tell you something," the man said. "We've all had bad travel experiences before where we've each ended up with unpleasant busmates. This time, we each booked an infinite number of tickets (each one on a different bus), so we could switch buses as much as we'd like. Now, I happened to have booked Bus 2, Seat 3; Bus 4, Seat 6; Bus 6, Seat 9; and many, many more. According to your program, I have rooms 8, 42, 97..."

"Don't worry," said the receptionist. "You may pick any room you like, and we won't charge you for the other rooms. It won't hurt our profits if we have some vacancies tonight. I recommend Room 8—it has a Jacuzzi."

If we take the ratio of bus number to seat number as a positive fraction (an element of $\mathbb{Q}^+$), this story suggests

Theorem 7.21 *$\mathbb{Q}^+$ has the same cardinality as $\mathbb{N}$.*

A formal proof of this is not easy, but the main idea is that we have a function (defined by the ML program) which is invertible, so it must be a one-to-one correspondence. We have one problem, though: Unlike the hotel receptionist, we care very much that the fractions $\frac{2}{3}$, $\frac{4}{6}$, $\frac{6}{9}$ ... are all different expressions for the same element of $\mathbb{Q}^+$, because that means the function from $\mathbb{N}$ to $\mathbb{Q}^+$ is no longer one-to-one. (Puzzlingly, this suggests that $\mathbb{Q}^+$ might be *smaller* than $\mathbb{N}$.) To fix this up, we would like a function that will assign natural numbers to rational numbers in a way that will skip over numbers we have already seen.

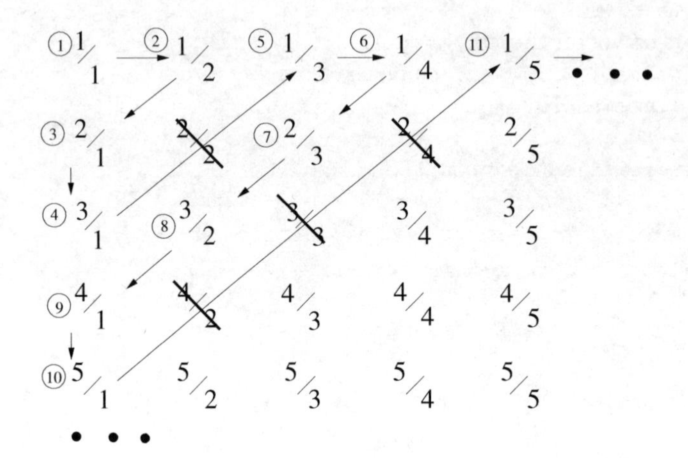

And a function like that exists and is computable. We just need to add a few contraptions to our program for reducing fractions to simplest form, maintaining a list of "already used" fractions in simplest form, and skipping any new fractions whose simplest form is in the list. This method is referred to as the *Cantor walk* or *Cantor diagonalization* of the rational numbers, invented by mathematician Georg Cantor.

```
- fun cantorDiag(n) =
=   let
=     fun gcd(a, 0) = a
=       | gcd(a, b) = gcd(b, a mod b);
=     fun reduce(a, b) =
=       let val comDenom = gcd (a, b);
=       in (a div comDenom, b div comDenom)
=       end;
=     fun nextRatio(a, b) =
=       if a = 1 andalso b mod 2 = 1 then (1, b + 1)
=       else if b = 1 andalso a mod 2 = 0 then (a + 1, 1)
=       else if (a + b) mod 2 = 1 then (a + 1, b - 1)
```

Biography: Georg Cantor, 1845-1918

Georg Cantor was born in Russia and spent most of his life in Germany. He led the way on new methods of proof and ways of thinking about the foundations of mathematics at a critical time in math history. He can be considered the founder of the theory of sets, and in particular he changed the way mathematicians think about the infinite. His ideas were controversial at the time, and it was not until late in life that his contributions began to be recognized for what they were. [2]

```
=        else (a - 1, b + 1);
=      fun contains((a, b), []) = false
=        | contains((a, b), (c, d)::rest) =
=          if a = c andalso b = d then true
=                                 else contains((a, b), rest);
=      fun nextNewRatio(oldRatio, usedList) =
=        let
=            val newRatio = nextRatio(oldRatio);
=        in
=            if contains(reduce(newRatio), usedList)
=            then nextNewRatio(newRatio, usedList)
=            else newRatio
=        end;
=      fun findRatio(i, currentRatio, usedList) =
=        if i < n
=        then findRatio(i + 1, nextNewRatio(currentRatio,
=                                            usedList),
=                          currentRatio :: usedList)
=        else currentRatio;
=    in
=      findRatio(1, (1, 1), [])
= end;

val cantorDiag = fn : int -> int * int

- map(cantorDiag, [1,2,3,4,5,6,7,8]);

val it = [(1,1),(1,2),(2,1),(3,1),(1,3),(1,4),(2,3),(3,2)]
     : (int * int) list
```

This function hits every positive rational exactly once, so it is a one-to-one correspondence. Thus at least $\mathbb{Q}^+$ is countably infinite, and by a process similar to the proof of Theorem 7.20, we can bring $\mathbb{Q}$ into the fold by showing its cardinal equivalence to $\mathbb{Q}^+$. Thus

$$|\mathbb{N}| = |\mathbb{W}| = |\mathbb{Z}| = |\mathbb{Q}|$$

This strongly suggests that all infinities are equal, but the tally is not finished. We still need to consider $\mathbb{R}$. To simplify things, we will restrict ourselves just to the set $(0, 1)$, that is, only the real numbers greater than 0 and less than 1. This might sound like cheating, but it is not.

Theorem 7.22 $(0, 1)$ *has the same cardinality as* $\mathbb{R}$.

This time we will use a geometric argument. Imagine taking the line segment $(0, 1)$ and rolling it up into a ball with one point missing where 0 and 1 would meet. Then place the ball on the real number line, so $.5$ on the ball is tangent with 0 on the line. Now define a function $f : (0, 1) \to \mathbb{R}$, so to find $f(x)$ we draw a line from the $0/1$ point on the ball through x on the ball; the value $f(x)$ is the point where the drawn line hits the real number line. Proving that this function is a one-to-one correspondence is a matter of using analytic geometry to find a formula for f and then showing that every $\mathbb{R}$ is hit from exactly one value on the ball.

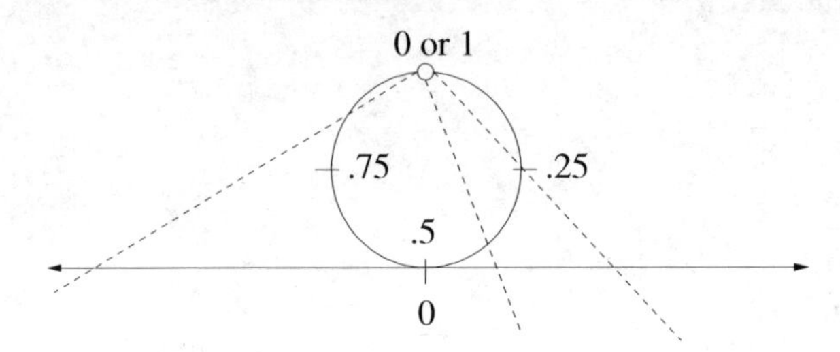

This appears to cut our task down immensely. To prove all infinities (that we know of) equal, all we need to show is that any of the sets already proven countable can be mapped one-to-one and onto the simple line segment $(0, 1)$. However,

Theorem 7.23 $(0, 1)$ *is uncountable.*

Since countability calls for an existence proof, uncountability requires a non-existence proof, for which we will need a proof by contradiction.

Proof. Suppose $(0, 1)$ is countable. Then there exists a one-to-one correspondence $f : \mathbb{N} \to (0, 1)$. We will use f to give names to the all the digits of all the numbers in $(0, 1)$, considering each number in its decimal expansion, where each $a_{i,j}$ stands for a digit.:

$$\begin{aligned} f(1) &= 0.a_{1,1}a_{1,2}a_{1,3}\dots a_{1,i}\dots \\ f(2) &= 0.a_{2,1}a_{2,2}a_{2,3}\dots a_{2,i}\dots \\ &\vdots \\ f(x) &= 0.a_{x,1}a_{x,2}a_{x,3}\dots a_{x,i}\dots \\ &\vdots \end{aligned}$$

Now construct a number $d = 0.d_1d_2d_3\dots d_i\dots$ as follows

$$d_i = \begin{cases} 1 & \text{if } a_{i,i} \neq 1 \\ 2 & \text{if } a_{i,i} = 1 \end{cases}$$

Since $d \in (0, 1)$ and f is onto, there exists an $x \in \mathbb{N}$ such that $f(x) = d$. Moreover,

$$f(x) = 0.a_{x,1}a_{x,2}a_{x,3}\ldots a_{x,x}\ldots$$

so

$$d = 0.a_{x,1}a_{x,2}a_{x,3}\ldots a_{x,x}\ldots$$

by substitution. In other words, $d_i = a_{x,i}$, and specifically $d_x = a_{x,x}$. However, by the way that we have defined d, we know that $d_x \neq a_{x,x}$, a contradiction.

Therefore $(0, 1)$ is not countable. □

Corollary 7.24 $\mathbb{R}$ *is uncountable.*

Proof. Theorems 7.22 and 7.23. □

Both our intuitions are wrong. There are just as many naturals as wholes as integers as rationals. But there are many, many more reals. E. T. Bell said, "Intuition (male, female, or mathematical) has been greatly overrated. Intuition is the root of all superstition" [2, pg 567].

The next day a truck pulled up to the hotel. Looking through the windows, the receptionist read "Alf and Juan's Floral Delivery." She saw a man jump out of the truck and enter the lobby.

"Delivery for you, ma'am," he said. "A travel agent wanted to thank you for helping him out yesterday, so he called to send you more flowers than you can count."

"That was sweet of him," said the receptionist. "But impractical. I can't possibly accommodate that many flowers, even if I were to put one in every room."

This chapter draws heavily from Epp[8].

Chapter summary

As with relations and sets, functions are a crucial tool for expressing mathematical ideas. By extension, they capture concepts in many other fields of study. Computer science is particularly dependent on the function concept as it is employed both in theoretical computer science and in describing the practical aspects of programming.

The connections between discrete mathematics and functional programming have come to the surface in this chapter. The ideas of images and functions as mathematical objects have direct analogs in functional programming. First-class functions, mapping, and folding form the main tools of more advanced programming in the functional style.

Key ideas:

A function is a relation where every element in the first set (the domain) is related to exactly one element in the second set (the codomain).

Two functions on the same domain are equal if for every element in the domain, the results of the functions are equal.

In ML, anonymous functions can be made using the `fn` construct. Moreover, functions are first class values, meaning they can be passed to functions, stored in variables, and returned from functions.

The image of a set (a subset of the domain) under a function is the set of items in the codomain mapped to from the set by the function.

The inverse image of a set (a subset of the codomain) under a function is the set of items in the domain that the function maps to that set.

The ML function `map` takes a function and a list and returns a list containing the results of applying that function to every item in the given list.

The ML function `foldl` takes a function, a list, and a seed value and computes the accumulation of applying the function to every item in the list.

A function is onto if every element of the codomain is hit by at least one element of the domain.

A function is one-to-one if every element of the domain hits a unique element of the codomain (or, every element of the codomain is hit by at most one element of the domain).

A one-to-one correspondence is a function that is onto and one-to-one.

The inverse relation of a one-to-one function is a function, referred to as the inverse of that function.

Functions can be composed to form a new function by feeding the output of one function as input into the other function.

Two sets have the same cardinality if there is a one-to-one correspondence between them. A finite set has cardinality n if there exists a one-to-one correspondence between that set and the set $\{1, 2, \ldots, n\}$.

A set with the same cardinality as $\mathbb{N}$ is countably infinite. $\mathbb{W}$, $\mathbb{Z}$, and $\mathbb{Q}$ are countably infinite.

An infinite set that that is not countably infinite is uncountable. $\mathbb{R}$ is uncountable.

Part III

Electives

Chapter 8 Graph

We commonly use the word *graph* to refer to a wide range of graphics and charts that provide a visual representation of information, particularly quantitative information. From your previous mathematics courses, you probably most closely associate graphs with illustrations of functions in the real-number plane. *Graph theory*, however, refers to something else.

A graph in this sense is a discrete structure composed of two kinds of entities: vertices and edges. The vertices are points and the edges are connections between them. From this simple foundation we can build structures for modeling systems and representing information, with countless applications throughout mathematics, computer science, and other fields.

Unfortunately, the beginning student of graph theory will likely feel intimidated by the horde of terminology required; some sources and textbooks use the terms differently from each other, and that aggravates the situation. The student is encouraged to take careful stock of the definitions in these sections, but also to enjoy the beauty of graphs and their uses. This chapter will present the basic vocabulary of graph theory including the various kinds of paths through graphs. From there we will use graph theory as a framework for discussing isomorphism, a central concept throughout mathematics. We also will look at a variety of kinds of graphs, their applications, and algorithms on graphs.

You will notice a difference in this chapter's proofs. Consider the proving style *more mature* than in earlier chapters. On one hand, that means that the proofs tend to be harder. It also means that they will be less rigid, though not less rigorous. Some aspects of graphs are tedious to reason about formally, and so our proofs will have some informality for convenience.

Chapter goals. The student should be able to

- use the basic definitions and terms for graphs, recognizing the common applications.
- prove propositions about graphs.
- prove isomorphic invariants.

- recognize various kinds of graphs.
- implement basic graph algorithms.

8.1 Definition and terms

Anna Karenina is a hefty book, but it quickly establishes its premise, its characters, and their relationships. Oblonsky and Dolly are in a marriage troubled by Oblonsky's infidelity. Oblonsky's friend Levin has come to town, intending to propose to Dolly's sister Kitty. Oblonsky warns Levin that another man, Vronsky, has been giving Kitty attention recently. Anna, Oblonsky's sister, also comes to town for a visit, and reconciles Oblonsky and Dolly. Levin proposes to Kitty, who rebuffs him, assuming Vronsky will soon propose to her. Later, at a ball, Vronsky ignores Kitty and dances with Anna, for whom he develops an infatuation. When the visit is over, Anna returns to her husband, Karenin, but Vronsky pursues her and eventually they begin an affair. (The rest of the book traces Levin's eventual re-proposal and marriage to Kitty and the consequences of Anna and Vronsky's affair.)

We can represent the connections between major characters this way:

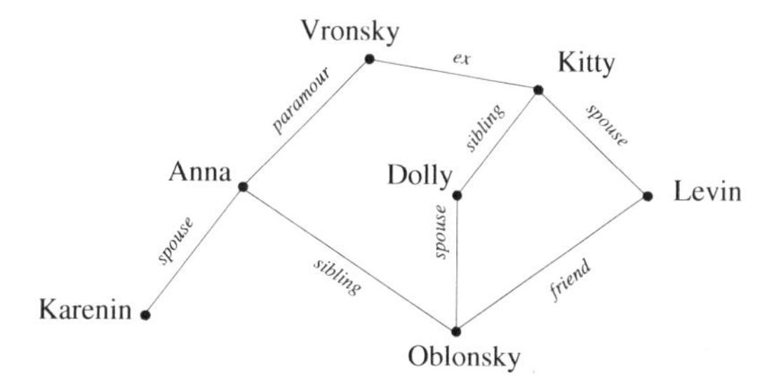

The information we are displaying has two kinds: characters and their relationships. Characters are independent entities, and each relationship connects exactly two characters. Accordingly, the figure displaying the information has two kinds of elements, dots that are labeled with character names and line segments between dots that are labeled with the nature of the relationship. This figure provides an abstraction for the relationships of the characters—it summarizes them, but hides the subtleties found in the narrative.

Now erase the labels—or replace them with dummy labels. It may seem like we are obliterating all the information when we do this, but we are not. This removes the *meaning*, but it leaves the *structure*. Now we have the sort of thing we study in discrete mathematics.

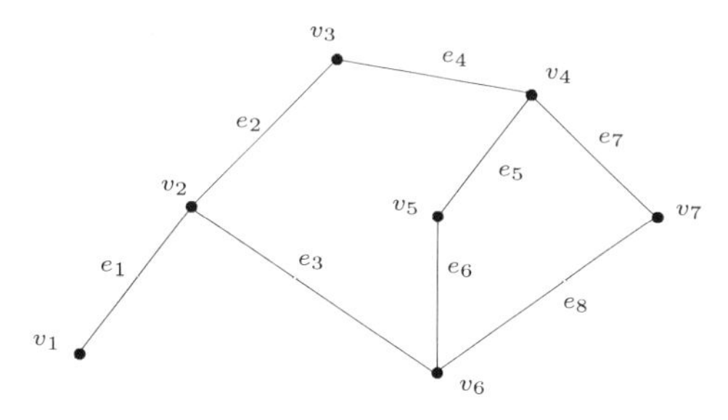

graph

vertex

edge

end points

A *graph* $G = (V, E)$ is a pair of finite sets, a set V of *vertices* (singular *vertex*) and a set E of pairs of vertices called *edges*. We will typically write $V = \{v_1, v_2, \ldots, v_n\}$ and $E = \{e_1, e_2, \ldots, e_m\}$ where each $e_k = (v_i, v_j)$ for some v_i, v_j; in that case, v_i and v_j are called *end points* of the edge e_k.

Graphs are drawn so that vertices are dots, and edges are line segments or curves connecting two dots. However, such drawings are mere representation—a graph is not a drawing but an abstract mathematical object. The graph above can be described just as well as $(\{v_1, v_2, v_3, v_4, v_5, v_6, v_7\}, \{e_1, e_2, e_3, e_4, e_5, e_6, e_7, e_8\})$ where $e_1 = (v_1, v_2)$, $e_2 = (v_2, v_3)$, etc.

The kinship between graphs and relations should be readily apparent, but graphs are more flexible. As we can see from this example, we can associate information with the edges that represent the relationships (close friendship, siblinghood, or romantic involvement—labeled on the drawing but not as names for the edges).

The definition above is not completely formal, as we will see later, but it will serve us for now as we consider the range of uses of graphs, that is, the kinds of information that graphs can capture.

Maps indicating routes among locations are a good place to start. Consider the cities of Naperville, Chicago, Champaign, Gary, Detroit, Lafayette, and Indianapolis. They are connected by several interstate highways. I57 can be taken between Champaign and Chicago. I90 and I80 are alternate routes between Chicago and Gary. I65 runs between Gary and Indianapolis, with Lafayette between the two. We have the following graph (with vertices and edges labeled). This is not quite a typical roadmap because it is more abstract—for example, it contains no accurate information about distance or direction.

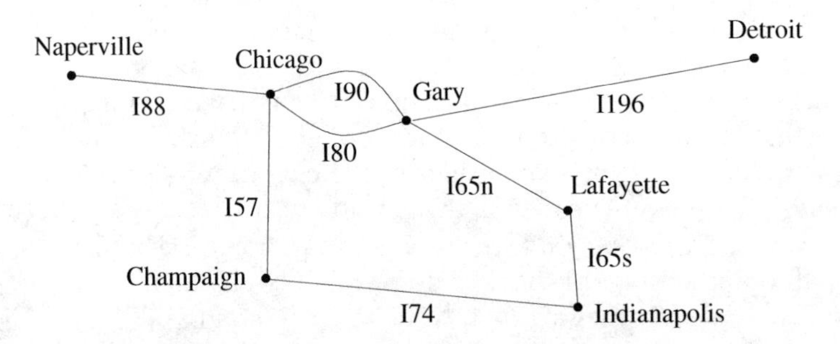

We call the edges *pairs* of vertices for lack of a better term. A pair is generally considered a two-tuple (in this case, it would be an element of $V \times V$). Moreover, we write edges with parentheses and a comma, just as we would write a tuple. However, we mean something slightly different. First, tuples are ordered. In our basic definition of graphs, we assume that the end points of an edge are *unordered*: we could write I57 as (Chicago, Champaign) or (Champaign, Chicago). Second, an edge as a pair of vertices is not unique. In the cities example, we have duplicate entries for (Chicago, Gary): both I90 and I88.

We can analyze many specialized maps as graphs. Consider how airports are vertices and flights are edges in Mesaba Airline's flight map for CRJ-200 aircraft, or, similarly, stations and parts of lines in the Lausanne Metro plan.

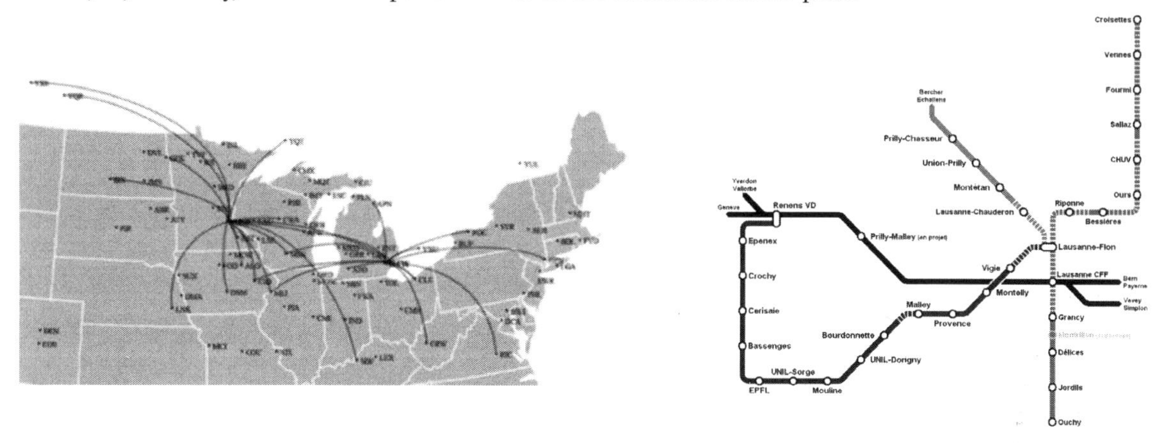

An edge (v_i, v_j) is *incident* on its end points v_i and v_j; we also say that it *connects* them. If vertices v_i and v_j are connected by an edge, they are *adjacent* to one another. See how the edges in these maps illustrate adjacency between vertices or the connections between them.

incident

connects

adjacent

Consider a maze, and then interpret that maze as a graph, where the vertices are crossroads and dead ends, and the edges represent halls between those points.

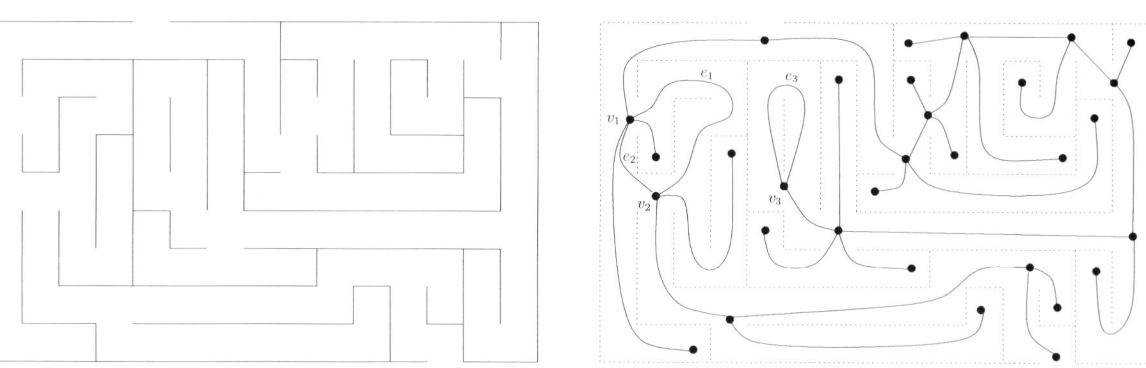

The vertices v_1 and v_2 above have two edges connecting them, e_1 and e_2, representing how there are two ways to go directly from one point to the other. Also, v_3 has an edge connected only to itself, e_3, since that hall does not lead to a crossroad or to a dead end, but only around in a loop.

If a vertex is adjacent to itself, that connecting edge is called a *self-loop*. If two edges connect the same two vertices, those edges are *parallel* to each other. Some authors use the term *multigraph* for graphs that allow parallel edges, and graphs that also may have self-loops they call *pseudographs*. We will not use those terms.

self-loop

parallel

As we observed earlier, if a graph has parallel edges like e_1 and e_2 above, then the description (v_1, v_2) is not unique—it could describe either e_1 or e_2. If a graph has no parallel edges then a description like (v_1, v_2) can be used as the name of an edge. A graph is *simple* if it contains no parallel edges or self-loops.

simple

The graphs described so far are called *undirected graphs* because the edges indicate a symmetric relationship between the two endpoints: if there is an edge $e = (v_1, v_2)$, then v_1 is adjacent to v_2 and v_2 is adjacent to v_1; we could have described this edge as (v_2, v_1). In a *directed graph*, sometimes abbreviated to *digraph*, the edges have a direction and are displayed with arrowheads.

undirected graphs

directed graph

Consider the board for the game Chutes and Ladders and its interpretation as a directed graph. Adjacency does not refer to physical proximity, because a graph is abstract. It refers only to the existence of a connection. However, in a directed graph each connection has only one direction.

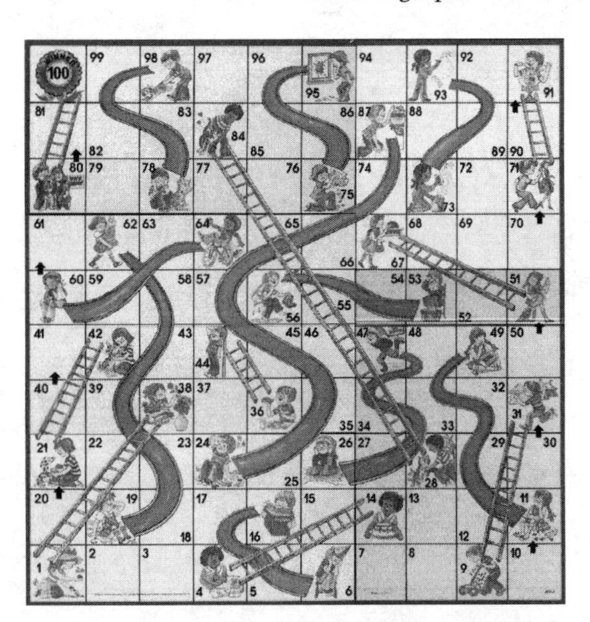

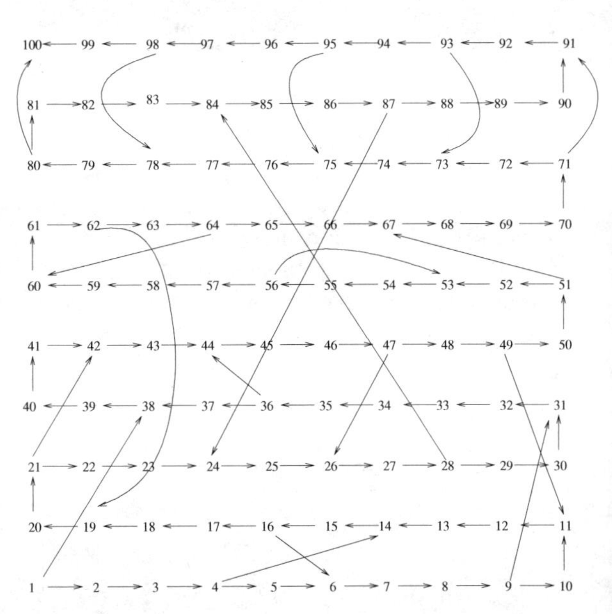

In Chapter 5 we used directed graphs to illustrate relations. In fact, a directed graph that has no parallel edges with the same direction is, formally, no different from a set together with a relation on that set (the original set being the vertices, and the relation being the set of edges). In this chapter we will be more interested in undirected graphs. (Notice that undirected graphs that have no parallel edges are essentially the same thing as symmetric relations.)

The *degree* $\deg(v)$ of a vertex v is the number of edges incident on the vertex, with self-loops counted twice. In the maze example, $\deg(v_1) = 5$, $\deg(v_2) = 4$, and $\deg(v_3) = 3$. In a directed graph, we would need to distinguish between the *in-degree* and *out-degree* of a vertex. In the Chutes-and-Ladders example, vertex 60 has in-degree 2 and out-degree 1, but vertex 80 has in-degree 1 and out-degree 2.

degree

in-degree

out-degree

A *subgraph* of a graph $G = (V, E)$ is a graph $H = (W, F)$ where $W \subseteq V$ and $F \subseteq E$ (and, by definition of graph, for any edge $(v_i, v_j) \in F$, $v_i, v_j \in W$). See below a simple graph and a subgraph of that graph, with v_5 and e_6, e_4, and e_2 removed.

subgraph

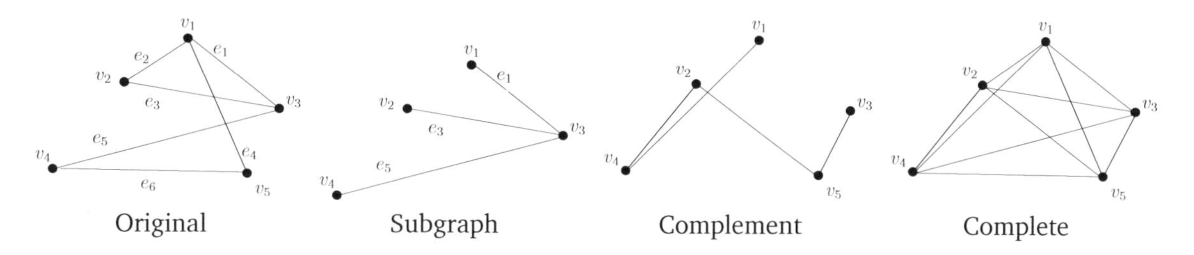

The *complement* of a simple graph $G = (V, E)$ is a graph $\overline{G} = (V, \overline{E})$ where for $v_i, v_j \in V$, $(v_i, v_j) \in \overline{E}$ if $(v_i, v_j) \notin E$; in other words, the complement has all the same vertices and all (and only) those possible edges that are not in G. See above the complement of the original graph, with edges $\{(v_1, v_4), (v_2, v_4), (v_2, v_5), (v_3, v_5)\}$. A simple graph $G = (V, E)$ is *complete* if for all distinct $v_i, v_j \in V$, the edge $(v_i, v_j) \in E$. See above for a complete graph on the vertices in the other examples.

complement

complete

Even though we tend to think of a graph as a picture, it is important to notice that a formal description of a graph is independent of the way it is drawn. Important to the essence of a graph is merely the names of vertices and edges and their connections (and in Section 8.4, we will see that even the names are not that important). The following two pictures show the same graph.

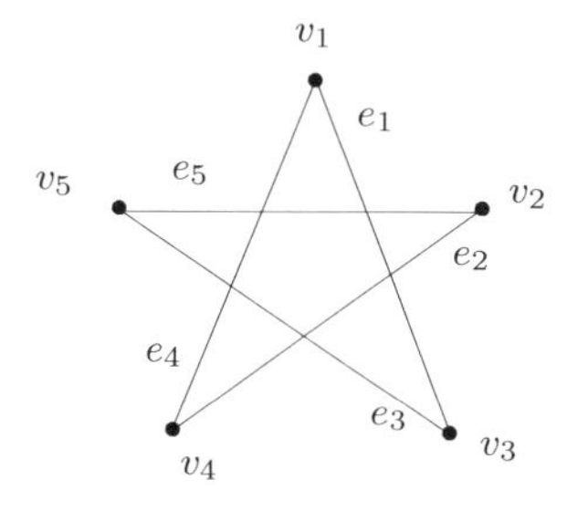

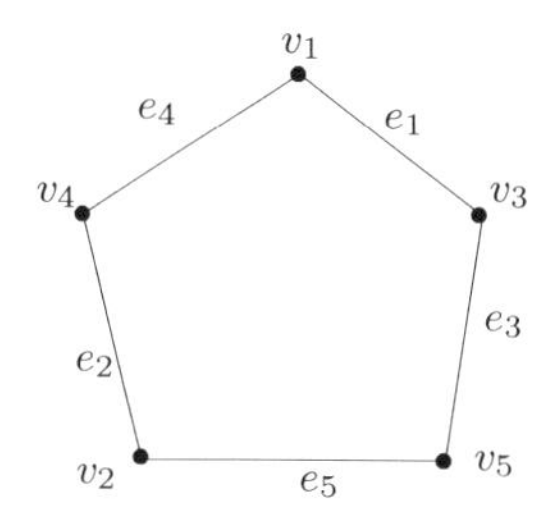

Finally, we revisit the definition of *graph*, which we earlier noted was informal. The definition, in fact, works for simple directed graphs if we interpret (v_i, v_j) in the ordinary way as an ordered pair, because then it is no more than a restatement of the definition of a relation. However, if we have undirected graphs in mind and want to allow for parallel edges, then (v_i, v_j) must be interpreted as an *unordered* pair (a set of cardinality 2) and that there may in fact be more than one distinct object labeled with such a pair.

Another way to state the problem is to ask what the primitive terms are. *Vertices* clearly are primitive—we cannot define what they are, they are simply elements, mathematical objects. But what about *edges*? Are they pairs or sets or something else definable from vertices, or are they also primitive, with pairs or sets serving merely as labels?

The best way to define them is to take both *vertex* and *edge* as primitive and, in our definition of graph, assert that the edges are *associated* with vertices. The way to formulate that association is with a function. So, a *graph* $G = (V, E, f)$ is a triple of a set of vertices V, a set of edges E, and a function $f : E \to \{\, \{u, v\} \mid u, v \in V\}$ (that is, the function f maps edges to sets of vertices, each set of cardinality 2 if $u \neq v$ or 1 if $u = v$). Since the definition is cumbersome, we mostly will use the informal one.

Exercises

For each of the following graphs, indicate which of the following terms apply: Directed, undirected, simple, complete. Also, find the greatest and least degree of any vertex, or greatest and least in-degree and out-degree, for directed graphs.

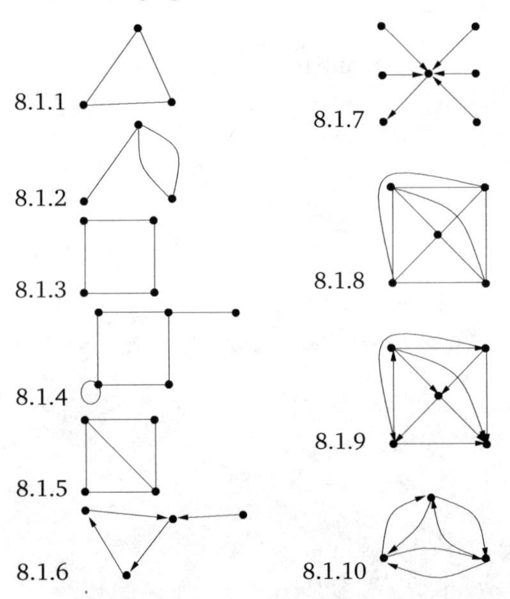

8.1.11 Find the complement of the graph in Exercise 8.1.5.

8.1.12 Find the complement of the graph in Exercise 8.1.6.

8.1.13 Find the complement of the graph in Exercise 8.1.9.

Based on our formal definition of a graph, formalize the definitions for the other graph terms in this section. (In Exercises 8.1.14–8.1.19, assume an undirected graph.)

8.1.14 Write a formal definition of *adjacent*, that is, "If $G = (V, E, f)$ is a graph, then $v, u \in V$ are adjacent if..."

8.1.15 Write a formal definition *simple*, that is, "A graph $G = (V, E, f)$ is simple if ..."

8.1.16 Write a formal definition of *degree*, that is, "If $G = (V, E, f)$ is a graph and $v \in V$, ..."

8.1.17 Write a formal definition of *subgraph*, that is, "If $G = (V, E, f)$ is a graph, then $H = (W, F, h)$ is a subgraph of G if ..."

8.1.18 Write a formal definition of *complement*, that is, "If $G = (V, E, f)$ is a simple graph, then $H = (W, F, h)$ is the complement of G if ..."

8.1.19 Write a formal definition for *complete*, that is, "If $G = (V, E, f)$ is a simple graph, then G is complete if ..."

8.1.20 Write a formal definition for *directed graph*, based on our formal definition for an undirected graph. There is more than one way to do this.

8.2 Propositions on graphs

We will ease slightly on the formality of proofs at this point. This is partially because by now you should have achieved a maturity for writing proofs that allows the privilege of writing with less rigidity, such as condensing a few steps. Moreover, completely formal proofs of graph propositions are inconvenient. Accordingly, we will work mainly with informal definitions of graphs and related terms.

Do not be misled, nevertheless, into thinking that this is lowering the standards for logic and rigor. We will merely be stepping over a few obvious details for the sake of notation, length, and readability. The proof of the following proposition shows what sort of argumentation is expected.

Theorem 8.1 (Handshake Theorem.) *If $G = (V, E)$ is an undirected graph with $V = \{v_1, v_2, \ldots, v_n\}$, then $\sum_{i=1}^{n} deg(v_i) = 2 \cdot |E|$.*

This is called the Handshake Theorem because we can imagine the vertices as people at a party and the edges as handshakes between people. If every person at the party were to tally up the number of times they shook hands with someone, the result would be twice the number of handshakes that occurred. Sound obvious? It still takes some work to prove.

First, consider why you think this is true. Every handshake event is counted twice, since there are two people involved in any handshake. (A person is unlikely to shake hands with herself, but if a vertex has a self-loop, then it is counted twice for the vertex's degree.) So, for our proof we need to find a way to formalize the counting. For any graph that has at least one edge, we can make a subgraph by taking away one edge. If we knew the proposition were true for that subgraph, then it is true for the original graph because it has one more edge and two more in the tally of degrees.

Do you recognize that this is a proof by mathematical induction? We need to rework the proposition so that it is quantified over the whole numbers.

Proof. By induction on the cardinality of E.

Base case. First, suppose that G has no edges, that is, $|E| = 0$. Then for any vertex $v \in V$, $\text{deg}(v) = 0$. Hence $\sum_{i=1}^{n} \text{deg}(v_i) = 0 = 2 \cdot 0 = 2 \cdot |E|$. Hence there exists an $N \geq 0$ such that for all $m \leq N$, if $|E| = m$ then $\sum_{i=1}^{n} \text{deg}(v_i) = 2 \cdot |E|$.

Inductive case. Now suppose $|E| = N + 1$, and suppose $e \in E$. Consider the subgraph of G, $G_1 = (V, E - \{e\})$. We will write the degree of $v \in V$ when it is being considered a vertex in G_1 instead

of G as $\deg'(v)$. $|E - \{e\}| = N$, so by our inductive hypothesis $\sum_{i=1}^{n} \deg'(v_i) = 2 \cdot |E - \{e\}|$. Suppose v_i, v_j are the end points of e. If $v_i = v_j$, then $\deg(v_j) = \deg(v_i) = \deg_1(v_i) + 2$; otherwise, $\deg(v_i) = \deg_1(v_i) + 1$ and $\deg(v_j) = \deg_1(v_j) + 1$; both by the definition of degree. For any other vertex $v \in V$, where $v \neq v_i$ and $v \neq v_j$, we have $\deg(v) = \deg_1(v)$. Hence $\sum_{i=1}^{n} \deg(v_i) = 2 + \sum_{i=1}^{n} \deg'(v_i) = 2 + 2 \cdot |E - \{e\}| = 2 + 2(|E| - 1) = 2 \cdot |E|$.
□

Here are the things left out of this proof.

- The third sentence claims but does not justify that having no edges implies every vertex has a degree of zero. This follows from the definition of degree, however.
- In the fourth sentence, substitution and rules of arithmetic are used without citation.
- The claim $|E - \{e\}| = N$ depends on Exercise 7.9.1 (the Difference Rule) and the facts that $E - \{e\}$ and $\{e\}$ are disjoint and that $|\{e\}| = 1$.
- The sixth sentence of the second paragraph together with the last sentence claims that whether or not e is a self loop, it contributes two to the total sum of degrees. It is difficult to state this more formally.
- The last sentence uses arithmetic, algebra, and substitution uncited.

total degree

Define the *total degree* of a graph to be the sum of the degrees of all the edges. Notice that the following corollary follows immediately from the Handshake Theorem.

Corollary 8.2 *The total degree of a graph is even.*

If eleven people are at a party, is it possible for each of them to shake hands with exactly five other people? Consider this result:

Theorem 8.3 *If $G = (V, E)$ is a graph, the total number of vertices with an odd degree are even.*

Proof. Let V_1 be the set of vertices of G that have an even degree and V_2 be the set of vertices of G that have an odd degree. Note that V_1 and V_2 make a partition of V.

For each $u_i \in V_1$, there exists a $d_i \in \mathbb{W}$ such that $\deg(u_i) = 2 \cdot d_i$ and for each $w_j \in V_2$, there exists a $c_j \in \mathbb{W}$ such that $\text{def}(w_j) = 2 \cdot c_j + 1$, by definition of even and odd.

By the Handshake Theorem, $\sum_{v \in V} \deg(v) = 2 \cdot |E|$. Since V_1 and V_2 make a partition of V, $\sum_{v \in V} \deg(v) = \sum_{u \in V_1} \deg(u) + \sum_{w \in V_2} \deg(w)$. So,

$$\begin{array}{rcll} \sum_{w \in V_2} \deg(w) & = & \sum_{v \in V} \deg(v) - \sum_{u \in V_1} \deg(u) & \text{by algebra} \\ & = & 2 \cdot |E| - (2d_1 + 2d_2 + \cdots + 2d_{|V_1|} & \text{by substitution} \\ & = & 2(|E| - d_1 - d_2 - \cdots - d_{|V_1|}) & \text{by algebra} \end{array}$$

Where $d_1, d_2, \ldots, d_{|V_1|}$ are some whole numbers such that $\deg(u_i) = 2d_i$. Moreover,

$$\begin{array}{rcll} \sum_{w \in V_2} \deg(w) & = & (2c_1 + 1) + (2c_2 + 1) + \cdots + (2c_{|V_2|} + 1) & \text{by substitution} \\ & = & 2(c_1 + c_2 + \cdots c_{|V_2|}) + |V_2| & \text{by algebra} \end{array}$$

Putting these together and doing algebraic rearranging, we get

$$|V_2| = 2(|E| - d_1 - d_2 - \cdots - d_{|V_1|} - c_1 - c_2 - \cdots c_{|V_2|})$$

And so $|V_2|$ is even by definition. □

Exercises

8.2.1 Prove that if a graph $G = (V, E)$ is complete, then $|E| = \frac{|V|(|V|-1)}{2}$.

8.2.2 Prove that the union of a simple graph and its complement is complete, that is, if $G = (V, E)$ is a graph, then $(V, E \cup \overline{E})$ is complete.

8.3 Strolling about a graph

The natural way to imagine a graph is to think of the vertices as places and the edges as connections or bridges between those places. Extending that, we can imagine ourselves or an object moving around in a graph from vertex to vertex. In fact, one of the earliest uses of graph theory is in a problem proposed by mathematician Leonhard Euler about crossing bridges while walking around town. More on that problem later.

In this chapter we categorize routes through graphs and prove propositions about them. Brace yourself for even more vocabulary—it will be a challenge to keep it all straight. Also beware that there is some variation in how different authors use these terms.

walk

A *walk* from vertex v to vertex w, where $v, w \in V$, is a sequence alternating between vertices in V and edges in E, written $v_0 e_1 v_1 e_2 \ldots v_{n-1} e_n v_n$ where $v_0 = v$ and $v_n = w$ and for all i, $1 \le i \le n$, $e_i = (v_{i-1}, v_i)$. v is called the *initial* vertex and w is called the *terminal* vertex.

initial

terminal

At first glance it might seem easier to think of a walk merely as a sequence of vertices, constrained in that vertices next to each other in the sequence must be adjacent in the graph. That would work in simple graphs, but not in general—if two vertices were connected by more than one edge, then such a description would be ambiguous about which edge was taken. It would not be ambiguous if we considered a walk to be a sequence of edges, but that makes it more difficult to describe the constraint of only moving to adjacent vertices.

When convenient, we will describe walks in reduced form, listing only edges. If the graph is simple, we may also refer to the walks by listing only vertices.

trivial

length

A walk is *trivial* if it contains only one vertex and no edges; otherwise it is *nontrivial*. The *length* of a walk is the number of edges (not necessarily distinct, since an edge may appear more than once). Consider the graph below, displayed twice, each with a walk highlighted.

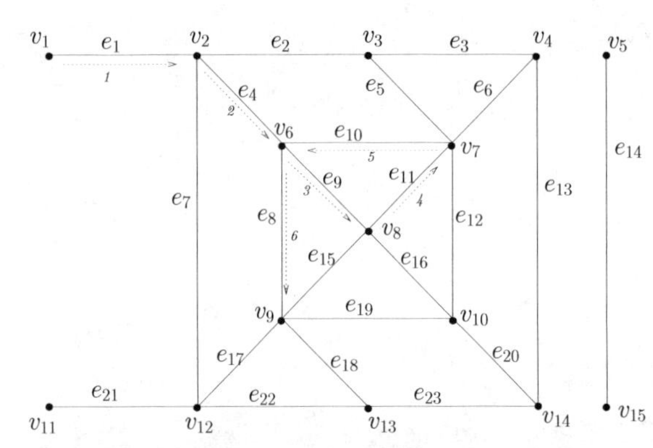

Walk $c_1 = v_1 e_1 v_2 e_4 v_6 e_9 v_8 e_{11} v_7 e_{10} v_6 e_8 v_9$, with length 6

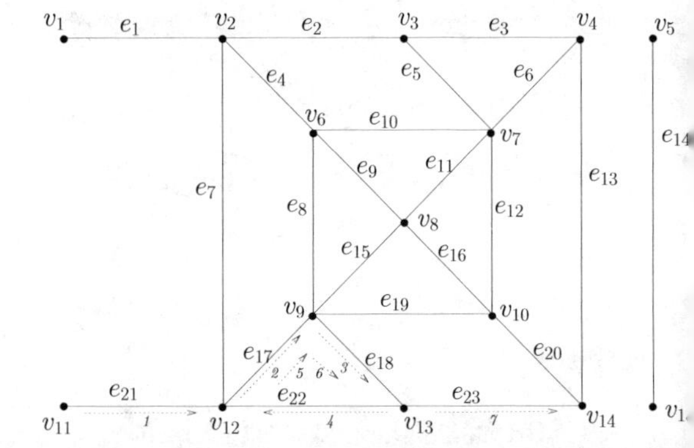

Walk $c_2 = v_{11} e_{21} v_{12} e_{17} v_9 e_{18} v_{13} e_{22} v_{12} e_{17} v_9 e_{18} v_{13} e_{23}$ with length 7

connected

A graph is *connected* if for all $v, w \in V$, there exists a walk in G from v to w. (Notice that for any vertex there exists a trivial walk to itself, so we do not need self loops for a graph to be connected.) This graph is not connected, since no walk exists from v_5 or v_{15} to any of the other vertices. However, the subgraph excluding v_5, v_{15}, and e_{14} is connected.

path

simple

A *path* is a walk that does not contain a repeated edge. c_1 above is a path, but c_2 is not, since e_{17} and e_{18} are repeated. If the walk contains no repeated vertices, except possibly the initial and terminal, then the walk is *simple*. Path c_1 is not simple because v_6 is visited twice, but if we take walk c_2 and cut out the trip around the triangle, we get the simple path $c_3 = v_{11} e_{21} v_{12} e_{17} v_9 e_{18} v_{13} e_{23} v_{14}$.

If $v = w$ (that is, the initial and terminal vertices are the same), then the walk is *closed*. A *circuit* is a closed path—no repeated edges, initial and terminal vertices the same. A *cycle* is a simple circuit Below on the left, $c_3 = v_9e_8v_6e_9v_8e_{11}v_7e_{12}v_{10}e_{16}v_8e_{15}v_9$ is a circuit, but not a cycle, since v_8 is repeated. Below on the right, $c_4 = v_2e_4v_6e_8v_9e_{17}v_{12}e_7v_2$ is a cycle.

closed

circuit

cycle

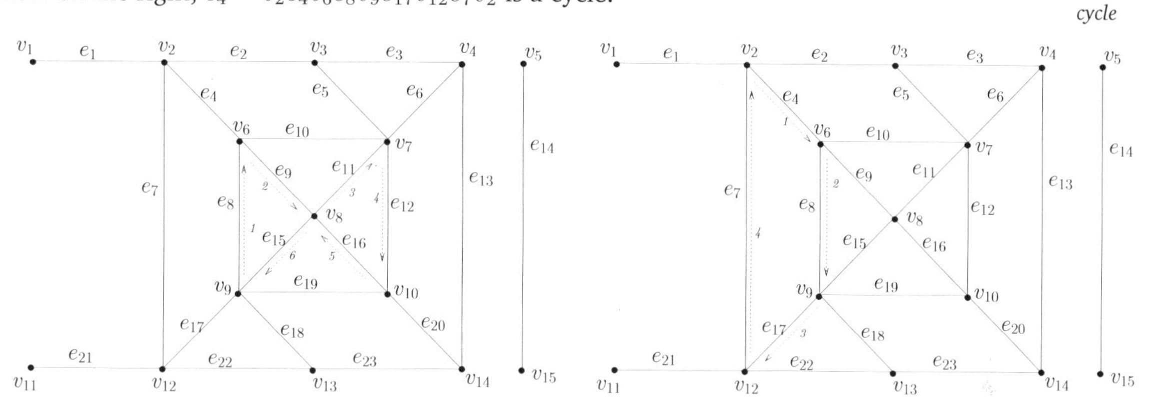

To help keep these terms straight, here is a summary:

Adjectives	
Trivial	Having only one vertex and no edges.
Simple	Having no repeated *vertices* (except, possibly, the initial and terminal).
Closed	Having the same vertex as initial and terminal.

Nouns	
Walk	An alternating sequence of vertices and edges, each edge coming between its end points.
Path	A walk with no repeated *edge* (repeated vertices are ok).
Circuit	A closed path (no repeated edges, initial and terminal the same).
Cycle	A simple circuit (no repeated edges or vertices, except the initial and terminal, which are the same).

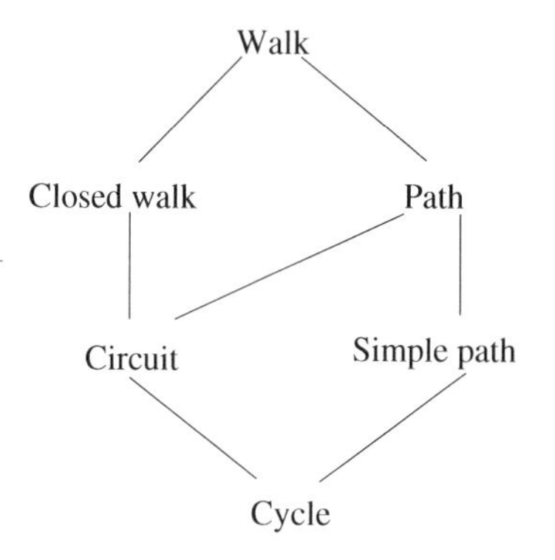

If you thought the earlier graph theory proofs were messy, walks only make things worse, with the notation to track edges and vertices. As before, we will countenance some informality to relieve the pain. This should allow us to focus on the core of the proof. Consider this one:

Theorem 8.4 *If $G = (V, E)$ is a connected graph and for all $v \in V$, deg$(v) = 2$, then G is a cycle.*

The important thing to think about is *What is the burden of this proposition? What do we need to show?* Identifying that is an exercise in applying the definitions listed above, and it provides a road map through the actual proof.

First of all, what we need to show is that G is a cycle. That means G *has* a cycle which happens to be all of G. This is our first step to unraveling what needs to be shown—it is a proof of existence. We must show *there exists* a cycle in G that constitutes all of G.

A cycle is a simple circuit. *Simple* means it has no repeated internal vertices. A circuit is a closed path. *Closed* means it has the same initial and terminal vertex. A *path* is a walk with no repeated edges.

Here is our proof outline or strategy:

1. Construct a walk.

2. Show that the walk has no repeated edges (so it is a path)

3. Show that it has the same first and last vertex (so it is closed—and it is also a circuit)

4. Show that it has no repeated internal vertices (so it is simple—and it is also a cycle)

5. Show that every vertex and edge in G is this cycle.

Now, why is this proposition true? Here is a a picture of a connected graph, all of whose vertices have degree 2. This makes the theorem almost obvious. All we need to do is pick any vertex to begin with and travel out by any edge. For every vertex we visit, we leave by the edge other than the one we entered.

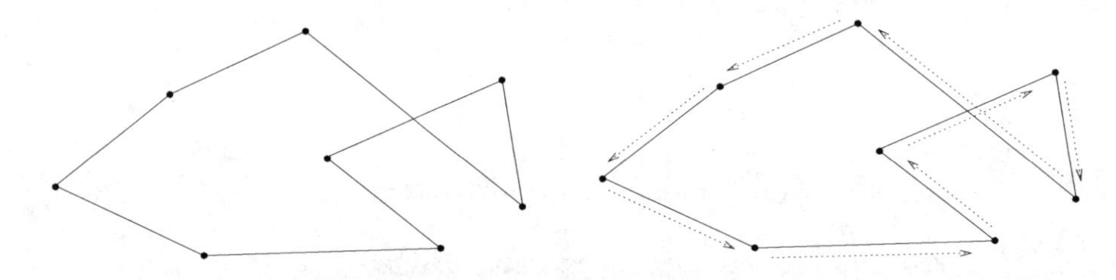

But do not forget a special case: . Ready to prove?

Proof. Suppose $G = (V, E)$ is a connected graph and for all $v \in V$, $\deg(v) = 2$. First suppose $|V| = 1$, that is, there is only one vertex, v. Since $\deg(v) = 2$, this implies that there is only one edge, $e = (v, v)$. Then the cycle vev constitutes the entire graph.

This looks like the beginning of a proof by induction, but actually it is a traditional division into cases. We are merely getting a special case out of the way. We want to use the fact that there can be no self-loops, but that is true only if there are more than one vertices.

Next suppose $|V| > 1$. By Exercise 8.3.5, G has no self-loops.

That part is left for you. What comes next is a constructive proof, as we saw in Section 7.9. We will build an appropriate walk by exploring the graph one edge at a time. In the figures below, the cloudy shape represents the unexplored portion of the graph, with the vertices discovered so far shown.

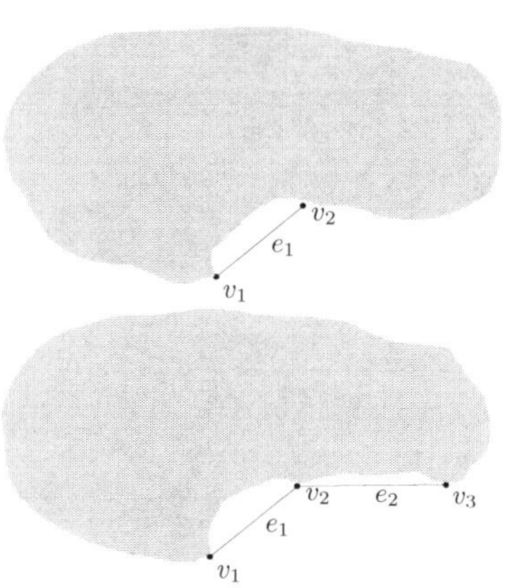

Then construct a walk c in this manner: Pick a vertex $v_1 \in V$ and an edge $e_1 = (v_1, v_2)$. Since $\deg(v_1) = 2$, e must exist, and since G contains no self-loops, $v_1 \neq v_2$.

Since $\deg(v_2) = 2$, there exists another edge, $e_2 = (v_2, v_3) \in E$.

Continue this process until we reach a vertex already visited, so that we can write $c = v_1 e_1 e_2 v_3 \ldots e_{x-1} v_x$ where $v_x = v_i$ for some i, $1 \leq i < x$. We will reach such a vertex eventually because V is finite.

Only one vertex in c is repeated, since reaching a vertex for the second time stops the building process. Hence c is simple.

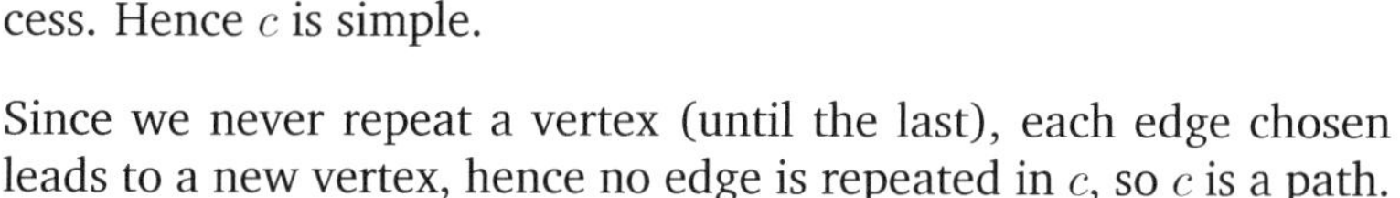

Since we never repeat a vertex (until the last), each edge chosen leads to a new vertex, hence no edge is repeated in c, so c is a path.

We are always choosing as the outgoing edge some edge other than the one we took into a vertex, so $i \neq x - 1$.

Suppose $i \neq 1$. Since no other vertex is repeated, v_{i-1}, v_{i+1}, and v_{x-1} are distinct. Therefore, distinct edges (v_{i-1}, v_i), (v_i, v_{i+1}), and (v_{x-1}, v_i) all exist, and so $\deg(v_i) \geq 3$. Since $\deg(v_i) = 2$, this is a contradiction. Hence $i = 1$. Moreover, $v_1 = v_x$ and c is closed.

As a closed, simple path, c is a cycle.

Suppose that a vertex $v \in V$ is not in c, and let v' be any vertex in c. Since G is connected, there must be a walk, c' from v to v', and let edge e' be the first edge in c' (starting from v') that is not in c, and let v'' be an endpoint in c' in c. Since two edges incident on v'' occur in c, accounting for e' means that $\deg(v'') \geq 3$. Since $\deg(v_i) = 2$, this is a contradiction. Hence there is no vertex not in c.

Suppose that an edge $e \in E$ is not in c, and let v be an endpoint of e. Since v is in the cycle, there exist distinct edges e_1 and e_2 in c that are incident on v, implying $\deg(v) \geq 3$. Since $\deg(v) = 2$, this is a contradiction. Hence there is no edge not in c.

Therefore, c is a cycle that comprises the entire graph, and G is a cycle. □

Leonhard Euler proposed a problem based on the bridges in the town of Königsberg, Prussia (now Kaliningrad, Russia). Two branches of the Pregel

River converge in the town, delineating it into a northern part, a southern part, an eastern part, and an island in the middle, connected with bridges, as shown below.

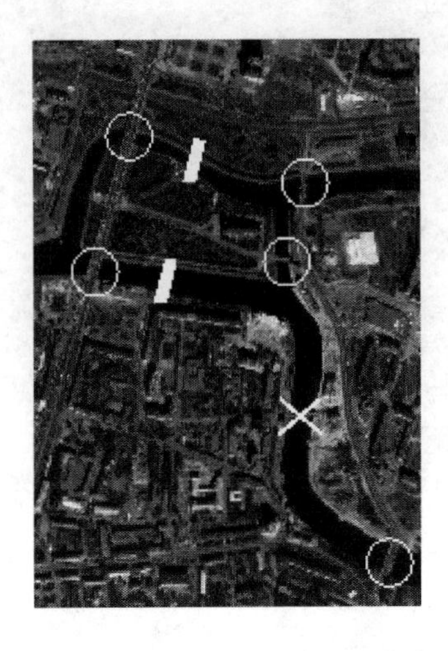

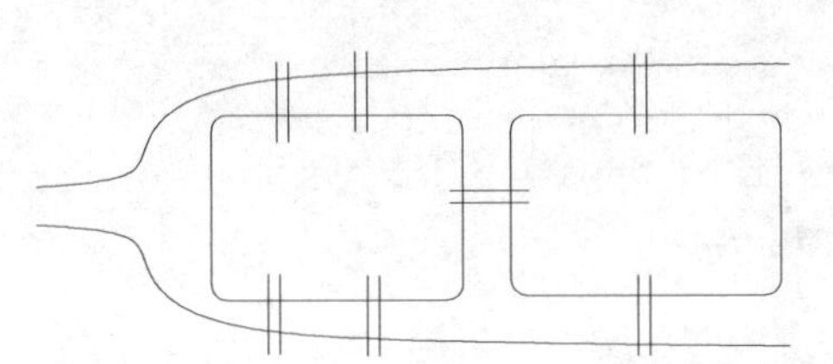

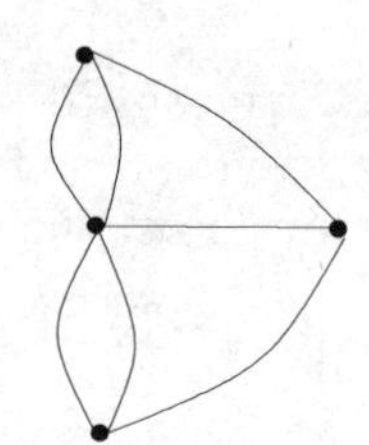

To the left is an aerial view from Google Maps of modern Kaliningrad showing five of the bridges still in approximately the same spot as in the eighteenth century. The locations of two older, no longer extant, bridges are indicated and a newer bridge exed out. Above is a more abstract map and a graph representation.

Supposing your house was in any of the four parts, is it possible to walk around town (beginning and ending at your house) and cross every bridge exactly once? We can turn this into a graph problem by representing the information with a graph whose vertices stand for the parts of town and whose edges stand for the bridges, as displayed above. Let $G = (V, E)$ be a graph. An *Euler circuit* of G is a circuit that contains every vertex and every edge. (Since it is a circuit, this also means that an Euler circuit contains every edge exactly once. Vertices, however, may be repeated.) The question now is whether or not this graph has an Euler circuit. We can prove that it does not, and so such a stroll about town is impossible.

Euler circuit

Biography: Leonhard Euler, 1707–1783

Euler was raised near Basel, Switzerland. His father was a pastor and amateur mathematician, and although the father wished the son to follow him into the ministry, the younger Euler's early mathematical talent attracted the attention of the Bernoulli family, who propelled him into a career in science.

Euler was the most prolific mathematician in history, producing results in many pure and applied areas of mathematics—including algorithms, a sort of computer-science-before-computers. He is most well known for his work in calculus and analysis, as he pioneered the use of Descartes's, Newton's, and Leibniz's tools to their full potential. In addition to ground-breaking research, Euler also wrote a textbook for children and popular works for the common person. Most impressively, his productivity did not slacken in face of such distractions as delicate political situations, devotion to his 13 children, and blindness in the last two decades of his life.

Theorem 8.5 *If a graph $G = (V, E)$ has an Euler circuit, then every vertex of G has an even degree.*

Proof. Suppose $G = (V, E)$ is a graph that has an Euler circuit $c = v_0 e_1 v_1 \ldots e_n v_n$, where $v_0 = v_n$. For all $e \in E$, e appears in c exactly once, and for all $v \in V$, v appears at least once, by definition of Euler circuit. Suppose $v_i \in V$, $v_i \neq v_1$, and let x be the number of times v_i appears in c. Since each appearance of a vertex corresponds to two edges, $\deg(v_i) = 2x$, which is even by definition.

For v_1, let y be the number of times it occurs in c besides as an initial or terminal vertex. These occurrences correspond to $2y$ incident edges. Moreover, e_1 and e_n are incident on v_1, and hence $\deg(v_i) = 2y + 1 + 1 = 2(y + 1)$, which is even by definition. (If either e_1 and e_n are self-loops, they have already been accounted for once, and we are rightly counting them a second time.)

Therefore, all vertices in G have an even degree. □

The northern, eastern, and southern parts of town each have odd degrees, so by the contrapositive of this theorem, no Euler circuit around town exists.

Hamiltonian cycle

Another interesting case is that of a *Hamiltonian cycle*, which for a graph $G = (V, E)$ is a cycle that includes every vertex in V. Since it is a cycle, this means that no vertex or edge is repeated; however, not all the edges need to be included. Here is a Hamiltonian cycle in a graph similar to the one at the beginning of this chapter (with the disconnected subgraph removed).

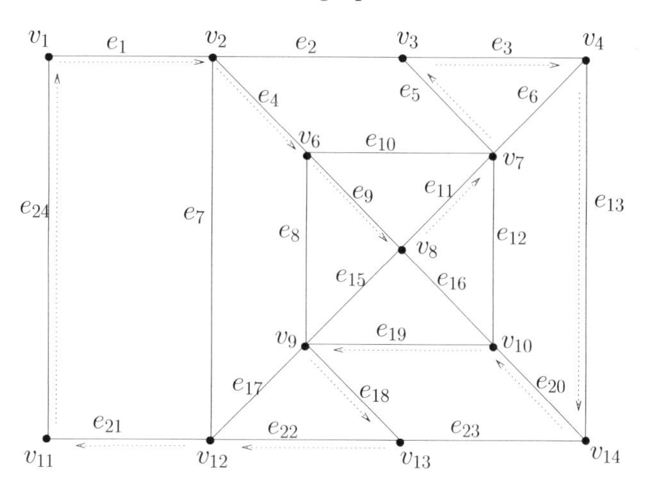

Exercises

8.3.1 Draw a graph that has a simple walk that is not a path.

Prove. Assume $G = (V, E)$ is a graph.

8.3.2 If G is connected, then $|E| \geq |V| - 1$.

8.3.3 If G is connected, $|V| \geq 2$, and $|V| > |E|$, then G has a vertex of degree 1.

8.3.4 If G is not connected, then $\overline{G}$ is connected.

8.3.5 If G is connected, for all $v \in V$, $\deg(v) = 2$, and $|V| > 1$, then G has no self-loops.

8.3.6 Every circuit in G contains a subwalk that is a cycle.

8.3.7 If for all $v \in V$, $\deg(v) \geq 2$, then G contains a cycle.

8.3.8 If $v, w \in V$ are part of a circuit c and G' is a subgraph of G formed by removing one edge of c, then there exists a path from v to w in G'.

8.3.9 If G has no cycles, then it has a vertex of degree 0 or 1.

8.3.10 If G has an Euler circuit, then it is connected.

In each of the following graphs, find an Euler circuit or explain why one does not exist.

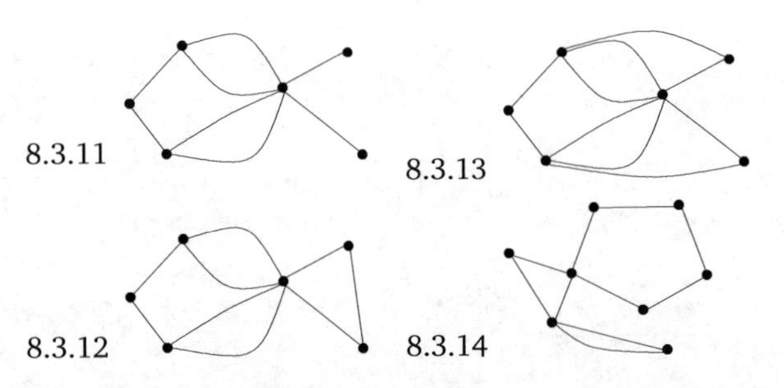

8.3.15 Make a graph of the modern bridges of Kaliningrad as shown from the aerial photo (include the exed out bridge, exclude the two old ones drawn back). Find an Euler circuit or explain why one does not exist.

8.3.16 Reinterpret the Königsburg Bridges Problem by making a graph whose vertices represent the bridges and whose edges represent land connections between bridges. What problem are you solving now?

8.3.17 Find a Hamiltonian cycle in the following graph (copy it and highlight the walk on your copy).

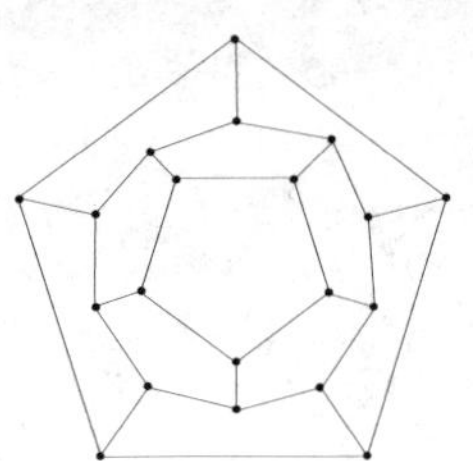

Prove.

8.3.18 If G has a non-trivial Hamiltonian cycle, then G has a subgraph $G' = (V', E')$ such that

- G' contains every vertex of G ($V' = V$)
- G' is connected,
- G' has the same number of edges as vertices ($|V'| = |E'|$), and
- every vertex of G' has degree 2.

8.3.19 If $G = (V, E)$ is a simple graph, $|V| \geq 3$, and for all $v \in V$, $\deg(v) \geq \frac{|V|}{2}$, then G has a Hamiltonian cycle. This is called Dirac's Theorem. To prove it, first draw an example or two by hand and find Hamiltonian cycles in them. Generalize how you found them into an algorithm. In your proof, state the algorithm and justify why each step is possible and why the steps converge to a Hamiltonian cycle.

8.3.20 If $G = (V, E)$ is a simple graph, $|V| \geq 3$, and for all $u, v \in V$ such that u and v are not adjacent, $\deg(u) + \deg(v) \geq |V|$, then G has a Hamiltonian cycle. This is called Ore's Theorem, and it is a corollary to Dirac's Theorem. Once you decipher what it is saying, it can be proven very quickly from the previous exercise.

8.4 Isomorphisms

We already have seen that the drawn shape of the graph—the placement of the dots, the resulting angles of the lines, any curvature of the lines—is not of the essence of the graph. The only things that count are the names of the vertices and edges and the abstract shape, that is, the connections that the edges define. However, consider the two graph representations below, which illustrate the graphs $G = (V = \{v_1, v_2, v_3, v_4\}, E = \{e_1 = (v_1, v_2), e_2 = (v_2, v_3), e_3 = (v_3, v_4), e_4 = (v_4, v_1), e_5 = (v_1, v_3)\})$ and $G' = (W = \{w_1, w_2, w_3, w_4, w_5\}, F = \{f_1 = (w_1, w_2), f_2 = (w_2, w_3), f_3 = (w_3, w_4), f_4 = (w_3, w_1), f_5 = (w_4, w_2)\})$.

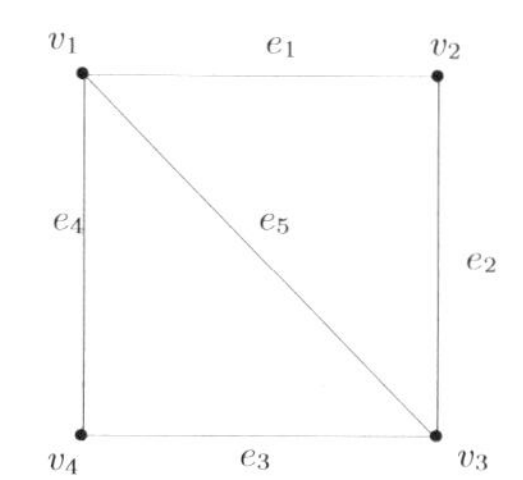

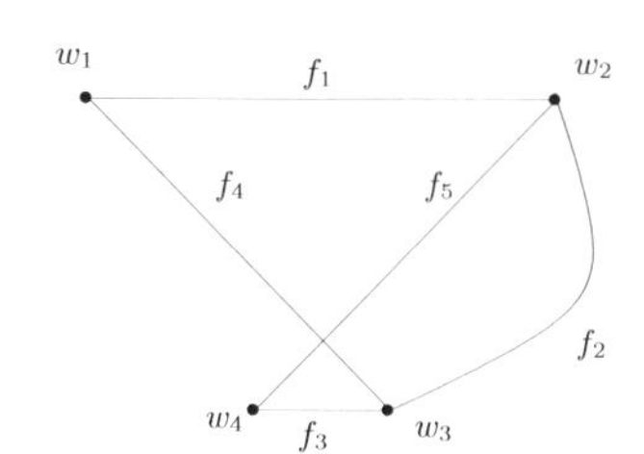

These graphs have much in common. Both have four vertices and five edges. Both have two vertices with degree two and two vertices with degree three. Both have a Hamiltonian cycle ($v_1 e_1 v_2 e_2 v_3 e_3 v_4 e_4 v_1$ and $w_1 f_1 w_2 f_5 w_4 f_3 w_3 f_4 w_1$, leaving out e_5 and f_2, respectively) and two other cycles (involving e_5 and f_2). In fact, if you imagine switching the positions of w_1 and w_2, with the edges sticking to the vertices as they move, and then doing a little stretching and squeezing, you could transform the second graph until it appears identical to the first.

In other words, these two really are the same graph, in a certain sense of sameness. The only difference is the arbitrary matter of names for the vertices and edges. We can formalize this with renaming functions, $\phi : V \to W$ and $\psi : E \to F$.

v	$\phi(v)$
v_1	w_2
v_2	w_4
v_3	w_3
v_4	w_1

e	$\psi(e)$
e_1	f_5
e_2	f_3
e_3	f_4
e_4	f_1
e_5	f_2

There is another way of looking at this. The practical use of graphs is that they model information. These two graphs each model the dummy information contained in the labels of the vertices and edges. Those labels are different, so in that sense the graphs are different. However, they have the same *structure*. To revise two earlier examples, compare the graph of the characters in *Anna Karenina* with a map of the connections among a few Midwestern cities.

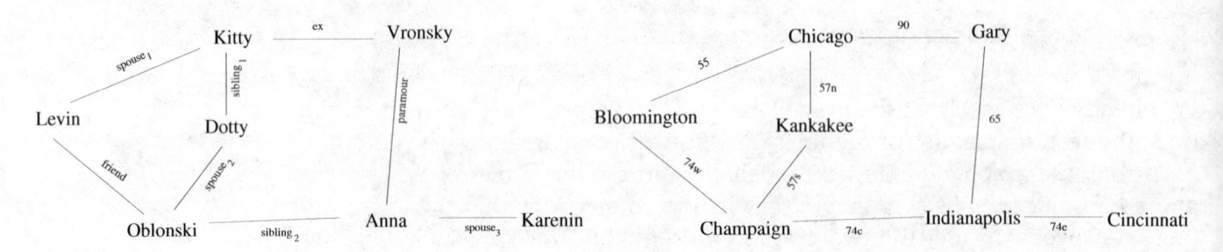

Again, the structure of these two graphs is the same. The difference is the meaning. Structure and meaning make an important distinction in many fields of study—for example, the difference between syntax (grammar) and semantics in the field of linguistics.

isomorphism

The term for this kind of equivalence between graphs is *isomorphism* from the Greek roots *iso* meaning "same" and *morphë* meaning "shape." This is a way to recognize identical abstract shapes of graphs, that two graphs are the same up to renaming.

What would a formal definition of isomorphism take? Rephrase the question this way: How would we define the relation "is isomorphic to" on graphs?

isomorphic

Let $G = (V, E)$ and $H = (W, F)$ be graphs. G is *isomorphic* to H if there exist one-to-one correspondences $\phi : V \to W$ and $\psi : E \to F$ such that for all $v \in V$ and $e \in E$, v is an endpoint of e iff $\phi(v)$ is an endpoint of $\psi(e)$. The two functions ϕ and ψ, taken together, are referred to as the *isomorphism* itself, that is, there exists an isomorphism, namely ϕ and ψ, between G and H.

What makes isomorphisms so important is that they preserve properties. That is, there are many properties of graphs that, if they are true for one graph, are true for any other graph isomorphic to that graph.

isomorphic invariant!for graphs

An *isomorphic invariant* is a property that is preserved through an isomorphism. Formally, a predicate P is an isomorphic invariant if for any graphs G and H, if $P(G)$ and G is isomorphic to H, then $P(H)$. Some isomorphic invariants are simple and therefore quite easy to prove; others require subtle and complex proofs. To illustrate what is meant, consider this theorem.

Theorem 8.6 *For any $k \in \mathbb{N}$, the proposition $P(G) =$ "G has a vertex of degree k" is an isomorphic invariant.*

This is not a difficult result to prove as long as one can identify what burden is required. Being an isomorphic invariant has significance only when two pieces are already in place: We have two graphs known to be isomorphic and that the proposition is true for one of those graphs.

> **Proof.** Suppose $k \in \mathbb{N}$, $G = (V, E)$ is a graph which has a vertex $v \in V$ with degree k, and H is a graph to which G is isomorphic.

Now we can assume and use the definition of isomorphic (those handy one-to-one correspondences must exist), and we must prove that H has a vertex of degree k. The definition of degree also comes into play, especially in distinguishing between self-loops and other edges.

By the definition of degree, there exist edges $e_1, e_2, \ldots, e_n \in E$, non self-loops, and $e'_1, e'_2, \ldots e'_m \in E$, self-loops, that are incident on v, such that $k = n + 2m$.

By the definition of isomorphism, there exist one-to-one correspondences ϕ and ψ with the isomorphic property.

Saying "with the isomorphic property" spares the trouble of writing out "for all $v' \in V$" etc, and instead more directly we claim

For each e_i, $1 \leq i \leq n$, $\psi(e_i)$ has $\phi(v)$ as one endpoint, and for each e'_j, $1 \leq j \leq m$, $\psi(e'_j)$ has $\phi(v)$ as both endpoints, and no other edge has $\phi(v)$ as an endpoint. Each $\psi(e_i)$ and $\psi(e'_j)$ is distinct since h is one-to-one. Hence

$$\deg(\phi(v)) = n + 2m = k$$

Therefore H has a vertex, namely $\phi(v)$, with degree k. □

The proof above contains the awkward clause "H is a graph to which G is isomorphic." This is necessary because our definitions directly allow us to talk only about one graph being isomorphic to another, not necessarily the same thing as, for example, "H is isomorphic to G." However, your intuition should suggest to you that these in fact are the same thing, that if G is isomorphic to H then H is also isomorphic to G, and we may as well say "G and H are isomorphic *to each other*."

If we think of isomorphism as a relation (the relation "is isomorphic to"), this means that the relation is symmetric. We can go further, and observe that isomorphism defines an equivalence class for graphs.

Theorem 8.7 *The relation R on graphs defined that $(G, H) \in R$ if G is isomorphic to H is an equivalence relation.*

Proof. Suppose R is a relation on graphs defined that $(G, H) \in R$ if G is isomorphic to H.

Reflexivity. Suppose $G = (V, E)$ is a graph. Let $\phi = i_V$ and $\psi = i_E$ (that is, the identity functions on V and E, respectively). By Exercise 7.6.7, ϕ and ψ are one-to-one correspondences. Now suppose $v \in V$ and $e \in E$. Then $\phi(v) = v$ and $\psi(e) = e$, so if v is an endpoint of e then $\phi(v)$ is an endpoint of $\psi(e)$, and if $\phi(v)$ is an endpoint of $\psi(e)$, then v is an endpoint of e. Hence ϕ and ψ are an isomorphism between G and itself, $(G, G) \in R$, and R is reflexive.

Symmetry. Suppose $G = (V, E)$ and $H = (W, F)$ are graphs and $(G, H) \in R$, that is, G is isomorphic to H. Then there exist one-to-one correspondences $\phi : V \to W$ and $\psi : E \to F$ with the isomorphic property. Since ϕ and ψ are one-to-one correspondences, their inverses, $\phi^{-1} : W \to V$ and $\psi^{-1} : F \to E$, respectively, exist. Suppose $w \in W$ and $f \in F$. By definition of inverse, $\phi(\phi^{-1}(w)) = w$ and $\psi(\psi^{-1}(f)) = f$. Suppose w is an endpoint of f. By definition of isomorphism, $\phi^{-1}(w)$ is an endpoint of $\psi^{-1}(f)$. Then suppose $\phi^{-1}(w)$ is an endpoint of $\psi^{-1}(f)$. Also by definition of isomorphism, w is an endpoint of f. Hence ϕ^{-1} and ψ^{-1} are an isomorphism from H to G, $(H, G) \in R$, and R is symmetric.

Transitivity. Exercise 8.4.4.

Therefore, R is an equivalence relation. □

There we have it—our "certain sense of sameness" is an example of mathematical equivalence. This is what abstract thinking is about, being able to ignore the details to identify what unifies. This proof, especially the part about symmetry, has much important review tucked into it. We had to notice that ϕ and ψ were one-to-one correspondences before we could assume inverse functions. We see again the pattern of analysis (taking apart what it means for G to be isomorphic to H) and synthesis (constructing what it means for H to be isomorphic to G). The isomorphic property is an "iff" property, so we were required to prove the matter of endpoints going either direction.

Now for a real workout—a theorem about paths and isomorphisms.

Theorem 8.8 *Having a Hamiltonian cycle is an isomorphic invariant.*

Proof. Suppose $G = (V, E)$ and $H = (W, F)$ are isomorphic graphs, and suppose that G has a Hamiltonian cycle, say $c = v_1 e_1 v_2 \ldots e_{n-1} v_n$ (where $v_1 = v_n$). By the definition of isomorphism, there exist one-to-one correspondences ϕ and ψ with the isomorphic property.

Suppose e_i is any edge in c, incident on v_i and v_{i+1}. By the definition of isomorphism, $\psi(e_i)$ is incident on $\phi(v_i)$ and $\phi(v_{i+1})$. This allows us to construct the walk $c' = \phi(v_1)\psi(e_1)\phi(v_2) \ldots \psi(e_{n-1})\phi(v_n)$.

By definition of Hamiltonian cycle, every $v \in V$ occurs in c. Since g is onto, every vertex $w \in W$ occurs in c', and since g is one-to-one, no vertex occurs more than once in c' except $\phi(v_1) = \phi(v_n)$. Also by definition of Hamiltonian cycle, no edge $e \in E$ occurs more than once in c. Since h is one-to-one, no edge $f \in F$ occurs more than once in c'.

Hence c' is a Hamiltonian cycle. Therefore, having a Hamiltonian cycle is an isomorphic invariant. □

Isomorphism is a crucial concept not only in graph theory. The definition of isomorphism is adapted to analyze similarities in many kinds of algebraic structures. In fact, discovering isomorphisms and their uses throughout mathematics is one of the principal goals of the second half of this text. The main idea of isomorphism is the independence of structure from meaning. If an isomorphism exists between two things, think of them as existing in parallel but equivalent universes, where everything that happens in one universe has an equivalent event in the other that keeps them in step.

Exercises

Find an isomorphism between the two graphs in each of the following pairs. Since the graphs are simple, it is enough to show a pairing of the vertices; a pairing of the edges will then be implied.

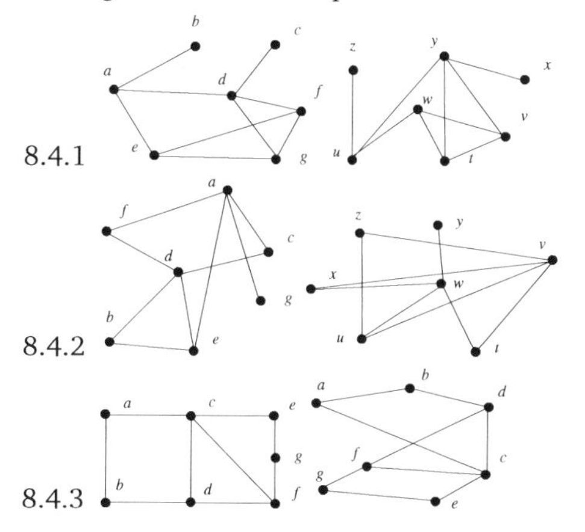

8.4.4 Prove that the relation "is isomorphic to" over graphs is transitive.

8.4.5 Prove that if simple graphs G and H are isomorphic, then their complements, $\overline{G}$ and $\overline{H}$ are isomorphic.

Prove that the following are isomorphic invariants.

8.4.6 G has n vertices, for $n \in \mathbb{N}$.

8.4.7 G has m edges, for $m \in \mathbb{N}$.

8.4.8 G has n vertices of degree k, for $k \in \mathbb{N}$.

8.4.9 G has a circuit of length k, for $k \in \mathbb{N}$.

8.4.10 G has a cycle of length k, for $k \in \mathbb{N}$.

8.4.11 G is connected.

8.4.12 G has an Euler circuit.

8.5 A garden of graphs

Graph theory is a vast subject. Entire books and courses are devoted to it. This text does not try to exhaust the topic, but this section presents a selection of the kinds of graphs often studied, as well as a few common applications.

Complete graphs. We earlier saw *complete* as an adjective describing graphs such that all vertices are adjacent to each other. We also can consider complete graphs as a species of graph, because all complete graphs of a given size are the same, up to isomorphism. The complete graph with n vertices is denoted K_n.

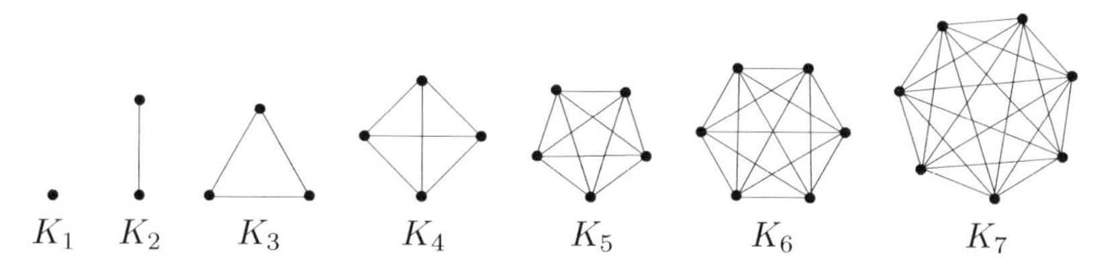

We have observed that complete graph K_n has $\frac{n(n-1)}{2}$ edges. A complete graph usually is a bad thing because the number of edges grows with the square of the number of vertices. No sane computer networking topology is set up so that every computer on the network has a direct connection to every other computer—not only would that be difficult to wire physically, but it would mean a large number of ports for any computer to poll. As another example, in any team larger than 4 or 5, communication overhead begins to push out all other work if every member of the team must talk to every other member. Imagine further if a college campus had sidewalks across a commons that made a complete graph among the entrances to the buildings. There would not be much green space left.

Cycles. Just as all complete graphs of a given size are identical up to isomorphism, we can identify a similar category of graphs: those of size n that contain only the edges that make a cycle of all the vertices (so, a Hamiltonian cycle). We simply call such graphs *cycles* and denote them as C_n, $n \geq 3$.

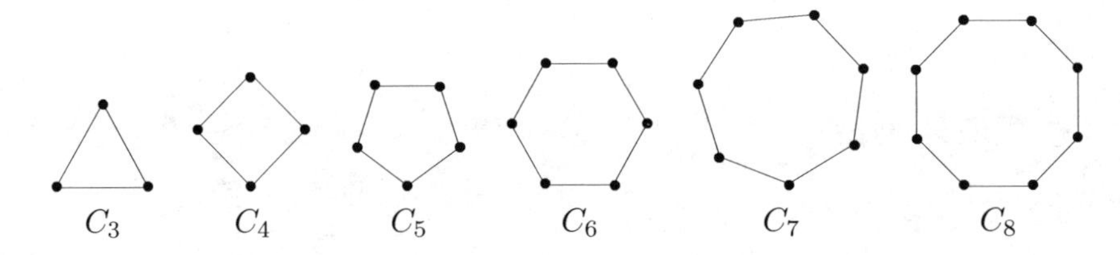

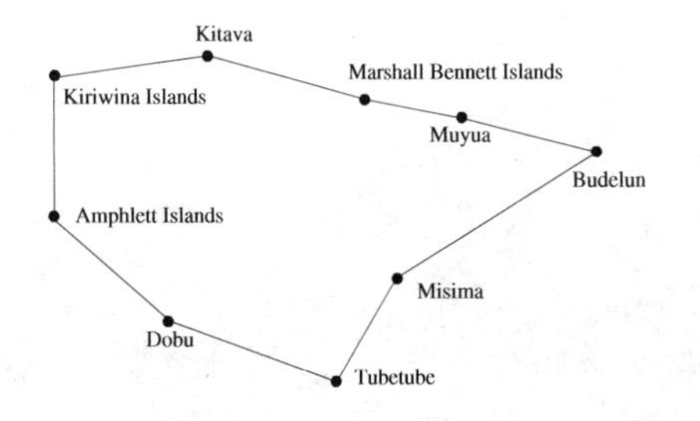

The Massim people of Papua New Guinea's Milne Bay Province (which comprises 160 inhabited islands and a section of New Guinea mainland) maintain social connections between settlements by exchanging gifts in a system called Kula. The gifts are arm bands and necklaces made from shells. Representatives from one community travel to another by canoe and exchange gifts, with the understanding that the gift will be enjoyed by the recipient only for a time and then be passed on as a Kula gift to another community. Each community participates in this exchange only with its immediate neighbor in what is essentially a cycle among the islands. The necklaces circulate clockwise and the arm bands circulate counter-clockwise. [33]

Some early work in computer networking used a technology where computers in a network were connected in a cycle. The computers would communicate by each passing a packet to its neighbor, which would then pass the packet on to the next computer until the packet arrived at the intended recipient. A special packet called a *token* would be passed continually around the cycle, indicating which computer's turn it was to initiate a packet. Such networks are called *token ring networks*, and they are still in use, though Ethernet has largely displaced them.

Wheels. A *wheel*, as a graph, is like a cycle except that it has one extra vertex which is adjacent to all other vertices. We denote a wheel of size n as W_n.

wheel

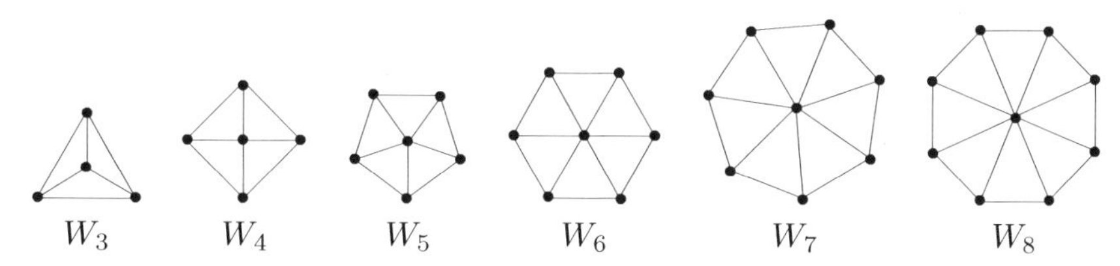

Graphs showing relationships among people often have subgraphs that are wheels or are similar to wheels, with a manager (or social butterfly) in the middle. Finding vertices with a high degree is also useful in criminal investigations and counter-terrorism efforts to identify ring-leaders.

n-cubes. Another identified variety of graph gets its name from the generalization of a cube. Notice that a cube can be thought of as two squares with the corresponding corners of each square connected with each other. (This is the natural technique for drawing a cube with pencil and paper.) In the same way we could draw two cubes and connect the corresponding corners to get a *hypercube*. In the other direction we can build a square (a *hypocube*?) by connecting the endpoints of two line segments.

In all of these cases, consider the corners of the shapes as vertices and the edges as graph edges. A graph in this family is called an n*-cube*, denoted Q_n. It has 2^n vertices and can be considered a n-dimensional object.

n-cube

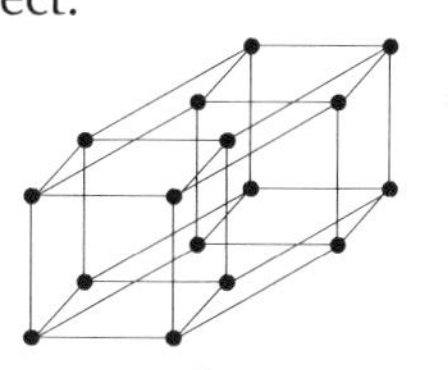

We can define n-cubes formally using recursion. Let $Q_1 = (\{v_1, v_2\}, \{(v_1, v_2)\})$. For $n \in \mathbb{N}$, $n \geq 1$, let $Q_n = (V \cup W, E \cup F \cup R)$ where

- $V \cap W = \emptyset$ and E, F, R are pairwise disjoint (so V and W make a partition of the vertices and E, F, and R make a partition of the edges).
- (V, E) and (W, F) each make a subgraph isomorphic to Q_{n-1}.
- There exists an isomorphism $(\phi : V \to W, \psi : E \to F)$ such that $R = \{(v, \phi(v)) \mid v \in V\}$.

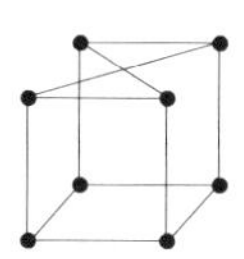

In other words, we match up the vertices of two copies of Q_{n-1} and draw edges between them to get Q_n. The fact that *phi* is part of an isomorphism excludes something like the graph on the right from being considered an n-cube, which is constructed from two copies of Q_2 but with twisted connections.

n-cubes are sometimes used for the network topology connecting nodes in parallel computers as they have some of the benefit of a complete graph (fast communication between nodes—any two nodes in Q_n are at most n links away) but without as many connections. Scaling is still a problem, however, as the number of vertices and edges doubles with each new dimension.

bipartite graph

Bipartite graphs. In a *bipartite graph*, the vertices can be partitioned into two sets such that all edges connect a vertex from one set to a vertex in the other; in other words, no edge exists between two vertices in the same set.

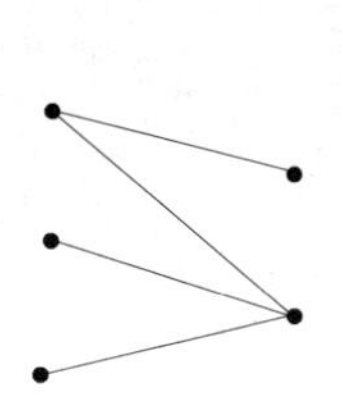
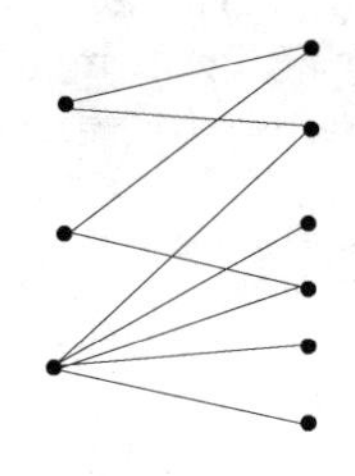
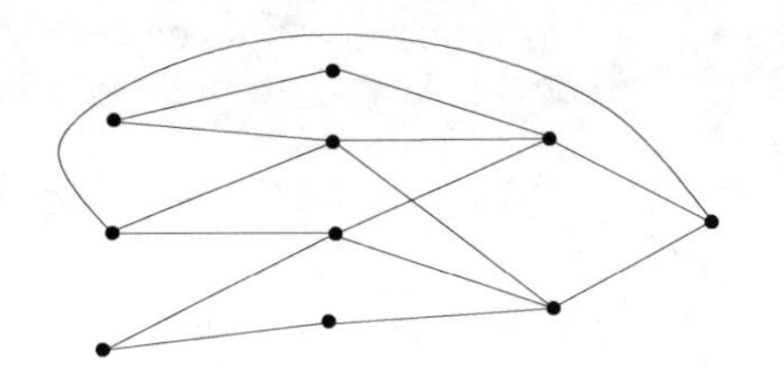
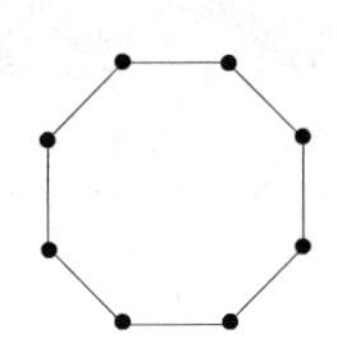

Imagine a graph modeling business relationships where each vendor and customer is a vertex and the edges indicate sales made by a vendor to a customer. Alternately, consider a singles mixer where the vertices stand for the men and women at the event and edges indicate phone numbers exchanged.

complete bipartite graph

A *complete bipartite graph* is a bipartite graph such that every pair of vertices from different sets in the partition are adjacent—that is, it is bipartite but has every possible edge it can have. (Note that although a complete bipartite graph is bipartite, it is not complete. This is a relaxing of the requirements for completeness.) Any complete bipartite graph is isomorphic to a graph in the family of graphs each denoted $K_{n,m}$, meaning a complete bipartite graph with one vertex set of size n and the other of size m.

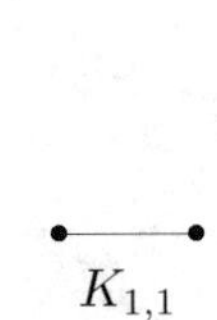
$K_{1,1}$

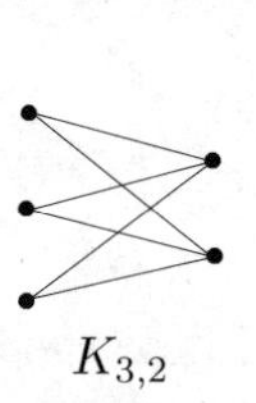
$K_{3,2}$

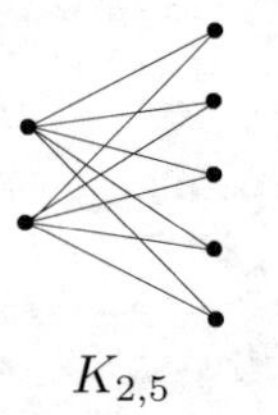
$K_{2,5}$

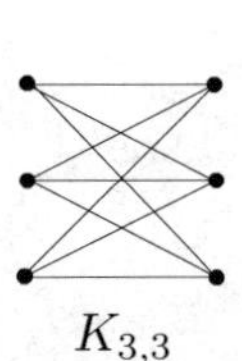
$K_{3,3}$

We have not given formal definitions for bipartite graph and complete bipartite graph. That is your task in Exercises 8.5.7 and 8.5.8.

Planar graphs. The complete bipartite graph $K_{3,3}$ leads to a problem that is a staple for graph theory courses: Suppose the graph represents three houses and three utilities (say, water, gas, and electricity). Is it possible for each utility to connect to each house without the lines crossing each other (if they were drawn in a two-dimensional schematic)? In other words, is it possible to draw $K_{3,3}$ in a plane with edges never crossing each other? The answer is *no*—try it.

A *planar graph* is a graph that can be drawn in a plane with no edges crossing. A formal definition takes some work; we present the following "semi-formal" definition. First, notice by inspection that when you draw a planar graph with no crossings, the area of the graph is broken up into regions that are bounded by cycles and have no overlap. We count the area around the graph to be a region.

planar graph

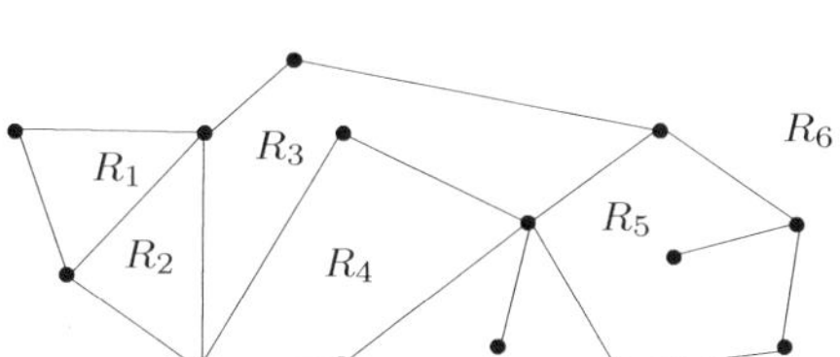

The edges in the cycles make up all the edges in the graph except for edges incident on vertices of degree 1. The main idea in our definition is the existence of such a set of cycles, which serve as a certificate of planarity.

Our definition is recursive. First, a graph is planar if it has exactly one edge; in that case it also has one region. (A graph with no edges is also planar, but not very interesting). A graph $G = (V, E)$ is planar also if for every edge $e \in E$ with endpoints v and w, we have that the graph with e taken away $(G_1 = (V, E - \{e\}))$ is itself planar and either v and w are disconnected or there exists for G_1 a certificate set of cycles R_1, R_2, $\ldots R_n$ such that v and w are both on some cycle R_i.

To make some sense of this thick definition, consider the graph below, both with and without the edge (c, f).

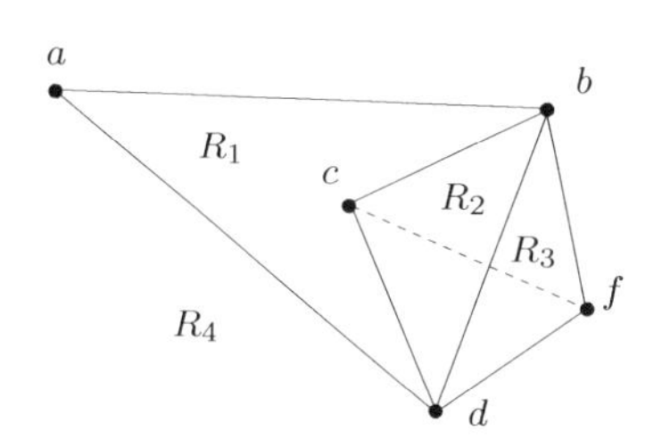

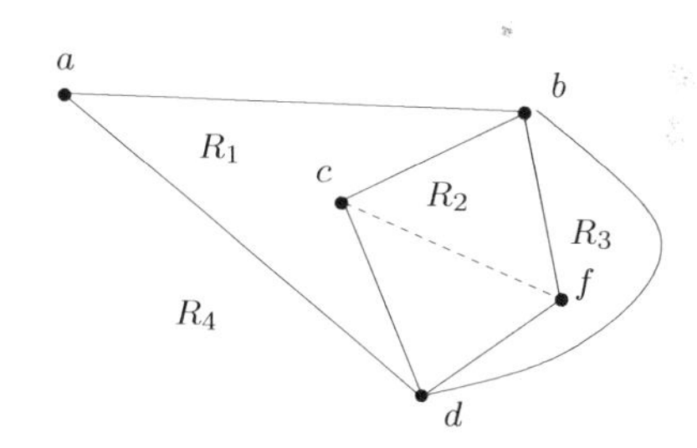

In the left drawing we can see that the graph without (c, f) is planar with certificate $\{R_1 = abcda, R_2 = bcdb, R_3 = bdfb, R_4 = abfda\}$. We cannot add (c, f) using this certificate because the endpoints are not on the same cycle. But that does not mean that the graph with (c, f) is not planar; we merely need a different certificate for the subgraph. On the right the same graph is drawn to suggest certificate $\{R_1 = abcda, R_2 = bcdfb, R_3 = bdfb, R_4 = abda\}$. Now c and f are on the same cycle, and indeed we can draw the edge without crossing any other. The graph is planar.

The process implicit in the definition is used to prove an interesting result discovered by Euler. A planar graph may have many possible certificates, but they each have the same number of cycles (or regions).

Theorem 8.9 (Euler's Formula) *If $G = (V, E)$ is a connected, simple, planar graph with certificate $\{R_1, R_2, \ldots, R_n\}$, then $n = |E| - |V| + 2$.*

Here is a sketch of the proof. Given a graph $G = (V, E)$, start with a subgraph $G_1 = (V_1, E_1)$ that contains only a single edge from E and its end points. It has one edge, two vertices, and one region, the one around the whole graph. So, $n = 1$ and $|E_1| - |V_1| + 2 = 1 - 2 + 2 = 1$.

Then, continually add edges from E one at a time to make larger (but still connected) subgraphs; add vertices from V as they become connected to the subgraphs. Each time we add an edge with no new vertex, we split a region, replacing an old cycle with two new cycles. This keeps the equation in balance. On the other hand, each time we add an edge with one new vertex, we do not add a region, but we increment the number of edges and vertices each by 1, again maintaining the formula.

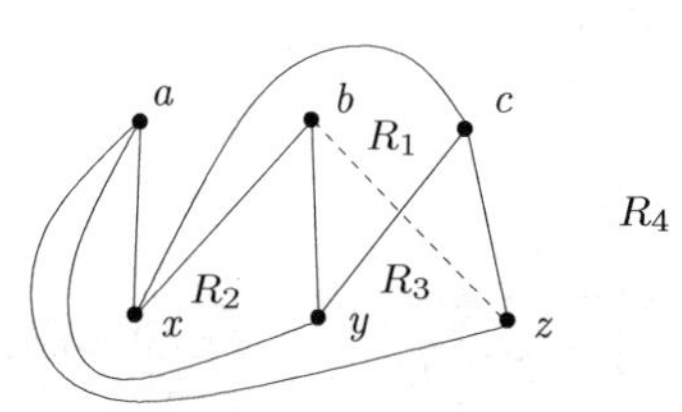

$K_{3,3}$ is not as lucky as the graphs already described. The diagram here shows at attempt to draw it in a plane. Without the edge (b, z), the subgraph is indeed planar with the regions shown, but b and z are not on the same cycle. We could redraw the edges to get a region bounded by a cycle including b and z, but it would no longer be a certificate of planarity because it would have its own crossings.

homeomorphic

$K_{3,3}$ leads to another interesting result. It turns out to be one of two poison pills that are the culprits of any failure of planarity. Two graphs G_1, G_2 are *homeomorphic* if they can each be derived from a third graph G_3 by repeatedly picking an edge from G_3 and breaking it with a new, intervening vertex.

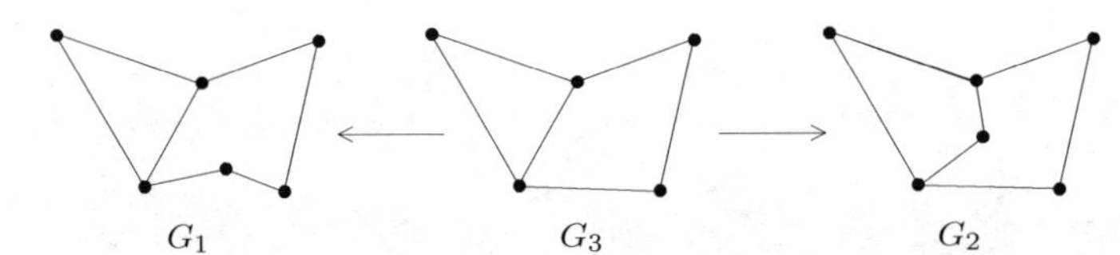

The following result was proven by Polish mathematician Kazimierz Kuratowski. Any nonplanar graph has a component derivable from $K_{3,3}$ or the other poison pill, K_5.

Theorem 8.10 (Kuratowski's Theorem) *A graph G is not planar iff there exists a subgraph of G that is homeomorphic to $K_{3,3}$ or K_5.*

Planar graphs are important for planning things that are embedded in real space and must appear on a surface, such as an electronic circuit on a single board with no crossings. (In fact circuit boards can be designed with crossings, but the number of planar subgraphs of a circuit can determine the number of layers needed.)

Subgraphs and connectivity. Some applications of graphs—whether real-world applications or just interesting problems—make use of subgraphs, sets of vertices, or just special vertices or edges. For example, in a simple, undirected graph, a *clique* is a complete subgraph, that is, a set of vertices that are pairwise adjacent to each other (together with the edges that connect them). As the name suggests, you can imagine a graph representing people and their friendships; the presence of a clique indicates a group of mutual friends. A similar idea for directed graphs is a *strongly connected component*, a set of vertices all of which are reachable from (not necessarily adjacent to) the others.

clique

strongly connected component

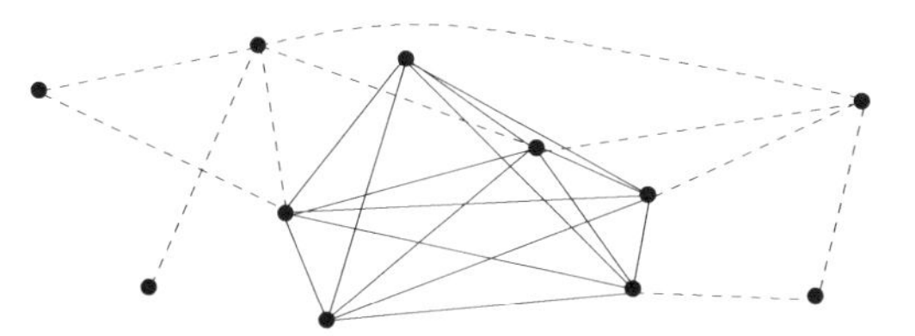

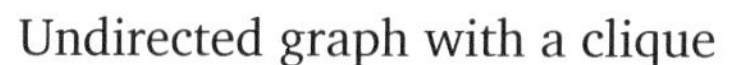

Undirected graph with a clique

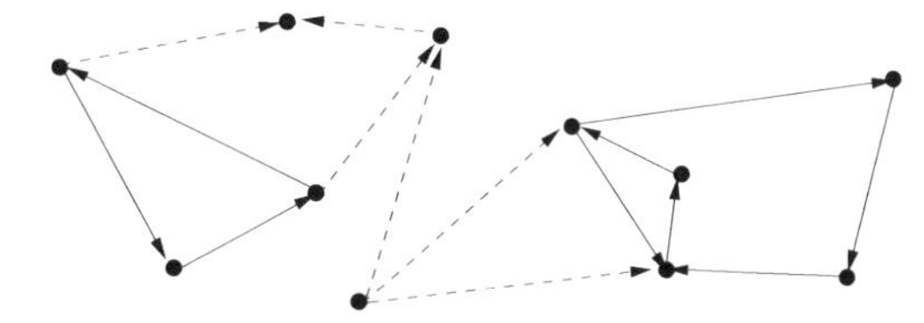

Directed graph with two strongly connected components

In a connected undirected graph, a *bridge* is an edge whose removal disconnects the graph. Similarly, an *articulation point* is a vertex whose removal (which implies the removal of all edges adjacent to it) disconnects the graph. Bridges and articulation points can be used to identify weaknesses in networks of any sort—computer, social, transportation, etc. In the graph below, a, b, and c are articulation points and the edge (a, b) is a bridge.

bridge

articulation point

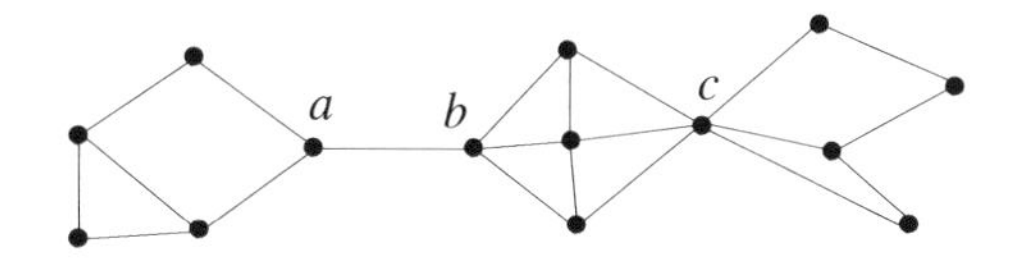

A *vertex cover* of an undirected graph is a set of vertices that includes at least one endpoint of every edge. The idea is that a vertex covers the edges incident on it, and we want to find a set that covers all the edges. Notice that the entire set of vertices is a trivial vertex cover. An interesting problem is finding a vertex cover with the smallest number of vertices. Below left is a graph with vertices in a vertex cover highlighted in white. It is not a minimal vertex cover, though—on the right is the same graph but with a smaller vertex cover.

vertex cover

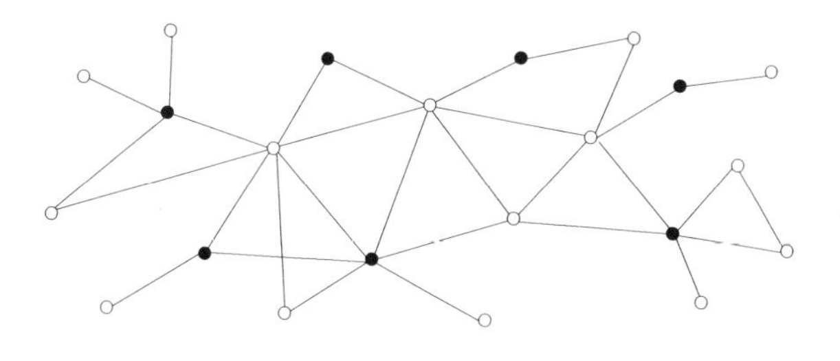

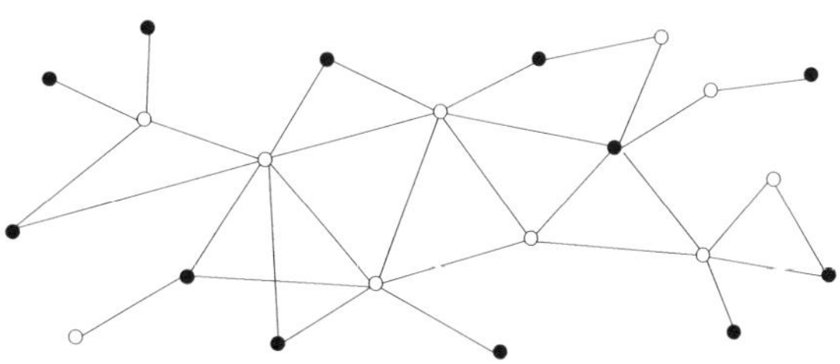

Trees. We met trees in Section 6.2 defined on their own terms:

> A *tree* is a collection of *nodes* and *links*. We will take *node* to be a primitive. A link is an ordered pair of nodes, and each node is the second node in exactly one link, except for one special node which is not the second node in any link.

tree, as a directed graph

However, if you replace the terms *node* and *link* with *vertex* and *edge*, we clearly have a (directed) graph. Formally, a directed graph $G = (V, E)$ with designated vertex $r \in V$ is a *tree* if for all $v \in V$ there exists exactly one path from r to v. We call r the *root*.

tree, as an undirected graph

Now, take a tree (as a directed graph) and erase the arrowheads, and you are left with another interesting structure, a special type of undirected graph, which we also call *tree*. The undirected-graph version of *tree* is in fact easier to define—it is a connected, simple graph with no cycles. (By the contrapositive of Exercise 8.3.6, it must not have any circuits, either.) Below left is an example of a directed-graph tree. Below center is a similar but undirected tree.

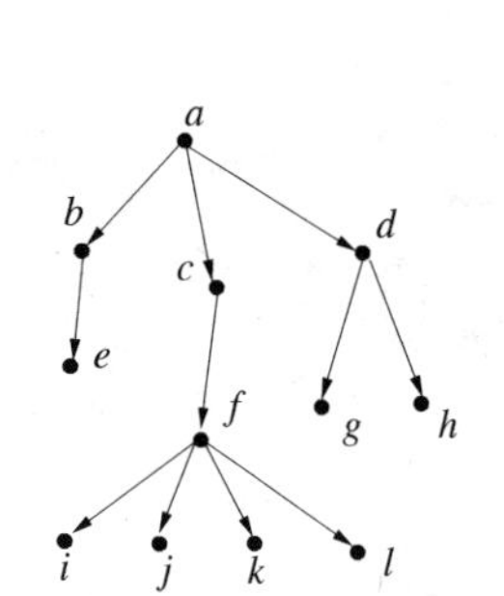

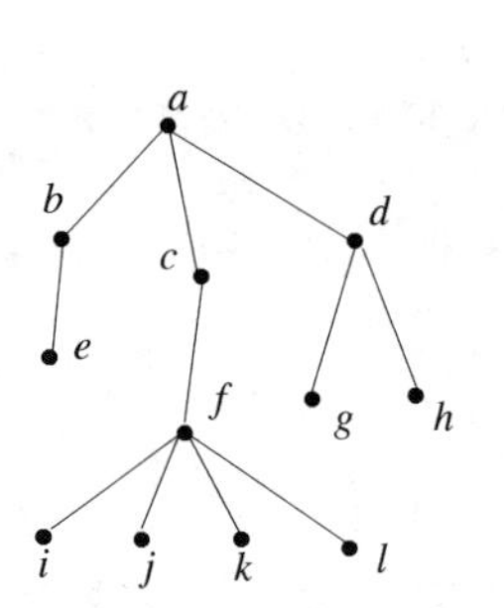

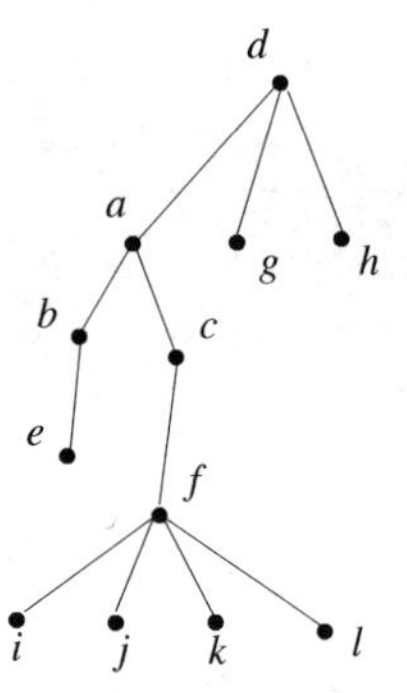

Notice that undirected-graph trees do not need a designated root. Any vertex can be a root. We conventionally draw a tree with the roots at the top, but imagine picking that tree up by any other vertex and allowing the other vertices and edges to flop down, as we have done above right. You have the same tree, but a different vertex is suggested for the root.

Game theory. The last few items in this section are not about varieties of graphs but applications of graphs. Game theory is a branch of mathematics for modeling games and puzzles (or social and economic phenomena) that have discrete states, such as the configuration of game pieces on a board, and where

Tangent: Tree roots

You know you are officially a computer scientist if you think trees grow down with their roots at the top and leaves at the bottom.

transitions between those states are the result of decisions made by players. Game theory makes extensive use of graphs.

Consider the following puzzle (which is a one-player game):

> You must transport a cabbage, a goat, and a wolf across a river using a boat. The boat has only enough room for you and one of the other objects. You cannot leave the goat and the cabbage together unsupervised, or the goat will eat the cabbage. Similarly, the wolf will eat the goat if you are not there to prevent it. How can you safely transport all of them to the other side?

We will solve this puzzle by analyzing the possible *states* of the situation, that is, the possible places you, the goat, the wolf, and the cabbage can be, relative to the river; and the *moves* that can be made between the states, that is, your rowing the boat across the river, possibly with one of the objects. Let f stand for you, g for the goat, w for the wolf, and c for the cabbage. / will show how the river separates all of these. For example, the initial state is $fgwc/$, indicating that you and all the objects are on one side of the river. If you were to row across the river by yourself, this would move the puzzle into the state gwc/f, which would be a failure. Our goal is to find a series of moves that will result in the state $/fgwc$.

First, enumerate all the states.

$$\begin{array}{cccc}
fcgw/ & fcg/w & fw/cg & c/fgw \\
cgw/f & gw/fc & fg/cw & g/fcw \\
fgw/c & cw/fg & fc/gw & w/fcg \\
fcw/g & cg/fw & g/cgw & /fcgw
\end{array}$$

Now, mark the starting state with a double circle, the winning state with a triple circle, each losing state with a square, and every other state with a single circle. These will be the vertices in our graph.

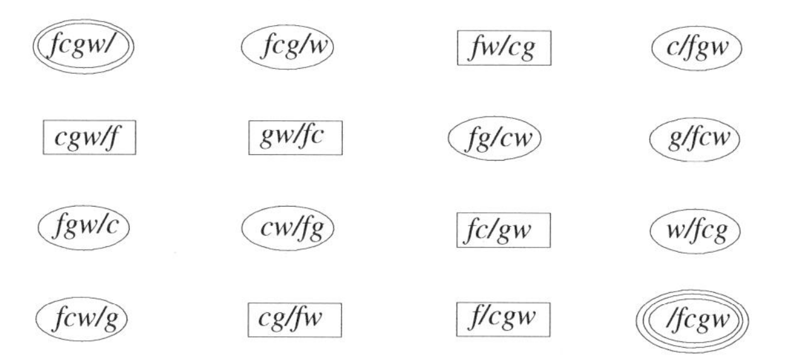

Finally, we draw edges between states to show what would happen if you cross the river carrying one or zero objects.

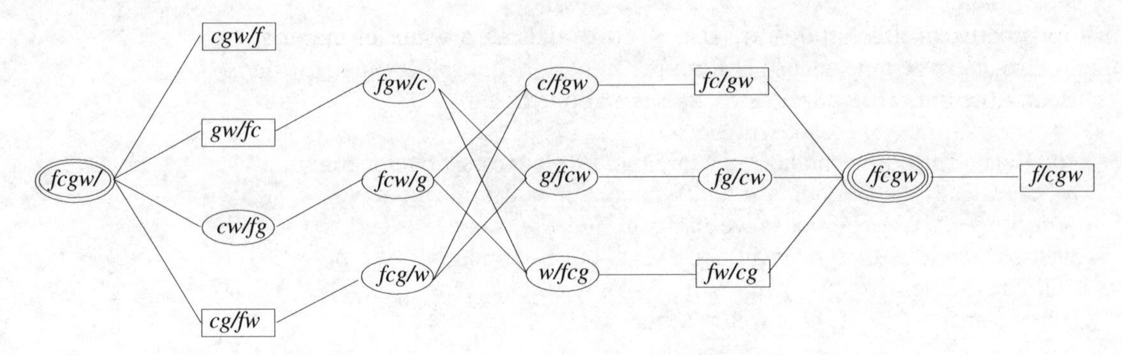

The puzzle is solved by finding a path from $fcgw/$ to $/fcgw$ that does not go through any losing state. One possible route suggests you to transport them all by first taking over the goat, coming back (alone), transporting the cabbage, coming back with the goat, transporting the wolf, coming back (alone), and transporting the goat again. (One could argue that f/cgw is an unreachable state, since you would first need to win in order to lose in that way.)

flow network

Flow networks. Some applications associate data not only with the vertices but also with the edges. In a *flow network* (also called a *transportation network*), a directed graph has two specially designated vertices, the *source* and the *sink*. The source has in-degree of 0, and all other vertices are reachable from the source; the sink has out-degree of 0, and it is reachable from all other vertices. The edges are labeled with natural-number valued *capacities*. Flow networks are used to model transportation of some material from the source to the sink. The edges can be thought of as pipes carrying fluid or routes for trucks carrying freight. The vertices besides the source and the sink are places where pipes join, or switching stations for the trucks. In the following example, the source is designated s and the sink t.

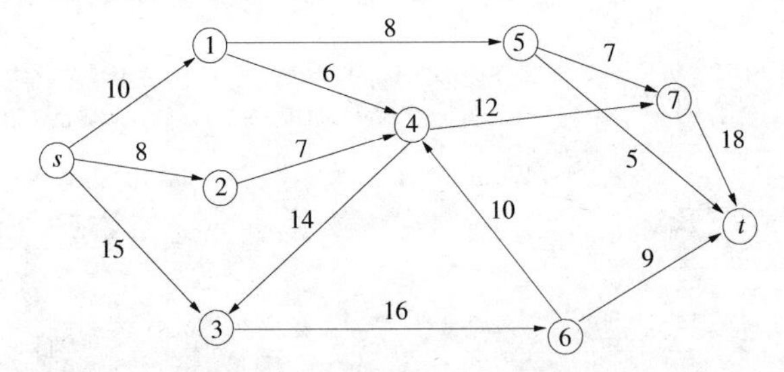

Formally, a flow network is a quintuple (V, E, s, t, c) where (V, E) is a simple directed graph, $s, t \in V$, the restrictions on s and t are as stated before, and $c : E \to \mathbb{N}$ is a function that assigns capacities to the edges. In the example above, $c(s, 2) = 8$.

flow

A *flow* on a network is another assignment to the edges (formally a function $f : E \to \mathbb{N}$) indicating how much material is actually flowing through the edge. A flow has a restriction that the assignment to any edge must be less than or equal to its capacity (for all $e \in E$, $f(e) \leq c(e)$) and the amount of material flowing into each vertex (except the source and sink) must be equal to the amount flowing out. For convenience, we can also define f to be $V \times V \to \mathbb{N}$ (that is, over pairs of vertices) so that for all $v, u \in V$ such that $(v, u) \notin E$, define $F(v, u)$ to be 0. Then we can define this second *flow conservation* constraint as

flow conservation

$$\forall\, v \in V - \{s, t\},\ \sum_{u \in V} f(u, v) = \sum_{w \in V} f(v, w)$$

Here is an example of a flow for the network shown above (the label 3/10 on an edge indicates 3 out of a possible 10 units of flow are being used, that is, $c(s, 1) = 10$ but $f(s, 1) = 3$.

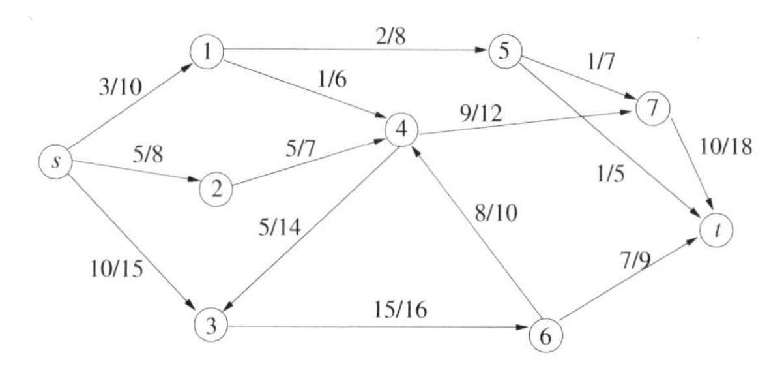

Notice that the amount flowing out of the source must be equal to the amount flowing into the sink. More generally, we define a *cut* of a flow network (V, E, s, t, c) to be any two-set partition of the vertices V, call it (S, T) where $s \in S$ and $t \in T$. A cut stands for a way to draw a dividing line among the vertices as long as the source and sink are on different sides. It must be the case that the net flow from S to T must be the same for any cut (S, T).

cut

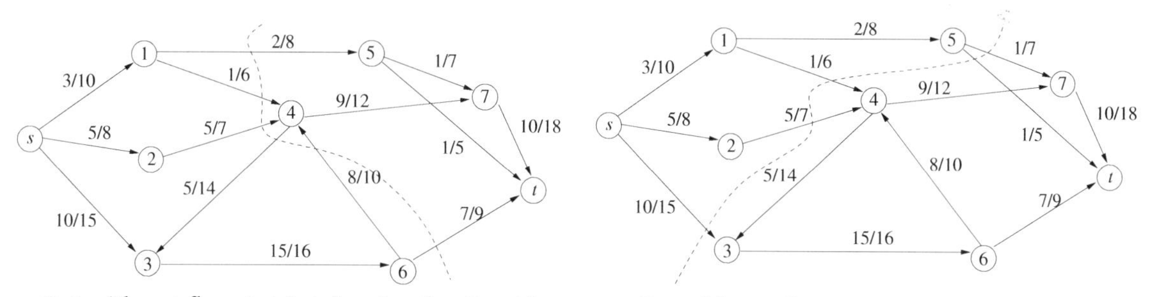

Cut with net flow $2 + 1 + 5 - 5 + 8 + 7 = 18$ Cut with net flow $1 + 1 + 1 + 5 + 10 = 18$

We can also extend the idea of capacity to cuts by saying that the capacity of a cut $c(S, T)$ is the sum of the capacities of all the edges from S to T (ignoring the edges from T to S). The cut above left has capacity $8 + 6 + 7 + 10 + 9 = 30$, the one above right has capacity $7 + 5 + 6 + 7 + 15 = 40$.

The central problem in flow networks is finding a maximum flow, an assignment to the edges that meets the constraints but has the most material flowing from source to sink of any legal flow. The main strategy for finding a maximum

augmenting path

flow involves looking for *augmenting paths*. An augmenting path for a flow is a path through a network on which every edge has a flow less than its capacity. What is helpful about an augmenting path is that we can safely increase the flow on each edge in the path (that is, increase the flow through that path) without disrupting any constraint.

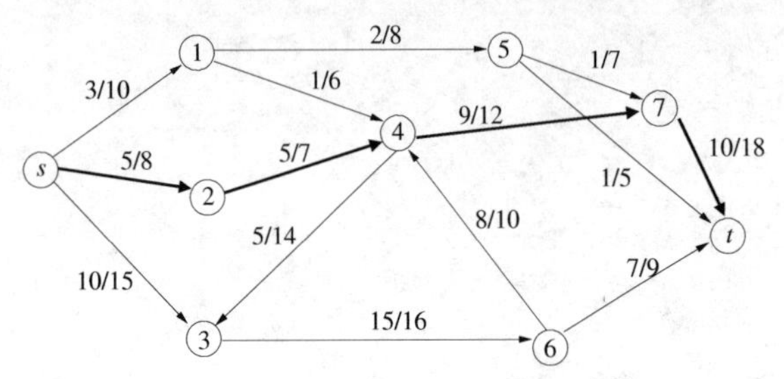

Edge (2, 4) has smallest cut/flow difference

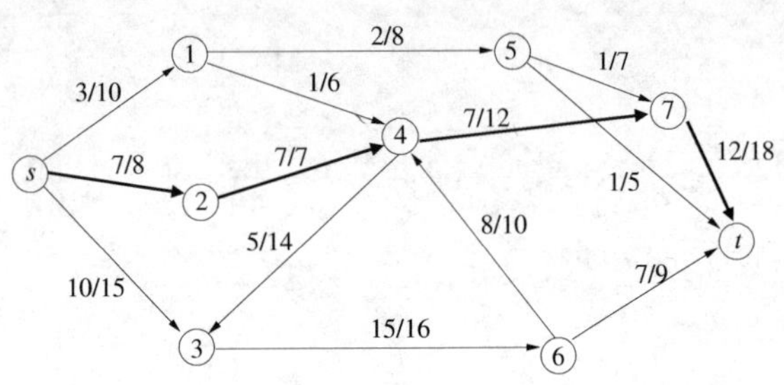

Adding 2 to each edge in path improves the flow.

The Ford-Fulkerson Method for finding a maximum flow is to take any flow (one can start with all edges having flow zero), and as long as an augmenting path exists, augment the flow along that path. When no more augmenting paths exists, the process converges to a maximum flow. This depends on the following theorem, which also states that with a maximum flow, there exists some cut with capacity equal to the flow (that is, some cut is flowing at maximum capacity).

Theorem 8.11 (Max-Flow Min-Cut Theorem) *If f is a flow in network (V, E, s, t, c), then f is a maximum flow iff f has no augmenting paths iff there exists a cut whose capacity is equal to the net flow.*

However, *finding* an augmenting path is a harder problem, which we do not discuss here.

Traveling Salesman Problem

The Traveling Salesman Problem. The *Traveling Salesman Problem* involves a graph with edges labeled not with capacities but by *weights* (although it might be easier to think of them as distances). Here we will define a complete, directed, weighted graph G as a tuple (V, w) where V is, as usual, the set of vertices and a function $w : V \times V \to \mathbb{N}$ is an assignment of natural-number valued weights for each edge. We do not include a set of edges because the graph is complete. Since the graph is undirected, assume that for all $v_1, v_2 \in V$, $w(v_1, v_2) = w(v_2, v_1)$ and for all $v \in V$, $w(v, v) = 0$.

Suppose a salesman needs to visit n cities on his itinerary. However, he wants to spend as little total time traveling as possible. Model the layout of the cities as a complete, directed, weighted graph with the vertices representing cities and the weights of the edges representing the distance between the cities or the time it takes to travel from one to another. The Traveling Salesman Problem is to find a Hamiltonian cycle (so, a route that hits every city exactly once and comes back to its starting point) with minimal total weight. Here is an example.

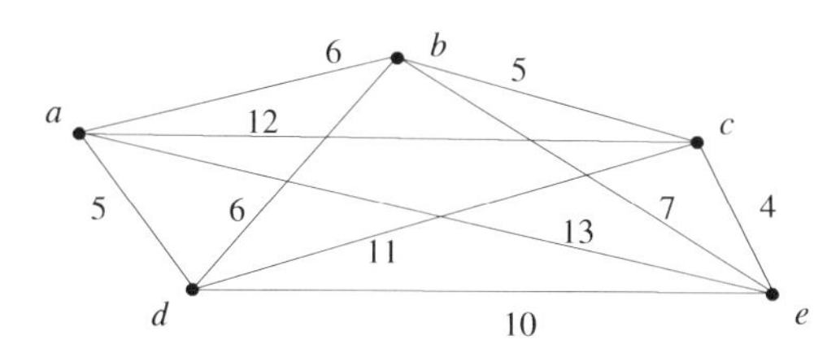

Since the graph is complete, finding a Hamiltonian cycle is simple. Finding one of minimal total weight is a different story. Suppose we take the approach of trying all possible routes (with a as our starting and ending point). We would have to check 24 routes.

$abcdea$	$abceda$	$abdcea$	$abdeca$	$abecda$	$abedca$	$acbdea$	$acbeda$
$acdbea$	$acdeba$	$acebda$	$acedba$	$adbcea$	$adbeca$	$adcbea$	$adceba$
$adebca$	$adecba$	$aebcda$	$aebdca$	$aecbda$	$aecdba$	$aedbca$	$aedcba$

There are only 12 distinct cycles, actually, since any route is essentially the same if taken in the opposite direction—$abcdea$ and $aedcba$, for example. This approach does not scale, because in general a graph with n vertices has $\frac{(n-1)!}{2}$ distinct Hamiltonian cycles.

In fact, there is no known efficient way to solve the Traveling Salesman Problem exactly, and most computer scientists believe that none exists (see Section 9.11). In practice approximation algorithms are used which are not guaranteed always to converge to the least weight route, but probabilistically will find a pretty good route most of the time.

Exercises

8.5.1 What complete graphs have Euler circuits?

8.5.2 What cycles are bipartite?

8.5.3 What wheels have Euler circuits?

8.5.4 Explain why every wheel has a Hamiltonian cycle.

8.5.5 What is the greatest n for which K_n is planar?

8.5.6 Explain how a bipartite graph and a symmetric relation R from a set A to a set B, where $A \cap B = \emptyset$, are the same thing.

8.5.7 Give a formal definition of a *bipartite* graph. Your definition must be existentially quantified—it depends on the existence of a partition.

8.5.8 Give a formal definition of a *complete bipartite* graph. Use your definition of bipartite from Exercise 8.5.7

8.5.9 Prove that a graph is bipartite iff all circuits have an even number of edges. (Informally, this means that in a bipartite graph it is impossible to go around the graph from a vertex and back again traversing an odd number of edges.) Use your formal definition from Exercise 8.5.7; note that the "only if" direction is a proof of existence.

8.5.10 How many vertices does an n-cube have, in terms of n? Prove your claim.

8.5.11 Prove that for all $n \in \mathbb{N}$, an n-cube is bipartite. (Hint: Use mathematical induction.)

Exercises, continued

8.5.12 Write a formal, non-recursive definition on an n-cube. (Hint: Label the vertices as shown below for Q_3. How do we determine what edges exist, based on the labels of the vertices?)

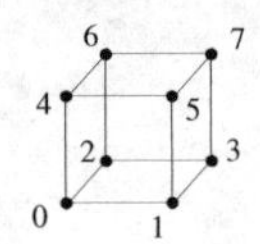

8.5.13 Explain why in W_3 any vertex could be considered the central one.

8.5.14 Explain, informally, why for all $n, m \in \mathbb{N}$, $K_{n,m}$ is isomorphic to $K_{m,n}$.

8.5.15 Describe, informally, an algorithm for determining if a graph is bipartite.

8.5.16 Prove that any tree is bipartite.

8.5.17 Show that K_4 is planar (by drawing it with no edge crossings). What other complete graphs are planar?

8.5.18 Show that Q_3 is planar by drawing it with no edge crossings. What other n-cubes are planar?

8.5.19 Give a formal definition of a clique.

8.5.20 If we include even very small strongly connected components (a single vertex is a strongly connected component by itself, as it is reachable from itself in 0 steps), then a graph is completely composed of strongly connected components. Explain how strongly connected components are equivalence classes (or a partition) for the vertices of a graph. Define an equivalence relation using strongly connected components. (Hint: Notice that many subgraphs of a strongly connected component are themselves strongly connected components. Hence we do not want to consider *all* of a graph's strongly connected components, but rather just the largest ones—those that are not subgraphs in other strongly connected components.)

8.5.21 Give a formal definition of a bridge.

8.5.22 Give a formal definition of an articulation point.

8.5.23 Explain why the end points of any bridge are articulation points.

8.5.24 Give a formal definition of a vertex cover.

8.5.25 In our definition of a tree as a directed graph, we did not assert that it must be simple or connected. We did not need to. Prove that any tree (as a directed graph) is simple and connected.

8.5.26 Why is it not enough to define a tree as a directed graph to be a directed graph that has no cycles? A directed graph with no cycles is called, not surprisingly, a *directed acyclic graph* or *dag*. Give an example of a directed acyclic graph that is not a tree.

8.5.27 Prove that an undirected graph $G = (V, E)$ is a tree iff for any $v, w \in V$ there exists a simple path in G from v to w.

Since simple directed graphs are equivalent to relations, we can think of a directed-graph tree as a relation. Exercises 8.5.28–8.5.32 ask you to define and reason about a tree as a kind of relation.

8.5.28 Formally define *path* in terms of a relation so that it is equivalent to our concept of path in a graph. (Hint: Use the transitive closure.)

8.5.29 Using your definition of path from Exercise 8.5.28, write a formal definition of *tree* as a relation (that is, "A relation R is a *tree* if . . . ").

8.5.30 From your definition in Exercise 8.5.29, prove that a tree is irreflexive.

8.5.31 Prove that a tree is asymmetric.

8.5.32 Prove that if R, a relation on a set A, is a tree, then for all $a, b, c \in A$, if $(a, b) \in R$ and $(b, c) \in R$, then $(a, c) \notin R$.

8.5.33 The following is known as the Jealous Husband Puzzle. Men a, b, and c and women x, y, z must cross a river using a two-person boat. a is married to x, b is married to y, and c is married to z. No woman may be on the same side of the river as any man unless her husband is also on that side. How can they all get across? Use a graph to solve this puzzle.

8.5.34 Find the maximum flow in the flow network given in this section.

8.5.35 Find the minimum Hamiltonian cycle in the Traveling Salesman graph given in this section.

8.6 Representing graphs

We have seen that graphs are indispensable tools for modeling information. Accordingly, many algorithms on graphs—algorithms for computing properties of graphs, finding paths in graphs, iterating over the vertices in a graph—are important problems in computer science.

Before we can formulate any graph algorithms, we need a representation for graphs, one that can be stored in computer memory. The natural way to represent a graph to a human is to do so visually, with a diagram. Clearly that will not do for a computer, however.

Since graphs have a lot in common with relations (a directed graph with no parallel edges, after all, is identical to a relation), we might look to our ways of representing relations for inspiration. Consider this graph from Section 8.1:

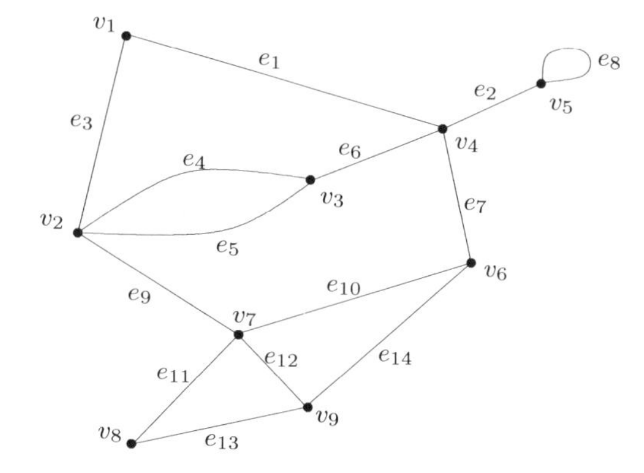

The straightforward way to interpret the definition of *graph* using set notation would be something like (V, E) where

$$V = \{v_1, v_2, v_3, v_4, v_5, v_6, v_7, v_8, v_9\}$$

$$E = \begin{array}{ll} \{ & e_1(v_1, v_4), e_2(v_4, v_5), e_3(v_1, v_2), e_4(v_2, v_3), e_5(v_2, v_3), \\ & e_6(v_3, v_4), e_7(v_4, v_6), e_8(v_5, v_5), e_9(v_2, v_7), e_{10}(v_6, v_7), \\ & e_{11}(v_7, v_8), e_{12}(v_7, v_9), e_{13}(v_8, v_9), e_{14}(v_6, v_9) \quad \} \end{array}$$

The elements of E appear as $e_1(v_1, v_4)$ and the like, with both a label and a description of the vertices they connect. If our graph had no parallel edges, we could list them only as a pair of vertices, without a label. For example, if the graph above had only one edge between v_2 and v_3, then we would write the edges as

$$E = \begin{array}{ll} \{ & (v_1, v_4), (v_4, v_5), (v_1, v_2), (v_2, v_3), \\ & (v_3, v_4), (v_4, v_6), (v_5, v_5), (v_2, v_7), (v_6, v_7), \\ & (v_7, v_8), (v_7, v_9), (v_8, v_9), (v_6, v_9) \quad \} \end{array}$$

This, then, is the basis for our first way to represent graphs on a computer: an explicit listing of both the vertices and the edges. Assume a vertex will be labeled with an int, and a vertex may also have data of some specified type associated with it at it.

```
- datatype 'a vertex = V of int * 'a;
```

The type parameter 'a stands for the type of the data, if any, stored at this vertex—call it the vertex's *payload*. An edge, then, will be represented as two ints (the labels of its end points) and its own payload of type 'b. The most common payload to store with an edge is a *weight*, as we saw with the Traveling Salesman Problem. Intuitively the weight can be thought of as something like the distance between the two vertices along that edge. (We assume no parallel edges—to allow for them, add a label to each edge.)

```
- datatype 'b edge = E of int * int * 'b;
```

It is important to distinguish between the *vertices themselves* as data objects and the *numerical labels* for vertices. In this representation, edges contain a pair of vertex labels, not a pair of vertices. It is most important for students who have programmed in another language such as Java to recognize this distinction, since in many languages it is natural to have references to data objects. In Java, classes for vertices and edges might look like this (T is the type parameter for vertex payloads, S for edge payloads):

```
public class Vertex<T> {
   private int label;
   private T payload;
   ...
}

public class Edge<T, S> {
   private Vertex<T> endpoint1, endpoint2;
   private S payload;
  ...
}
```

We could do the same thing in ML if we used reference variables, but that is not ML's natural style.

We finish off our ML interpretation by defining the graph type itself.

```
- datatype ('a, 'b) graph =
=       G of 'a vertex list * 'b edge list;
```

Keep in mind that this is our first attempt, a sort of literal interpretation of the definition of graph. There are several ways to represent graphs, each having their own advantages and liabilities. There are three factors for evaluating graph representations. First, how efficiently does the representation use computer memory—how compact is it? Second, how easy is it to store information at the vertices and edges? Third, how efficiently can information from the graph be read?

How do the data types defined above measure up? Space efficiency: The size of the representation of a particular graph would be proportional to the number of vertices plus the number of edges. It is difficult to imagine doing better than that. Payload convenience: We have a built-in way to store extra information at both the vertices and edges. Time efficiency: To find if an edge exists between two given vertices, we would have to search through the list of edges. We will see if other representations can improve on that.

One common representation is the *adjacency matrix*. In this scheme, each vertex is represented by a row and column in a matrix, and the entries in the matrix indicate whether there exists an edge between the two matrices—1 for the existence of an edge, 0 for its absence. (For a *directed* graph, we could decide that the row stands for the first vertex in the pair, the column for the second.) For the graph above (with the parallel edge e_5 removed),

adjacency matrix

$$
\begin{array}{c}
\\ 1 \\ 2 \\ 3 \\ 4 \\ 5 \\ 6 \\ 7 \\ 8 \\ 9
\end{array}
\begin{array}{c}
\begin{array}{ccccccccc} 1 & 2 & 3 & 4 & 5 & 6 & 7 & 8 & 9 \end{array} \\
\left(\begin{array}{ccccccccc}
0 & 1 & 0 & 1 & 0 & 0 & 0 & 0 & 0 \\
1 & 0 & 1 & 0 & 0 & 0 & 1 & 0 & 0 \\
0 & 1 & 0 & 1 & 0 & 0 & 0 & 0 & 0 \\
1 & 0 & 1 & 0 & 1 & 1 & 0 & 0 & 0 \\
0 & 0 & 0 & 1 & 1 & 0 & 0 & 0 & 0 \\
0 & 0 & 0 & 1 & 0 & 0 & 1 & 0 & 1 \\
0 & 1 & 0 & 0 & 0 & 1 & 0 & 1 & 1 \\
0 & 0 & 0 & 0 & 0 & 0 & 1 & 0 & 1 \\
0 & 0 & 0 & 0 & 0 & 0 & 1 & 1 & 0
\end{array}\right)
\end{array}
$$

Since there is an edge between v_6 and v_7, the values at positions $(6,7)$ and $(7,6)$ are 1. Adjacency matrix form makes looking up information about a graph very fast. If we want to know if two vertices are adjacent, we use the vertex labels as coordinates in the matrix; if we want to know all the vertices to which a certain vertex is adjacent, we can iterate through that vertex's row (or column). Notice that in a graph with n vertices they must be labeled 1 to n or 0 to $n-1$. Even though we so far have given vertices numerical labels, we have not assumed it needed to be that way.

This representation is less space-efficient, however, since the matrix must have an entry not only for existent edges but also for absent ones. (Notice that for undirected graphs, the matrix will always be symmetric about the upper-left to lower-right diagonal.) Unless the number of edges is proportional to the square of the number of vertices, this is a serious loss of efficiency, especially for large graphs with only a few edges (we would call such a graph *sparse*).

Instead of using the value 1 to indicate the presence of an edge, we could use the matrix to store extra information about edges, such as weight (the no-edge entries could be any special value indicating the lack of an edge, not necessarily 0). Any extra information for vertices would need to be stored in a separate structure. Note also that in the adjacency matrix representation we have almost no hope of representing parallel edges.

How convenient is using a matrix and what is the efficiency of the actual lookup? This depends on the programming language. In programming languages that provide direct support for multi-dimensional arrays in computer memory such as C and Java, this is easy. ML's support for arrays is poor, and for this reason alone we will not be using adjacency matrices for representing graphs.

adjacency list

An alternative to the adjacency matrix is the the *adjacency list* representation. In this case, we do not store edges separately, but instead our representation of each vertex contains a list of the vertices (or their labels) to which that vertex is adjacent.

```
- datatype 'a vertex = V of int * 'a * int list;
```

In this datatype, the first int is the vertex's label, the 'a is the payload, and the int list contains the labels of the adjacent vertices. (If we want a 'b payload for the edges, we could make the last component to be a (int * 'b) list. If we want parallel edges, we can further augment each entry in the adjacency list with a label for the edge.)

In this representation, edge lookup is less efficient because we must first find a vertex's adjacency list and then iterate through the list rather than jumping to an entry in an adjacency matrix. This is mitigated, however, by ML's being optimized for list processing. Moreover, the adjacency list representation is more space efficient because it uses space proportional to the number of edges. In fact, it is similar to our first attempt of storing a list of edges, with the main difference of not requiring a separate structure for the edges. In fact, we can now represent a graph as just a list of vertices (to simplify the following examples, we omit the payload):

```
- datatype vertex = V of int * int list;

- datatype graph = G of vertex list;
```

Now we are ready to consider operations on graphs. Finding the list of adjacent vertices for a given vertex is straightforward:

```
- exception NoSuchVertex;

- fun getAdjacent(G([]), v) = raise NoSuchVertex
=   | getAdjacent(G(V(u, adj)::rest), v) =
=       if u = v then adj else
=              getAdjacent(G(rest), v);
```

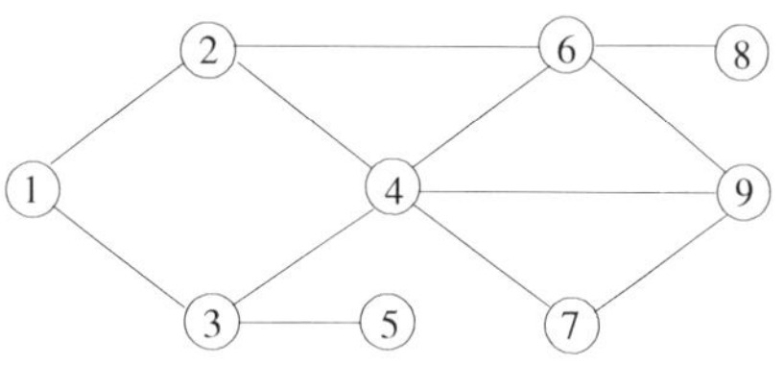

A key part to many tasks involving graphs is to search out a path from one vertex to another, that is, to search a graph starting from a vertex. Consider the graph on the right. We will find routes from vertex 1 to all other vertices (called *traversing* the graph or *visiting* the vertices).

There are two main strategies we can take. In the first strategy, we follow an edge from our starting point to another vertex, follow an edge from that one to a third vertex, and so on until we can go no further because we are at a vertex that is not adjacent to any vertex we have not yet visited. Then we retrace our steps, going back to the vertex most recently visited that has adjacent vertices we have not visited. Observe this process on this graph. Vertices that have been visited (or *discovered*) are shaded; edges that have been explored are thickened.

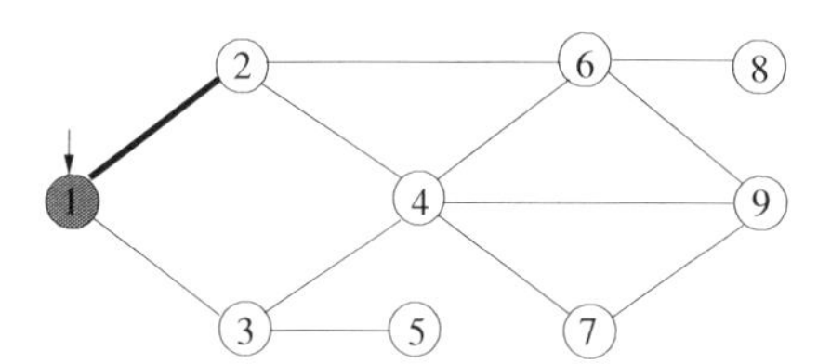

Starting from vertex 1, we explore the edge to vertex 2

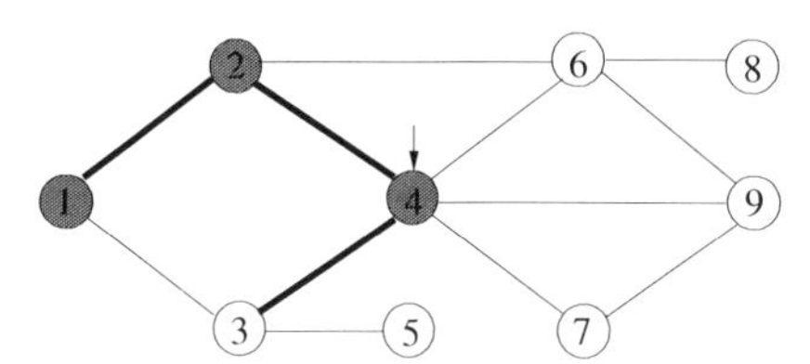

From vertex 2 (now discovered), we explore the edge to vertex 4

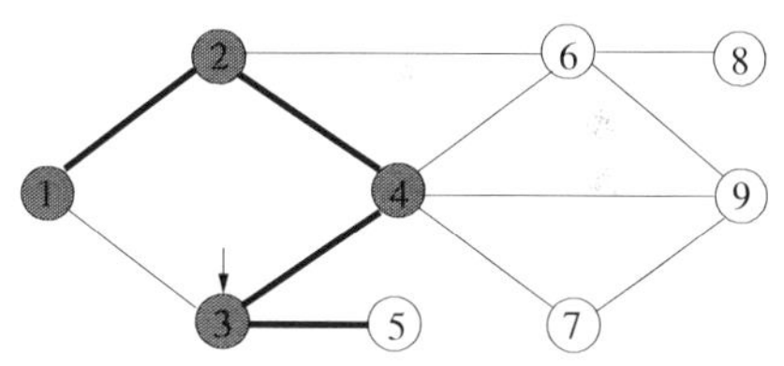

From vertex 4, we explore the edge to vertex 3

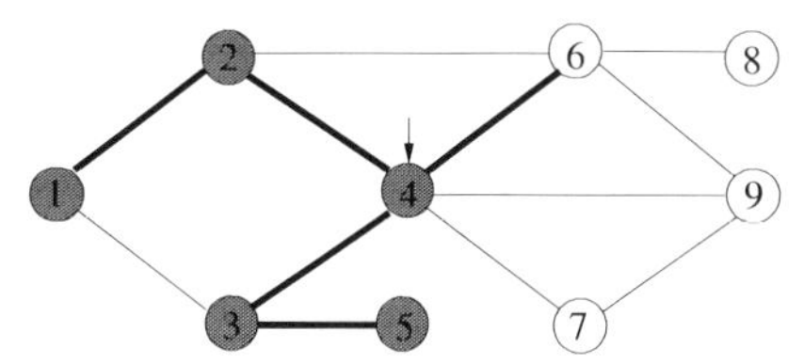

From vertex 3, we explore the edge to vertex 5.

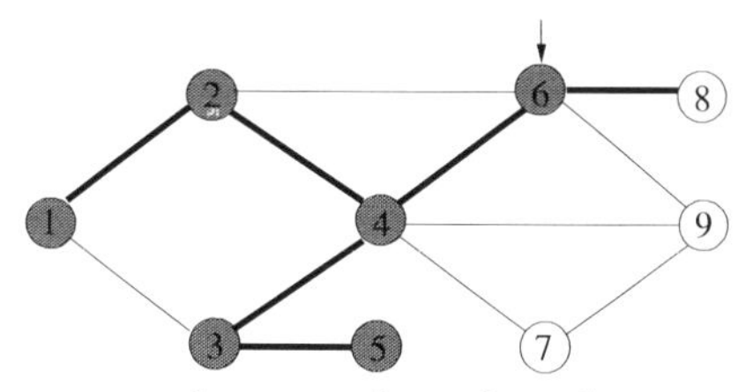

Having discovered vertex 5, we find it has no edges to undiscovered vertices. Vertex 3 has no more edges to undiscovered vertices either. So, we are back to vertex 4; we explore the edge to vertex 6.

From vertex 6, we explore the edge to vertex 8.

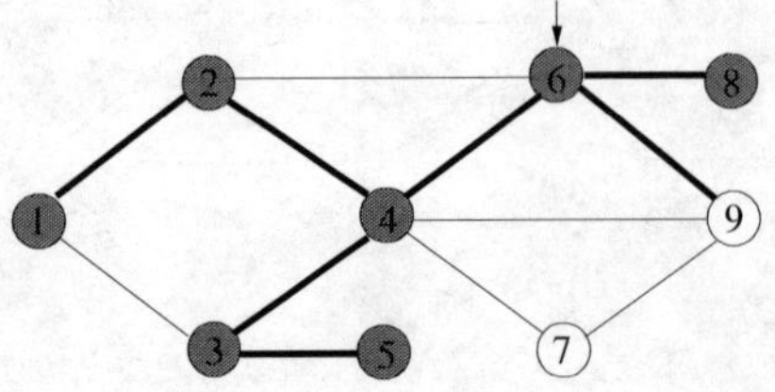

Vertex 8 has no edge to undiscovered vertices, so we are still working from vertex 6. We explore the edge to vertex 9.

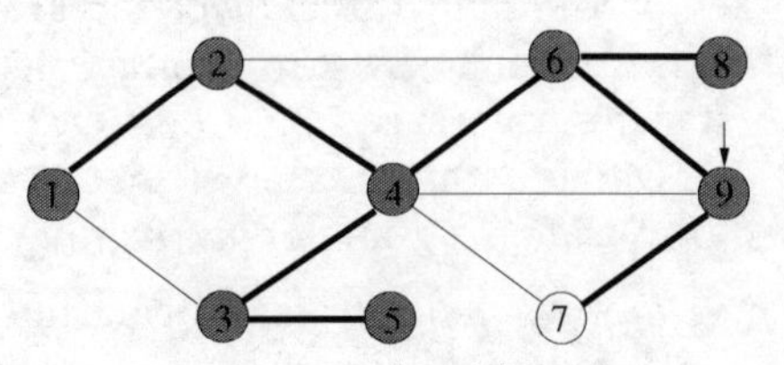

From vertex 9, we explore the edge to vertex 7.

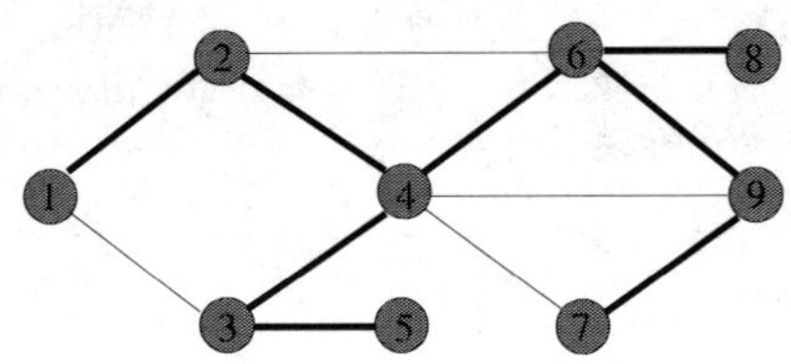

Vertex 7 has no edges to undiscovered vertices. Moreover, if we retrace our steps (vertices 9, 6, 4, 2, 1), no vertex has an edge to an undiscovered vertex. The traversal is complete.

depth-first search

breadth-first search

This strategy is called *depth-first search* (or *depth-first traversal*) because we always choose to go as deep in the graph as possible—every time we discover a new vertex we immediately move to it rather than looking at the other vertices adjacent to the current vertex. An alternative is *breadth-first search*. When visiting a vertex, we first discover all the (previously undiscovered) adjacent vertices and add them to a list of vertices to be visited. We visit vertices in the order they are discovered.

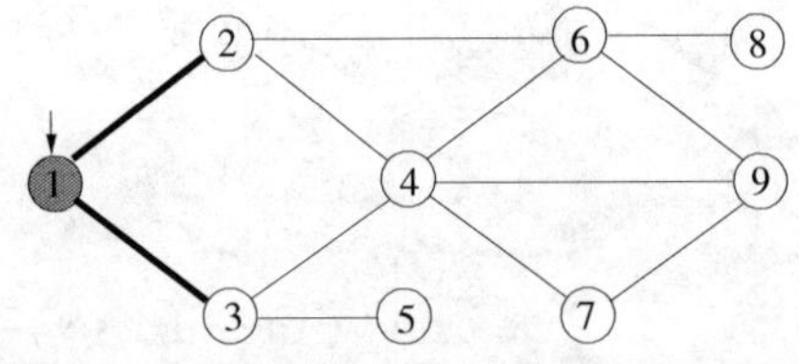

We begin at vertex 1 and explore the edges to vertex 2 and vertex 3. We add them to our list of discovered vertices.

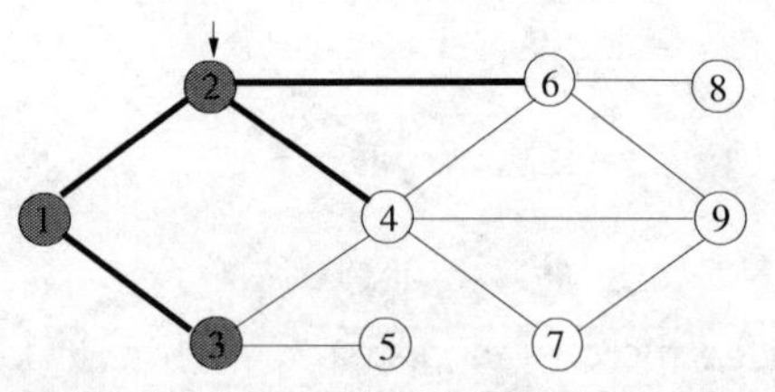

We visit vertex 2 and explore the edges to vertex 4 and vertex 6. Our current list of discovered-but-not-yet-visited vertices is 3, 4, 6.

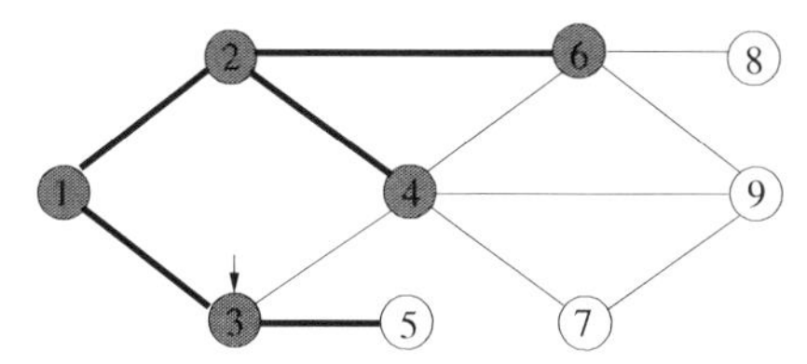

We visit vertex 3 and explore the edge to vertex 5. Our current list is 4, 6, 5.

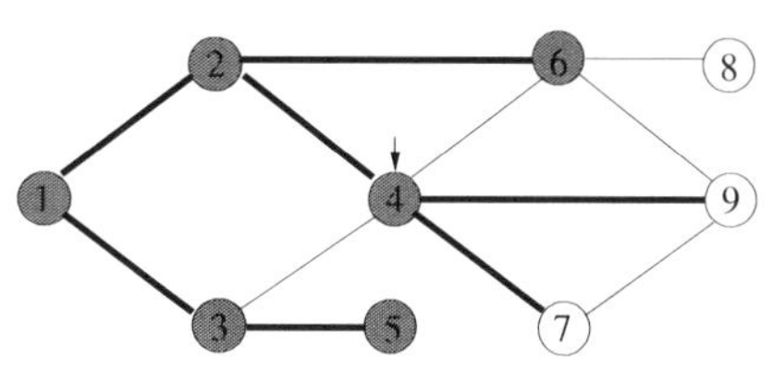

We visit vertex 4 and explore the edges to vertices 9 and 7. Our current list is 6, 5, 7, 9.

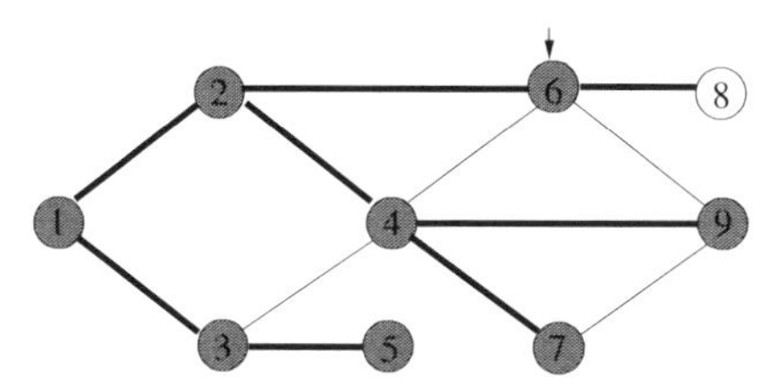

From vertex 6 we explore the edge to vertex 8. Our current list is 5, 7, 9, 8.

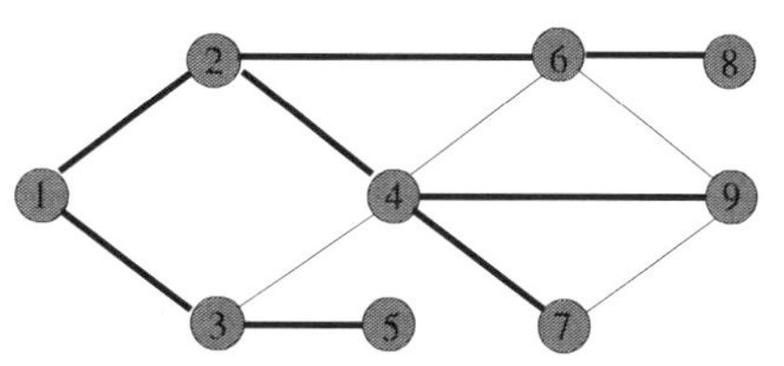

None of the vertices 5, 7, 9, or 8 have edges to undiscovered vertices, so the traversal is finished.

Our task is to implement these strategies in ML. First, what do they have in common? In either case, our goal is to find paths from the source to all the other vertices, and our output, then, is the set of edges on these paths. In fact, the set of edges we return defines a subgraph, and that subgraph happens to be a tree. Second, both strategies use the concepts of exploring an edge, discovering a vertex, and visiting a vertex, and this implies that in both algorithms vertices exist in three states: undiscovered, discovered but unvisited, and visited. Both strategies also visit vertices in the same order as they are discovered.

Looking at their similarities should clarify their differences. In depth-first search, when we discover a vertex, we visit it immediately—the other edges to be explored from the current vertex can wait. In breadth-first search, we add all the newly discovered adjacent vertices to a list to be visited later.

Depth-first search is naturally recursive, which makes it easier for us. We can sketch the algorithm as

To visit a vertex v (given also a graph and the set of discovered vertices so far
    and the set of explored edges so far):
  For each u in the adjacency list of v
    If u is undiscovered,
      Add (v, u) to the set of edges explored
      Mark u as "discovered"
      Visit u, getting back a new set of discovered vertices
        and the edges explored
      Union the returned set of edges to the set of edges explored here
  Return the set of discovered vertices and the set of edges explored

From this we will construct a helper function `dfsVisit` which will be called by our function `dfs`, to be written later. Notice that `dfsVisit` takes four things, all represented in the diagrams earlier: A representation of the entire graph, the vertex we are visiting (indicated in the diagrams by the small arrow), the set of discovered vertices (indicated by shading), and the set of explored edges (indicated by thickened edges). It will return an updated state of the computation: A new set of discovered vertices and a new set of explored edges.

We can figure from this that the type of `dfsVisit` will be

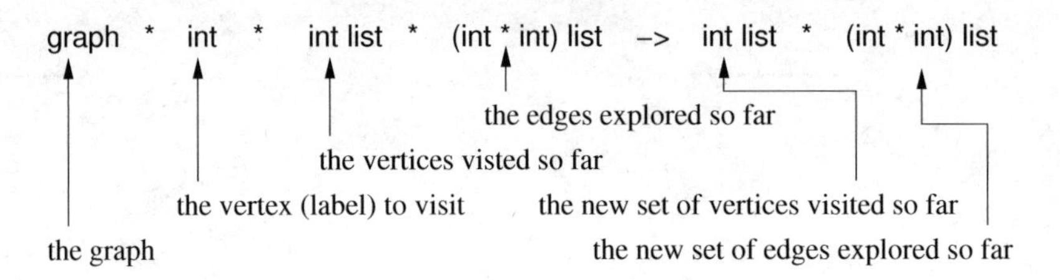

Keep in mind that `dfsVisit` is being called on a vertex. In the body of that function we need to consider each vertex adjacent to the vertex being visited—we will consider that a *step*. Each step may involve a recursive call to `dfsVisit`, if that vertex has not been discovered yet; it will also result in a new list of discovered vertices and explored edges. We will define `step` as a helper function within `dfsVisit`. Assume `v` is the vertex being visited. Then `step` takes current adjacent vertex `u` and current sets of discovered vertices `found2` and explored edges `dfsEdges2`:

```
- fun step(u, (found2, dfsEdges2)) =
=       if not (contains(found2, u))
=       then dfsVisit(g, u, u::found2, (v, u)::dfsEdges2)
=       else (found2, dfsEdges2);
```

We want to apply this to each vertex `u` in the adjacency list of `v`. We can find that adjacency list by `getAdjacent(g, v)` where `g` is the graph. Finally, we can use `foldl` to apply this function to each vertex in the list, accumulating the discovered vertices and explored edges as we go.

```
- fun dfsVisit(g, v, found, dfsEdges) =
=    let val adj = getAdjacent(g, v);
=        fun step(u, (found2, dfsEdges2)) =
=              if not (contains(found2, u))
=              then dfsVisit(g, u, u::found2,
=                                   (v, u)::dfsEdges2)
=              else (found2, dfsEdges2);
=    in
=       foldl(step)(found, dfsEdges)(adj)
=    end;
```

```
val dfsVisit =
    fn : graph * int * int list * (int * int) list ->
           int list * (int * int) list
```

The function dfs takes a graph and a starting point and returns a list of edges in the traversal tree. It is built simply from dfsVisit, starting the recursive calls at the given vertex and throwing away the final set of discovered vertices (which should, in the end, be all the vertices reachable from the given one). We test this on the graph diagramed earlier.

```
- fun dfs(g, s) =
=     #2(dfsVisit(g, s, [s], []));

val dfs = fn : graph * int -> (int * int) list

- val g = G([V(1, [2,3]), V(2, [1,4,6]), V(3,[1,4,5]),
=           V(4,[2,3,6,7]), V(5, [3]), V(6, [2,4,8,9]),
=           V(7,[4,9]), V(8,[6]), V(9,[4,6,7])]);

val g =  G[...] : graph

- dfs(g, 1);

val it = [(9,7),(6,9),(6,8),(4,6),(3,5),(4,3),(2,4),(1,2)]
    : (int * int) list
```

Breadth-first search is going to take more work. The difficulty—or just the new idea—is the list of discovered-but-not-yet-visited vertices. This fits into a family of algorithms that use a *worklist*, a data structure for recording information that needs to be processed and that is discovered during the processing of other information. The general outline of these algorithms is

worklist

Initialize the worklist.
As long as the worklist is not empty,
  Take the next item out of the worklist.
  Process it.
  If we discovered new items during the processing,
    Add them to the worklist.

Adapting this to breadth-first search,

To find paths to all other reachable vertices from a vertex s (given a graph);
  Make a worklist containing only s.
  As long as the worklist is not empty,
    Remove the next vertex v from the worklist.
    Visit v.
      (This might modify the worklist.)

Of course, we need to define how to visit a vertex.

To visit a vertex v (given a graph, the set of discovered vertices,
the set of explored edges, and the worklist),
For each u in the adjacency list of v,
If u is undiscovered,
Add (v, u) to the set of edges explored.
Mark u as discovered.
Add u to the worklist.

This prompts a helper function `bfsVisit` that takes a graph, an int vertex label, an int list set of discovered vertices, an (int * int) list set of explored edges, and an int list worklist. Programming this algorithm in ML is your task.

Exercises

8.6.1 The example in this section is an undirected graph. Explain why both depth-first search in general and our implementation would work also on directed graphs.

8.6.2 Finish the function `bfsVisit`. Like `dfsVisit`, it has a helper function `step` and uses `foldl` to apply it to every vertex in the adjacency list.

```
fun bfsVisit(g, v, found, bfsEdges,
                workList) =
   let val adj = getAdjacent(g, v);
       fun step(u, (found2, bfsEdges2,
                       workList2)) =
            if ??
            then ??
            else ??;
    in
       foldl(step)
         (found, bfsEdges, workList)(adj)
    end;
```

8.6.3 Finish the function `bfsLoop`, which repeatedly takes items from the worklist until it is empty. Notice the parameters `found`, `bfsEdges`, and `v::rest` are bound together as a tuple. This makes the interaction between `bfsLoop` and `bfsVisit` convenient. Pay careful attention to the types (parameter and return) of the two functions.

```
fun bfsLoop(g, (found, bfsEdges, [])) =
        bfsEdges
  | bfsLoop(g, (found, bfsEdges,
                     v::rest)) =
        bfsLoop(??, ??);
```

8.6.4 Finish the function `bfs`, which takes a graph `g` and a starting vertex `s`. If you have the other functions written correctly, then this is straightforward.

```
fun bfs(g, s) = bfsLoop(??, ??);
```

8.6.5 Will breadth-first search work on an undirected graph? Why or why not?

8.7 Extended example: Graph algorithms

The algorithms in the previous section found paths in graphs from a given vertex. No attempt was made, however, to discover paths that are shorter than all other paths between the vertices.

The problem of finding shortest paths is more interesting when considered on weighted graphs, where each edge has a weight or distance associated with it. The *shortest* path between vertices in a weighted graph is not necessarily the path with the fewest number of edges but the path having the smallest sum of the weights of its edges. Consider the graph on the right, with edges labeled with their weights.

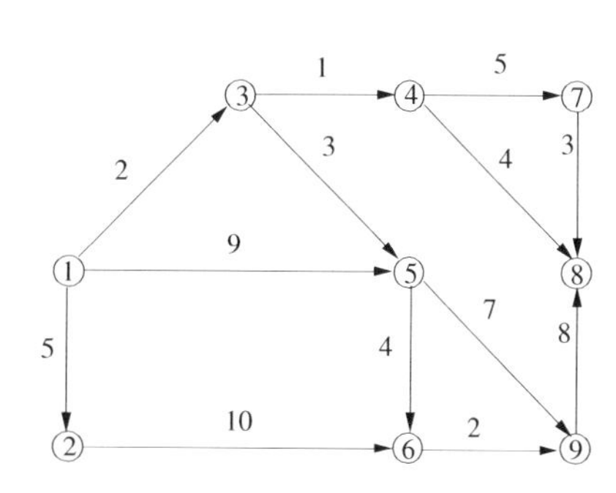

The path from vertex 1 to vertex 9 with the fewest edges is $v_1 v_5 v_9$. However, that path has total weight 16. The path $v_1 v_3 v_5 v_6 v_9$, even though it has twice as many edges, has total weight 11.

The **single-source shortest paths** problem assumes a weighted graph and a specified vertex called the *source* and asks, what are the shortest paths from that source to all other reachable vertices?

source

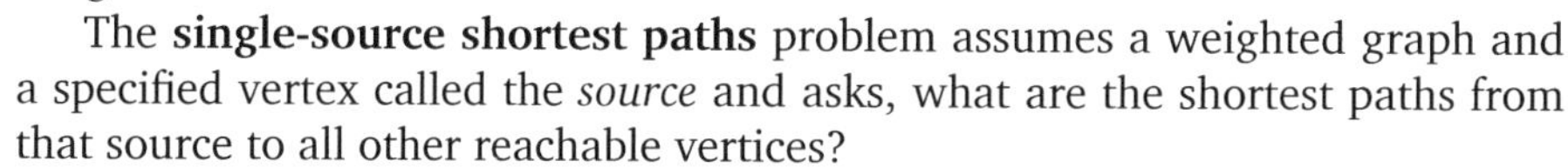

We will suppose that every edge has a positive weight and modify our graph representation from the previous section to store the weights for the edges. Specifically, we will represent the vertices with an int label and a list of int * int pairs. The first int in each pair is for the label of an adjacent vertex, and the second int is for the weight of the edge to that vertex. Assume in our examples that the graph is *directed*. In ML,

```
datatype vertex = V of int * (int * int) list;

datatype graph = G of vertex list;
```

For example, here is the ML representation of the graph above.

```
- val g = G([V(1, [(2, 5), (3, 2), (5, 9)]), V(2, [(6, 10)]),
=           V(3, [(4, 1), (5, 3)]), V(4, [(7, 5), (8, 4)]),
=           V(5, [(6, 4), (9, 7)]), V(6, [(9, 2)]),
=           V(7, [(8, 3)]), V(8, []), V(9, [(8,8)])]);


val g =  G [...] : graph
```

The algorithm will compute a set of edges that constitute the shortest paths. As in depth-first search and breadth-first search, the edges will form a tree. The total weight of the edges in the path from the source s to any vertex v is considered the *distance* from s to v.

distance

Here is how the algorithm works. In the end we want to find for each vertex v the distance from s to v. As we go along, each vertex will have a provisional, computed distance, a sort of best-distance-found-so far. This will always be an upper bound—in fact, we will initially give each vertex a computed distance of ∞ (except the source; we know right from the start that it has distance 0.) As the algorithm proceeds, we improve the computed distances until they converge on the true distance.

In addition to the computed distances, the algorithm will keep track of the last edge taken to each vertex on the shortest path discovered so far. Suppose v is a vertex and p is the shortest path yet discovered from s to v. Suppose u is the last vertex before v on path p. Then we keep track of edge (u, v) for vertex v—or, more precisely, we will keep track of vertex u as the *parent* of vertex v in the tree that the algorithm is building.

In our implementation, all this information will be stored in a list of `int` triples. Each element in the list is a record of what we know so far about a vertex; the three values in the triple are the label of the vertex, the computed distance, and the computed parent. Glancing at the graph we notice the edge (1, 5) with weight 9. From that we may guess that vertex 5 has computed distance 9 and parent 1, which would be the entry (5, 9, 1).

The list is sorted by computed distance from smallest to largest. As the algorithm starts, all computed distances are ∞ and all parents are unknown (except the source, which has distance 0 and no parent at all, which we symbolize with $\bot$). Our initial list for the graph above (picking vertex 1 as the source) is `[(1, 0, ⊥), (2, ∞, ?),(3, ∞, ?), (4, ∞, ?), (5, ∞, ?), (6, ∞, ?), (7, ∞, ?), (8, ∞, ?), (9, ∞, ?)]` This list is similar to the worklist from breadth-first search, although we will not be adding to it as we go along, only removing and rearranging it.

A high-level view of the algorithm is

Make a list of vertices together with computed distances and provisional parents
As long as the list is not empty,
  Remove the next vertex v with computed distance d and parent u.
  Mark d as the true distance and u as the true parent of v.
  For each outgoing edge (v, x),
    If edge (v, x) makes for a shorter path to x than previously known,
      Improve the computed distance and update the recorded parent of x.
  Rearrange the list to maintain it as sorted.

How can we mark the computed distance and parent of v as the true ones? Because by the time v is at the front of the list (that is, it has the shortest computed distance in the list), then the estimate must be right. More on that later. Notice that the core of this algorithm is the improving of the computed distances of all the vertices to which v has an outgoing edge. This is called *relaxing* the edges.

relaxing

Watch the algorithm as it processes this graph. Vertices are marked with (finite) computed distances in squares; when a vertex's true distance has been discovered, it is marked in a hexagon. Likewise, edges shown as dashed lines are the final edges in best-paths-discovered-so-far, but edges known to be in the shortest paths tree are thickened.

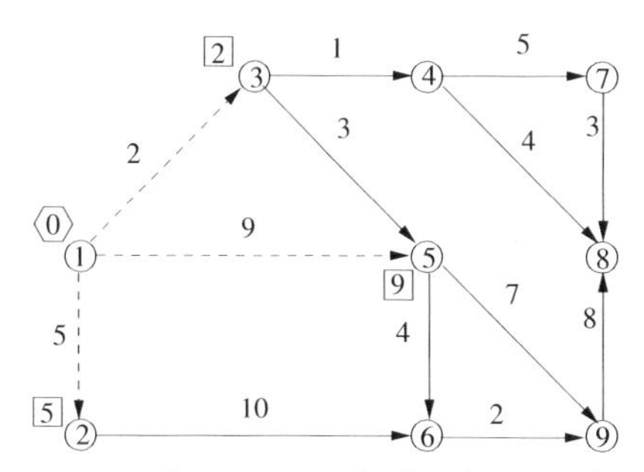

First we process the source, relaxing its outgoing edges. Vertices 2, 3, and 5 now have computed distances and the source as their provisional parent. New list: `[(3, 2, 1), (2, 5, 1), (5, 9, 1), (4, ∞, ?), (6, ∞, ?), (7, ∞, ?), (8, ∞, ?), (9, ∞, ?)]`

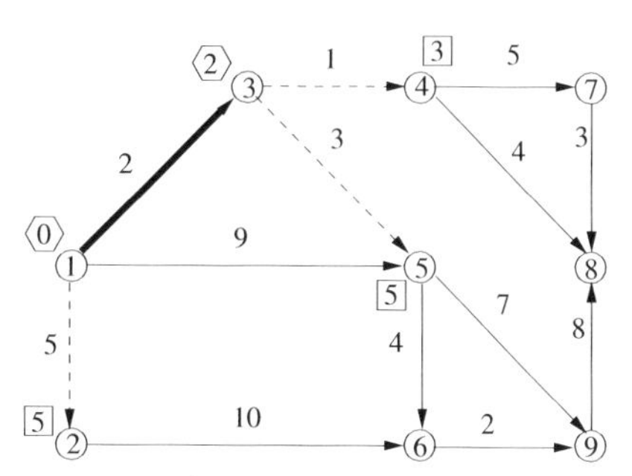

We remove the first entry in the list, vertex 3. This means its provisional distance, 2, is correct. Relax its outgoing edges to improve computed distances for vertex 4 and vertex 5. New list: `[(4, 3, 3), (2, 5, 1), (5, 5, 3), (6, ∞, ?), (7, ∞, ?), (8, ∞, ?), (9, ∞, ?)]`

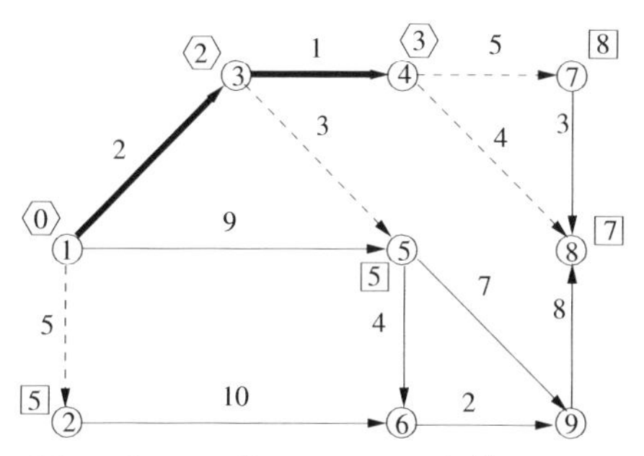

Vertex 4 has the smallest computed distance, so relax its edges, finding computed distances for vertex 7 and vertex 8. `[(2, 5, 1), (5, 5, 3), (8, 7, 4), (7, 8, 4), (6, ∞, ?), (9, ∞, ?)]`

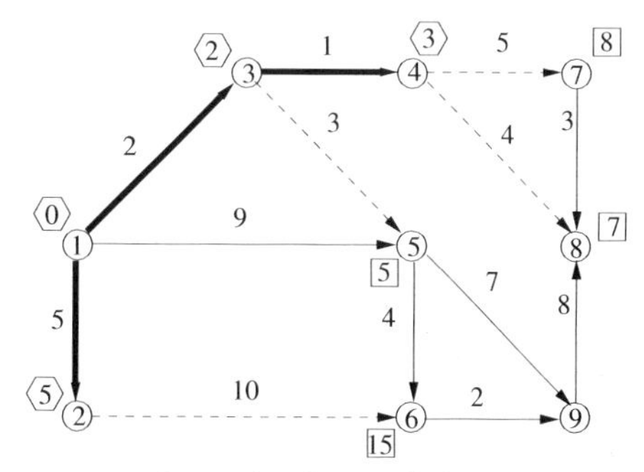

Since vertex 2 is at the front of the list, we confirm its computed distance as correct. Relax its edge to vertex 6. `[(5, 5, 3), (8, 7, 4), (7, 8, 4), (6, 15, 2), (9, ∞, ?)]`

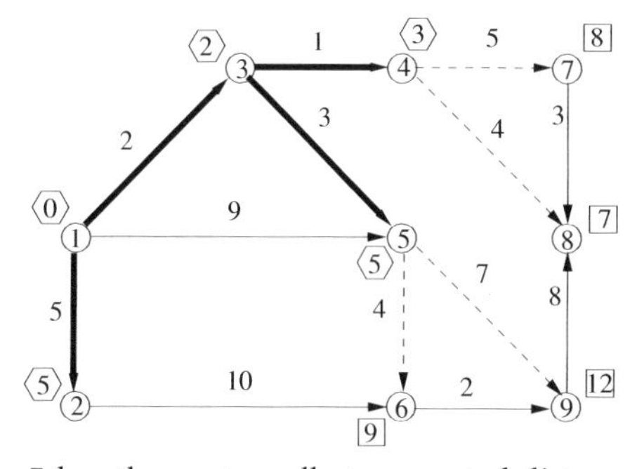

Vertex 5 has the next smallest computed distance, vertex 3 is its parent. Relax edges to vertex 6 and vertex 9. `[(8, 7, 4), (7, 8, 4), (6, 9, 5), (9, 12, 5)]`

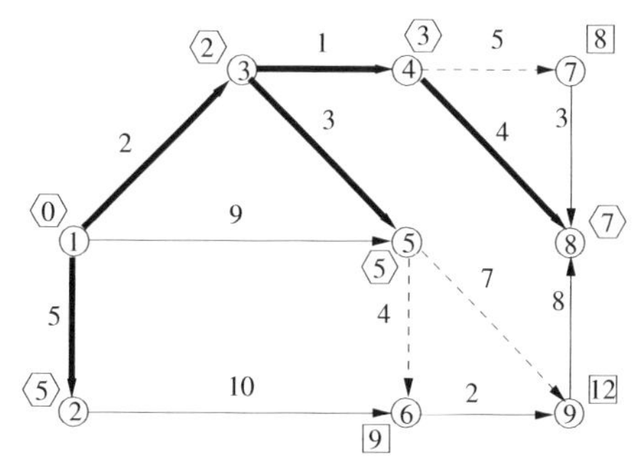

We now know the true distance to vertex 8, but there are no outgoing edges to relax. `[(7, 8, 4), (6, 9, 5), (9, 12, 5)]`

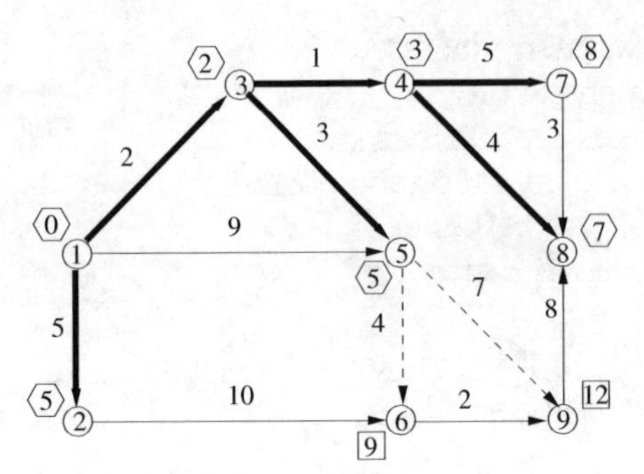

Vertex 7 is found to have distance 8, parent vertex 4. `[(6, 9, 5), (9, 12, 5)]`

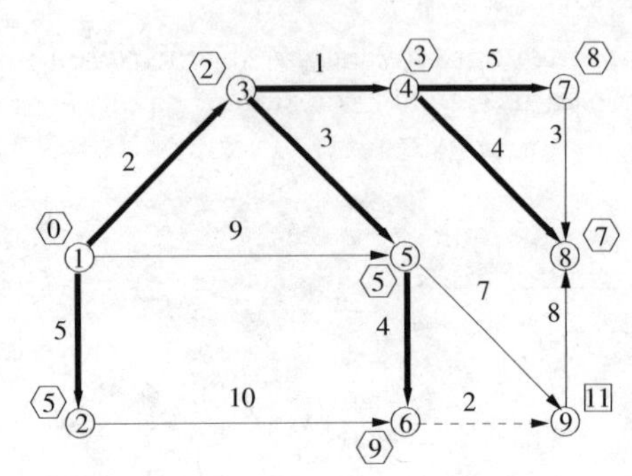

Relaxing edge (6, 9) improves the computed distance of vertex 9. `[(9, 12, 5)]`

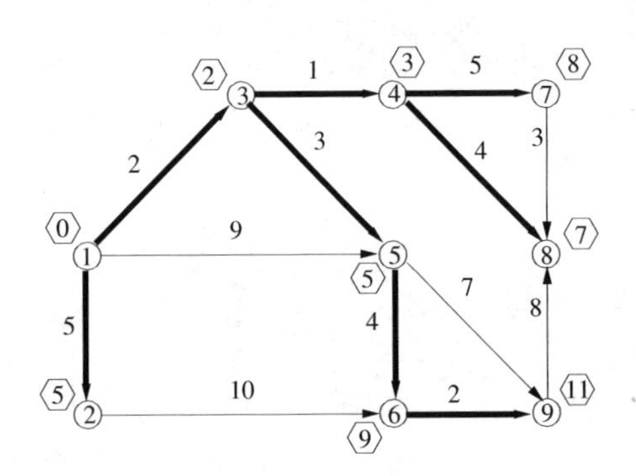

The graph with the final set of edges is shown on the left. The process we used is called *Dijkstra's Algorithm* after Edsger Dijkstra.

Dijkstra's Algorithm

optimal substructure

The correctness of Dijkstra's Algorithm is a standard topic in advanced algorithm design. We only sketch it here. How do we know that the paths in the resulting tree are truly the shortest? Take note of several assumptions hidden in our process—they are intuitively believable, but they would need to be shown to establish correctness formally. First, the shortest path to a vertex v (with u being the next-to-last vertex on the path) must have the shortest path to u as a subpath. This is called *optimal substructure*. For example, the shortest path to vertex 6 is $v_1v_3v_5v_6$, which includes the shortest path to vertex 5, v_1, v_3v_5.

Second, each time we remove the entry for a vertex v from the list, we assume that the computed distance must be the true distance, that no other shortest path will (or can) be found. We would prove this inductively, noticing that if this assumption held for the vertices already removed from the list, then

Biography: Edsger Dijkstra, 1930–2002

Edsger Dijkstra was born in Rotterdam, the Netherlands, and spent much of his career at the University of Texas at Austin. He made contributions to many areas of computer science, but his main interest was formal and mathematical methods for reasoning about computer programs. For Dijkstra, programming meant not only writing a program but proving its correctness—without a correctness proof, a program is just a conjecture. He proposed that this view shape even freshman programming courses, where students would write programs in a language with no implementation "so that students are protected from the temptation to test their programs." While admitting that his way of "really teaching computing science" was "cruel," he also claimed a student would "very soon make the exciting discovery that he is beginning to master the use of a tool that, in all its simplicity, gives him a power that far surpasses his wildest dreams." [8]

the parent of vertex v in the shortest path must already have been processed and its edges already relaxed. For a complete proof, see Cormen et al [5].

Project items 8.7.A–8.7.D walk you through an implementation of Dijkstra's algorithm to solve the single-source shortest paths problem. The output on the given graph would be a set of edges (the label `~1` stands in for the nonexistent parent of the root; hence the edge `(~1, 1)` indicates that 1 is the root or source).

```
- sssp(g, 1);

val it =
   [(~1,1),(1,3),(3,4),(1,2),(3,5),(4,8),(4,7),(5,6),(6,9)]
      : (int * int) list
```

The **minimum spanning tree** problem is related to single-source shortest paths. As before, we will be building a tree, but not necessarily one that contains the paths that are shortest from the source to each vertex. Instead we build a tree that connects ("spans") the entire graph but whose total weight (the sum of the weights of all of the tree's edges) is minimal. This time we will assume the graph is undirected, as in this example:

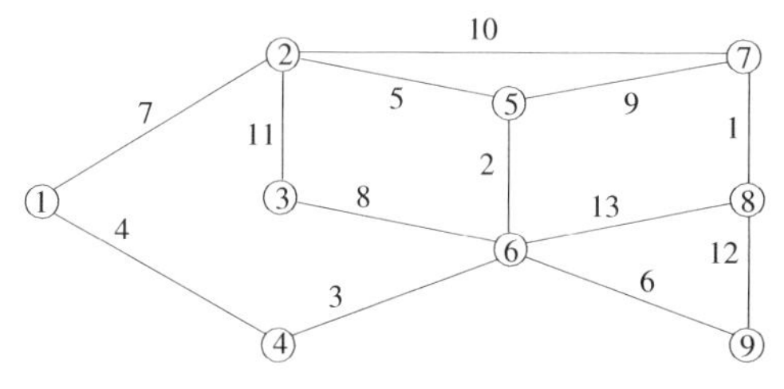

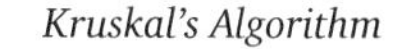
Kruskal's Algorithm

The desired result will be clear as we watch an algorithm for it unfold. In *Kruskal's Algorithm* (after Joseph Kruskal) like Dijkstra's Algorithm, we maintain a set of edges for the tree we are building, adding an edge at every step. In Kruskal's Algorithm, however, the set of edges does not always make a single (connected) tree as we go along. Every time we add an edge we connect two trees to make a bigger tree. Initially every vertex is its own, trivial tree. At the end, there is one tree spanning the entire graph—if that is possible. Also unlike Dijkstra's Algorithm, Kruskal's iterates over edges, not vertices—all edges in the graph, in fact, sorted from smallest to greatest weight.

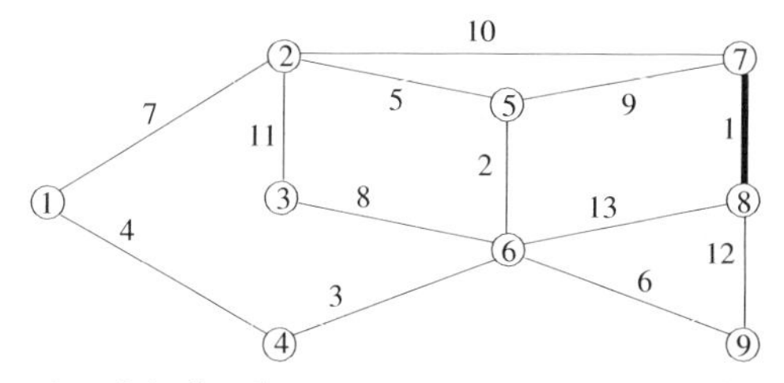

Edge (7, 8) is the shortest, so unite vertex 7 and vertex 8 into a tree.

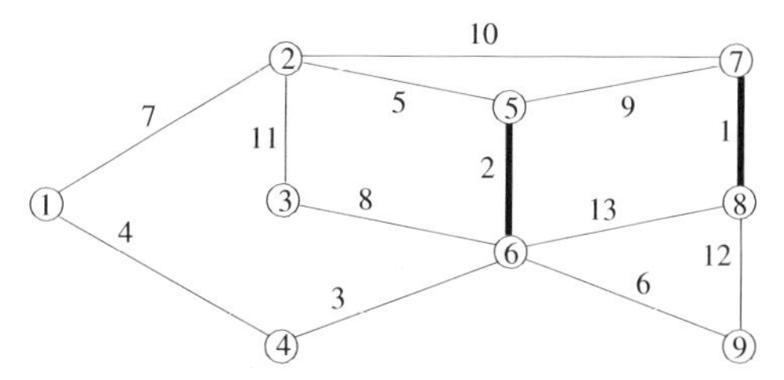

Edge (5, 6) is the next shortest, so similarly unite vertex 5 and vertex 6.

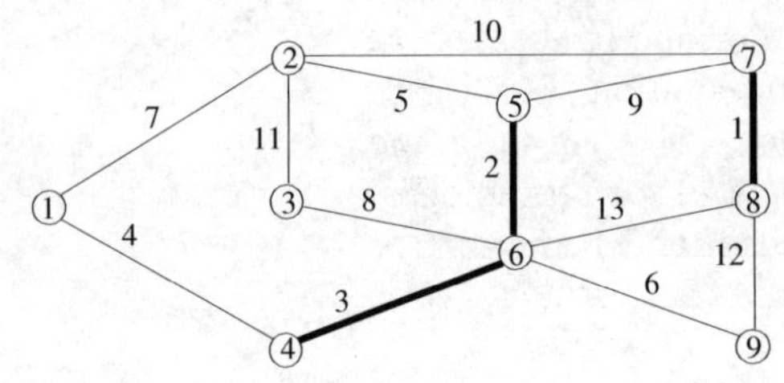

Edge (4, 6) is next.

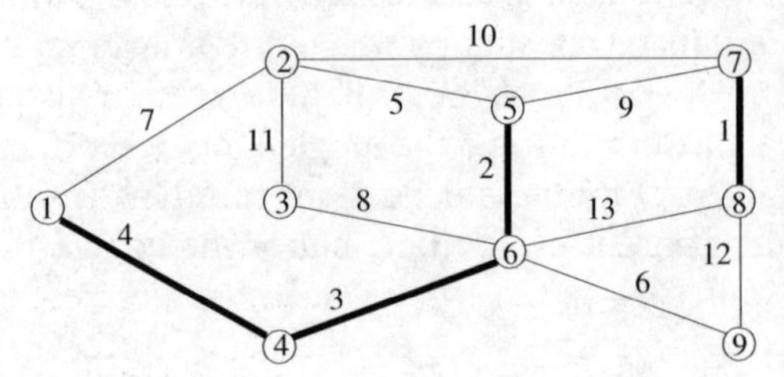

Then add edge (3, 4), as the 5–6–4 tree continu
to grow.

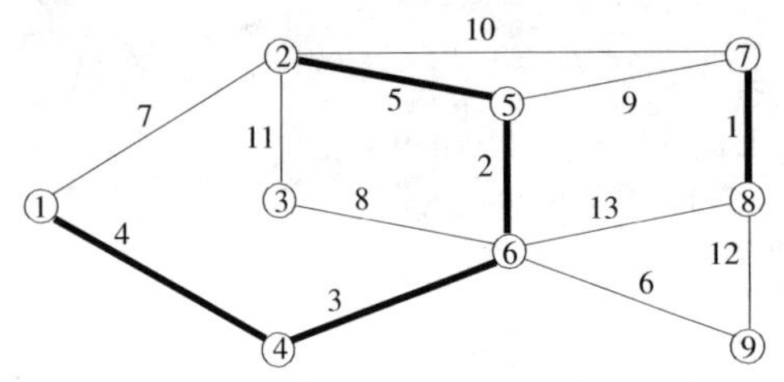

Edge (2, 5) unites the tree containing only vertex 2 to the big tree.

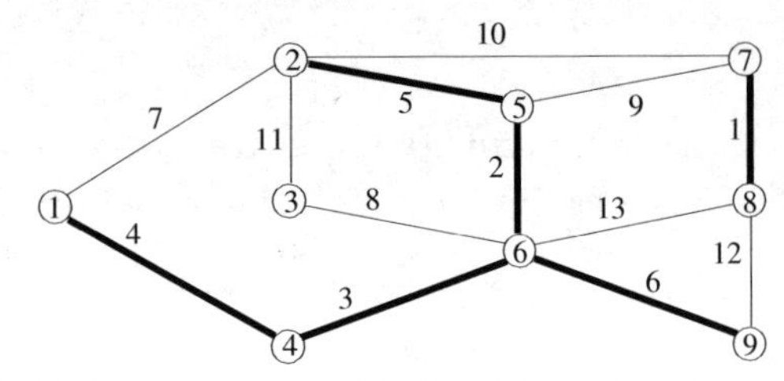

Similarly, vertex 9 is added on through edge (6,

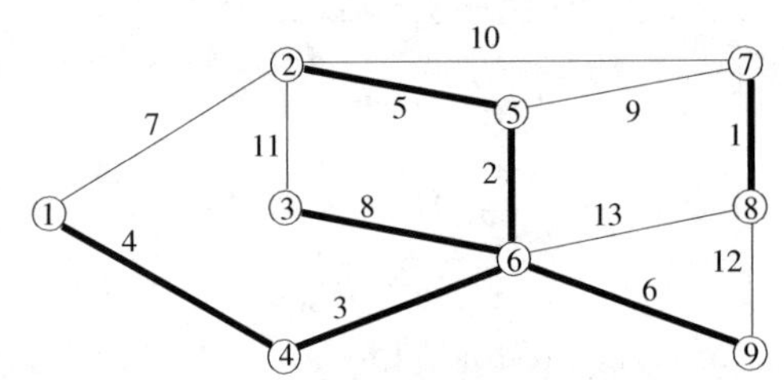

Edge (1, 2) is the next smallest, but it does not connect two distinct trees. Skip it. Instead unite vertex 3 to the big tree through edge (3, 6).

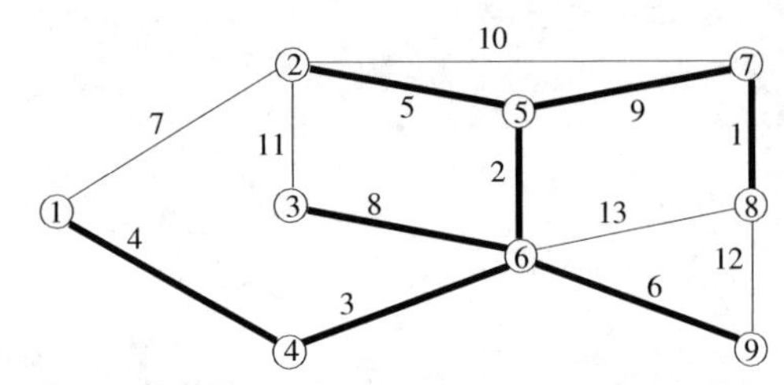

Finally we unite the two remaining trees throu
vertex (5, 7).

Notice that the individual paths through the tree are not necessarily the shortest between their respective initial and terminal vertices. For example, the path $v_1v_4v_6v_5v_7$ has total distance 18, but you can get from v_1 to v_7 with distance 17 in path $v_1v_2v_7$.

Why does this approach work? Suppose that at some point in the computation we have a set of trees such that each vertex is a part of exactly one tree. Suppose further that all the trees happen to be subgraphs of some minimum spanning tree T (notice we have not yet found T, but whatever such tree there is, what we are saying must be true for it). Now suppose edge (u, v) is the shortest edge connecting two distinct trees in our set. Let U be the tree containing vertex u, and similarly V. There must be an edge (x, y) in T that connects its subtrees U and V.

Since (u, v) is the shortest, then (x, y) cannot be shorter than (u, v)—it must be longer or equal. Either way, we could take tree T and break it apart by removing (x, y) and form a new tree, T' by reuniting the parts using edge (u, v). T' must have equal or smaller total weight than T, so T' is a minimum spanning tree. This means the tree that we get by connecting U and V using (u, v) is also a subtree of a minimum spanning tree.

For our implementation, consider the information that needs to be modeled and stored. First there is the work that still needs to be done, which is the list of edges, sorted from smallest to greatest weight. Second there is the portion of the solution we have built so far, that is, the trees.

We can model the work needing to be done by a simple list. Each item in the list needs to represent an edge, including its weight. We will use an ((int * int) * int) list, the first pair being the endpoints of the edge, the last int being the weight. In our example, `[((7,8),1),((5,6),2),((4,6),3),((1,4),4), ...]`.

For the solution-so-far, observe that not only will it be inconvenient to store the trees themselves (should they be graphs by themselves, or should we pick a root?), we also never need to treat them as trees. Rather we are interested in two aspects: The set of edges, since that will be our final answer, and the sets of vertices in each tree. In the second-to-last step, the set of sets of vertices in the same tree is originally `[[3], [5, 6, 4, 1, 2, 9], [7, 8]]`, but by taking edge (3, 6) we get `[[5, 6, 4, 1, 2, 9, 3], [7, 8]]`.

Most of the work in our implementation will be done by a helper function `kruskalBody` that takes the sorted list of edges, the set of connected-vertex sets, and the set of tree edges. If there are no graph edges left to process and there is only one set of connected vertices, then the set of tree edges is our final answer. If there are no graph edges left but there are zero or more than one sets of connected vertices, then we are in trouble. Otherwise, process the next smallest edge in the graph by asking, does it connect two distinct trees?

To answer that question, you will need to find the sets of the respective end points of the edge currently being processed. The function `getSet` takes an item and a list of lists and returns the list containing the given item.

```
- fun getSet(a, fstSet::rest) =
=       if contains(fstSet, a)
=       then fstSet else getSet(a, rest);

Warning: match nonexhaustive
val getSet = fn : ''a * ''a list list -> ''a list
```

We will need to grab all the edges in the graph and sort them by weight. Since the graph is undirected, each edge is represented twice, and we do not want duplicates in the list. Moreover, we want to grab all the vertices in a graph so we can make our initial set of connected-tree sets.

```
- fun compareEdges(((a, b), w1), ((c, d), w2)) = w1 < w2;
```

```
val compareEdges =
      fn : (('a * 'b) * int) * (('c * 'd) * int) -> bool

- fun getEdges(G([]), edges) = edges
=   | getEdges(G(V(v, []):: rest), edges) =
=          getEdges(G(rest), edges)
=   | getEdges(G(V(v, (u, w)::restA)::restB), edges) =
=          getEdges(G(V(v, restA)::restB),
=                    if contains(edges, ((u, v), w))
=                    then edges
=                    else insertSorted(((v, u), w),
=                                        compareEdges, edges));

val getEdges = fn
  : graph * ((int * int) * int) list ->
      ((int * int) * int) list

- fun getVertices(G[]) = []
=   | getVertices(G(V(a, uu)::rest)) =
=         a::getVertices(G(rest));

val getVertices = fn : graph -> int list
```

In Project 8.E–8.F, you will complete the implementation.

```
- val g2 =
=   G([V(1,[(2,7), (4,4)]),
=       V(2,[(1,7), (3,11), (5,5), (7,10)]),
=       V(3,[(2,11), (6,8)]), V(4,[(1,4), (6,3)]),
=       V(5,[(2,5), (6,2), (7,9)]),
=       V(6,[(3,8), (4,3), (5,2), (8,13), (9,6)]),
=       V(7,[(2,10), (5,9), (8,1)]),
=       V(8,[(6,13), (7,1), (9,12)]),
=       V(9,[(6,6), (8,12)])]);

val g2 = G [ ...]  : graph

- mst(g2);

val it = [(5,7),(3,6),(6,9),(2,5),(1,4),(4,6),(5,6),(7,8)]
    : (int * int) list
```

Project

8.A Write a general function `insertSorted` that takes a value, a comparison function, and a list (assumed sorted), and returns a new list like the one given except with the given value inserted into it. This will be similar to your function `putInPlace` from Exercise 6.8.15, but generalized for any type and sort order. The given comparison function, say `comp`, is used to determine what order the values in the list should be, with `comp(a, b)` returning `true` if `a` should come first. For example, `insertSorted(5, fn(a, b) => a > b, [8, 6, 3, 1])` would return `[8, 6, 5, 3, 1]`.

```
fun insertSorted(x, comp, []) = [x]
  | insertSorted(x, comp, a::rest) =  ??
```

8.B Write a function `makeInitialList` that takes a graph and a source and returns a list of triples: vertex (label), computed distance, and assumed parent vertex. The list should contain an entry for each vertex: for the source, the computed distance is 0 and parent vertex is a special value `~1` meaning no parent. For other vertices, the computed distance is some large number (say 1000) and the parent vertex is also the special value `~1`, this time meaning "unknown."

```
fun makeInitialList(g as G(vertices), s) =
  ??
```

8.C Write a function `dijkstraStep` that takes a vertex `v`, the computed (actually, *true*) distance for that vertex `dv`, the graph `g`, and the maintained sorted list of vertices `vv`; the one step that it does is to relax all the edges from `v` to any adjacent vertex in the maintained list. (Recall that the character `_` is a special identifier. It can be used many times in the same pattern to indicate a parameter we will ignore.)

```
fun dijkstraStep(v, dv, g, vv) =
    let val adj = getAdjacent(g, v);
        fun compare((a, x, _), (b, y, _)) =
                  x < y;
       (* u is the current adjacent vertex,
          du is u's old computed distance,
          p is u's old parent,
          and rest is the rest of the
          vertices in the maintained list *)
        fun relax((u, du, p), rest) = ??
    in
        foldr(relax)([])(vv)
    end;
```

8.D Write a function `sssp` that takes a graph and a source vertex and returns a list of edges in the shortest-path tree from that source to all reachable vertices, using Dijkstra's algorithm. Specifically, the helper function `dijkstraBody` processes the maintained list until it is empty.

```
fun sssp(g, s) =
  let val q = makeInitialList(g, s);
      fun dijkstraBody([]) = []
        | dijkstraBody((v, d, p)::rest) =
            ??
  in
      dijkstraBody(q)
  end;
```

8.E Finish the function `kruskalBody`. If the graph is not connected, we raise an exception. The only hard part is the recursive call in the then clause.

```
exception UnconnectedGraph;

fun kruskalBody([], [set], treeEdges) =
      ??
  | kruskalBody([], sets, treeEdges) =
      raise UnconnectedGraph
  | kruskalBody(((u, v), w)::rest, sets,
                treeEdges) =
      let val uSet = ??;
          val vSet = ??;
      in
          if uSet <> vSet
          then kruskalBody(rest, ??, ??)
          else kruskalBody(rest, sets,
                           treeEdges)
      end;
```

Project, continued

8.F Finish the function `mst`. This function is fairly straightforward, requiring you only to set up the initial conditions. Take note, however, that `getVertices` creates a flat list of all the vertices in the graph, whereas what you want is a list of lists, with each vertex in its own list. (Hint: Use your function `listify` from Exercise 2.2.10.)

```
fun mst(g) =
    let val allEdges = ??;
        val sets = ??;
    in
        kruskalBody(??, ??, ??)
    end;
```

8.G There is another algorithm used to compute minimum spanning trees, one that is almost identical to Dijkstra's Algorithm. Like Dijkstra's Algorithm, it maintains a single tree as a set of edges that grows as the algorithm progresses (in the end this tree is a minimum spanning tree, not a shortest-paths tree.) The difference from Dijkstra's algorithm is that instead of keeping track of a shortest known distance from the source for each edge, we keep track of the shortest known single-edge connection to the tree for each edge. This algorithm was discovered first by Vojtěch Jarník, but it often is called *Prim's Algorithm* after Robert Prim who independently rediscovered it.

Take the code you wrote for Dijkstra's algorithm (`dijkstraStep` and `sssp`, including the helper function `dijkstraBody`), rename them appropriately (`primStep`, `primBody`, `mst2`) and make one other small change to implement Prim's Algorithm for the minimum spanning tree.

8.8 Special topic: Graph coloring

In grade school you may have had an exercise for learning geography where you were given a map of the entities you were to learn—states or provinces of a country or countries of a continent—and instructed to color the entities. To make the distinctions between the entities clear, one naturally would not give entities with a common border the same color.

What is the smallest number of colors needed for such a coloring? This is a problem that can be studied with the tools of graph theory. Represent each geographic entity with a vertex and draw an edge between vertices for entities that share a boundary. Here it is done for the provinces of Australia.

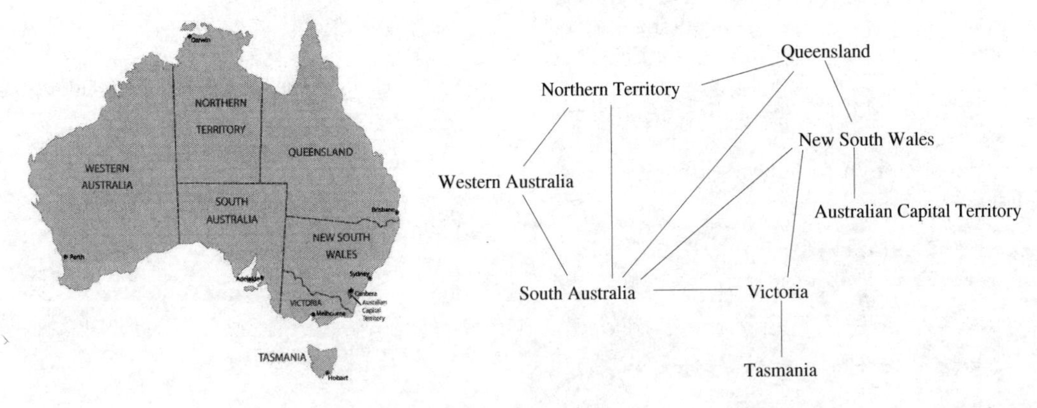

The Australian Capital Territory is an enclave within New South Wales, so its vertex has a degree of 1. Tasmania does not literally share a border with

Victoria, but its proximity means it is unlikely the two should have the same color.

Now we can define the problem formally. By "coloring" the graph what we really mean is a way to assign values (colors) to each vertex in the graph so that adjacent vertices do not have the same color. An assignment is really a function, and any set can represent the colors. Hence a *coloring* of a simple, undirected graph $G = (V, E)$ is a set C and function $f : V \to C$ such that for all $(u, v) \in E$, $f(u) \neq f(v)$.

coloring

The question of finding the minimum number of colors then is about the cardinality of the set C: What is the least $n \in \mathbb{N}$ such that there exists a coloring C, f where $|C| = n$. This is called the *chromatic number* of the graph.

chromatic number

Cartography is not the only context in which this question is important. Graph coloring can also model a variety of scheduling problems. Suppose the vertices stand for events and edges show which events must not happen at the same time, because they require the same people, location, or some other resource. The problem of finding a minimal coloring for a graph can now be interpreted as that of finding the fewest number of time slots needed to schedule every event. Alternately, suppose that an edge (u, v) rather represents the fact that events u and v might overlap in time and therefore should not be scheduled in the same location. The colors then represent room assignments for the events, as events whose durations do not overlap can be given the same room.

The problem is particularly important for constructing compilers, the programs that translate other programs from high-level programming languages to the machine language. One task a compiler must perform is assigning hardware registers (a small number of fast-access memory locations) to the variables used in the program. Since there are a limited number of registers, it is important for the program's performance that they be used as efficiently as possible. If we have two program variables each in use during different sections of the program that do not overlap, then we can use the same register to store both of them. The compiler constructs a graph with variables represented by vertices whose edges show which variables have overlapping "live ranges." The colors are registers.

Of course the minimal number of colors needed depends on properties of the graph. The most important of these is planarity, because most maps can be represented as planar graphs. More on that later. If a graph is planar, we can guarantee a limit on its chromatic number.

Recall Euler's Formula (Theorem 8.9), that the number of regions of a simple, planar graph $G = (V, E)$ is equal to $|E| - |V| + 2$. From this we get the following corollaries.

Corollary 8.12 *If $G = (V, E)$ is a simple, connected, planar graph with certificate $\{R_1, R_2, \ldots, R_n\}$ and $|V| \geq 3$, then $|E| \leq 3 \cdot |V| - 6$.*

Proof. G is divided into n regions. Each region has at least 3 edges on its border, and each edge occurs on the borders of two regions (or, if it is an edge to a vertex of degree 1 and juts into a region, then the edge occurs twice on that one region's border). So we have

$$\begin{aligned}
3n &\leq 2 \cdot |E| && \text{as stated above} \\
n &\leq \tfrac{2}{3} \cdot |E| && \\
|E| - |V| + 2 &\leq \tfrac{2}{3} \cdot |E| && \text{by Euler's Formula} \\
\tfrac{|E|}{3} &\leq |V| - 2 && \\
|E| &\leq 3 \cdot |V| - 6 \quad \square &&
\end{aligned}$$

Moreover, such a graph must have a vertex with a degree less than or equal to 5.

Lemma 8.13 *If $G = (V, E)$ is a simple, connected, planar graph, then there exists $v \in V$ such that $deg(v) \leq 5$.*

Proof. The result is trivial if $|V| = 1$ or $|V| = 2$. Suppose, then, that $|V| \geq 3$, there are at least 3 vertices. Suppose further that for all $v \in V$, $\deg(v) \geq 6$.

By Corollary 8.12, $|E| \leq 3 \cdot |V| - 6$, or $2 \cdot |E| \leq 6 \cdot |V| - 12$.

By Theorem 8.1 (the Handshake Theorem), $\sum_{v \in V} \deg(v_i) = 2 \cdot |E|$. By our supposition that each vertex has degree at least 6, $2 \cdot |E| \geq 6 \cdot |V|$. Contradiction.

Therefore, there is at least one vertex with degree 5 or less. □

(How do we decide whether to call something a corollary or a lemma? Remember that a theorem is a big result, interesting in its own right, whereas a lemma is a small result used to prove a theorem, and a corollary is a small result that follows from a theorem. In this case, Corollary 8.12 and Lemma 8.13 are both corollaries to Theorem 8.9 and lemmas to Theorem 8.14. We split the difference by calling the first a corollary and the second a lemma.)

Now an interesting result.

Theorem 8.14 (Six-color Theorem) *If $G = (V, E)$ is a simple, connected, planar graph, then its chromatic number is no greater than 6.*

Proof. By induction on the number of vertices.

Base case. Suppose $|V| = 1$. Then we can color that single vertex a single color, and so its chromatic number is $1 \leq 6$.

Inductive case. Suppose that for some $n \geq 1$, all simple, connected, planar graphs with n vertices have chromatic number no greater than 6. Suppose further that $|V| = n + 1$.

By Lemma 8.13, there exists a vertex $v \in V$ such that $\deg(v) \leq 5$. Construct graph H by removing v and its incident edges from G, that is, $H = (W, F)$ where $W = V - \{v\}$ and $F = \{(x, y) \mid (x, y) \in E \wedge x \neq v \wedge y \neq v\}$.

Since $|W| = |V - \{v\}| = |V| - 1 = n + 1 - 1 = n$, then by the inductive hypothesis, H has chromatic number no greater than 6, that is, there exists a set D and function $g : W \to D$ such that for all $(x, y) \in F$, $g(x) \neq g(y)$ and $|D| \leq 6$.

If D in fact has fewer than 6 colors, then add some colors to make a set C. Formally, let C be any set such that $D \subseteq C$ and $|C| = 6$.

Let $c \in C$ be some color such that for all $u \in V$ such that u is adjacent to v in G, $g(u) \neq c$. Such a c must exist because there are at most 5 vertices adjacent to v but there are 6 colors in C.

Now define $f : V \to C$ such that

$$f(x) = \begin{cases} g(x) & \text{if } x \neq v \\ c & \text{if } x = v \end{cases}$$

Set C and function f make a coloring for G because D and g make a coloring for H and v does not have the same color as any of its adjacent vertices.

Therefore the chromatic number of G is no greater than 6. □

A nifty proof, but this is far from the whole story. As it turns out,

Theorem 8.15 (Four-color Theorem) *If $G = (V, E)$ is a simple, connected, planar graph, then its chromatic number is no greater than 4.*

Do not try to find the difference between the premises of the Six-color Theorem and the Four-color Theorem, because there are none. Unlike the Six-color Theorem, however, the Four-color Theorem is extremely difficult to prove. Since a version of the problem was first proposed the 1850s, many mathematicians have worked on it, sometimes producing incorrect proofs that were accepted by the mathematical community for years before a flaw was discovered.

The Four-color Theorem was finally proved in 1976 with the help of a computer. The proof showed that any planar graph with chromatic number greater than 4 could be reduced to one of a finite set of planar graphs, which then could be checked by hand. However, there were thousands of such irreducible graphs in the set. The availability of computational power—and the expertise to write appropriate software—made such checking possible. Of course, that made a proof of the program's correctness necessary.

To see why this result about planarity means that most maps can be colored with only four colors, take a map and put a dot inside each entity for the vertices. Then for any entities that share a border you can draw the appropriate edge by starting at one dot, moving the pencil to the shared border, and then to the other dot. This can be done without any of the edges crossing, hence we have a planar graph.

Why, then, did we say that *most*, not *all* maps can be colored this way? Because the description above assumes that each entity is contiguous. Consider the map below where entity A has an exclave between D and B which would require the same color as the rest of A. In this case each of the 5 entities has a border with the other 4, making its graph representation isomorphic to K_5, which is not planar.

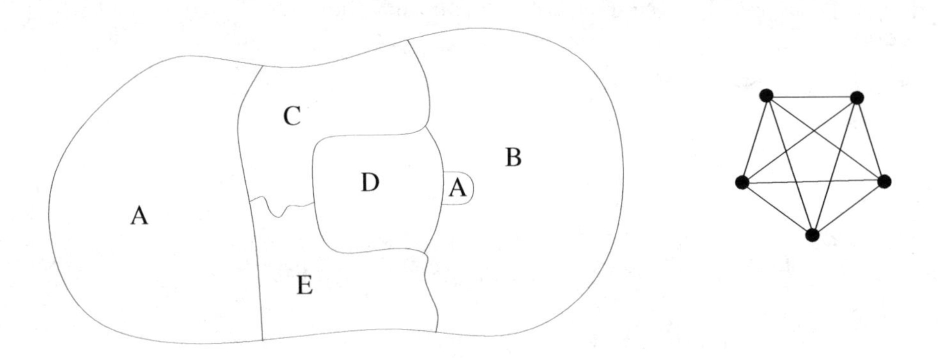

Exclaves or other non-contiguous entities occur in real maps: The United States has Alaska, Michigan has both a lower and upper peninsula, Labrador and Newfoundland are one province, Russia has an exclave including the city of Kaliningrad, Angola has an exclave of Cabinda, and West Berlin was an exclave of West Germany during the Cold War, to name a few. As it happens, all of these still can be modeled by planar graphs.

It is not only the proofs of graph-coloring results that are difficult. Assigning the actual colors to a graph is a real algorithmic challenge—perhaps not so much for cartography, but for problems like scheduling and register allocation, where the graphs are not necessarily planar.

In fact, there is no known efficient algorithm for finding minimal colorings for general graphs, and the prevailing opinion of computer scientists is that no such algorithm exists (see Section 9.11). Most effort in graph coloring applications is in algorithms that are not guaranteed to find a perfectly minimal coloring, but rather approximates such a solution with a coloring that uses as few colors as it is feasible to find.

Be glad that your grade school assignment did not require you to color a map of the Holy Roman Empire.

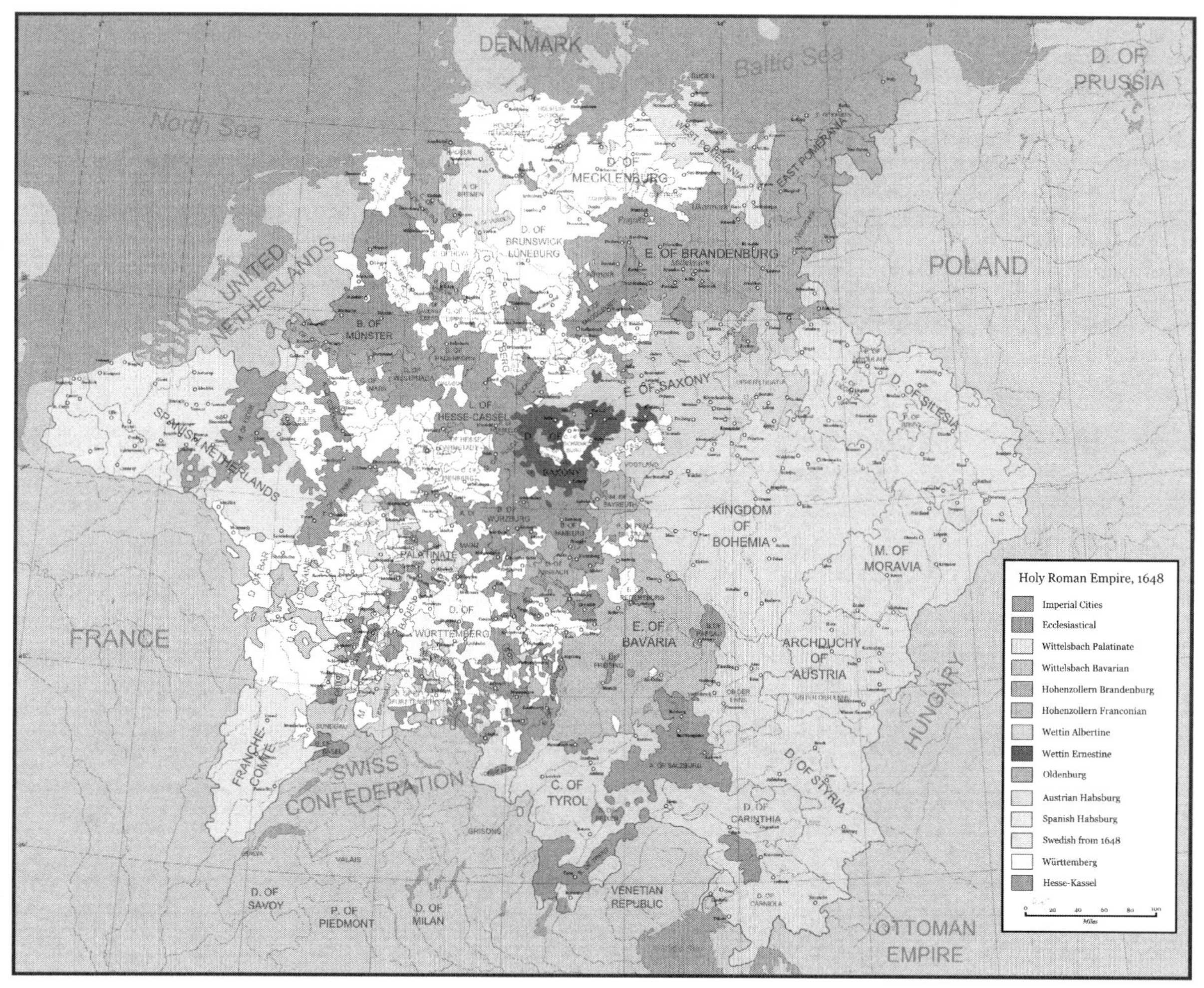

A color version of the map above can be found at `http://commons.wikimedia.org/wiki/File:Holy_Roman_Empire_1648.svg`. More information on the four color problem can be found in *Four Colors Suffice* by Robin Wilson [37].

Chapter summary

Graph theory is a vast subject and has been studied from many angles, both theoretical and applied. This chapter has presented a balanced summary of the most widely studied and used results. What is most interesting about graphs is how flexible they are for modeling information and hence a wide variety of real problems.

For the purposes of this course, graphs have served as a context for practicing proofs, especially for our first taste of the concept of isomorphism, which we will see again in later topics. Graph algorithms also form a core part of the field of algorithms in general, and the student will find the introduction given here as a foundation for future study, if he or she some day takes a course on data structures and algorithms or on the analysis of algorithms. Finally, the variety of graph applications have shown yet again how we can find discrete mathematics in the world around us.

Key ideas:

A graph is a pair of finite sets, a set V of vertices and a set E of edges. Graphs can be directed or undirected. Some graphs have self-loops and parallel edges, but those that do not are called *simple* graphs.

Proofs of propositions about graphs often require induction on the number of vertices in the graph.

Walks in a graph are series of vertices and the edges that connect adjacent vertices. There are many kinds of walks: simple walks, closed walks, paths, circuits, and cycles. Proofs of propositions about walks often require reasoning about the length or structure of the path.

Informally, two graphs are isomorphic if you can change one into the other by moving around the vertices and stretching or pulling the edges, but maintaining the connections between edges. Formally, an isomorphism from graph $G = (V, E)$ to graph $H = (W, F)$ is a pair of one-to-one correspondences $\phi : V \to W$ and $\psi : E \to W$ such that for all $v \in V$ and $e \in E$, v is an endpoint of e iff $\phi(v)$ is an endpoint of $\psi(e)$.

An isomorphic invariant is a predicate on a graph which, if it is true for a graph, is also true for all graphs isomorphic to that graph.

A large number of graph varieties have been studied: Complete graphs, cycles, wheels, n-cubes, bipartite graphs, planar graphs, graphs with strongly connected components and other features of connectivity, and trees. Graphs have been used in game theory and the optimization of transportation routes.

We can represent graphs in a programming language using an adjacency matrix or adjacency lists. Two main strategies for visiting all the vertices in a graph are depth-first search and breadth-first search.

For weighted graphs, algorithms exist for finding the shortest path from a source vertex to all other vertices or the smallest tree that connects or spans the entire graph.

Graph theory has been used to show that (almost) any map can be colored using only four colors. This line of research has other uses as well, such as scheduling problems.

Chapter 9 Complexity Class

A large part of programming is not only determining how to do something but also choosing the best way to do something, based on efficient use of resources. The resources may include time, computer memory, or network bandwidth, to name a few. Efficiency can be determined and predicted both theoretically and empirically.

In this chapter we consider mathematical tools for predicting the running times of algorithms—or, more specifically, how the running time of an algorithm depends on the algorithm's input. However, the core of this chapter looks at these mathematical tools as mathematical objects that are interesting in their own right. This chapter builds on both our study of functions in Chapter 7 and your previous experience with real-valued functions from algebra and precalculus. Specifically we will explore rigorous ways to categorize functions.

The sections of this chapter can be grouped into three parts: Sections 9.1–9.4 introduce the analysis of algorithms for time efficiency—though a separate course is necessary for mastering the analysis of algorithms. Sections 9.5–9.7 constitute the core of the chapter, investigating tools for categorizing functions as mathematical objects. Sections 9.8–9.10 explore a sequence of programming topics related to the rest of the chapter in that they can aid in improving the efficiency of algorithms.

The technical material is based on the presentation in Cormen et al [5].

Chapter goals. The student should be able to

- solve simple recurrence relations.
- determine formulas for the running time of simple ML functions and assign complexity classes to them.
- compare the expected running time of sorting algorithms.
- articulate and use the formal definitions of various complexity class families.

- prove propositions about the relationships among complexity classes.
- use tables and memoization to improve the efficiency of algorithms.

9.1 Recurrence relations

You may recall that a mathematical *sequence* is an ordered set of elements. We use the notation a_i to refer to the ith element in the sequence and $\{a_i\}$ to name the sequence as a whole. Simple sequences include:

sequence

$$a_i = i \quad \{a_i\} = 1, 2, 3, 4, \ldots \qquad b_i = 2^i \quad \{b_i\} = 2, 4, 8, 16, \ldots$$

$$c_i = \frac{1}{i} \quad \{c_i\} = 1, \frac{1}{2}, \frac{1}{3}, \frac{1}{4}, \ldots \qquad d_i = i^2 \quad \{d_i\} = 1, 4, 9, 16 \ldots$$

With each of these sequences, we have assumed that they start with $i = 1$ and go on forever. Sequences, however, can be either infinite or finite, and they can start at any integer index—0 and 1 being the most common "first" elements.

(By the way, what do we mean by defining a sequence as an *ordered set*? How do we order a set? By mapping a subset of the integers to it, of course. Formally, an infinite sequence with elements from a set X is a function $f : \mathbb{N} \to X$. If you want the sequence to begin with a zeroth element, replace $\mathbb{N}$ in the definition with $\mathbb{W}$. A finite sequence is a function $f : \{1, 2, \ldots, n\} \to X$ or $f : \{0, 1, \ldots, n\} \to X$. Finagle these definitions in the obvious way to start the sequence with an index other than 0 or 1.)

Sequences, like functions, can be defined recursively. Instead of giving a stand-alone formula for an arbitrary element, we can define a sequence by giving the value for the first one or more elements and then describing all other elements with a formula that refers to earlier elements in the sequence. There are two standard examples of recursively defined sequences: the Fibonacci sequence and the number of moves required in the Towers of Hanoi puzzle.

You know the Fibonacci sequence as the sequence of numbers where each number in the sequence is the sum of the two previous numbers, starting with 0 and 1. A recursive definition for the sequence can be formulated in the following way. Let f_n be the nth value in the sequence. We define

$$f_n = \begin{cases} 0 & \text{if } n = 0 \\ 1 & \text{if } n = 1 \\ f_{n-1} + f_{n-2} & \text{otherwise} \end{cases}$$

The Towers of Hanoi is a puzzle in which the player is given a set of three pegs and a certain number of disks, each of a different size. Initially all the disks are on one peg, stacked from the largest disk on bottom to the smallest on top. The player must move all the disks to the third peg by moving one disk at a time and never placing a larger disk on top of a smaller.

The key is to think of the problem recursively: If you have n disks, then first move the top $n - 1$ disks (in a series of single-disk moves) from the first peg to the second. Then move the bottom disk from the first peg to the third. Finally, move the $n - 1$ disks from the second peg to the third (again, by a series of moves). Thus we divide the problem of moving n disks into two subproblems, each for moving $n-1$ disks. Consider the problem with four disks:

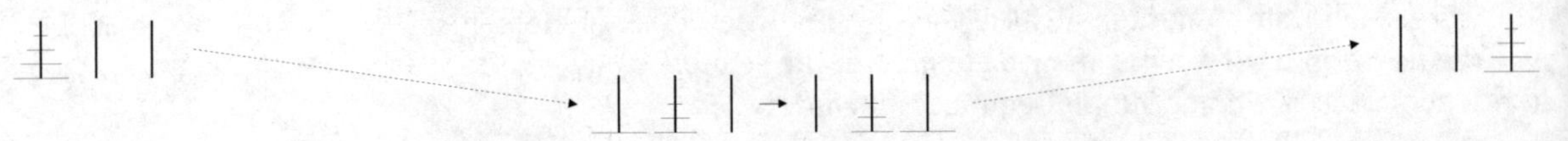

How do we move the three disks from the first to the second peg? First we move the top two disks from first peg to the third, move the second-largest disk from the first to the second, and then move the top two from the third to the second. A similar process moves the three disks from the second to the third in the second half of the solution.

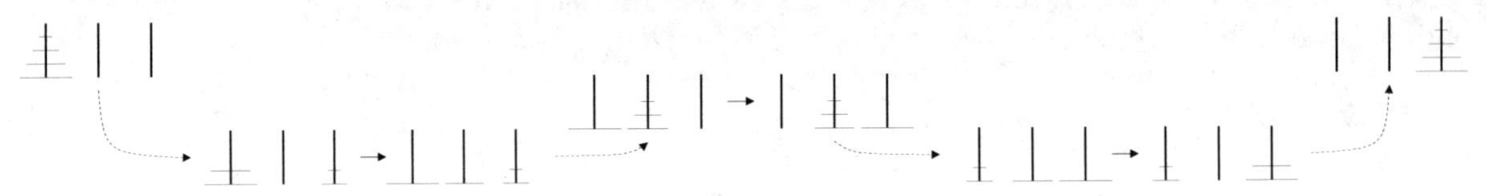

To generalize the algorithm, let us call the pegs *origin*, *destination*, and *scratch*. To move 1 disk from the origin to the destination, just move it. To move the top $n > 1$ disks from the origin to the destination using scratch,

- move the top $n-1$ disks from origin to scratch, using destination
- move the nth disk from origin to destination
- move the top $n-1$ disks from scratch to destination, using origin.

This generalized form also works for moving a subset of the disks. It directs us how to move the top n disks, but there might in fact be more than n disks on the origin peg, and there might be other disks already on the other pegs. The algorithm will leave those disks undisturbed.

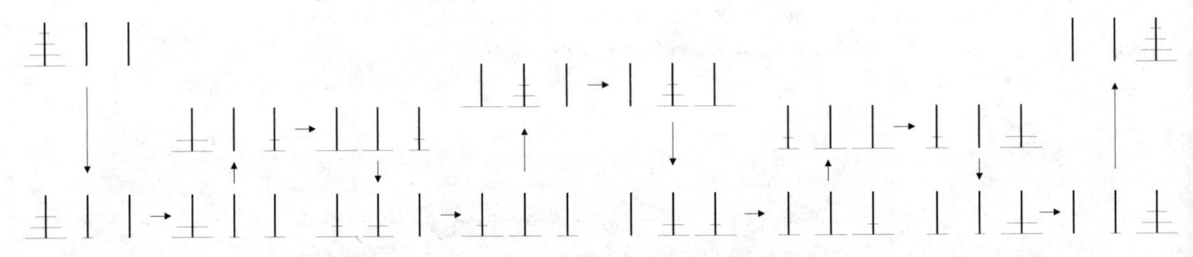

Let h_n be the number of individual moves this strategy would take for n disks. Then

$$h_n = \begin{cases} 1 & \text{if } n = 1 \\ 2 \cdot h_{n-1} + 1 & \text{otherwise} \end{cases}$$

recurrence relation

We now come to the main definition of this section. A *recurrence relation* for a sequence of numbers $\{a_i\}$ is a formula that expresses the values of the sequence in terms of earlier values. The first value in the sequence must be expressed as a constant—in fact, several values may be expressed as constants, just as the first two values in the Fibonacci sequence. But at least one rule in the formula for a_i must contain one or more term each in the form a_j where $j < i$.

To *solve* a recurrence relation means to find a non-recursive formula for the elements of the sequence. For example, consider the recurrence relation for the number of moves to solve Towers of Hanoi. Experiment by calculating a few terms

n	h_n
1	1
2	$2+1=3$
3	$2 \cdot 3+1=7$
4	$2 \cdot 7+1=15$
5	$2 \cdot 15+1=31$
6	$2 \cdot 31+1=63$

There is a clear pattern—it seems that each number is one less than a power of two. We conjecture that

$$h_n = 2^n - 1$$

is a solution to $h_n = 2 \cdot h_{n-1} + 1$. Since the sequence itself is defined recursively, it should come as no surprise that our proof will use induction.

Proof. By induction on n.

Base case. Suppose $n = 1$. Then $h_1 = 1 = 2 - 1 = 2^1 - 1$. Hence there exists $N \geq 1$, such that for all $n \leq N$, $h_n = 2^n - 1$.

Tangent: The origins of the Fibonacci Sequence and the Towers of Hanoi puzzle

The Fibonacci sequence is named after Leonardo Fibonacci, a medieval Italian mathematician who lived around 1170–1250. His most important contribution was introducing the Indian-Arabic number system to Europe (replacing Roman numerals) which he learned while working as a merchant in northern Africa.

Fibonacci did not discover the sequence that bears his name, but he wrote about it in his book *Liber Abaci* (*Book of Calculation*). In his presentation he described an island where two newborn rabbits, one male and one female, were introduced. Assume that this kind of rabbit reaches sexual maturity in two months, that adult rabbit couples each reproduce one male and one female child each month, and no rabbits ever die. How many rabbit pairs will there be after n months? During the first two months (month 0 and month 1), there is only the original pair. At any other point in time, however, the number of rabbit pairs in existence f_n is equal to the number in existence last month f_{n-1} plus the number born this month—equal to the number that are sexually mature, that is, were alive two months ago, f_{n-2}.

The Towers of Hanoi puzzle was invented in 1883 by French Mathematician Édouard Lucas, which he sold as a game. The instructions to the game mention a legend about Brahmin priests manipulating a version of the puzzle with 64 golden disks, and that the world would end when they complete it. (Moving one disk per second, that would take more than 5 billion centuries.) [31]

Inductive case. Suppose n is such that $h_n = 2^n - 1$. Then

$$\begin{aligned} h_{n+1} &= 2 \cdot h_n + 1 && \text{by how the recurrence relation is defined} \\ &= 2 \cdot (2^n - 1) + 1 && \text{by the inductive hypothesis} \\ &= 2^{n+1} - 2 + 1 \\ &= 2^{n+1} - 1 && \square \end{aligned}$$

The solution to the Fibonacci sequence is complicated (see Exercise 9.1.1). For a simpler example, we can give an upper bound on each value in the sequence: $f_n \leq (\frac{5}{3})^n$.

The recursive part of the definition of f_n has two previous terms in it, f_{n-1} and f_{n-2}. How will that affect our proof? A simple inductive step in the form of proving $I(n) \to I(n+1)$ will not work because the inductive hypothesis must hold for $n - 1$ also. Instead, we need to use so-called strong induction, where our inductive hypothesis supposes that the solution holds for all smaller values of n.

Proof. By induction on n.

Base case. For $n = 0$, $f_0 = 0 < 1 = (\frac{5}{3})^0$. For $n = 1$, $f_1 = 1 < \frac{5}{3} = (\frac{5}{3})^1$. Hence there exists $N \geq 1$ such that for all $n \leq N$, $f_n \leq (\frac{5}{3})^n$.

Inductive case. Suppose N such that for all $n \leq N$, $f_n \leq (\frac{5}{3})^n$. Then

$$\begin{aligned} f_{N+1} &= f_N + f_{N-1} && \text{by how the Fibonacci sequence is defined} \\ &\leq (\tfrac{5}{3})^N + (\tfrac{5}{3})^{N-1} && \text{by the inductive hypothesis} \\ &= (\tfrac{5}{3} + 1) \cdot (\tfrac{5}{3})^{N-1} && \text{by factoring} \\ &= \tfrac{8}{3} \cdot (\tfrac{5}{3})^{N-1} \\ &= \tfrac{24}{9} \cdot (\tfrac{5}{3})^{N-1} \\ &< \tfrac{25}{9} \cdot (\tfrac{5}{3})^{N-1} \\ &= (\tfrac{5}{3})^2 \cdot (\tfrac{5}{3})^{N-1} \\ &= (\tfrac{5}{3})^{N+1} && \square \end{aligned}$$

Exercises

9.1.1 Prove that the exact explicit formula for the Fibonacci series is

$$f_n = \frac{1}{\sqrt{5}} \cdot (\frac{1+\sqrt{5}}{2})^n - \frac{1}{\sqrt{5}} \cdot (\frac{1-\sqrt{5}}{2})^n$$

Find and prove an explicit formula for the following recurrence relations.

9.1.2

$$a_i = \begin{cases} 0 & \text{if } i = 0 \\ 3 + a_{i-1} & \text{otherwise} \end{cases}$$

9.1.3

$$b_i = \begin{cases} 1 & \text{if } i = 0 \\ 3 \cdot b_{i-1} & \text{otherwise} \end{cases}$$

9.1.4

$$a_i = \begin{cases} 0 & \text{if } i = 0 \\ i & \text{if } i \text{ is odd} \\ a_{i-2} + 2 & \text{otherwise} \end{cases}$$

9.1.5

$$b_i = \begin{cases} 1 & \text{if } i = 0 \\ 2 \cdot b_{i-1} & \text{otherwise} \end{cases}$$

9.1.6

$$c_i = \begin{cases} 0 & \text{if } i = 1 \\ 2 \cdot (b_{i-1} + 1) & \text{otherwise} \end{cases}$$

9.1.7

$$d_i = \begin{cases} 0 & \text{if } i = 0 \\ d_{i-1} + i & \text{otherwise} \end{cases}$$

9.1.8 Write an ML function `hanoiSolve` that will list the steps to solve a Towers of Hanoi puzzle of a certain size. The function should take four parameters: an `int` number of disks to move and three strings giving names to the pegs listed in origin, scratch, destination order. For example, to find the steps for moving 4 disks from peg 1 to peg 3 using peg 2, one would call `hanoiSolve(4, "1", "2", "3")`. The function should print the moves to the screen, in this case

```
from 1 to 2
from 1 to 3
from 2 to 3
from 1 to 2
...
```

9.2 Complexity of algorithms

Just as a recurrence relation helped us discover how many moves were involved in our strategy for the Tower of Hanoi puzzle, based on the size of the puzzle, we will use recurrence relations to determine the time required for computer algorithms.

How long does an ML program take? It is daunting to try to answer that question because of the number of things we do not know. Computer hardware differs in speed, and some ML interpreters may be implemented more efficiently than others. Even if we had a specific computer and interpreter in mind, how long does it take a computer to perform basic operations? Besides all these, a program is going to take longer for some inputs than others.

Even though we have all of these unknowns about the environment, they are unknown *constants*—it is a reasonable simplifying assumption that an operation like integer addition will take just as long every time. The running time still depends on the input to the program, and so we will represent the running time as a function of the attributes of the input, particularly the *size* of the input.

Start with a simple example from very early in this text:

```
- fun sum([]) = 0
=     | sum(a::rest) = a + sum(rest);
```

Think about what the computer is doing in all this. Break down the component steps.

- There is a test to see which pattern applies. This means we have to compare the input list with an empty list. We do not know how long this takes, but assume it is always the same and call it c_0, that is, it takes c_0 units of time to do the comparison and decide which pattern applies.

- If the first pattern applies, then the rest of the function is however long it takes simply to return 0. Call that c_1.

- If the second pattern applies, we need to break the list apart into a head `a` and a tail `rest`. This is a standard operation in ML, and any student who has programmed linked lists in another programming language such as Java can imagine what the ML interpreter is doing. Suppose it takes c_2 time.

- Also if the second pattern applies, we need to perform the recursive call. That takes as long as this entire process, except on a smaller list.

- Finally, we need to perform an addition (and return the result). int addition takes constant time, say c_3.

Let $T(n)$ be the function describing the running time of this ML function on a list of size n. (Notice that the running time does not depend on the contents of the list, just the size.)

$$T(n) = \begin{cases} c_0 + c_1 & \text{if } n = 0 \\ c_0 + c_2 + c_3 + T(n-1) & \text{otherwise} \end{cases}$$

It is a recurrence relation—just rename it T_n instead of $T(n)$. So how much is that, really? Can we solve this recurrence relation? Experiment first:

n	$T(n)$				
0	$c_0 + c_1$				
1	$c_0 + c_2 + c_3 + T(0)$	$=$	$c_0 + c_2 + c_3 + c_0 + c_1$	$=$	$2 \cdot c_0 + c_1 + c_2 + c_3$
2	$c_0 + c_2 + c_3 + T(1)$	$=$	$c_0 + c_2 + c_3 + 2 \cdot c_0 + c_1 + c_2 + c_3$	$=$	$3 \cdot c_0 + c_1 + 2 \cdot (c_2 + c_3)$
3	$c_0 + c_2 + c_3 + T(2)$	$=$	$c_0 + c_2 + c_3 + 3 \cdot c_0 + c_1 + 2 \cdot (c_2 + c_3)$	$=$	$4 \cdot c_0 + c_1 + 3 \cdot (c_2 + c_3)$

It is not hard to see from here that the solution is

$$\begin{aligned} T(n) &= (n+1) \cdot c_1 + c_1 + n \cdot (c_2 + c_3) \\ &= (c_0 + c_1) + (c_0 + c_2 + c_3) \cdot n \end{aligned}$$

Moreover, since we do not know what the constants are anyway, we will lose no information by a simple renaming: let $d_0 = c_0 + c_1$ and $d_1 = c_0 + c_2 + c_3$.

$$T(n) = d_0 + d_1 \cdot n$$

The main point is that the running time is some constant factor times the size of the list plus a constant. If we were to graph the function, it would be linear, looking something like this:

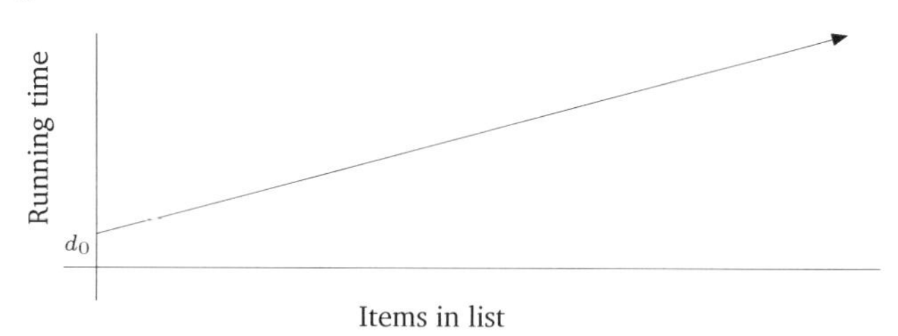

We made one more assumption we have not yet mentioned, and that is what the type of the list is. This function could be used on either a list of ints or a list of reals. Does it make a difference? One on hand, real addition is a completely different operation from int addition on computer hardware, and there is no reason to expect that the two operations would take the same amount of time. On the other hand, whichever type we are using, the running time for the operation is still constant for that type. We still would have a formula like the one above, it is only that the constants would be different.

Now consider how long this ML function would take:

```
- fun contains(x, []) = false
=   | contains(x, y::rest) = if x = y then true
=                            else contains(x, rest);
```

Of course this is structurally similar to sum. What stays the same and what is different?

- Again there is a test to see which pattern applies. Call that time c_0. (We are reusing names from the previous example. This is a different c_0.)
- If the first pattern applies, return `false`. Say that takes c_1 time.
- If the second pattern applies, we need to break the list apart—c_2 time.
- Now this example is a little different. We need to do a comparison. If x and y are ints, then we can assume every comparison takes the same amount of time, call it c_3 (let that also include the time it takes to execute the `if/then/else`). Although we will stick with that assumption because it is the best we can do at this point, we should note that it is far from being true in general. What if x and y were trees? The comparison of two trees itself depends on the size of the trees.

- Now the real kicker. What if the comparison is true? Then the process stops right then and there and we return an answer (which should take the same time as the other base case, c_1). Otherwise, it keeps going. In other words, the running time depends not just on the size of the data but the contents.

What do we do about that last point? We assume the worst. In this case, assume that the list does not contain the item we are searching for. That way we know the actual running time is no worse than what our analysis predicts. In this case we have

$$T(n) = \begin{cases} c_0 + c_1 & \text{if } n = 0 \\ c_0 + c_2 + c_3 + T(n-1) & \text{otherwise} \end{cases}$$

By analogy of the previous example, the solution is again $T(n) = d_0 + d_1 \cdot n$ for some constants d_0 and d_1—it is a linear function in n.

What about cases other than the worst? The best thing that could happen—for making the program run quickly, at least—is either that the list is empty or that the item we are looking for is the first one in the list. If the list is empty, then the whole thing takes c_0+c_1 time, using our original set of constants. If we find what we are looking for immediately, then it is $c_0 + c_2 + c_3$. The important thing is that those are just constants—every time the function is called with x at the front of the list, it will take the same amount of time, no matter how big the list is. We call this *constant time*, and you cannot get any faster than constant.

What is between best and worst? If we were to assume that the item searched for is in the list and that all positions are equally likely, then we could argue that on average the function will search through half of this list, that is, $T(n) = d_0 + d_1 \cdot \frac{n}{2}$. Half as much, however, is not a big enough difference for us to care about. n just has a different coefficient, $\frac{d_1}{2}$ instead of d_1. It is still a linear function, just like the worst case.

On to a more complicated example. Recall this function for testing if a relation is transitive:

```
- fun isTransitive(relation) =
=   let fun testOnePair((a, b), []) = true
=         | testOnePair((a, b), (c, d)::rest) =
=               ((not (b = c))
=                     orelse isRelatedTo(a, d, relation))
=               andalso testOnePair((a,b), rest);
=       fun test([]) = true
=         | test((a,b)::rest) =
=               testOnePair((a,b), relation)
=               andalso test(rest);
=   in
=     test(relation)
=   end;
```

This relies on a slew of helper functions, so we will pick at them one at a time. We do not show the code here for `isRelatedTo`, but it is essentially the same as `contains`, so assume its running time is described by a function $T_1(n) = a_0 + a_1 \cdot n$. For `testOnePair`, assume the base case takes b_0 time, and assume it takes b_1 time to perform the comparison and the boolean operators. Notice that `testOnePair` recursively calls itself with the rest of the given list, but calls `isRelatedTo` with the *original* list containing the entire relation. The deal with this complication, let m be the size of the sublist `testOnePair` is called on and let n be the size of the relation. Then for the running time of `testOnePair` we get

$$T_2(m) = \begin{cases} b_0 & \text{if } n = 0 \\ b_1 + T_1(n) + T_2(m-1) & \text{otherwise} \end{cases}$$

If we substitute our formula for $T_1(n)$ and regroup and rename the constants, we come up with

$$T_2(m) = \begin{cases} c_0 & \text{if } n = 0 \\ c_1 + c_2 \cdot n + T_2(m-1) & \text{otherwise} \end{cases}$$

The solution (see Exercise 9.2.1) is $T_2(n) = d_0 + d_1 \cdot n + d_2 \cdot n^2$ for some constants d_0, d_1, and d_2. For the function `test`, suppose e_0 accounts for the base case and e_1 for the overhead of each recursive case. Its running time is then

$$T_3(n) = \begin{cases} e_0 & \text{if } n = 0 \\ e_1 + T_2(n) + T_3(n-1) & \text{otherwise} \end{cases}$$

or

$$T_3(n) = \begin{cases} e_0 & \text{if } n = 0 \\ e_1 + d_0 + d_1 \cdot n + d_2 \cdot n^2 + T_3(n-1) & \text{otherwise} \end{cases}$$

The solution (see Exercise 9.2.2) is $T_3(n) = \alpha_0 + \alpha_1 \cdot n + \alpha_2 \cdot n^2 + \alpha_3 \cdot n^3$. Since the body of `isTransitive` simply calls `test` on the whole list, T_3 is also the running time for the whole function.

What is really going on? Cut through the symbols and look at it this way: `test` steps through each of the n items in the list. For each of those items it calls `testOnePair`, which also steps through the entire list, making n^2 calls to `testOnePair` in total. But each call to `testOnePair` calls `isRelatedTo`, which also steps through the entire list. That is where the term of n^3 comes from, and that is the big one.

Next example: We saw this "accumulate" version of `powerset` in Section 6.7.

```
- fun powerset([], yy) = [yy]
=   | powerset(x::rest, yy) =
=           powerset(rest, yy) @ powerset(rest, x::yy);
```

It is not easy to wrap your head around why it works, but you do not need to understand why it works to analyze its efficiency, which is our present concern. The number of recursive calls depends on the size of the first parameter, not the second, so assume the size of the input, n, is only the size of the first parameter. For the base case we make a new list with one item, which is the same as consing[1] an item to an empty list: a_0 time. In the recursive case we break the list `x::rest` apart and cons the list `x::yy`; let a_1 cover those steps together. We now have two recursive calls, both to lists of size $n - 1$.

How long does the concatenation take? At first glance we might suppose that it is a simple, constant-time operation like cons. Be careful. Do you remember Exercise 2.2.9 where you were asked to implement your own concatenation operation? Concatenating two lists, say `aa@bb`, requires breaking down the elements of `aa` and consing them in succession to a new list, with `bb` as our starting point.

$$[1, 2, 3, 4, 5]@[6, 7, 8] = 1 :: 2 :: 3 :: 4 :: 5 :: [6, 7, 8]$$

The point is that the time taken will be proportional to the size of the first list being concatenated. We know from the size of powersets (Theorem 6.5) that if `rest` has $n - 1$ elements, then `powerset(rest, yy)` will have 2^{n-1} elements. All told we have

$$T(n) = \begin{cases} a_0 & \text{if } n = 0 \\ a_1 + a_2 \cdot 2^{n-1} + 2 \cdot T(n-1) & \text{otherwise} \end{cases}$$

Does that look scary? In a way it should. How are we ever going to solve that recurrence? Try making a tree showing the work done for each call of the function. The original call of `powerset` on a list of size n takes $a_1 + a_2 \cdot 2^{n-1}$, discounting the recursive calls. Those two recursive calls each take $a_1 + a_2 \cdot 2^{n-2}$, discounting *their* recursive calls. Eventually we will get to lists of size 1—not the base cases, but the calls from which the base case is called—which each take $a_1 + a_2 \cdot 2^0$. The base cases, of which there are 2^n, take a_0 each.

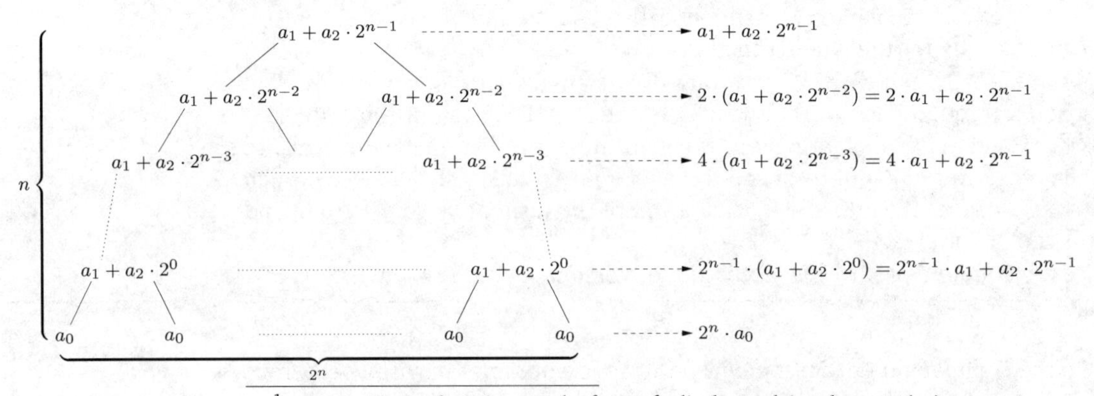

[1]**cons** • *v.* To attach an item to the front of a list by applying the cons (`::`) operator.

To the right of the tree you see the sum of of the work done at each level. There are n levels in the tree—$n-1$ of them follow a similar pattern, and the bottom level, for the base cases, is different. The pattern for the non-base-case levels is a term of a_1 times a power of 2 and a term of $a_2 \cdot 2^{n-1}$. Again, there are $n-1$ of these. Summing these we arrive at a grand total of

$$T(n) = a_0 \cdot 2^n + a_1 \cdot (1 + 2 + 4 + \cdots + 2^{n-1}) + a_2 \cdot n \cdot 2^{n-1}$$

The $(1+2+4+\cdots 2^{n-1})$ is a geometric series (see Exercise 6.5.4), and it is equal to 2^n. So, finally, we have

$$\begin{aligned}
T(n) &= (a_0 + a_1) \cdot 2^n + a_2 \cdot n \cdot 2^{n-1} \\
&= (a_0 + a_1) \cdot 2^n + \tfrac{a_1}{2} \cdot n \cdot 2^n \\
&= b_0 \cdot 2^n + b_1 \cdot n \cdot 2^n
\end{aligned}$$

Now why should we care? What useful information can we glean from all this? `sum` and `listMin` each have running times in the form $c_0 + c_1 \cdot n$. `isTransitive` takes $c_0 + c_1 \cdot n + c_2 \cdot n^2 + c_3 \cdot n^3$ time. For `powerset`, it is $c_0 \cdot 2^n + c_1 \cdot n \cdot 2^n$. How significant is that difference? Let us make three functions by assuming for each of these forms that the coefficient of the highest power is 1 and the other coefficients are 0:

$$\begin{aligned}
f(n) &= n \\
g(n) &= n^3 \\
h(n) &= n \cdot 2^n
\end{aligned}$$

Graph them, as you would have done in high school algebra. We show their graphs with two scales, a smaller one on the left.

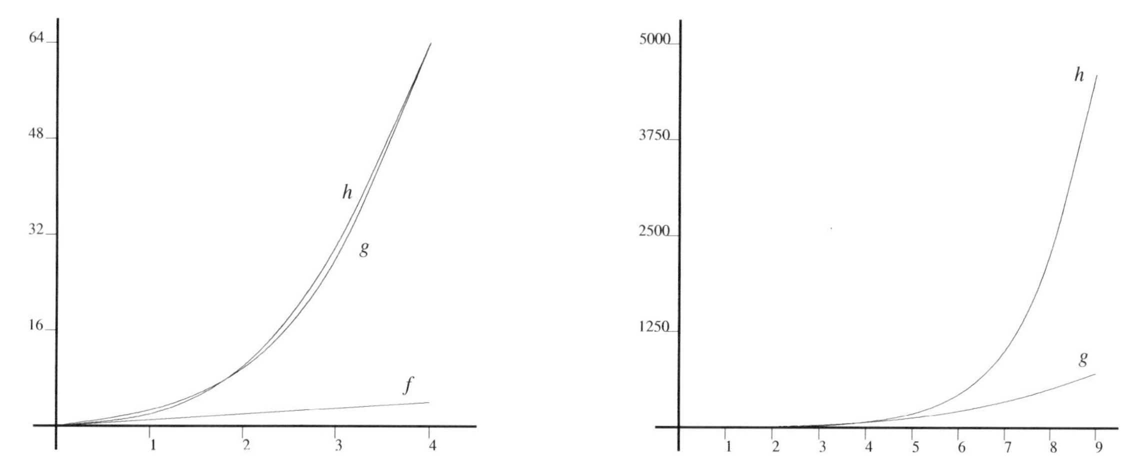

On the right we do not even plot f because it would not be visibly different from the x-axis. On the left h and g seem to track closely at first—indeed they are equal at 2 and 4. However, on the right we see that h dwarfs g soon after. Suppose we had algorithms whose running times were represented by these very functions and that the time unit was nanoseconds. Suppose we applied them each to a list of size 1000. The program with function f would complete in 1000 nanoseconds, which is 10^{-6} of a second. The program with function g would complete in 10^9 nanoseconds, that is, 1 second. The program with function h would take $1.07 \cdot 10^{304}$ nanoseconds, or $3.4 \cdot 10^{284}$ millennia.

This illustrates the dramatic difference in these functions. Moreover, this justifies our nonchalance about the constants. Lower ordered terms and coefficients may stretch or angle these functions a bit, but they will not change their fundamental shape. A function with an n^3 term with small constant factors eventually will be greater, much greater, than a linear function with huge constants. Likewise an exponential function eventually will overtake any polynomial function.

on the order of

big-oh

This leads to our rough way of categorizing functions: For any function standing for the running time of an algorithm, we take the dominant term—that is, the highest powered term in a polynomial function or whatever term eventually is the greatest—and ignore the constant. We then say that the original function is *on the order of* that term. In the examples above, the running time of `listMin` is on the order of n, written $O(n)$, which is often read "*big-oh* of n." The running time of `isTransitive` is $O(n^3)$, and the running time of `powerset` is $O(n \cdot 2^n)$.

Analyzing algorithms is difficult, and we are providing only a bare introduction here. It will take a course or two in data structures and algorithms to gain a real proficiency. In the next two sections we will consider a few other examples to show some of the range of problems in analyzing complexity.

Exercises

9.2.1 Show that the solution to the recurrence for `testOnePair` is $T_2(n) = d_0 + d_1 \cdot n + d_2 \cdot n^2$ for some constants d_0, d_1, and d_2.

9.2.2 Show that the solution to the recurrence for `isTransitive` is $T_3(n) = \alpha_1 + \alpha_1 \cdot n + \alpha_3 \cdot n^2 + \alpha_4 \cdot n^3$, for some constants α_0, α_1, α_2, and α_3.

Analyze the following functions to determine a recurrence relation expressing their running times as functions of the size of the list. Find a solution to the recurrence relation and a big-oh category.

9.2.3 Your function `doubler` from Exercise 2.2.3.

9.2.4 Your function `makeNoRepeats` from Exercise 3.7.9.

9.2.5 Your function `allLessThan` from Exercise 3.13.1.

9.2.6 Your function `allHaveOpposites` from Exercise 3.13.3.

9.3 Analyzing sorting algorithms

Sorting algorithms are a useful context for algorithmic analysis examples because we know several algorithms that all do the same thing—namely, sort—but that may have different efficiencies. Take merge sort, which we worked on in Section 6.8.

```
- fun split1([], aa, bb) = (aa, bb)
=   | split1(x::rest, aa, bb) = split2(rest, x::aa, bb)
= and split2([], aa, bb) = (aa, bb)
=   | split2(x::rest, aa, bb) = split1(rest, aa, x::bb);

- fun merge([], bb) = bb
=   | merge(aa, []) = aa
=   | merge(a::aRest, b::bRest) =
=     if a < b then a::merge(aRest, b::bRest)
=               else b::merge(a::aRest, bRest);

- fun mergeSort([]) = []
=   | mergeSort([a]) = [a]
=   | mergeSort(xx) =
=     let val (aa, bb) = split1(xx, [], []);
=     in merge(mergeSort(aa), mergeSort(bb))
=     end;
```

The ML functions `split1`, `split2`, and `merge` look hairy, but on closer inspection we can eyeball them as being similar to `sum` and `contains`. Let $T_1(n)$ and $T_2(n)$ be functions describing the running time for `split1` and `merge` (since `split2` is not called by `mergeSort`, we do not need a name for its running time). We claim

$$T_1(n) = a_0 + a_1 \cdot n$$

$$T_2(n) = b_0 + b_1 \cdot n$$

The parameter n in T_2 (for `merge`) is the total size of the two lists, combined.

Now we construct a recurrence relation for `mergeSort`, say T_3. Since `mergeSort` has two base cases, so will the recurrence relation. The terms of the recursive case correspond to the four function calls in the last pattern of `mergeSort`: `split1`, `merge`, and `mergeSort`, recursively, twice.

$$T_3(n) = \begin{cases} c_0 & \text{if } n = 0 \\ c_1 & \text{if } n = 1 \\ T_1(n) + T_2(n) + 2 \cdot T_3(\frac{n}{2}) & \text{otherwise} \end{cases}$$

Or, letting $c_2 = a_0 + b_0$ and $c_3 = a_1 + b_1$,

$$T_3(n) = \begin{cases} c_0 & \text{if } n = 0 \\ c_1 & \text{if } n = 1 \\ c_2 + c_3 \cdot n + 2 \cdot T_3(\frac{n}{2}) & \text{otherwise} \end{cases}$$

Of course that formula assumes n can be split evenly, that is, the two recursive calls work on equal portions of the list. If n is odd, then one recursive call will have one more element than the other. To simplify our analysis, we indeed will assume that n is always even, specifically that we are sorting a list of size 2^m for some m.

To get a handle on how much work is being done, we make a diagram showing how the problem is divided up into smaller subproblems, just as we did with `powerset` in the previous section, except this time we assume a specific size of the original list, $n = 16$ or $m = 4$, and show the entire tree. To see how much work is being done at each level, the splitting and merging, we take the formula but ignore the recursive call. At the top level we have $c_2 + c_3 \cdot 16$. For the second level we have two lists of size 8, and so we have $2 \cdot (c_2 + c_3 \cdot 8)$.

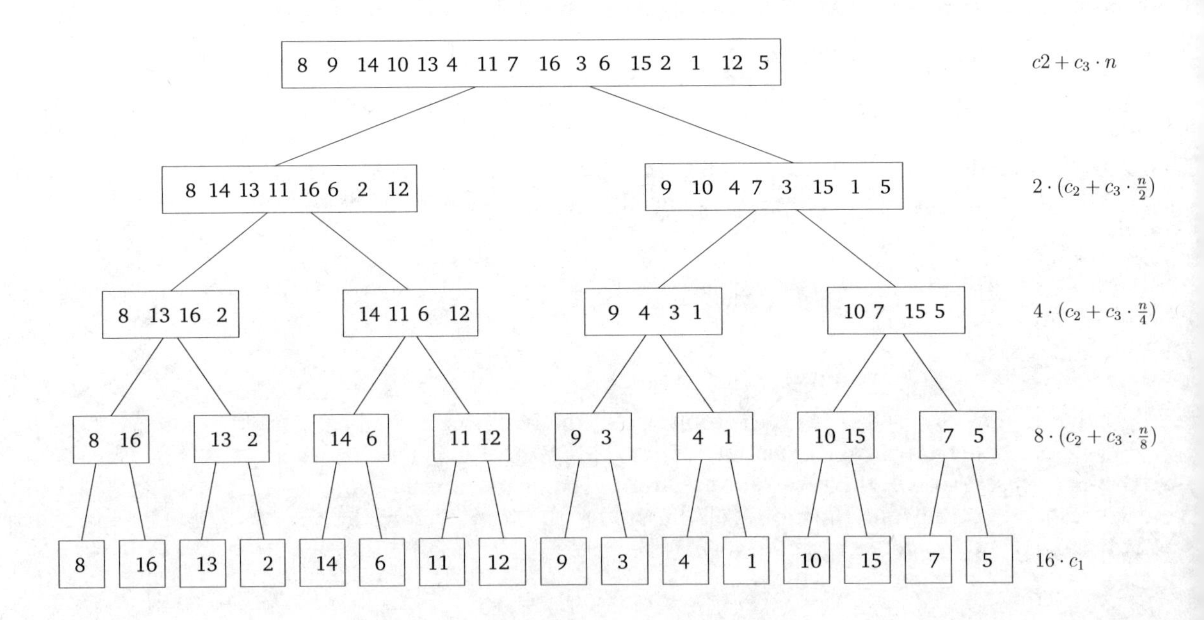

The total work done by the base cases is $c_3 \cdot 16$, or more generally $c_3 \cdot n$ or $c_3 \cdot 2^m$. There are $m = \lg n$ such levels. (lg stands for base-2 log.) We have

$$\begin{aligned}
T(n) &= 2^m \cdot c_1 + m \cdot c_3 \cdot 2^m + c_2 + 2 \cdot c_2 + 4 \cdot c_2 + \cdots {}_2^{m-1} c_2 \\
&= 2^m \cdot c_1 + m \cdot c_3 \cdot 2^m + c_2(1 + 2 + 4 + \cdots 2^{m-1}) \\
&= 2^m \cdot c_1 + m \cdot c_3 \cdot 2^m + c_2 \cdot \sum_{i=0}^{m-1} 2^i \\
&= 2^m \cdot c_1 + m \cdot c_3 \cdot 2^m + c_2 \cdot (2^m - 1) \\
&= d_0 + d_1 \cdot 2^m + d_2 \cdot m \cdot 2^m \\
&= d_0 + d_1 \cdot n + d_2 \cdot n \cdot \lg n
\end{aligned}$$

The second-to-last step was by renaming: $d_0 = c_3$, $d_1 = c_1 + c_2$ and $d_2 = c_1$. In short, merge sort has an efficiency function that is $O(n \cdot \lg n)$.

Compare that with selection sort, which you worked out in Exercises 3.7.11–3.7.13. Obviously we are not going to give the solution to those exercises here, but we can perform a back-of-the-envelope analysis by inspecting the problem. The version of selection sort described in those exercises uses helper functions `listMin` and `removeFirst`. The `sort` function takes in a list, finds the minimum element and removes it, sorts (by a recursive call) the list with the minimum removed, and concatenates the removed minimum to the front of the sorted sublist.

`listMin` examines every element of the list once, so like `sum` it will have a running time in the form of $T_1(n) = a_0 + a_1 \cdot n$. `removeFirst` examines every element in the list until it finds the one to remove, and then it does the same number of cons operations. Like `contains`, the worst case will have it search the entire list, so its running time is $T_2(n) = b_0 + b_1 \cdot n$.

For the `sort` function itself we have

$$T_3(n) = \begin{cases} c_0 & \text{if } n = 0 \\ T_1(n) + T_2(n) + T_3(n-1) & \text{otherwise} \end{cases}$$

Or, substituting in the bodies for $T_1(n)$ and $T_2(n)$ and renaming,

$$T_3(n) = \begin{cases} c_0 & \text{if } n = 0 \\ c_1 + c_2 \cdot n + T_3(n-1) & \text{otherwise} \end{cases}$$

To solve this recurrence, simply expand the terms.

$$
\begin{aligned}
T_3(n) &= c_1 + c_2 \cdot n + T_3(n-1) \\
&= c_1 + c_2 \cdot n + c_1 + c_2 \cdot (n-1) + T_3(n-2) \\
&= c_1 + c_2 \cdot n + c_1 + c_2 \cdot (n-1) + \cdots c_1 + c_2 \cdot 1 + c_0 \\
&= c_0 + c_1 \cdot n + c_2 \cdot \textstyle\sum_{i=1}^{n} i \\
&= c_0 + c_1 \cdot n + c_n \cdot \frac{(n-1)\cdot n}{2} \\
&= d_0 + d_1 \cdot n + d_2 \cdot n^2
\end{aligned}
$$

The second-to-last step is by the solution to arithmetic series, as seen in Exercise 6.5.1, which you recently used in Exercise 9.2.1. The last step is by renaming. This means that selection sort, under our "worst case" assumptions, is $O(n^2)$.

How do $O(n \cdot \lg n)$ and $O(n^2)$ compare? You know from high school algebra that n^2 looks like a parabola. A function with $n \cdot \lg n$ as its dominant term grows faster than n but more slowly than n^2. How much more slowly? See for yourself:

n	n^2	$n \cdot \lg n$
10	100	33.2
25	625	116
50	2500	282.2
75	5625	467.2
100	10000	664.4
125	15625	870.7
150	22500	1084.3
175	30625	1304
200	40000	1528.8

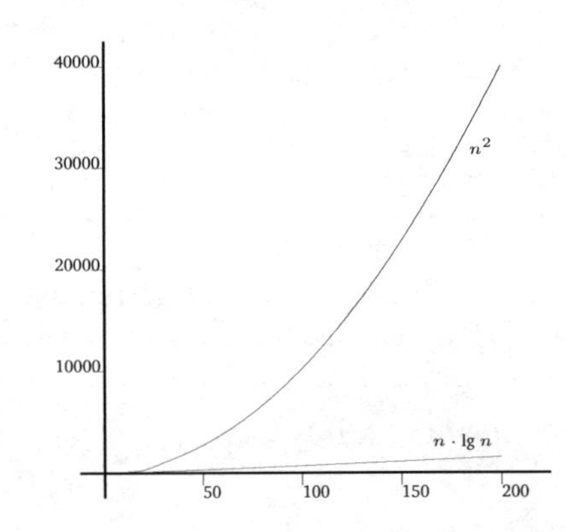

Exercises

9.3.1 Show carefully that `split1`, `split2`, and `merge` have the linear times we assert in this section.

9.3.2 What would a list look like that would cause the worst case behavior of selection sort that we assume? That is, can you say anything about how the elements are ordered?

9.3.3 On the other hand, describe a list that would result in best case time for a given n. Analyze selection sort under a best case assumption. What is its big-oh category?

9.3.4 Analyze the running time of bubble sort, as seen in Exercises 6.8.9–6.8.11.

9.3.5 Analyze the running time for your implementation of insertion sort from Exercises 6.8.15–6.8.19.

9.3.6 The analysis of quick sort from Section 6.8 is difficult. Its worst case time is $O(n^2)$ and best case $O(n \cdot \lg n)$. Can you give an intuitive explanation why this would be?

9.4 Alternative examples of analyzing algorithms

The examples shown so far have all been functions that operate recursively on lists and whose running times are functions of the number of elements in the list. Before moving on, we briefly examine examples that break this mold.

Iterative programming is not a major concern for this text, but students who go on to do more programming will likely need to reason about the complexity of iterative programs as well as recursive ones. Consider the iterative version of `listMin`.

For each line of the program—besides structural syntax like `let`—we consider what it does and assign a constant for the amount of time it would take to perform that step once: c_1 for finding the head of `L` and making a reference variable (similarly for c_2), c_3 for comparing the list referred to by `LL` with an empty list, etc. We multiply each constant by the number of times the step is performed—presumably based on the size of the list. Below we show our analysis for each line along the right.

```
- fun listMin(L) =
=   let
=       val min = ref (hd(L));               c1
=       val LL = ref (tl(L));                c2
=   in
=    (while !LL <> [] do                     c3 · n
=      (if hd(!LL) < !min                    c4 · (n − 1)
=        then min := hd(!LL) else ();        c5 · s
=       LL := tl(!LL));                      c6 · (n − 1)
=      !min)                                 c7
=   end;
```

Suppose `L` has n elements. In the declaration of `LL`, we remove the first, so the list we are working on now has $n-1$ elements. The guard of the while loop executes once for each of the $n-1$ remaining elements, plus when the list is empty, so that makes n times. The testing to see if the head of the current version of `LL` is the new minimum happens $n-1$ times—once for every element besides the first.

How many times do we assign to `min`—that is, how many times will the `then` clause be performed? We do not know. In the absence of knowledge, make a new variable, say s. Now at least we have a name for the unknown. We will worry about what s is later—for now, it stands for the number of times we change `min`.

Presumably the else clause takes no time, no matter how many times it happens. Finally, updating `LL` happens $n-1$ times. We have

$$T(n) = (c_1 + c_2 - c_4 - c_6 + c_7) + (c_3 + c_4 + c_6) \cdot n + c_5 \cdot s$$

What is s? It depends on how the list is ordered. If it goes from smallest to largest, then we will never update `min` after the original assignment to it. If the list is backwards sorted, then we will update `min` every time. Hence in the worst case, $s = n - 1$, and in the best, $s = 0$. Let $a_0 = c_1 + c_2 - c_4 - c_5 - c_6 + c_7$ and $a_1 = c_3 + c_4 + c_5 + c_6$; let $b_0 = c_1 + c_2 - c_4 - c_6 + c_7$ and $b_1 = c_3 + c_4 + c_6$. In the worst case we have

$$T_1(n) = a_0 + a_1 \cdot n$$

In the best

$$T_2(n) = b_0 + b_1 \cdot n$$

Only the constants differ. Both functions are $O(n)$.

Now for the second example. Consider our old friend `factorial`:

```
- fun factorial(0)  = 1
=    | factorial(x) = x * factorial(x-1);
```

Just by eyeballing it, it is pretty clear that the number of steps (subtractions, recursive calls, and multiplications) is going to be proportional not to the *size* of the input but the input value itself. `factorial(50)` will take twice as many steps as `factorial(25)`, but those values themselves take up the same amount of computer memory. It seems as though big-oh categories do not apply.

There is another way to look at this. Theoretically, larger numbers do take up more memory. 25 in binary is 11001, and 50 is 110010, requiring one bit more of memory. Let n be the size, in bits, of the input value x. Since the required number of bits doubles with every power of 2 that x is greater than (2 takes 2 bits, 4 takes 3 bits, 64 takes 7 bits), we have $n = \lg x$, or $x = 2^n$. Since the running time of `factorial` is linear with x, it is exponential with the size of x: $O(2^n)$.

In theory. On typical computer hardware, int and similar types have fixed length, so actually they would all have the same size. (If we did not consider the input to be of fixed size, we should not consider the output or intermediate representations as fixed size, either, and arithmetic operations would not be constant.)

Exercises

9.4.1 Determine how the running time of your function `digify` from Exercise 2.2.8 depends on the size of its input (theoretically, as though larger numbers take up more space). First consider how it depends on the size of the decimal representation of the number, then on the binary representation.

9.4.2 Analyze the running time of your iterative function `sum` from Exercise 6.9.9.

9.4.3 Analyze the running time of your iterative function `doubler` from Exercise 6.9.10.

9.5 Big-oh complexity classes

This section begins the second major part of this chapter. So far we have seen how to describe the running time of algorithms as functions of the size of the input. We also have noted informally that when we compare these functions, we do not care about constant coefficients or lower ordered terms but only the highest order term as a function of n, and with this we have have informally introduced big-oh notation.

This section will shift the course of our discussion in two ways. First, we will set our method of categorizing functions on rigorous grounds with a formal definition. Second, we will be interested in the functions in their own right, whether or not they represent the running time of any algorithm. In other words, the previous sections have been about distinctively computer science topics. The next sections will be on discrete math ideas, although the field of computer science uses these ideas very much.

Consider these two functions, based on `mergeSort` and `listMin`:

$$\begin{aligned} T_1(n) &= d_0 + d_1 \cdot n + d_2 \cdot n \cdot \lg n \\ T_2(n) &= b_0 + b_1 \cdot n \end{aligned}$$

We claim that no matter what the constants are, no matter which function is greater for small values of n, eventually T_1 will be much bigger than T_2 and stay that way. Formally, there exists some N such that for all $n \geq N$, $T_1(n) \geq T_2(n)$. We are going to prove that. It is existentially quantified, so we need to find an appropriate N and show that it works. The proof itself, however, is not very illuminating—it contains steps that seem to come out of thin air.

The whole process will make more sense if we do some scratch work first. We start with *what we want to show* and then figure out what N has to be.

$$\begin{aligned} d_0 + d_1 \cdot n + d_2 \cdot n \cdot \lg n &> b_0 + b_1 \cdot n && \text{(what we want)} \\ (d_1 - b_1) \cdot n + d_2 \cdot n \cdot \lg n &> b_0 - d_0 && \text{(isolate } n\text{)} \\ (d_1 - b_1 + d_2 \cdot \lg n) \cdot n &> b_0 - d_0 && \text{(pull } n \text{ out)} \\ n &> \tfrac{b_0 - d_0}{d_1 - b_1 + d_2 \cdot \lg n} && \text{(divide both sides)} \end{aligned}$$

What we are getting at is that if n is greater than what is on the right hand side, then the top inequality will hold. The whole thing is a little messy since we have not isolated n—there is still a term with $\lg n$ in the right hand side denominator. However, as n increases, so does both $\lg n$ and the whole denominator, which in turn means the entire right hand side gets smaller.

Based on our scratch work, we pick N so that

$$N \geq \frac{b_0 - d_0}{d_1 - b_1 + d_2}$$

We also need $\lg n > 1$, that is $n > 2$, so that we can replace the $d_2 \cdot \lg n$ term in the denominator with d_2. Finally, in order to divide both sides of the inequality by $d_1 - b_1 + d_2 \cdot \lg n$ as we did in our scratch work, we need

$$\begin{aligned} d_1 - b_1 + d_2 \cdot \lg n &> 0 \\ d_2 \cdot \lg n &> b_1 - d_1 \\ \lg n &> \tfrac{b_1 - d_1}{d_2} \\ n &> 2^{\frac{b_1 - d_1}{d_2}} \end{aligned}$$

Now we can choose our N. We will dress this up as a theorem:

Theorem 9.1 *For all $n \in \mathbb{N}$ such that $n > max(\frac{b_0 - d_0}{d_1 - b_1 + d_2}, 2, 2^{\frac{b_1 - d_1}{d_2}})$, we have $d_0 + d_1 \cdot n + d_2 \cdot n \cdot lg\, n > b_0 + b_1 \cdot n$.*

Proof. Suppose $n > \max(\frac{b_0 - d_0}{d_1 - b_1 + d_2}, 2, 2^{\frac{b_1 - d_1}{d_2}})$.
Then

$$\begin{aligned} \lg n &> 1 && \text{since } n > 2 \\ d_2 \cdot \lg n &> d_2 \\ d_1 - b_1 + d_2 \cdot \lg n &> d_1 - b_1 + d_2 \end{aligned}$$

So

$$\begin{aligned} n &> \tfrac{b_0 - d_0}{d_1 - b_1 + d_2} && \text{since } n > \tfrac{b_0 - d_0}{d_1 - b_2 + d_2} \\ &> \tfrac{b_0 - d_0}{d_1 - b_1 + d_2 \cdot \lg n} && \text{by what we showed previously} \end{aligned}$$

And finally

$$\begin{aligned} n \cdot (d_1 - b_1 + d_2 \cdot \lg n) &> b_0 - d_0 && \text{by multiplying } d_1 - b_1 + d_2 \cdot \lg \\ d_0 + d_1 \cdot n + d_2 \cdot n \cdot \lg n &> b_0 + b_1 \cdot n && \text{by algebra.} \end{aligned}$$

We can multiply both sides by $d_1 - b_1 + d_2 \cdot \lg n$ without disturbing the inequality because $n > 2^{\frac{b_1 - d_1}{d_2}}$ and so $d_1 - b_1 + d_2 \cdot \lg n > 0$. □

This is the essence of what it means that these two functions are in *different complexity classes*. The one is classified with other functions whose dominant term is in n, the other with functions whose dominant term is $n \cdot \lg n$. We write that as $O(n)$ and $O(n \cdot \lg n)$. Or goal is to understand what that means, and our tool will be that great tool which we always use to mold informal muck into mathematical rigor: the theory of sets.

The *complexity class on the order* of a function f is the set of functions that are asymptotically bounded by f scaled by some constant. Symbolically, if $f : \mathbb{W} \to \mathbb{R}^+$, then

complexity class on the order

$$O(f) = \{ \ g : \mathbb{W} \to \mathbb{R}^+ \mid \ \exists \ N \in \mathbb{W} \text{ and } c \in \mathbb{R}^+ \text{ such that } \forall \ n > N, \ g(n) \leq c \cdot f(n) \ \ \}$$

As we always say, do not let all those symbols intimidate you. If you ever picked apart the formal definition of a limit in a calculus or analysis class (we mentioned it in Section 3.12), you may notice some similarities. Here is what it all means:

- $O(f)$ is a set of functions; call an arbitrary one g.
- For g to be a member of this set, there must be a whole number N and positive real number c specific to g. N is the "eventually point" and c is the scaling factor.
- So any time n is larger than N, g will be less than or equal to f scaled by c.

Notation Note: What is $O(f)$?

There are several difficulties with big-oh notation. When we talk about a big-oh category, we mean the set of functions that are on the order of a specific function, which suggests we should write something like $O(f)$, using a function name like f. However, it is more common to use big-oh notation with part of a function description (or *rule* or *body*), things like $O(n^3)$ or $O(2^n)$. The class $O(n)$ is confusing because there is nothing in the notation to show that n is a function, not just a value or parameter. Most confusing of all is the set of constant functions, which traditionally is denoted $O(1)$.

What we really need is a way to refer to a complete description of a function, not just a name of a function or a function's body ripped from its definition. Such a notation does exist, the lambda calculus. In the lambda calculus, the expression $\lambda n.(3 \cdot n^2 + 5)$ means "the function that takes a parameter called n and returns $3 \cdot n^2 + 5$." This is exactly equivalent to anonymous functions in ML; compare with

```
fn n => 3 * n * n + 5
```

Thus it would be better to write $O(\lambda n.n)$. Unfortunately, lambda notation is not familiar enough for this to be practical and it is not as easy for algebraic manipulation in practice.

Another source of confusion is that big-oh categories are often used in ways that obscure the fact that they stand for sets. If a function f is quadratic, most authors would write $f = O(n^2)$ rather than $f \in O(n^2)$. We will stick with set notation for consistency, but students should recognize that they will likely see things like $f = O(n^2)$ in other texts.

A few simple examples will exercise this definition. Our intention for this definition is that a set like $O(n)$ will include all linear functions. Here we have it as a theorem:

Theorem 9.2 *For all* $d_0, d_1 \in \mathbb{R}$, $d_0 + d_1 \cdot n \in O(n)$.

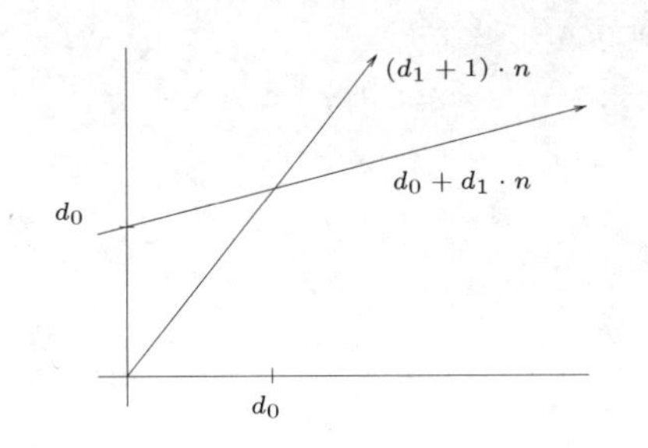

Proof. Pick $c = d_1 + 1$ and $N = d_0$. Then if $n > N$,

$$d_0 + d_1 \cdot n < n + d_1 \cdot n = (d_1 + 1) \cdot n = c \cdot n$$

□

And a similar result for all quadratic functions, that they all fit into the set $O(n^2)$:

Theorem 9.3 *For all* $d_0, d_1, d_2 \in \mathbb{R}$, $d_0 + d_1 \cdot n + d_2 \cdot n^2 \in O(n^2)$.

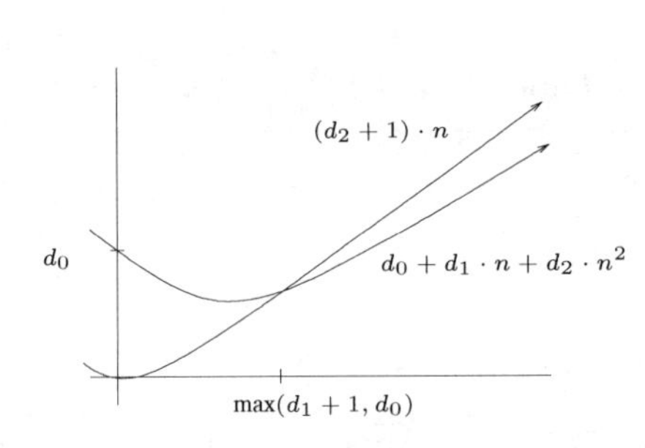

Proof. Pick $c = d_2 + 1$ and $N = \max(d_1 + 1, d_0)$. Then if $n > N$,

$$\begin{aligned}
\frac{d_0}{n} &< 1 && \text{since } n > d_0 \\
\frac{d_0}{n} + d_1 &< n && \text{since } n > d_1 + 1 \\
d_0 + d_1 \cdot n &< n^2 && \text{by multiplying by } n \\
d_0 + d_1 \cdot n + d_2 \cdot n^2 &< n^2 + d_2 \cdot n^2 && \text{by adding } d_2 \cdot n^2 \\
&= (d_2 + 1) \cdot n^2 \\
&= c \cdot n^2 && \square
\end{aligned}$$

These are pretty intimidating proofs given straight up, without the sort of scratch work we showed for Theorem 9.1. What you should notice is that it is much easier to reason our way in the reverse order from which we present them in a proof.

To derive the previous proof, start by saying "I want to find c and N so that $d_0 + d_1 \cdot n + d_2 \cdot n^2 < c \cdot n^2$." Since we know (intuitively) that the n^2 term is going to dominate, we have to pick a c that is a little bit bigger than d_2. Thus $c = d_2 + 1$ will work. Now we want to show

$$d_0 + d_1 \cdot n + d_2 \cdot n^2 < (d_2 + 1) \cdot n^2$$

Then, using algebraic manipulation, you can (almost) isolate n to the form

$$\frac{d_0}{n} + d_1 < n$$

Then what remains is the insight that if $n > d_0$, then $\frac{d_0}{n} < 1$, and thus $\frac{d_0}{n} + d_1 < d_1 + 1$. So if n is also greater than $d_1 + 1$, we have $\frac{d_0}{n} + d_1 < n$. This helps us choose $N = \max(d_1 + 1, d_0)$.

It is important that you recognize big-oh categories as being upper bounds. This means that a big-oh category is a superset of all categories of slower-growing functions. For example,

Theorem 9.4 *For all* $d_0, d_1 \in \mathbb{R}$, $d_0 + d_1 \cdot n \in O(n^2)$.

Proof. Pick $c = d_1 + 1$ and $N = \max(d_0, 1)$. Then if $n > N$,

$$d_0 + d_1 \cdot n < n + d_1 \cdot n = (d_1 + 1) \cdot n = c \cdot n < c \cdot n^2$$

We know the last part of that is true only because $n > N \geq 1$. □

And so

Corollary 9.5 $O(n) \subseteq O(n^2)$.

One more sample problem. When we were analyzing the accumulate version of `powerset`, we saw that we could turn a term with 2^{n-1} into a term with 2^n by pulling out a factor of $\frac{1}{2}$. Here is the same idea for 2^{n+1}.

Theorem 9.6 $2^{n+1} \in O(2^n)$.

First, we will do some scratch work. We want a c and N such that if $n > N$, $2^{n+1} < c \cdot 2^n$. Well, that is not so hard: $2^{n+1} = 2 \cdot 2^n$, so pick $c = 3$.

Proof. If $n > 0$, then $2^{n+1} = 2 \cdot 2^n < 3 \cdot 2^n$. □

Exercises

Prove.

9.5.1 $1000 + 1000 \cdot n \in O(.0001 \cdot n^2)$.

9.5.2 $1000 + 1000 \cdot n \in O(.0001 \cdot n)$.

9.5.3 For all $c \in \mathbb{R}$, $c \cdot n^3 \in O(n^3)$.

9.5.4 $O(n^2) \subseteq O(n^3)$. (You may find it easier to prove a proposition analogous to Theorem 9.4 first and then prove this as a corollary.)

9.5.5 $O(2 \cdot n + 5 \cdot n^2) = O(3 \cdot n^2 - 8)$

9.5.6 For all $c \in \mathbb{R}$, $c \cdot 2^n \in O(2^n)$.

9.5.7 For all $c \in \mathbb{R}$, $2^{c+n} \in O(2^n)$.

9.5.8 $n^2 \notin O(n)$.

9.5.9 $2^{2 \cdot n} \notin O(2^n)$.

9.5.10 For all $c \in \mathbb{N}$, $O(n^c) \subseteq O(2^n)$. (This says that any polynomial function is on the order of 2^n.)

9.5.11 If f and g are functions $\mathbb{W} \rightarrow \mathbb{R}^+$, then $\max(f(n), g(n)) \in O(f(n) + g(n))$.

9.5.12 If $f, g \in O(h)$, then $f(n) + g(n) \in O(h)$.

9.6 Big-theta and family

The previous section hit hard on the idea that big-oh categories are sets. When we use an idea like $O(f)$, we mean a set of functions bounded by a function f scaled by some constant, as though we were establishing some club that revolves around function f.

To explore the nature of these sets, it may help to think of them as a relation, say R defined so that $(f, g) \in R$ if $f \in O(g)$. Now we can ask questions about how big-oh categories relate to each other by phrasing them in terms of the relation R. For example, is R transitive? That is, is it the case that

Theorem 9.7 (Transitivity of O) *If $f \in O(g)$ and $g \in O(h)$, then $f \in O(h)$.*

Since we called this a theorem, you know the answer. Let us prove it.

Proof. Suppose $f \in O(g)$ and $g \in O(h)$. Then there exist c_1, c_2, N_1, and N_2 such that for all $n > N_1$, $f(n) < c_1 \cdot g(n)$ and for all $n > N_2$, $g(n) < c_2 \cdot h(n)$.

Let $c_3 = c_1 \cdot c_2$ and $N_3 = \max(N_1, N_2)$. So, if $n > N_3$,

$$\begin{array}{rcll} f(n) & < & c_1 \cdot g(n) & \text{since } n > N_3 \geq N_1 \\ & < & c_1 \cdot (c_2 \cdot h(n)) & \text{since } n > N_3 \geq N_2 \text{ and hence } g(n) < c_2 \cdot h(n) \\ & = & c_3 \cdot h(n) & \end{array}$$

Therefore $f \in O(h)$. □

Is R symmetric? That is, if $f \in O(g)$, is $g \in O(f)$? No. Notice that $n \in O(n^2)$ and $n^2 \notin O(n)$—as you proved in Exercise 9.5.8

Practically, this is a problem because if you say that an algorithm's running time is $O(f)$, it might be a very imprecise measure because the running time might be much less that f. If the running time is n nanoseconds, it is perfectly true to categorize it as $O(2^n)$, but not very helpful.

Big-oh notation essentially says "it is not worse than." We can make a new categorization that indicates that a function f can serve as a lower bound for g, that is, that g is no better than f. We call this *big-omega* notation and write $g \in \Omega(f)$.

big-omega

Formally, if $f : \mathbb{W} \to \mathbb{R}^+$, then

$$\Omega(f) = \{ \, g : \mathbb{W} \to \mathbb{R}^+ \mid \ \exists\, N \in \mathbb{W} \text{ and } c \in \mathbb{R}^+ \text{ such that } \forall\, n > N, c \cdot f(n) \leq g(n) \ \}$$

This is just like the definition of big-oh except that the scaled f is less than g. The quantification is the same.

The most useful categorization is when a function f can be scaled in two ways to provide both an upper bound and a lower bound for g. In other words, g tracks precisely with f, and f is a *tight bound* for g. We indicate this with *big-theta* notation and write $g \in \Theta(f)$.

tight bound

big-theta

$$\Theta(f) = \{\, g : \mathbb{W} \to \mathbb{R}^+ \mid \ \exists\, N \in \mathbb{W} \text{ and } c_1, c_2 \in \mathbb{R}^+, \text{ such that } \forall\, n > N, c_1 \cdot g(n) \le f(n) \le c_2 \cdot g(n) \,\}$$

The definition of big-theta simply combines the definitions of big-oh and big-omega. Notice that we need to assume two constants, one for the lower bound and one for the upper.

Now we explore the relationships among these categorizations. While we have seen before that the relation based on big-oh categories is not symmetric, it is clear intuitively that big-oh and big-omega have a sort of symmetry with each other, which we call *transpose symmetry*.

transpose symmetry

Theorem 9.8 (Transpose symmetry of O and Ω) *If f and g are functions $\mathbb{W} \to \mathbb{R}^+$, then $f \in O(g)$ iff $g \in \Omega(f)$.*

Proof. Suppose $f \in O(g)$. Then there exist N and c such that if $n > N$, then $f(n) \le c \cdot g(n)$. So, if $n > N$,

$$\frac{1}{c} \cdot f(n) \le g(n)$$

Hence $g \in \Omega(f)$. The converse is similar. □

We can take a similar approach to show that a relation based on big-theta—that is, R_1 defined so that $(f, g) \in R_1$ if $f \in \Theta(g)$ is symmetric, unlike big-oh and big-omega alone.

Theorem 9.9 (Symmetry of Θ) *$f \in \Theta(g)$ iff $g \in \Theta(f)$*

Proof. Suppose $f \in \Theta(g)$. By definition of Θ, there exist N, c_1, c_2, such that if $n > N$, then

$$c_1 \cdot g(n) \le f(n) \le c_2 \cdot g(n)$$

Thus $\frac{1}{c_1} f(n) \ge g(n)$ and $\frac{1}{c_2} \cdot f(n) \le g(n)$, that is,

$$\frac{1}{c_2} \cdot f(n) \le g(n) \le \frac{1}{c_1} \cdot f(n)$$

Therefore $g \in \Theta(f)$.

The converse is true by symmetry. □

Notice that O is analogous to $\leq$, Ω to $\geq$, and Θ to $=$. What if we would like things to correspond to $<$ and $>$? That is, an upper bound that is *not* tight: $f \in O(g)$ but $f \notin \Theta(g)$. Such things exist, called *little-oh* and *little-omega*.

little-oh

little-omega

$$o(f) = \{\, g : \mathbb{W} \to \mathbb{R}^+ \mid \forall\, c \in \mathbb{R}^+,\ \exists\, N \in \mathbb{W} \text{ such that } \forall\, n \geq N, g(n) < c \cdot f(n) \ \}$$

$$\omega(f) = \{\, g : \mathbb{W} \to \mathbb{R}^+ \mid \forall\, c \in \mathbb{R}^+,\ \exists\, N \in \mathbb{W} \text{ such that } \forall\, n \geq N, c \cdot f(n) < g(n) \ \}$$

These two are a little harder to wrap one's mind around. As is true with so many of the most difficult things in this book, the challenge is in the quantification. Like big-oh and big-omega notation, we assume a scaling factor c and an "eventually" point N. However, now the c is universally quantified and the N is existentially quantified but nested within the quantification of c. The intent for big-oh notation is that there exists a way to scale f to get an upper bound for g. In little-oh notation, we are saying that *any* scaling will work.

Unpacking the quantification, then, we get that $g \in o(f)$ if

- for any positive real number c you pick,
- if you scale f by c
- there is a whole number N
- so that after N, $g(n)$ is less than $f(n)$.

Little-omega is the same but for a lower bound. Watch how the quantification comes into play in this example.

Theorem 9.10 *For all* $d_0, d_1 \in \mathbb{R}$, $d_0 + d_1 \cdot n \in o(n^2)$.

Proof. Suppose $d_0, d_1 \in \mathbb{R}$. Suppose further that $c \in \mathbb{R}^+$.
Pick $N = \max(1, \frac{d_0}{c-d_1}, \frac{d_1}{c} + 1)$, and suppose $n \geq N$. Then

$$\begin{array}{rcll} n & \geq & \frac{d_0}{c-d_1} & \text{from how we chose } N \\ & \geq & \frac{d_0}{c \cdot n - d_1} & \text{since also } n \geq 1 \\ n \cdot (c \cdot n - d_1) & \geq & d_0 & \text{by multiplying by } c \cdot n - d_1 \\ c \cdot n^2 - d_1 \cdot n & \geq & d_0 & \\ d_0 + d_1 & \leq & c \cdot n^2 & \end{array}$$

Multiplying through by $c \cdot n - d_1$ works without disrupting the inequality because

$$\begin{aligned} n &\geq \tfrac{d_1}{c} + 1 && \text{by how we chose } N \\ &> \tfrac{d_1}{c} \\ c \cdot n &> d_1 && \text{since } c > 0 \\ c \cdot n - d_1 &> 0 && \square \end{aligned}$$

Our intention with little-oh is to make it analogous to $<$. Just as $x < y$ iff $x \leq y$ but $x \neq y$, we should have this result:

Theorem 9.11 *$g \in o(f)$ iff $g \in O(f)$ and $g \notin \Theta(f)$.*

> **Proof.** ($\Rightarrow$) Suppose $g \in o(f)$. Then pick any c and we immediately have $g \in O(f)$. Now suppose $g \in \Theta(f)$. Then there exists c_1 and N such that if $n > N$, $c_1 \cdot g(n) < f(n)$, contradicting $g \in o(f)$.
>
> ($\Leftarrow$) Suppose $g \in O(f)$ and $g \notin \Theta(f)$. Suppose there exists $c_1 \in \mathbb{R}^+$ such that for all $n > 0$, $g(n) \not< c_1 \cdot f(n)$, that is, $c_1 \cdot f(n) \leq g(n)$.
>
> Since $g \in O(f)$, there exists c_2 and N such that if $n > N$, $g(n) \leq c_2 \cdot f(n)$. Hence if $n > N$, $c_1 \cdot f(n) \leq g(n) \leq c_2 \cdot f(n)$, so $g \in \Theta(f)$. Contradiction.
>
> Hence there exists no such c_1, so for all c, there exists N such that if $n > N$, $g(n) < c \cdot f(n)$, and so $g \in o(f)$. $\square$

One last theorem. Part of the significance of a function that is $o(f)$ is that you can add that function to $f(n)$ without changing complexity class. That is,

Theorem 9.12 *If $g \in o(f)$, then $f(n) + g(n) \in \Theta(f)$.*

Think carefully about the quantification.

> **Proof.** Suppose $g \in o(f)$. Then for all $c > 0$, there exists $N > 0$ such that for all $n \geq N$, $g(n) < c \cdot f(n)$.

We want to show

$$c_1 \cdot f(n) < f(n) + g(n) < c_2 \cdot f(n)$$

Choose c_1 and c_2 to make this work.

> Let $c_1 = \frac{1}{2}$ and $c_2 = 2$.
>
> Since $g \in o(f)$, by definition there exist $N \in \mathbb{W}$ such that for all $n \geq N$, $g(n) < f(n)$.

N comes from our analytic use of the definition of little-oh. We will use that same N in our synthetic use of the definition of big-theta.

Suppose $n \geq N$. Then

$$\begin{array}{rcll} \frac{1}{2} \cdot f(n) & < & f(n) & \\ & < & f(n) + g(n) & \text{since } g(n) > 0 \\ & < & f(n) + f(n) & \text{since } g(n) < f(n) \text{ for all } n > N \\ & = & 2 \cdot f(n) & \square \end{array}$$

Exercises

Prove.

9.6.1 For all $c \in \mathbb{R}^+$, $n^2 \in \Omega(n)$.

9.6.2 For all $d_0, d_1 \in \mathbb{R}$, $d_0 + d_1 \cdot n \in \Omega(n)$.

9.6.3 If $f \in \omega(g(n))$ then $f \notin O(g)$.

9.6.4 For all $c \in \mathbb{R}^+$, $c \cdot n \in o(2^n)$.

9.6.5 $f \in \Theta(g)$ iff $f \in O(g)$ and $f \in \Omega(g)$.

9.6.6 If $f \in \Theta(g)$, then $f \notin \omega(g)$.

9.6.7 $o(f) \cap \omega(f) = \emptyset$

9.6.8 $\max(f(n), g(n)) \in \Theta(f(n) + g(n))$

9.7 Properties of complexity classes

We began the discussion in the previous section by talking about big-oh categorization as a relation. Having introduced big-theta, big-omega, little-oh, and little-omega, we now can talk about all of these as relations, specifically looking at what relation properties apply. These all follow from the definitions, and their proofs are reserved for exercises.

We already have seen that big-oh is transitive. It turns out that all five class families are also transitive.

Theorem 9.13 (Transitivity of all complexity class families) *If f, g, and h are all functions $\mathbb{W} \to R^+$, then the following hold:*

$$\begin{array}{llllll} \textit{if} & f \in O(g) & \textit{and} & g \in O(h) & \textit{then} & f \in O(h) \\ \textit{if} & f \in \Omega(g) & \textit{and} & g \in \Omega(h) & \textit{then} & f \in \Omega(h) \\ \textit{if} & f \in \Theta(g) & \textit{and} & g \in \Theta(h) & \textit{then} & f \in \Theta(h) \\ \textit{if} & f \in o(g) & \textit{and} & g \in o(h) & \textit{then} & f \in o(h) \\ \textit{if} & f \in \omega(g) & \textit{and} & g \in \omega(h) & \textit{then} & f \in \omega(h) \end{array}$$

Which ones are reflexive? Remember the analogy, that O, Ω, Θ, o, and ω are like $\leq$, $\geq$, $=$, $<$, and $>$. Reflexivity works across the analogy.

Theorem 9.14 (Reflexivity of O, Ω, and Θ) *If $f : \mathbb{W} \to R^+$, then $f \in O(f)$, $f \in \Omega(f)$, $f \in \Theta(f)$.*

o and ω are, in fact, *irreflexive*. This means no function is little-oh or little-omega of itself.

Theorem 9.15 (Irreflexivity of o and ω.) *If $f : \mathbb{W} \to R^+$, then $f \notin o(f)$ and $f \notin \omega(f)$.*

We have already seen that big-oh and big-omega are not symmetric and that big-theta is. We mentioned earlier the idea of *transpose symmetry*. Really what this means is that the two relations are inverses of each other. Just as O and Ω are transpose symmetric, so are o and ω.

transpose symmetry

Theorem 9.16 (Transpose symmetry of o and ω) *If f and g are functions $\mathbb{W} \to R^+$, then $f \in o(g)$ iff $g \in \omega(f)$.*

Exercises

Suppose f, g, and h are functions $\mathbb{W} \to \mathbb{R}^+$. Prove.

9.7.1 If $f \in \Omega(g)$ and $g \in \Omega(h)$, then $f \in \Omega(h)$.

9.7.2 If $f \in o(g)$ and $g \in o(h)$, then $f \in o(h)$.

9.7.3 $f \in O(f)$.

9.7.4 $f \in \Omega(f)$.

9.7.5 $f \notin \omega(f)$.

9.7.6 $f \in o(g)$ iff $g \in o(f)$.

9.7.7 Which complexity class families, if any make an equivalence relation?

9.7.8 Since big-oh is analogous to $\leq$, it is natural to ask, are all functions comparable, as all real numbers are with $\leq$? Prove, on the contrary, that the functions n and $n^{1+sin(2\cdot\pi\cdot n)}$ are not comparable, that is, neither $n \in O(n^{1+sin(2\cdot\pi\cdot n)})$ nor $n^{1+sin(2\cdot\pi\cdot n)} \in O(n)$.

9.8 Tables

This is our transition to the third portion of this chapter. The programming concepts we explore in these next two sections (and apply in the extended example) do not relate directly to the study of complexity classes—in fact, one could read these sections without having worked through the earlier material in this section. But the topics are valuable in their own right, and this is an appropriate enough place to discuss them because in many cases they are valuable for reducing the time complexity of an algorithm.

One of the most useful data structures in computer science is the *table*, also known as the *dictionary* or *map*. A table associates the elements in a set of *keys*

table

keys

values

each with an element in a set of *values*. The dictionary analogy is that the keys are like the words in the dictionary and the values like the definitions. Tables have three main operations: One can check if a potential key is associated with a value in the table, look up a value in a table using a key, and insert a new association between a key and value. You can see why the term *map* is used, since the structure acts something like a function.

For example, suppose we want a list of people in an organization with their office numbers.

Name	Office
Alice	8
Bob	12
Carol	3
Dave	5
Eve	15

The three operations, again, are to check if a person is on the list, to read the office number for a person, and to add (or overwrite) an entry for a person.

The description above implies that tables are inherently mutable—the operations can change them. That differs from typical use in functional programming where we tend to view any value as a simple value that cannot change. Nevertheless, we have a tool for mutability in ML: reference variables.

To implement our office directory example, we would want a table with string keys and int values, which we represent as a list of string * int pairs. Since the pair might change—we may update someone's number if he or she moves to a new office—we want the second item in the pair actually to be an int ref. Since the list itself might change—a new person might be added—the list too should be a reference value. Finally, we wrap all this in a datatype, both to give it a name and so there is a place to hang the list.

```
- datatype table = Table of (string * int ref) list ref;
```

However, we may have other uses for this besides associating strings and ints, so we use generic types instead. We rewrite our table datatype with 'k and 'v as our key and value types, respectively.

```
- datatype ('k, 'v)table = Table of ('k * 'v ref) list ref;


datatype ('a,'b) table = Table of ('a * 'b ref) list ref
```

Tangent: Hashtables

Sometimes you will hear the sort of structure we are building here referred to as a *hashtable* or *hashmap*. Technically, *hash* refers to a specific implementation strategy, where a number is calculated from the key to determine a point in the collection of values to jump to. It does not refer to the interface of the structure. Hashtables make the most sense in a language with good support for arrays. Since our table implementation here does not use hashing, it would be wrong for us to call it a hashtable.

So ML still prefers 'a and 'b to 'k and 'v. For the moment, let us forget about the datatype wrapper and think about how to look up a value or test if it is there in the list alone.

```
- fun hasAssoc(key, []) = false
=   | hasAssoc(key, (ke,va)::rest) =
=         key = ke orelse hasAssoc(key,rest);


Warning: calling polyEqual
val hasAssoc = fn : ''a * (''a * 'b) list -> bool

- fun assoc(key, (a,b)::rest) =
=    if key = a then (a,b) else assoc(key, rest);


Warning: calling polyEqual
Warning: match nonexhaustive
          (key,(a,b) :: rest) => ...
val assoc = fn : ''a * (''a * 'b) list -> ''a * 'b
```

We are playing a risky game. For any of this to work, the keys must be something we can compare. Moreover, assoc assumes we will find the key eventually. But now we can use these as helper functions for writing the functions that operate on the datatype.

```
- fun contains(key, Table(records)) = hasAssoc(key, !records);


val contains = fn : ''a * (''a,'b) table -> bool

- fun lookup(key, Table(records)) = !(#2(assoc(key, !records)));


val lookup = fn : ''a * (''a,'b) table -> 'b

- fun insert(key, value, tab as Table(records)) =
=    if contains(key, tab)
=    then let val (k, v) = assoc(key, !records) in v := value end
=    else records := (key, ref value)::(!records);


val insert = fn : ''a * 'b * (''a,'b) table -> unit
```

To use this, we need to make a new table to start with. If we make our empty table as `Table(ref [])`, the interpreter will be confused because it has no context for inferring the key and value types. Instead, we use explicit typing to specify our intent.

```
- val tab = Table(ref ([]:(string * int ref) list));

val tab = Table (ref []) : (string,int) table

- insert("Alice", 8, tab);

val it = () : unit

- insert("Bob", 12, tab);

val it = () : unit

- lookup("Alice", tab);

val it = 8 : int

- contains("Zeke", tab);

val it = false : bool
```

If you have any experience with object-oriented programming, you know that it is not desirable to have to pass the table to the function. The operations `insert`, `lookup`, and `contains` should be viewed as something the table provides, not operations performed on the table. Furthermore, the rest of the program does not even need a reference to the table itself (that is, the list of pairs). The rest of the program should interact with the table only through the three given operations. Can we make it so that the operations are the only things in scope, that the list itself is not even exposed to the rest of the program?

We can do this by wrapping those three functions in a let expression (making them local, or *private* in object-oriented terminology) and returning handles to the three operations.

```
- fun makeTable(kernel) =
=    let val tab = Table(ref kernel) in
=       (fn (k) => contains(k, tab),
=        fn (k) => lookup(k, tab),
=        fn (k, v) => insert(k, v, tab))
=     end;


val makeTable = fn
  : (''a * 'b ref) list ->
       (''a -> bool) * (''a -> 'b) * (''a * 'b -> unit)
```

Look carefully at what we have done here. If you do not have experience in a language like Java, then it is something completely novel. If you have programmed in a language like Java, however, you can compare this to a constructor or to a class. The function `makeTable` makes a new table value `tab`, but does not return it. Instead it returns a triple containing three anonymous functions like `contains`, `lookup`, and `insert`, but with the parameter for the table reference already filled in.

```
- val (has,get,put) = makeTable([]:(string * int ref) list);


val has = fn : string -> bool
val get = fn : string -> int
val put = fn : string * int -> unit

- put("Carol", 5);


val it = () : unit

- get("Carol");


val it = 5 : int
```

Exercises

9.8.1 Write a datatype and series of functions for a two-dimensional table. The table should associate two keys with a value. The first key should be used to find a table for all associations with that as the first key; the second key should be a key into that inner table.

9.9 Memoization

When we revisited the Fibonacci sequence at the beginning of this chapter, we discussed it as an example of a recurrence relation. We did not talk about the efficiency of the straightforward implementation of the sequence as function like the one you wrote in Exercise 1.12.8. (We reckon it is safe to give the answer now and hope you did not find this page at the time when you were working on that exercise.)

```
- fun fib(0) = 0
=   | fib(1) = 1
=   | fib(n) = fib(n-1) + fib(n-2);
```

A recurrence relation for the number of steps based on n (which is the parameter itself, not the size of the parameter) will be similar to the Fibonacci recurrence relation itself:

$$T(n) = \begin{cases} c_0 & \text{if } n = 0 \text{ or } n = 1 \\ c_1 + T(n-1) + T(n-2) & \text{otherwise} \end{cases}$$

Like the the solution to the sequence itself, this is exponential in n. Exponential is bad, and in this case it is unnecessarily so. A tracing of the recursive call will show us why. Take `fib(6)`.

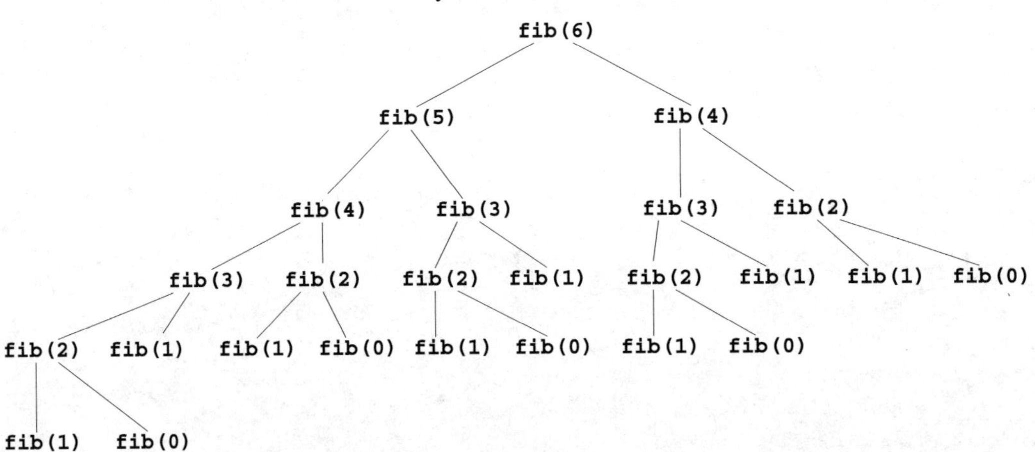

Because of the two recursive calls, much work is redone. The call `fib(2)` is itself made five times, and the base case `fib(1)` is hit seven times. The problem is that the subproblems of `fib(6)`—that is, `fib(5)` and `fib(4)`—have overlapping subproblems and therefore redundant work is done solving those subproblems.

At this point we beg forgiveness for teaching you such an obscenely inefficient way to compute the Fibonacci sequence. In this section we make amends. Specifically, we can fix this by making the function store its results in a table like we built in the previous section. Then when asked for a result it can check the table to see if it has computed the result already.

```
- val fib =
=   let
=      val (has, get, put) = makeTable([]);
=      fun memoFib(0) = 0
=        | memoFib(1) = 1
=        | memoFib(n) =
=           if has(n)
=           then get(n)
=           else let val result =
=                         memoFib(n-2) + memoFib(n-1);
=                in (put(n, result);
=                    result)
=                end;
=   in
=      memoFib
=   end;


val fib = fn : int -> int
```

This is a sophisticated piece of ML. Pick it apart carefully.

- It defines a function, but at the top level it is not a `fun` expression but a `let`, whose result we store in the variable `fib`. It defines the function `memoFib` locally and returns it as the result.
- It also defines a new table locally, though by calling the `makeTable` function it has access only to the table operations. Moreover, these are not returned, but used only locally.
- The function `memoFib` does the interesting stuff. It follows the basic structure of our old `fib` function and the base cases are identical. The recursive case, however, first checks if the result requested has been computed already by looking it up in the table. If we already did this one, then fetch the answer from the table. Otherwise, calculate it as we would in old `fib`, but put that result in the table before returning it.

Note that we did not need to give the type of the table's initial list explicitly. The outer let expression was enough context for the interpreter to infer that the list's pairs were int * int ref.

This approach, storing results for future retrieval and avoidance of redoing work, is called *memoization*. The name comes from the idea that we are creating a list of memos recording the cases we have already handled. It makes it possible to use algorithms that are intuitive but not naturally efficient.

memoization

To explore this process further, we might note that it is a lot of work to make a memoized version of a function. Our instincts urge the question, can this be automated? Could we write a function that simply takes another function and memoizes it? That function could clean up our mess after we carelessly write recursive functions with overlapping subproblems. Our attempt at this will be disappointing but interesting, instructive, and of some practical use.

The function we have in mind is fairly simple.

```
- fun memoize(f) =
=   let val (has, get, put) = makeTable([]) in
=     fn (x) => if has(x)
=               then get(x)
=               else let val y = f(x) in (put(x, y); y) end
=   end;


val memoize = fn : (''a -> 'b) -> ''a -> 'b
```

This is a fascinating function indeed. It takes another function, `f`. The declaration `val y = f(x)` tells us it should be considered to have one parameter, and it calls that type ''a (it is still possible that `f` has more than one parameter—then the type ''a would be a tuple type, and all this would still work); the result type is generalized to 'b. The anonymous function returned from `memoize`, then, must have type ''a -> 'b, and finally `memoize` has type (''a -> 'b) -> ''a -> 'b.

Now consider what happens when we apply this to the naïve `fib` function, as in

```
- val memoFib = memoize(fib);
```

This does not work as well as we hoped. Suppose we call `memofib(6)` at the prompt. The result will be stored in the table, so calling `memofib(6)` again will not rerun the process. However, an application of `memofib(7)` will be just as inefficient as before. Since it will not find 7 in the table, it will call `fib(7)` which in turn calls `fib(6)` and `fib(5)`. It will not call the memoized version. The duplicated work will cascade again, since it does not know to call `memofib` instead.

Another way to see it is that there are two sides to the inefficiency of overlapping subproblems. First, repeated calls from outside the function will request

***Tangent: Did you say* memoize?**

Teacher Elmer Fudd caught a student using a crib sheet on a test that required students to recall several formulas from class. When dragged before the principal, the student defended himself: "But Mr Fudd told us to *memoize* the formulas!"

repeated work if the same values are given to the parameters. This is true of all functions, not just functions with double recursive calls or even simple recursive functions. Second, the repeated recursive calls duplicate work. The second aspect is the more damaging. The function `memoize` addresses only the first.

What we would like is for the function to call the memoized version of itself instead of itself. Here is an attempt:

```
- val memofib =
=     memoize(fn (n) => if n = 0 then 1
=                       else if n = 1 then 1
=                       else memofib(n-1) + memofib(n-2));

Error: unbound variable or constructor: memofib
Error: unbound variable or constructor: memofib
```

No, that does not work because the scope of `memofib` does not include the definition. Only `fun`, not `val`, can handle a recursive use of the name being defined.

Try something more complicated. Instead of having `fib` call itself, it will receive a function as a second parameter to call instead. Then we memoize `fib` and call the memoized version instead, giving it the memoized version of itself to call recursively.

```
- val memofib =
=   let
=     fun fib(0, f) = 1
=       | fib(1, f) = 1
=       | fib(n, f) = f(n-1, f) + f(n-2, f);
=     val mfib = memoize(fib)
=   in
=     fn(n) => mfib(n, mfib)
=   end;

Error: operator is not a function [circularity]
  operator: 'Z
  in expression:  f (n - 1,f)
Error: operator is not a function [circularity]
  operator: 'Z
  in expression:  f (n - 2,f)
Error: operator and operand don't agree [equality type required]
  operator domain: int * ''Z
  operand:         int * (int * ''Z -> int)
  in expression:
    mfib (n,mfib)
```

If you could not follow that function, take comfort in the fact that the interpreter could not follow it either. The first problem is that an application like `f(n - 1, f)` requires the type of `f` to be recursive. Although a programmer may define a recursive type in ML using the `datatype` construct, ML's type inference is not powerful enough to discover it in cases like this. Another problem is that we would be making the key in the table to be pairs of numbers and functions. Functions are not equality types (that is, types that can be compared using =; real is not an equality type either) and so cannot be keys or parts of keys in the table.

One more try. Functions, like other values, can be stored in reference variables. This allows us to define a function to call a certain function by name but later to change what function is referred to by that name. In this case we will switch the function being called to a memoized version of the caller as soon as the memoization is complete.

```
- val memofib =
=     let
=       val f = ref (fn(n) => 0);
=       fun fib(0) = 1
=         | fib(1) = 1
=         | fib(n) = !f(n-1) + !f(n-2);
=     in
=       (f := memoize(fib);
=        !f)
=     end;

val memofib = fn : int -> int
```

Realistically, though, this is not any easier than memoizing it by hand.

Tangent: Caching

Memoization is similar to another computer science concept, *caching*, which is the storage of frequently-used data in a fast memory location. Suppose you are working on a research paper in a library. At any time you may have one or two books open in front of you as you read or take notes. Of all the information in the library, you have fastest access to the content of these books because all it takes is a glance or the turn of a page. You may, further, have a handful of books in a stack on the table that you are not using right at the moment but are handy in case you need them. If you discover you need a book that is still on the shelves then you must get up and find it, which takes the most time.

Similarly, computer memory has various levels of speed. Cache is a small area of memory that can be read from and written to faster than main memory or disk or a remote site. The operating system and hardware manage cache to make sure the most frequently read and changed data is stored there.

(If the library analogy does not work for you because you do all your research online, you can notice that in a Google search you have the option to see versions of the result pages that have been cached on Google's servers.)

Exercises

9.9.1 Consider the following solution to Exercise 4.10.2.

```
- fun exp(x, y) =
=     let
=         fun e(a, b, 0) = a
=           | e(a, b, n) =
=                 if n mod 2 = 0
=                 then e(a, b*b, n div 2)
=                 else e(a * b, b, n - 1);
=     in
=         e(1, x, y)
=     end;
```

Write two new versions of this: one in which you memoize the helper function `e` by hand, the other in which you use `memoize`.

9.9.2 The following function computes a summation, given a function to sum (which has the summation index as a parameter) and an upper and lower bounds:

```
- fun summation(f, b, n) =
=     if b > n
=     then 0
=     else f(b) + summation(f, b+1, n);
```

For example, the following functions are for arithmetic and geometric sums, respectively:

```
- fun arithmetic(d, b, n) =
=     summation(fn (x) => d * x, b, n);

- fun geometric(a, r,  b, n) =
=       summation(fn (x) => a * (exp(r, x)),
=                 b, n);
```

Write two new versions of `summation`, one that is memoized by hand, and one that is memoized using `memoize`. (In either case, you may use `memoize` on the parameter `f`; it is safe to assume `f` is not recursive.)

9.10 Extended example: The Knapsack Problem

Knapsack Problem

The *Knapsack Problem* is a well-known problem in computer science, and it has several variations. The general scenario is that a thief with a knapsack breaks into a building containing a variety of objects which he or she might wish to steal. Each object has a *weight* and a *value*. Likewise, the thief's knapsack has a finite *capacity*, meaning that it cannot contain items whose combined weight is too great. The problem is to determine which objects to take to maximize the total value without exceeding the knapsack's capacity. (It might make more sense to consider an object to have a volume rather than a weight which restricts whether it fits in the knapsack. The terms *weight* and *value*, however, allow us to use the variables w and v.)

Solutions to this problem are useful in several areas of computing, such as allocating resources in a system. A solution to the problem need not be put to such sinister ends as are implied by the problem statement.

The problem comes in a variety of flavors. Here are a few common versions:

- The thief may take any part of any object, receiving a proportional value and a proportional weight. For example, if the object is a wad of 1000 $100 bills, but the knapsack can hold only 750, the thief can take exactly 750 and leave the rest. This is called the *Fractional Knapsack Problem*.

- The thief must take an entire object or not take it at all. Assume all objects in the building either cannot be broken up or they will lose all value if they are. This is called the *0-1 Knapsack Problem.*

- The thief must take whole objects, but there happen to be an undetermined number of copies of each object, and the thief may take as many as desired. This is called the *Unbounded Knapsack Problem*.

Our interest is the last of these. Suppose the store being robbed sells items of weights 2, 3, 4, 7, 8, 13, 15 and values 1, 4, 6, 10, 11, 20, 21, as summarized below left, and the thief's knapsack has capacity 15. The thief can take five of item b for a value of 20 or one of item g for a value of 21. Each would fill the knapsack to capacity. Below right summarizes these and eight other ways to fill the knapsack to capacity.

	weight	*value*
a	2	1
b	3	4
c	4	6
d	7	10
e	8	11
f	13	20
g	15	21

	a	a	b	c	c	
w	2	2	3	4	4	15
v	1	1	4	6	6	18

	a	a	b	e	
w	2	2	3	8	15
v	1	1	4	11	17

	a	a	c	d	
w	2	2	4	7	15
v	1	1	6	10	18

	a	f	
w	2	13	15
w	20	1	21

	b	b	b	b	b	
w	3	3	3	3	3	15
v	4	4	4	4	4	20

	b	c	c	c	
w	3	4	4	4	15
v	4	6	6	6	22

	b	c	e	
w	3	4	8	15
v	4	6	11	21

	c	c	d	
w	4	4	7	15
v	6	6	10	22

	d	e	
w	8	7	15
v	11	10	21

	g	
w	15	15
v	21	21

Two arrangements are tied for the best (or, at least, the best that we have found so far): one b and three cs is worth 22, and so is two cs and one d. (Notice that a c and a b together are equivalent to one d in both weight and value.)

In this instance of the problem, it is not optimal to take the most valuable item, g weighing 15, since it yields a value of only 21 and leaves no room for any more. Item f happens to have the greatest value density ($\frac{13}{20} = .65$), but it still does not make sense to take it, since it leaves room only for one a, giving the knapsack a total value of 21.

In fact, the knapsack does not need to be filled to capacity. Taking two of item d would have weight 14, and since there are no items that weigh 1, we would not be able to put anything more in the knapsack. That would have value 20, so it would not be as good as our two configurations worth 22. But suppose there were an object of weight 14 with value 24. Taking that item would leave 1 unit of weight unused, but it would still be the best knapsack.

How can we find the best combination of items to take? The brute force way is to try every possible combination that is within capacity and keep track of the best so far. We can formulate this recursively by treating the room left in the knapsack after taking an item as its own knapsack of a smaller capacity. For example, in our knapsack of capacity 15, once we take one a we have enough room left over for items with combined weight 13. The subproblem is, what is the best way to fill the rest of the sack—that is, what is the best way to fill a knapsack of capacity 13? This relies on the problem's *optimal substructure*, the fact that best solutions to a subproblem can be used to construct a best solution to the whole problem, just as we saw with several graph algorithms in Section 8.7. For any best way to fill a knapsack, the sub-knapsacks must be filled in the best possible way.

optimal substructure

> To find the best way to fill a knapsack of capacity C choosing from items $\{(v_1, w_1), (v_2, w_2), \ldots\}$, try two possibilities:
>
> - Take (a copy) of item (v_1, w_1); find the best way to fill the remaining sub-knapsack (that is, the best way to fill a knapsack of capacity $C - w_1$ choosing from items $\{(v_1, w_1), (v_2, w_2), \ldots\}$); and form a new combination x by adding (v_1, w_1) to the best combination for the sub-knapsack.
> - Ignore item (v_1, w_1), but instead find the best combination y for filling the same knapsack (capacity C) by choosing only from the other items, $\{(v_2, w_2), \ldots\}$.
>
> Return x or y, whichever is better.

In addition to this description, we should note that if item (v_1, w_1) does not fit—that is, if $w_1 > C$—then we automatically choose the second option. Also, if C is 0 or the list of items is empty, then the best we can do is an empty knapsack; these are our base cases. This tree illustrates how subproblems are searched.

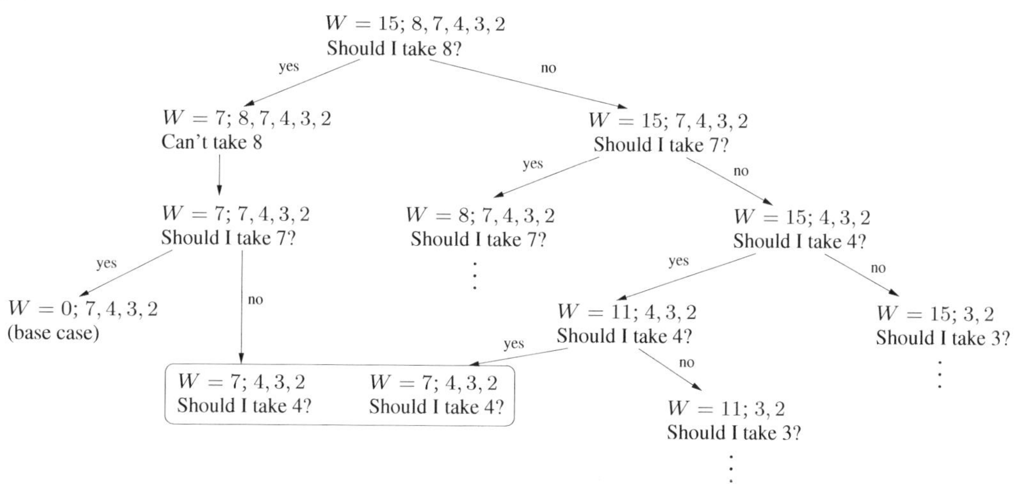

What we observe is that in more than one case, we arrive at the subproblem, "What is the best way to fill a knapsack of capacity 7 with only the items weighing 4, 3, and 2?" If we showed the entire tree, we would find large parts of it are redundant. This calls for memoization.

First, we need to distinguish between an *instance* of the problem (a capacity and a list of name/weight/value triples) and a *solution* to an instance of a problem, which would be a list of the names of items taken (the quantity of each item being represented by its frequency in the list), together with the total value, for convenience.

```
- datatype ksProb = Prob of int * (string * int * int) list;

- Prob(25, [("Rembrandt", 23, 1000), ("Diamond", 5, 500),
=           ("HulaHoop", 20, 2)]);

val it = Prob (25,[("Rembrandt",23,1000),("Diamond",5,500),
                   ("HulaHoop",20,2)])
  : ksProb

- datatype ksSoln = Soln of int * string list;
```

The use of the term "solution" is a little ambiguous here. A value of the type ksSoln is simply one way to fill a knapsack. It is not necessarily a solution in the sense of "the best way to fill the knapsack," which we are looking for. In the end, our program will be a function of type ksProb -> ksSoln, and what it should return is the best possible ksSoln.

At this point, there are some helper functions which will make parts of the program easier. For example, we may want a simple function to pick the better of two solutions, based on their value.

```
- fun maxSack(Soln(v1, items1), Soln(v2, items2)) =
=     if v1 > v2 then Soln(v1, items1) else Soln(v2, items2);
```

Do you remember ML's as construct from earlier examples? It allows us to refer to the parts of a complex parameter (as we can do through pattern-matching) and also to have a single variable to stand for the entire parameter. It makes this function a little shorter.

```
- fun maxSack(sack1 as Soln(v1, items1),
=             sack2 as Soln(v2, items2)) =
=     if v1 > v2 then sack1 else sack2;

val maxSack = fn : ksSoln * ksSoln -> ksSoln
```

The next function provides a simple way to add an item (with name n and value v) to a knapsack solution.

```
- fun addToSack(n, v, Soln(V, items)) = Soln(V+v, n::items);
```

```
val addToSack = fn : string * int * ksSoln -> ksSoln
```

Then, we need a table for memoization. Out of context, we need to tell ML the type of the seed list.

```
- val (has, get, put) =
=         makeTable([]:(ksProb * ksSoln ref) list);

val has = fn : ksProb -> bool
val get = fn : ksProb -> ksSoln
val put = fn : ksProb * ksSoln -> unit
```

Finally, here is a skeleton of the function—or, actually, functions. The function `bestSack` will handle the memoization, checking if it has seen the solution before. If not, it will call `checkEachItem`. However, `checkEachItem` will need to call `bestSack` for the subproblems. These two functions are mutually recursive—they call each other—see Section 6.3. Such functions can be written in ML by joining them with the keyword and. Note that and replaces the second `fun`:

```
- fun
=      bestSack(Prob(0, items)) = Soln(0, [])
=     | bestSack(Prob(W, [])) = Soln(0, [])
=     | bestSack(x) =   (* you won't need the components to *)
=       ??       (* the problem in this case--just call it x *)
=   and
=      checkEachItem(Prob(W, [])) = Soln(0, [])
=             (* technically, this base case is unnecessary;
=                it's just to avoid the warning *)
=     | checkEachItem(Prob(W, (n, w, v)::rest)) =
=          ??

val bestSack = fn : ksProb -> ksSoln
val checkEachItem = fn : ksProb -> ksSoln
```

The rest is up to you.

```
- val p = Prob(15, [("A", 2, 1),("B", 3, 4),("C", 4, 6),
=                    ("D",7,10),("E",8,11), ("F",13,20),
=                    ("G",15,21)]);

val p = Prob(15,
     [("A",2,1),("B",3,4),("C",4,6),("D",7,10),("E",8,11),
      ("F",13,20),("G",15,21)]) : ksProb
```

```
- bestSack(p);

val it = Soln (22,["C","C","D"]) : ksSoln
```

Project

9.A Finish the functions `bestSack` and `checkEachItem`. As you are debugging, you should reset the table (by re-pasting the call to `makeTable`) each time you make a change to your functions or start working on a different example; otherwise the table will still hold solutions from the previous (presumably erroneous) version of your functions and from a previous example.

9.B Resetting the table each time is annoying. Modify your program so that the table is local to a function to which you can just give a ksProb. You will no longer need to give an explicit type for the seed list. `bestSack` and `checkEachItem` will be local to your new function.

9.C Revise your program so that it handles the 0-1 Knapsack Problem.

9.D Write a solution to the Fractional Knapsack problem. This will involve using reals and conversion between int and real, but no memoization. Your implementation will be more efficient if you first sort the items by decreasing value density.

9.11 Special topic: P vs NP

As we have seen in this chapter, a complexity class, when talked about formally, is a set of functions, but complexity classes are often used as though they were sets of algorithms or even problems. When we say that an algorithm belongs to a complexity class, we simply mean that the function representing the running time of that algorithm belongs to the complexity class. But what would it mean for a *problem* to belong to a complexity class?

It is a question of how efficient an algorithm is possible for solving the problem. Specifically, does there exist an algorithm that solves this problem and whose running time is a function belonging to the complexity class?

Notice that this is existentially quantified. The best way to answer an existence question (if the proposition being asked is true) is to construct the object in question. To show that a problem belongs, say, to $O(n)$, we make an algorithm that solves the problem and prove that the algorithm's running time is a function that belongs to $O(n)$. Not only do we have an answer in the end, but we also have an algorithm.

Proving that a problem is not in a complexity class is harder. As we have seen in our other run-ins with proofs of nonexistence, we would have to prove this by contradiction. It is a crucial kind of question, though. Suppose you had an algorithm to solve a problem that ran in $O(n^2)$ time and needed better performance. You might try finding an algorithm that is $O(n \cdot \lg n)$, but if you knew no such algorithm exists, you would not waste your time looking for one.

How can we prove that a task cannot be accomplished in a certain amount of time—or, more accurately, how do we prove that there does not exist an algorithm whose running time *grows* with the size of the input within a certain complexity class? One example of such a result comes from the problem of sorting. It can be shown that if we rearrange elements in a list by comparing pairs of elements, no algorithm can avoid a minimum of $n \cdot \lg n$ comparisons in the worst case. Mind that that is only counting comparisons, not list manipulation or any other overhead. This means no comparison-based sorting algorithm is in a faster complexity class than merge sort is. (Nevertheless this still does not prove something inherent to the *problem* of sorting, as there do exist sorting algorithms that do not use pair-wise comparisons and are, under certain assumptions, linear in their running time.)

The most important distinction made among problems is whether or not there is an algorithm for them that can run in polynomial time. In other words, given a problem X, does there exist an algorithm A and a constant $c \in \mathbb{R}+$ such that the running time for A as a function of problem size n is $O(n^c)$. Notice that for something to be *polynomial*, there must be some exponent on n.

When we talk about algorithms that are worse than polynomial—what we would call *super*-polynomial—we mainly have in mind exponential time. There do exist functions that are super-polynomial and sub-exponential, like $n^{\lg n}$, and there do exist functions that are super-exponential, like 2^{2^n}, but algorithms in those complexity classes would be hard to come by.

Why does it make sense to draw the line here, between polynomials and exponentials? Because even really big polynomials get left in the dust by exponentials. Take for example n^{100} and 2^n. While an algorithm whose running time is $O(n^{100})$ is certainly not desirable, it is nothing like 2^n. To prove our point, we want to find an N such that for all $n > N$, $n^{100} < 2^n$. To simplify things, assume n is a multiple of 100, say $n = 100 \cdot x$.

$$
\begin{aligned}
n^{100} &< 2^n && \text{or so we claim} \\
\underbrace{n \cdot n \cdot \dots \cdot n}_{100} &< \underbrace{2 \cdot 2 \cdot \dots \cdot 2}_{n} \\
(100 \cdot x)^{100} &< 2^{100 \cdot x} && \text{by substitution} \\
&= (2^x)^{100} \\
100 \cdot x &< 2^x && \text{what we now want to show}
\end{aligned}
$$

...which is true when $x \geq 11$, since $1100 < 2^{11} = 2048$.

The set of all problems that can be solved by polynomial-time algorithms is called $\mathcal{P}$. It may have seemed strange to you earlier in this chapter when we began talking about analyzing algorithms for their complexity that we ignored constant coefficients, but now that we are talking about the theory of computational complexity we are going to lump all constant exponents into one category. A computer scientist does not care about constants, just exponents, but a *theoretical* computer scientist does not even care about constant exponents. It is only when there are variables in the exponents that we start to get worried.

As we talked about earlier, it is a straightforward thing to prove that a problem is in the class $\mathcal{P}$. Proving that a problem is *not* in $\mathcal{P}$ is a different matter. There are many problems that have been studied for years by computer scientists for which no polynomial-time algorithm has been found—but we have not been able to prove that no such algorithm exists, either.

Graph theory is a handy source for problems like this. Recall from Chapter 8 that a Hamiltonian cycle for a graph G is a cycle that visits every vertex exactly once. Consider the problem of finding a Hamiltonian cycle, if one exists, given a graph $G = (V, E)$. There are a few special cases that an algorithm can rule out—if any vertex in G has fewer than two non-loop edges incident on it, then there can be no Hamiltonian cycle—but in the general case the brute force approach is to try all orderings of the vertices, checking each ordering to determine whether edges exist to make them a cycle in the graph, checking until a cycle is found. There are $|V|!$ orderings, which is worse than $2^{|V|}$. It is impossible to make a polynomial time algorithm with that strategy.

(The preceding does not tell the whole story. The number of vertices is not the same as the size of the input. Not only does the input need to contain information about the edges, but different ways to represent the graph would take up different amounts of memory. Counterintuitively, a less space-efficient representation would mean the same algorithm's running time has a lower complexity class as a function of the input size. For example, suppose we used a bone-headed representation where the size of the input n was equal to $|V|!$. Then the number of combinations to check would be linear in the size of the input. Nevertheless, the brute force approach is provably super-polynomial for any reasonable representation of the graph.)

So far that is bad news for the Hamiltonian cycle problem. But who is to say that there is not some undiscovered algorithm that could find a Hamiltonian cycle faster? Here is one fact that dares us to look for such an algorithm: For each ordering of vertices, the brute force algorithm has $|V|$ edges to look for to check if the ordering makes a cycle, and in a reasonable representation of the graph this can be done in polynomial time. Thus if we have a potential solution to the problem, it can be checked or verified in polynomial time.

This qualifies the Hamiltonian cycle problem for a new classification. It is a member of the set of problems that can be *verified* in polynomial time, given a solution. This class is called $\mathcal{NP}$, which stands for *nondeterministic polynomial* time, because a nondeterministic algorithm could solve the problem in polynomial time. (Think, "nondeterministically guess the answer and check if it is right." If we happen to guess right, then we get the solution in polynomial time. The technical details are more complicated but not necessary for the present discussion.)

Consider another problem from graph theory, the Traveling Salesman Problem, which we introduced in Section 8.5. Given a complete, directed, weighted graph, find a Hamiltonian cycle of minimum weight. As we noted at the time, finding a Hamiltonian cycle is not the hard part, since the graph is complete. The challenge is finding the smallest such cycle. Here is one variation on the problem which will simplify our discussion. Suppose instead of finding the absolute smallest cycle, all we are looking for is a cycle no longer than some given k. Think of k as representing the travel budget for the salesman.

This budgeted version of the Traveling Salesman Problem is in the class $\mathcal{NP}$ since if we are given a route, all we need to do to verify it as a solution is see that every vertex appears exactly once (except that the first and last should be the same) and that the weights of the connecting vertices sums to no more than k.

But notice that an algorithm to solve the budgeted Traveling Salesman Problem can also be used to find a Hamiltonian cycle. Suppose we are given a graph $G_1 = (V_1, E_1)$ in which to find a Hamiltonian cycle. Construct an instance of the budgeted Traveling Salesman Problem $G_2 = (V_2, w, k)$ (we do not need an explicit set of edges E_2 because the graph is implicitly complete):

- For every vertex $v \in V_1$, make a corresponding vertex in V_2 (for convenience we will simply say $V_1 = V_2$, treating the vertices of the two graphs as identical).

- Define the weight function w such that for every edge (that is, every pair of vertices $v_i, v_j \in V_2$),

$$w(v_i, v_j) = \begin{cases} 1 & \text{if } (v_i, v_j) \text{ is an edge in } E_1 \\ 2 & \text{otherwise} \end{cases}$$

- Let $k = |V_1|$.

In other words, give every existing edge in G_1 a weight of 1 and every missing edge a wieght of 2. Then require the cycle to have a total weight of no more than k, which means it can use only edges of weight 1—in other words, it uses only edges that appear in the original graph and thus is a Hamiltonian cycle for G_1. We call this *reducing* the Hamiltonian Cycle Problem to the Traveling Salesman Problem.

reduction

Here is the big deal: Since the reduction itself takes polynomial time, if we had a polynomial-time algorithm to solve the Traveling Salesman Problem, we could build a polynomial-time algorithm to solve the Hamiltonian Cycle Problem. If the first is in the class $\mathcal{P}$, then so is the second.

Since the Hamiltonian Cycle Problem and the Traveling Salesman Problem are both from graph theory and both involve Hamiltonian cycles, the fact that one can be reduced to the other is not surprising. However, researchers have found a large number of polynomial-time reductions among $\mathcal{NP}$ problems from ostensibly dissimilar domains: problems involving logical formulas and circuits, problems for scheduling events, and problems similar to the Knapsack Problem.

One of the most marvelous results is identifying a set of problems to which *any* $\mathcal{NP}$ problem can be reduced in polynomial time. This class is called $\mathcal{NP}$-Complete, and it includes the Traveling Salesman Problem and the Hamiltonian Cycle Problem. The importance of the class is that if there exists an $\mathcal{NP}$-Complete problem that can be solved in polynomial time, then so can *every* $\mathcal{NP}$ problem. That is, $\mathcal{P} = \mathcal{NP}$. (Technically, it means only that $\mathcal{NP} \subseteq \mathcal{P}$, but $\mathcal{P} \subseteq \mathcal{NP}$ is obvious.)

The contrapositive is that if there is any problem in $\mathcal{NP}$ for which we can prove no polynomial time algorithm solving it exists, then we know that no $\mathcal{NP}$-Complete problem can be solved in polynomial time. In that sense, $\mathcal{NP}$-Complete problems are the hardest $\mathcal{NP}$ problems.

After decades of research, most computer scientists suspect $\mathcal{P} \neq \mathcal{NP}$ and therefore $\mathcal{NP}$-Complete problems are inherently intractable. (No one has been able to prove it, though.) If a problem is found to be $\mathcal{NP}$-Complete, that does not mean all hope is lost. It means that instead of trying to find a perfect algorithm that finds the best answer efficiently every time, effort would be spent better by developing an approximation algorithm or one that handles special cases of the problem.

Chapter summary

The complexity of algorithms had been studied even before the invention of machines to automate those algorithms. Analyzing algorithms for their efficiency has become crucial for judicious use of computational resources. The increase in computing speed and power that we have witnessed during the rise of the computer age has never alleviated the need to make software efficient. Rather, as more software is run on small devices like embedded computers, tablets, and personal electronics, there is all the greater concern for efficient use of constrained resources.

Even apart from their practical value, the formal definitions of complexity classes provide a domain in which to practice our proving techniques, especially as they exercise reasoning through quantification.

Key ideas:

A recurrence relation is a formula or rule for defining a sequence in which elements in the sequence are defined in terms of earlier elements.

Big-oh notation is used to categorize how a function's value grows. Formally, a big-oh class $O(f)$ is a set of functions for which we can scale the function f to make an upper bound.

A big-omega class $\Omega(f)$ is similar to a big-oh class, except that it denotes a set of functions for which we can scale f to be a lower bound. A big-theta class $\Theta(f)$ is used when f can be scaled for an upper and lower bound. Little-oh classes $o(f)$ and little-omega classes $\omega(f)$ are the sets of functions for which f is a strict upper and strict lower bound, respectively.

Big-oh, big-omega, big-theta, little-oh, and little-omega are analogous to $\leq$, $\geq$, $=$, $<$, and $>$, respectively.

A table is a data structure that associates keys with values. The basic operations on tables are testing to see if a key exists in the table, looking up the value associated with a key, and associating a new value with a key.

Memoization is the strategy of recording previously-computed results for a function for later retrieval, to avoid redundant computation. Memoization can make algorithms with overlapping subproblems more efficient.

The class of problems that can be computed in polynomial time is denoted $\mathcal{P}$, and the class of problems for which solutions can be verified in polynomial time is denoted $\mathcal{NP}$. If a problem in the class $\mathcal{NP}$ has the property that any $\mathcal{NP}$ problem can be reduced to it in polynomial, then that problem is $\mathcal{NP}$-complete. If any $\mathcal{NP}$-complete problem is found to be computable in polynomial time, then any $\mathcal{NP}$ problem can be computed in polynomial time. Contrapositively, if any $\mathcal{NP}$ problem is shown not to be computable in polynomial time, then no $\mathcal{NP}$-complete problem is computable in polynomial time.

Chapter 10 Lattice

Lattices are an example of an algebraic structure—a set together with a few operations. The power of mathematical structures—as we saw with graphs in Chapter 8 and will see with groups in Chapter 11—is their use in finding patterns across domains that do not immediately look similar. Lattices in particular are useful in generalizing many natural and human-built phenomena.

Lattices are an outgrowth of partially ordered sets, which we saw in Section 5.8. They provide an opportunity to practice proofs, to explore the concept and use of isomorphism, and to see how mathematical structures are applied in digital technology.

Chapter goals. The student should be able to

- use the basic definitions and terms about graphs, recognizing the common applications.
- prove theorems about lattices, including propositions about isomorphisms.
- understand and use the tests for distributivity and modularity in lattices.
- recognize the attributes and applications of boolean algebras.

10.1 Definition and terms

Section 5.8 introduced the idea of a partial order relation (*partial order* for short). As its name suggests, a partial order on a set proscribes a way to put elements in a set in a sequence—just as as we can organize a set of numbers from least to greatest, a set of events in the order they happened, a set of people by height, or a set of university courses by their prerequisite dependencies. What makes a partial order *partial* is that it does not necessarily disambiguate all pairs of elements. There may be some a and b in the set for which the partial order cannot tell you which comes first. In the examples already mentioned, there may be two events that occurred at the same time, two people with the same height, or courses that may be taken in any order.

What a partial order, say relation R, does provide is a guarantee of coherence. If a and b are distinct and $(a, b) \in R$, then $(b, a) \notin R$. Moreover if $(b, c) \in R$, then $(a, c) \in R$. If Betty is shorter than John and John is shorter than Sue, we know that John is not shorter than Betty and that Betty is shorter than Sue.

Do you remember the formal definition of *partial order*? If not, you should be able to infer it from what we have just said, because we have basically given the definition of antisymmetry and transitivity. Traditionally the definition of partial order also includes reflexivity (although most of the interesting properties of partial orders do not use it). Thus,

> A *partial order relation* is a relation R on a set X that is reflexive, transitive, and antisymmetric. A set X on which a partial order is defined is called a *partially ordered set* or a *poset*.

Transitivity and antisymmetry make it possible to represent posets with a form of a digraph modified to reduce clutter, called a Hasse diagram: arrange the items so that arrows always point up, then erase the arrow heads, self loops, and any arrow that is implied by transitivity. Here are the posets $\mathscr{P}(\{1, 2, 3\})$ on $\subseteq$, the set of divisors of 30 on $|$ (generally, let D_n be the set of divisors of $n \in \mathbb{N}$), and the set of tributary rivers to the Mississippi on "flows (eventually) into" (the "eventually" part is for transitivity).

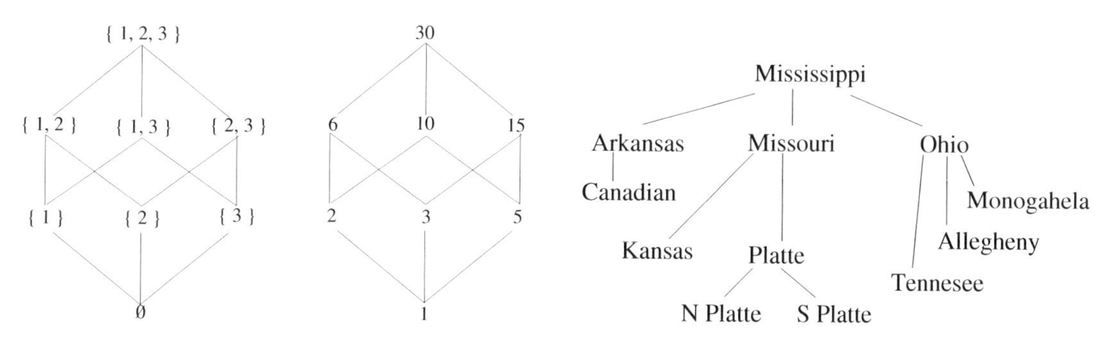

We use the symbol $\preceq$ for a generic partial order. For a partial order $\preceq$ on a set A, $a, b \in A$ are *comparable* if $a \preceq b$ or $b \preceq a$. 6 and 30 are comparable for $|$ $(6|30)$, but 6 and 15 are not comparable. For $\subseteq$, $\emptyset$ is comparable to everything; in fact, it is a subset of everything. The Mississippi is comparable to everything (since everything flows into it), but the Kansas and the Platte are not comparable.

In this chapter we explore a structure that is a refinement of the idea of a poset. Notice how $\mathscr{P}(\{1,2,3\})$ and the divisors of 30 have a common structure that vastly differs from the tributary system. How might we characterize the similarities and differences?

Notice in the tributary example that although Kansas and Platte are not comparable, they do both flow into Missouri. Moreover Missouri is the least (in the partial order) that they both flow into. Similarly Kansas and Allegheny have the Mississippi as the least element that they are both less than. None of these, however, have a common element that they are both greater than. If any two elements in a poset were to have a common element above and below them, then we could define other operations and make certain guarantees about their structure.

upper bound

Suppose A is a poset with relation $\preceq$, and suppose $B \subseteq A$. An element $a \in A$ is an *upper bound* of a subset B if for all $b \in B$, $b \preceq a$. Thus 5, 10, 15, and 30 are all upper bounds for $\{1, 5\}$. Mississippi, likewise, is an upper bound for { Kansas, Platte, Allegheny }.

least upper bound (LUB)

lower bound

greatest lower bound (GLB)

Furthermore, a is a *least upper bound (LUB)* of B if for all upper bounds c of B, $a \preceq c$. Similarly we can define *lower bound* and *greatest lower bound (GLB)*. 5 is the LUB for $\{1, 5\}$, and 1 is the GLB. Mississippi is the LUB of { Kansas, Platte, Allegheny }; as there is no lower bound, there certainly is no GLB.

lattice

Now for the main definition of this chapter. A poset A with relation $\preceq$ is a *lattice* if every pair of elements (or every subset with cardinality 2) has an LUB and a GLB. Suppose A is a lattice and $a, b \in A$. Then we define the *join* operator $\vee$ and the *meet* operator $\wedge$ to be the LUB and GLB, respectively, of a and b. These definitions immediately imply

Theorem 10.1 *If L is a lattice and $a, b \in L$, then $a \preceq a \vee b$ and $a \wedge b \preceq a$.*

$\mathscr{P}(\{1,2,3\})$ with $\subseteq$ is a lattice; here $\vee = \cup$ and $\wedge = \cap$. The set of divisors of 30 with $|$ is a lattice; here, as with any set D_n, $\vee$ is the least common multiple

Tangent: What is a lattice?

The word lattice primarily refers to building material with slats of wood arranged in a criss-cross pattern, as in a trellis or a screen to filter light from a window. The word is related to lath, a slat of wood as in a lattice or as used in a framework for tiles or plaster.

The mathematical structure is named simply from the appearance of the Hasse diagrams. The word is also applied to other things that in some way resemble the structure of lattice work: criss-cross patterns in heraldry, the arrangement of ions in a crystal, a part of a machine for processing cotton or wool, and a kind of electrical network.

(LCM) and $\wedge$ is the greatest common divisor (GCD). The set of tributaries is not a lattice because, for example, Kansas and Platte have no GLB.

The following Hasse diagrams are examples of structures that lattices may have.

Any total order is partial order, and we notice here that a total order is also a lattice. Any set of four numbers with $\leq$ would look like this.

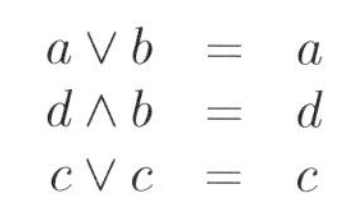

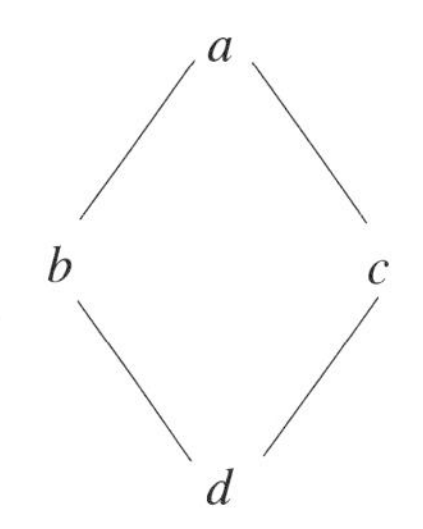

Any powerset of a set of cardinality 2 would have this structure.

$$b \vee c = a$$
$$b \wedge c = d$$
$$a \wedge b = b$$

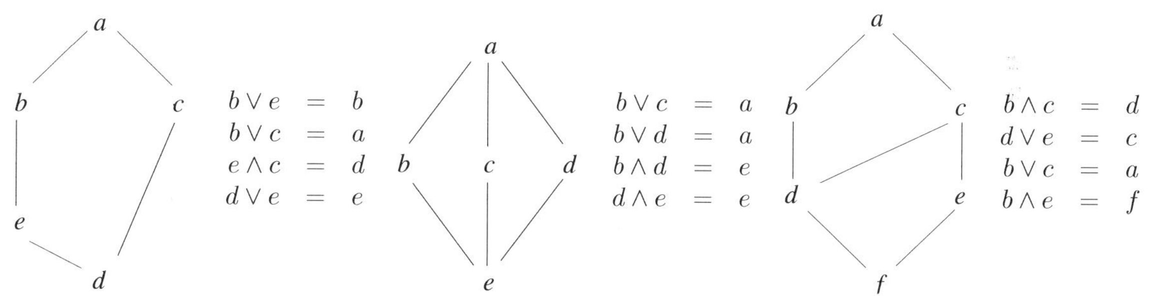

To gain an intuition for the structures of lattices it also helps to consider posets that are not lattices.

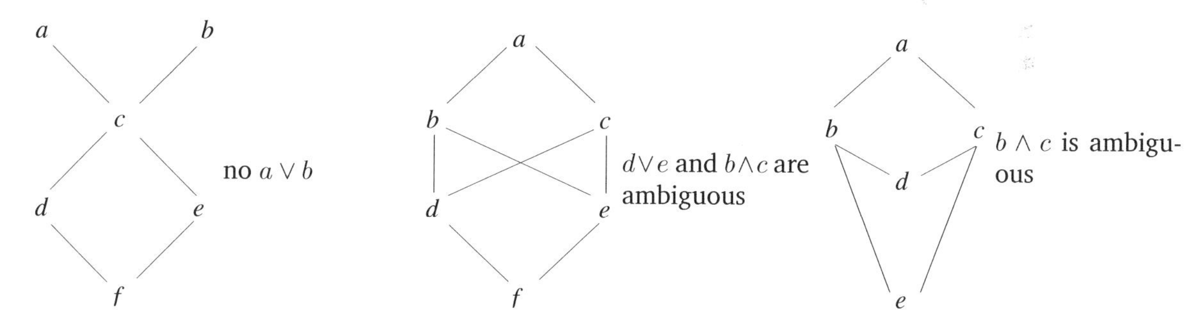

All the lattices in the diagrams above are finite. There do exist infinite lattices, though: $\mathbb{N}$ on $|$ and $\mathbb{R}$ on $\leq$ are both lattices and both infinite.

Another feature of the lattices we have drawn is that they all have a topmost element and a bottommost element. Formally, if L is a lattice, the LUB of the entire set, if it exists, is called the *top*, denoted $\top$; the GLB of the entire set, if it exists, is called the *bottom*, denoted $\bot$. If $\top$ and $\bot$ exist, then we say the lattice is *bounded*.

top

bottom

bounded

Not all lattices are bounded—looking again at $\mathbb{N}$ on $|$ and $\mathbb{R}$ on $\leq$, we see that $\mathbb{N}$ has 1 as $\bot$ but no $\top$ and $\mathbb{R}$ has neither $\bot$ nor $\top$. If the examples so far make you suspect that all finite lattices are bounded, you are correct. We will prove that in the next section. The converse is not true, however. We can construct a bounded, infinite lattice by taking the real interval $[0, 1]$ (that is, all real numbers between 0 and 1, inclusive) and defining a relation $\preceq$ so that

$$a \preceq b \text{ if } a = b \text{ or } \lceil a \rceil \leq b$$

Verify for yourself that 0 is $\bot$ and 1 is $\top$ but no two numbers between 0 and 1 are comparable.

Note that by definition, we have the following properties for $\top$ and $\bot$. If L is a bounded lattice, then for all $a \in L$,

$$\begin{array}{rclcrcl} a \vee \top & = & \top & \quad & a \vee \bot & = & a \\ a \wedge \top & = & a & \quad & a \wedge \bot & = & \bot \end{array}$$

This means that $\top$ is a universal bound for $\vee$ and an identity for $\wedge$; likewise $\bot$ is an identity for $\wedge$ and a universal bound for $\vee$.

sublattice

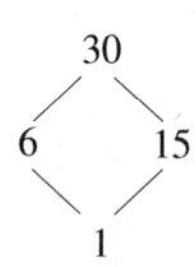

If L with $\preceq$ is a lattice, then $S \subseteq L$ is called a *sublattice* of L if for all $a, b \in S$, $a \vee b \in S$ and $a \wedge b \in S$. In that definition, it is important that $\vee$ and $\wedge$ are understood as the operators on L and not operators modified for S. In other words, to be a sublattice of L does not mean simply to be a subset that is itself a lattice—no, the meet and join operators must be the same. Consider the set $\{1, 6, 15, 30\}$ with $|$, shown on the left. This is indeed a subset of D_{30} and it is a lattice, but it is not a sublattice of D_{30} because $6 \wedge 15 = 1$, whereas in D_{30}, $6 \wedge 15 = 3$. For positive examples, D_{15} (the set of divisors of 15) is a sublattice of D_{30}, and $\mathscr{P}(\{1, 3\})$ is a sublattice of $\mathscr{P}(\{1, 2, 3\})$.

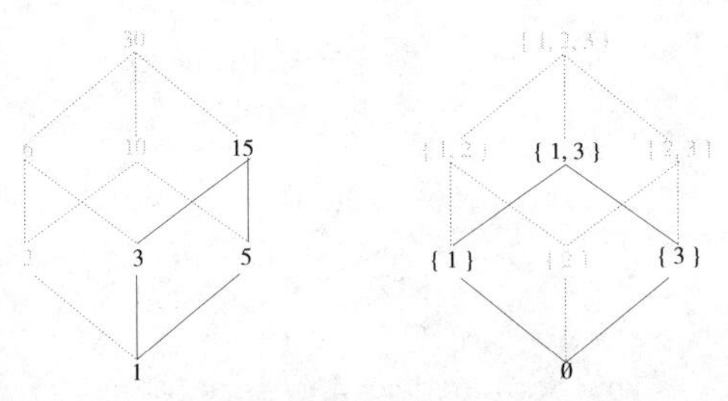

Tangent: Keeping lattice terms straight

Liz was enjoying a sunset from the roof of the building where she worked at the end of the day. She called Stephan, still working two stories down, to come up and see the view before they left for the day. They would not go back down together, however, because Liz takes the elevator and Stephan takes the stairs. Stephan said, "I'll join you at the top and then meet you at the bottom." (As we will see later, that is a complement.)

We can find lattices in everyday phenomena. Consider the water and sanitation system in a typical building. There are some pipes supplying clean tapwater and other pipes carrying away sewage. We can describe this situation as a small set { tapwater, sewage } with a relation isDirtierThan, which makes a poset. It is also a lattice, though not a very interesting one, where the $\wedge$ operator indicates the designation of mixing two flows of water. Sewage mixed with sewage (like when sewage pipes converge as they go to the city sewer) is of course sewage, and tapwater mixed with tapwater (as when hot and cold tapwater are mixed in a faucet) is still tapwater. If tapwater and sewage meet, the sewage contaminates the tapwater and the result is designated sewage. An expression like $a \vee b$ does not have as obvious an interpretation at this point, but we will return to it later.

One modification a building can have for environmental friendliness is a greywater system. Greywater refers to such things as the output of fixtures used for washing (sinks, baths, showers, dishwashers, and laundry machines) but not containing human waste such as from toilets. While greywater would not be potable or usable for further washing, it could be used for flushing toilets and irrigation. For our purposes, we note that greywater fits neatly into our lattice, as greywater mixed with tapwater is greywater and greywater mixed with sewage is sewage.

We can take this idea further. Some buildings are equipped with cisterns to collect rainwater, which would not be potable but could be used for the same purposes as greywater and more, such as some kinds of washing. In areas with hard water as from wells, buildings require water softening to reduce the concentration of rocky ions. The resulting water has a higher concentration of sodium and is less healthy to drink. Finally, suppose we wish to collect organic material such as food waste and waste from toilets for use as fertilizer, keeping it free from chemicals like salts and soaps. We could then add rainwater, softwater, and organic waste pipes, which would fit into our lattice as shown below (notice that either softwater or rainwater could flow into greywater, but mixing softwater and organic waste would result in sewage).

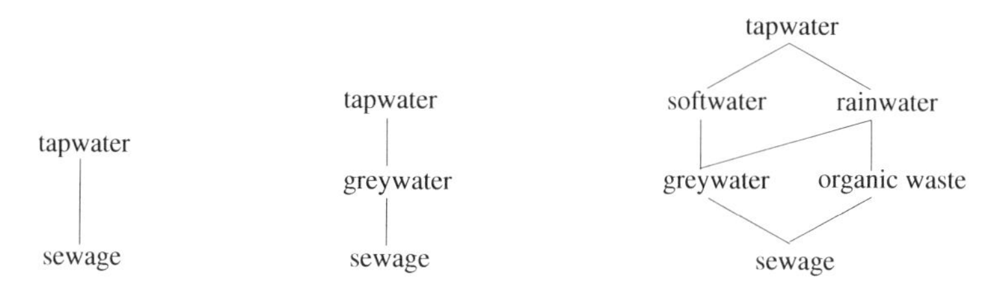

The usefulness of the lattice concept is its abstraction, as we see here: whatever the complexity of the actual system, we can reason about the joining of flows uniformly by using lattice operations. (From the last version we can also form an interpretation for the $\vee$ operator. The elements of the lattice can be used not only to categorize flows and pipes but also to categorize minimum sanitation requirements for a given purpose. For example, watering plants is a greywater need; softwater, rainwater, or tapwater would also do. If a and b are the sanitation requirements for two uses that need to be serviced by a single pipe, then that pipe requires $a \vee b$, that is, the least upper bound of the two requirements a and b.)

Lattices have proven useful in modeling information security to monitor whether an information management system will leak classified information for access at a lower level of classification. Let L be the set of security levels or clearances. The elements of L are used as classifications for documents, classification for access channels, and clearance levels for people. In a simplified version of the classification system used by American intelligence, $L = \{\text{Unclassified}, \text{Confidential}, \text{Secret}, \text{TopSecret}\}$. This set is a lattice with partial order $\preceq$ meaning "is less secret than." A person with clearance a may read any document with classification b as long as $b \preceq a$.

Moreover, if a person is given two clearances, say a and b, by different agencies, then that person's effective clearance is the upper bound of the two, $a \vee b$. If two classified documents are merged into a single document, then the resulting document must be as secret as its more secret component, which the $\vee$ operator also computes. If a document of classification a is published on access channel with classification b, then the resulting effective classification of the information is the lower of the two, $a \wedge b$.

The simplified classification system does not make for a very interesting lattice. An alternate system may categorize information by content. Suppose a database contains documents having some combination of medical, financial, and criminal records. A detective investigating securities fraud may have access documents with financial and criminal information, but a physician would have access to documents only with medical records.

In reality, intelligence agencies use a classification system with both levels and categories. A database containing information about cryptographic algorithms and potential ballistic targets may use a system like the one represented in the lattice below right.

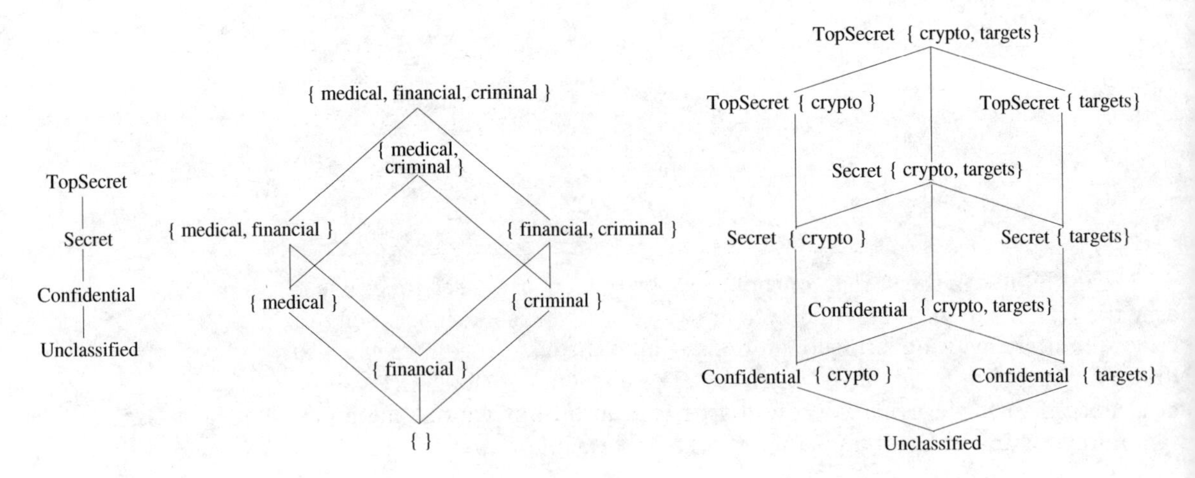

Exercises

10.1.1 Our example of an infinite, bounded lattice is much more complicated than necessary. Give a simpler example of an infinite, bounded lattice. (Hint: Use the same set, $[0, 1]$, but a more familiar partial order. Remember that a total order is considered a special case of a partial order.)

Determine which of the following are lattices.

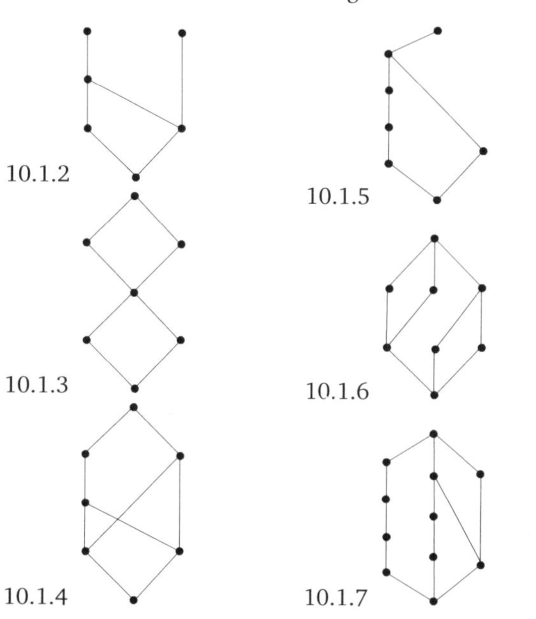

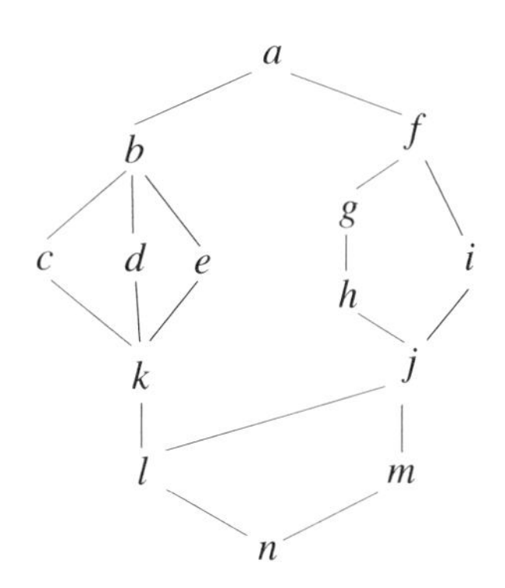

Using the lattice above, determine the results of the following operations.

10.1.8 $i \vee b$

10.1.9 $a \vee b$

10.1.10 $k \vee j$

10.1.11 $j \wedge f$

10.1.12 $m \vee b$

10.1.13 $n \wedge j$

10.1.14 $l \wedge g$

10.1.15 $l \wedge l$

10.1.16 $a \wedge l$

10.1.17 $h \wedge b$

10.1.18 $d \wedge g$

10.1.19 $j \vee a$

10.1.20 $j \wedge g$

10.1.21 $j \vee b$

10.1.22 $c \vee a$

10.1.23 $b \wedge d$

10.1.24 $f \vee c$

10.1.25 $d \wedge k$

10.2 Propositions on lattices

The previous section introduced you to the terms and concepts of lattices and built your intuition. In this section we apply those ideas to prove results about lattices. There are four kinds of results we may look for about lattices: that a set with its operation is a lattice, that a set is a sublattice of another set, that a certain property holds for all lattices, and that a certain property holds for lattices with extra assumptions.

First we take one of our examples from the previous section—or, a generalization of one—and prove that it is a lattice.

Theorem 10.2 *If A is a set, then $\mathscr{P}(A)$ with $\subseteq$ is a lattice.*

Proof. First we need to show that it is a poset, but we did that already back in Theorem 5.15.

Next suppose $B, C \in \mathscr{P}(A)$, that is, $B \subseteq A$ and $C \subseteq A$. Then $B \cup C$ is an upper bound for B and C (since $B \subseteq B \cup C$ and $C \subseteq B \cup C$). Now suppose $D \in \mathscr{P}(A)$ is an upper bound for C and B, in other words $B \subseteq D$ and $C \subseteq D$. Suppose $x \in B \cup C$. Then by definition of union, $x \in B$ or $x \in C$; either way, $x \in D$. Hence by definition of subset, $B \cup C \subseteq D$. Therefore $B \cup C$ is the LUB for B and C.

Similarly we can show that B and C have a GLB, and that shows that this is a lattice. □

In the previous section, we noted (or claimed) that $\mathscr{P}(\{1, 3\})$ is a sublattice of $\mathscr{P}(\{1, 2, 3\})$. More generally, here is a proof.

Theorem 10.3 *If A is a set and $B \subseteq A$, then $\mathscr{P}(B)$ is a sublattice of $\mathscr{P}(A)$.*

Proof. The meet and join operators are $\cap$ and $\cup$, respectively. Suppose $C, D \in \mathscr{P}(B)$. By definition of powerset, $C \subseteq B$ and $D \subseteq B$. We want to show that $C \cap D \in \mathscr{P}(B)$ and $C \cup D \in \mathscr{P}(B)$.

Suppose $x \in C \cap D$. By definition of intersection, $x \in C$ and $x \in D$. $x \in B$ by the definition of subset. By definition of subset, $C \cap D \subseteq B$, and $C \cap D \in \mathscr{P}(B)$ by definition of powerset.

Suppose $x \in C \cup D$. By definition of union, $x \in C$ or $x \in D$. Either way, $x \in B$ by definition of subset. By definition of subset, $C \cap D \subseteq B$, and $C \cap D \in \mathscr{P}(B)$ by definition of powerset.

Therefore $\mathscr{P}(B)$ is a sublattice of $\mathscr{P}(A)$. □

That proof was tedious. Fortunately many of these can be proven by referring to earlier results. Here is a sublattice proof for infinite lattices. In Exercise 10.2.1 you will show that $\mathbb{N}$ is a lattice with $|$. Moreover, $\vee$ and $\wedge$ are LCM and GCD, just as in D_n lattices. Let $\mathbb{N}_e$ be the set of even natural numbers.

Theorem 10.4 *$\mathbb{N}_e$ is a sublattice of $\mathbb{N}$ with $|$.*

Proof. Suppose $a, b \in \mathbb{N}_e$. By definition of even $2|a$ and $2|b$.

By definition of LCM, $a|a \vee b$ and $b|a \vee b$. By transitivity of $|$, $2|a \vee b$. By definition of even, $a \vee b \in \mathbb{N}_e$.

By definition of GCD, $a \wedge b|a$ and $a \wedge b|b$. Suppose $a \wedge b$ is not even. Then $a \wedge b < 2 \cdot (a \wedge b)$. Moreover, since a and b are even, $2 \cdot (a \wedge b)|a$ and $2 \cdot (a \wedge b)|b$. This contradiction $a \wedge b$ being the GCD. Hence $a \wedge b$ is even, and so $a \wedge b \in \mathbb{N}_e$. □

Next we consider properties of elements of any lattice.

Theorem 10.5 *If A is a lattice with $\preceq$, then for all $a, b, c \in A$,*

$$\begin{array}{rclrcl} a \vee b & = & b & \textit{iff} \quad a & \preceq & b \\ a \wedge b & = & a & \textit{iff} \quad a & \preceq & b \\ a \wedge b & = & a & \textit{iff} \quad a \vee b & \preceq & b \end{array}$$

Proof. (1) ($\Rightarrow$) Suppose $a \vee b = b$. That means b is the LUB of a and b. By definition of LUB, $a \preceq b$.

($\Leftarrow$) Suppose $a \preceq b$. Since $\preceq$ is reflexive, $b \preceq b$, so b is an upper bound for a and b. Now suppose c is an upper bound for a and b, that is $a \preceq c$ and $b \preceq c$. The fact that $b \preceq c$ is enough to show that b is the LUB. Hence $a \vee b = b$.

(2) and (3) are left as exercises. □

You may have noticed the similarities of the lattice operators $\vee$ and $\wedge$ to operators in other areas of mathematics—$+$ and $\times$ in arithmetic, $\cup$ and $\cap$ on sets, boolean operators $\vee$ and $\wedge$. As with those operators, we have a collection of handy and familiar properties. Compare this theorem with the table of logical equivalences in Theorem 3.1.

Theorem 10.6 *If L is a lattice, then for all $a, b, c \in L$, the following properties hold:*

Commutative laws:	$a \vee b$	$=$	$b \vee a$	$a \wedge b$	$=$	$b \wedge a$
Associative laws:	$a \vee (b \vee c)$	$=$	$(a \vee b) \vee c$	$a \wedge (b \wedge c)$	$=$	$(a \wedge b) \wedge c$
Absorption laws:	$a \vee (a \wedge b)$	$=$	a	$a \wedge (a \vee b)$	$=$	a
Idempotent laws:	$a \vee a$	$=$	a	$a \wedge a$	$=$	a
Universal bounds laws:	$a \vee \top$	$=$	$\top$	$a \wedge \bot$	$=$	$\bot$
Identity laws:	$a \wedge \top$	$=$	a	$a \vee \bot$	$=$	a

We omit the proof because most of these are pretty trivial in the definitions of $\vee$ and $\wedge$. You should note some absences, however. Since we do not have an analogue to logical negation (or set complement), we do not have laws corresponding to logical equivalences that involve negation, such as DeMorgan's laws. Another notable exception is distributivity. Not all lattices are distributive, and we will develop a test for distributivity in Section 10.4.

Before we close this section, however, we present one result about lattices with an extra assumption, a result we alluded to in the previous section.

Theorem 10.7 *Any non-empty finite lattice is bounded.*

Proof. Suppose L is a finite lattice. Let $n = |L|$. By definition of finite, $n \in \mathbb{N}$. Name the elements of L to be a_1, a_2, $\ldots a_n$. Consider the result of joining the entire set, $a_1 \vee a_2 \vee \ldots \vee a_n$.

Now suppose $a_i \in L$. We can rewrite the expression above as $a_1 \vee a_2 \vee \ldots \vee a_i \vee \ldots \vee a_n$. Then

$$\begin{aligned} & a_1 \vee a_2 \vee \ldots \vee a_i \vee \ldots \vee a_n \\ &\quad = a_1 \vee a_2 \vee \ldots \vee a_i \vee a_i \vee \ldots \vee a_n && \text{by idempotency} \\ &\quad = a_i \vee (a_1 \vee a_2 \vee \ldots \vee a_i \vee \ldots \vee a_n) && \text{by associativity and commutativity} \end{aligned}$$

Then by Theorem 10.5(1), $a_i \preceq (a_1 \vee a_2 \vee \ldots \vee a_i \vee \ldots \vee a_n)$. Hence $a_1 \vee a_2 \vee \ldots \vee a_n$ is an upper bound on the entire set.

Similarly, $a_1 \wedge a_2 \wedge \ldots \wedge a_n$ is a lower bound on the entire set. Therefore L is bounded. □

Exercises

10.2.1 Prove that $\mathbb{N}$ with $|$ is a lattice.

10.2.2 Prove that $\mathbb{B}$ (the set of boolean values) is a lattice with

- $T \preceq T$
- $F \preceq F$
- $F \preceq T$

Not surprisingly, $\vee$ and $\wedge$ align just right: logical *or* is join and logical *and* is meet.

10.2.3 Prove that for $n, k \in \mathbb{N}$, if $k|n$ then D_k is a sublattice of D_n.

10.2.4 Suppose A is a set, and let $\mathscr{R}$ be the set of all equivalence relations on A. (Think that one through—$\mathscr{R}$ is a set of *relations*, specifically all the equivalence relations on set A. For example, $i_A \in \mathscr{R}$.) Prove that $\mathscr{R}$ is a poset with $\subseteq$. Prove further that it is a lattice with join and meet defined as follows, for $R, S \in \mathscr{R}$:

- $R \wedge S = R \cap S$
- $R \vee S = (R \cup S)^T$, that is, the transitive closure of the union.

(The preceding exercise is from Kolman et al [18].)

10.2.5 Prove that if $L_1, \preceq_1$ and $L_2, \preceq_2$ are lattices, then $L_1 \times L_2, \preceq_3$ is a lattice, where $\preceq_3$ is defined so that $(a_1, a_2) \preceq_3 (b_1, b_2)$ if $a_1 \preceq_1 b_1$ and $a_2 \preceq_2 b_2$

Informally argue why each of the following must be true.

10.2.6 The commutative laws for lattices.

10.2.7 The associative laws for lattices.

10.2.8 The absorption laws for lattices.

10.2.9 The idempotent laws for lattices.

Suppose L is a lattice. Prove for all $a, b, c, d \in L$.

10.2.10 If $a \preceq b$ then $a \vee c \preceq b \vee c$ and $a \wedge c \preceq b \wedge c$.

10.2.11 $a \preceq c$ and $b \preceq c$ iff $a \vee b \preceq c$.

10.2.12 $c \preceq a$ and $c \preceq b$ iff $c \preceq a \wedge b$.

10.2.13 If $a \preceq b$ and $c \preceq d$, then $a \vee c \preceq b \vee d$ and $a \wedge c \preceq b \wedge d$.

10.2.14 If $a \preceq c$, $a \preceq d$, $b \preceq c$, and $b \preceq d$, then $a \vee b \preceq c \wedge d$.

10.3 Isomorphisms

In Chapter 8 we were introduced to the concept of graph isomorphism. Two graphs are isomorphic if they have the same essential structure even though their interpretations may differ. Take another look at D_{30} on $|$ and $\mathscr{P}(\{1,2,3\}$ on $\subseteq$.

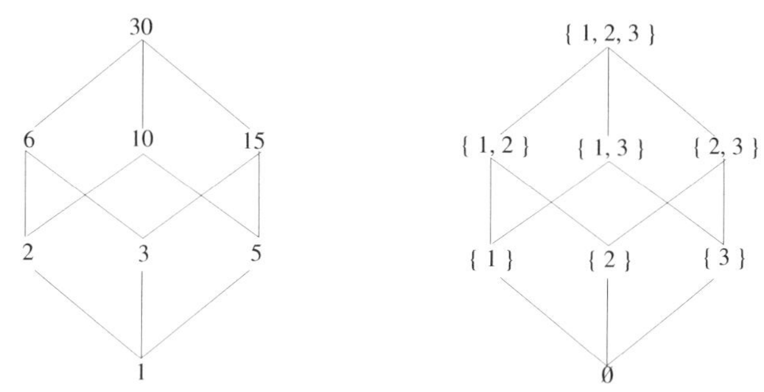

As with graphs, we clearly have corresponding structures. We want to capture this observation with a definition similar to that for graph isomorphisms. In this case we want a way to match up elements in two lattices in such a way that it preserves $\preceq$ and, along with that, $\vee$ and $\wedge$. In its formal definition, a lattice isomorphism is a bit simpler than a graph isomorphism.

isomorphism

If $L_1, \preceq_1$ and $L_2, \preceq_2$ are lattices, then L_1 is *isomorphic* to L_2 if there exists a one-to-one correspondence $f : L_1 \to L_2$ such that for all $a, b \in L_1$ $a \preceq_1 b$ iff $f(a) \preceq_2 f(b)$. f is called the isomorphism. (Notice this requires only one function, not two as with graphs.)

To show an isomorphism in this example, it might be easier first to show an isomorphism between D_{30} and $\mathscr{P}(\{2,3,5\})$.

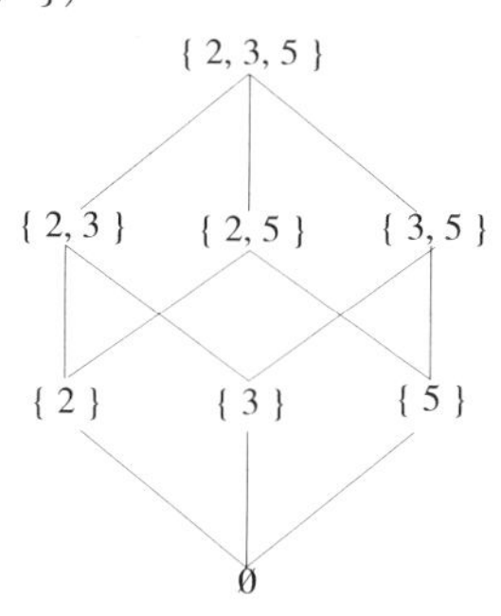

Theorem 10.8 *$\mathscr{P}(\{2,3,5\}), \subseteq$ is isomorphic to $D_{30}, |$.*

Remember that this is an existence proof. We need to construct an appropriate isomorphism (function) and show that it fits.

> **Proof.** Define $f : \mathscr{P}(\{2,3,5\}) \to D_{30}$ so that $f(A)$ is the product of the elements of A, $f(A) = \prod_{a \in A} a$. We will show first that f is a one-to-one correspondence:

First, suppose $A, B \in \mathscr{P}(\{2,3,5\})$ such that $f(A) = f(B)$. We want to show $A = B$. Suppose $a \in A$. Then $a|f(A)$, so $a|f(B)$ by substitution. Since the elements in B are relatively prime, $a \in B$, so $A \subseteq B$. By a symmetric argument, $B \subseteq A$ and $A = B$. Hence f is one-to-one.

Then suppose $x \in D_{30}$. Let A be the set of prime factors of x. Suppose $a \in A$. By definition of D_{30}, $a \in \{2,3,5\}$. By definition of subset, $A \subseteq \{2,3,5\}$, and by definition of powerset, $A \in \mathscr{P}(\{2,3,5\})$. Moreover $f(A) = x$—this happens to be true because no element in D_{30} has any number that is a prime factor more than once. Hence f is onto.

Now for the isomorphic property. Suppose $A, B \in \mathscr{P}(\{2,3,5\})$ such that $A \subseteq B$. Then, by definition of f, $f(B) = f(A) \cdot f(B - A)$. By the definition of divides, $f(A)|f(B)$.

Finally suppose $x, y \in D_{30}$ such that $x|y$. We want to show that $f^{-1}(x) \subseteq f^{-1}(y)$. Suppose $z \in f^{-1}(x)$. By definition of f, this means $z|x$. By the transitivity of $|$, we know that $z|y$. Again by definition of f, $z \in f^{-1}(y)$, so by definition of subset, $f^{-1}(x) \subseteq f^{-1}(y)$.

Therefore, by definition of lattice isomorphism□

See how even this late in our study so many proofs contain plain old Set Proof Form 1 at their heart.

Exercises

10.3.1 We defined *isomorphism* for lattices in terms of $\preceq$, not $\vee$ and $\wedge$, but some texts define it the other way. To show these two definitions are equivalent, assume our definition of isomorphism and show that if L_1 and L_2 are lattices with relations $\preceq_1$ and $\preceq_2$, respectively, and $f : L_1 \to L_2$ is an isomorphism, then for all $a, b \in L_1$, $f(a) \vee f(b) = f(a \vee b)$. ($f(a) \wedge f(b) = f(a \wedge b)$ is symmetric.) (Hint: You want to show that that $f(a \vee b)$ is the LUB of $f(a)$ and $f(b)$, that is, for all $x \in L_2$ such that $f(a) \preceq_2 x$ and $f(b) \preceq_2 x$, it is the case that $f(a \vee b) \preceq_2 x$. So, suppose such an x and think about the corresponding $f^{-1}(x)$ in L_1. It is legal to use f^{-1} since f is a one-to-one correspondence by definition of isomorphism and thus invertible.)

10.3.2 We have already seen that $\mathbb{N}_e$ is a sublattice of $\mathbb{N}$. Show that $\mathbb{N}_e$ is isomorphic to $\mathbb{N}$.

10.3.3 Prove that if A and B are sets and $|A| = |B|$, then the lattices $\mathscr{P}(A), \subseteq$ and $\mathscr{P}(B), \subseteq$ are isomorphic. (To do this formally, use the formal definition of *having the same cardinality as each other*. Notice that this does not require the sets to be finite.)

10.3.4 Prove that if L_1 is isomorphic to L_2 and L_3 is isomorphic to L_4 then $L_1 \times L_3$ is isomorphic to $L_2 \times L_4$.

10.3.5 Prove that if L_1 and L_2 are both bounded lattices, $f : L_1 \to L_2$ is an isomorphism, and $\top_1$ and $\top_2$ are the top elements in L_1 and L_2, respectively, then $f(\top_1) = \top_2$.

10.3.6 Prove that the relation "is isomorphic to" on lattices is transitive. (It is also reflexive and symmetric, but those are not interesting proofs.)

10.4 Modular and distributive lattices

Before you flip through this section—relax! The proofs may look monstrous. Indeed the section contains a romp through lattice-land with stops at eight lemmas and two theorems before arriving at our premiere result, Corollary 10.19. But this trip is worth it, and it can be read at different levels. For advanced students, the details of the proofs will provide a satisfying challenge. If you do not feel up to all that, no worries—skip the lemma proofs. The statements of the lemmas, the proofs of the theorems, and the discussion will provide the main ideas.

In the chapter on functions, Section 7.6 investigated properties that certain functions have, namely that some are onto, some are one-to-one, and some are both. In this section we consider two properties that some lattices have: modularity and distributivity. (A third property, complementedness, is coming up in Section 10.6.)

Distributivity is so familiar a property—you know it from arithmetic operations, set operations, and logical operations—that it might have surprised you not to find it in the lattice properties of Theorem 10.6. We have a definition ready to go: A lattice L is *distributive* if for all $a, b, c \in L$,

distributive

$$a \wedge (b \vee c) = (a \wedge b) \vee (a \wedge c)$$

Not all lattices are distributive. D_n on $|$ for all $n \in \mathrm{N}$ is, and $\mathscr{P}(A)$ on $\subseteq$ for all A is, but $\{1, 2, 3, 6, 10, 14, 28, 210, 420\}$ on $|$ is not:

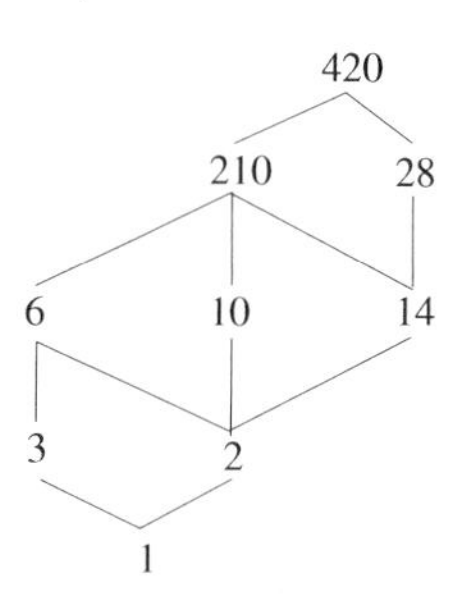

To see that this is not distributive, consider

$$\begin{aligned} 6 \wedge (10 \vee 14) &= 6 \wedge 210 &= 6 \\ (6 \wedge 10) \vee (6 \wedge 14) &= 2 \vee 2 &= 2 \end{aligned}$$

Keep in mind that in the arbitrary set $\{1, 2, 3, 6, 10, 14, 28, 210, 420\}$ things work differently from in, say, D_{420}, in which case $\wedge$ and $\vee$ are GCD and LCM. As already noted, D_{420} is distributive:

$$\begin{aligned} \gcd(6, \mathrm{lcm}(10, 14)) &= \gcd(6, 70) &= 2 \\ \mathrm{lcm}(\gcd(6, 10), \gcd(6, 14)) &= \mathrm{lcm}(2, 2) &= 2 \end{aligned}$$

modular

Modularity does not have such a familiar ring to it, but it does apply to most ready examples of lattices. A lattice L is *modular* if for all $a, b, c \in L$ such that $c \preceq a$,

$$(a \wedge b) \vee c = a \wedge (b \vee c)$$

Alternately you may read it as $(b \wedge a) \vee c = (b \vee c) \wedge a$. Modularity reads like a mixed-mode associativity, as though the two operations were associative with each other. D_n and $\mathscr{P}(A)$ are modular, and the example above is, too, but $\{1, 2, 3, 6, 12, 15, 24, 60, 120\}$ is not:

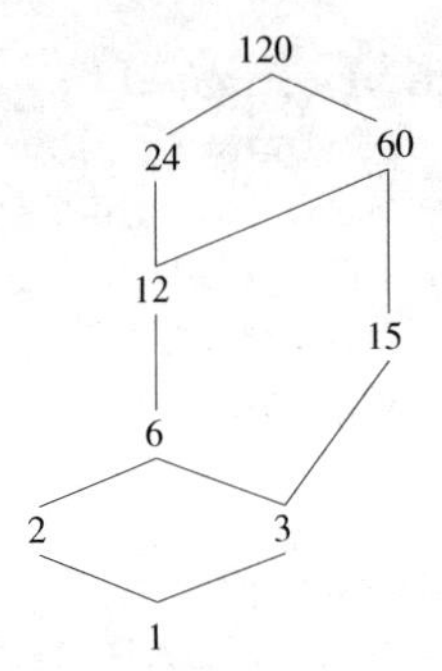

For example, $2|24$, but

$$\begin{aligned} (24 \wedge 15) \vee 2 &= 2 \vee 3 &= 1 \\ 24 \wedge (15 \vee 2) &= 24 \wedge 60 &= 12 \end{aligned}$$

Furthermore, this example is not distributive, either.

$$\begin{aligned} 12 \wedge (6 \vee 15) &= 12 \wedge 60 &= 12 \\ (12 \wedge 6) \vee (12 \wedge 15) &= 6 \vee 3 &= 6 \end{aligned}$$

It should also be clear that the lack of modularity and distributivity in these examples is reflected in the structure. We could replace the values to make lattices that are isomorphic to these and get the same result. Our main task in this section is to characterize lattices by their structure to determine whether they are modular and distributive.

To attack this problem, we search out examples from the smallest lattices up, looking for ones that are not distributive or not modular. There is one lattice each of size 1, 2, and 3, and two lattices of size 4 (up to isomorphism).

1-element 2-element 3-element 4-element

$\top$ $\top$ $\top$ $\top$

a a a b

b

$\top/\bot$ $\bot$ $\bot$ $\bot$ $\bot$

All of these are clearly distributive. For the four-element ones, notice that any expression involving three distinct items will include at least one of $\top$ and $\bot$, so, for example,

$$\begin{array}{rcccl} a \wedge (b \vee \top) & = & a \wedge b & = & a \\ (a \wedge b) \vee (a \wedge \top) & = & \bot \vee a & = & a \end{array}$$

Now consider five-element lattices. For convenience, we will name them by letter.

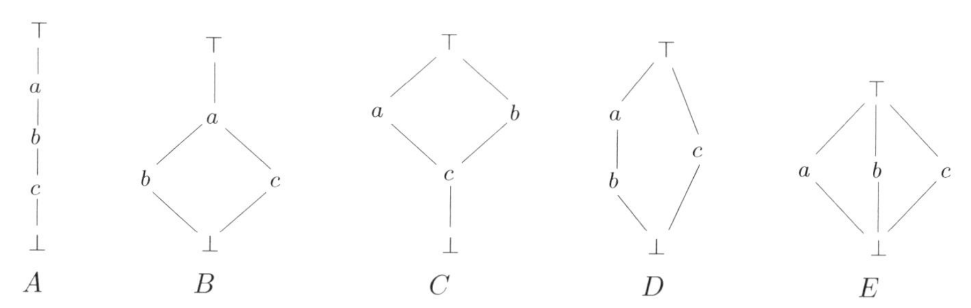

Lattice A, B, and C are distributive and modular by inspection. Lattice D is not modular or distributive, and E is modular but not distributive. The key observation is that the two lattice examples we looked at earlier each have a sublattice isomorphic to D or E:

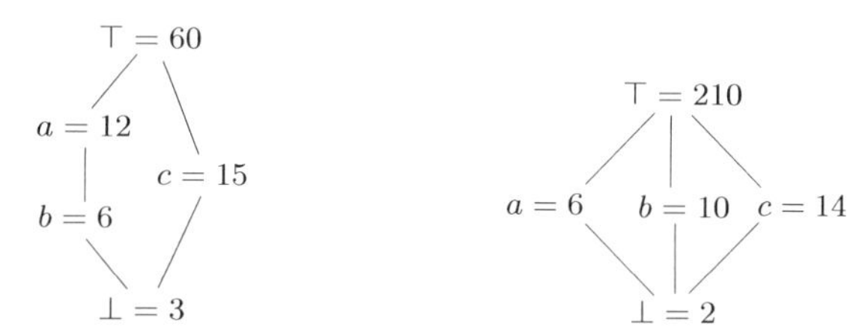

We now can conclude that if any lattice has a sublattice that looks like one of these two stinkers, then it is not distributive. If it is D that the sublattice looks like, then the lattice is not modular, either.

Now here is the marvelous result of this section. It turns out that the converse is also true: Any non-modular lattice (which is also non-distributive) has a sublattice isomorphic to E, and any lattice that is modular but not distributive has a sublattice isomorphic to D.

Consider what it will take to prove all this. First we would have to suppose that a lattice is non-modular or that it is modular but non-distributive. Then we are left with an existence proof, to show that it has an appropriate sublattice. We prove existence by identifying elements with which we can construct a sublattice isomorphic to E or D. That, in turn, requires three things to be shown: The elements are ordered correctly to match the structure of E or D, the meets and joins work out so that it is a sublattice, and the elements are in fact distinct elements.

For a warm-up, we prove a claim alluded to earlier: All distributive lattices are modular.

Lemma 10.9 *If a lattice is distributive, then it is modular.*

Proof. Suppose L is a distributive lattice. Suppose $a, b, c \in L$ such that $c \preceq a$. Then

$$\begin{aligned} (a \wedge b) \vee c &= (a \vee c) \wedge (b \vee c) && \text{by distributivity} \\ &= a \wedge (b \vee c) && \text{since } c \preceq a. \ \square \end{aligned}$$

We are more interested in the contrapositive, that is, if a lattice is non-modular, then it is also non-distributive. This establishes our two categories of non-distributive lattices, non-modular and modular-but-non-distributive, which we will consider under separate heads. The way forward, then, is shown by this map of our proofs, with lemmas at the bottom as a foundation and arrows showing which lemmas are used to prove what. "NM" and "M/ND" stand for "non-modular" and "modular but non-distributive," and "is sub-isomorphic to" is shorthand for "has a sublattice that is isomorphic to."

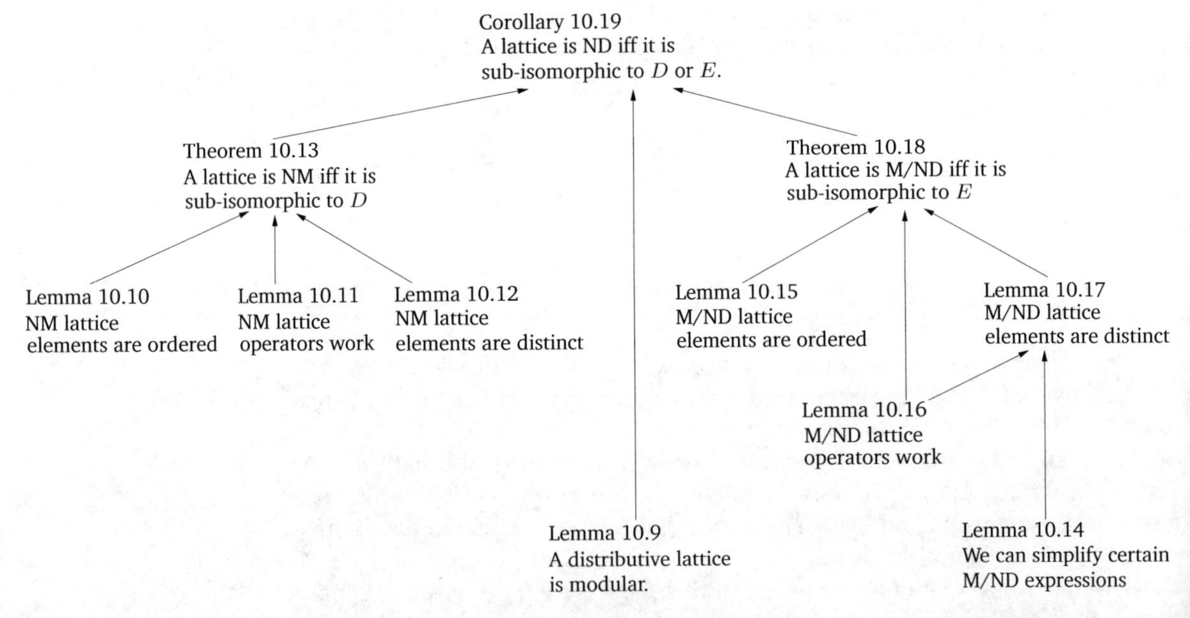

To start with modular lattices, what does it mean for a lattice not to be modular?

$$\sim (\forall\, a, b, c \in L, c \leq a \rightarrow (a \wedge b) \vee c = a \wedge (b \vee c))$$

$$\exists\, a, b, c \in L \mid c \leq a \text{ and } (a \wedge b) \vee c \neq a \wedge (b \vee c)$$

Thus for our lemmas about non-modular lattices, our hypothesis will always be that such a, b, and c exist. From those elements, we will construct an appropriate sublattice. The corner of the lattice it is in will look something like what appears on the right.

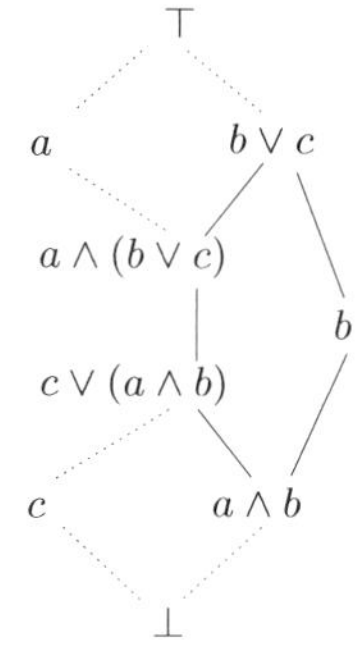

The elements $b \vee c$, $a \wedge (b \vee c)$, $c \vee (a \wedge b)$, b, and $a \wedge b$ are the ones that make a sublattice isomorphic to E, according to this picture. We prove that we have ordered them correctly:

Lemma 10.10 *If L is a lattice, $a, b, c \in L$, $c \leq a$, and $(a \wedge b) \vee c \neq a \wedge (b \vee c)$, then*

1. $a \wedge (b \vee c) \preceq b \vee c$
2. $b \preceq b \vee c$
3. $a \wedge b \preceq c \vee (a \wedge b)$
4. $a \wedge b \preceq b$
5. $c \vee (a \wedge b) \preceq a \wedge (b \vee c)$

Proof. (1) From Theorem 10.1, since $\wedge$ computes a lower bound of the two.

(2) Since $\vee$ computes an upper bound.

(3) Since $\vee$ computes an upper bound.

(4) Since $\wedge$ computes a lower bound.

(5) Note that $a \wedge b \preceq a$ and $c \preceq b \vee c$ by definition of $\wedge$ and $\vee$. Moreover $a \wedge b \preceq b \vee c$ by (2), (4), and transitivity. Since we are given $c \preceq a$, Exercise 10.2.14 tells us $c \vee (a \wedge b) \preceq a \wedge (b \vee c)$. □

Remember, to be a sublattice means that the meets and joins must "work out." There may not be any missing pieces from the larger lattice giving a least upper bound of b and $a \wedge (b \vee c)$ different from $b \vee c$, for example.

Lemma 10.11 *If L is a lattice, $a, b, c \in L$, $c \leq a$, and $(a \wedge b) \vee c \neq a \wedge (b \vee c)$, then*

1. $(a \wedge (b \vee c)) \wedge b = a \wedge b$
2. $(c \vee (a \wedge b)) \vee b = b \vee c$
3. $(a \wedge (b \vee c)) \vee b = b \vee c$
4. $(c \vee (a \wedge b)) \wedge b = a \wedge b$

Proof.

$$
\begin{array}{llcll}
(1) & (a \wedge (b \vee c)) \wedge b & = & a \wedge (b \wedge (b \vee c)) & \text{by associativity} \\
 & & = & a \wedge b & \text{by absorption}
\end{array}
$$

(2) Similar to (1).

$$
\begin{array}{llcll}
(3) & a \wedge (b \vee c) & \preceq & b \vee c & \text{by Lemma 10.10.1} \\
 & b & \preceq & b \vee c & \text{by Lemma 10.10.2} \\
 & (a \wedge (b \vee c)) \wedge b & \preceq & b \vee c & \text{by Exercise 10.2.11} \\
 & & & & \\
 & (c \vee (a \wedge b)) \vee b & \preceq & (a \wedge (b \vee c)) \vee b & \text{by Exercise 10.2.10} \\
 & b \vee c & \preceq & (a \wedge (b \vee c)) \vee b & \text{by substitution into (2)} \\
 & & & & \\
 & (a \wedge (b \vee c)) \vee b & = & b \vee c & \text{by antisymmetry}
\end{array}
$$

(4) Similar to (3). □

The really messy proof is in proving that these five elements ($b \vee c$, $a \wedge (b \vee c)$, b, $c \vee (a \wedge b)$,and $a \wedge b$) are distinct, which is necessary to establish isomorphism. Notice we are not claiming that a, c, $\top$ and $\bot$, are distinct, even though we drew them that way in the diagram. It is possible, for example, that $\top = a = b \vee c$.

Lemma 10.12 *If L is a lattice, $a, b, c \in L$, $c \leq a$, and $(a \wedge b) \vee c \neq a \wedge (b \vee c)$, then $b \vee c$, $a \wedge (b \vee c)$, b, $c \vee (a \wedge b)$ and $a \wedge b$ are distinct.*

Proof. [$a \wedge (b \vee c) \neq b \vee c$.] Suppose $a \wedge (b \vee c) = b \vee c$. Then

$$
\begin{array}{rcll}
(a \wedge (b \vee c)) \wedge b & = & (b \vee c) \wedge b & \\
a \wedge ((b \vee c) \wedge b) & = & b & \text{by associativity (left) and absorption (r} \\
a \wedge b & = & b & \text{by absorption}
\end{array}
$$

which we show to be impossible below.

[$b \neq b \vee c$.] Suppose $b = b \vee c$. Then

$$
\begin{array}{rcll}
a \wedge (b \vee c) & = & a \wedge b & \\
(a \wedge (b \vee c)) \vee c & = & (a \wedge b) \vee c & \\
 & \prec & a \wedge (b \vee c) & \text{by Lemma 10.10.3 and what is given}
\end{array}
$$

which is impossible because $\vee$ computes an upper bound.

[$(a \wedge b) \vee c \neq a \wedge (b \vee c)$.] Given.

[$c \vee (a \wedge b) \neq a \wedge b$.] Suppose $c \vee (a \wedge b) = a \wedge b$. Then

$$\begin{array}{rcll} c & \preceq & a \wedge b & \text{by Theorem 10.5} \\ c & \preceq & b & \text{by Exercise 10.2.12} \\ b \vee c & = & b & \text{by Theorem 10.5} \\ (a \wedge b) \vee c & \prec & a \wedge b & \text{by substitution into what is given} \end{array}$$

which is impossible because $\vee$ computes an upper bound.

[$a \wedge b = b$.] Suppose $a \wedge b = b$. Then $b \vee c \prec a \wedge (b \vee c)$ by substituting into what is given. However, that is impossible because $\wedge$ computes a lower bound.

[$a \wedge (b \vee c) \neq b$.] Suppose $a \wedge (b \vee c) = b$. Then

$$\begin{array}{rcll} a \wedge a \wedge (b \vee c) & = & a \wedge b & \text{by substitution} \\ a \wedge (b \vee c) & = & a \wedge b & \text{by idempotency} \\ c \vee (a \wedge b) & \prec & a \wedge b & \text{by substitution} \end{array}$$

The last part is because $c \vee (a \wedge b) \prec a \wedge (b \vee c)$. However, it is impossible because $\vee$ computes an upper bound.

[$c \vee (a \wedge b) \neq b$.] Suppose $c \vee (a \wedge b) = b$. Then

$$\begin{array}{rcll} c \vee (a \wedge b) \vee c & = & b \vee c & \text{by substitution} \\ c \vee (a \wedge b) & = & b \vee c & \text{by commutativity, associativity,} \\ & & & \text{and idempotency} \\ b \vee c & \prec & a \wedge (b \vee c) & \text{by substitution} \end{array}$$

which is impossible because $\wedge$ computes a lower bound. □

The work was in the lemmas. The theorem about non-modular lattices can be constructed quickly from them.

Theorem 10.13 *A lattice is not modular iff it has a sublattice isomorphic to D.*

Proof. ($\Leftarrow$) Suppose L is a lattice that has a sublattice isomorphic to D. Then we can choose elements from that sublattice that provide a counterexample to L being modular.

($\Rightarrow$) Suppose L is a modular lattice. This means there exist $a, b, c \in L$ where $c \preceq a$ and $(a \wedge b) \vee c \neq a \wedge (b \vee c)$. By Lemma 10.12, we know that the values $b \vee c$, $a \wedge (b \vee c)$, b, $c \vee (a \wedge b)$, and $a \wedge b$ are distinct. By Lemma 10.10, we know that as a poset they form the structure of D. By Lemma 10.11, the meets and joins work out so that they make a sublattice of L. □

The second leg of our journey lies with modular but non-distributive lattices. As with non-modular lattices, negating the definition of *distributive* will clarify our task.

$$\sim (\forall\, a, b, c \in L, a \wedge (b \vee c) = (a \wedge b) \vee (a \wedge c))$$

$$\exists\, a, b, c \in L \mid a \wedge (b \vee c) \neq (a \wedge b) \vee (a \wedge c)$$

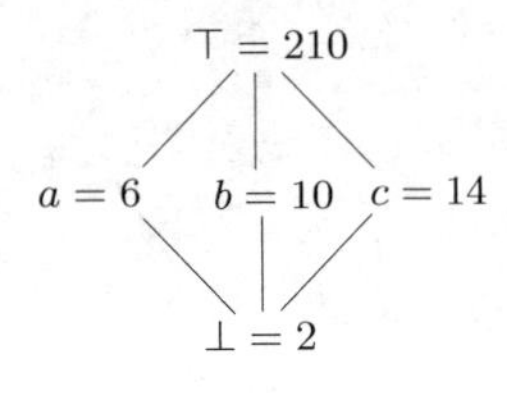

We have already seen an example of such a lattice, shown again at the left, but our task is to construct a sublattice that is isomorphic to that one from any three elements a, b, and c for which the inequality holds. Although we will construct a sublattice isomorphic to E, it is not necessarily the case that a, b, and c will be in that sublattice, as they are in the example on the right. We in fact know very little about the given a, b, and c, and it will take several steps to construct an appropriate sublattice.

We first find a top for the sublattice, which is found by joining each pair of the given three elements and then meeting those three. Call that value q.

$$q = (a \vee b) \wedge (a \vee c) \wedge (b \vee c)$$

Think of this as the greatest lower bound of the individual least upper bounds. For a bottom element of the sublattice, p, we construct the opposite

$$p = (a \wedge b) \vee (a \wedge c) \vee (b \wedge c)$$

The middle elements of the sublattice are $(a \wedge q) \vee p$, $(b \wedge q) \vee p$, and $(c \wedge q) \vee p$—see below left. If all of these values were distinct, then we might have a poset that looks like the one below right. Of course that is not a complete lattice. Moreover, some of them might be the same value—it might be, for example, that $q = a \vee b = a \vee c = b \vee c$ and $a = a \wedge q = (a \wedge q) \vee p$. That would be the case in the earlier example with 6, 10, and 14.

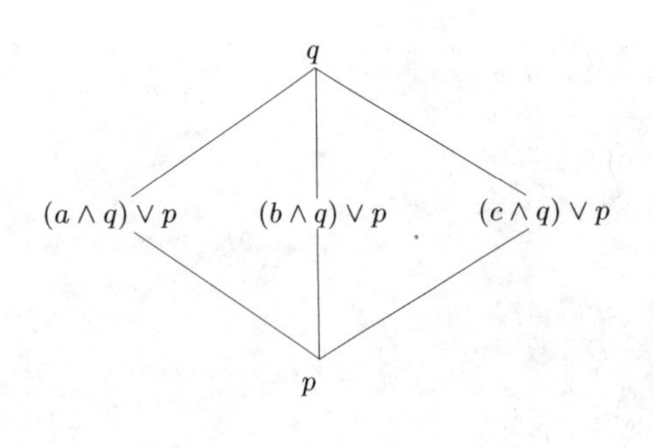

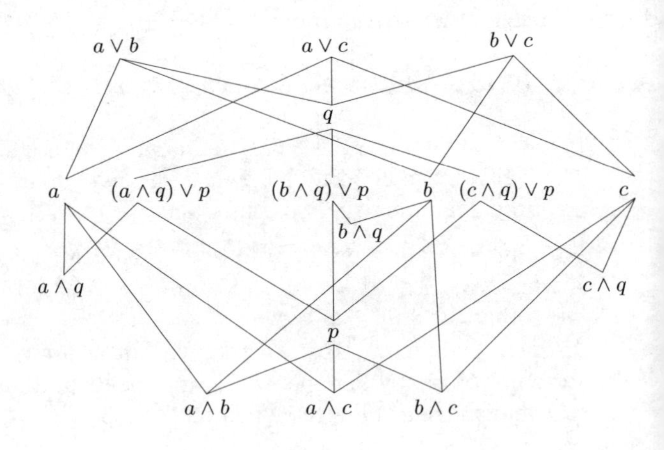

Our first lemma will give you a feel for working with these elements, p, q, $(a \wedge q) \vee p$, etc. It also provides simplifications that will make other lemmas easier.

Lemma 10.14 *If L is a modular lattice, $a, b, c, \in L$, and $(a \wedge b) \vee (a \wedge c) \neq a \wedge (b \vee c)$, then*

$$\begin{array}{rcl} a \wedge q & = & a \wedge (b \vee c) \\ a \wedge p & = & (a \wedge c) \vee (a \wedge b) \end{array}$$

where $p = (a \wedge b) \vee (b \wedge c) \vee (a \wedge c)$ and $q = (a \vee b) \wedge (b \vee c) \wedge (a \vee c)$.

Proof.

$$\begin{array}{rcll} a \wedge q & = & a \wedge ((a \vee b) \wedge (b \vee c) \wedge (a \vee c)) & \text{by substitution} \\ & = & (a \wedge (a \vee b)) \wedge (b \vee c) \wedge (a \vee c)) & \text{by commutativity} \\ & = & a \wedge (b \vee c) \wedge (a \vee c) & \text{by absorption} \\ & = & (a \wedge (a \vee c)) \wedge (b \vee c) & \text{by regrouping} \\ & = & a \wedge (b \vee c) & \text{by absorption} \end{array}$$

$$\begin{array}{rcll} a \wedge p & = & a \wedge ((a \wedge b) \vee (b \wedge c) \vee (a \wedge c)) & \text{by substitution} \\ & = & a \wedge ((b \wedge c) \vee ((a \wedge b) \vee (a \wedge c))) & \text{by regrouping} \\ & = & (a \wedge (b \wedge c)) \vee ((a \wedge b) \vee (a \wedge c)) & \text{by modularity} \\ & = & ((a \wedge c) \wedge b) \vee ((a \wedge b) \vee (a \wedge c)) & \text{by regrouping} \\ & = & (((a \wedge c) \wedge b) \vee (a \wedge c)) \vee (a \wedge b) & \text{by regrouping} \\ & = & (a \wedge c) \vee (a \wedge b) & \text{by absorption } \square \end{array}$$

"Regrouping" above is shorthand for "commutativity and associativity." As in the previous bunch of lemmas, we need to show that these elements are ordered as our diagrams show, in the same shape as E.

Lemma 10.15 *If L is a modular lattice, $a, b, c, \in L$, and $(a \wedge b) \vee (a \wedge c) \neq a \wedge (b \vee c)$, then*

$$\begin{array}{llcccc} 1. & p & \preceq & (a \wedge q) \vee p & \preceq & q \\ 2. & p & \preceq & (b \wedge q) \vee p & \preceq & q \\ 3. & p & \preceq & (c \wedge q) \vee p & \preceq & q \end{array}$$

where $p = (a \wedge b) \vee (b \wedge c) \vee (a \wedge c)$ and $q = (a \vee b) \wedge (b \vee c) \wedge (a \vee c)$.

Proof. Suppose L is a modular lattice, $a, b, c, \in L$, and

$$(a \wedge b) \vee (a \wedge c) \neq a \wedge (b \vee c)$$

Let p and q be as given in the lemma.

Sine $\vee$ computes an upper bound, $p \preceq (a \wedge q) \vee p$, and similarly if b or c is substituted for a.

By Exercise 10.4.1, $p \preceq q$. Hence

$$
\begin{aligned}
(a \wedge q) \vee p &\preceq (a \wedge q) \vee q && \text{by Exercise 10.2.10} \\
&= q && \text{by absorption}
\end{aligned}
$$

Similarly if we substitute b or c for a. □

Then we show that the meets and joins work. Fortunately all the parts of this proof are similar. Unfortunately they each are a mess.

Lemma 10.16 *If L is a modular lattice, $a, b, c, \in L$, and $(a \wedge b) \vee (a \wedge c) \neq a \wedge (b \vee c)$, then*

$$
\begin{aligned}
((a \wedge q) \vee p) \vee ((b \wedge q) \vee p) &= q & \qquad ((a \wedge q) \vee p) \wedge ((b \wedge q) \vee p) &= p \\
((b \wedge q) \vee p) \vee ((c \wedge q) \vee p) &= q & \qquad ((b \wedge q) \vee p) \wedge ((c \wedge q) \vee p) &= p \\
((a \wedge q) \vee p) \vee ((c \wedge q) \vee p) &= q & \qquad ((a \wedge q) \vee p) \wedge ((c \wedge q) \vee p) &= p
\end{aligned}
$$

where $p = (a \wedge b) \vee (b \wedge c) \vee (a \wedge c)$ and $q = (a \vee b) \wedge (b \vee c) \wedge (a \vee c)$.

We will give a careful proof that $((a \wedge q) \vee p) \wedge ((b \wedge q) \vee p) = p$. The rest are similar. We will use the modularity property several times. The underlining is to help you identify how it is applied. The underlined items are the ones that correspond to a and c in the definition of *modular*.

Proof.

$$((a \wedge q) \vee \underline{p}) \wedge \underline{((b \wedge q) \vee p)}$$

$$
\begin{aligned}
&= (((b \wedge q) \vee p) \wedge (a \wedge q)) \vee p && \text{by modularity, since } p \preceq (b \wedge q) \vee p \\
&= (((\underline{q} \wedge b) \vee \underline{p}) \wedge (a \wedge q)) \vee p && \text{by commutativity} \\
&= ((q \wedge (b \vee p)) \wedge (a \wedge q)) \vee p && \text{by modularity, since } p \preceq q \\
&= ((b \vee p) \wedge q \wedge q \wedge a) \vee p && \text{by commutativity and associativity} \\
&= ((b \vee p) \wedge (a \wedge q)) \vee p && \text{by idempotency and commutativity} \\
&= ((b \vee (c \wedge a)) \wedge (a \wedge (b \vee c))) \vee p && \text{by Lemma 10.14 and Exercise 10.4.2} \\
&= (a \wedge (\underline{(b \vee c)} \wedge (\underline{b} \vee (c \wedge a)))) \vee p && \text{by commutativity and associativity} \\
&= (a \wedge (((b \vee c) \wedge (c \wedge a)) \vee b)) \vee p && \text{by modularity, since } b \preceq b \vee c \\
&= (\underline{a} \wedge (\underline{(c \wedge a)} \vee b)) \vee p && \text{since } c \wedge a \preceq c \preceq b \vee c \\
&= ((a \wedge b) \vee (c \wedge a)) \vee p && \text{by modularity, since } c \wedge a \preceq a \\
&= p && \text{by idempotency (twice)}
\end{aligned}
$$

The last step is from the definition of p. □

While not all the elements in the big diagram earlier need to be distinct, the principle elements—p, q, $(a \wedge q) \vee p$, etc, are the ones in our constructed sublattice, and they must be distinct.

Lemma 10.17 *If L is a modular lattice, $a, b, c, \in L$, and $(a \wedge b) \vee (a \wedge c) \neq a \wedge (b \vee c)$, then p, q, $(a \wedge q) \vee p$, $(b \wedge q) \vee p$, and $(c \wedge q) \vee p$ are distinct elements, where $p = (a \wedge b) \vee (b \wedge c) \vee (a \wedge c)$ and $q = (a \vee b) \wedge (b \vee c) \wedge (a \vee c)$.*

Proof. We take on individual element pairs to show them to be unequal.

[$p \neq q$.] Suppose $p = q$. Then

$$\begin{array}{rcll} a \wedge p & = & a \wedge q & \text{by substitution} \\ a \wedge (b \vee c) & = & (a \wedge c) \vee (a \wedge b) & \text{by substitution and Lemma 10.14} \end{array}$$

which contradicts $a \wedge (b \vee c) \neq (a \wedge c) \vee (a \wedge b)$.

[$(a \wedge q) \vee p \neq (b \wedge q) \vee p$.]

$$\begin{array}{rcll} p & = & ((a \wedge q) \vee p) \wedge ((b \wedge q) \vee p) & \text{by Lemma 10.16} \\ q & = & ((a \wedge q) \vee p) \vee ((b \wedge q) \vee p) & \text{by Lemma 10.16} \\ p & \prec & q & \text{since } p \preceq q \text{ and } p \neq q \text{, above} \end{array}$$

Then $((a \wedge q) \vee p) \wedge ((b \wedge q) \vee p) \prec ((a \wedge q) \vee p) \vee ((b \wedge q) \vee p)$ by substitution which implies $(a \wedge q) \vee p \neq (b \wedge q) \vee p$.

[$(a \wedge q) \vee p \neq (c \wedge q) \vee p$.] Similar.

[$(b \wedge q) \vee p \neq (c \wedge q) \vee p$.] Similar.

[$(a \wedge q) \vee p \neq q$.] Suppose $(a \wedge q) \vee p = q$. Then

$$\begin{array}{rcll} p & = & ((a \wedge q) \vee p) \wedge ((b \wedge q) \vee p) & \text{by Lemma 10.16} \\ & = & q \wedge ((b \wedge q) \vee p) & \text{by substitution} \\ & = & (b \wedge q) \vee p & \text{since } (b \wedge q) \vee p \preceq q \text{ (Lemma 10.15)} \end{array}$$

Similarly we can show $p = (c \wedge q) \vee p$. Then $(c \wedge q) \vee p = (b \wedge q) \vee p$ by substitution, contradicting what we just showed.

[$(a \wedge q) \vee p \neq p$.] By a symmetric argument to the previous.

Similar proofs to the last two can be made for $(b \wedge q) \vee p$ and $(c \wedge q) \vee p$.
□

Finally we arrive at a theorem for modular, non-distributive lattices.

Theorem 10.18 *A modular lattice is not distributive iff it has a sublattice isomorphic to E.*

Proof. ($\Leftarrow$) Suppose L is a modular lattice that has a sublattice isomorphic to E. Then we can choose elements from that sublattice that provide a counterexample to L being distributive.

($\Rightarrow$) Suppose lattice L is modular but not distributive. By definition of distributivity, this means there exist $a, b, c \in L$ such that $(a \wedge b) \wedge (a \wedge c) \neq a \wedge (b \vee c)$. By Lemma 10.17, we know that the values p, q, $(a \wedge q) \vee p$, $(b \wedge q) \vee p$, and $(c \wedge q) \vee p$, as defined in earlier proofs, are distinct. By Lemma 10.15, we know that as a poset they form the structure of E. By Lemma 10.16, the meets and joins work out so that they make a sublattice of L. □

Since distributivity is the more well-known property, these results are usually summarized as

Corollary 10.19 *A lattice is not distributive iff it has a sublattice isomorphic to D or E.*

Proof. From Theorems 10.13 and 10.18. □

The proofs in this section are adapted from an exposition of lattices by Davey and Priestley [6].

Exercises

10.4.1 Prove that if L is a lattice and $a, b, c \in L$, then $(a \wedge b) \vee (b \wedge c) \vee (c \wedge a) \preceq (a \vee b) \wedge (b \vee c) \wedge (c \vee a)$.

10.4.2 Prove that in the context of Lemma 10.16, $b \vee p = b \vee (c \wedge a)$. It is similar to our proof of Lemma 10.14.

10.4.3 Explain why any sublattice of a distributive lattice must be distributive. (We would say "prove" if it were not so simple.)

10.4.4 Explain why modularity and distributivity are isomorphic invariants.

10.4.5 The "map" of our proofs on page 530 would be a lattice itself with relation "is used to prove," if we gave it a dummy bottom. Is this lattice modular? Is it distributive? How can you tell?

10.4.6 Prove that if L is a lattice and $a, b, c \in L$, then $(a \wedge b) \vee (a \wedge c) \preceq a \wedge (b \vee c)$. In other words, any lattice is "almost distributive," just with an inequality.

10.4.7 Logical operations and set operations each have two distributive laws, depending on which operator is being distributed. This is also true for distributive lattices. Show that if a lattice is distributive then for all $a, b, c \in L$, $a \vee (b \wedge c) = (a \vee b) \wedge (a \vee c)$.

10.4.8 Prove that for all $n \in \mathbb{N}$, D_n is distributive.

10.4.9 Prove that if $L_1, \preceq_1$ and $L_2, \preceq_2$ are distributive lattices, then $L_1 \times L_2$ with pairwise application of $\preceq_1$ and $\preceq_2$ is distributive.

10.4.10 Prove that if L is a distributive lattice and $a, x, y \in L$ such that $a \wedge x = a \wedge y$ and $a \vee x = a \vee y$, then $x = y$.

10.5 Implementing lattice operations

In this section we will take a break from proofs to do a little programming. We can illustrate lattices in ML by representing the set of lattice elements and implementing the operations. For common lattice examples, these are ready to go. For example, we implemented set operations in Exercises 3.7.7 and 3.7.9.

```
- val lat1 = powerset([1,2,3]);

val lat1 = [[1,2,3],[1,2],[1,3],[1],[2,3],[2],[3],[]]
    : int list list

- val join = union;

val join = fn : ''a list * ''a list -> ''a list

- val meet = intersection;

val meet = fn : ''a list * ''a list -> ''a list

- meet([1,3], [2, 3]);

val it = [3] : int list

- join([1,3], [2, 3]);

val it = [1,2,3] : int list
```

For lattices like D_n, meet (GCD) is known to us, and join (LCM) can be made easily.

```
- fun gcd(a, 0) = a
=   | gcd(a, b) = gcd(b, a mod b);

- fun lcm(a, b) =
=     let val x = gcd(a, b);
=     in x * (a * b div x) end;
```

How do we generate the set D_n? What we want are all the natural numbers that are divisors of n. Our strategy is to generate all the natural numbers less than n and then filter out the ones that do not divide n. With the function `enumerateInterval` below we can make the initial list, and `filter` from Exercise 7.3.9 will do the rest.

```
- fun enumerateInterval(low, high) =
=    if low > high
=    then []
=    else low::enumerateInterval(low+1, high);

val enumerateInterval = fn : int * int -> int list

- fun Dn(n) =
=       filter(fn(x) => n mod x = 0,
=              enumerateInterval(1, n));

val Dn = fn : int -> int list

- val lat2 = Dn(30);

val lat2 = [1,2,3,5,6,10,15,30] : int list
```

Now we can make functions to generate lattices (as set, meet, join triples) from two families:

```
- fun makePSLattice(A) = (powerset(A), intersection, union);

val makePSLattice =
  fn : 'a list ->
       'a list list *
       (''b list * ''b list -> ''b list) *
       (''c list * ''c list -> ''c list)

- fun makeDnLattice(n) = (Dn(n), gcd, lcm);

val makeDnLattice =
  fn : int ->
       int list * (int * int -> int) * (int * int -> int)
```

In general a lattice-making function returns 'a list * ('a * 'a →'a) * ('a * 'a →'a), and any function or system that worked with a generic lattice would take a value of that type. Our goal, however, is not to need to program the meet and join operations separately for every kind of lattice. Instead, it is possible to compute the LUB and GLB functions (for a finite lattice, at any rate), as long as we are given the lattice and the partial order relation.

For example, if we are given two lattice elements, the lattice itself, and the partial order (as a relation in predicate form), we can generate a list of elements that are upper bounds for two given elements (using `filter` again).

```
- fun upperBounds(a, b, latt, po) =
=       filter(fn (x) => po(a,x) andalso po(b,x),
=              latt);

val upperBounds =
    fn : 'a * 'a * 'b list * ('a * 'b -> bool) ->
       'b list

- upperBounds(2, 5, Dn(30), fn(a, b) => b mod a = 0);

val it = [10,30] : int list
```

The exercise below will walk you through writing a function `latticeOps` that produces a join and a meet for a lattice.

Exercises

10.5.1 Finish this function `latticeOps` which takes a lattice (as a set, that is, a list of elements) and a partial order relation (a function that takes two elements of the set and returns true or false, for example `subset` and `divides`) and returns a pair of functions, each taking two elements of the set and returning an element of the set, computing the LUB and GLB, respectively. Hints:

- `latticeOps` does not need to be recursive, though it may have recursive helper functions.
- Define a helper function `isLUB` that takes a "LUB candidate" (a member of a set that might be the LUB we are looking for) and a list of other upper bounds, and returns true or false, depending on whether the candidate is least of all others.
- Define a helper function `LUB` that computes the least upper bound; this function will be returned.
- Helper functions `isGLB` and `GLB` can be written similarly to `isLUB` and `LUB`.
- If you have worked through Section 9.9, then memoize `LUB` and `GLB`. Determine whether or not it is safe to use the function `memoize` to memoize your functions automatically; otherwise memoize them by hand.

```
fun latticeOps(latt, po) =
    let
      fun isLUB(x, []) = true
        | isLUB(x, y::rest) = ??
      fun LUB(a, b) =
         let val uBs = ??
             fun findLUB([bnd]) =  ??
               | findLUB(a::rest) = ??
         ..??...
    in
      (??, ??)
    end;
```

10.6 Boolean algebras

Take another look at our old standby examples of lattices:

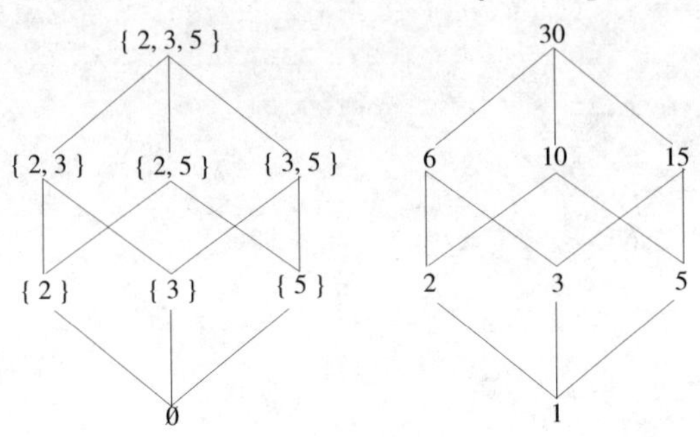

We know that $\{2,3\}$ and $\{5\}$ are complements of each other, $\overline{\{2,3\}} = \{5\}$ and $\{2,3\} = \overline{\{5\}}$, if $\{2,3,5\}$ is taken as the universal set. We should expect something similar for 6 and 5 in the other lattice because of the isomorphism. What exactly is the significance of being complements? Consider

$$\{2,3\} \cup \{5\} = \{2,3,5\} \qquad \{2,3\} \cap \{5\} = \emptyset$$

$$\mathrm{lcm}(6,5) = 30 \qquad \gcd(6,5) = 1$$

The same thing works on a small scale with boolean values: $T \vee F = T$, $T \wedge F = F$. In general, let a be an element in a lattice L. a' is the *complement* of a if $a \vee a' = \top$ and $a \wedge a' = \bot$. But do not get too comfortable with the idea of complement yet because it does not always work out as cleanly on some lattices as it does on the ones mentioned above. Consider these two.

complement

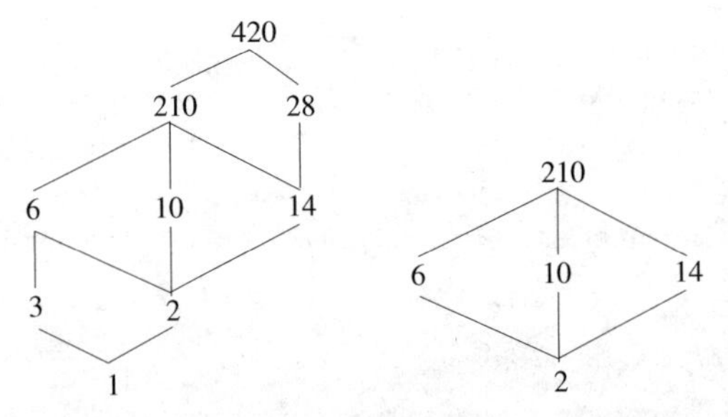

In the lattice on the left (an earlier example of a non-distributive lattice), 14 has no element that both joins it to the top and meets it to the bottom. If you take just the sublattice with 210 as the top and 2 as the bottom, as on the right (which is isomorphic to E, our prototypical modular/non-distributive lattice), both 10 and 6 join 14 to the top and meet it to the bottom. As we can see, it is not guaranteed that an element has a complement at all and, it if has one, that it is unique.

Our present task is to identify lattices and elements that do have complements, possibly unique. It turns out that distributivity helps.

Theorem 10.20 *If L is a bounded, distributive lattice and $a \in L$, then if a' exists, it is unique.*

Proof. Suppose L is a bounded, distributive lattice and $a \in L$. Suppose further $a', a'' \in L$ such that both are complements of a, that is

$$\begin{aligned} a \vee a' &= \top & \quad a \vee a'' &= \top \\ a \wedge a' &= \bot & \quad a \wedge a'' &= \bot \end{aligned}$$

Then,

$$\begin{aligned} a' = a' \vee \bot &= a' \vee (a \wedge a'') \\ &= (a' \vee a) \wedge (a' \vee a'') \quad \text{by distributivity} \\ &= (a'' \vee a) \wedge (a' \vee a'') \\ &= a'' \vee \bot \\ &= a'' \qquad \square \end{aligned}$$

complemented

Now we define a bounded lattice A to be *complemented* if for all $a \in A$, a' exists. Note this does not require the complement to be unique. E, for example (see the lattice $\{2, 6, 10, 14, 210\}$ above) is complemented. You should expect that complementedness is an isomorphic invariant—Exercise 10.6.13 asks you to prove it.

boolean algebra

Now for our main definition of this section. A *boolean algebra* is a complemented, distributive lattice. Note that being complemented implies being bounded, since if there are no $\top$ or $\bot$, then nothing can have a complement. Remember that all nonempty finite lattices are bounded, but not all bounded lattices are finite. Nevertheless, our discussion will be largely about *finite* boolean algebras. Both of our handy examples, D_{30} and $\mathscr{P}(2,3,5)$, are boolean algebras. Here is why boolean algebras are interesting:

Corollary 10.21 *If A is a boolean algebra, then for all $a \in A$, a has a unique complement.*

Proof. Since boolean algebras are complemented, at least one complement, a', exists. Since boolean algebras are distributive, Theorem 10.20 tells us that a' is unique. □

Tangent: Algebra

It might sound strange to refer to something as an algebra. In this sense, *algebra* simply means a system of symbols together with operators to combine those symbols. The word comes from Arabic *al-jabr*, which was part of the title of a medieval mathematical text, and it means restoration or reunion. For this reason algebra is also an obsolete English word for the surgery of setting fractured bones.

Recall the symbol $\mathbb{B}$ as set of boolean values, $\{T, F\}$. Now let $\mathbb{B}_n$ be the set of n-tuples over $\mathbb{B}$. For example, $\mathbb{B}_3 = \{000, 001, 010, 011, 100, 101, 110, 111\}$.

If $a \in \mathbb{B}_n$, then we refer to the components of a as $a_1, \ldots, a_n$. Define $\preceq$ so that for $a, b \in \mathbb{B}_n$, $a \preceq b$ if for all i, $1 \leq i \leq n$, $a_i \leq b_i$ (interpreting the components as numbers). It is straightforward to see that $\mathbb{B}_n, \preceq$ is a lattice.

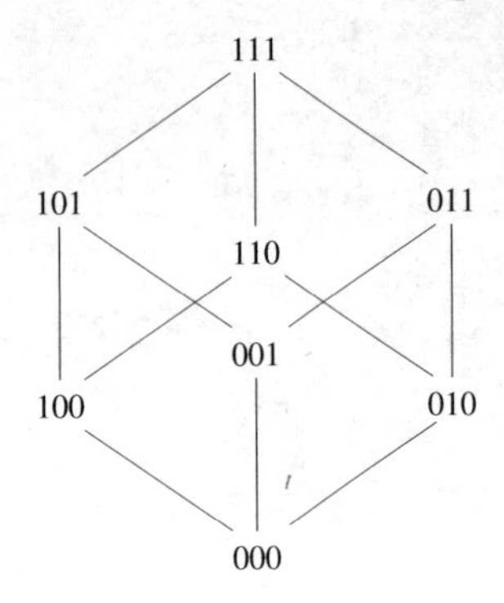

$\wedge$ and $\vee$ are just componentwise logical $\wedge$ and $\vee$, which should make it easy to remember.

$$110 \vee 101 = 111 = (T \vee T), (T \vee F), (F \vee T)$$

$$110 \wedge 101 = 100 = (T \wedge F), (T \wedge T), (F \wedge T)$$

The complement of an element in $\mathbb{B}_n$ is the componentwise negation of its components. Moreover, since $\mathbb{B}_n$ is a Cartesian product of distributive lattices, Exercise 10.4.10 tells us that $\mathbb{B}_n$ itself is distributive. $\mathbb{B}_n$ is thus a boolean algebra—in fact, hence the name *boolean algebra*.

$\mathbb{B}_n$ is going to be our beachhead for proving other things to be boolean algebras. Our tactic is to show conjectured boolean algebras to be isomorphic to $\mathbb{B}_n$ for some n.

Theorem 10.22 *If S is a finite set, then $\mathscr{P}(S), \subseteq$ is a boolean algebra.*

Proof. Let $n = |S|$. Name the elements of S to be $s_1, s_2, \ldots s_n$.

We can do this by the definition of *finite*. What we really are doing is supposing a one-to-one correspondence between S and $\{1, 2, \ldots n\}$.

Define a function $f : \mathscr{P}(S) \to \mathbb{B}_n \ldots$

As a side note, do you remember the cardinality of $\mathscr{P}(S)$? It is 2^n, which is also the cardinality of $\mathbb{B}_n$ and the number of vertices in the n-cube graph Q_n.

...such that if $A \in \mathscr{P}(S)$ (that is, $A \subseteq S$), then $f(A) = a_1 a_2 \ldots a_n$ where

Notation note: (F, F, T) ***and*** 001

We are writing 0 for *F* and 1 for *T* and omitting the commas and parentheses from tuple notation to make it easier to read. "001" would conventionally be written "(*F*, *F*, *T*)". But as anyone who ever tried to hedge a true/false test knows, *T* and *F* look alike at a quick glance. We will also sometimes treat the values as the numbers 0 and 1 rather than the boolean values *T* and *F*.

$$a_i = \begin{cases} 1 & \text{if } s_i \in A \\ 0 & \text{otherwise} \end{cases}$$

It is easy to see that f is a one-to-one correspondence (see Exercise 10.6.17).

Now suppose $A, B \in \mathscr{P}(S)$, $A \subseteq B$. Let $f(A) = a = a_1, a_2, \ldots a_n$ and $f(B) = b = b_1, b_2 \ldots b_n$.

Suppose $1 \leq i \leq n$. It must be that $a_i = 1$ or $a_i = 0$.

Case 1: Suppose $a_i = 1$. Then $s_i \subseteq A$. $s_i \subseteq B$ by definition of subset, so $b_i = 1$. Hence $a_i \leq b_i$.

Case 2: Suppose $a_i = 0$. Then $a_i \leq b_i$ no matter what.

Either way, $a_i \leq b_i$, so $f(A) \leq f(B)$.

Conversely, suppose $a, b \in \mathbb{B}_n$, $a = (a_1, a_2, \ldots a_n)$, $b = (b_1, b_2 \ldots b_n)$, and $a \leq b$. This means that for all i, $1 \leq i \leq n$, $a_i \leq b_i$.

Let $A = f^{-1}(a)$ and $B = f^{-1}(b)$. Suppose $s_i \in A$. By how we defined f, this means $a_i = 1$. Since $a_i \leq b_i$, it must be that $b_i = 1$, and so $s_i \in B$. Hence $A \subseteq B$. □

What about the D_n family of lattices? Some are boolean algebras, but not all. Consider this fact:

Theorem 10.23 *If $n \in \mathbb{N}$ is a product of unique primes $p_1, p_2, \ldots, p_k$, then D_n is a boolean algebra.*

"Product of unique primes" is saying that each prime number in the prime factorization of n occurs only once. Thus for every divisor x of n, the crucial question is, for each prime factor p_i, is p_i included in the x or not? Consider D_{2310} and its elements 14 and 154. We choose 2310 because its prime factorization is $2 \cdot 3 \cdot 5 \cdot 7 \cdot 11$.

	2?	3?	5?	7?	11?
2310	yes	yes	yes	yes	yes
154	yes	no	no	yes	yes
14	yes	no	no	yes	no

And now it is clear we can represent rows in this table with elements of $\mathbb{B}_5$: (11111), (10011), and (10010).

Proof. Define $f : D_n \to \mathbb{B}_k$ such that if $x \in D_n$, then $f(x) = x_1 x_2 \ldots x_k$ where

$$x_i = \begin{cases} 1 & \text{if } p_i | x \\ 0 & \text{otherwise} \end{cases}$$

We leave the rest as an exercise. □

atom

Now for our main result: Any (finite) boolean algebra is isomorphic to $\mathbb{B}_n$ for some n. To state this formally—and to prove it—we introduce the concept of an *atom*, which are elements that only $\bot$ is less than. Formally: If L is a boolean algebra, then $x \in L$ is an atom if for all $y \in L$, if $y \preceq x$ then $y = \bot$ or $y = x$.

Theorem 10.24 *If $L, \leq$ is a boolean algebra and there are n distinct atoms, then L is isomorphic to $\mathbb{B}_n$*

(The converse is also true, but straightforward; see Exercise 10.6.18.)

Do you suspect induction? Good call. We will prove the idea directly for $n = 0$, 1, and 2, even though that is not necessary, because it may help you understand how the proof works. By inspection we see that there is only one boolean algebra (up to isomorphism) with 0, 1, or 2 atoms.

Proof. By induction on n.

Base case 0. $n = 0$ isomorphic to $\mathbb{B}_0$

Base case 1. $n = 1$ isomorphic to $\mathbb{B}_1$

Note that  is not complemented.

Base case 2. $n = 2$ 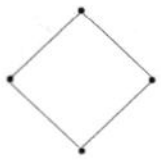 isomorphic to $\mathbb{B}_2$

Inductive case. Suppose this holds for $n - 1$, that is, all boolean algebras with $n - 1$ atoms, for some $n \geq 3$, are isomorphic to $\mathbb{B}_{n-1}$. Suppose L is a boolean algebra with n such atoms, $a_1, \ldots a_n$. Now we wish to make a sublattice L_1 that is made from all possible combinations of $\{a_1, \ldots a_{n-1}\}$ joined together. Formally, we define this recursively: Let L_1 be the smallest set such that

- $\{a_1, \ldots a_{n-1}\} \subseteq L_1$
- For all $b, c \in L_1$, $b \vee c \in L_1$

By Exercise 10.6.19, L_1 is a sublattice. By Exercise 10.4.3, L_1 is distributive. By Exercise 10.6.20, any element of L_1 can be written as $b_1 \vee b_2 \vee \ldots \vee b_m$ where b_1, b_2, $\ldots b_m$ are distinct elements of $\{a_1, \ldots a_{n-1}\}$. In that case, the complement of $b_1 \vee b_2 \vee \ldots \vee b_m$ is $c_1 \vee c_2 \vee \ldots \vee c_\ell$ where

$$\{c_1, c_2 \ldots c_\ell\} = \{a_1, \ldots a_{n-1}\} - \{b_1, b_2 \ldots b_m\}$$

Hence L_1 is complemented and, moreover, a boolean algebra.

By induction, L_1 is isomorphic to $\mathbb{B}_{n-1}$; let f_1 be the isomorphism.

Now define $f : L \to \mathbb{B}_n$ so that

$$f(z) = (z_1 z_2 \dots z_n)$$

where, if $z \in L_1$, then $(z_1, z_2, \dots z_{n-1}) = f_2(z)$ and $z_n = 0$; otherwise, if $z \notin L_1$, $(z_1, z_2, \dots z_{n-1}) = f_2(z \wedge a_n')$ and $z_n = 1$.

To see how this works, pick $z \in L$ such that $z \notin L_1$. Let a_n' be the complement of a_n and $w = z \wedge a_n'$. Note that $z = w \vee a_n$. For example, $\mathscr{P}(\{1, 2, 3, 4\})$ has atoms $\{1\}$, $\{2\}$, $\{3\}$, and $\{4\}$.

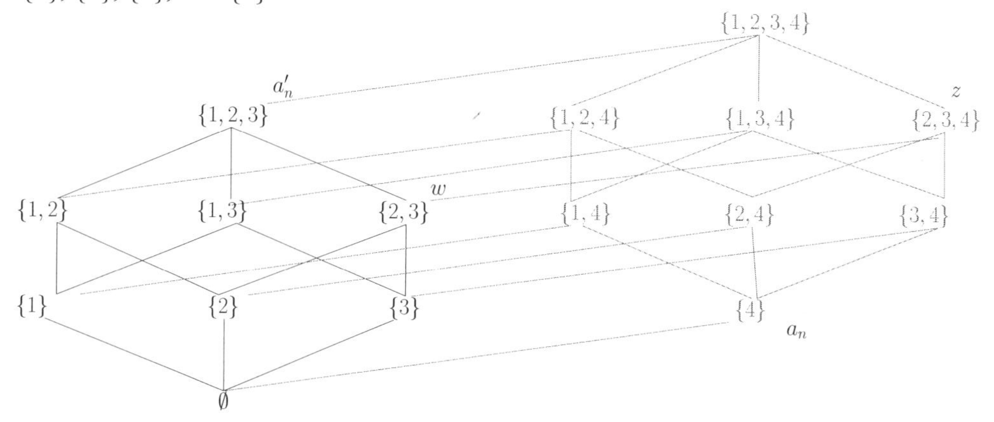

f is a one-to-one correspondence by Exercise 10.6.21.

Now suppose $a, b \in L$, $a \preceq b$. We need to show that $f(a) \preceq f(b)$. (For simplicity, we use the same partial order symbol $\preceq$ for L and $\mathbb{B}_n$; they are not really the same relation.) This depends on whether a and b are in L_1 or not.

Case 1. Suppose $a, b \in L_1$. Then by induction, $f_1(a) \preceq f_1(b)$. Applying the definition of f, we have $f(a) \preceq f(b)$, since $0 \preceq 0$ (for the last element).

Case 2. Suppose $a \in L_1$, $b \notin L_1$. Let $w = b \wedge a_n'$ We will show $a \leq w$, that is, $a \vee w = w$.

We have

$$\begin{aligned} a \vee w &= a \vee (b \wedge a_n') \\ &= (a \vee b) \wedge (a \vee a_n') \\ &= b \wedge a_n' && \text{because } a \preceq b \text{ and } a_n' \text{ is } \top \text{ in } L_1 \\ &= w \end{aligned}$$

So $a \preceq w$.

$f_2(a) \preceq f_2(w)$ by induction; apply the definition of f, noting that $0 \preceq 1$ (for the last element).

Case 3. Suppose $a \notin L_1$ and $b \in L_1$.

Time out for a lemma.

Lemma 10.25 *If $a \in L_1$, then $a_n \not\preceq a$ (with L_1, a_n, etc defined as they are in the proof of Theorem 10.24).*

Proof. Suppose $a \in L_1$ and $a_n \preceq a$. Then $a \wedge a_n = a_n$, so a is not an atom. By how we defined L_1, there must be at least two distinct atoms, say a_1 and a_2, such that $a_1 \preceq a$ and $a_2 \preceq a$.

$a_1 \vee a_2$ may or may not be distinct from a. If $a_1 \vee a_2 = a$, then we have a sublattice as seen on the left. If $a_1 \vee a_2 \neq a$, then we have a sublattice as seen on the right.

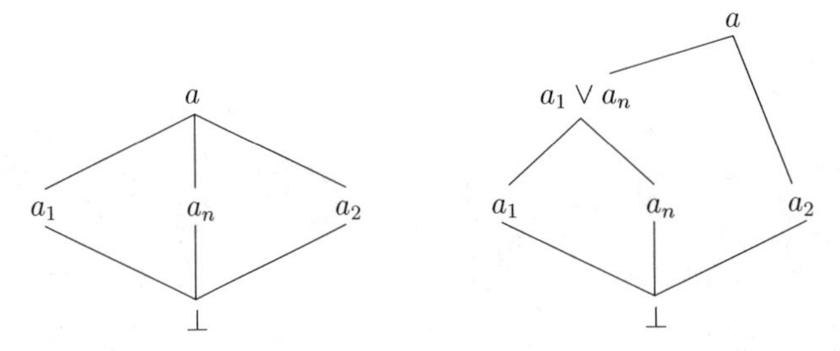

In the first case we have a lattice isomorphic to E. In the second case we can eliminate a_1 to get a lattice isomorphic to D. Either way, we have a sublattice of L that is not distributive, contradicting L being a boolean algebra. □

We now return to our regularly scheduled proof.

Proof (of Theorem 10.24, continued). Then $a_n \leq a$. By transitivity, $a_n \leq b$, contradicting the lemma. This case is eliminated.

Case 4. Suppose $a, b \notin L_1$. Then let $w = a \wedge a_n'$, $v = b \wedge a_n'$. Then $w \vee v = (a \wedge a_n') \vee (b \wedge a_n') = (a \vee b) \wedge a_n' = b \wedge a_n' = v$. Hence $w \preceq v$.

So $f_1(w) \preceq f_1(v)$ by induction. We can then apply the definition of f, noting that $1 \preceq 1$. □

We can pick up the following result almost for free.

Corollary 10.26 *If L is a boolean algebra with n atoms, then $|L| = 2^n$.*

Proof. By the same induction. Adding the nth atom doubles the size of the lattice.

Theorem 10.24 shows us that any boolean algebra is a set of the combinations of a handful of primitive elements. Each element in a boolean algebra can be interpreted as a subset of the available elements, or, alternately, as a tuple of booleans indicating which elements are included. Boolean algebras are simple and efficient to implement on computer hardware. Each element in the boolean algebra can be represented by a series of bits, each indicating the presence or absence of one of the atoms.

Finding boolean algebras in everyday life is easy and fun. The relationships between primary and secondary colors in a subtractive color model (as in pigments) form a boolean algebra, as seen below, with the atoms being the primary colors magenta, yellow, and cyan. Cyan, for example, is the complement of red. Notice that if we turn this lattice upside down we get a lattice representing the additive color model (as in light) with red, green, and blue as primary colors. As another example, consider the set of pizzas that can be made from a set of available toppings. Each topping (or single-topping pizza) appears as an atom, with cheese pizza being $\bot$ and supreme being $\top$.

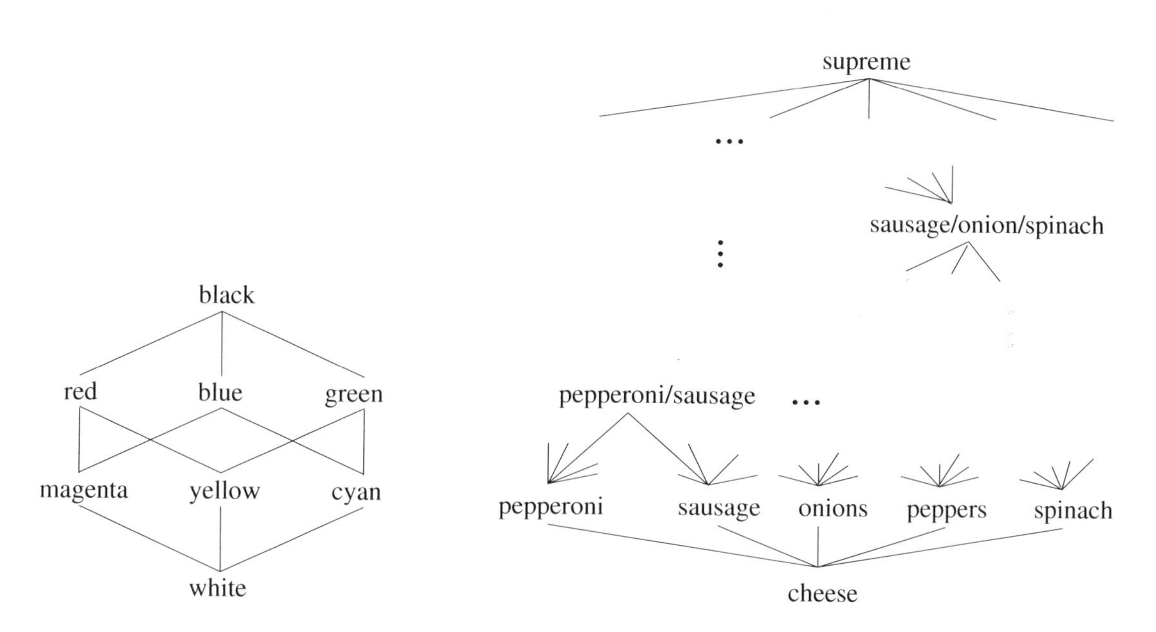

Boolean algebras are also the lattices that have all the nice properties that we have seen in other contexts. We have the following theorem, which expands Theorem 10.6 and parallels (actually, generalizes) Theorem 3.1.

Theorem 10.27 *If L is a boolean algebra and $a, b, c \in L$, then the following properties hold.*

Commutative laws:	$a \vee b = b \vee a$	$a \wedge b = b \wedge a$
Associative laws:	$a \vee (b \vee c) = (a \vee b) \vee c$	$a \wedge (b \wedge c) = (a \wedge b) \wedge c$
Distributive laws:	$a \vee (b \wedge c) = (a \vee b) \wedge (a \vee c)$	$a \wedge (b \vee c) = (a \wedge b) \vee (a \wedge c)$
Absorption laws:	$a \vee (a \wedge b) = a$	$a \wedge (a \vee b) = a$
Idempotent laws:	$a \vee a = a$	$a \wedge a = a$
Involution laws:	$(a')' = a$	
Identity laws:	$a \wedge \top = \top$	$a \vee \bot = \bot$
DeMorgan's laws:	$(a \vee b)' = a' \wedge b'$	$(a \wedge b)' = a' \vee b'$
Complement laws:	$a \vee a' = \top$	$a \wedge a' = \bot$
Universal bounds laws:	$a \vee \top = \top$	$a \wedge \bot = \bot$
Identity laws:	$a \wedge \top = a$	$a \vee \bot = a$
Top/bottom laws:	$\top' = \bot$	$\bot' = \top$

Exercises

Assume $A, \preceq$ is a boolean algebra and $a, b, c \in A$. Prove.

10.6.1 The involution law.

10.6.2 DeMorgan's law.

10.6.3 If $a \preceq b$, then $b' \preceq a'$.

10.6.4 $a = b$ iff $(a \wedge b') \vee (a' \wedge b) = \bot$.

10.6.5 If $a \preceq b$, then $a \vee c \preceq b \vee c$.

10.6.6 $((a \vee c) \wedge (b' \vee c))' = (a' \vee b) \wedge c'$.

10.6.7 If $a \preceq b$, then $a \vee (b \wedge c) = b \wedge (a \vee c)$.

10.6.8 $(a \wedge b) \vee (a \wedge b') = a$

10.6.9 $b \wedge (a \vee (a' \wedge (b \vee b'))) = b$

10.6.10 $(a \wedge b \wedge c) \vee (b \wedge c) = b \wedge c$

10.6.11 $((a \vee c) \wedge (b' \vee c))' = (a' \vee b) \wedge c'$

10.6.12 The following are equivalent

(a) $a \vee b = b$

(b) $a' \vee b = \top$

(c) $a \preceq b$

(d) $a \wedge b = a$

(e) $a \wedge b' = \bot$

(To prove that these are *equivalent*, you need to show that each one is true if and only if any of the others is. That does not mean that you need to write 20 proofs; just write a proof that a implies b, b implies c, c implies d, d implies e, and e implies a—or in a different order.)

Exercises, continued

10.6.13 Prove that complementedness is an isomorphic invariant.

10.6.14 Prove that if L is a bounded, distributed lattice and $a, b \in L$ and a has complement a', then $a \vee (a' \wedge b) = a \vee b$ and $a \wedge (a' \vee b) = a \wedge b$.

10.6.15 Prove that if L is a bounded lattice and $|L| \geq 2$, then for all $a \in L$, if a' exists, then $a \neq a'$. In other words, no element is its own complement.

10.6.16 It might be tempting to think that any sublattice of a boolean algebra is also a boolean algebra. Give a counterexample showing that this is not true.

10.6.17 Explain why (that is, informally prove that) the function we defined in the proof of Theorem 10.22 is a one-to-one correspondence.

10.6.18 Explain why the converse of Theorem 10.24 is true.

10.6.19 Prove that L_1 in the proof of Theorem 10.24 is a sublattice of L.

10.6.20 Prove that, in the context of the proof of Theorem 10.24, any element of L_1 can be written as $b_1 \vee b_2 \vee \ldots \vee b_m$ where $b_1, b_2, \ldots b_m$ are distinct elements of $\{a_1, \ldots a_{n-1}\}$.

10.6.21 Prove that, in the context of the proof of Theorem 10.24, f is a one-to-one correspondence. Use the proof that ψ is a one-to-one correspondence in the proof of Theorem 7.13 as a model, though your proof will not be quite so complicated.

10.6.22 The proof we gave for Corollary 10.26 is a hand-waving proof, suggesting we could just rewrite the induction, noting the size of each lattice as we go. However, we can prove it formally and succinctly by combining Theorem 10.24 with Theorem 6.5. Give the proof.

Some of the exercises in this section were taken from Kolman et al [18].

10.7 Special topic: Digital logic circuits

Our study of lattices brought us around to Cartesian products over the set of logical values. You should observe that propositional forms are really functions over $\mathbb{B}_n$ for some n. The *exclusive or* (true if exactly one operand is true) and conditional operators, for example, are both functions $\mathbb{B}_2 \to \mathbb{B}$:

exclusive or

$$\begin{aligned} \text{xor}(p, q) &= (p \vee q) \wedge \sim (p \wedge q) \\ \text{cond}(p, q) &= q \vee \sim p \end{aligned}$$

In ML,

```
- fun xor(a, b) = (a orelse b) andalso not (a andalso b);

- fun conditional(a, b) = not a andalso b;
```

Digital computing machinery is built on the insight that operations on boolean algebras can be implemented in electronic circuits. A *combinational circuit* is a system for modeling an electronic network with two kinds of components:

combinational circuit

- *wires*, each of which can be either on or off (representing the lattice values of $\top$ and $\bot$ or the logical values of T and F).

wires

- *gates*, which model functions on boolean algebras, specifically having wires as inputs and outputs.

gates

The three primitive logical operators are represented by three primitive gates.

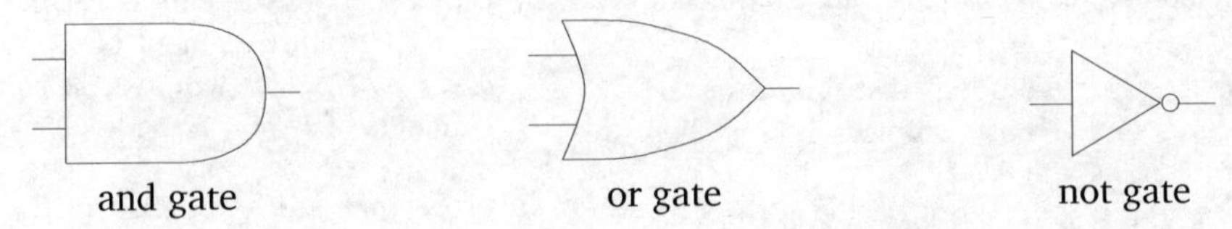

These gates can be combined, like the operators they represent, to build circuits to implement other boolean functions.

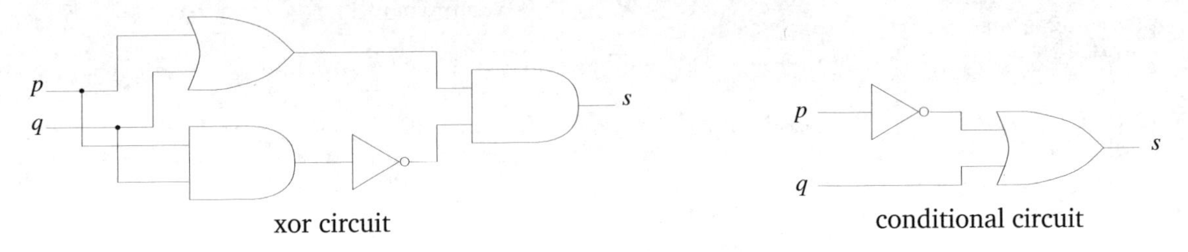

Notice that input into individual gates and into the entire circuit appear on the left, and output appears on the right. No gate's output can loop back to its input. Wires can branch, as indicated by the small dots in the *xor* circuit: both wire p and wire q are inputs into both the *and* and the *or* gates. When a wire branches, both resulting lines should be considered the same wire in the same way as when a variable appears more than once in the same formula.

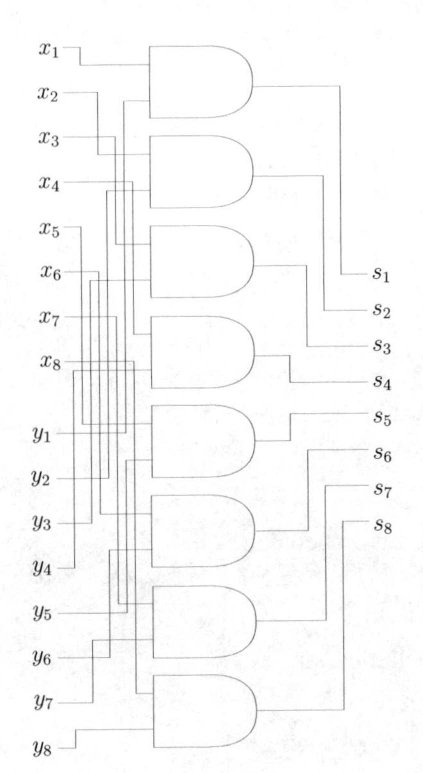

In the previous section we saw that the powerset of a set (so, a system of subsets) is a boolean algebra, and so such a system can be represented neatly and efficiently with wires in a circuit. Suppose we have a system of subsets of a set with eight elements, for example, pizzas with eight possible toppings (pepperoni, sausage, ham, onions, peppers, mushrooms, anchovies, and celery). We can represent this mathematically using elements of $\mathbb{B}_8$ or electronically using *bit vectors*. Thus 00000000 stands for cheese pizza, 00000001 for celery pizza, 10010000 for pepperoni and onion, and 11111111 for supreme.

We have long observed the correspondence between set operations and logical operations. Now we can see it in action as logic gates can be combined to implement set operations as in the circuit to the right that computes intersection. In the pizza scenario, this could find a set of pizza toppings acceptable to two people by taking the intersection of the toppings they each like.

More broadly, the digital age is built on the premise that almost any information can be represented as bit sequences. A simple but illuminating example is whole numbers. Instead of representing a whole number in our usual decimal (that is, base ten) system, we can use a binary (base two) system. Instead of columns standing for powers of ten, they stand for powers of two. Here we demonstrate how 115 in base ten is equivalently written in base two:

$$\begin{array}{rcl} 115_{10} & = & 1 \cdot 10^2 + 1 \cdot 10^1 + 5 \cdot 10^0 \\ & = & 64 \quad + \quad 32 \quad + \quad 16 \quad + \qquad\qquad\qquad\quad 2 \quad + \quad 1 \\ & = & 1 \cdot 2^6 \quad + \quad 1 \cdot 2^5 \quad + \quad 1 \cdot 2^4 \quad + \quad 0 \cdot 2^3 + 0 \cdot 2^2 + \quad 1 \cdot 2^1 \quad + \quad 1 \cdot 2^0 \\ & = & 1110011_2 \end{array}$$

How did we perform that conversion—that is, find powers of two (64, 32, 16, 2 and 1) that sum to 115? The same way your function `digify` broke a number up into a list of digits way back in Exercise 2.2.8. That is, we repeatedly mod the number by 2 to get the next bit and divide by 2 to get the next part of the number.

```
- fun intToBits(n) = if n = 0 then [] else
=                     (n mod 2)::intToBits(n div 2);

val intToBits = fn : int -> int list

- intToBits(115);

val it = [1,1,0,0,1,1,1] : int list
```

Wait—did you notice that `intToBits` put the digits from *least significant* to *most*, the opposite from how we would write it? The bits are "backwards." The computation is so much cleaner that way, however, that we will live with it. Deciding the direction to order the bits is an annoying problem. We will run into it again later.

One change to `intToBits` before we go on: Conceptually the list components are not ints as our ML code suggests, but bits. We should use the type system to disallow values other than 0 or 1. We could use our bit datatype from Section 6.12, but we also want to use ML's boolean operators. Accordingly, we rewrite the function to return a bool list. We also write a function `bitsToInt` to convert the other way. (Study carefully how this works; how would you rewrite it using `foldr`?)

```
- fun intToBits(n) = if n = 0 then [] else
=                     (n mod 2 = 1)::intToBits(n div 2);

val intToBits = fn : int -> bool list

- fun bitsToInt([]) = 0
=    | bitsToInt(b::rest) =
=      (case b of true => 1 | false => 0) + 2
=         * bitsToInt(rest);
```

```
val bitsToInt = fn : bool list -> int

- intToBits(115);

val it = [true,true,false,false,true,true,true] : bool list

- bitsToInt(it);

val it = 115 : int
```

The algorithm we use to do addition by hand generalizes to numbers denoted in bases other than ten. That is, we can simply line up the bits in binary numbers by columns, carrying numbers to the next column as necessary.

$$\begin{array}{rrrr} & & {\scriptstyle 1} & \\ & 1 & 1 & 5 \\ + & & 1 & 7 \\ \hline & 1 & 3 & 2 \end{array} \qquad \begin{array}{rrrrrrrr} & {\scriptstyle 1} & {\scriptstyle 1} & & & {\scriptstyle 1} & {\scriptstyle 1} & \\ & 1 & 1 & 1 & 0 & 0 & 1 & 1 \\ + & & & 1 & 0 & 0 & 0 & 1 \\ \hline 1 & 0 & 0 & 0 & 0 & 1 & 0 & 0 \end{array}$$

Consider the addition table we use for each step.

+	0	1
0	0	1
1	1	10

+	F	T
F	(F,F)	(F,T)
T	(F,T)	(T,F)

p	q	$p \wedge q$	$(p \vee q) \wedge \sim (p \wedge q)$
T	T	T	F
T	F	F	T
F	T	F	T
F	F	F	F

The middle table represents the same information as the left but with boolean values—the difference really is in the two ways to denote values in $\mathbb{B}$ and $\mathbb{B}_2$. In an expression like (F,T), the T is the immediate *sum* of the two addends that is the result for the current column; the F is the *carry* which is propagated to the next column. The most exciting fact, which we see by comparing entries in the middle and right, is that we can find boolean formulas that compute the sum and the carry.

half adder

If we have boolean formulas, then we can build digital circuits to implement them. The following circuit is called a *half adder*. It takes two wires representing one-bit addends and has two wires as output, representing the sum and carry.

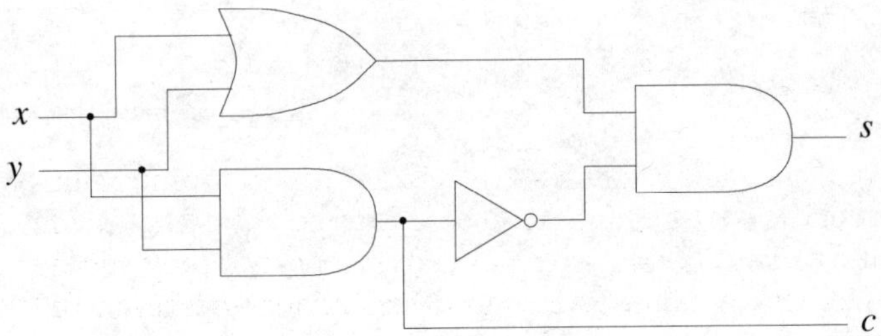

In ML:

```
- fun halfAdder(a, b) = ((a orelse b) andalso not (a andalso b),
=                         a andalso b);

val halfAdder = fn : bool * bool -> bool * bool

- map(halfAdder)
=     ([(false,false),(false,true),(true,false),(true,true)]);


val it = [(false,false),(true,false),(true,false),(false,true)]
  : (bool * bool) list
```

When interpreting the result of `halfAdder`, note that the first element is the sum, the carry second. The reason it is called a *half* adder is that in reality we also need to account for the carry of the previous column. A *full adder*, shown here, has three input wires (x and y for the current column entries for the addends, z for the previous carry) and two output wires, again modeling the sum and carry for the current column.

full adder

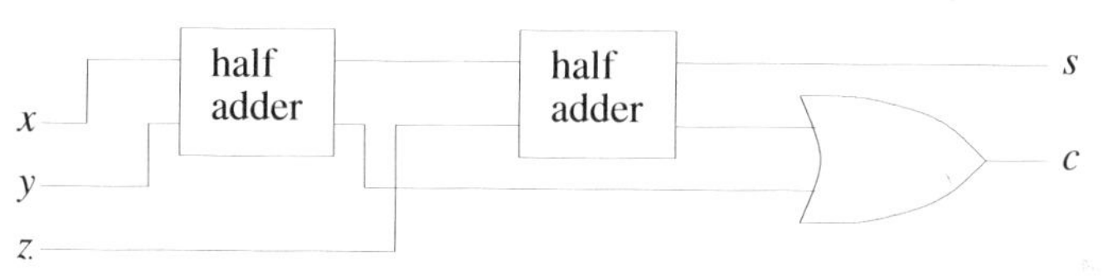

As the diagram shows, we do not have to compose all circuits from scratch. We can box up previously designed circuits and paste instances of them into new circuits just as we would with gates—and, just as we would make calls to other functions within a function. An ML implementation of a full adder:

```
- fun fullAdder(a, b, c) =
=     let val (s1, c1) = halfAdder(a, b);
=         val (s2, c2) = halfAdder(s1, c);
=     in
=         (s2, c1 orelse c2)
=     end;

val fullAdder = fn : bool * bool * bool -> bool * bool
```

Our final and biggest task is to use these as pieces in performing multi-bit addition on bit sequences representing numbers. Assume the addends will be whole numbers from 0 to 255, inclusive, so chosen because they can be represented in eight bits ($2^8 = 256$). We snap a half adder and seven full adders together, feeding the inputs and carries into the adders appropriately.

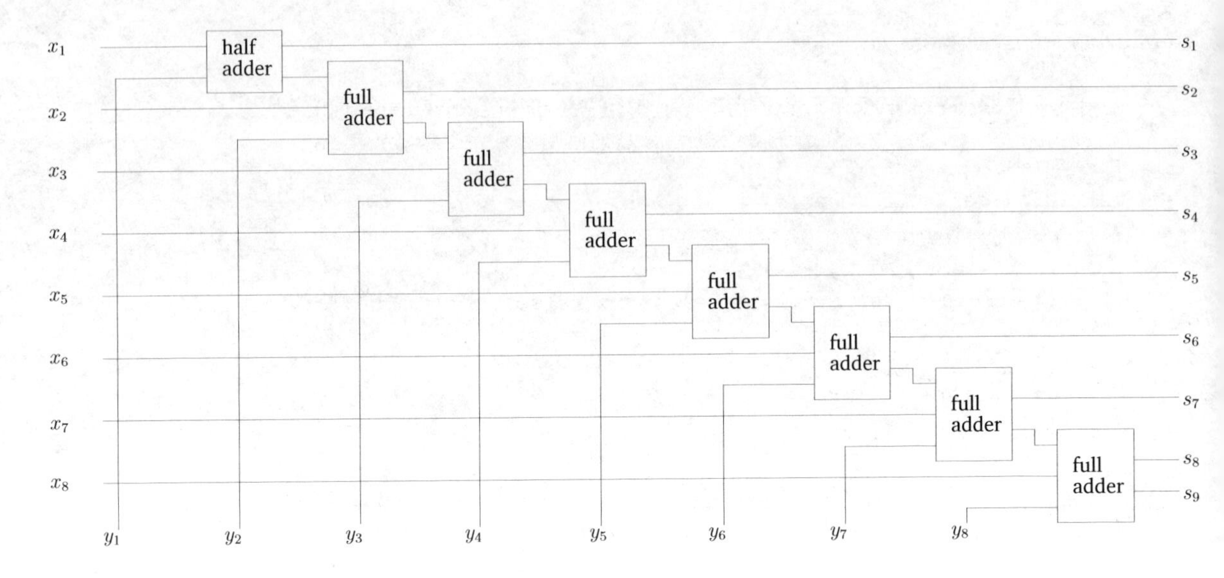

(The carry bit of the last full adder draws our attention to the fact that the sum may require more than eight bits. On most computer hardware this would be considered an *overflow*: the last carry bit is ignored and and the bits of the sum appear to roll over like an analog odometer: $255_{10} + 1_{10} = 11111111_2 + 00000001_2 = {}_{(1)}00000000_2 = 0_{10}$.)

Now we want to implement a binary adder like this in ML. In circuits we need to assume an arbitrary length—in our case we assumed eight bits, padding the number with zeros if necessary. We are not so constrained in ML. Since our binary numbers are lists of bits from least significant on, they can be as long or short as necessary, and we will write our `binaryAdder` function to generalize over lists of any size.

This is a tricky function. Approach it as you did your first ML list-processing problems. What happens at each step as we break down the lists? The tuples down the left represent the current carry and result-so-far. This is $21 + 5$:

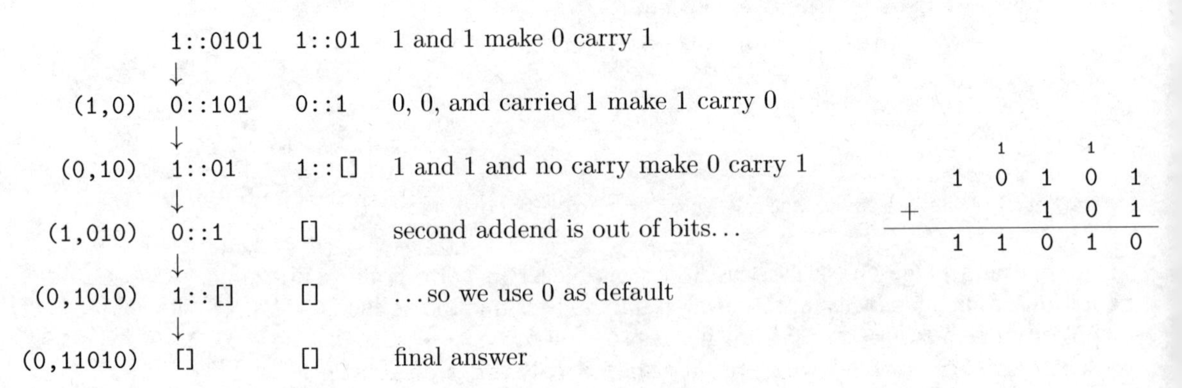

	`1::0101`	`1::01`	1 and 1 make 0 carry 1
	↓		
`(1,0)`	`0::101`	`0::1`	0, 0, and carried 1 make 1 carry 0
	↓		
`(0,10)`	`1::01`	`1::[]`	1 and 1 and no carry make 0 carry 1
	↓		
`(1,010)`	`0::1`	`[]`	second addend is out of bits. . .
	↓		
`(0,1010)`	`1::[]`	`[]`	. . . so we use 0 as default
	↓		
`(0,11010)`	`[]`	`[]`	final answer

```
      1   1
  1 0 1 0 1
+     1 0 1
-----------
1 1 0 1 0
```

In every step along the way

- we are given the carry bit and result-so-far (a list of bits) from the last step (on the first step, assume the carry is 0 and the result-so-far is an empty list)
- we grab the next element from each given list
- we apply `fullAdder` to the carry and two addend bits (remember that the carry to the first step is assumed 0, so we can replace the half adder in the diagram with a full adder with a dummy carry coming in)
- we cons the sum to the result-so-far list which together with the new carry makes the new result-so-far.

This process is like `foldl` where we apply a function to every element in a list, accumulating as we go. Remember

```
- foldl(fn(x,y) => x + y) (0)([4, 5, 6]);

val it = 15 : int
```

(We are using the standard, curried form of `foldl` introduced in Section 7.12.)

The difference is that while `foldl` accumulates over a single list, we need to step through two lists in lock-step. Since the two lists may not be the same length, we need a default value to use in case one list runs out before the other. To this end we write `foldlDouble`. In analogy with the curried form of `foldl`, it takes its parameters in three groups—the function to apply, a tuple containing the seed result and default, and a tuple containing the two lists.

```
- fun foldlDouble(f)(result, default)([], []) = result
=   | foldlDouble(f)(result, default)(a::aRest, []) =
=     foldlDouble(f)(f(a, default, result), default)
=                  (aRest, [])
=   | foldlDouble(f)(result, default)([], b::bRest) =
=     foldlDouble(f)(f(default, b, result), default)
=                  ([], bRest)
=   | foldlDouble(f)(result, default)(a::aRest, b::bRest) =
=     foldlDouble(f)(f(a, b, result), default)
=                  (aRest, bRest);

val foldlDouble = fn
  : ('a * 'a * 'b -> 'b) -> 'b * 'a -> 'a list * 'a list
      -> 'b
```

To try it out, suppose we have two lists of ints and we want to multiply them componentwise and sum the products (if you have studied vectors, you will remember this as the dot product). The function takes three parameters, one for each list and one for the result-so-far. In this case the seed result is 0, and we make the default 1. We test it with both equal-lengthed and unequal-lengthed lists:

```
- foldlDouble(fn(x, y, z) => (x * y) + z)(0, 1)
=             ([1, 3, 1],[5, 2, 4]);

val it = 15 : int

- foldlDouble(fn(x, y, z) => (x * y) + z)(0, 1)
=             ([1, 3, 1],[5, 2]);

val it = 12 : int
```

Now we put this to work in solving our `binaryAdder` problem. According to our tracing of a sample application, our result-so-far is a tuple containing the carry and the sum-so-far. We can unify those pieces in a single list, however, making the carry to be the head of the list of result bits. This is also convenient because if there is any carry left over from the last step, we would want to tack it on the end of the result as the next highest bit. The seed result is `[false]`, representing an initial carry of zero, and the default value is `false`, for the zeros with which we pad a smaller value so it would have the same number of bits.

```
- fun binaryAdder(x, y) =
=     foldlDouble(fn (a, b, c::rest) =>
=                        let
=                          val (s, c2) = fullAdder(a, b, c);
=                        in c2::s::rest end)
=                   ([false], false)(x, y);

Warning: match nonexhaustive
            (a,b,c :: rest) => ...

val binaryAdder = fn : bool list * bool list -> bool list

- binaryAdder(intToBits(21), intToBits(5));

val it = [false,true,true,false,true,false] : bool list

- bitsToInt(it);
```

```
val it = 22 : int
```

Huh? What went wrong?

It is bit-ordering problem again. Look carefully at our tracing of $21 + 5$. The addends 21 and 5 appear with the least significant bit first, as `intToBits` would compute them. We accumulate the result, however, by adding to the front of the list. We end up with the bits of 26 in the order we would normally write them, with most significant bit first.

This is unfortunate, but easy to fix. We can use your `reverse` function from Exercise 2.2.6.

```
- fun binaryAdder(x, y) =
=     reverse(foldlDouble(fn (a, b, c::rest) =>
=                              let val (s, c2) = fullAdder(a, b, c);
=                              in c2::s::rest end)
=                         ([false], false)(x, y));

Warning: match nonexhaustive
          (a,b,c :: rest) => ...

val binaryAdder = fn : bool list * bool list -> bool list

- fun adderTester(x, y) =
=     bitsToInt(binaryAdder(intToBits(x), intToBits(y)));

val adderTester = fn : int * int -> int

- adderTester(21, 5);

val it = 26 : int

- adderTester(115, 17);

val it = 132 : int
```

Chapter summary

Once you internalize a discrete mathematical structure, you begin to see it everywhere. Perhaps after working through the earlier chapters you began to see sets, Cartesian products, relations, functions, and graphs in your everyday life. Perhaps now when you consider ways to combine ingredients to make an omelet, you will see that the possible resulting omelets make a lattice, in fact, a boolean algebra.

Other sets we have seen before turn out to be lattices themselves. Lattices are more general, though, and not all lattices have the all of the nice properties that a system of sets or propositional values have. Boolean algebras are the lattices that have a full slate of expected properties. Boolean algebras also have proven useful in the design of electronics.

Key ideas:

A lattice is poset—a set together with a partial order relation—such that every pair of elements in the lattice has a least upper bound and a greatest lower bound in the partial order relation. Lattices have two binary operations: join (denoted $\vee$) and meet ($\wedge$), which compute the LUB and GLB, respectively, of the two operands.

The top element of a lattice (denoted $\top$), if it exists, is the least upper bound of all the elements in the lattice. Similarly the bottom ($\bot$) is the greatest lower bound of all the elements, if such a bound exists. A lattice having both a top and bottom is *bounded*.

All lattices have some of the properties we are familiar with from other mathematical structures: commutativity, associativity, absorption, idempotency, universal bounds, and identities (the last two only if $\top$ or $\bot$ exist.

Two lattices are isomorphic to each other if there exists a one-to-one correspondence between them that preserves the ordering relation.

For some lattices, a distributive property for the meet and join operators holds ($a \wedge (b \vee c) = (a \wedge b) \vee (a \wedge c)$). Some lattices are modular, meaning that for all $a, b, c \in L$, where $c \preceq a$, $(a \wedge b) \vee c = a \wedge (b \vee c)$. All distributive lattices are modular. A lattice is not modular iff it has a sublattice isomorphic to a certain five-element lattice, and a lattice is modular but not distributive iff it has a sublattice isomorphic to another certain five-element lattice. Hence a lattice is not distributive iff it has a sublattice isomorphic to one of two certain five-element lattices.

The complement of an element a in a bounded lattice is an element (say, a') such that $a \vee a' = \top$ and $a \wedge a' = \bot$. In a distributive lattice, any complement, if it exists, is unique (in non-distributive lattices, some elements may have more than one complement). A lattice for which all elements have a complement is called *complemented*. A boolean algebra is a complemented, distributive lattice. Powersets and sets of divisors of multiples of unique primes are boolean algebras, and all boolean algebras are isomorphic to $\mathbb{B}_n$ for some n.

Boolean algebras can be used to model digital circuits; alternately, digital circuits can be used to implement functions on boolean algebras.

Chapter 11 Group

The novel *The Princess Bride* purports to be the abridgement of a forgotten, voluminous adventure tale. Billed as the "good parts version," it presents all the action of the story with boring passages stripped away. Group theory is traditionally the first big unit in undergraduate courses that go by names like Abstract Algebra or Modern Algebra. The "full versions" of such courses are by no means boring, but few undergraduates besides math majors can make room in their schedules for one. In light of such constraints, this chapter presents the good-parts version of an introduction to group theory as a natural outgrowth of discrete mathematics.

This chapter follows a structure similar to Chapter 8 (Graph) and Chapter 10 (Lattice): a definition of terms, propositions and their proofs, isomorphisms, and a sampling of specific topics and applications. As with the other two chapters, one of the most important ideas is that of isomorphism, and the student would do well to compare group isomorphisms with those of graphs and lattices.

This chapter plays another role in the general plan of this text. It is a final round for the proof-writing skills so crucial to our goals. Not only will this chapter give you practice in working with new definitions (as the other chapters have done) but also reinforce the most important techniques from throughout the text. Many of the proofs are "back-to-basics" examples where proof patterns from sets, relations, and functions show up again. Use this chapter to test your mastery of these things, especially correct analysis and use of quantification.

How far should we go into group theory to give computer science majors, for example, a reasonable and balanced overview? This chapter provides enough to understand almost all of the mathematics behind the RSA encryption algorithm, which is our last stop on this journey. Since this chapter has less programming content than the other chapters, implementing RSA encryption demonstrates that this, too, has relevance to computer science.

The presentation of group theory in this chapter is influenced by that of Gallian [10] and Kolman et al [18].

Chapter goals. The student should be able to

- Prove propositions about binary operations, semigroups, and monoids.
- Describe the properties of a group, identify groups, and prove propositions about them.
- Prove that two groups are isomorphic and prove a proposition to be an isomorphic invariant.
- Identify and prove a subgroup of a group.
- Recognize dihedral groups, groups of relative primality, permutation groups and cyclic groups.
- Implement the RSA encryption algorithm.

11.1 Preliminary terms

Much of mathematics is the discovering of patterns and other similarities among things that do not seem to have much in common. The prevalence of the group pattern in mathematical and natural phenomena is one of the most remarkable examples of this at the undergraduate level, and it is the main topic of this chapter. In this section we observe other patterns that will prepare us for examining the definition of a group.

A closed *binary operation* on a set A is a function from $A \times A$ to A. The modifier *closed* indicates that every pair of elements in A maps to another element of A, as opposed to being undefined or having a result outside the set A.

binary operation

closed

This definition does not introduce any new idea. There are scads of examples of binary operations both in your prior mathematical experience and in this text. The most common ones are not usually written using function notation, though. Instead they often have their own operator symbol that is used in an infix style, meaning that you write the operator between the two operands.

- $+$, $\cdot$, and $-$ are closed binary operations on $\mathbb{R}$, $\mathbb{Q}$, and $\mathbb{Z}$. $+$ and $\cdot$ are also closed binary operations on $\mathbb{W}$ and $\mathbb{N}$, but $-$ is not, since $4-5 = -1 \notin \mathbb{N}$. $/$ is not closed on any of these number sets (except $\mathbb{N}$, interpreted as integer division) because division by zero is undefined.

- $\cup$, $\cap$, and $-$ are closed binary operations on the powerset of any set. In other words, if $X, Y \in \mathscr{P}(A)$, then $X \cup Y \in \mathscr{P}(A)$. $\subseteq$ is not a closed binary operation since the result of $X \subseteq Y$ is not a set but a truth value. Like other relations, we can think of $X \subseteq Y$ as a function $\mathscr{P}(A) \times \mathscr{P}(A) \to \mathbb{B}$.

- $\wedge$ and $\vee$ are closed binary operations on truth values. $\sim$ is a closed operation, but it is not binary since it has only one operand. $\wedge$ and $\vee$ on lattices are also closed binary operations.

- Concatenation is a closed binary operation on strings. If we have strings like `"disc"` and `"rete"`, we can concatenate them to get `"discrete"`.

We will use the symbol $*$ to denote a generic binary operation. (In this text we have used $\cdot$ for plain old arithmetic multiplication and $\times$ for Cartesian product.) We know a variety of properties that binary operations may or may not have:

- *Commutativity.* $\forall\ a, b \in A,\ a * b = b * a$. Examples of commutative operations include $+$ and $\cdot$ on the number sets, $\cup$ and $\cap$ on powersets, and $\wedge$ and $\vee$ on truth values and lattices. $-$ on number sets and powersets is not, and concatenation on strings is not.

- *Associativity.* $\forall\ a, b, c \in A,\ (a * b) * c = a * (b * c)$. All binary operations mentioned above except the two $-$s are associative.

- *Idempotency.* $\forall\, a \in A,\ a * a = a$. The only operations listed above that are idempotent are $\cup$, $\cap$, $\vee$ and $\wedge$

semigroup

We have a special name for a set with an associative, closed binary operation: *semigroup*. That is an odd name, but as you may guess, the concept is derived from that of a group, which gets top billing. The following is a diverse handful of example semigroups, some drawn from the examples above, others new.

- For a set A, $\mathscr{P}(A)$ with $\cup$.
- $\mathbb{B}$ (or any lattice) with $\vee$.
- For a set A, the set of functions $\{f : A \to A\}$ with $\circ$. (In Exercise 7.8.3 you proved the associativity of function composition.)
- $\mathbb{W}$, $\mathbb{Z}$, $\mathbb{Q}$, and $\mathbb{R}$ with $+$.
- $\mathbb{N}$, $\mathbb{W}$, $\mathbb{Z}$, $\mathbb{Q}$, and $\mathbb{R}$ with $*$.
- The set of strings on with concatenation.

Some semigroups—say, A with $*$—have a distinguished element e with the property that

$$\forall\, a \in A,\ a * e = e * a = a$$

identity

monoid

We call such an element an *identity*. A semigroup that has an identity is called a *monoid*. Not all semigroups have an identity, but all of the examples above do; respectively: $\emptyset$, F (or for general lattices, $\perp$), $f(x) = x$ (by freakish coincidence called the "identity function"), 0, 1, and the empty string.

Note that if we take the set $\mathbb{N}$ with $+$, we still have a semigroup, but not a monoid because it no longer has an identity. Note also that on number sets with multiplication, 1 is the identity. But is it right to say "*the* identity" or "*an* identity"? To justify the definite article, we should prove

Theorem 11.1 *If A with $*$ is a semigroup, then e, if it exists, is unique.*

This is an interesting result because it is a proof of uniqueness. Other proofs of uniqueness in this text have been the uniqueness of the transitive closure (Theorem 5.12) and proofs that a function is one-to-one. To prove that no more than one thing can have a certain property, suppose two things have that property and show that those two things in fact must be the same thing.

> **Proof.** Suppose A with $*$ is a semigroup, and suppose further that e_1 and e_2 are both identities. By definition of identity we have both $e_1 * e_2 = e_1$ and $e_1 * e_2 = e_2$. By substitution, $e_1 = e_2$. □

Monoids are useful in functional programming. Recall the function `foldl`:

```
- foldl;

val it = fn : ('a * 'b -> 'b) -> 'b -> 'a list -> 'b
```

The three parameters to `foldl` (in curried form) are a function, a seed value, and a list on which to apply the function. If 'a and 'b are the same type, then the function is a closed binary operation (as long as it does not raise exceptions for any combination). Consider two simple example uses:

```
- foldl(fn(x, y) => x + y)(0)([1, 4, 6, 3]);

val it = 14 : int

- foldl(fn(x, y) => x * y)(1)([2, 5, 7]);

val it = 70 : int
```

To sum a list, the seed value is the additive identity. To find the product of a list, the seed value is the multiplicative identity. Thus if a monoid's operation is being applied over a list of monoid values, the identity is the result over an empty list. The associativity of the monoid's operation enables efficiency: Since the individual operations may be done in any order, multiple processor cores can process parts of the folding simultaneously.

Exercises

11.1.1 Prove that $\mathbb{Z}$ is a monoid with the binary operation $*$ defined so that the result of $a*b$ is the lesser of a and b. (There is a long way to prove this and a short way that makes use of previous results).

For each of the following sets with an a binary operation, determine and prove whether (in decreasing order of specialization) the set is a monoid, the set is a semigroup, the operation is closed, or none.

11.1.2 $\mathbb{R}+$ with $a * b = \sqrt[a]{b}$.

11.1.3 $\mathbb{R}$ with $a * b = a^b$.

11.1.4 $\mathbb{Q} - \{0\}$ with $a * b = \frac{1}{a \cdot b}$.

11.1.5 $\mathbb{Z}$ with $a * b = 3 + a + b$.

11.1.6 $\mathbb{Z}$ with $a * b = 3 \cdot (a + b)$.

In Exercises 11.1.7–11.1.10, assume A_1 and A_2 are sets with closed binary operations $*_1$ and $*_2$, respectively and define $\circledast$ to be a binary operation on $A_1 \times A_2$ such that if $(a_1, a_2), (b_1, b_2) \in A_1 \times A_2$, then $(a_1, a_2) \circledast (b_1, b_2) = (a_1 *_1 b_1, a_2 *_2 b_2)$. Prove.

11.1.7 $\circledast$ is closed.

11.1.8 If $*_1$ and $*_2$ are associative, then $\circledast$ is associative.

11.1.9 If e_1 is the identity for A_1 and e_2 is the identity for A_2, then (e_1, e_2) is the identity for $A_1 \times A_2$.

11.1.10 If e_1 is the identity for A_1, e_2 is the identity for A_2, $a_1, b_1 \in A_1$ such that $a_1 *_1 b_1 = e_1$, and $a_2, b_2 \in A_2$ such that $a_2 *_2 b_2 = e_2$, then $(a_1, a_2) \circledast (b_1, b_2) = (e_1, e_2)$.

11.1.11 Write an ML function `catAll` that takes a list of strings and returns a single string that is the concatenation of all the strings in the list. The body of your function should be a `foldl` expression.

11.1.12 Similarly, write an ML function `composeAll` that takes a list of functions all of type 'a -> 'a for some type 'a. and returns a single function that is the composition of all the functions in the list.

11.2 Definition

Imagine an equilateral triangle cut out of some transparent material (like what long ago was used with overhead projectors) with its corners colored different shades of gray so we can track what corner is what. Suppose it is oriented with a corner at the top and a side at the bottom, the lightest shade at the top, the medium shade lower left, and the darkest shade lower right. Consider the ways we can manipulate the figure so that it remains oriented with a corner at the top.

We can rotate it clockwise 120° or 240°.

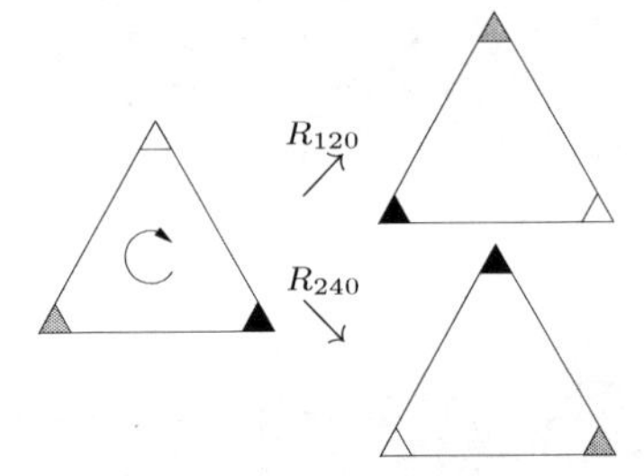

Or we can flip it about the vertical axis, or, likewise, about the axes through the other two corners.

We call the second two flips B and S for *bend* and *sinister bend*, the terms for shield-crossing stripes in heraldry.

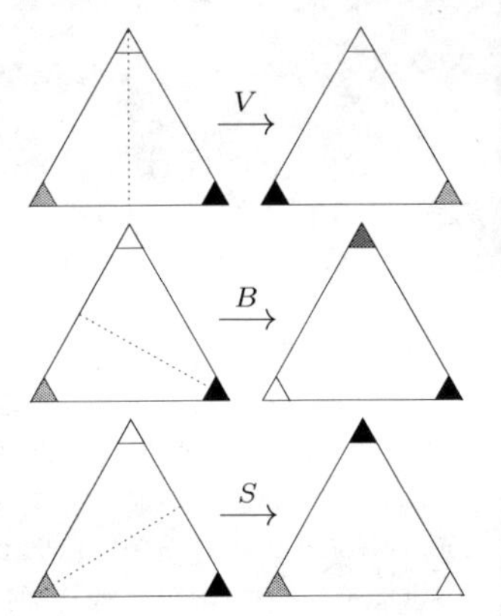

Since we have exhausted the ways we can arrange the shades among the corners (three items can be arranged six ways), there are no other manipulations that are not somehow a repeat of one of the manipulations we already have. For example, if we did one manipulation followed by another, we would get a resulting triangle that we could have gotten with only one of the other manipulations.

A vertical flip followed by 120° rotation is the same thing as a sinister-bend flip.

A bend flip followed by a sinister-bend flip is the same thing as a 240° rotation.

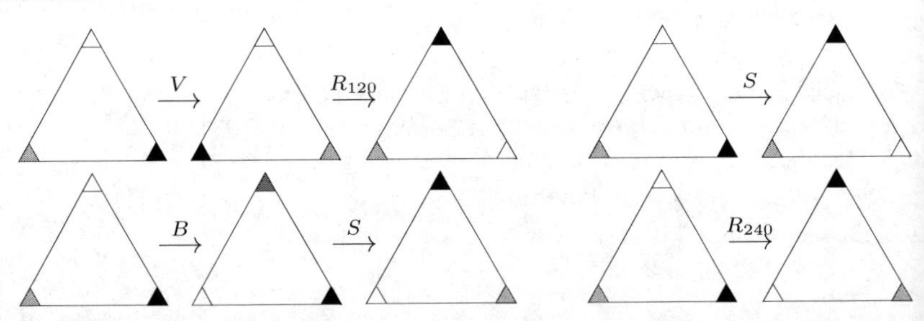

We have now a set of manipulations that, if we throw in R_0 for rotating 0° or doing nothing, comes to $\{R_0, R_{120}, R_{240}, V, B, S\}$. (Do not think of of this as a set of *triangles*. It is a set of *manipulations* to a triangle.) We also have a binary operation: combining two manipulations by doing one followed by the other. The "result" of this operation is the single manipulation we could have done for the same result. Call that operation $*$ (for example, $V * R_{120} = S$) and consider its summary in this table, analogous to an addition table from arithmetic (the value at the head of each row is the first operand, the head of the column the second).

$*$	R_0	R_{120}	R_{240}	V	B	S
R_0	R_0	R_{120}	R_{240}	V	B	S
R_{120}	R_{120}	R_{240}	R_0	B	S	V
R_{240}	R_{240}	R_0	R_{120}	S	V	B
V	V	S	B	R_0	R_{240}	R_{120}
B	B	V	S	R_{120}	R_0	R_{240}
S	S	B	V	R_{240}	R_{120}	R_0

Take all this in. The operation is closed since, as we observed already, every combination of manipulations is equivalent to a manipulation we already know. To see informally that the operation is associative, consider that if a, b, and c are manipulations, then in the expression $a * (b * c)$, the subexpression $(b * c)$ means "the manipulation equivalent to doing the b manipulation followed by the c manipulation." Thus $a*(b*c)$ would give you the same result as first doing the a manipulation, then the b manipulation, and finally the c manipulation, in other words, $(a * b) * c$.

We thus have established that this set is a semigroup. There is also a clear identity element, R_0, so it is furthermore a monoid. Notice one more thing:

$$V * V = R_0 \quad B * B = R_0 \quad S * S = R_0 \quad R_{120} * R_{240} = R_0 \quad R_{240} * R_{120} = R_0 \quad R_0 * R_0 = R_0$$

Every manipulation has another manipulation that, when they are combined, produces the identity manipulation. In other contexts we call such an element an *inverse* (4 and -4 are additive inverses; 4 and $.25$ are multiplicative inverses; T and F are inverses under both $\vee$ and $\wedge$). We will call it an inverse here, too, and denote the inverse of a given a as a^{-1}.

inverse

This last property is what makes the monoid a *group*, our main definition of this chapter. To put it all together, a group is a set A together with a closed binary operation $*$ such that

group

- for all $a, b, c \in A$, $(a * b) * c = a * (b * c)$ (associativity),
- there exists $e \in A$ such that $a * e = a = e * e$ (existence of an identity), and
- for all $a \in A$, there exists $a^{-1} \in A$ such that $a * a^{-1} = e$ (existence of inverses).

What is the big deal? The group concept's usefulness is its ability to summarize the similarities of different things. As an immediate example, instead of equilateral triangles, we could have considered the manipulations to a square (four rotations and four axes of symmetry) or a regular hexagon (six rotations, six axes of symmetry) and gotten larger sets than that of triangle manipulations but otherwise similar results. More strikingly, notice that $\mathbb{R}$ with the operation

0 is a group—we noticed in the previous section that it is a monoid and, just now, that a number's opposite is its inverse.

Consider sets with operations discussed in this book or elsewhere in your mathematical experience and look for ones that are groups and others that are almost groups, but not quite.

Set	*Operation*	*Identity*	*Inverses*
Flips and rotations of a regular n-gon.	combining manipulations in sequence	R_0	A flip is its own inverse; for a rotation R_α, $(R_\alpha)^{-1} = R_{360-\alpha}$.
$\mathbb{R}$	$+$	0	Opposites (for x, $x^{-1} = 0 - x$)
$\mathbb{R}$	$\cdot$	1	Reciprocals (for x, $x^{-1} = \frac{1}{x}$). Wait—**no**, that does not work for 0

We can fix that by excluding zero: $\mathbb{R}^+$ and $\mathbb{R} - \{0\}$ are groups. Also $\mathbb{Z}$ and $\mathbb{Q}$ are groups with $+$, and $\mathbb{Q}^+$ and $\mathbb{Q} - \{0\}$ are groups with $\cdot$.

Set	*Operation*	*Identity*	*Inverses*
A lattice L	$\vee$	$\bot$	The complement? **No.** $a \vee a' = \top$, not $\bot$. Lattices are not groups.
$\mathscr{P}(A)$	$\cup$	$\emptyset$	**None.** We already knew this would not work—powersets are lattices.
Strings	concatenation	the empty string	**None.** Concatenation can never get you back to the empty string.
Functions	$\circ$ (composition)	$f(x) = x$	The "inverse function." Wait—**no**, that does not always exist.

After three strike-outs, we can salvage this last example if we put it this way: For any set A, the set of *one-to-one* correspondences on A (that is, the set $\{f : A \to A \mid f \text{ is a one-to-one correspondence}\}$) is a group with composition.

Set	*Operation*	*Identity*	*Inverses*
Matrices	Component-wise addition: $\begin{bmatrix} a & b \\ c & d \end{bmatrix} + \begin{bmatrix} e & f \\ g & h \end{bmatrix} = \begin{bmatrix} a+e & b+f \\ c+g & d+h \end{bmatrix}$	$\begin{bmatrix} 0 & 0 \\ 0 & 0 \end{bmatrix}$	The matrix of opposites of the components.
$\mathbb{Z}_n$	addition modulo n	0	for $x \in \mathbb{Z}_n$, $x^{-1} = n - x$

If you have never seen modular arithmetic before, this one requires some explanation. For $n \geq 2$, $\mathbb{Z}_n$ is the set of whole numbers less than n, for example

$\mathbb{Z}_5 = \{0, 1, 2, 3, 4\}$. We can make the usual arithmetic operations closed on a set $\mathbb{Z}_n$ by "moding" the result of any operation by n—that is, the remainder of dividing by n, just like the `mod` operator in ML. Thus for $\mathbb{Z}_5$, modular arithmetic finds $2 + 3 = 0$ and $4 \cdot 3 = 2$. $\mathbb{Z}_n$ is a handy example of a finite group. Here is the addition table for $\mathbb{Z}_4$:

$+$	0	1	2	3
0	0	1	2	3
1	1	2	3	0
2	2	3	0	1
3	3	0	1	2

You can verify in this table that for any x, the inverse is $n - x$—for example, the inverse of 1 is 3.

Two more examples will take a little thought:

$\mathbb{R} - \{0\}$	$*$ where $a * b = \frac{a \cdot b}{2}$	2	for $x \in \mathbb{R} - \{0\}$, $\frac{4}{x}$
$\{1, -1, i, -i\}$	complex multiplication	1	$1' = 1$; $(-1)' = -1$; $i' = -i$

To understand this last example, recall that i is the "imaginary" unit in complex numbers: $i = \sqrt{-1}$, that is, $i \cdot i = -1$. If we take the set containing 1, i, and their opposites, multiplication turns out to be closed on that set:

$\cdot$	1	-1	i	$-i$
1	1	-1	i	$-i$
-1	-1	1	$-i$	i
i	i	$-i$	-1	1
$-i$	$-i$	i	1	-1

Notice also that the whole set of complex numbers excluding zero, $\mathbb{C} - \{0\}$, is a group with $\cdot$, like the other number sets.

Having seen a variety of examples we can now consider some properties of groups. We know from the previous section that the identity is unique.

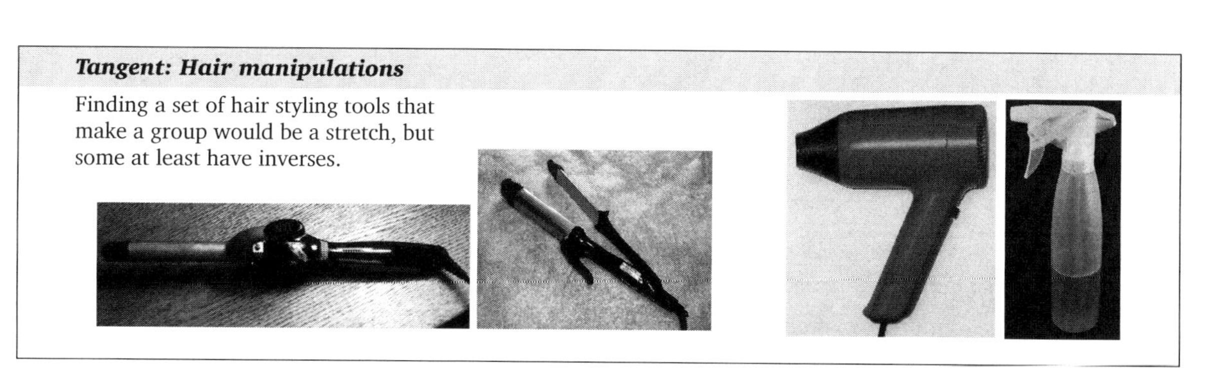

Tangent: Hair manipulations

Finding a set of hair styling tools that make a group would be a stretch, but some at least have inverses.

The definition we gave for group A stated that for all $a \in A$, there exists $a^{-1} \in A$ such that $a * a^{-1} = e$. Sometimes that part of the definition is stated as $a * a^{-1} = e = a^{-1} * a$. We do not need the last part of the equation in our definition because we can prove it.

Theorem 11.2 *If A is a group with operation $*$, then $(a^{-1})^{-1} = a$.*

This means that a is its inverse's inverse, or $a^{-1} * a = e$.

Proof. Suppose A is a group with operation $*$ and further suppose $a \in A$. By the definition of group, there exists $a^{-1} \in A$ such that $a * a^{-1} = e$. Then

$$\begin{array}{rcll} a^{-1} * a & = & (a^{-1} * e) * a & \text{by definition of identity} \\ & = & (a^{-1} * (a * a^{-1})) * a & \text{by substitution} \\ & = & (a^{-1} * a) * (a^{-1} * a) & \text{by associativity} \\ & = & e * e & \text{by definition of inverse} \\ & = & e & \text{by definition of identity} \end{array}$$

Therefore, by definition of inverse, $(a^{-1})^{-1} = e$. □

You may have noticed that one common property that is not part of the definition of a group is commutativity. Indeed, the group of manipulations to an equilateral triangle is not commutative:

$$V * R_{240} = B \neq S = R_{240} * V$$

Plenty of other groups, such as number sets with addition and multiplication, are commutative. A group with the commutative property is called an *Abelian group*.

Abelian group

We will use the notation of exponentiation to indicate that the operation is applied to the same element more than one time, for example $a^3 = a*a*a$. The notation a^{-1} to indicate the inverse of a is consistent with this, since $a^{-1} * a^3 = a^{-1} * (a * a * a) = (a^{-1} * a) * (a * a) = e * (a * a) = a^2$. We also can prove propositions like this:

Theorem 11.3 *If A is an Abelian group with $*$, then $(a * b)^n = a^n * b^n$.*

Proof. By induction on n.

Base case. Suppose $n = 1$. Then immediately $(a * b)^1 = a * b = a^1 * b^1$.

Inductive case. Suppose for some $n \geq 1$, $(a * b)^n = a^n * b^n$. Then

$$\begin{array}{rcll} (a * b)^{n+1} & = & (a * b)^n * (a * b) & \\ & = & a^n * b^n * (a * b) & \text{by the inductive hypothesis} \\ & = & a^n * a * b^n * b & \text{by associativity and commutativity} \\ & & & \text{(since } A \text{ is Abelian)} \\ & = & a^{n+1} * b^{n+1} & \square \end{array}$$

This last theorem was a result about a group known to be Abelian. We also may want to prove that a group *is* Abelian.

Theorem 11.4 *If A is a group with $*$ and for all $a, b, c \in A$, if $a * b = c * a$, then $b = c$, then A is Abelian.*

This is an awkward theorem to read since the hypothesis of the main conditional itself contains a conditional. Symbolically it is $(\forall\ a, b, c \in A, (a * b = c * a) \to (b = c)) \to (A \text{ is Abelian})$. In this proof the parts dealing with groups and Abelianness are not so bad, but dealing with the logic and the quantification can be tricky for even a moderately advanced prover such as yourself. Watch carefully.

> **Proof.** Suppose A is a group with $*$ and for all $a, b, c \in A$, if $a * b = c * a$, then $b = c$.

The first mistake one might make at this point is to start using a, b, and c. We have not supposed specific elements a, b, and c, but only some property that applies to any three elements we could pick. If one dodges that one, the next temptation would be to say

> Further suppose $a, b, c \in A$ such that $a * b = c * a \ldots$

No, no, no! To prove that A is Abelian you need to pick *two* arbitrary elements—call them x and y to avoid confusion with the earlier variables—and show that $x * y = y * x$.

> Further suppose $x, y \in A$. Then

$$\begin{array}{rcll} x^{-1} * (x * y) & = & (x^{-1} * x) * y & \text{by associativity} \\ & = & e * y & \text{by definition of inverse} \\ & = & y * e & \text{by definition of identity} \\ & = & y * (x * x^{-1}) & \text{by definition of inverse} \\ & = & (y * x) * x^{-1} & \text{by associativity} \end{array}$$

> By the given property (taking $a = x^{-1}$, $b = x * y$, and $c = y * x$), $x * y = y * x$. Hence A is Abelian by definition. □

Tangent: Group

Be wary of using the word *group* casually in the presence of a mathematician. It is interesting how everyday words can take on such strict technical meaning. The mathematical term *group*, like the plain English word, comes from the French *groupe*, as it was first used in its modern meaning in the writings of French mathematician Evariste Galois. *Abelian* groups are named after Norwegian mathematician Niels Abel. Both Galois and Abel died in their twenties, having already done ground-breaking work but before their work had received much recognition.

Make sure you understand what needs to be supposed and what needs to be shown—what we must pick blindly and what we may choose freely. To understand this is more important than all the specifics about groups.

Exercises

11.2.1 Find the set of manipulations (rotations and flips about axes of symmetry) on a square, analogous to the manipulations on an equilateral triangle described in this section. Make the operation table for combining manipulations in sequence.

Exercises 11.2.2–11.2.7 explore modular arithmetic and its relation to groups.

11.2.2 Make the addition table for $\mathbb{Z}_5$.

11.2.3 Make the addition table for $\mathbb{Z}_6$.

11.2.4 It is clear that for $n \in \mathbb{N}$, $n \geq 2$, addition modulo n is closed over $\mathbb{Z}_n$, addition is associative, and 0 is the identity. To show $\mathbb{Z}_n$ is a group, prove that for all $x \in \mathbb{Z}_n$, $n - x$ is the identity in $\mathbb{Z}_n$.

11.2.5 Consider the set $\mathbb{Z}_6 - \{0\}$ (that is, $\{1, 2, 3, 4, 5\}$). Make its *multiplication* table mod 6. Show that it is not a group (by counterexample: not all elements have inverses).

11.2.6 Do the same for $\mathbb{Z}_5 - \{0\}$, but show that it *is* a group (by exhaustion: every element has an inverse).

11.2.7 Prove that if p is prime, then $\mathbb{Z}_p - \{0\}$ is a group with multiplication mod p.

11.2.8 Prove that the set of even integers is a group with $+$.

11.2.9 Is the group of one-to-one correspondences on a set A ($f : A \to A$) with function composition Abelian? Prove or give a counterexample.

11.2.10 Recall the group $R - \{0\}$ with $*$ where $a * b = \frac{a \cdot b}{2}$. Show that 2 is the identity and that for $x \in \mathbb{R} - \{0\}$, the inverse of x is $\frac{4}{x}$. Is this group Abelian? Prove or give a counterexample.

11.2.11 In the section we glibly claimed that lattices are not groups. There is one exception, however. What lattice is also a group (there is only one such lattice, up to isomorphism)? This is trivial.

11.2.12 The set $\mathbb{B} = \{T, F\}$ is not a group with $\wedge$ or $\vee$, but it is with the exclusive or operation: $p * q = (p \vee q) \wedge \sim (p \wedge q)$. Write out the operation table and verify that it is a group.

In Exercises 11.2.13–11.2.21, assume A is a group with $*$. Prove.

11.2.13 For all $a, b, c \in A$, if $b*a = c*a$ or $a*b = a*c$, then $b = c$. This is called the *cancellation property*.

11.2.14 For all $a \in A$, a^{-1} is unique.

11.2.15 If $a, b \in A$, then $(a * b)^{-1} = b^{-1} * a^{-1}$. Is this also true for semigroups? Prove or give a counterexample.

11.2.16 A is Abelian iff for all $a, b \in A$, $(a * b)^{-1} = a^{-1} * b^{-1}$.

11.2.17 For all $a, b \in A$, if $(a * b)^2 = a^2 * b^2$, then $a * b = b * a$.

11.2.18 For all $a \in A$ and $n, m \in \mathbb{W}$, $a^n * a^m$. (Hint: Suppose these variables in three passes. First suppose $a \in A$, then $n \in \mathbb{W}$, and then finish the proof by induction on m.)

11.2.19 If for all $a \in A$, $a^2 = e$ (that is, every element is its own inverse), then A is Abelian.

11.2.20 For all $a \in A$ such that $a * a = a$, $a = e$. (This says that the identity is the only element that is idempotent.)

11.2.21 For all $a \in A$, $(a^n)^{-1} = (a^{-1})^n$. Note that this is proving that $(a^{-1})^n$ is the inverse of a^n. Use induction on n.

11.3 Isomorphisms

Recall the group $\{1, -1, i, -1\}$ and its multiplication table, below left, which we rearrange arbitrarily, below right.

$\cdot$	1	-1	i	$-i$
1	1	-1	i	$-i$
-1	-1	1	$-i$	i
i	i	$-i$	-1	1
$-i$	$-i$	i	1	-1

$\cdot$	1	i	-1	$-i$
1	1	i	-1	$-i$
i	i	-1	$-i$	1
-1	-1	$-i$	1	i
$-i$	$-i$	1	i	-1

Now compare that same table with $\mathbb{Z}_4$, shown below left.

$+$	0	1	2	3
0	0	1	2	3
1	1	2	3	0
2	2	3	0	1
3	3	0	1	2

$\cdot$	1	i	-1	$-i$
1	1	i	-1	$-i$
i	i	-1	$-i$	1
-1	-1	$-i$	1	i
$-i$	$-i$	1	i	-1

The thing to observe is that these tables are exactly the same, as long as you ignore the trifling detail of what the entries are called. A 0 in the left table appears in the same places as the 1 in the right table; 1 on the left is like i on the right; 2 on the left is like -1 on the right; and 3 on the left is like $-i$ on the right. After renaming, the groups $\mathbb{Z}_4$ and $\{1, -1, i, -i\}$ are identical.

But they mean different things, right? Possibly, but we do not study meaning, we study structure, and our discovery is the structural equivalence of these two groups. We have seen this pattern before in our study of graphs and lattices. The term for equivalence after renaming is *isomorphism*.

isomorphism

Formally, if A_1 with $*_1$ and A_2 with $*_2$ are groups, then an isomorphism from A_1 to A_2 is a one-to-one correspondence ϕ such that for all $a, b \in A_1$

$$\phi(a *_1 b) = \phi(a) *_2 \phi(b)$$

This last part of the definition is the *isomorphic property* .

isomorphic property

To get the intuition, think of A_1 and A_2 as parallel universes where any thing or action in one has its analogue in the other. $\phi(a)$ is a's doppelgänger in A_2-land. The effect of applying the operation $*_1$ to two elements of A_1 and then translating over to A_2 is the same as first translating to A_2 and then applying the operation $*_2$. The isomorphism is "operation-preserving."

Note that, as with graphs and lattices, an isomorphism *is* a function, strictly speaking. We say that two groups are *isomorphic* if there exists an isomorphism from one to the other.

isomorphic

As a second example, consider two groups each based on $\mathbb{R}$: the group $\mathbb{R}$ with $+$ and the group $\mathbb{R}^+$ with $\cdot$. Are we claiming addition and multiplication are structurally the same thing? Judge for yourself.

To prove that these two groups are isomorphic, we need to show that an isomorphism exists. For this proof of existence, we need to produce a function and show two things: that it is a one-to-one correspondence and that the isomorphic property holds for it.

Let $\phi : \mathbb{R} \to \mathbb{R}^+$ be defined so that $\phi(x) = 2^x$. The existence of the inverse function, $\phi^{-1}(y) = \lg y$, is a quick demonstration that ϕ is a one-to-one correspondence. Now suppose $a, b \in \mathbb{R}$. Then

$$\phi(a+b) = 2^{a+b} = 2^a \cdot 2^b = \phi(a) \cdot \phi(b)$$

As with graphs and lattices, the usefulness of the isomorphism concept is that if two groups are isomorphic, then almost any interesting property that one of them has the other has also. A predicate P on groups is called an *isomorphic invariant* if for any isomorphic groups A_1 and A_2, if $P(A_1)$ then $P(A_2)$. For example, being Abelian (its operation being commutative) is an isomorphic invariant. Spelled out explicitly:

isomorphic invariant

Theorem 11.5 *If ϕ is an isomorphism from $A_1, *_1$ to $A_2, *_2$ and $a, b \in A_1$, then $a *_1 b = b *_1 a$ iff $\phi(a) *_2 \phi(b) = \phi(b) *_2 \phi(b)$.*

(If you read this theorem carefully you will see that it actually makes a stronger claim than the isomorphic invariance of Abelianness. Even non-Abelian groups may have some elements that happen to "commute"—for example, everything is commutative with the identity element and inverses commute with each other. In this theorem A_1 does not have to be Abelian. On whatever pairs of elements on which the commutativity property holds, it will hold on their equivalents in A_2 also.)

Proof. Suppose ϕ is an isomorphism from A_1 to A_2 and $a, b \in A_1$ Further suppose $a *_1 b = b *_1 a$. Then

$$\begin{array}{rcll} \phi(a) *_2 \phi(b) & = & \phi(a *_1 b) & \text{by the isomorphic property} \\ & = & \phi(b *_1 a) & \text{as we supposed} \\ & = & \phi(b) *_2 \phi(a) & \text{by the isomorphic property} \end{array}$$

The converse is symmetric. □

In that last proof we never made use of the fact that ϕ was one-to-one or onto, or that $*_1$ or $*_2$ were associative, had inverses, or had an identity. This theorem would work just as well on monoids and semigroups. Moreover, we can loosen the requirement on an isomorphism being a one-to-one correspondence to get the definition of a *homomorphism*, that is, a function ϕ from a group A_1 to a group A_2 such that for all $a, b \in A_1$, $\phi(a *_1 b) = \phi(a) *_2 \phi(b)$.

homomorphism

For example, take the sets $\mathscr{P}(\{1,2,3\})$ and $\mathscr{P}(\{1,2\})$. As we know, these are monoids with $\cup$. Define $\phi : \mathscr{P}(\{1,2,3\}) \to \mathscr{P}(\{1,2\})$ so that

$$\phi(X) = X - \{3\}$$

So, for instance, $\phi(\{1,3\}) = \{1\}$ and $\phi(\{1,2\}) = \{1,2\}$. Notice

$$\begin{aligned}
\phi(\{1,3\} \cup \{1,2\}) &= \phi(\{1,2,3\}) \\
&= \{1,2\} \\
&= \{1\} \cup \{1,2\} \\
&= \phi(\{1,3\}) \cup \phi(\{1,2\})
\end{aligned}$$

This looks like a homomorphism. We will prove it: Suppose we have two subsets A and B of $\{1,2,3\}$. $\phi(A \cup B) = (A \cup B) - \{3\} = (A - \{3\}) \cup (B - \{3\}) = \phi(A) \cup \phi(B)$. The central step is by Exercise 4.3.18. ϕ is not an isomorphism because it is not one-to-one: $\phi(\{3\}) = \emptyset = \phi(\emptyset)$.

Of course all isomorphisms are homomorphisms. But with homomorphisms in general we are not guaranteed that everything in the codomain is hit, and it is possible that several elements of the domain are smooshed into the same element of the codomain. A homomorphism lets us preserve the structure of a semigroup even though we compress it into another semigroup that is smaller or stretch it over one that is bigger. Just as groups get top billing over semigroups and monoids, isomorphisms are more widely observed than homomorphisms. (Perhaps for fairness we should refer to the isomorphic property as the *'morphic* property.)

Another example:

Theorem 11.6 *If A with $*_1$ and B with $*_2$ are groups with identities e_1 and e_2, respectively, and ϕ is an isomorphism from A to B, then $\phi(e_1) = e_2$.*

This requires us to show that $\phi(e)$ is the identity in B. To do that, choose an arbitrary element from B and show that $\phi(e)$ works as an identity with it.

Proof. Suppose A and B are groups with identities e_1 and e_2 and isomorphism ϕ. Suppose further that $b \in B$. Since ϕ is onto, there exists an $a \in A$ such that $\phi(a) = b$. Now,

$$\begin{aligned}
\phi(e_1) *_2 b &= \phi(e_1) *_2 \phi(a) && \text{by substitution} \\
&= \phi(e_1 *_1 a) && \text{by the isomorphic property} \\
&= \phi(a) && \text{by the definition of identity} \\
&= b && \text{by substitution again}
\end{aligned}$$

Similarly $b *_2 \phi(e_1) = b$. Therefore, by definition of identity and the uniqueness of identity, $\phi(e_1) = e_2$. □

Did we use the fact that ϕ was a one-to-one correspondence? Well, we used onto, but not one-to-one. We also did not use the existence of inverses. So, we can say, more generally,

Corollary 11.7 *Let A with $*_1$ and B with $*_2$ be monoids with identities e_1 and e_2, respectively and ϕ be an onto homomorphism from A to B. Then $\phi(e_1) = e_2$*

(Since the theorem is a specialization of the corollary it would follow from it immediately. Perhaps Corollary 11.7 should be the theorem and Theorem 11.6 the corollary.)

Going back to the proof for moment, here is a friendly reminder to be vigilant against proof mistakes. What is wrong with the following proof attempt for Theorem 11.6?

> **Erroneous proof.** Suppose A and B are groups with identities e_1 and e_2 and isomorphism ϕ. Suppose further that $a \in A$. Then,
>
> $$\begin{aligned} \phi(e_1) *_2 \phi(a) &= \phi(e_1 *_1 a) && \text{by the isomorphic property} \\ &= \phi(a) && \text{by the definition of identity} \end{aligned}$$
>
> Similarly $\phi(a) *_2 \phi(e_1) = \phi(a)$. Therefore, by definition of identity and the uniqueness of identity, $\phi(e_1) = e_2$. ☐̸

$\phi(a)$ is an element of B and we showed $\phi(e_1) *_2 \phi(a) = \phi(a) = \phi(a) *_2 \phi(e_1)$ so $\phi(e_1)$ is the identity, right? **No**, we have shown only that $\phi(e_1)$ acts as an identity for *a certain* element called $\phi(a)$. To show that $\phi(e_1)$ acts as an identity for *all* elements of B we need to choose an element of B, not A, arbitrarily, as we did in the real proof.

Exercises

11.3.1 Prove that the relation “is isomorphic to” on groups is an equivalence relation.

11.3.2 Prove that the group of even integers with $+$ is isomorphic to $\mathbb{Z}$ with $+$. (Remember that to prove two groups to be isomorphic is a proof of existence. You need to come up with *an isomorphism*. Define a function, show that it is a one-to-one correspondence, and show that the isomorphism property holds. Of course the hard part is finding an appropriate function.)

11.3.3 Prove that $\mathbb{Z}_2$ with $+$ `mod` 2 is isomorphic to $\mathbb{B}$ with exclusive or (see Exercise 11.2.12). (Hint: Compare their operation tables.)

Tangent: Isomorphism and analogy

We reason by analogy all the time: *A* is a like *B*, *p* is true for *A*, and *q* is to *B* as *p* is to *A*; hence *q* is true for *B*. But this is logically dangerous and can lead to fallacies. Just because *A* and *B* have recognizable similarities does not mean that they match at every point. How do we know that the analogy holds for *q*?

An isomorphism is a rigorous, mathematical version of an analogy. In proving isomorphic invariants we are discovering the places where the analogy holds and, therefore, what things that are true for *A* are automatically, analogously true for *B*.

Exercises, continued

11.3.4 Prove that $\phi(x) = \sqrt{x}$ is an isomorphism from $\mathbb{R}^+$ to itself with $\cdot$. An isomorphism from a group to itself is called an *automorphism* and it suggests a self-similarity within a group.

11.3.5 Prove that if A is an Abelian group with $*$, then the function $\phi(x) = x * x$ is a homomorphism from A to itself.

11.3.6 Prove the converse of the previous exercise: If A is a group with $*$ and $\phi(x) = x * x$ is a homomorphism, then A is Abelian.

In Exercises 11.3.7–11.3.11, suppose A with $*_1$ and B with $*_2$ are groups and $\phi : A \to B$ is an isomorphism. Prove the following to be isomorphic invariants and, for each, determine whether they are also homomorphic invariants and whether they also hold for monoids or semigroups.

11.3.7 For all $a \in A$, if $a = a^{-1}$, then $\phi(a) = \phi(a)^{-1}$. (This means that the element a is its own inverse. The flips in the group of manipulations to an equilateral triangle are their own inverses.)

11.3.8 For all $a \in A$, $\phi(a^{-1}) = (\phi(a))^{-1}$.

11.3.9 For all $n \in \mathbb{N}$, for all $a \in A$, $\phi(a^n) = (\phi(a))^n$.

11.3.10 For all $a \in A$, if n is the smallest natural number such that $a^n = e$, then n is the smallest natural number such that $\phi(a)^n = e_2$.

11.3.11 The *center* of a group C, $\mathcal{Z}(C)$, is the largest subset of C of elements that commute with every other element in C. (Formally, $\mathcal{Z}(C) = \{c \in C \mid \forall\, x \in C, c * x = x * c\}$). The image of the center of A under ϕ is the center of B. (Formally, $\{\phi(x) \mid x \in \mathcal{Z}(A)\} = \mathcal{Z}(B)\}$.) Note that this requires a lot more from you than just proving something about isomorphisms. You also need remember the basics of proving set equality and using the definition of image, not to mention dealing with the new definition of *center*.

11.4 Subgroups

You probably have noticed the similar themes among our study of graphs and lattices in Chapter 8 and Chapter 10 and our present study of groups. Just as we had subgraphs and sublattices, the analogous concept of a *subgroup* is not surprising. If A is a group with operation $*$, then $B \subseteq A$ is a subgroup of A if it also is a group with $*$.

subgroup

The definition sounds simple, but consider what it entails. Since A is a group, we already know $*$ is associative; that part comes for free. We already know $*$ is closed, right? No, for B to be a group means that $*$ must be closed in B—for no two elements $b_1, b_2 \in B$ can $b_1 * b_2$ go outside of B into $A - B$. Moreover, the identity of A must be an element of B, and the inverses of all the elements of B must be in B.

Consider some examples:

- The set consisting only of rotations of an equilateral triangle, $\{R_0, R_{120}, R_{240}\}$, is a subgroup of the set of all manipulations. No combination of rotations will have the same affect as any flip. This set has the identity R_0, and R_{120} and R_{240} are inverses of each other.

- With $+$, the set of even integers is a subgroup of $\mathbb{Z}$, which is a subgroup of $\mathbb{Q}$, which is a subgroup of $\mathbb{R}$.

- The group $\{1, -1, i, -i\}$ is a subgroup of $\mathbb{C} - \{0\}$ with $\cdot$.

Consider also some subsets that are *not* subgroups.

- The set of only flips of an equilateral triangle, $\{V, B, S\}$, is not a subgroup of the set of all manipulations. Although it contains all the inverses of its elements (since a flip is its own inverse), it does not contain the identity R_0. Proposing $\{R_0, V, B, S\}$ as a fix does not work either because $V * B = R_{240}$, so the operation is not closed in the subset.

- The set $\{-1, 0, 1\}$ is not a subgroup of $\mathbb{Z}$ with $+$ because the operation is not closed on the set: $1 + 1 = 2$.

- The group $\mathbb{Z}_4$ with $+ \texttt{ mod } 4$ is not a subgroup of $\mathbb{Z}$ with $+$ because $+ \texttt{ mod } 4$ and regular $+$ are not the same operation.

Does the prospect of proving a subset of a group is a subgroup sound tedious? Fortunately we need to do the whole routine only once—in the proof of the following theorem which gives us a shortcut for proving a set to be a subgroup.

Theorem 11.8 (One-step subgroup test) *If A is a group with $*$ and $B \subseteq A$, $B \neq \emptyset$, then B is a subgroup of A if for all $a, b \in B$, $a * b^{-1} \in B$.*

What do we need to show?

- For all $x, y \in B$, $x * y \in B$. ($*$ is closed in B.)

- $e \in B$. (B has the identity.)

- For all $x \in B$, $x^{-1} \in B$. (Inverses are included.)

The second can be proven from the other two since $x * x^{-1} = e$, but it will be just as easy to prove that directly.

Proof. Suppose $B \subseteq A$, $B \neq \emptyset$, and for all $a, b \in B$, $a * b^{-1} \in B$.

Further suppose $x \in B$. First, by our supposition $x * x^{-1} \in B$, so $e \in B$. Moreover, since we have $x, e \in B$, we know from our supposition that $e * x^{-1} \in B$, so $x^{-1} \in B$.

Next suppose $x, y \in B$. From what we said earlier, $y^{-1} \in B$. Appealing to our supposition yet again, $x * (y^{-1})^{-1} \in B$, and so $x * y \in B$.

Since B meets all the criteria for being a group, B is a subgroup of A. □

Try it out:

Theorem 11.9 *If $n, x \in \mathbb{N}$, such that $n \geq 2$ and $x|n$, then the set of multiples of x in $\mathbb{Z}_n$ (formally, $\{a \in \mathbb{Z}_n \mid x|a\}$) is a subgroup of $\mathbb{Z}_n$ with $+$.*

Proof. Suppose $n, x \in \mathbb{N}$, $n \geq 2$, and $x|n$. Suppose further that $a, b \in \mathbb{Z}_n$ such that $x|a$ and $x|b$.

By Exercise 11.2.4, we know that $b^{-1} = n - b$. By definition of divides, there exist $i, j, k \in \mathbb{Z}$ such that $a = i \cdot x$, $b = j \cdot x$, and $n = k \cdot x$. Then $a + b^{-1} = i \cdot x + (k \cdot x - j \cdot x) = (i + k - j) \cdot x$ by substitution and algebra.

Now, what is $(i + k - j) \cdot x \texttt{ mod } n$? It depends on how i and j are related.

Case 1: Suppose $i < j$. Then $i+k-j < k$ and $(i+k-j)\cdot x < k\cdot x = n$. and so $a + b^{-1} = (i + k - j) \cdot x$ in arithmetic mod n just as in regular arithmetic.

Case 2: Suppose $i = j$. Then $i+k-j = k$ and $(i+k-j)\cdot x = k\cdot x = n$. In arithmetic mod n, $a + b^{-1} = 0$.

Case 3: Suppose $i > j$. Then $(i + k - j) \cdot x = (i - j) \cdot x + k \cdot x = (i - j) \cdot x + n$ which, in arithmetic mod n, equals $(i - j) \cdot x$. (Note that i and j must each be less than k, so $i - j$ must be less than k.)

In each case, $a + b^{-1}$ using arithmetic mod n results in a multiple of x. Hence, by the one-step subgroup test, $\{a \in \mathbb{Z}_n \mid x|a\}$ is a subgroup of $\mathbb{Z}_n$. □

Exercises

Prove.

11.4.1 If $n \in \mathbb{N}$, then the set of multiples of n is a subgroup of $\mathbb{Z}$ with $+$.

11.4.2 If $n, m \in \mathbb{N}$ and $n|m$, then $\mathbb{Z}_n$ is isomorphic to a subgroup of $\mathbb{Z}_m$. (This requires a double proof of existence. You need to find an appropriate subgroup of $\mathbb{Z}_m$ and an isomorphism from $\mathbb{Z}_n$ to that subgroup.)

Assume A is a group with $*$. Prove.

11.4.3 If $a \in A$ and $a = a^{-1}$, then $\{a, e\}$ is a subgroup of A.

11.4.4 If A is Abelian, then $\{a \in A \mid a = a^{-1}\}$ (the set of all things that are their own inverse, which would include e) is a subgroup of A.

11.4.5 If A is Abelian, then $\{a^2 \mid a \in A\}$ (the set of elements that are the square of another element) is a subgroup of A.

11.4.6 $\mathcal{Z}(A)$ is a subgroup of A. (Recall from Exercise 11.3.11 that the center of a group $\mathcal{Z}(A)$ is the subset of A that commutes with every element in A, $\mathcal{Z}(A) = \{a \in A \mid \forall\, x \in A, a * x = x * a\}$.)

11.4.7 If B is a subgroup of A and ϕ is an isomorphism from A to C, a group with $*_2$, then $\Phi(B)$ (the image of B under ϕ) is a subgroup of C. (In other words, "having a subgroup" is an isomorphic invariant. Note that B is isomorphic to $\Phi(B)$.)

11.4.8 If B is a subgroup of A and $a \in A$, then the set $\{a * b * a^{-1} \mid b \in B\}$, called the *conjugate* of B with a, is a subgroup of A.

11.4.9 If B is a subgroup of A, $a \in A$, and B is Abelian, then the conjugate of B with a is Abelian.

11.5 A garden of groups

We make no attempt to comprehend the vast terrain of group theory. This section surveys some of the prominent and more accessible features. Think of it as a one-day guided bus tour. If you want to rough it in the wild, take a course in modern or abstract algebra.

dihedral group

Dihedral groups. Our first example of a group, the set of manipulations to an equilateral triangle, is one of a family of groups called the *dihedral groups*. The group of rotations and symmetric flips of a regular n-gon is called the *dihedral group of order* $2n$ and written D_n. (One could say that D_n is an abstract group that the set of manipulations of a regular n-gon is isomorphic to. We will not distinguish between isomorphic groups. Any group isomorphic to D_n we will simply regard as D_n, even if the names are different.)

The difference between the written name and the pronounced name is confusing. We call the set we studied at the beginning of Section 11.2 "the dihedral group of order 6" and denote it D_3. The subscript in the written form indicates the sides of the shape. The *order* of a group is the same as the set's cardinality, which for these groups is twice the number of sides of the shape.

order

Dihedral groups are interesting because they come up wherever we find structural symmetry all around us. Various molecules, crystals, and flowers exhibit the rotational symmetries of dihedral groups and knowledge about how these rotations and flips combine is valuable to chemists, geologists, and botanists. Sand dollars have symmetry group D_5, snowflakes D_6. Rotational symmetries also abound in symbols, artwork, and other human designs.

relatively prime

Groups of relative primes. Two natural numbers x and y are *relatively prime* if they have no common factors; consider 1 to be relatively prime to anything. Let $\mathcal{U}(n)$ be the set of all positive integers less than n and relatively prime to n. For example, $\mathcal{U}(5) = \{1, 2, 3, 4\}$ and $\mathcal{U}(8) = \{1, 3, 5, 7\}$. These sets make interesting and useful groups with multiplication mod n.

Theorem 11.10 *For all $n \in \mathbb{N}$ such that $n \geq 2$, $\mathcal{U}(n)$ with $\cdot \texttt{ mod } n$ is a group.*

Before we prove this, get a feel for what these alleged groups look like. Take $\mathcal{U}(8)$:

$\cdot$	1	3	5	7
1	1	3	5	7
3	3	1	7	5
5	5	7	1	3
7	7	5	3	1

By inspection we see that the operation is closed. We know it is associative already, and, as we would expect, 1 is the identity. Checking for inverses, we notice that everything is its own inverse. Is that always the case? Try $\mathcal{U}(5)$.

$\cdot$	1	2	3	4
1	1	2	3	4
2	2	4	1	3
3	3	1	4	2
4	4	3	2	1

No, but at least this second example is still a group. Time to prove the theorem.

Proof. As mentioned already, we know that multiplication is associative and that 1 will be the identity for any kind of multiplication. We need to prove closure and inverses.

Suppose $a, b \in \mathcal{U}(n)$. The Quotient-Remainder Theorem tells us that there exist $q, r \in \mathbb{N}$ such that $a \cdot b = n \cdot q + r$, where $0 < r \leq n$. In arithmetic mod n, $a \cdot b = r$. What we need to show for $r \in \mathcal{U}(n)$ is that r is relatively prime with n.

Suppose r is not relatively prime with n. That means there exists an $x \in \mathbb{N}$ such that x is a common factor of r and n (that is, $x|r$ and $x|n$). That would mean $x|(n \cdot q + r)$, and hence $x|(a \cdot b)$. Then x is a factor of either a or b, and thus either a or b is not relatively prime with n; either $a \notin \mathcal{U}(n)$ or $b \notin \mathcal{U}(n)$. Contradiction. Hence r is relatively prime with n, and multiplication mod n is closed on $\mathcal{U}(n)$.

Showing inverses is a bit more complicated. First, a lemma:

Lemma 11.11 *If $a, b, c \in \mathcal{U}(n)$ and $b \neq c$, then $a \cdot b \neq a \cdot c$ (using modular arithmetic).*

Proof (of lemma). Suppose $a, b, c \in \mathcal{U}(n)$ and $b \neq c$. (Notice that it could be that $a = b$ or $a = c$.)

Suppose further that $a \cdot b = a \cdot c$. Then there exist q_1, q_2, and r such that $a \cdot b = q_1 \cdot n + r$ and $a \cdot c = q_2 \cdot n + r$.

Suppose (without loss of generality) b is the greater of the two, that is, $b > c$. Then we can subtract equations

$$\begin{array}{rrcl} & a \cdot b & = & q_1 \cdot n + r \\ - & a \cdot c & = & q_2 \cdot n + r \\ \hline & a \cdot (b - c) & = & (q_1 - q_2) \cdot n \end{array}$$

Since a is relatively prime with n, a cannot divide n, so it must divide $q_1 - q_2$. Now, solving for b:

$$b = \frac{q_1 - q_2}{a} \cdot n + c$$

Since we said $a|(q_1 - q_2)$, then $\frac{q_1 - q_2}{a} \geq 1$, and so $b > n$. This is a contradiction because we assumed $b \in \mathcal{U}(n)$. □

What this lemma says is that given $a \in \mathcal{U}(n)$, every element in $\mathcal{U}(n)$ must take a to something different. This further means that for every element in $\mathcal{U}(n)$, something must take a to it, simply because otherwise we would run out of elements (compare the Pigeonhole Principle in Section 7.9). This has to include 1, the identity, therefore a's inverse must exist in $\mathcal{U}(n)$. Getting technical:

Proof (of Theorem 11.10, continued). Suppose $a \in \mathcal{U}(n)$. Define $f_a : \mathcal{U}(n) \to \mathcal{U}(n)$ such that $f_a(x) = a \cdot x$. By the lemma, f_a is one-to-one. By Exercise 7.9.4, f_a is onto.

Since f_a is onto, there exists and $b \in \mathcal{U}(n)$ such that $f_a(b) = 1$. In other words, $a \cdot b = 1$, or $b = a^{-1}$.

This accounts for all the requirements for $\mathcal{U}(n)$ to be a group. □

There is one very frustrating part of the proof of the existence of inverses—we never actually showed how to find an inverse for a given $a \in \mathcal{U}(n)$, only that such an inverse exists. This is a *non-constructive* proof of existence, where we prove something must exist by deriving a contradiction from its non-existence (in this case the contradiction was in the lemma).

non-constructive

To find the inverse of a given a, we need to find $b \in \mathcal{U}(n)$ such that $a \cdot b = i \cdot n + 1$ for some $i \in \mathbb{N}$. Rewriting this as $a \cdot b + i \cdot n = 1$, we can use the Extended Euclidean Algorithm to find b. See Exercise 4.10.7.

Permutation groups. In combinatorics, we think of a permutation of a set as simply a (re)arrangement of the elements in the set. We looked at this briefly in Section 7.11. It is like a way to shuffle the cards. Thus, for the set $\{1, 2, 3, 4\}$, the permutations are

$$\begin{array}{cccccc} 1,2,3,4 & 1,2,4,3 & 1,3,2,4 & 1,3,4,2 & 1,4,3,2 & 1,4,2,3 \\ 2,1,3,4 & 2,1,4,3 & 2,3,1,4 & 2,3,4,1 & 2,4,1,3 & 2,4,3,1 \\ 3,1,2,4 & 3,1,4,2 & 3,2,1,4 & 3,2,4,1 & 3,4,1,2 & 3,4,2,1 \\ 4,1,2,3 & 4,1,3,2 & 4,2,1,3 & 4,2,3,1 & 4,3,1,2 & 4,3,2,1 \end{array}$$

We will forge a new definition for our present purposes. We will say that a *permutation* of a finite set A is a one-to-one correspondence from A to A.

permutation

What fellowship does that definition have with our intuitive understanding of permutations? Consider an example. Define the following one-to-one correspondence, α, on $\{1, 2, 3, 4\}$:

x	$\alpha(x)$
1	2
2	1
3	3
4	4

The second column looks just like one of the "permutations" we listed above. Moreover, if we extend our notion of α so that it can be applied to lists of elements of A (sort of like the image of a set under a function, except the elements or ordered; more like the `map` function in ML), then

$$\alpha([1, 2, 3, 4]) = [2, 1, 3, 4]$$

The standard way to represent a permutation uses a $|A| \times 2$ matrix with the input to the function as the top row and the result for each input on the bottom. The one above (α) would be written

$$\begin{bmatrix} 1 & 2 & 3 & 4 \\ 2 & 1 & 3 & 4 \end{bmatrix}$$

Again, read that by finding the input on top and the corresponding output on the bottom: 1 maps to 2, 2 maps to 1, 3 maps to 3, 4 maps to 4. We also have a ready binary operation to apply to permutations: function composition. Let β be the permutation listed originally as 3, 4, 1, 2. Then

$$\alpha \circ \beta = \begin{bmatrix} 1 & 2 & 3 & 4 \\ 2 & 1 & 3 & 4 \end{bmatrix} \circ \begin{bmatrix} 1 & 2 & 3 & 4 \\ 3 & 4 & 1 & 2 \end{bmatrix} = \begin{bmatrix} 1 & 2 & 3 & 4 \\ 3 & 4 & 2 & 1 \end{bmatrix}$$

To get your mind around this, you need to read from right to left. What is $\alpha \circ \beta(1)$? We feed 1 into β , which gets 3; feed 3 into α, and we still get 3. Hence $\alpha \circ \beta(1) = 3$.

To gain intuition, consider a larger example:

$$\alpha = \begin{bmatrix} 1 & 2 & 3 & 4 & 5 \\ 2 & 4 & 3 & 5 & 1 \end{bmatrix}, \quad \beta = \begin{bmatrix} 1 & 2 & 3 & 4 & 5 \\ 5 & 4 & 1 & 2 & 3 \end{bmatrix}$$

$$\alpha \circ \beta = \begin{bmatrix} 1 & 2 & 3 & 4 & 5 \\ 2 & 4 & 3 & 5 & 1 \end{bmatrix} \circ \begin{bmatrix} 1 & 2 & 3 & 4 & 5 \\ 5 & 4 & 1 & 2 & 3 \end{bmatrix} = \begin{bmatrix} 1 & 2 & 3 & 4 & 5 \\ 1 & 5 & 2 & 4 & 3 \end{bmatrix}$$

Notice some things.

- In α, $1 \to 2$, $2 \to 4$, $4 \to 5$, $5 \to 1$; $\quad 3 \to 3$.
- In β, $1 \to 5$, $5 \to 3$, $3 \to 1$; $\quad 2 \to 4$, $4 \to 2$.
- In $\alpha \cdot \beta$, $1 \to 1$; $\quad 2 \to 5$, $5 \to 3$, $3 \to 2$; $\quad 4 \to 4$.

Visualize these with digraphs:

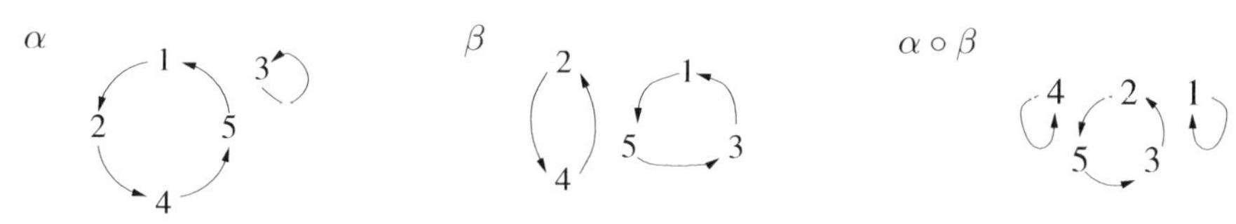

Within a permutation we have *cycles* (not to be confused with *cyclic groups*, discussed below). This leads to a new, compact notation where we list the values in each cycle.

$$\alpha = (1245)(3), \beta = (153)(24), \alpha \cdot \beta = (1)(253)(4)$$

Here is how to read this notation: (1245) means that this is a cycle in α; again $\alpha(1) = 2$ etc. To find $\alpha(x)$, find x in one of the cycles and see what number comes next. If x is the last number in the cycle, then go to the beginning of the cycle. Clearly there are several ways to describe any permutation in this notation. α could also be written (3)(5124). Exercise 11.5.1 asks you to prove that this notation is well founded.

Since a permutation is simply a one-to-one correspondence on a set, we know from earlier discussion that the set of *all* permutations of a set A is a group with function composition. But many smaller sets of of permutations turn out to be groups—subgroups of the group of all permutations. A set of permutations that forms a group under function composition is called a *permutation group*.

permutation group

Consider a set of permutations on the set $\{1, 2, 3\}$: $\{(1)(2)(3), (123), (132), (1)(23), (13)(2)(12)(3)\}$. Notice that $(1)(2)(3)$ is the identity function, so that must be in any group. For an arbitrary set like this, we would need to use exhaustion to verify closure and the presence of inverses, for example $(123) \circ (1)(23) = (12)(3)$, $(123) \circ (132) = (1)(2)(3)$, $(1)(23) \circ (1)(23) = (1)(2)(3)$—that makes three successful tests for closure and two inverses. Here is the whole operation table, along with the operation table for D_3, for comparison:

$\circ$	(1)(2)(3)	(123)	(132)	(1)(23)	(13)(2)	(12)(3)
(1)(2)(3)	(1)(2)(3)	(123)	(132)	(1)(23)	(13)(2)	(12)(3)
(123)	(123)	(132)	(1)(2)(3)	(13)(2)	(12)(3)	(1)(23)
(132)	(132)	(1)(2)(3)	(123)	(12)(3)	(1)(23)	(13)(2)
(1)(23)	(1)(23)	(12)(3)	(13)(2)	(1)(2)(3)	(132)	(123)
(13)(2)	(13)(2)	(1)(23)	(12)(3)	(123)	(1)(2)(3)	(132)
(12)(3)	(12)(3)	(13)(2)	(1)(23)	(132)	(123)	(1)(2)(3)

$*$	R_0	R_{120}	R_{240}	V	B	S
R_0	R_0	R_{120}	R_{240}	V	B	S
R_{120}	R_{120}	R_{240}	R_0	B	S	V
R_{240}	R_{240}	R_0	R_{120}	S	V	B
V	V	S	B	R_0	R_{240}	R_{120}
B	B	V	S	R_{120}	R_0	R_{240}
S	S	B	V	R_{240}	R_{120}	R_0

Why, they are isomorphic! There is a simple way to understand this. If we label the corners of the triangle clockwise from the top as 1, 2, and 3, then every flip or rotation is just a function of how to reorder those corners.

We cheated on this example, though—this is in fact all the possible permutations on $\{1, 2, 3\}$, not a strict subset. D_4, the set of flips and rotations on a square, however, is isomorphic to a set of permutations on $\{1, 2, 3, 4\}$ that is a strict subgroup of all possible permutations. In fact, we have the following profound result:

Theorem 11.12 (Cayley's Theorem) *Every group is isomorphic to a group of permutations.*

"A group of permutations" means, of course, a subgroup of the set of all permutations on some set, with function composition.

Proof. Suppose A is a group with $*$. Suppose $a \in A$. Define a function $f_a : A \to A$ such that $f_a(x) = a * x$. (We defined a similar function in the proof of Theorem 11.10.) This function merely applies a on the left.

Suppose $b \in A$. Then $f_a(a^{-1} * b) = a * a^{-1} * b = b$, so f_a is onto.

Suppose $b, c \in A$ such that $f_a(b) = f_a(c)$. Then $a * b = a * c$. By the cancellation property (Exercise 11.2.13), $b = c$, so f_a is one-to-one.

Now let $B = \{f_a \mid a \in A\}$, that is, the set of all such functions for all elements of A. The preceding showed that all such functions are one-to-one correspondences. We will show that B is a group with $\circ$.

Suppose $f_a, f_b \in B$ (this implicitly picks a and b arbitrarily from A). Suppose $x \in A$. Then

$$\begin{array}{rcll} f_a \circ f_b(x) & = & f_a(f_b(x)) & \text{by definition of function composition} \\ & = & f_a(b * x) & \text{by how we defined our set of functions} \\ & = & a * (b * x) & \text{similarly} \\ & = & (a * b) * x & \text{by associativity} \\ & = & f_{a*b}(x) & \text{again by how we defined our set} \end{array}$$

By definition of function equality, $f_a \circ f_b = f_{a*b}$, so the operation is closed on B Moreover, it is easy to see that f_e is the identity and for $f_a \in B$, $f_a^{-1} = f_{a^{-1}}$. We already know that function composition is associative. Hence B is a group.

Define $\phi : A \to B$ such that $\phi(a) = f_a$.

Read that carefully. ϕ is a function whose codomain is a set of functions. That should be easy for an ML programmer to understand, since it is just like an ML function that returns a function.

Suppose $a, b \in A$. Then $\phi(a * b) = f_{a*b} = f_a \circ f_b = \phi(a) * \phi(b)$. Hence ϕ is an isomorphism.

Therefore A is isomorphic to B, a group of permutations. □

Cyclic groups. Suppose A with $*$ is a group, and $a \in A$. Let $\langle a \rangle$ be the set $\{a^n \mid n \in \mathbb{Z}\}$ For example, if the group is $\mathbb{Q}$ with addition and $a = \frac{1}{2}$, then $\langle \frac{1}{2} \rangle$ is

$$\{\ldots \ \frac{1}{2}^{-2} = -1, \ \frac{1}{2}^{-1} = -\frac{1}{2}, \ \frac{1}{2}^{0} = 0, \ \frac{1}{2}^{1} = \frac{1}{2}, \ \frac{1}{2}^{2} = 1, \frac{1}{2}^{3} = \frac{3}{2}, \ \frac{1}{2}^{4} = 2 \ \ldots\}$$

In other words, this is the set of things we can make with repeated applications of $\frac{1}{2}$ and its inverse $-\frac{1}{2}$. Note that we do not assume the exponent is positive. We interpret a^{-n} as n terms of the inverse of a, $(a^{-1})^n$.

For an example on a finite group, consider $\mathbb{Z}_{32}$ with $+$ mod 8. $\langle 4 \rangle = \{0, 4, 8, 12, 16, 20, 24, 28\}$ These are handy examples of subgroups.

Theorem 11.13 *If A with $*$ is a group and $a \in A$, then $\langle a \rangle$ is a subgroup of A.*

Proof. Suppose A with $*$ is a group and $a \in A$.

Suppose $x, y \in \langle a \rangle$. By definition, there exist $n, m \in \mathbb{Z}$ such that $x = a^n$ and $y = a^m$. Recall that $(a^m)^{-1} = (a^{-1})^m$ by Exercise 11.2.21.

Then $x * y^{-1} = a^n * (a^m)^{-1} = a^{n-m} \in \langle a \rangle$.

Therefore $\langle a \rangle$ is a subgroup of A. □

To understand the expression a^{n-m}, consider whether $n - m$ is positive or negative, that is, which is bigger, n or m. For the earlier expression $a^n * (a^m)^{-1} = a^n * (a^{-1})^m$ imagine a sequence of as battling a sequence of a^{-1}s.

$$\underbrace{a * a \cdots a * a * a}_{n} * \underbrace{a^{-1} * a^{-1} * a^{-1} \cdot a^{-1}}_{m}$$

Associate from the inside out.

$$a * a \cdots (a * (a * (a * a^{-1}) * a^{-1}) * a^{-1}) \cdot a^{-1}$$

Elements cancel each other out at the middle until all that is left are as from the left or a^{-1}s from the right, not both. Hence a^{n-m} will be a power of a or of a^{-1}.

cyclic subgroup

We call $\langle a \rangle$ the *cyclic subgroup* of A generated by a. We deferred giving that definition because we needed to prove it was a subgroup before we could use that word in the name.

cyclic group

generator

If it so happens that $A = \langle a \rangle$ for some a, then A is called a *cyclic group* and a is called the *generator* of A. For example, 1 is the generator of $\mathbb{Z}$ with $+$. It is possible that a cyclic group has more than one generator—in the case of $\mathbb{Z}$, -1 is also a generator. $\mathbb{R}$ with $+$ is not a cyclic group.

Our foray into cyclic groups will focus on one big result:

Theorem 11.14 (The Fundamental Theorem of Cyclic Groups) *Every subgroup of a cyclic group is cyclic.*

Here is what that means. We know already that for any group we can find a subgroup—not necessarily a strict subgroup—by taking any element a and treating that element as a generator to find $\langle a \rangle$. But if the original group itself has a generator, then *these are the only subgroups*.

To get some intuition for this, consider $\mathbb{Z}$ with $+$. The set of even integers $\mathbb{Z}^e$ is a subgroup (if you add two evens, you get an even; all inverses of evens are even; 0 is even). While 1 generates $\mathbb{Z}$, 2 generates $\mathbb{Z}^e$. On to the proof.

Proof. Suppose A is a cyclic group with generator $a \in A$, that is, $A = \langle a \rangle$, and B is a subgroup of A.

We need to show that B is cyclic. To do that, we need to find a $b \in B$ that generates B. We can knock off an easy case to begin with.

First suppose $B = \{e\} = \langle e \rangle$, which is trivially a subgroup.

Next suppose $b \in B$ such that $b \neq e$. (Contrary to the previous supposition, we are assuming that B has at least one element besides e.) Since A is a cyclic group, $x = a^i$ for some $i \in \mathbb{Z}$. Since B is a subgroup, $x^{-1} \in B$, and by substitution, $a^{-i} \in B$. Since either i or $-i$ must be positive, B contains at least one element with a positive exponent on a.

Take an intuition break. In $\mathbb{Z}$ with $+$, $2 = 1^2$. (That looks funny, but for us exponentiation means many applications of the group's operation, in this case $+$.) Moreover, 2 is the element in $\mathbb{Z}^e$ with the smallest positive exponent, since $1^1 \notin \mathbb{Z}^e$. Just as 2 is a generator for $\mathbb{Z}^e$, the next step in the proof is to show that the element in B that is the smallest positive power of A's generator is a generator for B.

Let $m \in \mathbb{N}$ be the smallest natural number such that $a^m \in B$.

We are going to show $\langle a^m \rangle = B$. What kind of proposition is that? It is our good old friend set equality. We must show $\langle a^m \rangle \subseteq B$ and $B \subseteq \langle a^m \rangle$. Which do you think is easier?

$\langle a^m \rangle \subseteq B$ by closure, since B is a subgroup and $a^m \in B$.

Now suppose $b \in B$.

Showing $b \in \langle a^m \rangle$ is the heart of the matter.

Since $b \in \langle a \rangle$, $b = a^j$ for some $j \in \mathbb{Z}$. We will show $m | j$.

The Quotient Remainder Theorem (Theorem 4.21, used in Exercise 4.10.5) says there exist $q, r \in \mathbb{W}$ such that $j = m \cdot q + r$ and $0 \leq r < m$. We will show that $r = 0$. Now,

$$
\begin{array}{rcll}
a^j & = & a^{m \cdot q + r} = a^{m \cdot q} * a^r & \\
a^r & = & a^{-m \cdot q} * a^j & \text{by cancellation} \\
 & = & a^{-m \cdot q} * b & \text{by substitution} \\
(a^m)^{-1} & \in & B & \text{since } a^m \in B \\
((a^m)^{-1})^q & \in & B & \text{by closure} \\
a^{-m \cdot q} & \in & B & \text{by substitution}
\end{array}
$$

This means $a^r \in B$ by closure. Moreover, $0 \leq r < m$. But we supposed that m is the smallest natural number such that $a^m \in B$. It must be, then, that $r = 0$.

Hence $b = a^j = a^{m \cdot q} \in \langle a^m \rangle$, and $B \subseteq \langle a^m \rangle$.

Therefore $B = \langle a^m \rangle$; B is a cyclic subgroup. □

Here are a few other interesting facts about cyclic groups and subgroups. See the exercises for proofs.

- If $\langle a \rangle$ is finite and $|\langle a \rangle| = n$, then the order of any subgroup of $\langle a \rangle$ divides n.
- For every divisor k of n, there is exactly one subgroup of order k, namely $\langle a^{\frac{n}{k}} \rangle$.
- For $\mathbb{Z}_n$ with $+ \texttt{ mod } n$, $\langle \frac{n}{k} \rangle$ is the unique subgroup of order k, and these are the only subgroups.

To see this in action, consider $\mathbb{Z}_{24}$ the set of whose subgroups makes a nice lattice with the relation *is subgroup of*.

$$\begin{array}{rrcll}
 & & & & \textit{order} \\
\mathbb{Z}_{24} = & \langle 1 \rangle & = & \{0,1,2,\ldots,23\} & 24 \\
 & \langle 2 \rangle & = & \{0,2,4,\ldots 22\} & 12 \\
 & \langle 3 \rangle & = & \{0,3,6,\ldots 21\} & 8 \\
 & \langle 4 \rangle & = & \{0,4,8,\ldots 20\} & 6 \\
 & \langle 6 \rangle & = & \{0,6,12,18\} & 4 \\
 & \langle 8 \rangle & = & \{0,8,16\} & 3 \\
 & \langle 12 \rangle & = & \{0,12\} & 2 \\
\langle 0 \rangle = & \langle 24 \rangle & = & \{0\} & 1
\end{array}$$

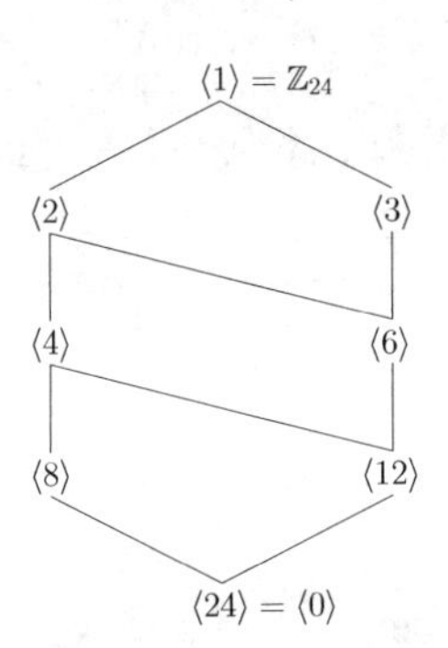

Exercises

11.5.1 Prove that every permutation α of a finite set A can be written as a cycle or a product of disjoint cycles. To do this, describe an algorithm for making the cycles (pick an element $a \in A$ and compute $\alpha(a)$, $\alpha(\alpha(a)), \ldots$ until you hit a repeat, etc.), prove that the sequence generated this way is a cycle, and derive a contradiction from supposing the cycles have an overlap.

11.5.2 Prove that disjoint cycles commute, that is, if A is a permutation group and $a, b \in A$ are in distinct cycles, then $a * b = b * a$.

In Exercises 11.5.3–11.5.5 prove the corollaries of Theorem 11.14 mentioned in the section.

11.5.3 If $\langle a \rangle$ is finite and $|\langle a \rangle| = n$, then the order of any subgroup of $\langle a \rangle$ divides n. (Hint: Suppose a subgroup B of $\langle a \rangle$. The theorem tells you $B = \langle a^m \rangle$ for some m.)

11.5.4 For every divisor k of n, there is exactly one subgroup of order k, namely $\langle a^{\frac{n}{k}} \rangle$. This is a proof of *unique existence* and requires a proof of existence followed by a proof of uniqueness. Once you have found an appropriate subgroup, suppose there exists another subgroup of the the same order. Prove these two are in fact the same.

11.5.5 For $\mathbb{Z}_n$ with $+ \texttt{ mod } n$, $\langle \frac{n}{k} \rangle$ is the unique subgroup of order k, and these are the only subgroups. To prove this, simply show how the theorem and other corollaries are relevant to $\mathbb{Z}_n$.

11.5.6 Prove that if A is a group, B is a subgroup of A, $a \in A$, and B is cyclic, then the conjugate (see Exercise 11.4.8) is cyclic.

11.5.7 Prove that if A is a finite cyclic group with generator a (that is, $A = \langle a \rangle$) and $|A| = n$, then $a^n = e$. (For example, think about 1 on $\mathbb{N}$ with $+ \texttt{ mod } n$.)

11.5.8 Prove that if A is a cyclic group with generator a, B is a group, and $\phi : A \to B$ is an isomorphism, then $\phi(a)$ is a generator of B. (In other words, isomorphisms preserve generators. This also implies that being cyclic is an isomorphic invariant.)

Exercises, continued

11.5.9 Prove that if A is a finite cyclic group and $|A| = n$, then A is isomorphic to $\mathbb{Z}_n$. (Hint: Think about what the previous exercise says.)

11.5.10 Prove that if A is a finite cyclic group with generator a, $|A| = n$ and $k \in \mathbb{N}$, then a^k is a generator of A (that is, $\langle a^k \rangle = A$) iff $\gcd(k, n) = 1$ (that is, k and n are relatively prime). (You may use the fact that for $x, y \in \mathbb{N}$, if $\gcd(x, y) = 1$ then $x \cdot u + y \cdot v = 1$ for some $u, v \in \mathbb{Z}$.)

11.5.11 You can use the result of the previous exercise to write a program that determines the generators of $\mathcal{U}(n)$, given n:

- First find all the elements of $\mathcal{U}(n)$. You can do this by going through the integers less than n and picking out the ones whose gcd with n is 1. (Hint: Use `filter` from Exercise 7.3.9.)
- Find the order of $\mathcal{U}(n)$, call it m. This is just the number of elements in the list.
- Next find one of its generators, say a. A test for whether a is a generator is to count how many times you have to apply it to itself before you get the identity. Say $a^x = 1$. If $x = n$ (or, if $x|n$), then a is a generator.
- Find all other generators by raising a to every power relatively prime with m—that is, all elements in $\mathcal{U}(m)$.

If there are no generators, then $\mathcal{U}(n)$ is not cyclic. For example, take $\mathcal{U}(50)$. The set itself is $\{1, 3, 7, 9, 11, 13, 17, 19, 21, 23, 27, 29, 31, 33, 37, 39, 41, 43, 47, 49\}$. Its order is 20 and 3 is a generator. All together, the set of generators is $\{3, 27, 37, 22, 47, 23, 13, 17\}$.

Write an ML program that computes the generators of $\mathcal{U}(n)$, given n. Its output should be a list (an empty list if there are no generators).

11.6 Extended example: RSA encryption

Most extended examples in this text are all about programming, explaining a large system of code. This chapter's extended example is a romp through group theory with some programming for you at the end. We meander through a forest of obscure and seemingly unconnected group theory results finally to arrive at one of the most important cryptographic algorithms.

This section will have less commentary than other similar sections, being mostly a series of theorems and proofs. On one hand, less commentary is required to understand the theorems, as long as you remember what the terms and notations mean from previous sections. Moreover, the relevance of the theorems is found in the theorems that follow them.

We begin, then, with a result, following on the heels of some topics in the previous section. Recall that $\mathcal{U}(n)$ is the set of numbers relatively prime with n, and that it is a group with multiplication mod n.

Theorem 11.15 *If $p \in \mathbb{N}$ is prime, then $\mathcal{U}(p)$ is cyclic.*

Remember that to be cyclic, a group must have a generator—there must be some $a \in \mathcal{U}(p)$ such that $\langle a \rangle = \mathcal{U}(p)$. The proof requires ideas just beyond what we have covered in this chapter, but here are a few examples, where we display $\langle a \rangle$ as $\{a, a^2, \ldots, a^{n-1} = 1\}$:

$$\begin{aligned} \mathcal{U}(3): \ \langle 2 \rangle &= \{2,1\} &&= \mathcal{U}(3) \\ \mathcal{U}(5): \ \langle 2 \rangle &= \{2,4,3,1\} &&= \mathcal{U}(5) \\ \mathcal{U}(7): \ \langle 3 \rangle &= \{3,2,6,4,5,1\} &&= \mathcal{U}(7) \end{aligned}$$

Actually we are really interested in the next result, which follows from the previous theorem.

Corollary 11.16 *If p is prime, then $\mathcal{U}(p)$ isomorphic to $\mathbb{Z}_{p-1}$.*

> **Proof.** Suppose p is prime. By Theorem 11.15, $\mathcal{U}(p)$ is cyclic. Note that $|\mathcal{U}(p)| = p - 1$ since everything less than p is relatively prime with it. By Exercise 11.5.9, $\mathcal{U}(p) = \mathbb{Z}_{p-1}$. □

Keep that result under your hat. Now we are going to move on to a seemingly unrelated topic.

Theorem 11.17 *If $A_1, *_1$ and $A_2, *_2$ are groups, then $A_1 \times A_2$ is a group with component-wise application of the operations (symbolized as $\circledast$).*

external direct products

By "component-wise application of the operations" we mean that if we have two tuples $(a, b), (c, d) \in A_1 \times A_2$, then $(a, b) \circledast (c, d) = (a *_1 c, b *_2 d)$. We call these groups *external direct products*. The proof is straightforward, and you have already done each part of it.

> **Proof.** Closure, associativity, identity, and inverses are verified by Exercises 11.1.7, 11.1.8, 11.1.9, and 11.1.10, respectively. □

Moreover, external direct products are isomorphic to external direct products of groups to which their component groups are isomorphic.

Theorem 11.18 *If A_1, A_2, B_1, and B_2 are groups, A_1 is isomorphic to B_1, and A_2 is isomorphic to B_2, then $A_1 \times A_2$ is isomorphic to $B_1 \times B_2$.*

To reduce the proliferation of symbols in the proof, let $*$ refer to the operation on A_1, A_2, B_1, or B_2, depending on the context. Similarly let $\circledast$ stand for the operation on $A_1 \times A_2$ or $B_1, \times B_2$, as appropriate.

> **Proof.** Suppose $\phi_1 : A_1 \to B_1$ and $\phi_2 : A_2 \to B_2$ are isomorphisms. Define $\psi : A_1 \times A_2 \to B_1 \times B_2$ so that $\psi(a, b) = (\phi_1(a), \phi_2(b))$.
>
> Note that ψ is a one-to-one correspondence.
>
> Suppose $(a, b), (c, d) \in A_1 \times A_2$. Then

$$
\begin{array}{rcll}
\psi((a,b) \circledast (c,d)) & = & \psi(a * c, b * d) & \text{by definition of } \circledast \\
& = & (\phi_1(a * c), \phi_2(b * d)) & \text{by definition of } \psi \\
& = & (\phi_1(a) * \phi_1(c), \phi_2(b) * \phi_2(d)) & \text{by definition of isomorphism} \\
& = & (\phi_1(a), \phi_2(b)) \circledast (\phi_1(c), \phi_2(d)) & \text{by definition of } \circledast \\
& = & \psi(a,b) \circledast \psi(c,d) & \text{by definition of } \psi
\end{array}
$$

Therefore ψ is an isomorphism. □

All right, then, apply direct product groups to $\mathcal{U}$-groups:

Theorem 11.19 *If $s, t \in \mathbb{N} - \{1\}$, such that s and t are relatively prime, then $\mathcal{U}(s \circ t)$ is isomorphic to $\mathcal{U}(s) \times \mathcal{U}(t)$.*

First, consider a few examples:

$\mathcal{U}(3)$:

$\circ$	1	2
1	1	2
2	2	1

$\mathcal{U}(4)$:

$\circ$	1	3
1	1	3
3	3	1

$\mathcal{U}(12)$:

$\circ$	1	5	7	11
1	1	5	7	11
5	5	1	11	7
7	7	11	1	5
11	11	7	5	1

$\mathcal{U}(3) \times \mathcal{U}(4)$:

		1	5	7	11
	$\circledast$	(1, 1)	(2, 1)	(1, 3)	(2, 3)
1	(1, 1)	(1, 1)	(2, 1)	(1, 3)	(2, 3)
5	(2, 1)	(2, 1)	(1, 1)	(2, 3)	(1, 3)
7	(1, 3)	(1, 3)	(2, 3)	(1, 1)	(2, 1)
11	(2, 3)	(2, 3)	(1, 3)	(2, 1)	(1, 1)

Our proof that these are isomorphic will construct an isomorphism. Demonstrating that the function we will construct is a one-to-one correspondence requires more work with various kinds of groups than we have given in this chapter, so for our purposes we will prove only that it meets the isomorphic property.

Proof (of homomorphism). Let $\phi : \mathcal{U}(s \cdot t) \to \mathcal{U}(s) \times \mathcal{U}(t)$ be defined so that $\phi(x) = (x \texttt{ mod } s, x \texttt{ mod } t)$.

Suppose $a, b \in \mathcal{U}(s \cdot t)$. Then

$$
\begin{array}{rcl}
\phi(a \cdot b) & = & (a \cdot b \texttt{ mod } s, a \cdot b \texttt{ mod } t) \\
& = & (a \texttt{ mod } s, a \texttt{ mod } t) \circledast (b \texttt{ mod } s, b \texttt{ mod } t) \\
& = & \phi(a) \circledast \phi(b) \; \square
\end{array}
$$

Remember that one. We again turn to a seemingly new topic.

Theorem 11.20 *If p and q are prime, $r, M \in \mathbb{N}$ such that $r \bmod lcm(p-1, q-1) = 1$ and $M < p \cdot q$, then $M^r \bmod p \cdot q = M$.*

Recall LCM is *least common multiple.*

Proof. Suppose p and q are prime. Let $m = \text{lcm}(p-1, q-1)$. Suppose $r, M \in \mathbb{N}$ such that $r \bmod m = 1$ and $M < p \cdot q$.

p and q are relatively prime, so by Theorem 11.19, $\mathcal{U}(p \cdot q)$ is isomorphic to $\mathcal{U}(p) \times \mathcal{U}(q)$. By Corollary 11.16, $\mathcal{U}(p)$ and $\mathcal{U}(q)$ are isomorphic to $\mathbb{Z}_{p-1}$ and $\mathbb{Z}_{q-1}$. By Theorem 11.17, $\mathcal{U}(p) \times \mathcal{U}(q)$ is isomorphic to $\mathbb{Z}_{p-1} \times \mathbb{Z}_{q-1}$. By the transitivity of isomorphism (Exercise 11.3.1), $\mathcal{U}(p \cdot q)$ is isomorphic to $\mathbb{Z}_{p-1} \times \mathbb{Z}_{q-1}$.

Thus there is some element in the form x^m in $\mathcal{U}(p \cdot q)$ that corresponds to some element $(x_1, x_2)^m$ in $\mathcal{U}(p) \times \mathcal{U}(q)$ and similarly to some element $(x_3, x_4)^m = (m \cdot x_3, m \cdot x_4)$ in $\mathbb{Z}_{p-1} \times \mathbb{Z}_{q-1}$.

Since m is the least common multiple of $p-1$ and $q-1$, we have

$$\begin{aligned} m &= u \cdot (p-1) \\ m &= v \cdot (q-1) \end{aligned}$$

Then in $\mathbb{Z}_{p-1} \times \mathbb{Z}_{q-1}$,

$$\begin{aligned} (m \cdot x_3, m \cdot x_4) &= (u \cdot (p-1) \cdot x_3, v \cdot (q-1) \cdot x_4) \\ &= (0, 0) \end{aligned}$$

Thus $x^m = 1$ in $\mathcal{U}(p \cdot q)$ since isomorphism preserves identity. Since $M < p \cdot q$, $M \in \mathcal{U}(p \cdot q)$ and $r = t \cdot m + 1$ for some $t \in \mathbb{W}$, then, using arithmetic mod $p \cdot q$,

$$\begin{aligned} M^r &= M^{t \cdot m 1} = M^{t \cdot m} \cdot M = (M^m)^t \cdot M \\ &= 1^t \cdot M = M \end{aligned}$$

□

One more result, from the previous.

Corollary 11.21 *If p and q are prime and $r, s \in \mathbb{W}$ such that both r and s are less than lcm$(p-1, q-1)$ and $r = s^{-1}$ in multiplication modulo lcm$(p-1, q-1)$, then $(M^r \bmod p \cdot q)^s \bmod p \cdot q = M$.*

Proof. In addition to what we said in the proof of Theorem 11.20, $M^{r \cdot s} = M^{t \cdot m + 1} = M$. □

So these esoteric group theory results all pull together. But so what? We change the topic one more time. *Cryptography* is an action of data security involving a *plaintext*, some information to be encrypted; a *ciphertext*, the scrambled version of the information; and a *key*, some piece of information used to generate the ciphertext from the plaintext or recover the plaintext from the ciphertext. We call the generation of ciphertext *encryption* and the recovery of plaintext *decryption*.

plaintext

ciphertext

key

encryption

decryption

More specifically, *public key cryptography* is distinguished by requiring two keys, a *public key* for encryption and a *private key* for decryption. It uses the following protocol:

public key cryptography

- A person named Alice wants to allow people to send private messages to her. She generates two keys, k_1 and k_2. She makes k_1 public, but k_2 is her secret.
- A second person, Bob, wants to send Alice message M and to ensure that any potential interceptor of the message will not be able to read it. He derives the cipher text M' using k_1 and sends it to Alice.
- Alice recovers M from M' using k_2.

A cryptographic protocol like this explains only how people can use an encryption and decryption scheme. It does not explain how the keys encrypt or decrypt the message or why it is secure. In this case, whatever encryption scheme is used, it must have the attribute that k_2 can be used to undo what k_1 does—and, importantly, that k_1 itself *cannot* be used to undo what it does.

If, in such an encryption scheme, k_1 also can undo what k_2 does (if the keys are *symmetric*), then an extra feature is that Alice can "sign" documents using k_2. If Alice wants to send M along with proof that she sent it, she can generate M' using k_2 and send both M and M'. Any receivers of the messages can verify the authenticity by deciphering M' with k_1 and checking that the result matches M. Since only Alice with k_2 can generate such an encryption M', this proves that she signed the message.

The *RSA encryption algorithm*, named after its inventors Rivest, Shamir, and Adleman, uses the results presented in this section for secure, symmetric public key encryption. Specifically, to be able to receive private messages,

RSA encryption algorithm

- Alice picks two large primes, p and q. In real life, these large primes should be greater than 100 digits long. These two primes must be kept secret. She computes $n = p \cdot q$.
- Alice also computes $m = \text{lcm}(p - 1, q - 1)$.
- She finds some r relatively prime to m.
- She computes the inverse of r for multiplication mod m, that is $s = r^{-1}$ or $r \cdot s \texttt{ mod } m = 1$.

- Finally, she announces r and n as the public keys, but keeps s as her private key. (She does not need p and q any more, but they must be kept secret, since knowing them would allow someone to calculate s.)

To send a message to Alice,

- Bob converts message M into blocks of digits $M_1, M_2 \ldots M_i \ldots$, each being less than p and q.
- For each block M_i, Bob calculates $M_i' = M_i^r \texttt{ mod } n$.
- Bob sends the blocks $M_1', M_2', \ldots M_i', \ldots$.

For each block, Alice then calculates $M_i = (M_i')^s \texttt{ mod } n$. The correctness of this depends on what we saw in Corollary 11.21. The security depends on the difficulty of factoring n. An efficient algorithm for factoring numbers—something other than trying all prime numbers less than the number—is unknown. The size of the prime numbers used to generate n make factoring n infeasible.

RSA is a crucial algorithm for secure communication on the Internet. The larger lesson here is that seemingly abstract and useless mathematical results may end up providing surprising and useful applications.

Project

This project will walk you through an implementation of the RSA algorithm.

11.A Write a function `lcm` that takes two positive integers and computes their least common multiple.

11.B Write a function `findRelativelyPrime` that takes a positive integer and finds another integer that is relatively prime to it. (You may use "brute force," searching through the integers until one is found with a gcd of 0 with the given integer.)

11.C Write a function `findModInverse` that takes two positive integers, r and m and finds r's inverse mod m, that is, finds s such that $r \cdot s = 1 \mod m$. Do not compute this with brute force—instead use the Extended Euclidean Algorithm (See Exercise 4.10.7).

11.D Using your solutions to the previous parts, write a function `findKeys` that takes two prime numbers p and q and returns a triple (r, n, s) where r and n constitute the RSA public key and s is the private key.

11.E Write a function `cryptChar` that takes a character (or, rather, the integer ASCII representation of a character), and an RSA public key (r and n) and encrypts the character. Write it so that it can be used by the following two functions to encrypt and decrypt a message using RSA (`Char.ord` converts a character to an int; `Char.chr` converts an int to a character; `explode` converts a string to a list of characters; `implode` converts a list of characters to a string):

```
fun encrypt(M, r, n) =
  map(fn x => cryptChar(Char.ord(x), r, n)
      explode(M));

fun decrypt(M, s, n) =
  implode(map(fn x =>
              Char.chr(cryptChar(x, s, n))
          M));
```

To avoid overflow, you will need to create your keys using small (two-digit) primes. This is not secure for real purposes but still proves the concept.

Chapter summary

Group theory is generally considered an advanced topic that is reserved for math majors in their junior or senior year, yet its applications pervade the sciences and other fields, and the fundamentals are accessible to anyone with a general competence in the earlier topics in this text. This chapter has showcased some of the important results of the field.

The most important concept presented here is the idea of *isomorphism*, that two sets with their operations can be in some sense "the same," structurally identical with only their names and meanings taken to be different. The idea of *isomorphism* helps us see patterns and other similarities in many different places.

Key ideas:

A binary operation on a set A is a function from $A \times A$ to A. A semigroup is a set together with a closed, associative binary operation on that set. A semigroup with an identity element is called a monoid.

A group is a set together with a closed, associative binary operation such that an identity element exists and every element has an inverse. Not all groups are commutative, but those that are are called Abelian.

An isomorphism is an operation-preserving one-to-one correspondence between two groups. If an isomorphism exists between two groups, we say those groups are isomorphic to each other, which suggest that they are structurally identical, the same up to renaming. An isomorphic invariant is a proposition such that if it is true for one group, then an analogous proposition is true for any group isomorphic to the first.

A subgroup of a group is a subset that is also a group under the same operation.

A permutation is a one-to-one correspondence on a set to itself. A set of permutations that is a group with function composition is called a permutation group. Every group is isomorphic to some permutation group.

A generator of a group is an element such that every other element of the group can be computed by applying the generator to itself a certain number of times. A group that has a generator is called a cyclic group. Every subgroup of a cyclic group is cyclic.

The RSA encryption scheme is used for public key encryption. The correctness and secrecy of RSA can be demonstrated by results from group theory.

Chapter 12 Automaton

We often think of computer science as the field studying what we can do with computing machinery. Yet the principles of the field and its theoretical foundations preceded the invention of the electronic digital computers we know today. This was possible by the discovery of formal models of computation or *automata*. These models helped researchers to study puzzling mathematical problems, foresee what physical computers could be used for, and discover the limitations of computation.

The problems which many classes of automata address are in the form of languages, specifically the question of whether or not a given string is a legal part of a language. This has led to applications in the processing of natural and artificial languages.

Chapter goals. The student should be able to

- Define a language using set notation.
- Design a DFA to accept a given language and implement that DFA in ML.
- Implement an NFA in ML.
- Define a language using a regular expression.
- Explain and demonstrate the equivalence among DFAs, NFAs, and regular expressions.
- Define a language using a CFG.
- Implement a PDA in ML.
- Reduce an expression in the lambda calculus.

12.1 Alphabets and languages

Formal models of computation are our ways to simplify what algorithms and computers are doing. These simplifications make it easier to reason about computation and provide a context in which to discover and articulate deep facts about the nature of computation. Before we build abstract machines themselves, we must define a way to talk about the input to these machines that is just as simple.

alphabet

symbol

An *alphabet* is a set of symbols. We will not define the term *symbol* because it is a primitive term, but the idea of a *character*, such as represented by the ML type `char`, is an intuitive approximation. Traditionally Σ is used to stand for an alphabet, so we could say, for example, $\Sigma = \{\mathtt{a}, \mathtt{b}, \mathtt{c}\}$ or $\Sigma = \{A, B, \Gamma, \Delta\}$ or $\Sigma = \{$ ☎, ⬢, ☕, ⏢ $\}$.

string

A *string* over an alphabet is a sequence of symbols of the alphabet. We use $\Sigma*$ to stand for the set of all strings over an alphabet. The symbol $*$ here suggests "zero or more things drawn from..." Since symbols may be repeated and since strings may be arbitrarily long, $\Sigma*$ is necessarily an infinite set, unless Σ itself is empty. We use strings in this context much as we do in a computer program that processes text—operations like string concatenation, reversal, length, and substring are as you would expect, and we also have the concept of the *empty string*.

We do not denote the empty string as $\emptyset$, because a string is a sequence, not a set. The symbol we use for the empty string is ε. (If ε happens to be a symbol in the alphabet, we are in trouble.)

alphabet	$\{\mathtt{a}, \mathtt{b}, \mathtt{c}\}$	$\{A, B, \Gamma, \Delta\}$	{ ☎, ⬢, ☕, ⏢ }
some strings	ε	$AA\Gamma\Delta BAA$	☎☎
over the	aaaabaaabcaaaa	$\Gamma\Gamma\Gamma$	☎⏢⬢⏢
alphabet	bbabc	$\Delta\Gamma AA\Delta$	☕☕☎☕⬢☕
	ccccc	ε	⬢⬢⏢☕⏢⬢⬢

language

A *language* over an alphabet is a set of strings over that alphabet. Read that carefully. We said before that $\Sigma*$ is *the* set of all strings over Σ, whereas a language is *any* set of strings over Σ. Thus as alternate definition is, a language over Σ is any subset of $\Sigma*$. Let $\Sigma = \{\mathtt{a}, \mathtt{b}, \mathtt{c}\}$, L_1 be the language of strings of size 3, L_2 the language of strings with only occurrences of a, and L_3 be the language of palindromes (strings the same backwards and forwards), all over the alphabet Σ.

$$
\begin{array}{lcl}
\mathtt{aaa} & \in & L_1 \\
\mathtt{aba} & \in & L_1 \\
\mathtt{abc} & \in & L_1 \\
\mathtt{ccb} & \in & L_1 \\
\mathtt{cbb} & \in & L_1
\end{array}
\qquad
\begin{array}{rcl}
\mathtt{aaa} & \in & L_2 \\
\mathtt{a} & \in & L_2 \\
\mathtt{aaaaaaaaa} & \in & L_2 \\
\mathtt{aa} & \in & L_2 \\
\varepsilon & \in & L_2
\end{array}
\qquad
\begin{array}{rcl}
\mathtt{aaa} & \in & L_3 \\
\mathtt{aba} & \in & L_3 \\
\mathtt{aabaa} & \in & L_3 \\
\mathtt{aaccaa} & \in & L_3 \\
\varepsilon & \in & L_3
\end{array}
$$

Notice also that $L_2 \subseteq L_3$.

This may sound like an odd way to define what a language is. The set of strings of size 3 or an arbitrary set like $\{\texttt{a}, \texttt{abaa}, \texttt{abcaa}, \texttt{caaaaab}\}$ appear to have little in common with what we usually call "languages," whether natural languages or programming languages. Recognize, however, that our normal understanding of languages fits under this definition: The string "*Je voudrais du café*" is in the language of French sentences and the string "`fun f(0) = 1 | f(n) = n * f(n-1);`" is in the language of ML functions. Spoken languages work, too, in which case the alphabet is the set of phones (individual speech sounds). For sign languages, the alphabet is either the set of individual signs or the cheremes (hand positions and movements) that make them up.

One difference between our earlier "language" examples and what we normally consider to be languages lies in the difficulty of defining these languages. While specifications exist for programming languages, it is debatable whether a language like French can be defined at all. Another difference is that for both natural languages and programming languages we pay a lot of attention to what those languages *mean*. We are not concerned at all about semantics in the languages we study here—hence they are *formal*.

In this chapter we will see various systems to define specific languages. For now we can define languages using basic set notation, plus a few basic string operations. Let $|x|$ be the length (number of characters) of string x, let xy be the concatenation of strings x and y, let x^n be the string x concatenated to itself n times, and let x^R be the reverse of string x. A single character can also be considered a string of length 1. Then the three languages mentioned earlier can be specified as

$$L_1 = \{x \in \{\texttt{a}, \texttt{b}, \texttt{c}\} * \mid |x| = 3\}$$

$$L_2 = \{\texttt{a}^n \mid n \geq 0\} \text{ or simply } \{a\}*$$

$$L_3 = \{xyx^R \mid x, y \in \{\texttt{a}, \texttt{b}, \texttt{c}\}*,\ |y| = 0 \text{ or } |y| = 1\}$$

In the last of those, note that x is any string in the language and y is either a single character or empty. Odd-lengthed palindromes have a middle character that is not repeated, but even-lengthed palindromes do not.

Exercises

List all the strings in these languages.

12.1.1 $\{x \in \{\texttt{a}, \texttt{b}\} * \mid |x| = 2\}$

12.1.2 $\{\texttt{a}^n\texttt{b}^m \mid 2 \leq n \leq m \leq 3\}$

12.1.3 $\{xx^Rx \mid x \in \{\texttt{a}, \texttt{b}\}*,\ |x| = 2\}$

Use set notation to define the following languages.

12.1.4 The language of seven-digit phone numbers. Include the `-` character. Ignore the fact that North American phone numbers do not begin with 0, 1, or 555.

12.1.5 The language of strings with a certain number of `a`s followed by the same number of `b`s.

12.1.6 The language of strings with alternating `a` and `b`. Be sure to cover strings like `aba` and `baba`.

12.1.7 The language of strings of three letters in the pattern consonant, vowel, consonant.

12.2 Deterministic finite automata

An *automaton* is any model of a machine that works on its own (that is, it works *autonomously* or *automatically*). For our purposes, an automaton is an abstract device or a model of computation. It is defined by a set of rules and, when given input, processes the input according to those rules.

automaton

The simplest kind of automaton is a *deterministic finite automaton* (DFA). The abbreviation DFA is sometimes interpreted to mean *deterministic finite acceptor* and such a model is also called a *finite state machine* (FSM). It is called this because the machine has a finite number of possible states and it transitions among these as it processes input. Its input consists in symbols from an alphabet, processed one at at time.

deterministic finite automaton

finite state machine

To get your mind around this idea, try some real world analogies. A ceiling fan has four *states*: off, slow, medium, and fast. Its "input" consists of pulls of the chain. For each unit of input—in this case, each pull—the fan transitions from one state to the next.

Let $\{O, S, M, F\}$ be the set of the fan's states. Then we can describe the relationship among the states and the input using this diagram:

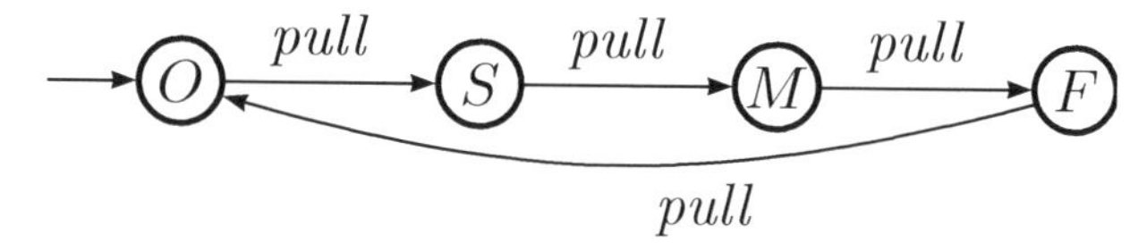

Each state is represented by a circle. The transitions are represented by arrows between circles labeled with the input that effects the transition. A special state called the *start state*, the state of the automaton before processing any input, is indicated by a small, unlabeled arrow with nothing at its tail.

start state

As a second real-world example, consider mechanical parking meters. It might be hard to find any today, but old-fashioned parking meters had *state* internally represented by the positions of gears that modeled the amount of money that had been put into the meter to purchase time, or the remaining time that had been purchased. The weight and dimensions of different coin denominations would cause levers to advance the gears in appropriate ways. Here is how the different coins—the input symbols—cause the machine to change state. To keep this example of manageable size, assume that the meter accepts

Tangent: Automata

Historically automaton has been used to refer to gadgets that run on their own (once power has been supplied to them) such as wind-up toys or clockwork that use spring-power or devices that use hydraulics. Legend and history record ancient and medieval automata made in centers of learning, some more plausible than others. These include mechanical birds that would flap wings and chirp and mechanical musical devices. There also have been famous hoax automata such as an 18th-century chess-playing device purported to be an automaton but in fact was controlled by a human chess master from a hidden compartment.

only nickels and dimes, counts only up to 25¢, and that any money put in after that is ignored. (Also this model does not account for the elapse of time.)

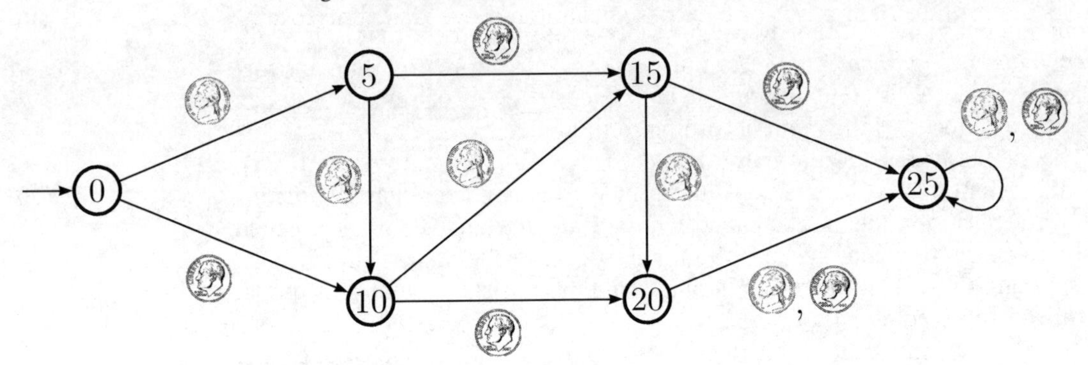

In the state where the meter has 25¢, any further input does not change the state. This is indicated by the loop. Notice that the edge from 20¢ to 25¢ and from 25¢ to itself are labeled with a list of the appropriate tokens since both tokens have the same effect.

For our purposes, a DFA is used to identify whether a string is in a language. Consider a string to be the DFA's input. Each character in the string is a token that causes the DFA to transition to a new state. The DFA has a special state called the *accept state*, and if the DFA is in the accept state when reaching the end of the string, then the DFA has accepted that string as being in the language. If the DFA is in any other state, then we say that it rejects the string. Put more carefully, we say that a DFA *accepts* a language L if, after processing a string s, the DFA is in the accept state iff $s \in L$.

accept state

Recall two of the languages mentioned in the previous section. Let the alphabet be $\Sigma = \{\mathtt{a}, \mathtt{b}, \mathtt{c}\}$. Let L_1 be the language over Σ of strings with length 3. Let L_2 be the language over Σ of string with only occurrences of a. The DFA for L_1 is shown on the left, L_2 on the right.

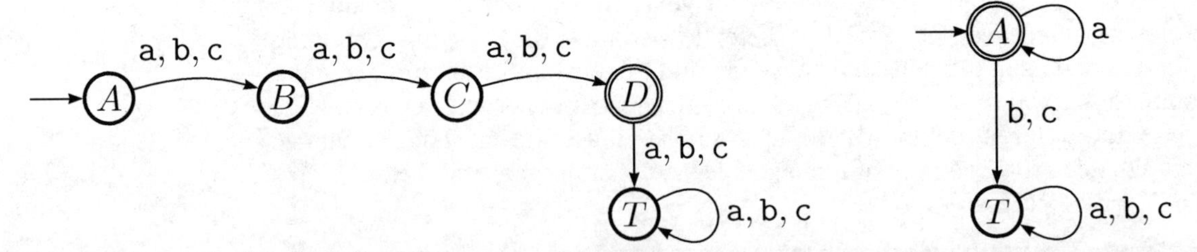

For L_1, the four states A, B, C, and D count how many characters have been seen, 0 through 3. If the string ends after three characters, we are in state D and the string is accepted. The accept state is shown with a double circle. If there is a fourth character, the DFA transitions to state T—and it will stay there no mater how many more characters are in the string. If, when the entire string has been read, the DFA is in state A, B, or C (too few characters) or T (too many), then the string is rejected.

For L_2, the start state A is also the accept state, because, as we saw before, the empty string trivially has only the a character. As long as all we see are as, we remain in state A. If any b or c ever is encountered, we transition to state T and never get out.

The states labeled T in each diagram are *trap states*. If the automaton sees a symbol in a place where that symbol cannot be—in other words, no string in the language has such a symbol there—then we transition to the trap state. After that, no other symbols encountered matter. Notice that no DFA needs more than one trap state.

trap state

How would we capture the concept of a DFA formally? The components of a DFA are the states and the transitions among states. If you have gotten this far in this book, you should say "set of states" without thinking about it. How do we model the transitions? Our formal definition needs to address the question, given a state A and symbol x, what should the next state be? We turn again to the function—our second favorite tool, after the set.

Formally, a DFA is a set of states Q together with an alphabet Σ and a transition function $\delta : \Sigma \times Q \to Q$. The transition function takes a (current) state and an input symbol and returns a new state. Moreover, one state $q_0 \in Q$ is designated the start state (or *initial state*) and the states in set $F \subseteq Q$ are designated accept states (or *final states*). Note that this definition allows for there to be more than one accept state. The trap state does not need any special designation—if there is a trap state, it acts as a trap simply by the way it is used in the transition function, δ.

Altogether, this makes a quintuple, $(Q, \Sigma, \delta, q_0, F)$. For our DFA accepting L_1,

$$
\begin{aligned}
Q &= \{A, B, C, D, T\} \\
\Sigma &= \{\mathtt{a}, \mathtt{b}, \mathtt{c}\} \\
\delta &= \{((\mathtt{a}, A), B), (\mathtt{b}, A), B), (\mathtt{c}, A), B), \\
&\quad ((\mathtt{a}), B, C), (\mathtt{b}, B), C), (\mathtt{c}, B), C), \\
&\quad ((\mathtt{a}, C), D), (\mathtt{b}, C), D), (\mathtt{c}, C), D), \\
&\quad ((\mathtt{a}, D), T), (\mathtt{b}, D), T), (\mathtt{c}, D), T), \\
&\quad ((\mathtt{a}, T), T), (\mathtt{b}, T), T), (\mathtt{c}, T), T)\} \\
q_0 &= A \\
F &= \{D\}
\end{aligned}
$$

It need not even be said that the diagrams are a preferable way to express this information.

Learning to compose DFAs for given languages requires some practice and observation of examples. Take the alphabet $\Sigma = \{0, 1\}$ and the language of strings that begin with 00 and end with 11, such as 0011, 00011, 00111, 0001001010110011, etc. In other words, the strings must start with at least two 0s and end with at least two 1s. The middle can be anything.

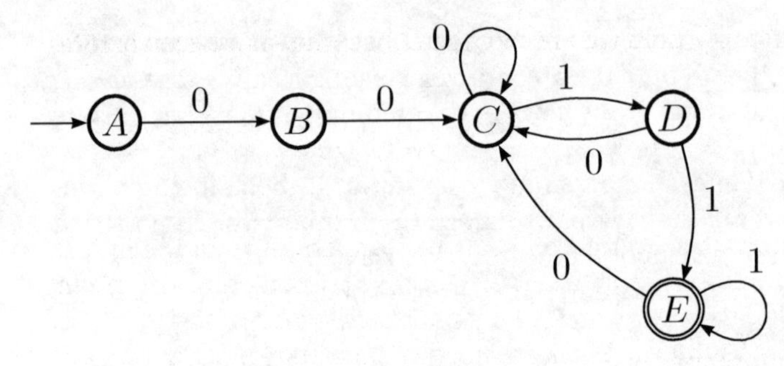

The way to read this is that the left "branch" shows the entry into the system—a string must begin with two 0s. State A means "no 0s have been seen," and state B means "one 0 has been seen." The route from C to D to E—transitions labeled with 1—needs to be matched at the end of the string. Thus D means "one 1 has just been seen" and E means "at least two consecutive 1s have been seen—ready to accept." If there are more than two 1s at the end of the string, the loop on state E handles it, just as the loop on state C handles extra 0s at the beginning. If we ever see a 0 after seeing one or more 1, we go back to C. Thus a rather wordy interpretation of state C would be, "At least two 0s at the beginning have been seen, plus any number of more 0s and 1s; the most recent symbol was 0, and we need two consecutive 1s before we can accept."

The diagram does not show any transition from states A or B for the symbol 1. If we do see a 1 as the first or second symbol in the string, we should reject, that is, transition to a trap state. We have omitted the trap state and any transitions to it in order to reduce clutter. In all subsequent transition diagrams, assume that any missing transition goes to an implicit trap state.

Again with $\Sigma = \{0, 1\}$, consider the language of strings with an even number of 0s and an even number of 1s. ε is in the language, as are 00, 1111 0111100011. This DFA accepts the language:

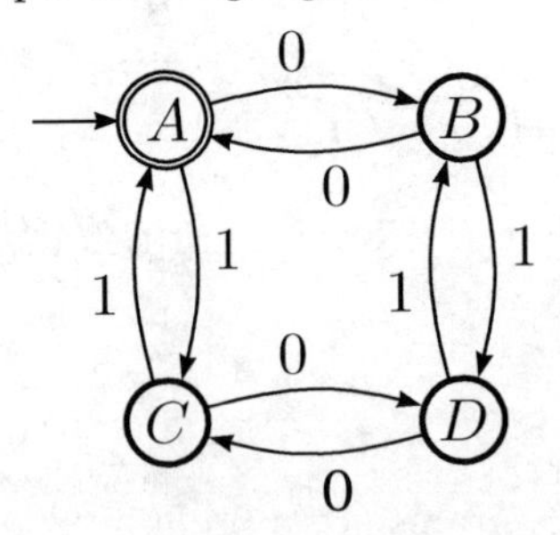

Here the states have a straightforward interpretation. Since 0 is an even number, we start out having seen an even number of each symbol. Any time we see a 0 when in a left state (A or C), we move to the right, and any time we see a 0 in a right state (B or D), we move to the left. Thus the left states stand for having seen an even number of 0s, the right states for having seen an odd number of 0s. Similarly, the top states (A and B) stand for having seen an even number of 1s, and any time we are in C or D, we must have seen an odd number of 1s.

A	even 0s, even 1s	B	odd 0s, even 1s
C	even 0s, odd 1s	D	odd 0s, odd 1s

This also explains why A is the accept state. There is no trap.

The formal definition of a DFA readily suggests an implementation in ML. Ultimately we want to build a function that can take a string and determine whether it is in the language or not. A DFA will need a set of states, which we can model using a datatype. Then we need a transition function, a start state, and a list of accept states. We will use the char type for our alphabet, even though the alphabet has been much smaller in the examples given. Any character not in the alphabet will cause the DFA to go to the trap state.

The DFA datatype, then, is parameterized with 'a as the state type and contains the transition function, start state, and accept states:

```
- datatype 'a dfa =
=      DFA of ((char * 'a) -> 'a) * 'a * 'a list;
```

Now we can write a function check that takes a dfa and a string and determines if the string is in the language. The process we want is to apply the transition function to the first character in the string and the start state, then apply the transition function to the next character and the result of the first application, and so on.

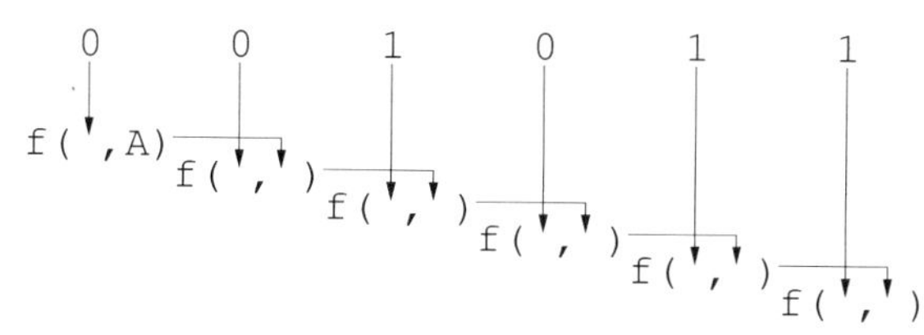

Does this process look familiar? It is generalized by the standard ML function foldl, which we first saw in Section 7.5. The start state is the seed value and an exploded version of the string is the list. The result from foldl will be a state, and we then need to check that the state exists in the list of accept states. We can use contains for that.

```
- fun check(DFA(trans, startState, acceptStates), str) =
=      contains(foldl(trans)(startState)(explode(str)),
=                acceptStates);

val check = fn : ''a dfa * string -> bool
```

Now we define a datatype, transition function, and string-testing function (using check) for the language of strings over $\Sigma = \{\mathtt{a}, \mathtt{b}, \mathtt{c}\}$ with exactly three characters. (We show the interpreter's response only for the `testTrans1` function and our trying it out. Recall that in a pattern we can use _ in place of a variable for a value we want to ignore more than once in the same pattern.)

```
- datatype States1 = A | B | C | D | Trap;

- fun trans1(#"a", A) = B
=   | trans1(#"b", A) = B
=   | trans1(#"c", A) = B
=   | trans1(#"a", B) = C
=   | trans1(#"b", B) = C
=   | trans1(#"c", B) = C
=   | trans1(#"a", C) = D
=   | trans1(#"b", C) = D
=   | trans1(#"c", C) = D
=   | trans1(_,_) = Trap;

- fun testTrans1(str) = check(DFA(trans1, A, [D]), str);

val testTrans1 = fn : string -> bool

- testTrans1("ab");

val it = false : bool

- testTrans1("abc");

val it = true : bool
```

Observe the code for the two examples on $\Sigma = \{0, 1\}$—the language of strings beginning with 00 and ending with 11 and the language of strings with an even number of 0s and even number of 1s:

```
- datatype States2 = A | B | C | D
=                  | E | Trap;
- fun trans2(#"0", A) = B
=   | trans2(#"0", B) = C
=   | trans2(#"0", C) = C
=   | trans2(#"1", C) = D
=   | trans2(#"0", D) = C
=   | trans2(#"1", D) = E
=   | trans2(#"0", E) = C
=   | trans2(#"1", E) = E
=   | trans2(_,_) = Trap;
- fun testTrans2(str) =
=       check(DFA(trans2, A, [E]), str);
```

```
- datatype States3 = A | B | C | D | Trap;
- fun trans3(#"0", A) = B
=   | trans3(#"0", B) = A
=   | trans3(#"1", B) = D
=   | trans3(#"1", D) = B
=   | trans3(#"0", D) = C
=   | trans3(#"0", C) = D
=   | trans3(#"1", C) = A
=   | trans3(#"1", A) = C
=   | trans3(_, _) = Trap;
- fun testTrans3(str) =
=       check(DFA(trans3, A, [A]), str);
```

Of course it is a tedious task writing DFAs this way, specifying each transition explicitly with a pattern in a function. A realistic, scalable approach to implementing DFAs would need a more succinct representation of the transition function, such as a table. Given a current state and next character, we would look up the appropriate transition in the table. Moreover we could automate this process with a DFA-generator, a program that reads a succinct description of the language and produces code equivalent to a DFA.

Exercises

Draw a DFA for each of the following languages. In Exercises 12.2.1–12.2.4, let $\Sigma = \{0, 1\}$.

12.2.1 The language of strings with alternating `0` and `1` for example `010101`, `010`, `10101` and `10`.

12.2.2 The language of strings composed of two to four `0`s followed by zero to three `1`s, for example `00111`, `00011`, and `0000`.

12.2.3 The language of strings containing the substring 111, for example 111, 1011100, 1110000010, and 110110111.

12.2.4 The language of strings containing the substring 01001. Be careful of how an occurrence of 01001 might overlap with a "false prefix"—for example, in the string 1101010010, the substring 1101010010 looks like it is the beginning of the substring we are looking for, but fails when we hit the 1 that follows it. However, it overlaps with the actual occurrence of the substring: 1101010010.

In Exercises 12.2.5–12.2.10, let $\Sigma = \{\texttt{a}, \texttt{b}, \texttt{c}\}$.

12.2.5 The language of strings beginning and ending with `a` and containing exactly one of either `b` or `c` (not both), for example `aba` and `aaacaa`.

12.2.6 The language of strings with an even number of `a`s and any number of `b`s and `c`s, for example `aabacaabcbcbcbca`.

12.2.7 The language of strings with an even number of `a`s, an even number of `b`s, and an odd number of `c`s.

12.2.8 The language of strings that begin with `aaa` or `acb`, followed by anything.

12.2.9 The language of strings with either exactly two `a`s or exactly one `b` (not both) and any number of `c`s, anywhere.

12.2.10 The language of strings containing any number of the following substrings (an nothing else): `abbc`, `acb`, `aabacc`.

12.2.11 Implement the DFA from Exercise 12.2.1 in ML.

12.2.12 Implement the DFA from Exercise 12.2.2 in ML.

12.2.13 Implement the DFA from Exercise 12.2.5 in ML.

Instead of writing an ML function for δ, the transition function, we could represent the same information using a list of pairs—the first item in the pair itself being a (symbol, state) pair and the second item being the state to transition to. For example, the list version of `trans1` would be

```
val trans1 =
    [((#"a", A), B), ((#"b", A), B),
     ((#"c", A), B), ((#"a", B), C),
     ((#"b", B), C), ((#"c", B), C),
     ((#"a", C), D), ((#"b", C), D),
     ((#"c", C), D)];
```

We would rewrite the dfa datatype to include that list:

```
datatype 'a dfa =
   DFA of ((char * 'a) * 'a) list *
          'a * 'a list * 'a;
```

The last `'a` is to indicate the *trap state*.

Exercises, continued

12.2.14 Finish the function `makeChecker` below. It takes a list description of a transition function and returns a function like the transition functions in this section (`trans1` etc). That is, the function it returns should take a `char` and state and return a next state. Notice that the list will not include transitions to the trap state. If the function returned by `makeChecker` cannot find an appropriate next state for a given symbol and state, it should return the trap state, which is a parameter to `makeChecker`.

For example, `makeChecker(trans1,Trap)(#"a", B)` would return `B` and `makeChecker(trans1, Trap)(#"a", D)` would return `Trap`.

```
fun makeChecker(DFA(trans, startState,
                    acceptStates), trap) =
  let fun checker(s, q, []) = ??
        | checker(s, q,
                  ((ss, qq1), qq2)::rest) =
            ??
  in fn (s, q) => ??
  end;
```

12.2.15 Finish the function `check` below. It is analogous to `check` in this section (taking a dfa and a string and determining whether the string is in the language) but for our new dfa datatype. Use your `makeChecker` function from Exercise 12.2.14.

```
fun check(d as DFA(trans, startState,
           acceptStates), trap, str) =
  ??
```

12.2.16 Implement the DFA from Exercise 12.2.1 and Exercise 12.2.11 for the version of `check` in Exercise 12.2.15.

12.3 Nondeterminism

The fact that DFAs are called deterministic finite automata raises the question, why do we emphasize that this model is *deterministic*? What would nondeterminism look like?

nondeterministic finite automaton

A *nondeterministic finite automaton* (NFA) is an automaton largely like a DFA except that in some states it may have more than one possible transition for the same symbol. Whenever it is ambiguous what the next state should be, then an arbitrary choice is made. That is the nondeterminism. (Usually NFAs are defined so that they also may transition to another state without consuming a symbol of input, in which case a transition would be labeled ε in the diagram. We omit that from our discussion for simplicity.)

Recall the language from the previous section that contains strings of 0s and 1s such that the first two characters are 0, the last two are 1, and any characters in between are either 0 or 1. We constructed a DFA to accept that language. Here is an NFA that also accepts the language:

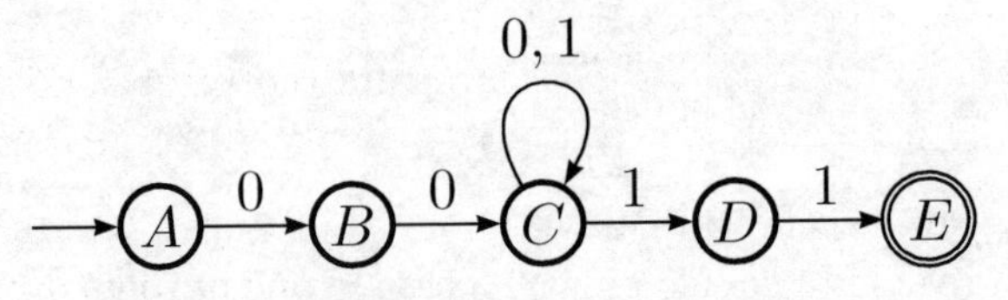

After this NFA sees two 0s, it will be in state C. If it then sees a 1, it will transition either to state D or stay in state C. Nondeterminism is unintuitive. Since we think of computers doing exactly what they are told without variation, a model of computation whose steps are not fully determined is strange. Seeing what the model means and why it is useful will take some careful defining of what is going on, but first keep in mind that it is a model, not an actual computer—it is an abstract machine, not a real one.

First, what does it mean for an NFA to accept a string or a language? Take the string 0011 as an example. It is clearly part of the language, and the NFA would end up in the accept state E after reading the string if the NFA happened to transition to state D when seeing the first 1, that is

character read		0		0		1		1	
NFA state	A		B		C		D		E

On the other hand, what if the NFA stays in state C when it sees the first 1, as the transition diagram shows is possible?

character read		0		0		1		1	
NFA state	A		B		C		C		D

Then the NFA ends up in state D, which is not an accept state, and so rejects the string. Or does it? In fact we will define *acceptance* in the case of an NFA by saying an NFA accepts a string if there exists any path through the transition diagram following the characters of the string that leads to an accept state. Accordingly, because of the existence of the first scenario, illustrated above, we say this NFA accepts the string `0011`.

There are several ways to think about this. Here are a few:

- When reaching a choice point in its operation, the automaton simply knows, oracularly, the right state to which it should transition in order to reach an accept state, if any.

- When the automaton reaches a choice point, the universe diverges into two (or more) parallel universes, each representing the universe in which the automaton has made one of the possible choices. Those universes themselves may diverge. If in any universe the automaton reaches an accept state, then that universe is the one where existence continues.

- Rather than splitting the universe, suppose that when a choice point is reached the automaton itself splits into two or more automata, and each goes its own way. Imagine the water-carrying brooms in "The Sorcerer's Apprentice." Soon you will have a herd of automata that keep forking more. If any of the herd reaches an accept state, then the string is accepted.

- When the automaton reaches a choice point, it chooses one of them arbitrarily, and if it leads to accepting the string, then fine. If the automaton reaches the end of the string but is not in an accept state, then it retraces its steps back to the choice point and makes a different choice. When all *back-tracking*

choices have been exhausted without reaching an accept state, the string is finally rejected. This approach is called *back-tracking*. It is the most realistic (or least abstract) approach and is the basis of the NFA implementation we will make later.

Here are a few more examples:

$\Sigma = \{0, 1\}$. Strings beginning with 1 followed by occurrences of 001, with the special case that string 101 is in the language. Examples: 101, 1001, 1001001.

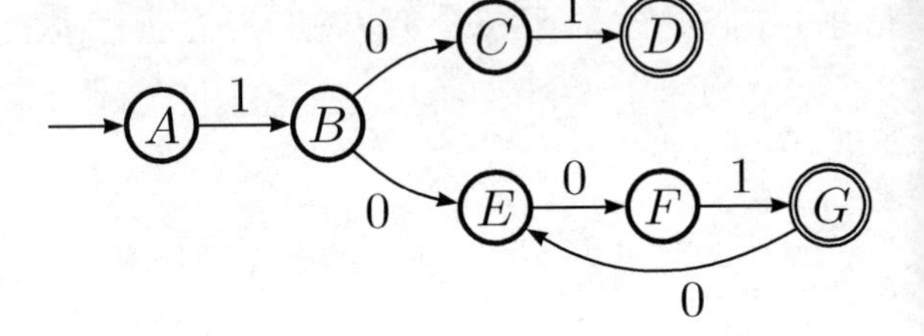

$\Sigma = \{0, 1\}$. Strings containing the substring 0101. Examples: 01011111, 000101001, 0010111010100.

$\Sigma = \{\mathtt{a}, \mathtt{b}, \mathtt{c}\}$. The language of just the strings `abba` and `abcb`.

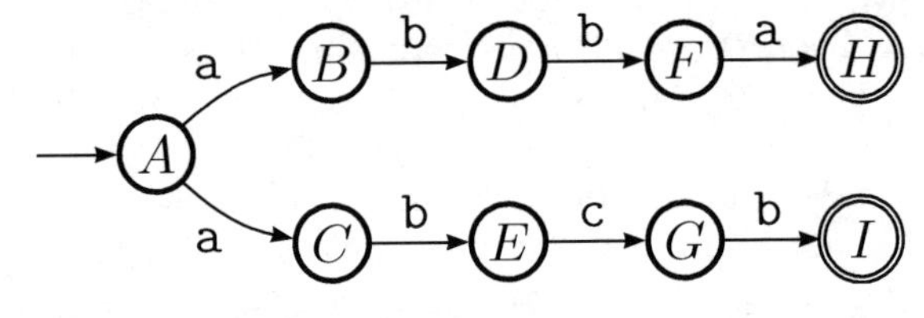

Now we are ready to construct a formal definition. Recall that a DFA is defined as $(Q, \Sigma, \delta, q_0, F)$ where Q is the set of states, Σ is the alphabet, $\delta : \Sigma \times Q \rightarrow Q$ is the transition function, $q_0 \in Q$ is the start state, and $F \subseteq Q$ is the set of final states. The formal definition of an NFA is similar except that we no longer consider δ to be a function but a *relation*, possibly associating a given pair (c, q_1) with many other states. Formally δ is a relation from $\Sigma \times Q$ to Q.

(Alternately we could keep δ a function but make its codomain $\mathscr{P}(Q)$. In other words, we could could consider δ to be a function to sets of states.)

An NFA is said to *accept* a string $s_1 s_2 \ldots s_n$ if there exists a sequence of states $q_0 q_2 \ldots q_n$ such that for all i, $1 \leq i \leq n$, $((q_{i-1}, s_i), q_i) \in \delta$ and $q_n \in F$. Read that notation as meaning that if the machine is in state q_{i-1} and sees symbol s_i, then (q_{i-1}, s_i) must be related to q_i in δ. As stated earlier, this is the requirement that there be a sequence of transitions that could be made while reading the input and that will result in an accept state. Notice that this definition is existentially quantified.

Now consider how to implement an NFA. Exercises 12.2.14–12.2.16 in the previous section developed an alternative implementation for a DFA where the transition function was represented using a list rather than an ML function. Since we have long used lists of tuples to represent relations, that implementation will suit NFAs as well.

```
- datatype 'a nfa =
=     NFA of ((char*'a)*'a) list * 'a * 'a list;
```

(The alternate DFA datatype in the exercises stored the trap state as well as the transition function, start state, and final state. That will not be necessary in our NFA implementation.)

For example, an implementation of the NFA at the beginning of this section.

```
- datatype nState1 = A | B | C | D | E;

- val nfa1 =
=    NFA([((#"0",A),B),((#"0",B),C),((#"0",C),C),
=          ((#"1",C),C),((#"1",C),D),((#"1",D),E)],
=        A, [E]);
```

It is more of a challenge to write a function for checking that a string is accepted by an NFA. As with DFAs, we will simulate the stepping through of the input symbols, transitioning to a new state each time. The difficulty is what to do when there is more than one possible next state. Our function should find a string to be accepted if there exists a sequence of choices that could lead to an accept state. The central idea to our strategy is

> *Try one of the possible states and test if the remaining substring is accepted when starting from that state. If not, try the next state. When all states are exhausted, reject the string.*

To understand this strategy, you must see the recursion in it. It is more complicated than most, so we will trace examples before we code it up. Consider how this works on the string 00100111, which is in the language. The following tracing shows the transitions from one state to another, with speculative transitions marked with a question mark and the alternatives indicated by *or*. The symbol seen in that state is indicated by a superscript. Reaching a non-accept state with no more symbols or a state and symbol for which there is no next state results in `false`. Reaching an accept state and the end of the string results in `true`.

$A^0 \rightarrow B^0 \rightarrow C^1 \quad \rightarrow ?D^0$ `false`

or $\rightarrow C^0 \rightarrow C^0 \rightarrow C^1 \quad \rightarrow ?D^1 \rightarrow E^1$ `false`

or $\rightarrow C^1 \rightarrow ?D^1 \rightarrow E^1$ `true`

Now consider the string 00010, which should be rejected because it does not end with 11. After seeing 000, the NFA is in state C. The 1 means it can transition to D or stay in C. If it transitions to D, then the NFA is stuck on seeing the 0 that follows. On the other hand, if the NFA stayed in C it will stay in C again on seeing the final character 0, but then will be in non-accept state C at the end of the string. In either case, the string is rejected, hence the string is rejected overall.

$$\begin{array}{rl} A^0 \to B^0 \to C^0 \to C^1 & \to ?D^0 \ \texttt{false} \\ \text{or} & \to C^0 \to C \ \texttt{false} \end{array}$$

Here is the implementation of a `check` function, the analogue of `check` from Section 12.2 and Exercise 12.2.15.

```
- fun check(NFA(trans, startState, acceptStates), str) =
=     let
=         fun checkHelper(currentState, [], transRest) =
=                 contains(currentState, acceptStates)
=           | checkHelper(currentState, strRest, []) = false
=           | checkHelper(currentState, currentSym::strRest,
=                         ((s, q1), q2)::transRest) =
=                 (currentState = q1 andalso currentSym = s
=                  andalso checkHelper(q2, strRest, trans))
=                 orelse
=                   checkHelper(currentState,
=                               currentSym::strRest,
=                               transRest);
=     in
=         checkHelper(startState, explode(str), trans)
=     end;

Warning: calling polyEqual
val check = fn : ''a nfa * string -> bool
```

As the interpreter's response shows, this function takes an NFA and a string and returns `true` or `false`, depending on whether the NFA accepts the string. The NFA component `trans` is the transition relation, and it is a list of items like `((#"0", A), B)`, indicating that when we see `0` in state A, we can transition to B. The main part of the function is a call to a helper function with the start state, the input string as a list of chars, and the transition relation. Thus `checkHelper` takes a current state, the remaining symbols to be processed, and (a portion of) the transition relation, and it primarily works by iterating through the transition relation looking for appropriate transitions to try. Dissecting `checkHelper`:

```
fun checkHelper(currentState, [], transRest) =
        contains(currentState, acceptStates)
```

If we are at the end of the input sequence, then accept if our current state is an accept state.

```
  | checkHelper(currentState, strRest, []) = false
```

If we have no more transitions in the relation to try, then we are stuck. Report a failure to accept.

```
  | checkHelper(currentState, currentSym::strRest,
              ((s, q1), q2)::transRest) =
        (currentState = q1 andalso currentSym = s
         andalso checkHelper(q2, strRest, trans))
        orelse
          checkHelper(currentState,
                      currentSym::strRest,
                      transRest);
```

The head of the list for the transition relation indicates we can transition to state q_2 from state q_1 if we see symbol s. If q_1 and s are indeed our current state and symbol, then try transitioning to q_2 and looking for a path to acceptance. If any of that fails—wrong state, wrong symbol, or the call `checkHelper(q2, stRest, trans)` return false—then look for some other transition from our current state and current symbol to try.

To test how well you follow this, explain to yourself why the first recursive call to `checkHelper` uses the original transition relation list `trans` given to `check` but the second uses only `transRest` given to `checkHelper`. Trying this function out on the string discussed earlier:

```
- check(nfa1, "00100111");

val it = true : bool

- check(nfa1, "000100");

val it = false : bool
```

Exercises

Draw NFAs for the following languages. Even though you have already drawn a DFA for these, draw a new automaton that uses nondeterminism at some point.

12.3.1 The language described in Exercise 12.2.1.

12.3.2 The language described in Exercise 12.2.2.

12.3.3 The language described in Exercise 12.2.3.

12.3.4 The language described in Exercise 12.2.4.

12.3.5 The language described in Exercise 12.2.5.

12.3.6 The language described in Exercise 12.2.6.

12.3.7 The language described in Exercise 12.2.7.

12.3.8 The language described in Exercise 12.2.8.

12.3.9 The language described in Exercise 12.2.9.

12.3.10 The language described in Exercise 12.2.10.

12.4 Regular expressions

The previous two sections presented DFAs and NFAs as implementations of languages. In the examples we used, we specified languages informally, such as "the language of strings of alternating as and bs," and constructed DFAs or NFAs to accept the languages. On the other hand, we could think of an automaton as a *specification* of a language, as in "Let L be the language accepted by the following DFA..."

regular expression

In this section we consider another way to specify a language—or implement one, depending on your perspective. A *regular expression* is an expression that specifies a language using the rules given below. A language is a set, so keep in mind that a regular expression, then, is a description of a set. Given an alphabet Σ, a regular expression is one of the following:

- $\emptyset$, which denotes the set $\emptyset$, that is, the set/language of no strings.
- ε, which denotes the set $\{\varepsilon\}$, that is, the set/language containing only the empty string.
- a, where $\mathtt{a} \in \Sigma$, which denotes the set $\{\mathtt{a}\}$, that is, the set/language containing only the string with only the symbol a.
- $r|s$, where r and s are regular expressions, which denotes the set/language $r \cup s$ (remember that r and s each denote a set). For example, $\mathtt{a}|\varepsilon$ represents the language $\{\mathtt{a}, \varepsilon\}$, and $\mathtt{a}|\mathtt{b}|\mathtt{c}$ represents the language $\{\mathtt{a}, \mathtt{b}, \mathtt{c}\}$.
- rs, where r and s are regular expressions, which denotes the set/language of strings each composed of a string from r concatenated with a string from s; formally, $\{RS \mid R \in r \text{ and } S \in s\}$. For example, $(\mathtt{a}|\mathtt{b}|\varepsilon)\mathtt{c}(\mathtt{d}|\mathtt{e})$ represents the language $\{\mathtt{acd}, \mathtt{ace}, \mathtt{bcd}, \mathtt{bce}, \mathtt{cd}, \mathtt{ce}\}$.
- $r*$, where r is a regular expression, denoting the set of strings which are composed of the concatenation of zero or more strings, each from r; formally, $\{R_0R_1 \ldots R_n \mid n \in \mathbb{W} \text{ and } \forall\, i, 0 \le i \le n, R_i \in r\}$. For example, $\mathtt{a}(\mathtt{bc}) * \mathtt{d}$ represents the language $\{\mathtt{ad}, \mathtt{abcd}, \mathtt{abcbcd}, \mathtt{abcbcbcd}, \ldots\}$.

To combine these in a big example, consider the language of strings beginning with two 0s, ending with two 1s, and having anything in the middle, which we have seen in the two previous sections. It is described by the regular expression `00(0|1)*11`.

Note that parentheses are also legal, used for grouping. There are also a variety of extensions to regular expressions that are popularly used to make them more concise. For example, $r+$ means "one ore more occurrences of..."; thus $r+$ is equivalent to $rr*$. These are notationally convenient but do not add to the "power" of regular expressions, which we discuss in the next section.

Suppose we have the regular expression `(abc|de)(f|g)*(h|i|`ε`)`. What strings does it produce? Here are some:

```
abcfh       abcfffffh   abcffi      abcgg       abcgggh
abcgfggfh   abcgggf     abcgffffi   abcffffg    abcffffffffffffi
deh         de          deffffi     degfgh      deggggg
```

Regular expressions are a very important tool for processing text because they can be used to define kinds of strings to search for. Almost any modern text-processing tool or programming language will have some facility for finding occurrences of strings in a text that match a regular expression. (For programming languages, this is often a library, such as the RegExp package in ML and Java's `java.util.regex` package.)

Suppose we had a large text such as a book or paper or computer program or email archive, and suppose we wanted to find all occurrences of DNA sequence descriptions or phone numbers or the like that occur in them. We can describe what such strings look like with a regular expression, and then a regular expression tool can find and return all the strings or locations in the text that match that description. Some examples:

- *DNA sequences:* `(A|C|G|T)*`. This defines strings containing only the letters C, A, T, and G. When we have a list of alternative symbols like this, a common abbreviation is simply to list them without the | separator inside of square brackets, that is `[ACGT]*`. If we want this not to be case sensitive, we would write `[ACGTacgt]*`.

- *Identifiers:* ((’|ε)[A-Za-z][A-Za-z0-9_])|_. Recall the rules for forming identifiers in ML: a sequence of letters, digits, and underscores, beginning with a letter (optionally with an apostrophe before the first letter), or just an underscore. Since we do not want to list every letter in the alphabet, we introduce another abbreviation to indicate a *range* of characters. `A-Z` indicates any capital letter and `a-z` indicates any lowercase letter.

  Moreover, it is fairly common to have a regular expression in the form of $(r|\varepsilon)$ as in (’|ε). This indicates that a component to the string is optional. Use the symbol ? to abbreviate this, as in `(’?[A-Za-z][A-Za-z0-9_])|_`.

- *Phone numbers:* [2-9][0-9]^2–[2-9][0-9]^2–[0-9]^4. In North America, phone numbers are composed of a 3-digit area code (not beginning with 0 or 1), a 3-digit exchange code (also not beginning with 0 or 1), and a 4-digit subscriber number, the three parts separated by hyphens, such as 630-555-3097. To avoid writing a sequence like `[0-9][0-9][0-9][0-9]`, we use the abbreviation [0-9]^4, indicating 4 occurrences of strings from the given regular expression.

- *Dates:* ((1[0-2])|[1-9])/(30|31|([12][0-9])|[1-9])/[1-9][0-9]^{0,3}. In the month/day/year formatting of calendar days, months can be named by 10, 11, 12, or a one-digit number besides 0; days can be named 30, 31, a two-digit number starting with 1 or 2, or a one-digit number besides 0; and years must begin with a digit from 1 to 9 and may have up to three more digits. For example, 11/1/2011. The superscript $0,3$ indicates between 0 and 3 occurrences. (This works for dates up to the year 9999. To fix this Y10K problem, we can change the last part of the regular expression to `[1-9][0-9]*`.)

- *US Postal Addresses:* `[0-9]+ [NSEW]`^{0,2} `[A-Z][a-z]* (St|Ave|Rd|Ln|Dr|Blvd),([A-Z][a-z]*)*, [A-Z]`^{2}`[0-9]`^{5}. Here we use +, mentioned earlier. The fragment `([A-Z][a-z]*)+` indicates one or more occurrences of words beginning with a capital letter followed by zero or more lowercase letters. Clearly we had to make many assumptions here about how the address is formated, even apart from narrowing ourselves to the USA. This regular expression will accept many addresses, such as

  930 N President St, Wheaton, IL 60187
  1 Windsor Dr, Oak Brook, IL 60523
  1600 NW Pennsylvania Ave, Washington, DC 20500
  22462 SW Washington St, Sherwood, OR 97140

  It breaks on many others (PO boxes, rural routes, apartment and suite numbers, streets with more than one word in the name ...).

Take note of the regular expression abbreviations used in these examples. Here is a summary:

Abbreviation	*Meaning*	*Equivalence*
`[abc]`	One occurrence of any of these symbols	`(a|b|c)`
`[a-c]`	One occurrence of any symbol in this range	`(a|b|c)`
$r?$	Optionally an occurrence of a string defined by r	$(r\|\varepsilon)$
r^n	n occurrences of a string defined by r	$rrrrr$ (if $n = 5$)
$r^{m,n}$	Between m and n occurrences of a string defined by r	$(rrr\|rrrr\|rrrrr)$ (if $m = 3$, $n = 5$)
$r+$	One or more occurrences of a string defined by r	$rr*$

Since for all of our abbreviations there are ways to write equivalent regular expressions that do not use the abbreviations, this demonstrates that the abbreviations do not add any *power* to regular expressions in the sense that they do not allow you to define a language you could not define otherwise. It is important to understand that the n and m must be constants. Regular expressions do not contain variables.

Exercises

Write a regular expression specifying each of the following languages.

12.4.1 The language described in Exercise 12.2.1.

12.4.2 The language described in Exercise 12.2.2.

12.4.3 The language described in Exercise 12.2.3.

12.4.4 The language described in Exercise 12.2.4.

12.4.5 The language described in Exercise 12.2.5.

12.4.6 The language described in Exercise 12.2.6.

12.4.7 The language described in Exercise 12.2.7.

12.4.8 The language described in Exercise 12.2.8.

12.4.9 The language described in Exercise 12.2.9.

12.4.10 The language described in Exercise 12.2.10.

12.4.11 The set of words that begin with a capital letter, other letters lowercase.

12.4.12 The set of words that end with *ment*.

12.4.13 The language of Dewey Decimal Classification numbers (or the system used at your school's library).

12.4.14 The set of course program and number (for example, CSCI 243), given the following rules:

- Programs are CSCI, MATH, PHYS, CHEM, BIOL.
- Course numbers are three digits long.
- Course numbers begin with 1, 2, 3, or 4.
- No course number ends in a 0.
- If 0 or 9 is the middle number, then only 1, 2, 3, or 4 can be the last number.

12.4.15 The language of numbers in scientific notation, for example -7.2365E27.

12.5 Language model equivalence and limitations

DFAs, NFAs, and regular expressions can each be used to define sets of strings, that is, languages. We have not yet asked the question of *what range* of languages any model can be used to define, but you probably sense that we are unlikely able to make a DFA that can determine whether a sentence is legal French or an expression is legal ML. Intuitively, these languages are too complicated for such a simple model. For each model there is a set of languages for which it is possible to make an instance of the model (a DFA, an NFA, or a regular expression) that can determine whether a given string is in the language.

On the other hand, you may have noticed that each of the previous sections contains the example of the language of strings each beginning with 00, ending with 11, and having any sequence of 0s and 1s in the middle.

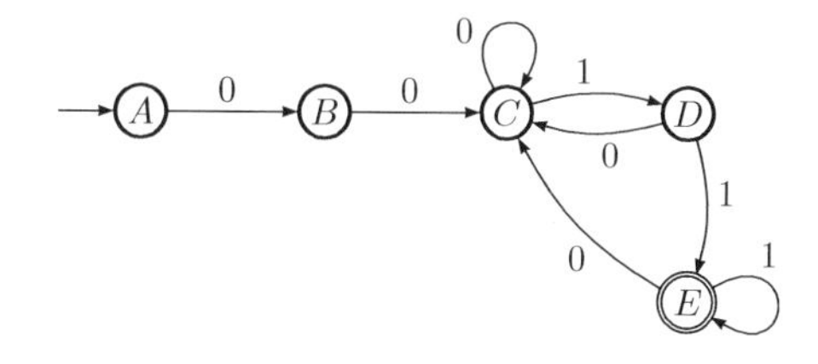

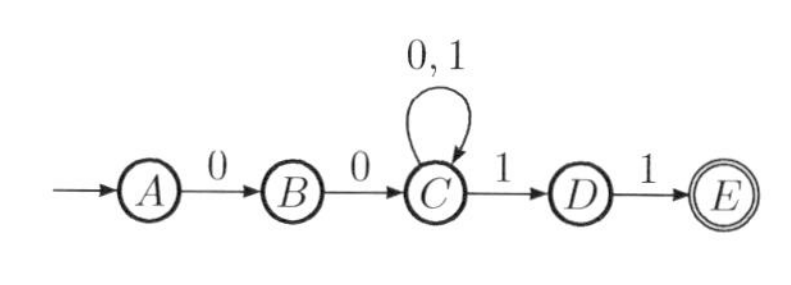

`00(0|1)*11`

So there is some overlap among the languages defined by these models. In fact, since any DFA is inherently a degenerate NFA (a DFA's transition function δ is, after all, a relation), the set of languages definable by a DFA is a subset of those definable by an NFA. In other words, it is a reasonable guess that the NFA model is *more powerful* than the DFA model. But how much more powerful, and how are regular expressions related in terms of power?

The big news of this section is that *all three models are equivalent.* Any language that can be accepted by an NFA also can be accepted by a DFA. Any language that can be defined by a regular expression can be accepted by a DFA or an NFA, and vice versa. Two parts of this are surprising or interesting: first, adding nondeterminism adds no power to the DFA model; second, despite how different regular expressions and finite automata seem, they are capturing the same set of languages.

equivalent

To put this in technical terms, we say that two automata are *equivalent* if they accept the same language—if the set of strings accepted by one is equal to the set of strings accepted by the other. We claim that for any NFA, there exists an equivalent DFA, and vice versa. A regular expression is equivalent to an automaton if the language defined by the regular expression is equal to the language accepted by the automaton. We claim that for any regular expression, there exists an equivalent DFA and NFA, and that for any DFA or NFA, there exists an equivalent regular expression.

We will tackle the DFA/NFA equivalence first. Make sure you have the quantification right.

Theorem 12.1 *For any language L over alphabet Σ, there exists a DFA that accepts L iff there exists an NFA that accepts L.*

We noted earlier that the "only if" part is obvious. The real work is in this lemma:

Lemma 12.2 *Every NFA has an equivalent DFA.*

Proof. Suppose $M_1 = (Q, \Sigma, \delta, q_0, F)$ is an NFA, where Q is a set of states, Σ is an alphabet, δ is a transition relation from $\Sigma \times Q$ to Q, q_0 is the start state, and F is the set of final states.

Before we go any further, we should think about where we are heading. The phrasing "has an..." in the lemma means that what follows is existentially quantified—it is a proof of existence. There are two approaches—prove it directly by coming up with such a DFA, or use proof by contradiction to show it is impossible for there not to be such a DFA. Proving that a DFA cannot not exist sounds hard and it would not be very helpful. Instead we will pursue a *constructive proof*, as we did to prove facts about cardinality in Section 7.9 and facts about graph walks in Section 8.3.

Specifically, we will give an algorithm for constructing a DFA equivalent to a given NFA, $(Q, \Sigma, \delta, q_0, F)$. Each state in the DFA will represent *the set of NFA states* that the NFA could be in at an equivalent step in the computation. Thus each state in Q_1 (the set of states in the DFA we are constructing) is an element of $\mathscr{P}(Q)$, and $Q_1 \subseteq \mathscr{P}(Q)$.

There is only one place to start the computation in the NFA—the start state, q_0. The DFA start state, then, is $\{q_0\}$—the DFA state corresponding to when the only state the NFA could be in is q_0. Consider our earlier example again, where $q_0 = A$:

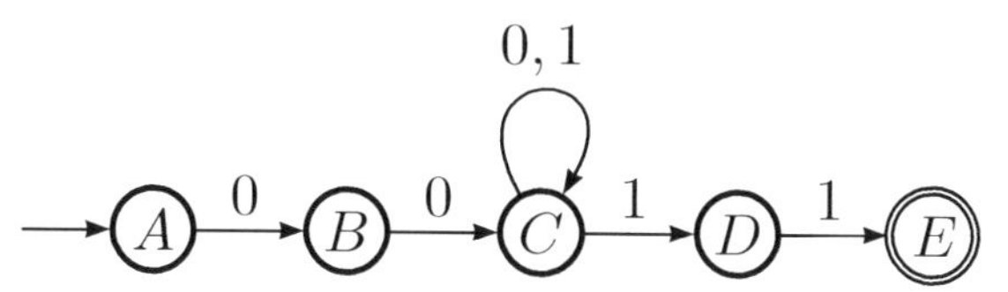

The start state in the corresponding DFA is $\{A\}$. On input, the only place we can go from A in this NFA is B, that is, the set of states we can transition to on 0 is $\{B\}$. Likewise, the set of states we can transition to from B on 0 is $\{C\}$, and the set of states we can transition to from C on 0 is $\{C\}$. So far we can construct the following DFA:

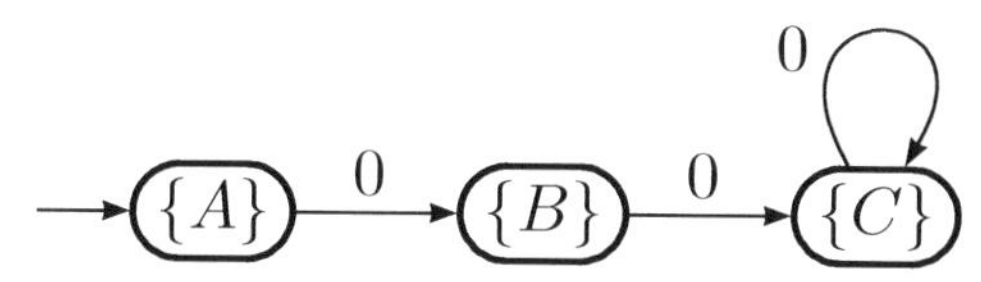

But on seeing 1 in state C in the NFA we can either transition to D or stay in C. Thus the set of states we can transition to from the C on 1 is $\{D, C\}$.

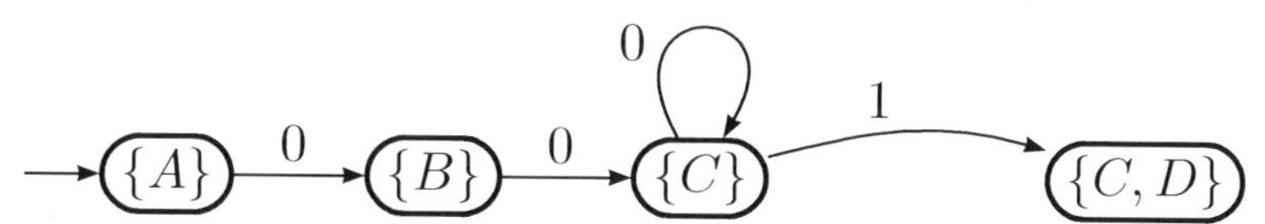

Now what happens in the DFA state $\{C, D\}$ if we see 1? Since in the NFA, C could lead C or D and D can lead only to E, the set of NFA states we can go to from the states $\{C, D\}$ is $\{C, D, E\}$. Note also that since D and E both go only to C on 0 in the NFA, the DFA states $\{C, D\}$ and $\{C, D, E\}$ go to $\{C\}$ on 0.

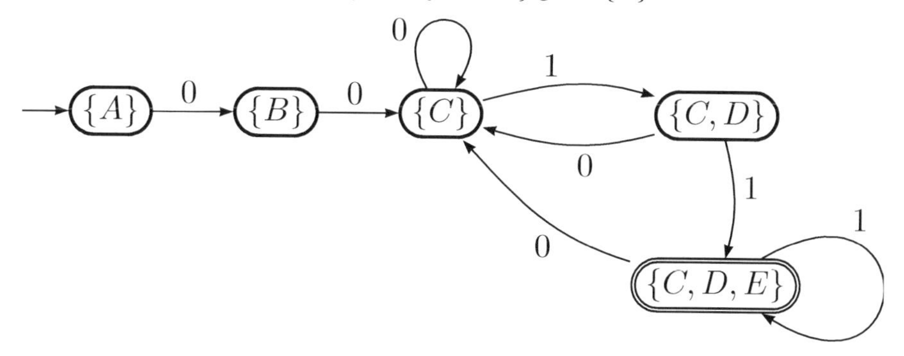

Finally you see that if the DFA is in state $\{C, D, E\}$ at the end of the string, then the NFA *could* have been in state E, which is our definition of acceptance. Thus $\{C, D, E\}$ is the accept state. We are left with a DFA that accepts the same language as the NFA, and incidentally it is structurally identical (*isomorphic*) to the DFA for this language from earlier in this chapter.

Summing up, here is how we form an equivalent DFA given an NFA:

- $\{q_0\}$, the set containing only the start state of the NFA, is a state in the DFA, indeed, the start state of the DFA.
- For any set X of NFA states already found to be a state in the DFA, and for any character s in the alphabet, the set of states that any state in X transitions to on a in the NFA is a state in the DFA, and X transitions to it. Formally, X, s transitions to $\{q \in Q \mid \exists\, q_1 \in X$ such that $((q_1, s), q) \in \delta\}$
- If X is a state in the DFA and contains an accept state of the NFA, then X is an accept state of the DFA.

What we have, implicitly, is an algorithm for making a DFA, given an NFA. Fully formed, here is the algorithm in ML, which takes a value of the nfa datatype from Section 12.3 and produces a value in the dfa datatype from Exercises 12.2.14–12.2.16. It works by building a list of DFA states and transitions. As it explores edges from DFA states it discovers new DFA states whose edges must be explored. To handle this it maintains a list of discovered but unexplored DFA states—a *worklist*, similar to breadth first search in Section 8.6.

worklist

```
- datatype 'a nfa =
=     NFA of ((char * 'a) * 'a) list * 'a * 'a list;

- datatype 'a dfa =
=     DFA of ((char * 'a) * 'a) list * 'a * 'a list * 'a;

- fun nfa2dfa(NFA(nfaTrans,nfaStartState,nfaAcceptStates)) =
=   let
=      (* Discover new transitions for the DFA:
=          Given a DFA state and a list of NFA transitions,
=          find the adjacent NFA state sets *)
=     fun exploreDfaEdges(dfaState, newDfaTrans, []) =
=                newDfaTrans (* no more NFA transitions *)
=       | exploreDfaEdges(dfaState, newDfaTrans,
=                              ((a,q),p)::rest) =
=                                   (* q transitions to p on a *)
=       (* if this DFA state contains q *)
=       if contains(q, dfaState) then
=         let   (* add p as a destination to the existing
=                    transitions on a *)
=           fun addDestinationState([]) =
=                 [((a, dfaState), [p])]
=             | addDestinationState(((b,_),dest)::rest) =
=                 if a = b
=                 then ((a, dfaState),
```

```
=                       if contains(p, dest)
=                       then dest else p::dest)::rest
=                 else ((b, dfaState), dest)
=                      ::addDestinationState(rest);
=         in
=           exploreDfaEdges(dfaState,
=                          addDestinationState(newDfaTrans),
=                          rest) (* explore other edges *)
=         end
=       else exploreDfaEdges(dfaState, newDfaTrans, rest);
=     (* Given a list of newly discovered DFA transitions, find
=        newly discovered DFA states in them *)
=     fun collectNewDfaStates(newTrans) =
=       map(fn((b, x), y) => y, newTrans);
=     (* Build the DFA. First parameter is a list of DFA states
=        whose edges still need to be explored (worklist);
=        second parameter is a list of DFA states that have
=        been processed; third parameter is our running
=        collection of DFA transitions *)
=     fun processDfaState([], dfaStates, dfaTrans) =
=         (dfaStates, dfaTrans)  (* worklist empty, all done *)
=       | processDfaState(nextState::rest,  (* worklist head *)
=                         dfaStates, dfaTrans) =
=       if contains(nextState, dfaStates)
=                  (* already did that one *)
=       then processDfaState(rest, dfaStates, dfaTrans)
=       else
=         let
=           val newDfaTrans = (* find transitions from here *)
=               exploreDfaEdges(nextState, dfaTrans, nfaTrans);
=                  (* any new states discovered? *)
=           val newDfaStates = collectNewDfaStates(newDfaTrans)
=         in           (* add new states to back of worklist *)
=           processDfaState(rest@newDfaStates,
=                          (* finished nextState *)
=                          nextState::dfaStates,
=                          (* add new transitions *)
=                          newDfaTrans@dfaTrans)
=             end;
=     (* find DFA states and transitions *)
=     val (dfaStates, dfaTrans) =
=           processDfaState([[nfaStartState]], [], []);
=     in
=     (* package them up in DFA data constructor *)
=     DFA(dfaTrans,
```

```
=                [nfaStartState],   (* dfa start state *)
=                (* dfa final states *)
=                filter(fn (state) =>
=                        intersection(state, nfaAcceptStates)
=                        <> [],
=                     dfaStates),
=                [])  (* dfa trap *)
=         end;

val nfa2dfa = fn : ''a nfa -> ''a list dfa
```

At long last, back to the proof. Remember that $M_1 = (Q, \Sigma, \delta, q_0, F)$ is our NFA.

> Let $M_2 = (Q_2, \Sigma, \delta_2, \{q_0\}, \{Z \in Q_2 \mid Z \cap F \neq \emptyset\})$ where Q_2 and δ_2 are calculated using the algorithm described above. We claim M_2 is a DFA equivalent to M_1.

The $\{Z \in Q_2 \mid Z \cap F \neq \emptyset\}$ business just means the set of all states Z that have some overlap with the set of NFA accept states F. What remains to be proven? That M_2 *is* (exists), that it is a *DFA*, and that it is *equivalent* to M_1. As for existence, M_2 must be well defined: Does our algorithm terminate? As for being a DFA, is M_2 deterministic? As for being equivalent, do M_1 and M_2 accept the same language?

> **That the algorithm terminates:** Each element of $\mathscr{P}(Q)$ could be processed at most once, and $\mathscr{P}(Q)$ is finite. Every time we add something to the worklist, we also add something to the list that becomes Q_2. Either the worklist will become empty (in which case the algorithm terminates) or, in the worst case, Q_2 becomes $\mathscr{P}(Q)$. If the worklist indeed became $\mathscr{P}(Q)$, then from then on we would remove from it without adding, and so eventually the worklist will become empty and terminate.
>
> **That M_2 is deterministic:** The way we have written the algorithm ensures that no DFA state (element of $\mathscr{P}(Q)$) will have more than one transition for a single alphabet symbol. Thus δ_2 will have at most one entry for any $(X \in \mathscr{P}(Q), s \in \Sigma)$ pair.
>
> **That M_2 is equivalent to M_1:** Suppose the string $s_1, s_2, \ldots s_n$ is accepted by M_1. Then M_1 has states $q_0, q_1, \ldots q_n$ such that for all i, $0 < i \leq n$, $q_i \in \delta(q_{i-1}, s_i)$.
>
> Our proof that M_2 accepts this string will be by structural induction on the length of the string. Restating our claim: If $s_1, s_2, \ldots s_n$ can put M_1 into a state q_n, then it puts M_2 into a state X_n such that $q_n \in X_n$.

Suppose $n = 1$. Then $q_1 \in \delta(q_0, s)$, and in our algorithm, $\delta(q_0, s_q) \subseteq \delta_1(\{q_0\}, s_1)$.

Next, suppose this holds for some $n \geq 1$, that is $s_1, s_2, \ldots s_n$ can put M_1 into state q_n and also puts M_2 into state X_n such that $q_n \in X_n$, and suppose some $q_{n+1} \in \delta(q_n, s_{n+1})$. By our algorithm, $\delta(q_n, s_{n+1}) \subseteq \delta_1(X_n, s_{n+1})$, and so $q_{n+1} \in \delta_1(X_n, s_{n+1})$.

Then we can apply this to the case where the last NFA state q_n is an accept state.

Conversely, if M_2 accepts a string, a similar process can construct a path in M_1 that accepts the string. □

The exercises will walk you through a result that regular expressions are equivalent to NFAs and DFAs—the set of languages that can be specified by regular expressions is equal to the set of languages that can be accepted by an NFA or DFA. For now we will demonstrate this intuitively with an example of a language none of these can handle.

Consider the language of strings that begin with a certain number of 1s followed by the same number of 0s: $\{\varepsilon, 10, 1100, 111000, 11110000, \ldots\}$. The regular expression `1 * 0*`—any number of 1s followed by any number of 0s—will accept every string in this language, but it will also accept strings like 110. The pseudo-regular expression 1^n0^n, taken to mean for all n, n instances of 1 followed by n instances of 0, would indeed accept this language, but it is not a regular expression. The "exponent notation" in regular expressions is shorthand for a specific number of copies of a regular expression and can be used only with constant superscripts, not variables.

If we were to test if a string were in the language by hand or with a computer program, we would either first count the 1s, then the 0s, and compare, or test it recursively, seeing if the string has a 1 at front, a 0 at back, and another string in the language in the middle. In short, regular expressions have no way of counting and no mechanism for recursion.

The following DFA attempts to accept this language:

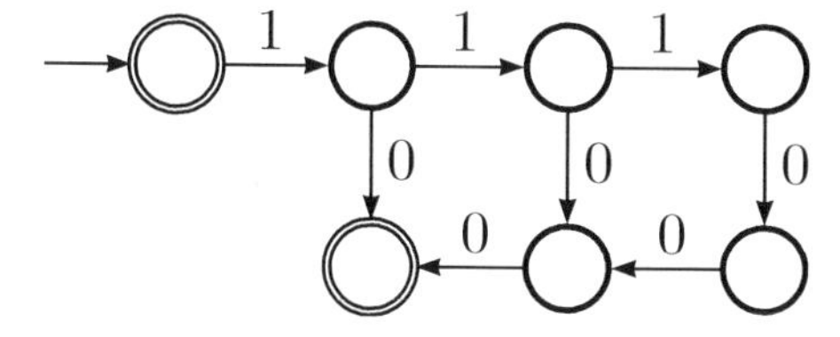

However, this works only for series of 1s and 0s no longer than 3. To handle arbitrarily long strings in this language, the DFA itself would need to be arbitrarily big—in which case it no longer would be *finite*.

Exercises

These exercises will walk you through a proof of the following theorem:

Theorem 12.3 *For any regular expression r, there exists an NFA M such that M accepts the language defined by r.*

The proof is by induction on the structure of r. Recall that r is either $\emptyset$, ε, a for some $\mathtt{a} \in \Sigma$, $r_1|r_2$, r_1r_2, or r_1*, for some regular expressions r_1 and r_2. Thus the set of regular expressions is recursively defined. The first three forms are simple and are our base cases; the last three are inductive cases.
For example, if $r = \emptyset$, then that language is "accepted" by an NFA having one state that is both the initial state and the trap state, with no accept states. It rejects all strings.

12.5.1 Suppose $r = \varepsilon$. Prove that there exists an NFA that accepts the language containing only the empty string by describing that NFA. This is a simple base case.

12.5.2 Suppose $r = \mathtt{a}$ for some $\mathtt{a} \in \Sigma$. Prove that there exists an NFA that accepts the language containing only the string containing only the character a once. This also is a very simple base case.

12.5.3 Suppose $r = r_1|r_2$ for some regular expressions r_1 and r_2. Prove that there exists an NFA that accepts the language specified by r. (Hint: Using structural induction, you know there exist NFAs accepting the languages defined by r_1 and r_2. Use those NFAs to construct a new NFA for r.)

12.5.4 Suppose $r = r_1r_2$ for some regular expressions r_1 and r_2. Prove that there exists an NFA that accepts the language specified by r. This will be similar to the previous exercise.

12.5.5 Suppose $r = r_1*$ for some regular expression r_1. Prove that there exists an NFA that accepts the language specified by r.

Of course that is only half of equivalence. We also have this theorem:

Theorem 12.4 *For any NFA M, there exists a regular expression r such that r defines the language accepted by M.*

Suppose $M = (Q, \Sigma, \delta, q_1, F)$ and let $Q = \{q_1, q_2, \ldots q_n\}$ (this merely gives names to all the states of M; notice we name the start state q_1 this time, not q_0). We will prove this using the following definition:
Let $R(i, j, k)$ be the set of strings which, if M is in state q_i, can bring M into q_j passing through only states $\{q_1, q_2, \ldots q_k\}$ in between. In other words, suppose the front part of a certain string fed into M can get M into state q_i. What substrings can come next that can get M into state q_j without using any state numbered greater than k? (i and j themselves may be greater than k.)
We will prove that a regular expression exists for any such set $R(i, j, k)$ by induction on k.

12.5.6 Prove that for any i and j where $0 \leq i, j \leq n$, there exists a regular expression defining the set $R(i, j, 0)$. This is our base case, and since there is no state q_0, it means the string must bring M directly from i to j. (Hint: Consider two subcases: $i = j$ and $i \neq j$.)

12.5.7 Next suppose that $R(i, j, k)$ is regular for some k—that is, for any i and j, there exists a regular expression that defines the set of strings taking M from q_i to q_j without using any state numbered greater than k. Prove, then, that for any i and j, there exists a regular expression defining the set $R(i, j, k+1)$. (Hint: Consider, among other things, the regular expressions defining $R(i, k+1, k)$, $R(k+1, k+1, k)$, and $R(k+1, j, k)$. Form a new regular expression from these from the rules of regular expressions.)

12.5.8 By the two previous exercises, for all accept states $q_a \in F$, there exists a regular expression defining the set of strings $R(1, a, n)$. Show that this implies there exists a regular expression defining the language accepted by M.

12.5.9 All of the previous work established the equivalence between regular expressions and *NFAs*. Prove the following corollary. (This is a simple composition of previous results.)

Corollary 12.5 *Let L be a language over alphabet Σ. There exists a regular expression r specifying L iff there exists a DFA M accepting L.*

12.6 Context-free grammars

The previous section ended with the example $\{\varepsilon, 10, 1100, 111000, \ldots\}$, the language of strings made up of a series of 1s followed by the same number of 0s, which no regular expression can specify and no DFA or NFA accept. The following notation can define it, though:

$$S \quad \rightarrow \quad \varepsilon \mid 1\,S\,0$$

The symbol S ranges over strings in the language. The notation means that S can either be the empty string or a 1 followed by another string in the language followed by a 0. Note the similarity between this and the recursively-defined sets we saw in Chapter 6 like Peano numbers and trees. We can use this to prove, for example, that 1111100000 is a string in the language as seen below on the left. Alternately, we can analyze how this notation generates the string using a tree, seen on the right, with each S having children either ε or $1\ S\ 0$ and the string itself being a concatenation of the leaves.

1111100000	is in the language because
11110000	is in the language because
111000	is in the language because
1100	is in the language because
10	is in the language because
ε	is in the language.

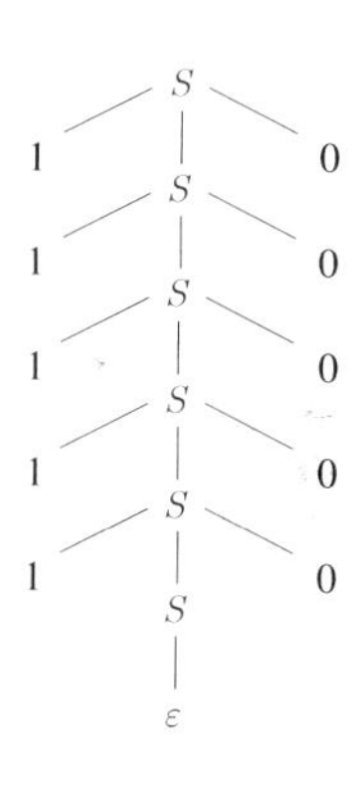

This form of specifying a language is called a *context-free grammar* (CFG), and it is composed of *productions* in the form $S \rightarrow x$ which indicate that the idea symbolized by S can be expanded using the rule x. Specifically, S is called a *nonterminal symbol*, meaning it is not a literal part of the language but something that can be expanded into a string or substring in the language. In the example above symbols like ε, 0, and 1 are called *terminal symbols* because they are literal parts of the string being defined an cannot be expanded. In the form $S \rightarrow x$, x is a string of terminal and nonterminal symbols. Two productions $S \rightarrow x_1$ and $S \rightarrow x_2$ can be abbreviated as $S \rightarrow x_1 \mid x_2$.

context-free grammar

production

nonterminal symbol

terminal symbol

Consider other examples of languages defined using CFGs. First, the language of strings containing the symbols { `(`, `)`, `[`, `]`, `{`, `}` }, properly nested, for example `(({[]}[]))[()]`.

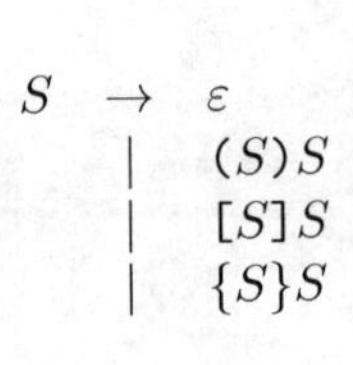

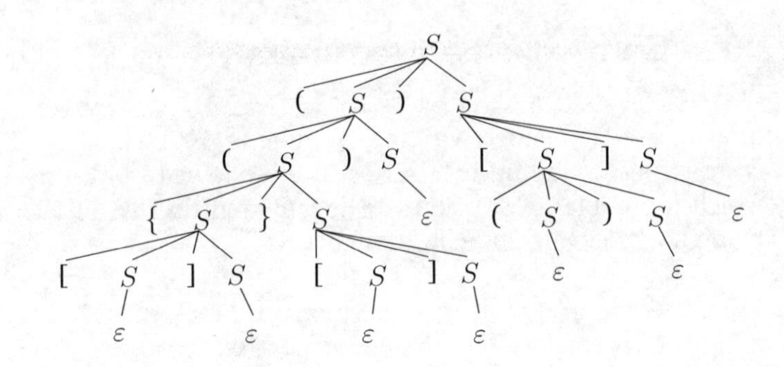

parse tree

A tree like this displaying how nonterminals are expanded into a string is called a *parse tree*. For a second example, consider the language of fully-parenthesized arithmetic expressions, such as `((2+9)*(3/(14-7)))`. This language has three nonterminals including N for numbers and O for operators, but only S stands for complete strings in the language.

$$\begin{array}{lcl} S & \rightarrow & N \mid (\, S\, O\, S\,) \\ N & \rightarrow & 0 \mid 1 \mid 2 \mid 3 \mid 4 \mid 5 \mid 6 \mid 7 \mid 8 \mid 9 \\ & \mid & 1N \mid 2N \mid 3N \mid 4N \mid 5N \\ & \mid & 6N \mid 7N \mid 8N \mid 9N \\ O & \rightarrow & + \mid - \mid * \mid / \mid \end{array}$$

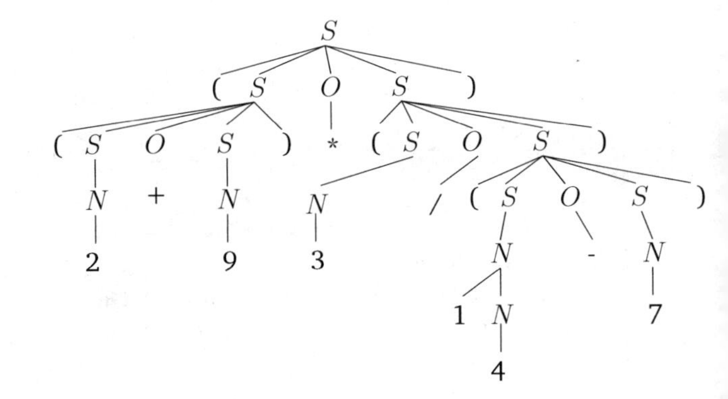

start symbol

Formally, a *context-free grammar* is a quintuple (T, NT, S, P) where T is a set of terminal symbols, NT is a set of nonterminal symbols, $S \in NT$ is a *start symbol* (standing for a complete string in the language), and P is a set of *productions*.

The set T is roughly equivalent to Σ, the alphabet which the language defined by the CFG is over. However, sometimes in our examples it will be convenient to think of T not necessarily consisting only of single symbols but also of substrings that appear atomically in the strings of the language.

Moreover, the productions (elements of P) are in the form $A \rightarrow x$ where $A \in NT$ and $x \in (T \cup NT)*$, that is, x is a string of terminals and nonterminals. Notice that even though we have a specialized notation for productions, a production is really just a tuple; that is, $P \subseteq NT \times (T \cup NT)*$. What makes this a *context-free* grammar is the way productions are defined. A *grammar* more generally in formal languages is just like our definition of CFG except that the specification of how productions are formed is broader.

You should notice how similar CFGs are to ML's data types. The datatype expression from Exercises 6.2.14–6.2.17 is conceptually the same as the language of fully parenthesized mathematical expressions above:

```
- datatype operation = Plus | Minus | Mul | Div;

- datatype expression =
=         Internal of operation * expression * expression
=       | Leaf of int;
```

What you can observe from these examples is that CFGs are more powerful than regular expressions and are especially convenient for defining languages with nested—that is, recursive—structure. In the Exercises, you will prove that CFGs are *strictly* more powerful than regular expressions: any language that can be specified with a regular expression can be specified with a CFG.

CFGs also capture most of the grammatical phenomena in natural languages. The subset of English described in the second version of the language processor (Section 6.3) is

$$\begin{array}{rcl}
T & = & \{\mathtt{man}, \mathtt{woman}, \mathtt{dog}, \mathtt{unicorm}, \mathtt{ball}, \mathtt{field}, \mathtt{flea}, \mathtt{tree}, \mathtt{sky}, \mathtt{a}, \mathtt{the}, \mathtt{big}, \\
 & & \;\mathtt{bright}, \mathtt{fast}, \mathtt{beautiful}, \mathtt{smart}, \mathtt{red}, \mathtt{smelly}, \mathtt{in}, \mathtt{on}, \mathtt{through}, \mathtt{with}, \\
 & & \;\mathtt{quickly}, \mathtt{slowly}, \mathtt{dreamily}, \mathtt{happily}, \mathtt{chased}, \mathtt{saw}, \mathtt{bit}, \mathtt{loved}, \mathtt{ran}, \\
 & & \;\mathtt{slept}, \mathtt{sang}, \mathtt{was}, \mathtt{felt}, \mathtt{seemed}, \mathtt{knew}, \mathtt{believed}, \mathtt{proved}\} \\
NT & = & \{\mathit{Sentence}, \mathit{NounPhrase}, \mathit{Predicate}, \mathit{VerbPhrase}, \mathit{PrepPhrase}, \mathit{Noun}, \\
 & & \;\mathit{Article}, \mathit{Adjective}, \mathit{Preposition}, \mathit{Adverb}, \mathit{TransitiveVerb}, \\
 & & \;\mathit{IntransitiveVerb}, \mathit{LinkingVerb}, \mathit{TransitivePropObjVerb}\}
\end{array}$$

$$\begin{array}{rcl}
\mathit{Sentence} & \rightarrow & \mathit{NounPhrase}\ \mathit{Predicate}\ \mathit{PrepPhrase} \\
 & | & \mathit{NounPhrase}\ \mathit{Predicate} \\
\mathit{NounPhrase} & \rightarrow & \mathit{Article}\ \mathit{Adjective}\ \mathit{Noun} \\
 & | & \mathit{Article}\ \mathit{Noun} \\
\mathit{Predicate} & \rightarrow & \mathit{Adverb}\ \mathit{VerbPhrase} \\
\mathit{VerbPrhase} & \rightarrow & \mathit{TransitiveVerb}\ \mathit{NounPhrase} \\
 & | & \mathit{IntransitiveVerb} \\
 & | & \mathit{LinkingVerb}\ \mathit{Adjective} \\
 & | & \mathit{TransitivePropObjVerb}\ \mathtt{that}\ \mathit{Sentence} \\
\mathit{PrepPhrase} & \rightarrow & \mathit{Preposition}\ \mathit{NounPhrase} \\
\mathit{Noun} & \rightarrow & \mathtt{man} \mid \mathtt{woman} \mid \mathtt{dog} \mid \mathtt{unicorm} \mid \mathtt{ball} \mid \mathtt{field} \mid \mathtt{flea} \mid \mathtt{tree} \mid \mathtt{sky} \\
\mathit{Article} & \rightarrow & \mathtt{a} \mid \mathtt{the} \\
\mathit{Adjective} & \rightarrow & \mathtt{big} \mid \mathtt{bright} \mid \mathtt{fast} \mid \mathtt{beautiful} \mid \mathtt{smart} \mid \mathtt{red} \mid \mathtt{smelly} \\
\mathit{Preposition} & \rightarrow & \mathtt{in} \mid \mathtt{on} \mid \mathtt{through} \mid \mathtt{with} \\
\mathit{Adverb} & \rightarrow & \mathtt{quickly} \mid \mathtt{slowly} \mid \mathtt{dreamily} \mid \mathtt{happily} \\
\mathit{TransitiveVerb} & \rightarrow & \mathtt{chased} \mid \mathtt{saw} \mid \mathtt{bit} \mid \mathtt{loved} \\
\mathit{IntransitiveVerb} & \rightarrow & \mathtt{ran} \mid \mathtt{slept} \mid \mathtt{sang} \\
\mathit{LinkingVerb} & \rightarrow & \mathtt{was} \mid \mathtt{felt} \mid \mathtt{seemed} \\
\mathit{TransitivePropObjVerb} & \rightarrow & \mathtt{knew} \mid \mathtt{believed} \mid \mathtt{proved}
\end{array}$$

Grammars like this are sometimes called *generative grammars* because the

generative grammar

rules indicate how to build or generate a string. The language, then, is generated from the productions.

It is possible that a grammar can generate a string in more than one way. Consider the following grammar for *unparenthesized* arithmetic expressions, such as 2*9-4. A grammar for this language can be found below left. However, this grammar can generate 2*9-4 in two ways, either interpreting 9-4 as a subexpression as in the parse tree below center, or interpreting 2*9 as a subexpression as seen below right.

$$
\begin{array}{rcl}
S & \rightarrow & N \mid S\,O\,S \\
N & \rightarrow & 0 \mid 1 \mid 2 \mid 3 \mid 4 \mid 5 \mid 6 \mid 7 \mid 8 \mid 9 \\
 & \mid & 1N \mid 2N \mid 3N \mid 4N \mid 5N \\
 & \mid & 6N \mid 7N \mid 8N \mid 9N \\
O & \rightarrow & + \mid - \mid * \mid / \mid
\end{array}
$$

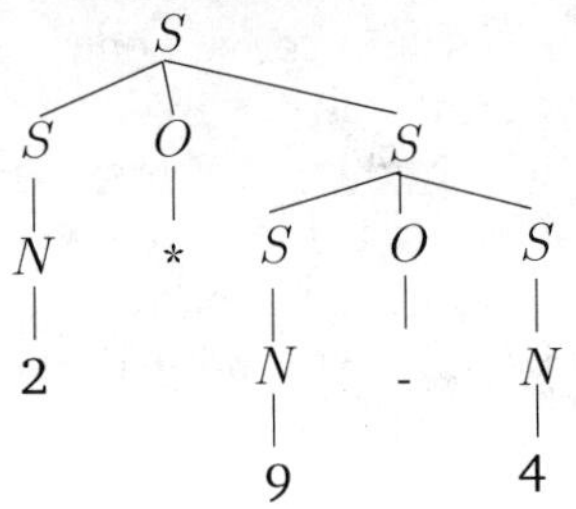

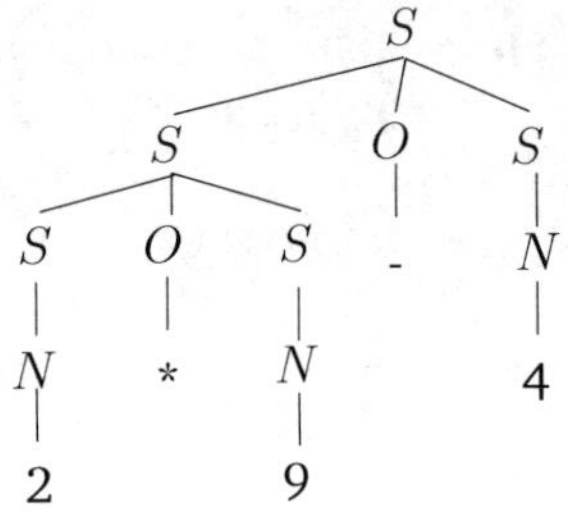

The parse tree on the right makes more sense in terms of how we normally read 2*9-4: since order of operations says we perform the multiplication first, 2*9 is the subexpression. That is a matter of meaning, however. It is not a relevant concern if all we want to do is define a set of strings. When the same string can be generated by a grammar in more than one way, we say the parse is *ambiguous*. That does not mean there is ambiguity about *whether* the string is in the language, only *how* it is generated.

ambiguous parse

For some languages it is possible to write several strikingly different CFGs, each generating the same language. The language we just considered is also generated by the language below, with a parse tree for 2*9-4+3.

$$
\begin{array}{rcl}
S & \rightarrow & N\,M \\
M & \rightarrow & \varepsilon \mid O\,N\,M \\
N & \rightarrow & 0 \mid 1 \mid 2 \mid 3 \mid 4 \mid 5 \mid 6 \mid 7 \mid 8 \mid 9 \\
 & \mid & 1N \mid 2N \mid 3N \mid 4N \mid 5N \\
 & \mid & 6N \mid 7N \mid 8N \mid 9N \\
O & \rightarrow & + \mid - \mid * \mid / \mid
\end{array}
$$

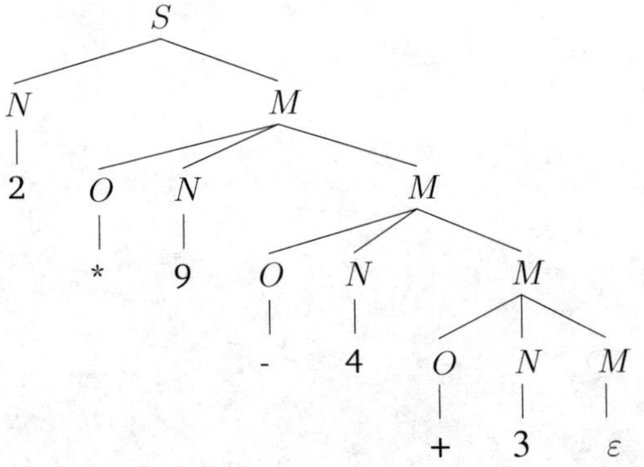

Interpret the nonterminal M to mean "more." $S \rightarrow N\ M$ means an expression is a number followed by more. The production $M \rightarrow \varepsilon$ implies that the "more" can be empty—perhaps a more intuitive reading is that M means "optionally more." Otherwise it is $M \rightarrow O\ N\ M$, an operator, a number, and yet more.

We can use this "optionally more" strategy to write a grammar for the language of ML lists of bool literals, such as `[]`, `[true]`, or `[true,true,false]`. This is tricky because we need to account for empty lists, lists with one item, and lists with many items. This does not scale easily because of how the commas work—if there are any items, then there are one fewer commas than items. The following grammar generates this language; parse trees for the three examples mentioned are shown. (L, for *list*, is the start symbol.)

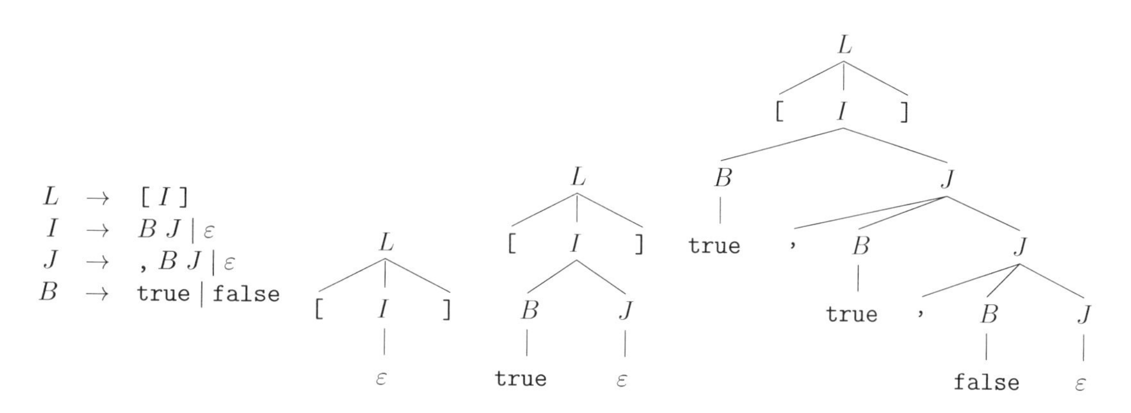

I stands for the content of the list, which may be empty ($I \to \varepsilon$). Otherwise I is a bool literal followed by an "optional more," that is, J. J is either a comma, a bool literal, and more, or it is nothing. Alternately, we could restructure this grammar so that the base case for J is a single bool literal rather than empty:

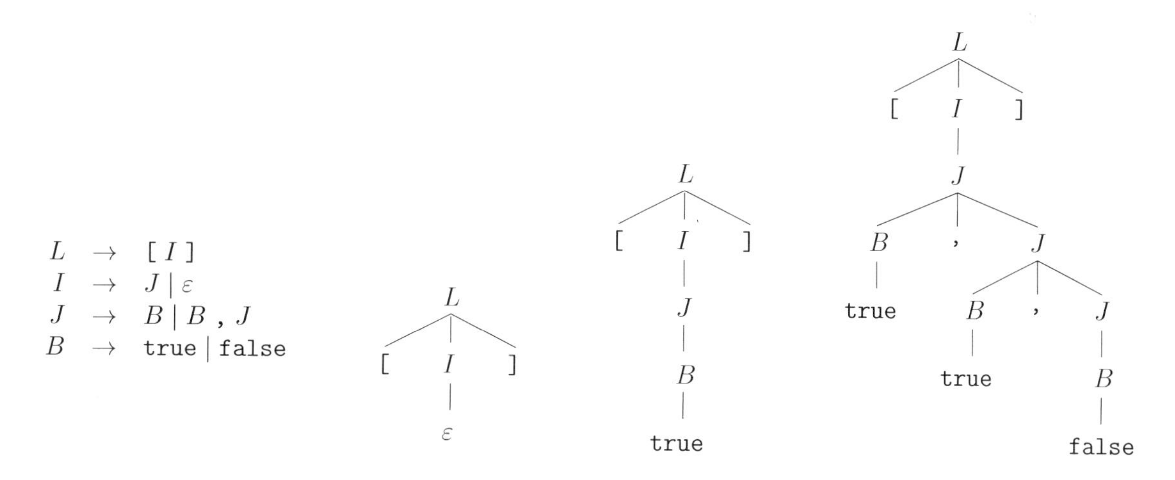

Here J is the content of a *nonempty* list—a single bool literal or a bool literal followed by a comma and the rest-of-the-list. The symbol I serves to allow the list to be empty.

Exercises

Write a context-free grammar for each of the following languages.

12.6.1 The language described in Exercise 12.2.1.

12.6.2 The language described in Exercise 12.2.2.

12.6.3 The language described in Exercise 12.2.5.

12.6.4 The language described in Exercise 12.2.6.

12.6.5 The language of palindromes (strings that are the same backwards and forwards) over $\Sigma = \{0, 1\}$. For example, `0010100`. Be careful that your language accounts for both even-lengthed and odd-lengthed strings.

12.6.6 The language of ML tuples with bool literals as their most basic component. That is, the tuples may be indefinitely nested, but the innermost tuples have bool literals. For example, `(true, false)` or `((true, false), false, true)`. Assume that any tuple has at least two elements—exclude trivial tuples like `()` or `(true)`.

12.6.7 The language of ML lists and tuples with `bool` literals as their most basic component. For example, `(true, [true, true, false])` and `[(true, (true, true), [true]), (false, (true, false), [])]`. Both tuples and lists may be arbitrarily nested within each other, and the topmost part may be either a list or tuple. Lists may have zero, one, or many items, but tuples must have two or more. Your grammar may allow expressions that are not type correct (in fact, it is impossible to make a CFG that accepts all such expressions that are type correct and excludes all those that are incorrect).

Exercises 12.6.8–12.6.12 will walk you through a proof that any language that can be specified with a regular expression also can be specified with a CFG. Stated formally:

Theorem 12.6 *If r is a regular expression, then there exists a CFG (T, NT, S, P) that generates the language specified by r.*

The proof, like that done in the exercises of Section 12.5, is by induction on the structure of r.

12.6.8 Suppose $r = \epsilon$. Give a CFG that generates the language consisting only of the empty string. This is trivial—your CFG needs only one nonterminal (which is then the start state), no terminals, and one production.

12.6.9 Suppose r = `a` for some `a` $\in \Sigma$. Give a CFG that generates the language consisting only of the string containing only `a`. This also is trivial.

12.6.10 Suppose $r = r_1|r_2$ for some regular expressions r_1 and r_2. By structural induction, assume there exist CFGs (T_1, NT_1, S_1, P_1) and (T_2, NT_2, S_2, P_2). Give a CFG that generates the language specified by r.

12.6.11 Suppose $r = r_1 r_2$ for some regular expressions r_1 and r_2. Give a CFG that generates the language specified by r.

12.6.12 Suppose $r = r_1*$ for some regular expression r_1. Give a CFG that generates the language specified by r.

12.6.13 Why do we not ask you to prove the converse (given a CFG, there exists a regular expression...)?

Tangent: Context-free grammars

The modern understanding of context-free grammars comes primarily from the work of linguist Noam Chomsky beginning in the 1950s. He used them to describe syntactic features in natural language. This coincided with the development of the first high-level programming languages. John Backus was the first to use CFGs to describe the syntax of a programming language in his work on the committee defining the programming language ALGOL 58. A few years later Peter Naur improved on this in his work defining ALGOL 60. The concept of a CFG, however, is much older. An ancient Indian grammarian named Panini used essentially the same idea to describe the grammar of Sanskrit in the fourth century BC.

12.7 Push-down automata

A language that can be specified by a regular expression is called a *regular language*. A language for which there exists a CFG that generates it is called a *context-free language*. Exercises 12.6.8–12.6.12 proved that any regular language is context-free. We also have seen that any regular language can be accepted by an NFA or DFA—in fact, the set of languages for which one can write an NFA or DFA is equal to the set of regular languages. Not all context-free languages are regular—we have seen CFGs that generate languages that no regular expression can specify and no NFA or DFA can accept.

regular language

context-free language

Just as every villain needs a hero to be defeated by, every set of languages needs a class of automata that will accept those languages. In this section we will consider a new kind of automaton for accepting context-free languages. Recall that the problem with DFAs and NFAs is that they do not have any mechanism for indefinite counting or recursion, which is also to say that they do not have any memory except for the the implicit memory about what state they currently are in. The sequence of states transitioned through to get to the current state, for example, is not remembered. To build a kind of automaton that can accept context-free languages, we will take our idea of a DFA and equip it with a minimal memory.

A *stack* is a mutable, ordered, finite collection of data on which certain operations, described below, can be performed. The *mutable* part means that the collection can be modified—we can add and remove things from the collection. In ML, reference variables are mutable, but other variables are not. When we say that a stack is *ordered* we simply mean that the elements exist in a sequence: there is a first, a second, and so on.

stack

mutable

To think about a stack we use a real-world analogy of a stack of books or plates, or a Pez-dispenser. We refer to the first item in the sequence as the *top* of the stack, and we have access only to the top of the stack. This means we can retrieve, remove, or add to the stack only at the top, not in the middle of the stack or at the other end. Accordingly, a stack conceptually has four operations:

- We can add a new item to the top of the stack. This is called *pushing* an item on the stack.
- We can remove the top item in the stack, assuming the stack is not empty. This is called *popping* an item from the stack.
- We can retrieve—that is, read but not remove—the top item in the stack (again assuming it is not empty).
- We can test if the stack is empty.

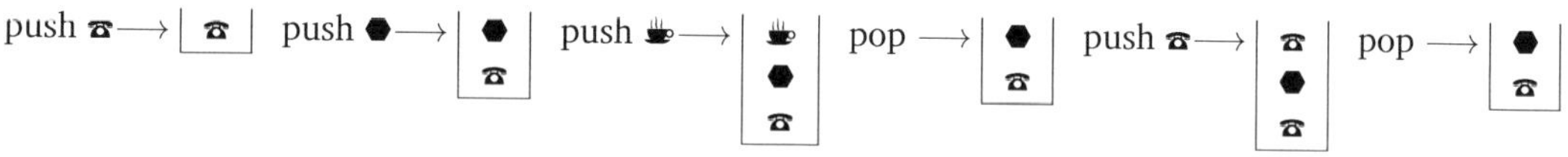

push ☎ ⟶ [☎] push ⬢ ⟶ [⬢ ☎] push ☕ ⟶ [☕ ⬢ ☎] pop ⟶ [⬢ ☎] push ☎ ⟶ [☎ ⬢ ☎] pop ⟶ [⬢ ☎]

Do you immediately recognize how easily a stack can be implemented in ML? ML lists seem ready-made to be used as stacks—except for not being mutable. If we consider the head of the list to be the top of the stack, then we can retrieve the top element with `hd` (or pattern-matching) and test for emptiness by comparing the list with `[]`.

We can simulate the changing of the stack in the first two operations by deriving a new list from the old. The new stack (after pushing) is constructed using cons (`::`). The stack-after-popping is found by applying `tl` to the list (or, again, by using pattern-matching).

push-down automaton

A *push-down automaton* (PDA) is a model of computation similar to a DFA except that in addition to having a state the machine also has a stack of symbols. Like a DFA, a PDA processes one input symbol per step in the computation, and it accepts or rejects the string based on what state it is in after all the input symbols have been read. Unlike a DFA, what to do at each step is determined not only by the current state and current input symbol but also by the top item on the stack. Moreover, when a PDA takes a step, it not only consumes an input symbol and changes to a new state, but it also modifies the stack by popping a symbol, pushing one or more symbols, neither, or both.

Consider an example before we define PDAs formally. We will build a PDA to accept the language of strings composed of a certain number of 1s followed by the same number of 0s, which, we have seen, is generated by the CFG $S \rightarrow \varepsilon \mid 1\ S\ 0$. The PDA will have four states (besides a trap): an initial state A, a state B for while it is reading 1s, a state C for while it is reading 0s, and an accept state D. Every time we see a 1, we will push some indicator on the stack—say, stack symbol X. As we see 0s, we pop these symbols from the stack. For the first 1 we see, we will push stack symbol Y instead of X; that way when we see Y again we will know that we have come to the bottom of the stack. To summarize this PDA's behavior:

State A	(Before we see anything)		
	if the input symbol is 0,	push Y	
		transition to state B	
State B	(Reading/counting more 0s)		
	if the input symbol is 0,	push X	
		stay in state B	
	if the input symbol is 1,	if the top is X,	pop
			transition to C
		if the top is Y,	pop
			transition to D
State C	(Reading/uncounting 1s)		
	if the input symbol is 1,	if the top is X,	pop
			stay in state C
		if the top is Y,	pop
			transition to D

In any case not covered above, transition to the trap. If the PDA is in state

A or D at the end of the string, then the string is accepted (we include A as an accept state to allow empty strings). Watch this PDA in action on the string 111000. The current input symbol at each step is underlined.

$\underline{1}11000$	$\longrightarrow$	$1\underline{1}1000$	$\longrightarrow$	$11\underline{1}000$	$\longrightarrow$	$111\underline{0}00$	$\longrightarrow$	$1110\underline{0}0$	$\longrightarrow$	$11100\underline{0}$	$\longrightarrow$	$111000.$
A ___		B [Y]		B [X, Y]		B [X, X, Y]		C [X, Y]		C [Y]		D ___

Now we are ready to define the PDA model formally. A *deterministic push-down automaton* is a sextuple $(Q, \Sigma, \Gamma, \delta, q_0, F)$ where

- Q is a set of states,
- Σ is an alphabet,
- Γ is a set of stack symbols,
- $\delta : Q \times \Sigma \times (\Gamma \cup \{\varepsilon\}) \to Q \times \Gamma *$ is a transition function, explained below,
- q_0 is the start state, and
- $F \subseteq Q$ is the set of accept states.

The transition function δ has three parameters. The first two—the current state and the current input symbol—are clear. The $\Gamma \cup \{\varepsilon\}$ business means that we are "unioning" the empty string to the set of stack symbols to get the set of possible values for the third parameter. This allows us to specify what to do when the stack is empty[1]. The codomain of δ (what the transition function returns) is a new state and a *string* of stack symbols.

This means the tuples in δ can be in any of the forms below. Remember that as a function, δ is a set of pairs—in this case, the first element in the pair is itself a triple and the second element is itself a pair, standing for the parameters and function value, respectively. In each form, assume we are in state A, see input symbol c, and transition to state B (possibly $B = A$).

$((A, c, \varepsilon), (B, \varepsilon))$	The stack is empty; leave it empty.
$((A, c, \varepsilon), (B, X))$	The stack is empty; push X.
$((A, c, X), (B, X))$	Leave the top stack symbol X unchanged.
$((A, c, X), (B, \varepsilon))$	Pop the top stack symbol X.
$((A, c, X), (B, YX))$	Leave the top stack symbol X; push Y.
$((A, c, X), (B, Y))$	Pop the top stack symbol X *and then* push Y.
$((A, c, X), (B, ZYX))$	Leave X; push Y; push Z.
$((A, c, X), (B, ZY))$	Pop X; push Y; push Z.

[1]Most descriptions of deterministic PDAs allow the transition function to take ε as the third parameter even if the stack is not empty, essentially allowing it to ignore the top stack element. To simplify our discussion, we discount that possibility.

So you see that in one step of the computation, the PDA may inspect at most one top stack symbol, may pop at most that one top symbol, and may push any number of symbols.

The transition function for the PDA described earlier is

$$\{ ((A,1,\varepsilon),(B,Y)), ((B,0,Y),(B,XY)), ((B,0,X),(B,XX)), ((B,1,X),(C,\varepsilon)), ((B,1,Y),(D,\varepsilon)), ((C,1,X),(C,\varepsilon)), ((C,1,Y),(D,\varepsilon)) \}$$

Consider how we would implement a PDA. In an approach similar to our original implementation of a DFA in Section 12.2 (not the alternate version defined in the exercises of that section), we can define a PDA as a transition function, a start state, and a set of accept states. In our DFA implementation, the transition function naturally took a char and current state as parameters and returned the next state. The configuration of a PDA includes not only the state but also the current stack. Accordingly we define the transition function to take a char and a tuple containing a state and a list of stack symbols. (Note that while the transition function in the formal definition takes as arguments the next input symbol, current state, and *top* stack symbol, our implementation of the transition function will take *the entire stack*, although it will modify at most the top stack symbol.)

```
- datatype ('a,'b) pda =
=       PDA of ((char * ('a * 'b list)) -> ('a * 'b list))
=                * 'a * 'a list;
```

This datatype has two type parameters—'a is the state type and 'b is the stack symbol type. We then can implement the transition function for the PDA described above as

```
- datatype state = A | B | C | D | T;

- datatype stkSymb = X | Y;

- fun m1(#"1", (A, [])) = (B, [Y])
=    | m1(#"1", (B, Y::stk)) = (B, X::Y::stk)
=    | m1(#"1", (B, X::stk)) = (B, X::X::stk)
=    | m1(#"0", (B, X::stk)) = (C, stk)
=    | m1(#"0", (B, Y::stk)) = (D, stk)
=    | m1(#"0", (C, X::stk)) = (C, stk)
=    | m1(#"0", (C, Y::stk)) = (D, stk)
=    | m1(_, _) = (T, []);

val m1 = fn : char * (state * stkSymb list) ->
     state * stkSymb list
```

The check function for PDAs will be largely similar to the check function for DFAs. The difference stems from the transition function's having a tuple for its second parameter. First, the "seed value" to foldl needs to be the start state *and an empty list* as a pair, the latter standing for the initial, empty stack. The result from foldl will likewise be a state and a stack as a pair, of which we are interested only in the state. Thus we need to grab the first item in that pair and test if that state is contained in the list of accept states.

```
- fun check(machine as PDA(trans, startState, acceptStates),
=           str) =
=     contains(#1(foldl(trans)(startState, [])(explode(str))),
=              acceptStates);

val check = fn : (''a,'b) pda * string -> bool

- fun checkM1(s) = check(PDA(m1, A, [A, D]), s);

val checkM1 = fn : string -> bool

- checkM1("111000");

val it = true : bool

- checkM1("11100");

val it = false : bool

- checkM1("111001");

val it = false : bool
```

As a second example, consider the alphabet of parentheses, square brackets, and curly braces, $\Sigma = \{(,),\{,\},[,]\}$, and the language of strings over this language such that the delimiters are balanced and nested properly, for example ([]{([])}). We saw a grammar for this language in Section 12.6.

The stack's natural use here is to keep track of what opening delimiters we have seen that still need to be matched. Whenever we see an opening delimiter, we push it on the stack; whenever we see a closing delimiter, we compare it to the top of the stack to see if it matches the most recent unmatched opening delimiter. Here we show the progress of a computation to accept the sample string above by showing only the stack—we will reason about what states we need later.

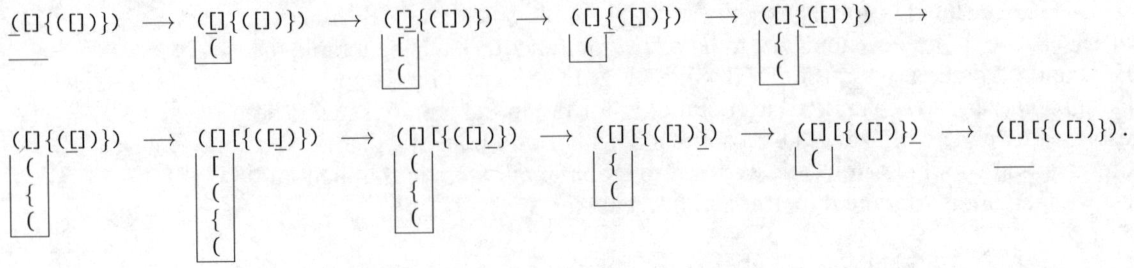

```
([]{([])}) ⟶ ([]{([])}) ⟶ ([]{([])}) ⟶ ([]{([])}) ⟶ ([]{([])}) ⟶
              (             [             (             {
                            (                           (

([]{([])}) ⟶ ([][{([])}) ⟶ ([][{([])}) ⟶ ([][{([])}) ⟶ ([][{([])}) ⟶ ([][{([])}).
 (             [              (              {              (
 {             (              {              (
 (             {              (
               (
```

If we were to make a PDA for performing this computation, what states must it have? Notice that we were able to describe the steps merely by looking at the stack. If the stack tells the whole story, then the set of states should be pretty simple. We will need an accept state, and that will correspond to when there are no opening delimiters left to be matched. Call that state *Matched*. Notice that this condition is met before we look at any input symbols, so *Matched* is also our start state. It makes sense for our start state to be an accept state since the empty string is vacuously part of the language.

A string fails to be in the language if we see a closing delimiter that does not match the opening delimiter on top of the stack. To reject a string, we need a *Trap* state. All other conditions—not yet matched, but not yet mismatched—we can subsume under a single remaining state, call it *Wait*.

As for the stack symbols, notice that we need to know if we are looking at the *bottom* stack symbol, because if we see a matching, closing delimiter at that step then we will transition into the *Matched* state. Thus we need six stack symbols, a bottom and non-bottom symbol for each kind of delimiter.

```
- datatype state = Matched | Wait | Trap;

- datatype stkSymb = Paren | Curly | Square |
=                    ParenBot | CurlyBot | SquareBot;

- fun m2(#"(", (Matched, stk)) = (Wait, ParenBot::stk)
=   | m2(#"{", (Matched, stk)) = (Wait, CurlyBot::stk)
=   | m2(#"[", (Matched, stk)) = (Wait, SquareBot::stk)
=   | m2(#"(", (Wait, stk)) = (Wait, Paren::stk)
=   | m2(#"{", (Wait, stk)) = (Wait, Curly::stk)
=   | m2(#"[", (Wait, stk)) = (Wait, Square::stk)
=   | m2(#")", (Wait, Paren::stk)) = (Wait, stk)
=   | m2(#"}", (Wait, Curly::stk)) = (Wait, stk)
=   | m2(#"]", (Wait, Square::stk)) = (Wait, stk)
=   | m2(#")", (Wait, ParenBot::stk)) = (Matched, stk)
=   | m2(#"}", (Wait, CurlyBot::stk)) = (Matched, stk)
=   | m2(#"]", (Wait, SquareBot::stk)) = (Matched, stk)
=   | m2(_,_) = (Trap, []);

- fun checkM2(s) = check(PDA(m2, Matched, [Matched]), s);
```

```
- checkM2("([]{([])})");

val it = true : bool

- checkM2("([]{([])}");

val it = false : bool
```

This discussion of PDAs has suggested that the PDA computational model corresponds to the CFG language model. We have not told you the whole story, however. In Exercise 12.6.5 we asked you to write a CFG for the language of palindromes over $\Sigma = \{0, 1\}$. Imagine a PDA to accept a language of palindromes (say, over the English alphabet, to make it more interesting). For each character in the first half of the string we would push a stack symbol indicating which input symbol we had seen, and then we would process the second half of the string by popping those symbols and checking that they match the input symbols.

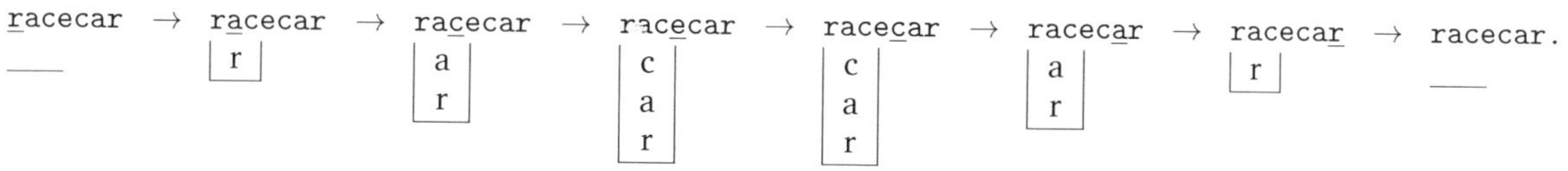

$\underline{\text{r}}$acecar →	r$\underline{\text{a}}$cecar →	ra$\underline{\text{c}}$ecar →	rac$\underline{\text{e}}$car →	race$\underline{\text{c}}$ar →	racec$\underline{\text{a}}$r →	raceca$\underline{\text{r}}$ →	racecar.
—	r	a	c	c	a	r	—
		r	a	a	r		
			r	r			

But—how can the PDA tell when it is halfway through the input string? In the case of a string with an odd number of characters, like *racecar*, we also need to know to skip the middle letter *e* altogether. So far we have been thinking only about *deterministic* PDAs. If we define a *nondeterministic* PDA in analogy to an NFA, then we would let it nondeterministically guess when it has reached the middle point and start popping instead of pushing (and, possibly, skip a middle character). If there exists a correct midpoint choice, then the string is accepted.

The bad news is *this cannot be done with a deterministic PDA*. Note several implications. Although deterministic PDAs are more powerful than DFAs and NFAs in that there exist languages that are not accepted by a DFA or NFA but are accepted by a deterministic PDA, nevertheless deterministic PDAs are not powerful enough to accept all context-free languages. Instead, the nondeterministic PDA model is the one that corresponds completely to CFGs. Finally, although nondeterminism does not increase the power of finite automata (the set of languages accepted by NFAs is equal to the set of those accepted by DFAs), nondeterministic PDAs are strictly more powerful than deterministic PDAs.

The PDA model, nevertheless, is a crucial tool for processing languages, both natural and artificial. Algorithms based on PDAs—often fudged to make them a little more powerful—form the basis for *parsing* programs in a programming language, that is, analyzing their structure.

parsing

Exercises

Implement a PDA for each of the following languages by making datatypes for the states and stack symbols and writing a transition function in ML.

12.7.1 The language of ML lists of bool literals, as described on page 629.

12.7.2 The language described in Exercise 12.6.6.

12.7.3 The language described in Exercise 12.6.7.

12.8 The lambda calculus

This chapter has been about models of computation. The models presented so far have been abstract machines, but a programming language also can model computation by defining how to express algorithmic ideas and what the result of those expressions should be. Most programming languages are too complex to be useful *models*. The *lambda calculus* is a minimal language that models computation.

lambda calculus

Think back to your first encounter with ML. The basic pieces introduced first were values, types, and expressions. We will start with a taxonomy of expressions—meanwhile, you observe what the values and types are. The lambda calculus has only three kinds of expressions. First there are variables, and we will assume that any mathematical variable is a legal identifier in the lambda calculus.

The second kind of expression is like the anonymous functions in ML we first saw in Section 7.3, called a *function abstraction*. It is formed by the Greek letter λ, the name of the parameter, a period, and the body of the function. For example, the following abstraction takes a value with parameter x and returns that same value:

function abstraction

$$\lambda x.x$$

It is equivalent to the ML expression

```
- fn (x) => x;

val it = fn : 'a -> 'a
```

Tangent: Calculus

The lambda calculus has nothing to do with derivatives and integrals. In Latin the word *calculus* means "little stone," such as pebbles used in abacus-like counting devices. By extension *calculus* came to refer to the calculation itself with verb form *calculare*, to calculate. The general meaning of *calculus* in English is "a system or method of calculation" (OED). Since differential calculus and integral calculus have played such an important role in science and engineering, they became referred to as "*the* calculus," and eventually simply as "calculus." The word has another relative in English: *chalk*, also a stone we use for calculation.

The third kind of expression is the *application* of a function. If two expressions occur one after the other, then the first is interpreted as a function and the second as the actual parameter to that function. If the first expression is an abstraction, then the whole expression is evaluated by taking the second expression and replacing every occurrence of the first expression's formal parameter with the second expression. In addition to these expressions, parentheses can disambiguate the order of operations. We will parenthesize the two expressions in an application for clarity, except when the parentheses become overbearing.

application

As an example, consider the following expression:

$$(\lambda x.x\ \lambda z.z)$$

Evaluating this results in $\lambda z.z$ because that is the actual parameter to the function $\lambda x.x$, which simply returns the value it is given.

The lambda calculus's very simplicity makes it hard to understand. Here is the language's grammar formally:

$$\begin{array}{rcl} \mathit{Expression} & \rightarrow & \mathit{Variable} \mid \mathit{Application} \mid \mathit{Abstraction} \\ \mathit{Variable} & \rightarrow & \mathit{Identifier} \\ \mathit{Application} & \rightarrow & (\mathit{Expression}\ \mathit{Expression}) \\ \mathit{Abstraction} & \rightarrow & \lambda\ \mathit{Identifier}\ .\ \mathit{Expression} \end{array}$$

Here is its single evaluation rule, where x stands for any variable and M and N stand for any expressions:

$$(\lambda x.M\ N) \longrightarrow M[N/x]$$

What this means: Suppose we have an expression with two subexpressions, the first $\lambda x.M$ and the second N, where M is any expression and N is any expression. The computation takes a step by reducing this expression to a new one—"$M[N/x$," meaning M but with every occurrence of x in M replaced with N.

Consider this expression and the three steps it takes to reduce it all the way. In each step we will underline the function being applied and the actual parameter to that function.

$$\begin{array}{rcl} ((\underline{\lambda x.\lambda y.(y\ x)}\ \underline{\lambda z.z})\ \lambda w.w) & \longrightarrow & (\underline{\lambda y.(y\ \lambda z.z)}\ \underline{\lambda w.w}) \\ & \longrightarrow & (\underline{\lambda w.w}\ \underline{\lambda z.z}) \\ & \longrightarrow & \lambda z.z \end{array}$$

What do you notice about the values and types in the lambda calculus? It turns out, all values are functions. There are no integers, reals, or even booleans. It is as if in ML every value had the type `'a`, where `'a = 'a -> 'a`. This means that `'a` is a function type for functions that also take `'a` parameters and return `'a` results. While this seems difficult to imagine, we shall see that we can construct things we expect in a programming language from these simple blocks.

Just as finite automata have their trap states, things can go wrong when a lambda calculus expression is being evaluated. If the first subexpression in an application is not an abstraction and cannot be reduced to one, then the expression itself cannot be reduced, as in this example:

$$(\lambda x.(y\ x)\ \lambda z.z) \longrightarrow (y\ \lambda z.z)$$

We can take this one step, but then we have $(y\ \lambda z.z)$. This happened because the variable y was free in the first abstraction—it was not the formal parameter to any enclosing abstraction. After reducing the first application, we were left with just y as the function to be applied to $\lambda z.z$. The expression is *stuck*, which is what we call it when an expression is not a value but cannot be reduced. (Note that an expression is a *value* if it is a function abstraction whose body cannot be reduced.)

stuck

Some expressions neither reduce to a value nor get stuck. Notice how the following is equivalent to infinite recursion or an infinite loop:

$$(\lambda x.(x\ x)\ \lambda x.(x\ x)) \longrightarrow (\lambda x.(x\ x)\ \lambda x.(x\ x))$$

It may seem unduly restrictive for functions to have only one parameter. In fact, we can simulate functions with more parameters by currying, which we saw in Section 7.12. In the following expression, the first subexpression (parenthesized) is a triply-nested abstraction, but it works like a function with three parameters x, y, and z. (M, N, O, and P could be any expressions.)

$$\begin{array}{rcl}(\lambda x.\lambda y.\lambda z.M)\ N\ O\ P & \longrightarrow & \lambda y.\lambda z.(M[N/x])\ O\ P \\ & \longrightarrow & \lambda z.(M[N/x][O/y])\ P \\ & \longrightarrow & M[N/x][O/y][P/z]\end{array}$$

Fully parenthesized, the original expression is $(((\lambda x.\lambda y.\lambda z.M\ N)\ O)\ P)$. Notice that without parentheses the applications are evaluated from left to right. The result means "M with occurrences of x replaced with N, then occurrences of y replaced with O, and finally occurrences of z replaced with P."

Now, what good is all this? As a first demonstration of the lambda calculus's power, we will use it to implement booleans and conditionals. Since "everything is a function," we need to find a way to think of `true` and `false` and conditional expressions as functions. This is not too bad for conditionals—it is natural to think of an if/then/else as a function that takes three parameters, returning the second if the first is true, the third otherwise.

```
- fun ifthenelse(c, t, e) = if c then t else e;


val ifthenelse = fn : bool * 'a * 'a -> 'a
```

But what about `true` and `false`? How can those be imagined as functions? Their main purpose is for use in a conditional, so we can think of them as helper functions for our conditional function. Here they are:

$$\begin{array}{rcl} \textit{True} & = & \lambda t.\lambda f.t \\ \textit{False} & = & \lambda t.\lambda f.f \\ \textit{If} & = & \lambda \ell.\lambda m.\lambda n.((\ell\ m)\ n) \end{array}$$

We are simply declaring *True*, *False*, and *If* as mathematical variables for our discussion, standing for these lambda calculus expressions. This sort of declaration is not part of the language. *True* essentially takes two things—a "true" thing and a "false" thing—and chooses the true. *False* chooses the false thing. *If* is harder to understand, but the intent is that if M and N are expressions, then

$$\textit{If True } M\ N \longrightarrow_* M$$

and

$$\textit{If False } M\ N \longrightarrow_* N$$

The symbol $\longrightarrow_*$ means "reduces eventually to," that is, after some number of steps. It is based on the $*$ from regular expressions.

Try this out:

$$\begin{array}{rcl} \textit{If True } M\ N & = & (\lambda \ell.\lambda m.\lambda n.((\ell\ m)\ n)\ \textit{True})\ M\ N \\ & \longrightarrow & \lambda m.\lambda n.((\textit{True}\ m)\ n)\ M\ N \\ & \longrightarrow & \lambda n.((\textit{True}\ M)\ n)\ N \\ & \longrightarrow & (\textit{True}\ M)\ N \\ & = & (\lambda t.\lambda f.t\ M)\ N \\ & \longrightarrow & \lambda f.M\ N \\ & \longrightarrow & M \end{array}$$

Notice that we use $\longrightarrow$ only when an actual reduction is made. When we simply replace a symbol like *True* with what we defined for it previously, we use $=$ to indicate these expressions are the same "by substitution."

Remember the Peano axioms for whole numbers? We used them in Section 6.1 to implement whole numbers in ML using a simple, recursive datatype. The main idea is that a whole number is either zero or one more than a whole number. We can use a similar approach to encode whole numbers in the lambda calculus. In this context they are called *Church numerals* after Alonzo Church, the inventor of the lambda calculus.

Church numerals

A whole number is represented as a function (of course). It has two parts, a "zero" part and a "successor" part, and the parts are represented as parameters. Numbers differ from each other by how they treat the successor and zero.

$$\begin{array}{rcl} c_0 & = & \lambda z.\lambda s.z \\ c_1 & = & \lambda z.\lambda s.(s\ z) \\ c_2 & = & \lambda z.\lambda s.(s\ (s\ z)) \\ c_n & = & \lambda z.\lambda s.\underbrace{(s\ (s\ \ldots (s}_{n \text{ applications of } s}\ z) \ldots)) \end{array}$$

The interpretation is that to represent n we apply s to z n times. Just as booleans are designed to be used in conditionals, whole numbers are for repetition. Now we can define simple operations. Define a "boolean" function to test if a value, presumably a whole number, is zero:

$$\mathit{IsZero} = \lambda m.((m\ \mathit{True})\ (\lambda x.\mathit{False}))$$

Try it:

$$\begin{aligned}
\mathit{IsZero}\ c_0 &= \lambda m.((m\ \mathit{True})\ (\lambda x.\mathit{False}))\ c_0 \\
&\longrightarrow (c_0\ \mathit{True})\ (\lambda x.\mathit{False}) \\
&= (\lambda z.\lambda s.z\ \mathit{True})\ (\lambda x.\mathit{False}) \\
&\longrightarrow \lambda s.\mathit{True}\ \lambda x.\mathit{False} \\
&\longrightarrow \mathit{True}
\end{aligned}$$

$$\begin{aligned}
\mathit{IsZero}\ c_1 &\longrightarrow (c_1\ \mathit{True})\ (\lambda x.\mathit{False}) \\
&= (\lambda z.\lambda s.(s\ z)\ \mathit{True})\ (\lambda x.\mathit{False}) \\
&\longrightarrow \lambda s.(s\ \mathit{True})\ (\lambda x.\mathit{False}) \\
&\longrightarrow \lambda x.\mathit{False}\ \mathit{True} \\
&\longrightarrow \mathit{False}
\end{aligned}$$

Try to see behind the magic. *IsZero* has m *True* at its core. m is a function that determines how many times we apply $\lambda x.\mathit{False}$ to *True*: if zero times, then our answer is *True*; if one or more times, then we get *False*.

Here is a harder one, addition:

$$\mathit{Plus} = \underbrace{\lambda m.\lambda\ n.}_{\text{take two numbers}}\ \overbrace{\lambda z.\lambda s.}^{\text{construct a number}}\ \underbrace{(m\ \ \overbrace{(n\ z\ s)}^{\text{apply } s \text{ to } z\ n \text{ times}}\ \ s)}_{\text{take that result and apply } s \text{ to it } m \text{ more times}}$$

Try it:

$$\begin{aligned}
\mathit{Plus}\ c_1\ c_2 &= (\lambda m.\lambda n.\lambda z.\lambda s.m\ (n\ z\ s)\ s)\ c_1\ c_2 \\
&\longrightarrow_2 \lambda\ z.\ \lambda\ s.(c_1\ (c_2\ z\ s)\ s)
\end{aligned}$$

The $\longrightarrow_2$ means we just made two steps.

$$= \lambda z.\lambda s.(c_1\ ((\lambda z'.\lambda s'.(s'(s'\ z')))\ z\ s)\ s)$$

c_2 has variables s and z just like c_1 does, but we intend them to be different variables. To keep them straight, we renamed them s' and z'.

$$\begin{aligned}
&\longrightarrow_2 \lambda z.\lambda s.(c_1\ (s\ (s\ z))\ s) \\
&= \lambda z.\lambda s.((\lambda z''.\lambda s''.(s''\ z''))(s\ (s\ z))\ s) \\
&\longrightarrow \lambda z.\lambda s.((\lambda s''.s''(s\ (s\ z)))\ s) \\
&\longrightarrow \lambda z.\lambda s.s(s\ (s\ z)) = c_3
\end{aligned}$$

Addition and multiplication may seem like paltry results, but consider how basic our foundation is. In the next section we will discuss the comparative power of the lambda calculus with other models of computation.

This presentation of the lambda calculus is based on that of Benjamin Pierce [26].

Exercises

Reduce the following expressions to values.

12.8.1 $(\lambda x.(x\ \lambda y.\lambda z.y)\ \lambda w.\lambda t.t)$

12.8.2 $(\lambda x.(x\ \lambda y.\lambda z.y)\ \lambda w.\lambda t.w)$

12.8.3 $((\lambda x.(x\ \lambda y.\lambda z.y)\ \lambda w.w)\ \lambda t.t)$

12.8.4 $(\lambda x.\lambda y.(x\ x)\ \lambda z.z)\ \lambda w.w)$

12.8.5 $\mathit{Plus}\ c_2\ c_0$

12.8.6 $\mathit{Plus}\ c_2\ c_2$

We can encode tuples and extraction of tuples with the Lambda Calculus in the following manner:

$$\begin{aligned} \mathit{Pair} &= \lambda f.\lambda s.\lambda b.((b\ f)s) \\ \mathit{Fst} &= \lambda p.(p\ \mathit{True}) \\ \mathit{Snd} &= \lambda p.(p\ \mathit{False}) \end{aligned}$$

The way to read the above is that Pair takes two parameters (f and s) and makes an "object" (actually a function, as everything is) which is like an ML tuple containing f and s. Fst and Snd are functions that take a pair and return the first and second item, respectively. It is equivalent to

```
fun Pair(f, s) = (f, s);
fun Fst(p) = #1(p);
fun Snd(p) = #2(p);
```

12.8.7 Translate the lambda calculus versions of *Pair*, *Fst* and *Snd* to ML (using just anonymous functions, not ML's tuples). Note this should start as `val Pair = ...`, not `fun Pair(f, s) = ...` Parts of your solution might not be typable in ML.

12.8.8 Confirm that $\mathit{Fst}\ (\mathit{Pair}\ M\ N) \longrightarrow_* M$. (You may assume pieces like $\mathit{True}\ X\ Y \longrightarrow_* X$ without showing the steps.)

We can define multiplication with

$$\mathit{Times} = \lambda m.\lambda n.(m\ c_0(\mathit{Plus}\ n))$$

Reduce the following expressions.

12.8.9 $\mathit{Times}\ c_0\ c_2$

12.8.10 $\mathit{Times}\ c_2\ c_1$

12.8.11 $\mathit{Times}\ c_2\ c_2$

Biography: Alonzo Church, 1903–1995

The lambda calculus was invented by Alonzo Church. Born in 1903 in Washington, DC, he studied at Princeton University where he also served on the faculty. Later in his career he moved to UCLA. Many important contributions to logic and the foundations of mathematics and computation were being made at the time, and the lambda calculus was built on earlier and contemporary systems. The use of the Greek letter λ for function abstraction was derived from the earlier use of $\hat{x}$ to indicate x as a parameter by Alfred Whitehead and Bertrand Russell. Changing the "hat" to a lambda may have been simply because of printing constraints. [4]

Exercises, continued

We can represent lambda calculus expressions in ML with the following datatype

```
datatype expression =
      Var of string
    | App of expression * expression
    | Abs of string * expression;
```

(Notice that the Var data constructor is for variables acting as expressions. The parameter to a function abstraction is just a string.) To display an expression as a string, use a function like this:

```
fun toString(Var(x)) = x
  | toString(App(m, n)) =
    "(" ^ toString(m) ^ " " ^ toString(n) ^ ")"
  | toString(Abs(x, e)) =
      "^" ^ x ^ "." ^ toString(e);
```

12.8.12 Finish the function replace below, which takes an expression, a string a (interpreted as variable names), and a second expression b and replaces all the occurrences of variable a with b. If a is the parameter to an abstraction in the expression, then no replacements are made within that abstraction (for example, replacing x with $(y\ y)$ in $\lambda z.(x\ \lambda x.x)$ results in $\lambda z.((y\ y)\ \lambda x.x)$).

```
fun replace(Var(x), a, b) = ??
  | replace(App(m, n), a, b) = ??
  | replace(Abs(x, e), a, b) =
        if x <> a then ?? else ??;
```

12.8.13 Finish the following function reduce which reduces an expression to a value or a stuck expression, printing the result. Specifically, finish the helper function red which takes an expression and either returns a value or raises an exception indicating a stuck expression (together with the expression that is stuck).

```
exception Stuck of expression;

fun reduce(e) =
 let
    fun red(Var(x)) = ??
      | red(Abs(x, e)) = (Abs(x, red(e))
        handle Stuck(g) => Abs(x, g))
     | red(App(a, b)) =
       let
       (* reduce a and b all the way *)
         val redA =
           (red(a) handle Stuck(g) => g);
         val redB =
           (red(b) handle Stuck(g) => g);
       in
          (* redA is either a value or
             is stuck *)
           case redA of
                Abs(x, e) => ??
              | Var(x) => ??
              | App(m, n) => ??
       end;
 in "value: " ^ toString(red(e))
      handle Stuck(ee) =>
          "stuck: " ^ toString(ee)
 end;
```

You can test this on the following expressions:

```
val firstExample =
App(App(Abs("x",Abs("y",App(Var("y"),
 Var("x")))),Abs("z",Var("z"))), Abs("w",
 Var("w")));

val True = Abs("t",Abs("f", Var("t")));
val False = Abs("t",Abs("f", Var("f")));
val If = Abs("l",Abs("m",Abs("n",
 App(App(Var("l"),Var("m")),Var("n")))));

val M = Abs("q",Var("q"));
val N = Abs("w",Var("w"));

val bTest = App(App(App(If,True),M),N);

val zero = Abs("z", Abs("s",Var("z")));
val one =
   Abs("z",Abs("s",App(Var("s"),Var("z"))));
val two = Abs("z",Abs("s",App(Var("s"),
   App(Var("s"),Var("z")))));

val Plus = Abs("m",Abs("n",Abs("z",Abs("s",
  App(App(Var("m"),App(App(Var("n"),
  Var("z")),Var("s"))),Var("s"))))));

val plusTest = App(App(Plus,one),two);
```

12.9 Hierarchies of computational models

We have seen that the set of languages that can be specified by regular expressions and accepted by finite automata is a strict subset of the set of languages that can be generated by CFGs and accepted by nondeterministic PDAs. The natural question to ask next is whether there exist more powerful kinds of grammars to specify larger sets of languages and more powerful automata to accept them. In this section we comment about other languages and classes of automata in this hierarchy.

We briefly mentioned earlier that what makes a grammar *context free* is that productions can be used to expand a nonterminal no matter what context the nonterminal is found in. This is why the productions are in the form $S \to x$ where S is a single nonterminal and x is a string of terminals and nonterminals. In a *context-sensitive grammar* (CSG), the productions may restrict the expansion to certain situations. Specifically, in a CSG, productions are in the form $y \to x$, where both x and y are strings of terminals and nonterminals, as long as the length of x is greater than or equal to the length of y—in other words, applying a production will never make the working string smaller.

context-sensitive grammar

Consider a language over the set of delimiters, $\Sigma = \{\texttt{(}, \texttt{)}, \texttt{\{}, \texttt{\}}, \texttt{[}, \texttt{]}\}$, similar to our earlier language except having first a series of opening delimiters followed by matching closing delimiters (so, no more opening delimiters once the closing delimiters have started) and with the restriction that no two delimiters in a row can be the same kind—for example, `({[{[([])]}]})`. This can be generated by the following CSG:

$$
\begin{array}{rcl}
S & \to & \varepsilon \mid (T) \mid [T] \mid \{T\} \\
(T) & \to & (\{T\}) \mid ([T]) \mid (R) \\
[T] & \to & [(T)] \mid [\{T\}] \mid [R] \\
\{T\} & \to & \{(T)\} \mid \{[T]\} \mid \{R\} \\
R & \to & \varepsilon
\end{array}
$$

When (T) appears on the left hand side of the production, it indicates the production is applicable if T appears between two parentheses. Although the language above can also be generated by a CFG, there are languages for which there are CSGs but no CFG. Note that any CFG is also a CSG. CSGs are a strictly more powerful language model than CFGs.

When we introduced PDAs, we claimed we were defining a new class of automata by equipping finite automata with *minimal* memory. A PDA's stack in fact can grow indefinitely—how can we call that minimal? Although that stack is very extravagant in terms of the amount of memory, the point was the severe restrictions in what could be done with that memory—only the top symbol can be read or removed, and information can be added only at the top.

(Note also that we can predict how big the stack will grow for a given PDA and its input. Suppose that of all the tuples in the PDA's transition function, no tuple adds more than k symbols to the stack. Suppose further that the input string has length n. Since the PDA takes one step for each input symbol, the stack can never grow to more than $k \cdot n$ symbols.)

linear bounded automaton

tape

An alternative is to equip an automaton with memory that can be used more flexibly. A *linear bounded automaton* (LBA) is automaton equipped with memory not described as a stack but as a *tape*. A tape is a sequence of mutable cells, each of which can hold a symbol. The metaphor is based on the idea of magnetic tapes that were used for data storage in old-time computers or in audio cassettes. The LBA looks at only one cell in the tape at a time, so during a computation, part of the LBA's configuration at each step is its current position in the tape. Unlike a PDA, an LBA can move back and forth within its memory, reading and writing arbitrarily. Moreover, in an LBA, the input string and the memory are unified: we assume that the tape is initially populated with the input string. This means that an LBA can make several passes over the input string. One more restriction gives it the name *linear bounded* automaton: The tape is not of indefinite length. Instead, every LBA has a linear function $f(x)$ such that if the input string is of length n then the tape is no larger than $f(n)$ cells.

unrestricted grammars

We can push this further. If we take the definition of CSGs and remove the restriction that the right hand side of a production must not be shorter than the left hand side (thus allowing a production like $T\ x\ S\ y\ z \rightarrow z\ S$), then we have a new class of languages defined by *unrestricted grammars*. The set of unrestricted languages is a strict superset of context-sensitive languages.

Turing machine

Likewise we can remove the restriction that the length of an LBA's tape be no longer than a linear function of length of the LBA's input. This defines a new kind a automaton called a *Turing machine* named after Alan Turing who invented this model of computation. For any language that can be defined by an unrestricted grammar, there exists a Turing machine that accepts that language. Thus we have the following hierarchy of language classes and automaton classes with power and generality increasing as you go up:

Language class...	*...generated by*	*...accepted by*
Unrestricted languages	Unrestricted grammars	Turing machines
Context-sensitive languages	Context-sensitive grammars	Linear bounded automata
Context-free languages	Context-free grammars	Push-down automata
Regular languages	Regular expressions	Finite automata

We do not cover Turing machines (or the related LBA class) in detail here, but they play a crucial role in the theory of computation. Although the accepting of languages is a handy domain in which to discuss the relative complexity of computational tasks, Turing machines also are used to talk about computation more generally. The "result" returned by a finite automaton or PDA is simply the state it is in when it has processed the entire output—a "yes" or "no" about whether the given string is in the language. A Turing machine, however, can use its tape to print a result. Thus we can think of a Turing machine as modeling a function or computer program.

On the other hand, the previous section did cover another way to model computation—the lambda calculus. The great result at the end of this discussion is that the lambda calculus and the Turing machine model are *equivalent*. Anything that can be computed using one model can be computed using the other. This is proven by designing a Turing machine that can evaluate lambda calculus expressions and by writing a lambda calculus expression that can simulate the execution of a Turing machine. Since these two models can simulate each other, we know that they are equal in computational power.

Since Turing's work on Turing machines and Church's work on the lambda calculus in the 1930s and 1940s, many other models of computation have been proposed. Some, like a *random access machine*, are based on how real computers work. Others, like Conway's Game of Life, do not immediately seem related to computation. But all this has led to an even greater observation: *They are all equivalent to Turing machines—and, thus, to the lambda calculus and each other.* (This excludes those that are less powerful, of course. A model equivalent to Turing machines is *Turing complete*.)

Turing complete

The fact that so many strikingly different models have been shown to be equivalent and that no model that is more powerful has been discovered suggests that these models capture something fundamental about computation, that these models are the most powerful computational models possible. Stated another way, if some problem can be computed at all, it can be computed using a Turing machine, the lambda calculus, or any of the other Turing-complete models. This is called the *Church-Turing Thesis*. Notice that it is not a theorem. It cannot be proven deductively or even stated formally. It is more like a scientific theory that becomes increasingly strong as more Turing-complete models are discovered and we continue to fail to find a more powerful model. The thesis is falsifiable in the sense that the discovery of a more powerful model would disprove it—that would involve new ideas fundamentally different from how anyone has ever thought of computation.

Church-Turing Thesis

Biography: Alan Turing: 1912–1954

Alan Turing is often regarded as the most important figure in theoretical computer science. He studied at the University of Cambridge and Princeton University. Even before coming to Cambridge he had published his paper describing what would later be called a Turing machine. It was proposed to solve the Halting Problem (see Section 12.10), one of Hilbert's 23 problems (see page 384). This theoretical machine was particularly remarkable in light of the fact that the first recognizable electronic computer was several years away. During the Second World War he played a crucial role in the deciphering of German military communications. After the war he was involved with the development of early computing machinery and in contemplating the things that nascent computers would make possible. [16]

12.10 Special topic: Computability

In Section 7.15, we noted the amazing property that natural numbers, integers, and rationals all have the same cardinality, being countably infinite, but that real numbers are more infinitely numerous. That discussion considered comparing sizes only of number sets. What about infinite sets of other things?

Let us take for example computer programs. The finite nature of any given computer necessitates that the set of computer programs is not even infinite. However, suppose we even allow for a computer with an arbitrary amount of memory, where more always could be added if need be. Since computer programs are stored in memory, and memory is a series of bits, we can interpret the bit-representation of a program as one large natural number in binary. Thus we have a function from computer programs to natural numbers. This function is not necessarily a one-to-one correspondence (it certainly is not, if we exclude from the domain bit representations of invalid programs), but it is one-to-one, since every natural number has only one binary representation. This means that there are at least as many natural numbers as programs, perhaps more. The set of computer programs is therefore countable.

In ML programming, we think of a program as something that represents and computes a function. How many functions are there? That is not a particularly fair question, almost like asking how many sets there are. It would make more sense to limit ourselves to a specific domain and codomain and consider how many distinct functions exist between them. (Our concept of *distinct functions* is based on our definition of function equality. Two functions $f, g : X \to Y$ are distinct if there exists $x \in X$ such that $f(x) \neq g(x)$.)

Let us pick functions with domain $\mathbb{N}$ and codomain $\{0, 1, 2, \ldots, 9\}$. This is an arbitrary choice except that it is a fair realm to consider for our countably infinite computer. Since the domain is countable, all possible inputs are representable in the computer; the codomain is even finite. If either input or output involved a set like $\mathbb{R}$, our computer would not even be able to represent all the elements. So, we are considering the set of functions between these sets:

$$T = \{f : \mathbb{N} \to \{0, 1, 2, \ldots, 9\}\}$$

This set is uncountable. To see why, we will need to come up with a one-to-one correspondence between it and another uncountable set. In this case, we will define a function $h : (0, 1) \to T$. Note that this function takes a real number between 0 and 1 and returns a function in the set T. Now, suppose we represent a number in $(0, 1)$ as an infinite sequence of digits, $0.a_1a_2a_3 \ldots a_n \ldots$. Then define the function so that

$$h(0.a_1a_2a_3 \ldots a_n \ldots) = \text{ the function } f \text{ defined by } f(n) = a_n$$

Every number in $(0, 1)$ will produce a unique function. Moreover, every distinct function in T defines a unique real number if one simply feeds the natural numbers into it and interprets the output as a decimal expansion. Hence h is one-to-one and onto, and so T is uncountable.

If T is uncountable and the set of possible programs is countable, this means there are more functions than there are computer programs, even if we consider a (countably) infinite computer working on (countably) infinite data. It follows, then, that there must exist some functions for which there do not exist computer programs to compute them. *Some functions cannot be computed.* This is a very humbling thought about the limitations of the machines that we build.

The set of functions that map natural numbers to digits is still fairly unrealistic. Are any of these uncomputable functions something that we would actually want to compute?

One of the most common errors a beginning programmer can make is a program with infinite recursion or a loop that will not terminate. It would be a valuable tool for computer programming instruction if someone would write a program that would scan another program and determine, without running that other program, whether or not it will ever finish. Nevertheless, a program like this is impossible; the function we intend to implement is not computable.

To clarify what we are talking about, let us put it this way. We want a program/function that takes as its input (1) another program/function and (2) input for that program/function, and returns true if the given program halts (that is, does not loop forever or have infinite recursion) on that input, false otherwise. The following program does half of the intended work:

```
- fun halfHalt(f, x) = (f(x); true);

val halfHalt = fn : ('a -> 'b) * 'a -> bool
```

This program will indeed return true if the given program halts on the given input, but rather than returning false if it does not, it will go on forever itself. We cannot just let the program run for a while and, if it does not end on its own, break the execution and conclude it loops forever because we would never know whether or not we had waited long enough before breaking execution.

To prove that the intended function is not computable, we need to show that there does not exist a program that computes it. A proposition of *non*-existence requires a proof by contradiction. So, suppose we had such a program; call it `halt`. Supposing this program does exist, we can then use it to write the following:

```
- fun d(m) =
=   if halt(m, m) then while true do (); else  ();

val d = fn : ('a -> 'b) -> unit
```
[2]

[2]This actually would not type in ML because the application `halt(m, m)` requires the equation 'a = ('a -> 'b) to hold. ML cannot handle recursive types unless datatypes are used. It is still valid for our reasoning, though, because one could write a new language like ML except that it is less strict about types.

This means that given a function `m`, this program feeds `m` into `halt` as both the *function to run* and the *input to run it on*. If `m` does not halt (when running on itself), then this function finishes (and returns `unit ()`). If it does halt, then this program loops forever.

What would be the result of running `d(d)`? If `d` halts when it is applied to itself, then `halt` will return true, so then `d` will loop forever. If `d` loops forever when it is applied to itself, then `halt` will return false, so then `d` halts. If it will halt, then it will not; if it will not, then it will. This contradiction means that the function `halt` cannot exist.

halting problem

This, the unsolvability of the *halting problem*, is a fundamental result in the theory of computation since it defines a boundary of what can and cannot be computed. We can write a program that *accepts* this problem, that is, that will answer true if the program halts but does not answer at all if it does not. What we cannot write is a program that *decides* this problem. Research has shown many problems to be undecidable by reducing them to the halting problem. If we had a model of computation—something much different from anything we have imagined so far—that could decide the halting problem, then all problems we can currently accept would then also be decidable.

Chapter summary

Formal languages and automata form the basis for theoretical computer science since they are used to model computational problems and the processes that solve them. A thorough study of languages and automata would fill at least an entire semester at the undergraduate level and the related ideas of computational complexity are among the most difficult that one finds in computer science graduate programs. The results are profound, though: that all sufficiently powerful models of computation are equivalent and that real limitations to our computational models can be discovered and articulated. As the power of computational machinery has changed life and society so greatly, we are humbled by its clear limitations.

Key ideas:

An alphabet is a set of symbols. A string over an alphabet is a sequence of those symbols. A language over an alphabet is a set of strings.

A deterministic finite automaton (DFA) is a formal machine with a set of states and a transition function. When given an input string, it begins in the special start state and for each symbol in the input string it applies the transition function to the current state and current symbol to determine its next state. If, when the entire input string has been processed, the machine's current state is one of the specially designated accepts states, then we say that the DFA accepts the string.

A nondeterministic finite automaton (NFA) is like a DFA except that at some points the next state, based on the current state and input symbol, is ambiguous, and the NFA will then choose nondeterministically between more than one option. Formally this is modeled by a transition relation rather than transition function. A given string is accepted by the NFA if there exists a sequence of such choices that lead the NFA to be in an accept state at the end.

A regular expression is a way to specify a language using a set of simple rules. There are three primitive rules: the empty set of strings, the empty string, and strings containing a single symbol; and there are three recursive rules: alternation (the choice between two regular expressions to generate a string), concatenation (a string generated by one regular expression followed by a string generated by another), and closure (the concatenation of zero or more strings generated by the same regular expression).

Two models—such as two automata classes or two ways to specify a language—are equivalent if the sets of languages that can be accepted by some automaton in that class or specified by the form of expression are the same. The class of DFAs, the class of NFAs, and regular expressions are equivalent.

Languages with nested structure or that require some form of counting to specify or compute cannot be accepted by NFAs and DFAs or generated by regular expressions. Context-free grammars (CFGs) can generate them, however. A CFG specifies a language with a set of productions which define how nonterminal symbols can be expanded into strings of nonterminal and terminal symbols.

A push-down automaton (PDA) is like a finite automaton except that it also has a stack of symbols, acting as a memory, that can be manipulated at each step. Transitions are determined not only by the current state and input symbol but also by the symbol on the top of the stack. Deterministic PDAs can accept more languages than NFAs and DFAs, but nondeterministic PDAs can accept all languages that can be generated by CFGs.

The lambda calculus is a minimal, formal language that models computation as a collection of functions. The three kinds of expression in the language are variables, function applications, and function abstractions. Execution is governed by one rule for applying functions.

Language classes and classes of automata to accept them are arranged in a hierarchy of increasing power. The largest class of languages is that of unrestricted languages, which are accepted by the most powerful class of machine, Turing machines. As a model of computation, Turing machines represent the most powerful category, as all other sufficiently powerful models are equivalent to that of Turing machines. Such models, including the lambda calculus, are called Turing complete. The claim that these are the most powerful models is the Church-Turing Thesis.

Appendix A Patterns for proofs

This appendix summarizes the kinds of proofs used in this text. The student can refer here to review proof techniques from previous chapters or to make comparisons.

A.1 Set proofs

The most fundamental proofs are those for propositions from basic set theory. In Chapter 1, these were organized into General Forms 1–3 and Set Proposition Forms 1–3.

Subset proofs (Set Form 1). This proof pattern is known as the *element argument*. See Section 4.2.

> **To prove** $X \subseteq Y$:
>
> Pick an element from X. *Suppose* $x \in X$.
>
> Demonstrate that this element also is in Y. $x \in Y$ *by* . . .
>
> Conclude that X is fully contained in Y. *Therefore* $X \subseteq Y$ *by the definition of subset.*

Proofs of set equality (Set Form 2). This is merely two proofs of Set Form 1. See Section 4.3.

> **To prove** $X = Y$:
>
> Prove $X \subseteq Y$.
>
> Next, prove $Y \subseteq X$.
>
> Conclude that X and Y have exactly the same elements. *Therefore* $X = Y$ *by the definition of set equality.*

Proofs of set emptiness (Set Form 3). This is our first and most common example of proof by contradiction, based on the argument form $p \rightarrow F, \therefore\ p$. See Sections 3.8 and 4.4

To prove $X = \emptyset$:

Pick a (hypothetical) element from X. *Suppose* $x \in X$. This implicitly assumes X is not empty, since it means, "suppose there is an element (namely, x) in X."

Derive a proposition known to be false from this fact. *Contradiction.*

Conclude that X is empty. *Therefore, by contradiction,* $X = \emptyset$.

Proofs of conditional propositions (General Form 2). Many more complicated theorems are conditional propositions at the surface level. See Section 4.5.

To prove *If* p *then* q:

Suppose p is true. *Suppose* $p \ldots$

Derive the proposition q. *Therefore* q *by* ...

Technically we are concluding that *If* p *then* q, but conventionally we do not state this at the end. It is enough to treat q as our conclusion.

Proofs of biconditional propositions (General Form 3). This is merely two proofs of General Form 2. See Section 4.7.

To prove p *iff* q:

Suppose p is true. *Suppose* $p \ldots$

Derive the proposition q. *Therefore* q *by* ...

Clear all assumptions and start from the top. Suppose q is true. *Conversely, suppose* $q \ldots$

Derive the proposition p. *Therefore* p *by* ...

As with proofs of conditionals, there is no need to conclude with the original proposition. That conclusion is implied.

This is not the only way to prove biconditionals. A biconditional also could be proven by a series of other biconditionals previously known to be true. Suppose we knew that p iff a, a iff b, b iff c, and c iff q. In that case we can conclude p iff q by the transitivity argument form (see Section 3.8). However, interesting propositions are rarely proven that easily, and beginners are not recommended to try.

A.2 Proofs with quantification

For propositions of any significant complexity, unravelling the quantification and analyzing what it demands is one of the most crucial parts of the proof. The patterns presented here inevitably make it look easier than it is. The trick is applying these patterns multiple times, according to the layers of quantification often found in complicated propositions.

Proofs of universally quantified propositions. To prove that every element in a set has some property, we invite the skeptic to pick an element at random, and then we show the property for that element. This should look similar to the element argument, which is essentially a proof over universal quantification.

> **To prove** $\forall\, x \in X, P(x)$:
>
> Pick an element in X. *Suppose* $x \in X$. Technically the x appearing in the proposition is different from the x appearing in the proof. This rarely leads to confusion, though.
>
> Show that the predicate or property holds for that element. *Hence* $P(x)$ *by ...*
>
> Conclude the property holds for all. No explicit justification is necessary. *Therefore* $\forall\, x \in X, P(x)$. The conclusion itself, in fact, may be omitted.

Proofs of existence. There are two ways to prove that something exists. The first and direct way is first to describe the thing, or tell how to construct one, or where to find one, and then to demonstrate that this thing fits the description of what we are looking for. This would be as if someone doubted whether any student on campus could solve the Rubik's Cube, but you knew that your friend Billy can. You would bring Billy into the room (or give the doubter Billy's address) and have Billy perform for the doubter.

> **To prove** $\exists\, x \in X \mid P(x)$:
>
> Describe a specific element in X or give an algorithm or the equivalent for constructing the element. *Let x be the element in X that ...* We prefer to use the word *let* rather than *suppose* because we are only giving a name, x, to a specific object. We are not picking that object at random.
>
> Demonstrate that x meets the qualifications. *Then* $P(x)$ *by ...*
>
> Conclude that such an element exists. In many cases, this conclusion does not need to be stated explicitly.

This is what would be called a *constructive proof* because it gives you not only a verification of the proposition but also an algorithm for making or finding the desired object. For this reason, the pattern above is the preferred way to prove an existence proposition. In not every case is such a proof possible, though. Suppose you left a scrambled Rubik's Cube in the dorm lobby overnight and in the morning found it solved. There must be someone on campus capable of solving the puzzle, but you do not know who it is. Similarly, some existence propositions can (and some must) be proven by contradiction.

> **To prove** $\exists\, x \in X \mid P(x)$:
>
> Assume the opposite—there is no item like that. *Suppose* $\forall\, x \in X$, $\sim P(x)$.
>
> Derive a contradiction from this. *Contradiction.*
>
> Conclude that that an item like this must exist. *Therefore, by contradiction,* $\exists\, x \in X \mid P(x)$.

An example of a non-constructive proof of existence is the proof of Theorem 11.10. Constructive or not, proofs of existence are among the hardest in a course like this one. One difficulty is that the statement of the theorem rarely contains the existential quantifier explicitly. Instead, definitions used by the proposition will assert that something exists. See Section 7.9 for examples of proofs like this.

Proofs of nonexistence. In principle, it is very difficult to prove that something does not exist. Just because you have not found one yet, how can you be sure that one is not lurking out there somewhere? Still, we know there is no integer that is both even and odd, there are no married bachelors, and there does not exist an algorithm that decides the halting problem—so do not bother looking for one. How do we know? Each of them would involve a contradiction if they were true. Hence proofs of nonexistence mirror the second (non-constructive) proof of existence.

> **To prove** $\sim \exists\, x \in X \mid P(x)$:
>
> Assume the opposite. *Suppose* $x \in X \mid P(x)$.
>
> Derive a contradiction from this. *Contradiction.*
>
> Conclude that no item like this exists. *Therefore, by contradiction,* $\sim \exists\, x \mid P(x)$.

This should look similar to the proofs of set emptiness. Note also that in some cases it may be easier to prove the equivalent proposition, $\forall\, x \in X$, $\sim P(x)$.

Proofs of uniqueness. Occasionally when we know an element exists we also want to show that no other element is like it—say, x is the only element in X such that $P(x)$. We could turn this into a proof of nonexistence to show that no element in $X - \{x\}$ has this property, but here is a more straightforward way:

To prove *if* $y \in X$ *such that* $P(y)$*, then* $x = y$:

Actually, all you need to do is look carefully at how we stated the uniqueness property formally. The pattern is just that of a conditional proof.

Proofs of unique existence. We can add yet another category: Prove that *exactly one* of certain thing exists. Notice the subtle difference from the previous kind of proof: we are not proving that *if* one exists that it is the only one, but that it *does* exists *and* it is the only one.

We do not need to spell this pattern out explicitly because it is the combination of a proof of existence and a proof of uniqueness. First prove that such an item exists, then prove that it is unique.

A.3 Proofs of relation properties

There is nothing fundamentally different about proving propositions involving relation properties. It is a matter of understanding and applying the relevant definitions. These properties are described starting in Section 5.4. In all of these, assume R is a relation over A.

Proving reflexivity. The definition of reflexivity is universally quantified.

To prove R *is reflexive*:

Pick any element from the domain. *Suppose* $x \in A$.

Show it is related to itself. *Hence* $(x, x) \in R$ *by ...*

Conclude that R is reflexive. *Therefore, by definition of reflexivity,* R *is reflexive.*

Proving symmetry. The definition is universally quantified but with a restriction.

To prove R *is symmetric*:

Pick two elements where one is related to the other; equivalently, pick a tuple in the relation. *Suppose* $a, b \in A$ *such that* $(a, b) \in R$. Or, *Suppose* $(a, b) \in R$. The latter option is more succinct, but the former is analogous to proofs of reflexivity and transitivity.

Show the tuple can be inverted. *Hence* $(b, a) \in R$ *by ...*

Conclude that R is symmetric. *Therefore, by definition of symmetry,* R *is symmetric.*

Proving transitivity. Similar principles apply, just in a more complex way.

> **To prove** *R is transitive*:
>
> Pick three elements known to be related. *Suppose $a, b, c \in A$ such that $(a, b), (b, c) \in R$.*
>
> Show that a is related to b. *Hence $(a, b) \in R$ by ...*
>
> Conclude that R is transitive. *Therefore, by definition of transitivity, R is transitive.*

Proving antisymmetry. Here we are picking two elements again but with a more complicated restriction.

> **To prove** *R is antisymmetric*:
>
> Pick two elements mutually related. *Suppose $a, b \in A$ such that $(a, b), (b, a) \in R$.*
>
> Show the two are actually one and the same. *Hence $a = b$ by ...*
>
> Conclude that R is antisymmetric. *Therefore, by definition of antisymmetry, R is antisymmetric.*

A.4 Inductive proofs

We have seen the general form of structural induction and the specific form on natural numbers, known as mathematical induction.

Structural induction. This is for sets that are recursively defined; that is, some elements of the set are described or listed simply and explicitly, but others are specified in terms of other elements in the set. We saw structural induction beginning with Section 6.4.

> **To prove** $\forall\, x \in X, P(x)$ where X has partition $\{Y, Z\}$, Y being the set of elements described directly and Z being the set of elements described by a recursive rule:
>
> Suppose $x \in Y$, that is, suppose we are in the base case of the recursive definition of X. ***Base case.*** *Suppose $x \in Y$ ...*
>
> Demonstrate the property for the simple elements. *Hence $P(x)$ by...* It is possible there will be several elements or sets of elements described directly, in which case you may need to repeat these two steps for each set.
>
> Next pick a complicated element. ***Inductive case.*** *Suppose $x \in Z$...*
>
> Assume that the property holds true for the other elements in X from which x is built. *By structural induction ...*

Demonstrate the property holds for the complicated elements. *Hence* $P(x)$ *by ...*

Conclude the property holds true for all.

Mathematical induction. Mathematical induction is simply structural induction applied to whole numbers or natural numbers. Proofs by mathematical induction can take slightly different forms depending on such things as whether we start from 0 (to prove a proposition quantified over whole numbers) or from 1 (to prove a proposition quantified over natural numbers) and whether the inductive step is formulated as $I(n) \rightarrow I(n+1)$ or $I(n-1) \rightarrow I(n)$. One also may find a varying level of explicitness in these proofs. See Section 6.5.

To prove $\forall\, n \in \mathbb{W}, I(n)$, explicit version:

Show that it works for 0. ***Base case.*** *Suppose* $n = 0$*...* *So* $I(n)$ *by ...* *Hence there exists* $N \in \mathbb{W}$ *such that* $I(N)$.

Show that if it works for one whole number, it works for the next. ***Inductive case.*** *Suppose* $n \in \mathbb{W}$ *such that* $I(n)$*...* *Hence* $I(n+1)$ *by ...*

Conclude that it works for the whole set. *Therefore, by mathematical induction,* $\forall\, n \in \mathbb{W}, I(n)$.

To prove $\forall\, n \in \mathbb{W}, I(n)$, concise version:

Show that it works for 0. ***Base case.*** *... Hence* $I(0)$ *by ...*

Show that if it works for one whole number, it works for the next. ***Inductive case.*** *Suppose* $I(n)$*...* *Hence* $I(n+1)$ *by ...*

Conclude that it works for the whole set. *Therefore, by mathematical induction,* $\forall\, n \in \mathbb{W}, I(n)$.

In either version, you can refer to the proposition $I(n)$ as the *inductive hypothesis*.

A.5 Proofs of function properties

As with relation properties, these proofs amount to exercises of your skill at applying definitions. These patterns (and the patterns of relation proofs) show how to use definitions synthetically. The justification in the proofs may require their analytic use. The properties are described in Section 7.6. Assume $f : X \rightarrow Y$.

That a function is onto. At the heart, this is a proof of existence, but it is wrapped in a for-all proof.

> **To prove** f *is onto*:
>
> Pick an element in the codomain. *Suppose* $y \in Y$.
>
> Find an appropriate element in the domain which you will later show hits y. *Let* x *be the element in* X *that ...*
>
> Demonstrate that indeed x hits y. *Hence* $f(x) = y$ *by ...*
>
> Conclude that all elements in the codomain are hit. *Therefore,* f *is onto by definition.*

That a function is one-to-one. At its heart, this is a proof of uniqueness, expressed as a conditional, and wrapped in a for-all proof.

> **To prove:** f *is one-to-one*:
>
> Pick two elements in the domain. *Suppose* $x_1, x_2 \in X$.
>
> Prove the conditional; suppose they hit the same thing. *Suppose further that* $f(x_1) = f(x_2)$.
>
> Demonstrate that these two elements are in fact the same element. *Hence* $x_1 = x_2$ by ...
>
> Conclude that each element in the codomain is hit at most once. *Therefore,* f *is one-to-one by definition.*

That two sets have the same cardinality. This is a hefty example of an existence proof. See Section 7.9.

> **To prove** $|X| = |Y|$:
>
> Show there exists a one-to-one correspondence between them. That is, first describe a function. *Let* f *be the function defined by ...*
>
> Then, show that the function is one-to-one and onto. These two steps can overlap by proving that each element in the codomain is uniquely hit. *Suppose* $y \in Y$. *Let* x *be the element in* X *that ... So* $f(x) = y$ *by ... Suppose* $x' \in X$ *such that* $f(x') = y$*... So* $x' = x$ *by ... Hence* f *is a one-to-one correspondence by the definitions of onto and one-to-one.*
>
> Conclude that they have the same cardinality. *Therefore,* $|X| = |Y|$ *by the definition of cardinality.*

Bibliography

[1] Harold Abelson and Gerald Jay Sussman. *Structure and Interpretation of Computer Programs*. McGraw Hill and the MIT Press, Cambridge, MA, second edition, 1996.

[2] E. T. Bell. *Men of Mathematics*. Simon & Schuster, New York, 1986. Originally published 1937.

[3] Felice Cardone and J Roger Hindley. Lambda-calculus and combinators in the 20th century. In Dov Gabbay and John Woods, editors, *Logic from Russell to Church*, volume 5 of *Handbook of the History of Logic*, chapter 14, pages 723–817. Elsevier, Amsterdam, 2009.

[4] G.K. Chesterton. *Orthodoxy*. Image Books, Garden City, NY, 1959.

[5] Thomas H. Cormen, Charles E. Leiserson, Ronald L. Rivest, and Clifford Stein. *Introduction to Algorithms*. MIT Press, Cambridge, MA, third edition, 2001.

[6] B. A. Davey and H. A. Priestley. *Introduction to Lattices and Order*. Cambridge University Press, Cambridge, UK, second edition, 2002.

[7] Edsger W Dijkstra. On the cruelty of really teaching computing science, December 1988. `http://userweb.cs.utexas.edu/users/EWD/transcriptions/EWD10xx/EWD1036.html`.

[8] Sussana S. Epp. *Discrete Mathematics with Applications*. Thomson Brooks/Cole, Belmont, CA, third edition, 2004.

[9] H. W. Fowler. *A Dictionary of Modern English Usage*. Oxford University Press, Oxford, 1965. Originally published 1926. Revised by Ernest Gowers.

[10] Joseph A Gallian. *Contemporary Abstract Algebra*. Brooks Cole, Belmont, CA, seventh edition, 2009.

[11] Erich Gamma, Richard Helm, Ralph Johnson, and John Vlissides. *Design Patterns: Elements of Reusable Object-Oriented Software*. Addison-Wesley, Reading, MA, 1995.

[12] G. H. Hardy. *A Mathematician's Apology*. Cambridge University Press, Cambridge, UK, 1940.

[13] John Harrison. *Handbook of Practical Logic and Automated Reasoning*. Cambridge University Press, Cambridge, UK, 2009.

[14] Tony Hoare (professional biography page). `http://research.microsoft.com/en-us/people/thoare/`, October 1999.

[15] Andrew Hodges. Alan Turing. In *Oxford Dictionary of National Biography*. Oxford University Press, 2004. Online version available at `http://www.turing.org.uk/bio/part1.html`.

[16] Leah Hoffman. Robin Milner: The elegant pragmatist. *Communications of the ACM*, 53(6):21, June 2010.

[17] Karel Hrbacek and Thomas Jech. *Introduction to Set Theory*. Marcel Dekker, New York, 1978. Reprinted by University Microfilms International, 1991.

[18] Bernard Kolman, Robert Busby, and Sharon Ross. *Discrete Mathematical Structures*. Prentice Hall, Upper Saddle River, NJ, fifth edition, 2008.

[19] Alexander MacFarlane. *Lectures on Ten British Mathematicians of the Nineteenth Century*. Number 17 in Mathematical Monographs. John Wiley and Sons, New York, 1916. Available at `http://www.gutenberg.org/ebooks/9942`.

[20] Merriam-Webster Online. `http://www.m-w.com`. Based on *Merriam-Webster's Collegiate Dictionary, Eleventh Edition*.

[21] J J O'Connor and E F Robertson. Helmut Hasse, 1996. `http://www-history.mcs.st-andrews.ac.uk/Biographies/Hasse.html`.

[22] J J O'Connor and E F Robertson. Giuseppe Peano, 1997. `http://www-history.mcs.st-andrews.ac.uk/Biographies/Peano.html`.

[23] J J O'Connor and E F Robertson. David Hilbert, 1999. `http://www-history.mcs.st-and.ac.uk/Biographies/Hilbert.html`.

[24] Ovid. *Amores*, 15 BC. Translated by A. S. Kline, 2001. Available at `http://www.poetryintranslation.com/PITBR/Latin/Amoreshome.htm`.

[25] Lawrence C Paulson. *ML for the Working Programmer*. Cambridge University Press, Cambridge, UK, second edition, 1996.

[26] Benjamin C. Pierce. Foundational calculi for programming languages. In Allen B. Tucker, editor, *Handbook of Computer Science and Engineering*, chapter 139. CRC Press, 1996.

[27] George Pólya. *Induction and Analogy in Mathematics*. Princeton University Press, Princeton, NJ, 1954. Volume I of *Mathematics and Plausible Reasoning*.

[28] Stuart Russell and Peter Norvig. *Artificial Intelligence: A Modern Approach*. Prentice Hall, Upper Saddle, NJ, 3rd edition, 2009.

[29] Barbara W Sarnecka and Susan Carey. How counting represents number: What children must learn and when they learn it. *Cognition*, 108(3):662–674, 2008.

[30] Gary Stix. Profile: David A. Huffman. *Scientific American*, pages 54, 58, Sept 1991. Available at `http://www.huffmancoding.com/my-family/my-uncle/scientific-american`.

[31] Paul Stockmeyer. The tower of hanoi. `http://www.cs.wm.edu/~pkstoc/toh.html`.

[32] William Strunk and E. B. White. *The Elements of Style*. Macmillan, New York, 1972.

[33] Nick Szabo. An explanation of the kula ring. `http://szabo.best.vwh.net/kula.html`, 2005.

[34] Jeffrey D Ullman. *Elements of ML Programming*. Prentice Hall, Upper Saddle River, NJ, ML97 edition, 1998.

[35] Mary-Claire van Leunen. *A Handbook for Scholars*. Knopf, New York, 1978.

[36] Obituary of John Venn. *The Times* (of London), April 1923. Available at `http://www-groups.dcs.st-and.ac.uk/~history/Obits/Venn.html`.

[37] Robin Wilson. *Four Colors Suffice*. Princeton University Press, Princeton, NJ, 2002.

Index

D

E

T

U

V